IRELAND'S GENEROUS NATURE

THE PAST AND PRESENT USES OF WILD PLANTS IN IRELAND

IRELAND'S GENEROUS NATURE

THE PAST AND PRESENT USES OF WILD PLANTS IN IRELAND

PETER WYSE JACKSON

MISSOURI BOTANICAL GARDEN
ST LOUIS, MISSOURI, USA

and formerly

NATIONAL BOTANIC GARDENS OF IRELAND
GLASNEVIN, DUBLIN, IRELAND

Missouri Botanical Garden

national botanic gardens of Ireland

Cloth edition ISBN: 978-0-915279-78-4
Leather edition ISBN: 978-0-915279-90-6
Library of Congress Control Number: 2013953830

Scientific Editor and Head: Victoria C. Hollowell
Managing Editor: Allison M. Brock
Associate Editor: Tammy M. Charron
Associate Editor: Lisa J. Pepper
Associate Editor: Laura Slown
Press Coordinator: Cirri R. Moran
Editorial Assistant: Maryjane M. Hotaling
Assistance with editing and compositing: Kay Yatskievych
Design and typesetting: Adrianna Sutton

P.O. Box 299, St. Louis, Missouri 63166-0299, U.S.A.
www.mbgpress.info

Front cover images:
Top: *Erica cinerea* growing among bog pine (*Pinus sylvestris*) in a cut-over bog in Co. Mayo. Bottom, left to right: Making a lobster pot on the Blasket Islands, Dingle Peninsula, Co. Kerry (image courtesy of The Blasket Centre/Ionad an Bhlascaoid Mhóir). Crushing the shoots of gorse (*Ulex europaeus*) for animal fodder. Preparing the fruits of hawthorn (*Crataegus monogyna*) for jelly-making. Rock samphire (*Crithmum maritimum*) can be boiled or steamed as a vegetable or pickled.

Back cover images:
Left to right: 'Strawboys' during the annual 'Wran' parade on 26 December 2009, Dingle, Co. Kerry; the costumes are made from oat straw (photo courtesy of Tom Fox). Foxglove (*Digitalis purpurea*). A simple thatched stone house in Keel Village, Achill Island, Co. Mayo, from a postcard, ca. 1910. Irish saxifrage (*Saxifraga rosacea*) growing on the limestone of the Burren, Co. Clare.

All photos taken by Dr Peter Wyse Jackson unless otherwise noted.

CONTENTS

FOREWORD

Set offshore at the far fringe of Europe, its historic subsistence rural culture so long remote from industrial development or bourgeois mores, Ireland should offer an encyclopedic heritage of the use of wild plants as food and medicine, for everyday utility or symbolic ritual. For early centuries, indeed, the importance of plant knowledge was tempered as much by a continuing pagan respect for nature as by the herb gardens of the many monasteries.

How this knowledge was lost or rejected after the Famine and in the drive for modernity and prosperity is just one thread of Peter Wyse Jackson's epic exploration of Ireland's ethnobotanical heritage. It has not been an easy venture, pursued over years and to many sources. One of the richest has been the remarkable folklore collection held in University College, Dublin, much of it in the Irish language. Loss of Irish in the island as a whole, and hostile attitudes bred through clumsy schooling, have helped to rob modern rural culture even of names of plants, let alone their many virtues and uses.

In an increasingly synthetic and digital world, however, there remains, or has revived, an intuitive need for reconnection with plants. From the greener reaches of the internet to the new allotments and communal city gardens of Dublin—even, indeed, the current fascination with culinary herbs and spices—this renewal of curiosity adds much value to this book, alongside its obvious beauty and great ethnobotanical worth. More than half of the 925 native plants of this island have been useful to its people, but nowhere, until now, have their stories been gathered together so systematically, both for Irish readers and the wider cultural and scientific audience.

Michael Viney
Thallabawn, Westport, Co. Mayo,
January 2014

ACKNOWLEDGEMENTS

There are very many people I want to acknowledge for their kindnesses and generous support during the preparation of this book. To the countless people who also helped me but do not get an acknowledgement by name, I hope that you will accept these thanks.

Formal acknowledgement is due to the following institutions and individuals for use of text or images:

National Botanic Gardens, Glasnevin, for granting permission to reproduce watercolours by Lydia Shackleton.

Ionad an Bhlascaoid Mhóir/The Blasket Centre for use of images.

Mr Standish Mason for permission to reproduce the images from his grandfather's book, *The Islands of Ireland* by Thomas H. Mason.

Tom Fox allowed reproduction of several of his excellent photographs of the Dingle 'Wran'.

Desmond Kavanagh, whose father 'interviewed' Charles McGlinchey, for granting permission to use textual passages from *The Last of the Name.*

The Mercier Press, Ltd. for granting permission to reproduce textual passages from *Róise Rua: An Island Memoir* by Pádraig Ua Cnáimhsí.

Dublin Institute for Advanced Studies for granting permission to reproduce textual passages from *Early Irish Farming* by Fergus Kelly.

Ed Greenwood for use of information from Botanical.com.

My personal thanks I extend to:

My colleagues at the National Botanic Gardens, Glasnevin, Dublin, for their general encouragement and support. In particular, I would like to thank Glynn Anderson, Alexandra Caccamo, Colette Edwards, Howard Fox, Felicity Gaffney, Matthew Jebb, Edel McDonald, Paul Maher, Maurice Maxwell, Seamus O'Brien, Myles O'Riada and Brendan Sayers. All helped in many ways, pointing me in the direction of interesting records of plant usage and references.

Mícheál de Mórdha and Dáithí de Mórdha of The Blasket Centre (OPW), Dunquin, Co. Kerry, for information on the usage of plants in West Kerry, and in particular on the Blasket Islands.

Robin Bryan, for useful information on medicinal and herbal plants in Ireland.

Eugene Deane and Bernie Goggin, for their rich store of knowledge about the Dingle Peninsula.

Sue Redican, the Blasket Island weaver, for valuable information on plants used for dyes.

Tom Fox and Fergus O'Flaherty, for information on the Dingle 'Wran' customs and for providing me with a magnificent strawboy costume.

All my friends and colleagues in the Irish botanical community, for suggestions, advice, support and information, in particular, Declan Doogue, Paddy Reilly, Gerry Sharkey, Micheline Sheehy Skeffington and Nick Scott. The late Maura Scannell, an inspiration for several generations of Irish botanists, was very generous in sending me snippets of information, most of which I would never have tracked down on my own.

Joe Hogan, for generously allowing me to quiz him about native plants used in Irish basketry and for making some wonderful baskets that are now a treasured part of my personal collection and those of the National Botanic Gardens.

Séamas MacPhilib of the National Museum of Ireland, Turlough Park, Castlebar, Co. Mayo, for providing me with information on objects and artefacts in the Museum's collection.

My ethnobotanical colleagues around the world who I also have the pleasure of calling friends—Glenn Wightman, Edelmira Linares, Robert Bye and Brien Meilleur—for sharing their knowledge, advice and guidance.

Patricia Moloney of the Department of Folklore, University College, Dublin, for pointing me in the direction of several valuable references.

David Simpson, Royal Botanic Gardens, Kew, United Kingdom, for information on the use of sedges.

Maírín Uí Chonchubhair, for help with the Irish names of plants, particularly from Co. Kerry.

Roy Vickery, for allowing me access to the wealth of information and lore he has gathered on plant uses in Britain and Ireland.

John McLoughlin, for information on the uses of Irish trees and timbers.

Michael and Ethna Viney, for their encouragement and information on plant uses in Co. Mayo, and to Michael for writing the Foreword to this book.

All my colleagues at the Missouri Botanical Garden, especially Rainer Bussmann, Kathy Farris, Doug Holland, Bob Magill, Linda Oestry and Beverly Wilcox.

I cannot adequately express how grateful I am to all the staff of the Missouri Botanical Garden Press, in particular Victoria Hollowell, Tammy Charron, Laura Slown, Allison M. Brock, Maryjane M. Hotaling, and Kay Yatskievych, who have dedicated so much time and effort to editing and preparing this book for publication. Without their expertise and assistance this work would have been a shabby affair. Nevertheless, I should point out that any errors are my responsibility alone, which I hope can be corrected in (any) future editions of this work.

My brother John for suggestions and help given in many ways, including translations of various Irish texts, all my extended Wyse Jackson family for advice and support, and especially my brother Michael for valuable botanical advice and other information.

My mother, Lois Wyse Jackson, with love, for all her support and wonderful stories and from whom I have learnt so much.

My children, Alice, Robert and Thomas, for their constant support and love, and for listening patiently whenever I told them, at length, about some fascinating (to me) new piece of information I had learnt about yet another plant.

My dear wife, Diane, for her support, love and for lending me to this project, which she could not have known would take so many years to complete. I dedicate this book to you, with my love.

Hippuris vulgaris. Watercolour by Lydia Shackleton. Courtesy of the National Botanic Gardens, Glasnevin.

The watercolours of native Irish plants included in this book are by Lydia Shackleton, the originals of which are part of the collections of the National Botanic Gardens, Glasnevin, Dublin. Most of those included here have never before been published. Shackleton began painting for the National Botanic Gardens (then the Royal Botanic Gardens) in about 1884, at the request of the director, Frederick Moore. She continued to paint at Glasnevin for 23 years, until 1907, when her failing eyesight made it impossible for her to continue. She was a Quaker, a native of Ballitore, Co. Kildare (Nelson & McCracken, 1987). During her career she did at least 1500 paintings at Glasnevin, most of which were watercolours of exotic orchids.

Erica cinerea (heather). Watercolour by Lydia Shackleton. Courtesy of the National Botanic Gardens, Glasnevin.

INTRODUCTION

ETHNOBOTANY is the study of the links and relationships between people, cultures and plants. Those who study ethnobotany aim to research and document not only the varied uses of plants, but also how they are thought of culturally. We focus on the huge range of purposes humans have found for plants—as food, medicine, textiles and fibres; for construction and shelter; as tools themselves; in rituals; as cosmetics, poisons and dyes; even as currency—and the broader literature and lore surrounding them. This book documents such uses of wild plants in Ireland.

In 2006, while writing text to be included in some interpretative labels for plants on display in the National Botanic Gardens, Glasnevin, I wanted to include some information on how native plants were used in the past. Looking through the great range and diversity of botanical literature in Ireland, I was surprised to find that although it is rich in information on the distribution, status and taxonomy of Irish plants, I could find no comprehensive scientific account of Irish ethnobotany that examined the past and present uses of each wild plant species (other than a number of publications that looked at plants for medicinal purposes). Of course much has been written about Irish folklore, and this literature is rich in references to the use of plants for many purposes, particularly for medicine, but little has been published from the botanical point of view. For that reason, I decided to compile a simple checklist that would list the known uses of wild plants in Ireland, past and present. However, as I delved deeper into the subject, I found that although much had been written in numerous publications related to the topic, there was no single work that provided an overview of the value of wild plants on this island for the human population. Although I had intended to prepare only a brief checklist for my own use, I found that the work expanded so much that a book was necessary if I wanted to review the published literature and to include whatever I could find on Irish ethnobotany.

Other countries are well served with literature on the uses of wild plants. For example, William Milliken and Sam Bridgewater have recently documented the uses of plants in Scotland down through the ages in their excellent *Flora Celtica: Plants and People in Scotland* (2004). Many of the uses of plants that they have outlined in Scotland can be applied equally to Ireland, but not all of them by any means. There are very significant differences between the usage of plants in these two countries, despite their extensive cultural linkages for thousands of years and very similar floras.

As I carried out the research for this book, I became increasingly impressed not only by the diversity of uses to which wild plants have been put, but also by the high percentage of plants that have documented uses. The native flora of Ireland is regarded as including in the region of 900 to 1000 plant species. Of those, I was astonished to find that approximately half (53%, see Tables 1, 2) are known to have been used for some specific purpose, particularly for medicine and for food. Table 1 lists the number of wild plant species (native and non-native) that are known to have been used (generally in Ireland) for various purposes. From this work it is clear that wild plants have been a hugely important natural resource throughout Ireland's history, freely providing materials that meet the needs of every part of our lives, suggesting and justifying the title that I have chosen for this book: *Ireland's Generous Nature.*

One cannot help but be impressed by the ingenuity of the people of Ireland in using the wild plant resources for a multitude of purposes to support their daily lives. The wide suite of uses represents a remarkable and sophisticated understanding of the physical and sometimes chemical properties of each plant species, as well as skill in adapting particular attributes to meet a specific need. Moreover, one can see that a high percentage of the native plants were used for some purpose or other. Yet it is equally remarkable, and somewhat of a tragedy, that not only has most of this usage slipped into obscurity, but most of the traditional knowledge of how plants were used has also disappeared.

Table 1. Statistics on the historical uses of wild plants in Ireland.

Use category	Flowering plants	Conifers	Ferns and fern allies	Total taxa	Percentage of total taxa number
Aphrodisiacs					
Native[1]	13	—	—	13	1.4%
Introduced[2]	2	—	—	2	
Aromatics					
Native	13	—	—	13	1.4%
Introduced	4	—	—	4	
Beverages (teas and infusions—non-alcoholic)					
Native	56	—	1	57	6.2%
Introduced	5	—	—	5	
Beverages (used in making wines, beers, liqueurs)					
Native	56	—	—	56	6.1%
Introduced	2	—	—	2	
Children's games					
Native	41	—	—	41	4.4%
Introduced	9	—	—	9	
Cosmetics (ingredients of)					
Native	21	—	—	21	2.3%
Introduced	5	—	—	5	
Dyes					
Native	83	1	3	87	9.4%
Introduced	4	—	—	4	
Fibres[3]					
Native	66	1	2	69	7.5%
Introduced	8	—	—	8	
Foods (used for food or edible)					
Native	113	2	4	119	12.9%
Introduced	50	—	—	50	
Insecticides, pesticides or insect repellents					
Native	10	—	—	10	1.1%
Introduced	3	—	—		
Medicinals					
Native	287	2	17	306	33.1%
Introduced	63	—	—	63	
Musical instruments (mainly timbers)					
Native	12	1	1	14	1.5%
Introduced	2	—	—	2	
Narcotics, hallucinogens and intoxicants[4]					
Native	6	—	—	6	0.6%
Introduced	6	—	—	6	

Table 1. Statistics on the historical uses of wild plants in Ireland. *(continued)*

Use category	Flowering plants	Conifers	Ferns and fern allies	Total taxa	Percentage of total taxa number
Ornamentals					
Native	127	3	6	136	14.7%
Introduced	47	15	—	62	
Religious and symbolic					
Native	11[5]	—	—	11	1.2%
Introduced	1	—	—	1	
Smoked					
Native	29	—	—	29	3.1%
Introduced	2	—	—	2	
Tanning leather					
Native	19	—	—	19	2.1%
Introduced	2	—	—	2	
Thatch					
Native	21	—	1	22	2.4%
Introduced	1	—	—	1	
Timbers					
Native	34	3	—	37	4.0%
Introduced	10	20	—	30	
Veterinary medicines					
Native	92	—	2	94	10.2%
Introduced	6	—	—	6	
Poisonous plants					
Native	54	1	10	65	7.0%
Introduced	18	—	—	18	
TOTALS	**Flowering plants**	**Conifers**	**Ferns and fern allies**	**Total number**	**Percentage total number**
Total number of native taxa	**865**	**3**	**57**	**925**	
Total number of native taxa known to have past/ present ethnobotanical uses	**462**	**3**	**25**	**490**	**53%**
Total number of native taxa with no known ethnobotanical uses, past or present	**403**	**—**	**32**	**435**	**47%**
Total number of introduced taxa naturalised in the wild with known ethnobotanical uses	**125**	**20**	**—**	**145**	

1. Those taxa which are 'Possibly Introduced' to Ireland are included in the totals given for 'Native' taxa.
2. Those taxa which are 'Probably Introduced' to Ireland are included in the totals given for 'Introduced' taxa.
3. Used for ropes, stuffing, threads, string, fabrics, basketry, etc., excluding thatch use.
4. In natural unfermented state.
5. This is certainly an underestimate of the total. There are many taxa which were used for folklore and ritual purposes that have not been documented in this book, e.g., for magical purposes, which are outside the scope of this work.

Note: Table 1 has been compiled through an analysis on the information given on the usage of plant species included in the Systematic List of Wild Plants. The numbers given for 'Introduced' (including 'Probably Introduced') taxa represent introduced taxa that are known to have been collected for various purposes in the wild or those that were originally and deliberately introduced for some purpose and have now gone wild (become naturalised) in Ireland. Trees grown for forestry purposes or for their timber are also included in these totals, despite the fact that they are not strictly 'naturalised'.

The figures and percentages given in Table 1 should be taken as indicative, rather than definitive. For example, we cannot be certain that all native willow species were used to obtain fibres, or that all sedges (*Carex* species) listed as 'Ornamental' in the Systematic List have actually been grown in an Irish garden. Nevertheless, they are included in the totals.

Other categories of use and uses not mentioned in Table 1 include plants for fodder (numerous taxa), those used for fuel (many timbers as well as other species with woody twigs and/or roots), those taxa used for binding soils (several taxa) and other species used for one-off or diverse purposes.

Table 2. Statistics on the use of algae, fungi, lichens, mosses and liverworts.

Use category	Algae	Fungi	Lichens	Mosses and liverworts
Aphrodisiacs	1	—	—	—
Cosmetics (ingredients of)	3	—	—	—
Dyes	1	—	3	—
Fibres[1]	—	—	—	1
Foods (known to have been used for food or edible)	13	13	—	—
Medicinals	7	2	3	3
Narcotics, hallucinogens and intoxicants[2]	—	2	—	—
Poisonous plants	—	1	—	—
Total number of taxa included with known past or present ethnobotanical uses	**462**	**3**	**25**	**490**

1. For ropes, stuffing, threads, string, textiles, basketry, etc., excluding for thatch.
2. In natural unfermented state.
Note: The percentages of species of present and former ethnobotanical use in the native flora of Ireland for algae, fungi, lichens and bryophytes (mosses and liverworts) have not been calculated.

The loss of traditional knowledge about plants in Ireland is well illustrated today by the relatively small number of words that survive in Irish that are associated with woods and woodland crafts. Most Irish language dictionaries contain a wealth of vocabulary about the use of peat, turf for fuel and its harvesting, bogs, and bog timber and its extraction. Many fewer words are found about woodland management, such as coppicing, felling timber, woodworking tools, charcoal and charcoal making. I imagine that this is, in part, due to the disappearance of such words and associated knowledge as the bulk of Ireland's great woodlands were cleared from the earliest times up until the 1600s. However, it must be acknowledged that, in part, this may also be because the Irish language survived best in regions of Ireland (the north-west, west and south-west in particular) where woodlands were more scarce and where the local economy had less reliance on growing trees than for peatland products (including bogwood).

As so much of this traditional knowledge on plants and their uses was handed down from generation to generation without ever being written down, many pieces of information are missing today. Indeed, as Edmund Lenihan has described, putting together an account of Ireland's ethnobotany is like completing a jigsaw puzzle composed of tens of thousands of pieces. Describing efforts to uncover the story and folklore of Biddy Early, the so-called wise woman of Co. Clare, Lenihan wrote, 'Now one has a sense of collecting the fragments of what was once vital, pieces that no longer make up a complete picture. One is fishing in the puddles of an all but dried-up channel, the river having long flowed in another direction' (1987: 11). Nevertheless, like an incomplete jigsaw puzzle, it is possible to gain an impression of the picture, which I hope that I have achieved in this book. I have no doubt that over years to come others will fill in the gaps I have left and correct the errors or misinterpretations I have made. If my book stimulates more interest and research in this subject, then I will have succeeded in what I set out to achieve.

Fifty years ago Irish folklorist Kevin Danaher also lamented the loss of traditional knowledge of plants in Ireland when he wrote, 'many … plants had their uses, as food or drink, as medicines, as material for this and that. Nowadays all such things come from the shops, but rural life is so much less interesting for the loss of the old knowledge of plants and their uses' (1964b: 78). The trend that Danaher identified and referred to has continued to the present day to such an extent that now most young people have very little knowledge about or interest in wild plants and their past or present uses, and traditional knowledge continues to decline.

E. Estyn Evans (1967 [first published in 1942]) wrote that 'The Ireland I am writing about is one which is passing away, and some of the customs and tools described … have almost entirely gone'. Nevertheless, with some optimism, he added, 'but I have learnt that it is rarely safe to use the past tense in writing about Irish matters'. Today, we may note with regret how little traditional knowledge of the uses of plants seems to survive in the memories of our population in Ireland. Despite this, I am occasionally surprised and delighted even today to find traces of living practices or memories handed down through an oral tradition, sometimes with people who are completely unaware how rare, valuable or unique their knowledge is. Ireland can be grateful for the efforts made to document and record Irish folklore and folklife before it disappeared. Most of these efforts were made by the State and by a small group of visionary people, particularly through the work of the Irish Folklore Commission, founded in 1935. So much has been recorded of the customs, music, traditions and stories of ancient Ireland, and the collection is hugely important to the national heritage. Less successful have been efforts to conserve traditional practices and crafts. We may know, for example, that many particular plants were used for a particular purpose. Nevertheless, in many cases, we really don't know with any certainly *how* they were collected, prepared, processed or applied. Much research needs to be undertaken to fill in the gaps in our remaining knowledge before we can be satisfied that we have completed a comprehensive Irish ethnobotany. In recent years there has been a significant resurgence in interest in country crafts. This is encouraging, and it may result in a rediscovery of traditional knowledge as people experiment to see how particular tasks were performed in former times and how historic artefacts were constructed. An excellent example is the construction of country furniture, such as greenwood chairs, which is becoming a popular craft in Ireland, and is championed by such craftspeople as Alison Ospina of West Cork (Ospina, 2009).

Many of the plant resources used in the past are now close to extinction, or at the very least, no longer available in the quantity that would support any exploitation. Plants will be lost—not only plants that once were important, but also plants that may be important in the future—as natural habitats are damaged or lost through unsustainable development, agricultural intensification, urbanization, pollution and many other human-induced factors. Threats to Irish plant species from climate change have also been documented (Wyse Jackson, 2007). Over the coming decades many of the plants listed as native in this book may disappear, their legacy becoming but a footnote in history, discarded artefacts of times past. I, for one, believe that this would be a great tragedy. I hope that the seeming inevitability of this loss can be avoided and that the contributions of plants will be recognised as just one reason why we need to make great efforts to conserve them and our natural environment for the future.

Urtica dioica (nettle). Watercolour by Lydia Shackleton. Courtesy of the National Botanic Gardens, Glasnevin.

HISTORICAL PLANT USE IN IRELAND

The tide gone out for good,
Thirty-one words for seaweed
Whiten on the foreshore.

"Death of Irish", Aidan Carl Mathews (1956–)

I HAVE FOUND it difficult and in many cases impossible to come up with a historical timeline for the ethnobotanical usage of particular plants in Ireland. Plant usage most likely evolved over time and changed significantly. I have therefore not tried to outline the different historic periods in which plants were used, other than in very general terms.

In ancient Ireland, in the pre-history before any written records were kept, the first settlers would have used plants as fundamental resources to support their lives—for food, shelter, medicines, fibres, weapons and many other purposes. The first arrivals would have found a rich nature, abundant with wild resources for life—woodlands full of game; rivers plenty with fish; seashores brimming with shellfish; timber for tools, weapons, shelter and boats; and plants for food and medicines. These people were hunter–gatherers subsisting on the plants and animals they could catch or collect. Not all writers have imagined these early inhabitants had an easy or comfortable lifestyle, though. Wood (1821: 145), for example, wrote:

> [The first Irish] differed but little from that of the wild animals which they chased. Clad in skins they lived in a filthy state of poverty, devoid of houses, horses, and even of every kind of arms except arrows, tipped with bone. They lived upon herbs and the produce of the chase, in which they were accompanied by their wives, as they had no fixed place of abode. They slept on the ground, and the only protection from old age and infancy, from the attack of wild beasts, and the inclemency of the weather, was a sort of wattled work made of interwoven branches. . . . In Ireland their huts or tents must have been of a very temporary nature, as they were erected in the evening, and after they had dined. The materials of their bedding were the branches of trees, placed next the ground, upon which they laid a quantity of dry moss, and upon the top of all a bundle of green rushes.

We cannot say whether Wood's description is accurate, but I suspect that the miserable existence described in this passage is probably far from the truth. Although people did not generally live long and only a few survived into old age, the richness of Ireland's nature would probably have provided them with a comfortable and relatively prosperous life with shelter from the cold, wind, and rain and enough food to eat, at least throughout much of the year.

As agriculture spread into Ireland and the hunter–gatherer lifestyle faded, wild plants would have remained and continued to be important for food, particularly in times of famine and shortage and to supplement food obtained through agriculture. However, by Neolithic times, about 4000 BC, the majority of food consumed in Ireland was probably derived from agriculture, in particular, cultivated cereal crops and cattle. Kelly (1997) observed that Old Irish texts from the 7th and 8th centuries make it clear that while most food came from agriculture, there was still significant wild-source food, with mentions of deer-hunting, fishing, and the collecting of nuts, wild fruits, herbs and seaweed.

Attitudes towards wild plants as food apparently shifted mid-19th century after the Irish Potato Famine. The shift in attitude was possibly due, in part, because foraging became associated with poverty. Danaher (1964b: 75) says very few wild plants were considered as food:

> Once in a while a person might pick up a piece of wood sorrel or water-cress and chew it, but they never appeared on the table. Some of the older people kept up the custom of taking a dish of boiled nettles three times in the spring when the nettles were young and green; they had quite a passable taste, rather like spinach, but it was for health reasons, as a renowned blood purifier, that the old people took them. They knew, too, that the *praiseach bhuí*, the charlock, is edible and told us how the poor people gathered it in the bad times and boiled it for food. The wild fruits were, of course, eaten. The favourite was the *fraochán*, the blue whortleberry that grew in profusion in the bogs, but the blackberry, the crab apple, the sloe and the hazelnut were gathered and eaten, although there were doubts about the elderberry and nobody dreamed of tasting rowan berries. And of all the fungus growths, only the mushroom was eaten.

Despite Danaher's reservations (which in any case probably refer to the period in Ireland after the potato famine), I think that there can be little doubt that in earliest times in Ireland wild plants were extensively used for food and only came to be significantly replaced as the dominant non-meat food source when Ireland ceased to have a hunter–gatherer society and moved towards a society dominated by agriculture.

We know little about the plants that were used by Ireland's early settlers or for what purposes they were used. The first hunter–gatherers in Ireland clearly used plants for food, but other than a handful of plants for which there is archaeological evidence of their use, we can do little more today than speculate as to the species that were important. However, it is reasonable to suggest that some of the ethnobotanical uses of native Irish plants developed in such times and continued and evolved until the present day. For example, undoubtedly much of the rich folklore on the use of timbers and woodcraft has disappeared with the clearance of the native woods. By way of an example, Danaher (1978) pointed out that there is archaeological and written evidence of numerous examples of timber used in frame construction of buildings in ancient and medieval times, but subsequently these disappeared more or less entirely, no doubt largely due to the destruction of forests in the 16th and 17th centuries. The surviving examples of vernacular architecture are generally stone-built houses that post-date the loss of forests. Folklore related to such building traditions in Ireland has now disappeared. But folklore related to use of, for example, bog timber has survived because it is more recent, as people may have increasingly switched to using bog timber when trees became scarce.

I have no doubt that the enormous social disruption and change in Ireland during the 18th and 19th centuries radically and permanently altered traditional folk knowledge of plants and disrupted the long-established ways that such knowledge was passed down through generations. So much of this change can be attributed to the potato, both its intensive cultivation as a monoculture and its failure that so tragically led to the Great Famine during the 1840s. In about 1780, the Irish population was ca. 3.5 million and began to grow rapidly. Population increase between 1779 and 1841 has been estimated at 172% (Woodham-Smith, 1962). Following this growth, the huge crash in population caused by the potato famine from 1845 to 1852 took with it much of the knowledge of

On the Blasket Islands. Courtesy Ionad an Bhlascaoid Mhóir/The Blasket Centre.

the 'old ways' in Ireland. During this period approximately one million people died and a further million people left Ireland. By 1860, the population was under 6 million.

It is also very possible, too, that after the famine passed, people were reluctant to continue harvesting the wild foods they had turned to when their crops failed. Usage of wild plants, such as charlock (*Sinapis arvensis*), sorrel (*Rumex acetosa*), nettles (*Urtica dioica*) and most hedgerow fruits, would have reminded them of poverty, hardship and starvation. Even during the early part of the 20th century, hedgerow produce was generally little regarded as a valuable resource for home consumption, despite the widespread and free availability and abundance of fruits, such as rosehips, sloes, wild cherries, rowan berries and haws. In a leaflet detailing home preservation of fruit (Department of Agriculture, 1933), the ways to preserve a wide variety of fruits are described, including apples, pears, plums, raspberries, gooseberries, rhubarb, loganberries, apricots, damsons, cherries, and currants and all garden fruit crops. The only wild fruit listed was the blackberry.

Evans (1957: 11) has described the donkey and goat as 'symbols of poverty' that were neglected and despised in many parts of Ireland after the famine. To such symbols, I am sure various wild plant foods could be added that were similarly despised. Up until quite recently, even people foraging on the seashore for seaweeds and shellfish would have been watched with sympathy as people would imagine that they must be going through hard times. After the famine years the majority of new generations in Ireland would quite simply have forgotten how plants were used by the generations that came before them.

One group of people who remained intimately connected with the Irish countryside, both after the Great Famine and to some extent even up to the present day, are the Irish Travellers. Irish Travellers[1] are a traditionally nomadic people of ethnic Irish origin, who maintain a separate language (Shelta) and set of traditions. They live predominantly in Ireland and the United Kingdom, although some Traveller communities can also be found in the United States. In 2009 a valuable and interesting book of Irish Travellers'

1. In Irish, *an lucht siúil*; in Shelta, *pavee.*

recipes using traditional ingredients was published. The book described Travellers' food in rural Ireland (Virmond, 2009). Virmond wrote that one of the factors that made Travellers' food so exciting was that they could find their ingredients along the sides of the roads during their travels. When I obtained the book I expected that it would be rich in lore about edible wild plants because of their nomadic lifestyle and long oral tradition. Although wild animals such as hares, pheasants, eels and salmon were mentioned as having been important for Travellers' food, only blackberries were included amongst the wild plants. Potatoes, carrots, turnips, parsnips, cabbage, onions, apples and rhubarb were listed as some of the commonest plant ingredients used, all of them cultivated plants. In the same book in a glossary of food-related words given in Traveller cant, or Shelta, no words are given for wild plant species used for food (although the following Shelta words for various food and other plants are given: *groola*—'apple'; *gritchoon*—'onion'; *kawb*—'cabbage'; *kogi*—'turnip' and *shkrawkh* or *skrakho*—'tree'). This suggests to me that a presumed ancient traditional use of wild plants for food amongst the Travellers' community has either disappeared or is buried deep and lost from everyday practice and knowledge. In another important work on languages in Ireland, by R. A. Stewart Macalister (1937), no words associated with wild plant foods are listed in Shelta, or in another Irish dialect, Bēarlagair na Sāer, an old vernacular language used by the Freemasons.

The use of plants in traditional medicine would have declined as well, as modern pharmaceutical medicine developed in the 20th century. Traditional medicine in Ireland largely relied on an unbroken oral tradition of healers passing on their skills and knowledge to the next generation. An interesting folktale told by a Dingle fisherman to John Millington Synge, whose *Travels in Wicklow, West Kerry and Connemara* was first published in 1911, details rural life in Ireland at the turn of the 20th century. Clearly, local rural people were already aware of and lamented the loss of traditional knowledge of plants in Ireland. Recounting the conversation, Synge says, 'We met Maurice, the fisherman I have spoken of, and we sat down under a hedge to shelter from a shower. We began to talk of fevers and sicknesses and doctors—these little villages are often infested with typhus—and Maurice spoke about the traditional cures':

> There is a plant, he said, which is the richest that is growing out of the ground, and in the old times the women used to be giving it to their children till they'd be growing up seven feet maybe in height. Then the priests and doctors began taking everything to themselves and destroyed the old knowledge, and that is a poor thing; for you know well it was the Holy Mother of God who cured her own Son with plants the like of that, and said that after that no mother should be without a plant for ever to cure her child. Then she threw out the seeds of it over the whole world, so that it's growing every place from that day to this. (Synge, 2009: 86)

In modern-day Ireland the connection between Ireland's people and its wild plants has grown even more distant. For most children food is something that comes from a supermarket, and this is not only for children living in urban situations in Ireland's towns and cities. Growing vegetables and fruits in gardens and orchards went through a massive decline during the latter part of the 20th century. Explaining this trend is neither straightforward nor easy. Undoubtedly the ready availability of food in shops and supermarkets has made it easier for people to obtain all the food they need, without having to go to the bother of sowing, tending, storing and harvesting their own produce. It also made the availability of many foods much less seasonally dependent, through imports and new varieties.

The trend towards urbanization in Ireland certainly contributed to the decline in home production of food, as did the greater affluence and increase in the disposable income of Irish citizens. Subsistence agriculture and small-scale farming was never an easy life in Ireland, and it is not surprising that when new opportunities presented themselves through education and in the growing industrial, public and commercial sectors that people should embrace these options. O'Súilleabháin wrote in his 1942 *Handbook of Irish Folklore* that 'most country people have some knowledge of plant lore, and in each district there is (or was) a person with a special store of it' (1942: 283). Sadly, this is probably no longer the case.

There are hopeful signs, though, of a revival in interest in wild plants as a source of nutrition. For example, *The Sunday Times* (6 March 2011) recently included an advertisement from an organic restaurant in Co. Wicklow looking to hire a 'forager' of wild foods, with the ideal person possessing an understanding of wild herbs and leaves, wild game and fish, fungi, nuts and berries. Provided such an interest does not result in the loss of species through over-harvesting, connecting people once again to the wild plant resources that grow around them can only be a good thing. For centuries, everyday life and home industries have depended upon wood and other plant products, but the craft and traditional knowledge involved are almost entirely gone. Evans in his book *Irish Folk Ways* (1957: 200) observed the essential relationship between knowledge of natural materials and the craft of making everyday objects:

> Most of the home industries depended on wood and other plant products, natural or cultivated, and on the skin and hair of animals . . . . Nearly all these crafts are now dead or dying and their products perishable by nature, are either gone or fast disappearing. A few skills live on, others survive in children's pastimes or in the memories of old people. The unwritten lore of the countryside included a vast store of information on all the varied trees, plants and animals, birds and fishes which in times past played their part in the lives of people. Nearly every species had its practical, medicinal or magical uses, and through careful observation, empirical knowledge and a wealth of poetic fancy, a vivid interest in the world of nature was maintained and passed on from one generation to another. Much of this lore has been lost with the decline of the Gaelic tongue and the spread of formal book-learning. The country school must conform to a clerk-breeding curriculum designed to meet urban standards. The destruction of woodland and the debasement of standards of living which followed the adoption of the alien potato and the consequent famines had serious consequences on home crafts.

Ethnobotanical knowledge and plant lore are an important part of Ireland's history and culture. But beyond the cultural importance, information on former uses of plants often points the way towards finding new uses in the future, for medicine, food and many other purposes. Our native flora is a vital resource for the present day and for the future. We are too well aware that dozens of native species are being lost in Ireland. Native plants are increasingly endangered or driven to extinction as our impact on the landscape becomes more and more significant and wild habitats are destroyed or damaged. In addition, human-driven climate change is altering the environment to such an extent that plants that have thrived in Ireland for more than 10,000 years will no longer be able to survive. I hope that in a small way this book will be a useful reference, not only to document the fascinating and rich role that plants have played in human history in Ireland, but also to stimulate the interest of people today in wild plants and Irish ethnobotany. Wild plants are a resource that should be cherished and nurtured for the future, and traditional knowledge, which is a vital part of that resource, should also be documented and safeguarded.

Historical Sources of Information on Plant Use

In compiling this work, I have based the information included, by and large, from published sources, supplemented with my own personal experiences, as well as stories, information and anecdotes relayed to me by various people around Ireland, to all of whom I am extremely grateful. I have had neither the time nor the opportunity to delve deeply into the remarkable manuscript resources that exist related to folklore and Irish folk tradition in Ireland, much of it held by the School of Irish, Celtic Studies, Irish Folklore and Linguistics in University College Dublin (UCD), or into many of the early manuscripts or other publications written in the Irish language that are sadly inaccessible to me. Such tasks are for others who will follow me in the field of Irish ethnobotany, mining these potentially rich resources of information on the uses of plants in Ireland.

One of the richest sources of manuscript information on the usage of plants in Ireland during the last century is the remarkable Schools' Folklore Scheme, now known as the Schools' Manuscript Collection, which ran from July 1937 until December 1938. This scheme was operated by National Schools in the Republic of Ireland where children throughout the 26 counties acted as folklore collectors, gathering information and stories from their parents, grandparents and neighbours. The results of the scheme included about half a million pages of manuscript, which are now bound into 1126 volumes housed in the library of the National Folklore Collection at UCD. In the preparation of this book I have not researched these manuscript sources, although information from them has been included when I have come across it published or referred to by others. Most notable amongst previous researchers who have used the results of the Schools' Manuscript Collection are Allen and Hatfield (2004) for their work on medicinal plants in folk tradition in Britain and Ireland. Much of the work undertaken for their book on plants in Ireland was carried out by Mrs Sylvia Reynolds who sifted through tens of thousands of pages extracting information on the medicinal uses of Irish plant species, much of which I have drawn on in summary from their publication, and which I am very pleased to acknowledge.

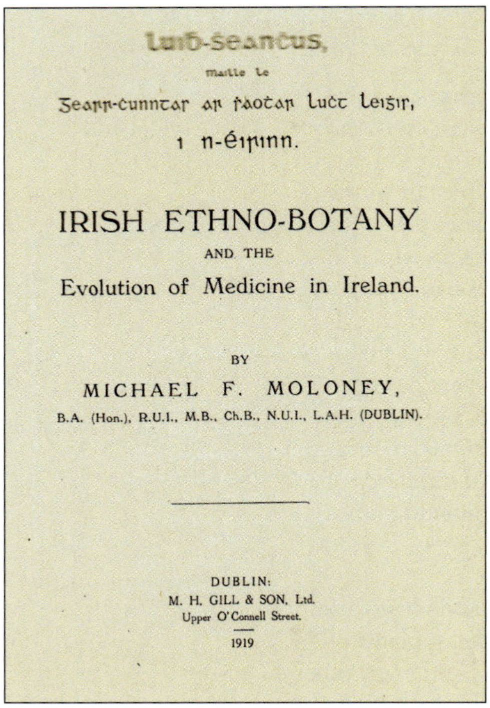
Luiḃ-Seanċus,
maille le
Gearr-ċunntas ar ṡaoṫar Luċt Leiġir,
i n-Éirinn.

IRISH ETHNO-BOTANY
AND THE
Evolution of Medicine in Ireland.

BY
MICHAEL F. MOLONEY,
B.A. (Hon.), R.U.I., M.B., Ch.B., N.U.I., L.A.H. (DUBLIN).

DUBLIN:
M. H. GILL & SON, Ltd.
Upper O'Connell Street.
1919

Moloney's *Irish Ethno-Botany* from 1919.

Another important source of information that I have drawn on is *Irish Ethno-botany and the Evolution of Medicine in Ireland* by Dr Michael F. Moloney, who published this slim book in 1919. Moloney was a medical doctor from Dungarvan, Co. Waterford, who had a great interest in Irish botany, the herbal use of plants for medicine and the Irish language. His book not only provides an extremely useful starting point for anyone interested in the use of wild plants for medicine in Ireland but also includes extensive lists of Irish plant names. Moloney lived in Youghal Road in Dungarvan where I am told he had a herb garden similar to a monastic garden at Clonmacnoise in the 9th century.

Amongst the published sources available on Irish ethnobotany, information is widely scattered and sometimes a single, almost throwaway, remark occurs about plant use. In some cases it is difficult to be certain what specific species is being referred to when a particular use is mentioned. For example, a wide range of *Juncus* species (rushes) grow in Ireland. There is little or no information available on whether a particular *Juncus* species is preferred for particular usage. Information on such 'folk taxonomy' is now unavailable, and one can only guess at which species might have been most appropriate for a particular use. I have also undertaken very little evaluation of the ethnobotanical information included in these published sources but have had to take much of it at face value. In only a few instances I have found ethnobotanical information that I regard as an error, perhaps due to the misapplication of a common name to a particular plant species. In these cases I have either indicated what I believe is a mistake, or omitted the wrong information entirely (such as many references to the use of particular medicinal plants in Ireland contained in an unreliable work by Jude, 1933). In most cases it has not been possible to verify the information given on folk uses of plants in Ireland, and I have had to presume that the previous authors have been correct both in their identification of plants and in their descriptions of usage. Much of what is reported by me is already from secondary sources, and it is always possible that previous authors have repeated or compounded earlier errors. In some cultures local people have been reluctant to share information on plant uses, in particular for medicinal purposes. Indeed, there are said to be instances where local people will provide information that is deliberately wrong or misleading simply to get rid of an annoying researcher. However, I know of no such cases in Ireland, and in general it seems that collectors of folklore have found those with whom they worked to be open and very helpful in passing on information.

I have tried to avoid including information in this book on the supposed magical properties of particular plants or the superstitions that surround them. Of course this has sometimes been difficult, particularly when there is such a rich folklore in Ireland about myths and legends, many of which have plants as central 'characters'. In some instances superstitions about use may be based on genuine reasons why a plant has not been or should not be used ethnobotanically because it is poisonous, for example, yew (*Taxus baccata*). In those cases it has been important to include myths, superstitions and magical stories in order to provide insight into a plant's folklore. Furthermore, it is difficult to ascertain whether the mythology surrounding a particular plant has been determined by its ethnobotanical use or is simply a result of other reasons.

Disclaimer and a Warning

This book is not meant to be a **manual** for the use of wild plants in Ireland, nor should it be used as a manual for edible plants or medical self-treatment. Many wild plants, fungi and the substances they contain are toxic and can cause illness, blindness or even death if used improperly. Any usage of wild plants, for whatever purpose, should be approached with both caution and common sense.

Some species of wild plants and fungi are scarce in Ireland, or even rare and endangered, and harvesting them may result in threatening their ultimate survival in the wild. Particular species are protected by law in Ireland and should not be gathered at all. When collecting wild plants for use, care should be taken not to unduly deplete wild populations. Remember to ask permission from the owners of property before collecting on their land.

Remember, also, that two plants may superficially appear to be the same but are actually different species, one of which may be good to eat and the other deadly poisonous. In some cases collecting plants in the wild should be avoided entirely because they may be toxic, poisonous, dangerous or potentially harmful. Some plants growing on toxic soils or in polluted places should be avoided as they may concentrate poisons in their tissues, taken up from the soil. Residues of herbicides and pesticides can remain in or on plants, too, so if you do use any wild crop, make sure you gather it from a 'clean' location where you know that it will not have been sprayed. Plants should be washed well before use wherever they are gathered. Moreover, some plants may cause an allergic reaction in particularly sensitive people, individuals or children, so test them with a small portion first.

Plants are not always easy to identify with confidence, so it is strongly recommended that you do not experiment with the use of a particular plant unless you are entirely confident of its identity. In using wild plants for food or medicine, it is best to follow the simple maxim—*if in doubt, leave alone and don't use!*

Plants Not Used Ethnobotanically

As the Systematic List of Wild Plants (see p. 151) indicates, I have been unable to find any information on the use of several hundred wild plant species in Ireland. Although I have probably failed to track down information on the use of some of them, or information on their use has not been recorded up to now, there is a clear set of categories of plants that seem to have been little used ethnobotanically in Ireland:

1. Rare plants—species that are now rare and were probably always rare in Ireland and plants occurring in inaccessible habitats.

   These include plants that are found only in small populations or in isolated localities, such as on high mountain or cliff locations. This group includes several species that belong to a category of plants referred to as Arctic–Alpine plants, some of which are geographically isolated remnants of when Ireland had a colder climate and significant ice cover during the Ice Age. Examples include several Saxifrage species such as *Saxifraga nivalis*, *Poa alpina*, *Polygonum viviparum* and *Silene acaulis*.

2. Plants that are poisonous, unpalatable or are lacking in known active medicinal compounds.

   Ironically, most of the notoriously poisonous plants in Ireland have been used medicinally at one time or the other and possibly used in such small quantities that they may have been regarded as 'safe'. This category also includes species that have no known active ingredients and can't be used medicinally or those plants that do not have significant wood, fibre or dye uses and cannot be eaten because they are poisonous, unpalatable, or simply look unappetizing.

3. Species which are difficult to distinguish taxonomically from each other.

   These include a wide range of species within the genus *Carex* (sedges) and *Potamogeton* (pondweeds), as well as some grasses. While some are widespread and sometimes even common or abundant, the difficulty of distinguishing them from each other at a superficial level has not helped to ensure that their individual usages have been well understood or documented.

4. Plants that have only been recently discovered in Ireland (most of these are rare).

   In these cases there has been insufficient time for the development of folk uses to have occurred. However, such species are rare anyway. Examples include species such as *Trifolium occidentale* and *Parapholis incurva*, which are rare species discovered in Co. Dublin during recent decades.

Plants in Place Names

A relatively small, but significant, range of plant species feature in Irish place names. The most frequently featured are the common tree species of alder, ash, elm, hazel, holly, oak and yew, as well as species that make up scrub thickets, such as hawthorn and blackthorn. In general, plants that appear in place names are ones that are either dominant in a particular habitat or particularly conspicuous. Examples might be a large yew tree on a hillside, a woodland coloured blue with bluebells in spring, or a large boulder covered with foxgloves. Plants grown as crops (such as flax, wheat, barley and oats) or wild plants collected for food (such as blackberries and gooseberries) also feature prominently. Probably the commonest plant that occurs in place names in Ireland is the oak (*doire*, an oak wood). Joyce (1870) recorded over 1300 names that begin with this word in various forms, and seemingly innumerable instances where the word is contained as a termination. These names present us with a surviving shadow of a time when Ireland was well wooded, rather than the image we have today of the few scraps of native oak woodland that remain.

The common English names of some plants are also used in some place names, such as Ashford in Co. Wicklow, or The Furry Glen (from furze, or gorse) in Phoenix Park in Co. Dublin. Common English names of plants, mainly trees, have been used in the names of roads, particularly in modern housing estates and house names, such as 'Oak Downs', 'Sycamore Road', 'Magnolia Terrace' and 'The Elms' (sometimes even in estates that are entirely devoid of any mature trees!). I have not attempted to provide information on such usage in Ireland, much of which is recent in origin. I have concentrated instead on providing details of the use of Irish plant names in local place names as this may give better insights into plants either formerly or currently found or prominent in those places or cultivated.

The most useful source of these names is the work of Patrick Weston Joyce (1827–1914), an Irish historian and writer, who particularly contributed to knowledge on Irish place names (see Plant Names in Irish Local Names). Of course not all of the interpretations of the meanings of these place names will be correct, but, nevertheless, it is clear that plants that were conspicuous or useful for various purposes often provided the names for places throughout Ireland, and many of them have survived up to modern times.

Corylus avellana (hazel). Watercolour by Lydia Shackleton. Courtesy of the National Botanic Gardens, Glasnevin.

NATURE'S MATERIALS: WILD PLANTS USED FOR CONSTRUCTION / HANDICRAFTS

Trees and Timber

At the end of the last Ice Age, about 13,000 years ago, Ireland's tundra-type vegetation began to form. By 12,000 to 11,000 years ago the dominant trees were Scots pine, birch, and hazel, which were gradually supplemented by other trees as the climate warmed and they migrated from the warmer lands further south. By about 8000 years ago oak and elm had established themselves as widespread species. Alder, yew and ash were also present (Mitchell & Ryan, 1997). The first human settlers arrived about 9000 years ago and were hunter–gatherers, living off the wild plants, animals and fish they found. Wild pigs, red deer, hares, salmon, trout, eels, pigeons, duck, geese and capercaillie[2] were all species that were hunted and eaten (Mitchell, 1986). Shellfish, including limpets, oysters, periwinkles and other species, were also gathered. Trees would have been important to them, for fuel, as a raw material for making tools (along with bone, antlers and flints) and for many other purposes.

About 6000 years ago, the first signs of agriculture in Ireland were emerging, characterised by an increase in the numbers of pollen grains of grasses, herbs and weed species found in sediments dating to that period (Mitchell & Ryan, 1997). Elms declined (probably due to the combination of clearance for agriculture and disease). From about 6000 years ago, there is evidence of increasing forest clearance associated with the arrival of the first Neolithic farmers. Woodland patches would have been cleared for cultivating crops and grazing stock. Timber remained a vital resource for these farmers, and ancient examples are known of houses constructed from timber as well as fences, palisades and trackways (Byrnes & Little, n.d., ca. 1990–2010). Domesticated animals had to be protected from predators, wolves, lynx and fox, and so stockades built from timber were constructed to protect animals and settlements. This general trend continued for many centuries.

While the development of early agriculture clearly resulted in the loss of significant tracts of ancient woodland, the main period of clearance was during the 16th and 17th centuries. Woodlands were exploited for charcoal production (for iron-ore smelting), ship building and barrel-making. A significant quantity of Irish timber was exported too. Agriculture, too, expanded into areas and onto land that had not previously been used for such purposes. A driving force for this would have been the four-fold increase in population in Ireland that occurred from 1700 to 1840, resulting in even more loss of woodlands. As a result most of the large areas of woodland that still survived in Ireland by 1600 were gone by 1800 (Neeson, 1991), and timber became an increasingly scarce and valuable resource for the Irish population.

Ireland has very little ancient woodland left that can be linked back to the prehistoric wildwoods. Perrin and Daly (2010) define 'ancient woodland' as woodlands with continuous historical cover since 1660, before the period when planting and afforestation became a common practice, that is, in about the mid-1600s. They identified in the Republic of Ireland 123 sites with possible or certainly 'ancient woodland', as well as a

2. Capercaillie is a bird of pine forests and is not now found in Ireland.

further 416 sites with long-established woodland (i.e., sites where there is no evidence available to determine whether they are ancient or not). Most of these sites are mere fragments of woodland, sometimes little more than a few hectares in size, and few are more than 100 ha. Rackham (1995) estimated that by the time of the Civil Survey in Ireland (1654–1656)[3] only 2.1% of Ireland was wooded. Furthermore, Rackham goes on to say these surviving woodlands had been reduced by as much as 90% by 1830–1844, when the 1st edition of the Ordinance Survey maps at 6 in. to the mile was completed. Therefore, by the mid-1800s only about 0.2% of Ireland was covered with ancient woodland. Rackham (1995) has used a wide range of evidence to determine whether woodland may be ancient or not, including pollen analysis, place names, historical documents, woodland structure, indicator species, archaeology and earthworks. Archaeological evidence of traditional management practices include coppiced and pollarded trees, wood-banks[4], charcoal hearths, limekilns, sawpits and quarries (Byrne, 2004), and structures associated with iron-ore smelting.

In Britain there is excellent documentation available on woodlands and their history, and on woodland management and crafts (e.g., Edlin, 1973 and Oliver Rackham's work [e.g., Rackham, 1976, 1985, 1995, 2003]). There is much less information available on the early usage of woodland products and on woodland management in Ireland. Nevertheless, Perrin and Daly (2010) noted that woodland archaeology in Ireland seems to indicate that Irish woodlands were managed similarly as those in Britain, with the majority of sites they studied containing earthworks and remnants of former coppice.

Coppicing is a woodland management system that maintains tree stock through regular cuttings to produce crops of long, straight rods, used for such purposes as fencing posts, rods for basketry and hurdles, hoops for barrels, spars for thatching and timber for

Rods harvested from coppiced hazel trees (*Corylus avellana*) in St John's Wood, Co. Roscommon.

3. The Civil Survey was established to profile the Irish lands that were due to be confiscated and redistributed to the English under the Cromwellian 'Act of Settlement', 1652 (Perrin & Daly, 2010).

4. Usually an earthen bank, sometimes with an associated ditch, often on the woodland boundary (Ardron et al., 2008). Banks surrounding woodlands would have been used to define ownership and to exclude stock and grazing animals who would otherwise browse the young coppice shoots.

charcoal production. Commonly coppiced species included hazel (*Corylus avellana*), ash (*Fraxinus excelsior*) and willows (*Salix* species). The length of time left between each coppice harvest depended on the size of rods required and the species used. Typically, 4–20 years is left between cutting a coppice. Pollarding is a similar process but involves cutting the trees at a height of 2–3 m, where the young shoots will be out of reach of grazing animals such as cattle, goats and deer. Pollarding is much used today to prevent street trees in towns and cities from becoming hazardous for pedestrians and cars and interfering with power lines and buildings.

In early Ireland trees were clearly regarded as important resources, and some species were afforded greater importance and status than others, depending on their relative value and uses. One example of this is the well-known Old Irish tree list, which dates back to early Christian times. It was included in the early 8th-century Irish law-text *Bretha Comaithchesa* and classified trees into four categories for protection: (1) the 'nobles of the wood' (*Airig Fedo*); (2) the 'commoners of the wood' (*Aithig Fedo*); (3) the 'lower divisions of the wood' (*Fodla Fedo*) and (4) the 'bushes of the wood' (*Losa Fedo*). Kelly (1997) provided a copy of this list in Irish and its translation into English in his book *Early Irish Farming*. There were penalties for people who felled or damaged trees from the different categories if they did so without permission. The penalties differed depending on the kind of tree.

There is a strong old Irish connection between the ancient Irish alphabet Ogham and Irish tree folklore. Ogham came into use in Ireland about the 4th century and consisted of a series of markings that survive today cut into the edges of standing stones. Ogham is read from the bottom up. Each letter consists of a series of markings and was assigned a name, often associated with an Irish tree. For example, the letter 'B' was *beith* (birch) and 'D' was *dair* (oak). Originally, probably only eight letters were named after trees (birch,

OLD IRISH TREE LIST

1. 'nobles of the wood' *(Airig Fedo)*
   - *daur* 'oak' (*Quercus robur, Quercus petraea*)
   - *coll* 'hazel' (*Corylus avellana*)
   - *cuileann* 'holly' (*Ilex aquifolium*)
   - *ibar* 'yew' (*Taxus baccata*)
   - *uinnius* 'ash' (*Fraxinus excelsior*)
   - *ochtach* 'Scots pine' (*Pinus sylvestris*)
   - *aball* 'wild apple' (*Malus sylvestris*)
2. 'commoners of the wood' *(Aithig Fedo)*
   - *fern* 'alder' (*Alnus glutinosa*)
   - *sail* 'willow, sally' (*Salix caprea, Salix cinerea* and other willows)
   - *scé* 'whitethorn' (*Crataegus monogyna*)
   - *cáerthann* 'rowan' (*Sorbus aucuparia*)
   - *beithe* 'birch' (*Betula pubescens, Betula pendula*)
   - *lem* 'elm' (*Ulmus glabra*)
   - *idath* 'wild cherry' (?) (*Prunus avium*)
3. 'lower divisions of the wood' *(Fodla Fedo)*
   - *draigen* 'blackthorn' (*Prunus spinosa*)
   - *trom* 'elder' (*Sambucus nigra*)
   - *féorus* 'spindle-tree' (*Euonymus europaeus*)
   - *findcholl* 'whitebeam' (*Sorbus aria*)
   - *caithne* 'arbutus' (*Arbutus unedo*)
   - *crithach* 'aspen' (*Populus tremula*)
   - *crann fir* 'juniper' (?) (*Juniperus communis*)
4. 'bushes of the wood' *(Losa Fedo)*
   - *raith* 'bracken' (*Pteridium aquilinum*)
   - *rait* 'bog-myrtle' (*Myrica gale*)
   - *aiteann* 'furze, gorse, whin' (*Ulex europaeus, Ulex gallii*)
   - *dris* 'bramble' (*Rubus fruticosus* agg.)
   - *fróech* 'heather' (*Calluna vulgaris, Erica cinerea*, etc.)
   - *gilcach* 'broom' (*Cytisus scoparius*)
   - *spín*[1] 'wild rose' (?) (*Rosa canina*, etc.)

Source: Adapted from Kelly (1997) and MacCoitir (2003).

1. MacCoitir (2003) suggested that *spín* is 'gooseberry' (*Ribes uva-crispa*), but gooseberry is an introduced plant in Ireland. There is no evidence of such an early introduction of this species. Kelly (1997) suggested 'wild rose'.

THE OLD IRISH TREE ALPHABET

I	Yew	*Iodha*
E	Aspen	*Eadha*
U	Heather	*Ura*
O	Gorse	*Onn*
A	Scots pine	*Ailm*
R	Elder	*Ruis*
Z	Blackthorn	*Straif*
Ng	Reed	*Ngetal*
G	Ivy	*Gort*
M	Bramble	*Muin*
Q	Apple	*Quert*
C	Hazel	*Coll*
T	Holly	*Tinne*
D	Oak	*Dair*
H	Hawthorn	*Huath*
N	Ash	*Nin* (or *Nuin*[1])
S	Willow	*Saille*
F	Alder	*Fearn*
L	Rowan	*Luis*
B	Birch	*Beith*

Source: Coll (2000).
1. Uí Chanainn (1962).

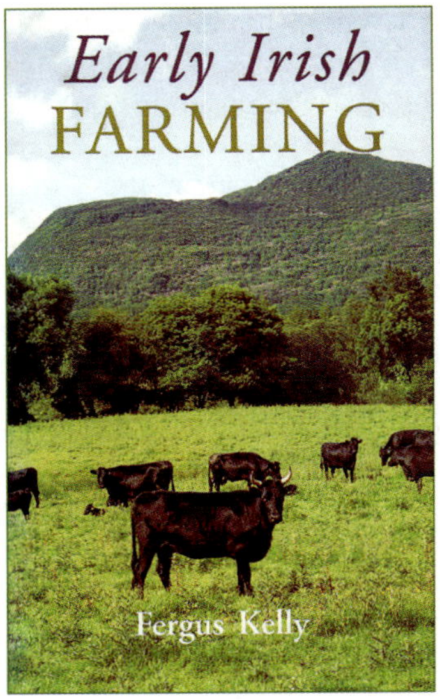

Fergus Kelly's *Early Irish Farming*. Reprinted with permission. The Governing Board of the School of Celtic Studies of the Dublin Institute for Advanced Studies.

alder, willow, oak, hazel, ash, yew and perhaps pine). In the Middle Ages other tree names were added to the remaining letters, resulting in a full tree alphabet (Coll, 2000). Both the old Irish name and its equivalent in the modern alphabet are given here. Today, there are under 500 known surviving examples of Ogham writing in Ireland and western Britain. Most are personal names, perhaps marking ownership or boundaries of lands.

The source of a modern Irish tree alphabet given here is as provided in Rev Patrick S. Dinneen's *Foclóir Gaedhilge agus Béarla: An Irish–English Dictionary* (1927). In the dictionary it is worth noting that Dinneen does not list a 10th letter and gave H as an auxiliary letter. Is this an error[5]? It is also notable that the 11th and 14th letters, vine and dwarf elder, are plants that are not native to Ireland. Perhaps 'vine' refers not to the grape vine but instead to another vine, such as the common, wild Irish woody climbing shrub, woodbine or honeysuckle (*Lonicera periclymenum*).

Trees were indeed one of Ireland's greatest natural resources from ancient times, and they provided shelter, timber for building houses and cabins, for furniture (chairs, tables, settlebeds, dressers, storage presses for flour and meal, barrels, churns, noggins, kitchen implements, etc.), rafters, floors, doors and window frames. Down through the ages in Ireland, craftsmen, including carpenters, coopers and turners, made many of the objects from wood that were an essential part of everyday life: boats, carts, ploughs, barrows and handles of implements such as spades, slauns, flails, scythes, walking sticks and much more. The properties of specific timbers were important, too, and so were used for specific purposes when water-resistant, flexible, tough, long-lasting or soft timbers were needed. As the basis of the original main natural habitats in Ireland, woodland trees also provided a home for a diversity of plants and animals that provided the natural resources

5. Uí Chanainn (1962) listed a 10th letter (M), ending on the 17th letter (U).

An Irish home with a wooden spinning wheel. Reprinted with permission from Mason, Thomas H. 1936. *The Islands of Ireland.* Batsford, London.

needed for food, medicines and many other purposes to sustain human life in Ireland, particularly before the development of agriculture.

Timber for furniture was and remains one of the most important uses for native timbers. Even today, our homes are generally furnished with wooden furniture. Different native trees have timber with differing properties, and through traditional usage the timber best suited for a variety of specific purposes was widely known. For example, woods that don't rot quickly when wet, such as alder and elm, were used for purposes such as making water pipes, clogs, bridge piles, etc. Fine-grained timbers needed for making household utensils and tools were chosen from such species as sycamore and ash. Long-lasting hard timbers, such as oak, were chosen for purposes such as making furniture, where the timber needs to be strong. Species where the timber can be polished to yield attractive colours and finishes, such as arbutus, yew and holly, were also chosen for use in making furniture (Table 3). Dinneen (1927) gave the Irish name *radhbhóg* for an unidentified wood or tree that was suitable for making spinning-wheel parts, suggesting that specific timbers had well-known and well-defined uses.

In the use of timber for furniture in Ireland, though, one should remember that houses in ancient and historic Ireland were generally very sparsely furnished, right up to the 20th century. For example, Woodham-Smith (1962) mentioned that furniture

MODERN IRISH TREE ALPHABET

A *Ailm*—the 1st letter: Elm tree (*Ulmus glabra*)
B *Beith*—the 2nd letter: Birch tree (*Betula* species)
C *Coll*—the 3rd letter: Hazel tree (*Corylus avellana*)
D *Dair*—the 4th letter: Oak tree (*Quercus* species)
E *Eadhadh*, or *Eabhadh*—the 5th letter: Aspen tree (*Populus tremula*)
F *Fearn*—the 6th letter: Alder tree (*Alnus glutinosa*)
G *Gath*, or *Gort*—the 7th letter: Ivy (*Hedera helix*)
H *Uath*—auxiliary letter: Whitethorn tree (or Hawthorn) (*Crataegus monogyna*)
I *Íodha*—the 8th letter: Ivy (*Hedera helix*), and/or Yew tree (*Taxus baccata*)
L *Luis*—the 9th letter: Quicken tree (or Mountain ash, or Rowan) (*Sorbus aucuparia*)
M *Muin*—the 11th letter: Vine (? *Vitis vinifera*[1]).
N *Nuim*—the 12th letter: Ash tree (*Fraxinus excelsior*)
O *Oir*—the 13th letter: Broom (*Cytisus scoparius*)
P *Peith*—the 14th letter: Dwarf elder tree (*Sambucus ebulus*)[2]
R *Ruis*—the 15th letter: Elder tree (*Sambucus nigra*)
S *Sail*—the 16th letter: Willow tree (*Salix* species)
T *Teithne*—the 17th letter: Furze (or Gorse) (*Ulex* species)
U *Ur*, or *Oir*—the 18th letter: Heath (or Heather) (*Erica* species, or *Calluna vulgaris*)

Source: Adapted from Dinneen (1927).

1. The question mark is added because *Vitis vinifera* is not a native plant and it is perhaps unlikely that this is the vine intended in this alphabet. The native honeysuckle or woodbine (*Lonicera periclymenum*) is perhaps more likely.

2. *Sambucus ebulus* is not native to Ireland but has probably occurred in Ireland for many centuries, either cultivated or established in the wild.

Table 3. Irish timbers: colour, characteristics and uses.

Scientific name	Common name	Wood colour*	Wood textures and features*	Uses	Availability from home-grown sources
Native trees					
Alnus glutinosa	Alder	Pale pinkish brown, darkens with age to light reddish brown.	Indistinct grain. Closely resembles birch in grain and texture but is lighter weight. Broad rays appear as dark lines or streaks on longitudinal surfaces. Little distinction between sapwood and heartwood. The wood is light and not very durable, although it is good for use in damp places where it is likely to stay wet.	A general purpose wood used for furniture, plywood, fencing stakes and craft purposes.	Occasionally, in small quantities
Betula species	Birch	Pale or light brown, often with dark streaks or flecks.	High-density wood but not very durable.	Veneers and plywood, floors, furniture, turnery and kitchen utensils.	Rarely
Fraxinus excelsior	Ash	White to light brown in colour and usually little difference between sapwood and heartwood. Heartwood can be dark brown.	Growth rings are distinct with a bank of obvious pores. Typically straight grained. Tough and flexible but perishable.	Furniture, including bentwood furniture, sports goods such as hurleys, ladders, boat building, tool handles, plywood, floors and decorative veneers for panelling.	Quite rare with limited supply
Ilex aquifolium	Holly	Greenish white or grey.	A dense and heavy wood with a fine, even texture, with no distinct heartwood. The grain can be irregular.	Mainly for small turner objects and inlays. It was sometimes stained black and used as a substitute for ebony.	Very rarely
Prunus avium	Wild cherry	Pale sapwood and pinkish brown heartwood.	Fine texture; annual growth rings indistinct.	Furniture, veneer, turnery, carving and decorative joinery.	Rarely

Table 3. Irish timbers: colour, characteristics and uses. *(continued)*

Scientific name	Common name	Wood colour*	Wood textures and features*	Uses	Availability from home-grown sources
Native trees *(cont'd)*					
Quercus species	Oak	Cream coloured sapwood; heartwood is yellowish brown to light tan to deep brown.	Obvious growth rings as alternating banks of large pored early wood and late dense wood. Grain is normally straight but can be irregular. Tough and hard timber.	Furniture, joinery, panels, building construction, floors, veneer, cladding, boat building, fences and railway sleepers.	Quite widely available
Taxus baccata	Yew	Heartwood varies from rich red to pale brown. Sapwood lighter in colour to almost yellow.	Can have attractive grain patterns. Durable wood that is very hard and heavy but very elastic. The growth rings are narrow but clearly defined.	Furniture, joinery, veneer, turnery, carving, tool handles. Formerly used for weapons and longbows.	Rarely
Ulmus species	Elm	The heartwood is beige-brown.	Coarse, textured timber with irregular growth rings. Not very durable.	Cabinet-making and furniture, including bentwood and the seats of Windsor chairs, turnery, boat-building, veneer.	Very rare now due to Dutch elm disease.
Non-native broadleaved trees					
Acer pseudoplatanus	Sycamore	Creamy-white when freshly sawn, which is retained if the wood is dried quickly, otherwise it can darken to pinky brown. Heartwood varies from rich red to pale brown.	Moderately hard, with a fine, even texture.	Furniture, veneer, carving, fine domestic items such as kitchen utensils, tables, bread boards and rolling pins, violin backs, floors.	Rarely
Castanea sativa	Spanish chestnut	Similar to oak in colour.	Moderately coarse grained and textured, similar to oak but not as hard and therefore easier to work. Straight grained when young but gains a spiral twist when mature. Durable.	Floors and furniture, veneer, coffins, fence posts, poles and stakes.	Rarely

Table 3. Irish timbers: colour, characteristics and uses. *(continued)*

Scientific name	Common name	Wood colour*	Wood textures and features*	Uses	Availability from home-grown sources
Fagus sylvatica	Beech	Whitish or pale brown, darkening over time to pale reddish brown or deep reddish brown after steaming.	Straight grained with a fine even texture. Growth rings clear; rays showing up as small spindle-shaped marks or dark irregular flecks; perishable. No strong difference between sapwood and heartwood.	Furniture and cabinet-making, including bentwood furniture, interior joinery, construction purposes, floors, domestic implements such as tool handles, turnery, veneer and plywood.	Rarely
Juglans species	Walnut	Sapwood is whitish to yellow brown, and the heartwood varies from light greyish brown to deep chocolate brown or almost a black-brown.	Grain is slightly open, usually straight but can be wavy or irregular and so is famous for its wavy, curly or mottled figures, from burls, crotches and stumpwood. Texture coarse and uniform. Moderately durable.	Fine furniture and cabinet-making, gun stocks, musical instruments, carving, panels, plywood and decorative veneer.	Only very rarely, no commercial production in Ireland
Non-native conifer forestry trees					
Abies alba	Silver fir	Pale cream coloured wood. Very similar to *Picea abies* timber.	Straight grain and fine texture. Non-resinous. Not very durable. Can be knotty.	Joinery, construction of building, poles.	Not commonly available
Cedrus libani	Cedar of Lebanon	Light brown heartwood	Straight grained with a medium to fine texture. Durable.	Indoor and outdoor furniture. Building purposes.	Rarely available
Chamaecyparis lawsoniana	Lawson's cypress	Pale pinkish-brown heartwood and slightly lighter sapwood.	Fine, even-textured timber with a characteristic spicy fragrance. Typically non-resinous. Very durable.	Indoor and outdoor furniture, fences and poles. Building purposes.	Rarely available
Larix species	Larch	Sapwood is light coloured to pale yellow, and heartwood is pale reddish brown to brick red.	Sapwood and heartwood clearly differentiated. Clear growth rings and a straight grain, resinous and relatively durable and harder than many softwoods. Contains hard knots.	Joinery, including door and window frames, floor planks, boat building, pit props, poles, cladding, fencing and posts.	Quite widely available

Table 3. Irish timbers: colour, characteristics and uses. *(continued)*

Scientific name	Common name	Wood colour*	Wood textures and features*	Uses	Availability from home-grown sources
Non-native conifer forestry trees *(cont'd)*					
Picea abies	Norway spruce	Almost white in colour to pale yellowish brown, little if any difference between sapwood and heartwood. Similar to *Pinus sylvestris*.	Straight grained and fine textured with clear growth rings. Non-durable.	Joinery, interior use in building construction, floors, plywood, pulpwood, veneer and packaging. Soundboard for musical instruments including violins, guitars and pianos.	Widely available
Picea sitchensis	Sitka spruce	Whitish or pale yellow with little distinction between heartwood and sapwood.	Straight even grained. Non-durable. Non-resinous. Relatively light and strong timber.	Interior use in building and joinery, musical instruments, oars, plywood, pallets and fibres for manufactured panel boards.	Widely available
Pinus contorta	Lodgepole pine	Pale yellow colour, sometimes with a brownish tinge; little difference between sapwood and heartwood.	Wood can have large knots. Grain is straight and wood is fine textured and fairly even. Growth rings are distinct. With darker latewood. The resin canals show as faint streaks on the side grain. Non-durable.	Joinery, building construction, fences, pallets, pulpwood and fibre wood for panel manufacture.	Widely available
Pinus sylvestris	Scots pine	Straw-coloured sapwood and pale reddish brown heartwood.	Clearly marked annual growth rings. Non-durable.	Furniture, joinery, interior construction and building purposes, plywood, telegraph poles, railway sleepers, turnery, piles and pit props, wood pulp.	Quite widely available
Pseudotsuga menziesii	Douglas fir	Sapwood whitish to pale yellow, or reddish white. Heartwood is variable but generally pale reddish brown. But can be yellowish to deep red.	Slow-growing material with a narrow band of sapwood. Growth rings are distinct. Uniform texture in slow-growing trees but uneven in fast-grown trees. Moderately durable.	Building construction, joinery, flooring, veneer, packaging, poles, cladding and plywood.	Quite widely available

Table 3. Irish timbers: colour, characteristics and uses. *(continued)*

Scientific name	Common name	Wood colour*	Wood textures and features*	Uses	Availability from home-grown sources
Thuja plicata	Western red cedar	Whitish sapwood and dark chocolate-brown to salmon-coloured heartwood.	Straight, even grained with a medium texture. Very durable. Non-resinous but aromatic.	Shingles and exterior boarding, cladding and decking, fences, greenhouses and sheds, doors and window frames, posts and poles, boat building, veneer and packaging, garden furniture.	Occasionally available
Tsuga heterophylla	Western hemlock	Whitish or pale yellow wood with little difference between heartwood and sapwood.	Straight, even grained. Non-durable and non-resinous.	Building construction, plywood, pallets and internal joinery.	Quite widely available

Sources: *Adapted from Anonymous (1941), Jackson and Day (1991) and Magner (2007).

Wine stoppers made from native Irish timber and turned by Kieran Reynolds. Top (L-R): Burr yew, ash, yew and oak. Bottom (L-R): bog oak, cherry, elm, Scots pine, and spalted birch.

was a luxury in 17th-century Ireland. She quoted a survey conducted in 1837 just before the famine which said that in Tullaghobegley, Co. Donegal, there were only 10 beds, 93 chairs and 243 stools for a population of about 9000 people. McGlinchey (2007: 16), whose autobiography *The Last of a Name* details life in Co. Donegal (1861–1954), mentioned that most homes in the 1800s were sparsely furnished:

> The carpenters made the chairs and tables and dressers and wooden beds. People made three-legged stools, too, out of bog fir, and a house with children would have a supply of creepies or low stools that anybody could make. A lot of the old people never used chairs or stools but would sit on a bag at the fireplace with their back to the hob or backstone.

The following page includes a list of 19 native trees and shrubs that have been used in Ireland for their timber as well as for making stakes.

The destruction of Ireland's native woodlands is outlined briefly later in this book, mainly under the entry for oaks (*Quercus* species). However, it is notable to record that following the large-scale clearance of woodlands in the Middle Ages and earlier, significant efforts were made by various governmental and other bodies and private individuals beginning particularly in the 1700s to plant trees in a landscape largely depleted of trees. These trees were primarily planted for forestry—the use of woodlands for amenity purposes is a 20th-century development. The area of woodlands planted up to 1841 is documented in a publication titled *Farming Since the Famine: Irish Farm Statistics 1847–1996*, prepared by the Irish

NATIVE TREES AND SHRUBS USED FOR TIMBER

Alnus glutinosa—Alder
Arbutus unedo—Strawberry-tree
Betula pendula—Silver birch
Betula pubescens—Downy birch
Corylus avellana—Hazel
Fraxinus excelsior—Ash
Ilex aquifolium—Holly
Populus tremula—Aspen
Quercus petraea—Sessile oak; Irish oak
Quercus robur—Pedunculate oak; English oak
Prunus avium—Wild cherry
Prunus padus—Bird cherry
Prunus spinosa—Sloe; Blackthorn
Salix aurita—Eared willow
Salix caprea—Goat willow
Salix cinerea—Sally; Rusty willow
Salix pentandra—Bay-leaved withy
Taxus baccata—Yew
Ulmus glabra—Wych elm; Irish elm

Ulmus species (elm) was used for cabinet-making and furniture.

Government's Central Statistics Office (Anonymous, 1997). It recorded that 345,604 acres (139,861 ha) of woodlands and tree plantations were planted prior to 1841 (see Tables 4–6).

The Irish Forestry & Timber Yearbook for 2007 states that 10.4% of the Republic of Ireland was covered by forest in 2006 (Anonymous, 2007), in total 717,300 ha. Of that, 397,700 ha are in state ownership and 319,600 ha are privately owned. However, the vast majority of this woodland is made up of non-native conifer plantations, and only 25% of the existing woodlands are broadleaf woodlands (i.e., <2.5%), mainly mixed woodlands. Of the remaining woodlands, little if any is pristine native woodland derived from the ancient woodland cover of the island that would have predominated before human colonization and settlement.

Coillte, the forestry industry body in the Republic, a semi-state company, owns more than 1 million acres* (= 404,685 ha) of land in Ireland, most of which is forested. It grows a range of conifer and broadleaved species for forestry purposes, the majority of which are non-native conifer species. Of these, by far the greatest acreage is planted with Sitka spruce (*Picea sitchensis*).

Apart from the conifers mentioned above, other commonly planted non-native conifers in Ireland include cedars (*Cedrus* species), monkey-puzzle[6] (*Araucaria araucana*), hemlock spruce (*Tsuga* species), Corsican pine (*Pinus nigra*), redwoods (*Sequoia sempervirens* and *Sequoiadendron giganteum*) and cypress (in Irish, *cufróg*) (*Cupressus* species and *Chamaecyparis* species). Most are grown for decorative and amenity purposes in parks, estates and larger gardens throughout Ireland and little used for timber.

6. Ó Dónaill (1977) gave the Irish word *arócar* for 'monkey-puzzle'. Hart (1898: 370) gave a common name for this tree recorded in Co. Donegal as 'crazzool', apparently a corruption of *Crassula*, from the resemblance in the leaf arrangement.

Quercus robur (oak). Watercolour by Lydia Shackleton. Courtesy of the National Botanic Gardens, Glasnevin.

Ireland has the lowest percentage of forest cover in Europe[7] (with the possible exception of Iceland), and until recently, good accurate information on the composition of the forests in the Republic of Ireland was inadequate. However, in recent years a National Forestry Inventory was undertaken by the Forest Service in the Department of Agriculture, Fisheries and Food to make more information available on the composition of forests, in relation to species, timber volumes, biodiversity and the status of woodlands. The Inventory is the first purely statistical approach to forest inventory undertaken in Ireland to provide an assessment of growing stock volume in both public and private national forest estates. The results of the Inventory were based on a detailed nationwide field survey that began in November 2004. The results of the Inventory confirm that the total forest area in Ireland now stands at 10% of the total land area, of which 57% is in public ownership and 43% in private ownership. Almost two-thirds of national forest estate has been planted in the past 20 years.

The Inventory also concluded that forests cover nearly 700,000 ha of the Republic (Table 4). Figures from 1995 to 1999 indicate that 85,000 ha of woodland exists in Northern Ireland (Table 5). In 2009, land under its control amounted to 75,500 ha, of which 61,200 ha was forested (Forest Service of Northern Ireland, 2009). Also see Table 6 with corresponding data for the Republic.

In 1891 a total of 1,351,775 trees were felled in Ireland, both for clearance and for thinning plantations (Table 7), which is fewer than the 4 million trees planted during the same year (Table 8). The majority of trees used for paling were larches, and a significant percentage of the trees felled during that period was for 'proppings'.

In western parts of Ireland, especially near the coasts and islands, trees and timber were often in short supply and local people made good use of any driftwood washed up on shore. Sometimes this was derived from the lost cargos of wrecked ships. Other times, it was logs and branches that had drifted sometimes for thousands of miles in the oceans. Milliken and Bridgewater (2004) mention that in excavations of Neolithic settlements at Skara Brae in Orkney, Scotland, charcoal fragments have been found of the non-native trees of spruce, larch and Weymouth pine (*Pinus strobus*), which can really only have been derived from driftwood. In some pieces, shipworm boreholes have been found, indicating that the wood must have been floating in the sea for many

Table 4. Republic of Ireland: Total stocked forest area by species type (conifer/broadleaved).

Species (conifer/broadleaved)	Area (ha)	Total area of Republic of Ireland
Conifer	462,580	
Broadleaved	151,950	
Temporarily unstocked	11,220	
Total	**625,750**	**69,000 km²**

Sources: Department of Agriculture, Fisheries and Food (2007).

Table 5. Northern Ireland: Total stocked forest area by species type (conifer/broadleaved).

Species (conifer/broadleaved)	Area (ha)	Total area of Northern Ireland
Conifer	66,000	
Broadleaved	19,000	
Total	**85,000**	**14,000 km²**

Sources: Forestry Commission, Forest Service and 1995–1999 National Inventory of Woodland and Trees.

Table 6. Total area of the Republic of Ireland with woodland and tree land-use types.

Land-use type	Area (ha)
Forest	697,730
Hedgerow	272,360
Other woodland	49,270
Individual trees	5610
Scrub	88,120

Sources: Department of Agriculture, Fisheries and Food (2007).

7. Ireland has 10.15% forest cover compared to a European average of over 30%. Government policy is to bring the national forest cover to 17% by 2030. Source: Teagasc, <http://teagasc.ie/forestry/technical_info/forestry_history.asp>, accessed 21 August 2011.

Table 7. Uses of felled trees in Ireland 1891–1917.

Uses	1891	1900	1910	1917
Propping	324,340	458,089	489,206	838,308
Sleepers	15,768	2715	31,935	54,885
Paling	88,028	29,277	30,298	93,046
Spools	8056	—	2054	790
Fuel	29,446	15,588	12,268	14,159
Furniture and bedding	8577	10,411	34,476	13,286
Carts and wagons	7702	7488	14,303	17,205
Clog soles	293	2170	8820	4001
Ship building	5245	—	3760	21,405
Telegraph/telephone poles	—	—	672	—
Boxes and packing cases	—	—	53,043	34,269
Other specified uses	24,609	56,880	1300	657,182
Unspecified	839,711	574,341	73,728	21,267
Total felled	**1,351,775**	**1,156,959**	**755,863**	**1,769,803**

Source: Adapted from Anonymous (1997).

Table 8. Number of trees planted by species and related area.

Species	1891	1900	1910	1917
Larch	692,111	583,685	1,287,584	644,701
Fir	306,281	225,467	372,352	503,496
Spruce	144,471	115,718	364,590	239,700
Pine	161,100	56,871	101,310	113,175
Oak	69,964	10,023	22,575	12,100
Ash	52,290	19,470	25,406	40,425
Beech	73,557	25,053	26,524	28,170
Sycamore	49,805	21,376	12,164	5290
Elm	18,596	1750	1084	7700
Other trees	145,022	22,397	13,304	33,255
Mixed trees	2,304,279	161,734	331,020	89,580
Number of trees	**4,017,476**	**1,243,544**	**2,557,913**	**1,717,592**
Area planted (acres)	**1886**	**629**	**1174**	**1134**
Area planted (hectares)	**763**	**255**	**475**	**459**

Source: Adapted from Anonymous (1997).

Timber being transported. A postcard from 1909 from Clonmel, Co. Tipperary.

AREA OF FORESTRY LAND OWNED AND PLANTED BY COILLTE IN THE REPUBLIC OF IRELAND

Sitka spruce (*Picea sitchensis*)[1]	231,395 ha
Norway spruce (*Picea abies*)	17,886 ha
Lodgepole pine (*Pinus contorta*)	51,015 ha
Douglas fir (*Pseudotsuga menziesii*)	8220 ha
Larches (Larix species)	13,306 ha
Scots pine (*Pinus sylvestris*)	4926 ha
Other conifers	4752 ha
Broadleaved species	18,852 ha
Total	**350,352**

Source: <http://www.coillte.ie>, November 2008 figures.
1. One hectare (ha) = 2.471 acres.

months. Weymouth pine is native to the eastern part of North America and is also called the eastern white pine. Accounts from people living on the islands of Ireland, such as the Blaskets in Co. Kerry, the Aran Islands in Co. Galway and Arranmore Island in Co. Donegal, mention the use of driftwood timber for construction purposes, making furniture, rafters and beams of houses, and so on.

Basketry

Baskets have been a fundamental and essential part of daily life in Ireland since the earliest times, used for constructing houses and shelters and for everyday items and containers. They were straightforward to make and could be constructed in a wide variety of shapes and sizes using materials that could be gathered freely from the wild from woods and hedgerows. While today most of us would struggle to make a basket from hazel, willows or rushes, there was a time in Ireland when most country people could have made simple baskets and other objects for use in their daily lives.

While writing this piece I looked around my home to survey what baskets I have in everyday use, most of which I realised that I take completely for granted. I was surprised to find that even in my relatively modern home I had more than 30 different baskets in use, of various shapes and sizes, and I prize the few that were made in Ireland from Irish-grown materials. Most of the baskets available in the shops in Ireland today are cheap imports made in China, the Philippines and other parts of south-east Asia (made from a wide variety of plant materials). Homemade Irish baskets are now few and far between. Indeed, basketry in Ireland has been transformed from being a widespread craft supporting everyday life to an art form, where the objects made are frequently expensive, albeit usually very well made, and often as not have become objects for display rather than for use.

A remarkable array of objects and articles have been made in Ireland from wickerwork or basketry (in Irish, *caoladóireacht* or *cliabhadóireacht*). These include baskets (in many shapes and sizes), potato skibs (in Irish, *ciseog*), bowls, coffins, hampers, turf creels, chimneys, panniers, cradles, bee hives, chairs, lobster pots, fish traps, eel traps, calf-muzzles, partitions, doors, cart-sides, palings, children's rattles and a wide variety of other household, farmyard and everyday objects. Dinneen (1927: 954) gave the Irish word *scalpán* as meaning 'a bundle of rods or sticks, used as a door for cattle sheds, etc.'

Some of the earliest known dwellings in Ireland were made from wickerwork. A ring of posts set in the ground, surrounded by a double wall of hazel wicker (wattle) stuffed with a variety of materials, such as heather or grass for insulation, were amongst the first houses. Other wattle houses were constructed with wicker plastered with mud or cow-dung ('wattle and daub'), sometimes mixed with a variety of plant fibres to help bind the mud together. Such methods of construction were common in Ireland right up to the 1700s (Hogan, 2001). According to Nelson and Walsh (1993), the medieval wattle and daub houses of Wood Quay in Dublin were constructed using laths of hazel (*Corylus avellana*) and ash (*Fraxinus excelsior*), and hazel would have been a common species used throughout Ireland for wattle and simple houses. The construction of houses using wattle and daub is described further later in this book. Wickerwork was also sometimes used on the roofs of houses, onto which the thatch was laid (*faoi chaol-ach an tí*, meaning 'under the roof of, within the house' in Irish and *cró caolaigh* is a 'wattle hut' (Ó Dónaill, 1977).

Basket-making in Ireland has had a long history. Ancient objects woven from plant materials have been discovered in archaeological sites, although baskets are generally not well preserved in bogs or other deposits so our record of the types and construction of baskets in Ireland long ago is not well known.

Lobster pots. Images courtesy Ionad an Bhlascaoid Mhóir/The Blasket Centre. (above left and middle)

Making a skib on the Aran Islands probably from willow rods. Reprinted with permission from Mason, Thomas H. 1936. *The Islands of Ireland.* Batsford, London. (above right)

Panniers, probably made from willow and hazel rods, used astride a donkey. (left)

During recent centuries itinerant basket-makers who would travel the land, stopping for a few days in each area to make or repair baskets to order were common. They would cut the materials they needed on the roadsides and hedgerows to make or repair any baskets required, such as creels, panniers, skibs and lobster pots. A wide variety of baskets were also sold at markets in the country towns.

We are fortunate to have Joe Hogan's important, authoritative and comprehensive book, *Basketmaking in Ireland* (2001). It describes not only the history, significance and variety of basketry in Ireland up to the present day, but also provides practical guidance on methods and techniques by which many traditional objects were and still can be made. Hogan documents this traditional craft, from the perspective of someone who has worked in basketry all of his professional life. I am fortunate to have met and watched him at work in his studio in Loch na Fooey, Co. Galway. I have in my own collection one of his beautifully and skilfully made baskets, which can be used and enjoyed not only for my own lifetime, but if treated with care, for generations to come. The Economic Botany Collection at the National Botanical Gardens, Glasnevin, includes examples of Joe Hogan's work too, such as a traditional potato skib and a lobster pot. The Economic Botany Collection also includes an interesting homemade tool from Connemara, used to strip the bark off willow rods as they were prepared for basketry.

A tool used to strip the bark from willow rods, which were then used for basketmaking. From the Economic Botany Collection at the National Botanic Gardens, Glasnevin.

A famous basket-making industry was established in Letterfrack, Co. Galway, during the late 1800s. In the museum collections of the National Botanic Gardens there is an example of a woman's bicycle basket made from willow labelled 'Connemara Basket Industry, Letterfrack, Co. Galway'. This is a product of the famous cottage industry started in 1888 by Miss Sophia (or Sophy) Sturge, a Quaker philanthropist from Birmingham who came to live in Connemara, who was horrified by the extent of poverty there and began a basket-making industry. Miss Sturge's basket-makers made baskets, hampers, bookstands, drawing-room kennels, doll's furniture, full-sized furniture (including wickerwork beds, chairs and tables) and other objects. The industry died out in the 1990s but was later revived by the Congested Districts Board. Miss Sturge settled in Letterfrack and brought a skilful French wickerworker with her to train the local people. She started working in the local court house, which she borrowed, and then moved on to a large iron building that later became a technical school and factory.

An interesting letter in 1891 to the editor of the *New York Times* described Sturge's humanitarian efforts and sought financial contributions for her work:

Joe Hogan's *Basketmaking in Ireland* documents the historical significance of the trade. (below left)

Hogan at work and his finished product: a willow shopping basket. (below middle and right)

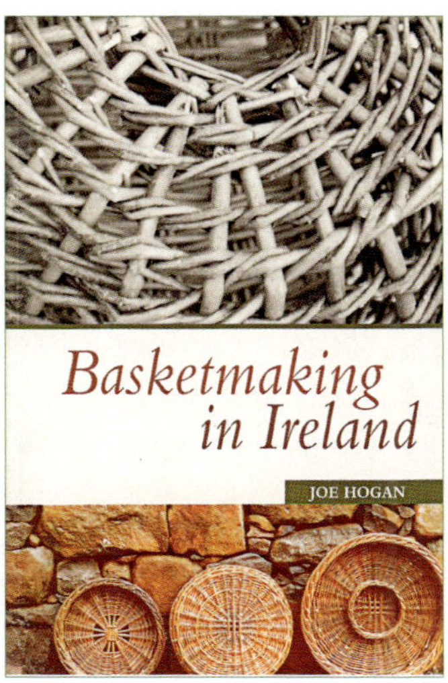

> [She] devotes herself to teaching the women and children the art of fancy basket making, the baskets finding a ready market in England. Miss Sturge is a believer in work, and not in almsgiving, but the present misery of the people on the west coast of Ireland, in consequence of the failure of the potato crop, forces her to appeal for immediate aid to friends in England and America (Garrison, 1891).

Webb and Scannell (1983) mention Miss Sturge's industry in their *Flora of Connemara and the Burren* but say that almost no trace remains of the willow (*Salix triandra*) and perhaps other species planted for this industry, as they fell victim to a severe fungal attack, perhaps with the exception of a single bush that was identified as a *S. triandra* hybrid.

Basket-making was widely practiced in local communities throughout Ireland. Róise Rua wrote in her memoir of life in Arranmore, Co. Donegal, in the early 1900s that her husband Séimí MacGrianna would make creels and baskets for local people during the wintertime during cold and wet days when outdoor work was unpopular (Ua Cnáimhsí, 2009). She said that there were plenty of sally rods (willow) on the island and his customers would bring him as many as he needed. When he first started making baskets he charged three shillings but eventually rose the price to five and then 10 shillings. She said that baskets and creels could also be bought at the fair in Dungloe, but it became an important source of income for her husband when he stopped being a migrant worker who had travelled to Scotland as an agricultural labourer each year:

> He used to make three types of creel—shoulder creels, donkey creels and panniers. The panniers were loaded with manure from the midden, and then it was brought to the grass fields and the fields that were being tilled. The bottom of the pannier was much broader than the mouth, and when you released the hinge on the base, it emptied easily. Whoever led the donkey or pony pulled the bar that kept the hinge closed; the hinge opened, the manure fell on the ground, the hinge was closed with the same bar, and back they'd go for another load. People needed panniers for drawing loads on their donkeys; Séimí wove panniers from a big sheaf of straw (Ua Cnáimhsí, 2009: 161).

PLANTS USED FOR BASKETRY

- *Ammophila arenaria*—Marram
- *Betula pendula*—Silver birch
- *Betula pubescens*—Downy birch
- *Calluna vulgaris*—Heather; Ling
- *Clematis vitalba*—Traveller's joy
- *Corylus avellana*—Hazel
- *Cynosurus cristatus*—Dog's tail; Thraneen grass
- *Cytisus scoparius*—Broom
- *Ilex aquifolium*—Holly
- *Juncus* species—Rush
- *Ligustrum vulgare*—Privet
- *Quercus petraea*—Sessile oak; Irish oak
- *Quercus robur*—Pedunculate oak; English oak
- *Rosa* species—Rose
- *Rubus fruticosus* agg.—Bramble; Blackberry
- *Rumex obtusifolius*—Broad-leaved dock
- *Salix* species, varieties and hybrids, including:
  - *Salix alba*—White willow
  - *Salix fragilis*—Withy
  - *Salix purpurea*—Purple osier
  - *Salix triandra*—Almond-leaved willow or French willow
  - *Salix viminalis*—Osier
- *Sambucus nigra*—Elder
- *Schoenoplectus* lacustris—Common club-rush
- *Sorbus aucuparia*—Mountain ash; Rowan

In Scotland there are records of some other species used for basketry, including grasses, crowberry (*Empetrum nigrum*), sedges (*Carex* species) and juniper (*Juniperus communis*) (Milliken & Bridgewater, 2004), but I have found no records of the use of such plants for basketry in Ireland.

Growing Willows for Basketry

Willows are not difficult to cultivate, and it is easy to grow your own willow rods for basketry. Hedgerows around Ireland provide a rich source of material that can be propagated so that you can start your own willow garden quickly and without cost. Cuttings are taken during the winter months when the plant is leafless. Each cutting should be about 25 cm long. These are placed in the ground up to about two-thirds of their length, placed about 15–20 cm apart. They are best grown in a sunny spot and in damp (rather than waterlogged) soil. The next spring most of the cuttings should have rooted, and within a year you will have rods worth cutting. One-year-old rods are ideal for basketry, but they can be left longer if thicker rods are needed for other purposes. Once cut, the rods should be tied in bundles, dried and stored until they are needed. Before use, the rods should be soaked for a few days until they are supple. In the past many basket-makers would have cultivated their willows in hedgerows, banks and field margins to save valuable land needed for other purposes. Many of these plants survive today, and my guess is that old examples of osiers (*Salix viminalis*) found throughout Ireland in particular are often derived from their former use as a source of rods for making baskets.

A rope winder.

Ropes

Ropes made from straw, hay and other plant-derived materials were used in Ireland for a wide variety of purposes (tying and holding down thatch, holding heavy loads, clothes lines, tethers and hobbles for animals, torches, belts for clothes, crafts such as baskets, cradles, footballs, bee-hives, armchairs and seats for chairs, bed bases, mattresses, mats, harnesses and reigns for horses and donkeys, and many other uses). Straw ropes were also used in building walls, instead of wattle and daub, where the rope would be wound between a series of upright posts and then plastered with mud. In Cos. Mayo and Antrim fowlers would lower themselves down cliffs on ropes made of straw, bog fir or horse hide to collect sea birds or their eggs (Evans, 1957).

Ropes were also made from animal-derived materials, including horsehide, cattle and horse hair. In terms of strength, Evans (1957) wrote that the strongest ropes were made of rawhide, horse hair, willow bark, birch twigs, bog fir and heather, followed by ropes made from rushes or sedges, and finally, ropes made from hay and straw. The following is a list of wild plant species that have been used in Ireland to make rope.

WILD PLANT SPECIES AND GROUPS USED TO MAKE ROPE

Ammophila arenaria—Marram
Betula pendula—Silver birch (twigs)
Betula pubescens—Downy birch (twigs)
Calluna vulgaris—Heather; Ling (twigs)
Corylus avellana—Hazel (twigs)
Juncus species—rushes
Molinia caerulea—Purple moor-grass
Pinus sylvestris—Scots pine; bog pine (bark, wood and root fibres)
Salix species—Willows (bark and twigs)
Sedges (various species including *Scirpus* and probably some *Carex* species)
Ulmus glabra—Wych elm; Irish elm (bark)
Peat (preserved fibres of sedges and rushes mainly)
Various species of grass (particularly in the form of hay)
Straw (dried stalks of barley, oats, rye[1] and wheat)
Flax
Seaweed
Sorbus aucuparia—Mountain ash; Rowan

1. Rye straw is longer, softer and more flexible than any other straw (Evans, 1957).

Linum usitatissimum (flax). Watercolour by Lydia Shackleton. Courtesy of the National Botanic Gardens, Glasnevin.

Materials Used to Make Irish Musical Instruments

Over 200 different woods are used worldwide to make musical instruments. Tropical hardwoods, such as ebonies, rosewoods and mahoganies, are the most popular and widely used, both for their beauty and the quality of the sound they produce. Some of the most notable and celebrated tropical timbers used in the manufacture of musical instruments include African blackwood (*Dalbergia melanoxylon*), used in the making of clarinets, oboes and other woodwind instruments, and Pau-Brasil (*Caesalpinia echinata*) from Brazil, used in the making of violin bows.

Irish musical instruments using native timbers have included the violin, Irish harp[8], bodhrán, Irish flute, bagpipes (sometimes called 'Great Irish Warpipes') and uilleann pipes. Although most of the timbers used today in the manufacture of Irish musical instruments are imported in Ireland, Irish timbers, particularly hardwoods from trees such as alder, willow, birch, arbutus and yew, were used in the past.

Various timbers have been used to make the wooden hoop of the traditional bodhráns, which is then covered with a treated goatskin or sheepskin. I have come across records of ash (*Fraxinus excelsior*), which is very flexible, and beech (*Fagus sylvatica*) being used to make these hoops.

TIMBERS USED IN MAKING IRISH HARPS

Soundbox
- Alder
- Hornbeam
- Sycamore
- Willow

Forepillar
- Alder
- Birch
- Hornbeam
- Spindle
- Sycamore
- Willow

Neck
- Alder
- Hornbeam
- Pear
- Sycamore

Source: Rimmer (1969)

Tanning

Tanning is an ancient process whereby animal skins (hides) are preserved using extracts from plants, mainly from the bark of trees. Fishing nets were also tanned, to help preserve them during the long periods immersed in sea water.

The need for bark for tanning in Ireland contributed in no small part towards the destruction of Ireland's native woodlands. The tanning process goes back at least 5000 years, and Egyptian tanning using the bark of a species of oak has even been recorded in an Egyptian tomb painting 3000 years old (Ciesla, 2002). The same author recorded that Native Americans used several plant species to tan the hides of buffalos (American bison). About 500 species worldwide are known to contain tannins that can be used for tanning. Tannins are acidic and astringent and combine with the collagen in animal skins to produce leather. Amongst temperate trees, members of the Fagaceae family, including species of *Castanea* (chestnuts) and *Quercus* (oaks), are important sources.

In Ireland, the bark of oak trees has been the main source of plant material for tanning, but other species are recorded as having also been used. I have recorded at least 15 native species as having been used for tanning purposes in Ireland (see Plants Used for Tanning Leather).

Samuel Johnson wrote in *A Journey to the Western Islands of Scotland* (1775) that brogues made from hides (with the fur side inwards) were tanned in Skye, mainly with oak bark, but also with birch bark and the roots of tormentil (*Potentilla erecta*).

PLANTS USED FOR TANNING LEATHER

- *Alnus glutinosa*—Alder
- *Arctostaphylos uva-ursi*—Bearberry
- *Berberis vulgaris*—Barberry
- *Betula pendula*—Silver birch
- *Betula pubescens*—Downy birch
- *Calluna vulgaris*—Heather; Ling
- *Castanea sativa*—Spanish chestnut, sweet chestnut
- *Fraxinus excelsior*—Ash
- *Myrica gale*—Bog myrtle
- *Potentilla erecta*—Tormentil
- *Quercus petraea*—Sessile oak; Irish oak
- *Quercus robur*—Pedunculate oak; English oak
- *Salix caprea*—Goat willow
- *Salix viminalis*—Osier
- *Sorbus aucuparia*—Mountain ash; Rowan

8. The Irish word is *cláirseach*.

A thatched cottage in Inisheer shows the intricate system of ropes used to tie down the straw thatch.

A modern thatched house in Annascaul, Co. Kerry.

Thatching

Thatched dwellings were probably constructed on the island from the very earliest of times. When the first settlers arrived on Ireland's shores they built simple houses of timber and covered them with leaves, branches, grasses or reeds to shelter themselves from the rain, wind and weather. The aim of a thatching is to produce a layer of vegetation that protects the inner roof of a house from rain, hail, sleet and snow. For this reason, most thatched roofs are pitched at various angles to encourage rain-water run-off. For many centuries, thatching remained the only roofing option for homes in Ireland.

Elsewhere in the world, thatching has been and remains an important means for roofing houses, and a variety of plant materials are used, such as bamboo, timber and palm and other leaves. Today, thatched houses are relatively rare in Ireland but are still part of the tourist folk image of Ireland and popular with residents and visitors alike. Ironically, although thatching was carried out in the past by the poorest people for simple housing, it is today affluent people who aspire to own, restore or live in a thatched house. A significant proportion of contemporary thatched houses in Ireland are maintained for tourism purposes and some of the commonest thatched houses are pubs, restaurants, rental cottages, guest houses, souvenir shops and the like.

At least 17 native plant species have been used for thatching purposes in Ireland[9]. The choice of one species or another was largely determined by availability. Thatch made from reeds (*Phragmites australis*) lasts longer than thatch made from rushes (*Juncus* species); therefore, if the former was easily available it would have been chosen before the latter. Some species used for thatching were available only in particular parts of Ireland (such as marram, *Ammophila arenaria*, on sandy coasts). Simple thatching was probably carried out by the poorest people and would have been undertaken with grasses, sedges, heathers and other species. I have been able to find very little documentation on what specific plants were used in this way.

A variety of non-native plant species have also been used for thatching in Ireland, such as straw and flax. Contemporary thatching is done with straw or imported reeds. Specific methods and uses of particular species are described in full in the Systematic List of Wild Plants, and for straw, under the section on straw in this Introduction.

9. In Irish, 'thatching' is *tuigheadóireacht* (Dinneen, 1927) or *tuíodóireacht* (Ó Dónaill, 1977), and 'a thatcher' is *tuigheadóir* (Dinneen, 1927) or *tuíodóir* (Ó Dónaill, 1977).

NATIVE AND OTHER PLANTS USED FOR THATCHING

Agrostis stolonifera—Creeping bent; Fiorin grass
Ammophila arenaria—Marram
Calluna vulgaris—Heather; Ling
Cladium mariscus—Saw sedge
Cytisus scoparius (*Sarothamnus scoparius*)—Broom
Eleocharis palustris—Common spike-rush—this may have been used for thatch but no evidence has been found
Iris pseudacorus—Yellow flag (thatching haycocks)
Juncus species (The following species are probably suitable for use in thatching and may have been used. Most grow from 50 to 100 cm tall, in tufts and have rigid shiny stems.)
- *Juncus acutiflorus*—Sharp-flowered rush
- *Juncus acutus*—Sharp rush
- *Juncus articulatus*—Jointed rush
- *Juncus conglomeratus*—Compact rush
- *Juncus effusus*—Common rush; Soft rush
- *Juncus inflexus*—Hard rush
- *Juncus maritimus*—Sea rush
- *Juncus subnodulosus*—Blunt-flowered rush

Molinia caerulea—Purple moor-grass (used in making ropes to tie down the thatch)
Phragmites australis—Reed
Pteridium aquilinum—Bracken
Schoenoplectus lacustris—Common club-rush
Schoenus nigricans—Black bog-rush
Typha angustifolia—Lesser bulrush
Typha latifolia—Bulrush; Reed mace
Ulex europaeus—Gorse; Whin; Furze
Ulex gallii—Western gorse
Zostera species (underlay for thatch)
- *Zostera angustifolia*—Narrow-leaved eelgrass
- *Zostera marina*—Common eelgrass
- *Zostera noltei*—Dwarf eelgrass

Non-native plants

Straw from oats, barley, rye and wheat
Rye straw was formerly the most popular straw used for thatching but declined everywhere in Ireland as a crop, lasting longest in the west of Ireland. It is longer, softer and more flexible than any other straw (Evans, 1957). Wheat straw was used instead as second best, but in fact, most farmers used oat straw instead.
Flax
Potato stalks
'Bean Haums' (stalks) is mentioned in *A Society of Gentlemen* (1780).
According to Rutty (1772: 123), '*Pisum vulg.* Common Pease' stalks are used in thatching houses, 'being put under the straw'.

Straw: An Important Resource Through the Ages

A wide range of straw artefacts and uses have been recorded in Ireland. The collections of the Museum of Country Life, National Museum of Ireland in Castlebar, Co. Mayo, include many straw objects, such as mattresses, various forms of baskets, stools and chairs, ropes, sun hats, hens nests, horse collars, riding harnesses, doors, child's rattles, spindle holders, pot scrubbers, horse saddles and Wrenboy costumes.

Thatching and Rope-Making

Straw has been used widely for thatching houses, sheds and animal shelters. Thatching with straw was generally preferred in Ireland because of its long life span and because it was clean, easy to prepare and of uniform length. Wheat and rye straw were more commonly used for thatching because they are more durable than oat straw (Evans, 1957). Expert thatchers preferred to use straw that had been hand-cut with a sickle rather than machine cut and mechanically threshed straw, which was often broken and damaged. The harvesting of wheat by the combine harvester has encouraged thatchers to return to using (mainly imported) reeds for thatching in Ireland. This lack of available hand-cut straw has resulted in the renewed exploitation of some reed beds in recent years, which were formerly heavily harvested, such as on the River Shannon south of Limerick city, to provide material for local thatchers (McInerney, 2005).

However, Kelly (1997) reports that in Old Irish texts, reeds are generally mentioned as having been used for thatching, floor covering and withies (*Salix* species) for basketry and rope-making, and it was not until the 10th to 12th centuries that straw was mainly used for such purposes. This may have been because during the Old Irish period most cereal crops were harvested close to the tops of the stalks and the straw was not saved.

Ropes[10] made from straw were widely used for many purposes as well, including thatching, making straw furniture such as chairs and beds, and tethers for animals, etc. A simple tool was used to twist the straw into ropes. It consisted of a hook on a crank that was operated by two people, one to twist the crank and the other to feed straw from a stack (Staniforth, 2008). (In Britain this object is called a 'whimmer' or 'whimble', but I have not found the Irish equivalent name.)

Simple straw ropes have been made in the west of Ireland right up to the present day by using a broad hay fork; a loop of straw is wrapped around the fork, then, as the rake is turned around and around by one person, the straw is fed into it by another to make the rope (Brendan Kelly, pers. comm., 2010). John Millington Synge (1979: 105) included an account of straw rope-making in his book *The Aran Islands*, first published in 1907:

> In our cottage the thatching—it is done every year—has just been carried out. The rope-twisting was done partly in the lane, partly in the kitchen when the weather was uncertain. Two men usually sit together at this work, one of them hammering the straw with a heavy block of wood, the other forming the rope, the main body of which is twisted by a boy or girl with a bent stick specially formed for this employment. In wet weather, when the work must be done indoors, the person who is twisting recedes gradually out of the door, across the lane, and sometimes across a field or two beyond it. A great length is needed to form the close network which is spread over the thatch, as each piece measures about fifty yards. When this work is in progress in half the cottages in the village, the road has a curious look, and one has to pick one's steps through a maze of twisting ropes that pass from the dark doorways on either side into the fields. When four or five immense balls of rope have been completed, a thatching party is arranged, and before dawn some morning they come down to the house, and the work is taken in hand with such energy that it is usually ended within the day.

Household and Other Items

Straw was formerly used as bedding in the home, particularly in winter in poorer households. It was commonly used for stuffing mattresses, cushions and upholstered furniture and was woven into comfortable armchairs used in country cottages and to make seats for wooden chairs. Before glass windows and wooden doors became standard in Irish cottages, straw mats were used as draft excluders and to stop the wind from whistling under the doors. Straw was also woven to make a variety of traditional farm items, such as horse harnesses and saddles, and was used as bedding for livestock. In his book *Cappaghglass*, on the experiences of people in that west Cork village, Somerville-Large (1985: 168) wrote that Timmy Mahon, a local craftsman, reported that straw horse 'collars were the most difficult thing to make, for it all came out of your head and there were no patterns or anything'. Most collars were made to fit a particular horse. Straw was also used as animal feed, but because it is low in nutrients, it would have formed only a small part of the diet of

10. Straw ropes are called *súgán* in Irish and *strat* in Yola.

farm animals and generally was used only when more nutrient-rich fodder was scarce. The use of hay in Irish ethnobotany is described under *Lolium perenne* (perennial rye-grass) in the Systematic List.

In several parts of Ireland, straw was used to weave special covered baskets for laying chickens. Bee-hives were also constructed of straw and generally made of coiled straw ropes, or 'lips'. The craft of making 'lips' was widespread in the United Kingdom in the 19th century. They were formed by passing straw through a hollow section of cow horn, about 30 mm (1.5 in.) in diameter, to form a long, untwisted roll of straw. These rolls could then be bound together with other materials such as split bramble stems (*Rubus fruitcosus* was frequently used) and made into such objects as bee-hives (or 'bee skeps') (Staniforth, 2008). Lip work was also practiced in Ireland. In the early 1900s, a small rural industry of straw hat-making was centred around Wexford and the Wexford Straw Hat Co., which operated at least during the year 1906. In the early part of the 20th century, a school for straw-hat manufacturing was established in Wexford. Examples of the products produced by the Wexford straw industry were exhibited in the Irish Industrial Exhibition at the World's Fair held in St Louis, Missouri, in the United States in 1904 (Deptartment of Agriculture and Technical Instruction, 1904).

A traditional chair with a straw rope seat.

Straw has also traditionally been used as packing material and to line baskets in which eggs were carried. Throughout Europe, straw was used as an important packing material for glass bottles. A straw envelope was made from strands of straw laid side by side and then sewn together. In 1913, Britain imported about 150 million straw envelopes for wrapping bottles, mainly from the Netherlands and France (Staniforth, 2008). I don't know whether such straw envelopes were ever made in Ireland, although they would certainly have occurred in Ireland, particularly imported with wine bottles from continental Europe. Most, if not all, wine bottles are now imported into Ireland, packed in a protective cardboard wrapping instead of straw.

Straw harvest in Inchigeela, Co. Cork. A tourist postcard dated 1910.

Straw has also been used under strawberries to prevent the ripening fruits from touching the soil. It is also used in horticulture in other ways, such as to protect plants in winter from frost damage.

Another by-product of cereal crops is the chaff (*cáith*), or the dry bracts that enclose the mature cereal grains that are left over after the crop is winnowed. In ancient Ireland this was generally regarded as worthless but may have occasionally have been stored for use as a fire-lighter (Kelly, 1997). Mattresses were also sometimes stuffed with chaff. Kelly described a useful food eaten in ancient Ireland, though, called 'sowens', made from a cereal by-product. The husks of bran from oats or barley were soaked for a few days to allow some fermentation to take place. Then the resulting liquid was strained and drunk, or boiled until it became a jelly.

Cultural Artefacts

As an alternative to rushes, straw was used in the weaving of St Brigid's Crosses, and for toys, as well as masks and costumes associated with particular traditional festivals. For example, on St Stephen's Day, 26th December (*Lá an Dreoilín*, 'Wran's Day' and *lucht Dreoilín*, 'Wran-boys'), 'Wrenboys' dressed in full-length straw costumes and hats were a frequent sight. On that day, crowds of people dress in a wide variety of colourful and motley costumes, usually wearing masks and often straw suits, keeping alive a tradition that has flourished for centuries, and one that probably pre-dates the arrival of Christianity in Ireland. In some parts of Ireland the Wrenboys are called 'Mummers', and there are strong English influences in the festival also. The straw suits were generally woven from oat straw and came in three parts: a hat or a mask to cover the face, a cape to cover the chest and a skirt to cover the lower part of the body. In earlier times the small wren, a native bird, was hunted and attached to a holly bush and carried at the head of a Wrenboys procession. Today, the wren is no longer hunted and is remembered only as part of the festival in the traditional rhyme about St Stephen's Day:

Illustration of the Wrenboys from Hall and Hall (1841).

The wran, the wran, the king of all birds,
St Stephen's Day was cot in the furze,
Although he was little his family's grate,
Put yer hand in yer pocket and give us a trate.

Sing holly, sing ivy—sing ivy, sing holly,
A drop just to drink it would drown melancholy.

And if you dhraw it ov the best,
I hope in heaven yer sowl will rest,
But if you dhraw it ov the small
It won't agree wid de wran boys at all.

(Hall and Hall, 1841: 24).

Various versions of this rhyme exist. The following version of the second half of the rhyme is well known in Ireland:

Up with the kettle and down with the pan,
And give us a penny to bury the wren.
Droolin, Droolin, where's your nest?
Tis in the bush that I love best,
In the tree, the holly tree,
Where all the boys do follow me.

The wren is said to have woken St Stephen's jailers while he was escaping just before his execution, which accounts for why wrens were killed on St Stephen's Day (Kinahan, 1881). Several regions and towns in Ireland still have active wren celebrations. In Dingle, Co. Kerry, about eight costumes are made of oat straw each year for the 'wran' (as it is known in Dingle). The elaborately woven straw hat from one of these costumes, made by Fergus O'Flaithbheartaigh (O'Flaherty) and Kevin O'Connor of the Green and Gold Wren, in December 2009, is preserved in the museum collections of the National Botanic Gardens, presented to the author in 2010 by Dingleman Tom Fox.

Alfred C. Haddon, the distinguished British anthropologist who worked in Ireland for some years from 1883 (while in Dublin he helped to found the Dublin Naturalists' Field Club in 1885/86), described the use of straw to make a wedding dance-mask in Erris, Co.

Mayo, in the 1890s (Haddon, 1893). There, whenever a wedding would take place, gangs of men and boys would appear uninvited at the wedding dressed up in women's clothes and wearing straw masks that completely covered their faces and heads. The 'captain' of the gang would ask the bride to dance with him and it was bad luck to refuse. The masks were conical in shape and topped with three rings of straw. The mask was 21 in. high and 11½ in. wide at the extreme base. The National Museum of Ireland (Museum of Country Life) in Castlebar, Co. Mayo, includes several Wrenboy costumes made of straw. Strawboys' (*glainairí*) costumes were also used on St Brigid's Day, as well as on St Stephen's Day. The Strawboys' tradition lives on today in several parts of Ireland. For example, I heard of straw costumes being made for a wedding on the Dingle Peninsula in 2011. In his autobiography of his life as a matchmaker in Co. Clare, Willie Daly (Daly, 2010: 127–131) wrote:

> The strawboys are local folk who disguise themselves by dressing up in costumes made from straw, and turn up at the wedding celebrations. . . . It's supposedly very lucky for you if they do come, and a little bit unlucky if they don't come or if you don't let them into the house. . . . There'd be many strawboys in elaborate costumes that were awfully well done, a hat like a crown with four prongs coming up off it into a point with a ball on top. But most times as youngsters we just stuck straw into our normal clothes. . . . The tradition is not so common now, although it carries on in the hotels, in a more organised way, for wedding parties, more of a performance and more impersonal, but you can still have great fun.

The green and gold wren (or wran), 2009, Dingle, Co. Kerry. Courtesy Tom Fox.

Other straw objects held in the Castlebar Museum include a club made from a straw rope, a *harra*, used to enforce the rules and inflict punishment during wakes games. 'Wakes' are an Irish funeral tradition during which the recently departed person's life is celebrated by family and friends, in the form of a party. The *harra* example on display was made in Claregalway, Co. Galway (Ref. No: F1957.55). Another straw object in the museum is a holy water sprinkler, about 1 ft long and originally from Ballygar, Co. Galway (Ref. No: F1967.97). The museum also has a pot scrubber made of straw, from Erris, Co. Mayo (Ref. No: F1997.583).

Wren costumes. Courtesy Tom Fox.

The museum includes a good collection of harvest knots, too, some of them 'female', woven with the ears of corn still attached and some of them 'male', without the ears of corn. These were woven and exchanged as love tokens as part of harvest celebrations and during the Lúnasa (Lughnasadh) festival held on 1st August. Examples from throughout different parts of Ireland are known. The last sheaf of corn cut as part of the harvest was of special significance and sometimes woven into a traditional object called a corn dolly, or *cailleach* in Irish.

One such example is presented in the museum's collections from Raheenduff, Co. Louth (Ref. No: 1949.389). These last sheaves of corn were said to hold the captured corn spirit, which guaranteed the fertility of the crop. The *cailleach* would be the home for the spirit for the winter. The name 'dolly' is probably not derived from a child's doll toy, but instead is perhaps a corruption of the word 'idol'. The following season when the land was being furrowed in preparation for the next year's crop, the *cailleach* was taken back to the field and placed in the earth, symbolically returning the spirit and the corn fertility to the land to ensure a good harvest for the coming year.

Vickery (2010: 16) mentioned that there were various traditions associated with cutting the last sheaf of the harvest in Ireland: 'In County Carlow each of the unmarried girls would have a stroke at the last sheaf with a reaping hook and the one who succeeded in cutting it down would marry within a year. Elsewhere in Ireland it was believed that whoever cut or bound the last sheaf would die unmarried'. The last sheaf was also thought to have curative powers and was sometimes fed to sick animals or calving cows. It was also burnt and the ashes used for an ointment for skin ailments (Vickery, 2010).

A variety of corn dollies from throughout Ireland.

Corn dollies or other objects woven from straw have been important talismans or charms in several parts of the world. For example, in India, they were hung in the doorways of newly married couples about to take up residence, and in Japan figures were made of warriors. Many countries have traditions associated with cutting the last sheaf of corn in the field, and in Ireland making corn dollies was a way of honouring, capturing and marking the harvest (Lambeth, 2008). Although a range of different traditional dollies have been documented in Britain and other countries, in Ireland the range of corn dollies seems to have been less diverse, but perhaps we are simply seeing the last small fragments of a lost tradition.

A wide variety of weaves and pleats have been used in making corn dollies, such as two-, three- and four-straw pleats. Different types of straw (from wheat, barley, rye and oats) have different properties for weaving. Wheat straw is probably the best and easiest for beginners to use. Undamaged straw should be chosen of even lengths, and old straw should be avoided. Wheat varieties with hollow stems are best. The straw needs to be damp before use and can be soaked in water for about 15 minutes and then stood upright to drain off the excess water. This prevents the straw from being hard and brittle.

Wood-Martin (1902: 376) described an interesting former folklore use of straw in his remarkable book, *Traces of the Elder Faiths of Ireland.* In it, he said that in Ireland a widespread superstitious custom is that the body of a drowned person can be discovered by floating a bundle of straw on the surface of the water and 'it will stop and quiver over the spot where the corpse is lying'.

In Britain, archery targets were made from coiled, heavy-gauge straw ropes sewn tightly together. These were made up to World War II and were most notably manufactured by the Egerton Company near Guildford, Surrey (Staniforth, 2008). Most archery targets used in Ireland were probably imported, and I have found no records of their manufacture in this country.

Straw was also widely used in the construction of cob walls in Britain. The cob was made up of a mixture of straw and clay. Cob houses and walls were most frequent in the West Country. They were long-lasting, provided they were constructed on dry foundations and protected from the rain on top by a roof or thatch (Staniforth, 2008). I suspect that the weather in Ireland is too wet for cob walls to have been frequently constructed, except perhaps as internal house walls.

Textiles and Dyes

Dyes made from plant extracts have been used throughout the world since the earliest of times. In Europe, dyes extracted from plants such as woad, saffron and madder were commonly used, mainly to colour textiles, and occasionally to dye the skin. These dyes have been used in Ireland to colour wool, silk, cotton and linen.

In Ireland, wool is derived from sheep (elsewhere in the world, other animals provide wool, such as alpaca, vicuna and llama in South America, and mohair from the Angora goat, Kashmir wool from the Tibetan goat, and camel hair from the soft underwool that is shed annually by the camel). Wool from sheep comes in three main forms: lamb's wool, which is soft and fine and cut from lambs that are under six months old; hogs' and tegs' wool, which is the first shearing of sheep that were not shorn as lambs; and wethers' wool, which is clipped from sheep after their first shearing (Mairet, 1916).

Linen is made from the stem fibres of flax (*Linum usitatissimum*), which was formerly widely grown in Ireland, particularly in the north-east, and used for the manufacture of textiles. After harvesting, the plants are decomposed (called 'retting') by placing them in a pond, pool or drain so that the fibres can be separated from the other decayed parts of the plant.

Water-soluble dyes were extracted from the dye plant and added to the materials that were to be dyed. The essence of a successful dye was its ability to then be fixed (i.e., made insoluble in water). Various chemical compounds, called mordants, were used for such purposes, such as the salts of various metals, including iron, aluminium, tin and chrome, which combined with the dye to form an insoluble compound. A wide variety of native plants were used in Ireland as dye to yield most colours. Dyes derived from lichens were especially popular and widely used.

Lichens have been used in Ireland as dyestuffs for centuries. They generally require no chemical to make them permanent, unlike dyes made from decoctions of leaves, roots, flowers and fruit juices that require a mordant to fix them permanently to the textile or other material they are being used to dye (Mahon, 1983). Bríd Mahon provided a very valuable overview of traditional dyestuffs used in Ireland in a volume of essays dedicated to the great Irish folklorist Kevin Danaher (Caoimhín Ó Danachair), and listed most of the native plants formerly used for dyeing purposes in Ireland. Mahon also mentioned what is perhaps the earliest reference to dyeing in Irish literature, when in the *Book of Leinster*, King Tighernmas, the son of Ollaig, dyed cloth brown, red and crimson with ornamental borders.

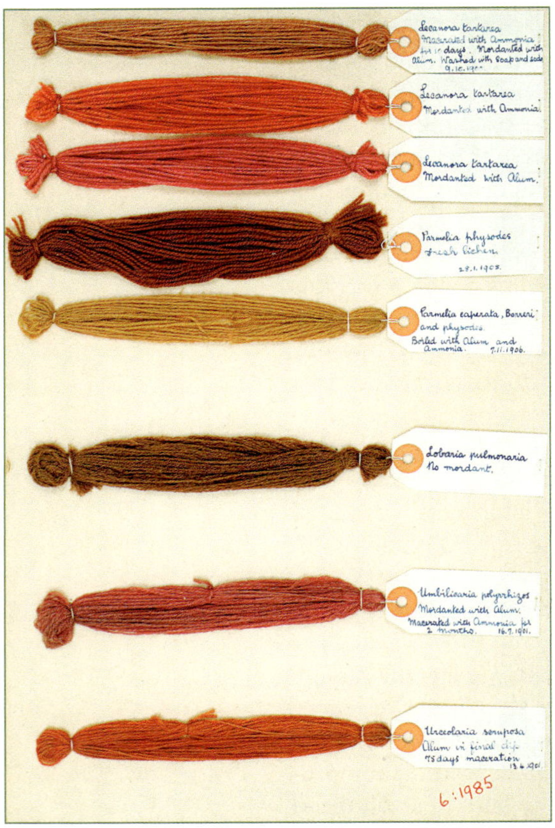

Examples of dye colours produced by lichens. From the National Botanic Gardens, Glasnevin, Dublin, Ireland, Museum objects collection. © Department of Environment, Heritage and Local Government.

Danaher (1964b: 77) documented quite a few native plants used for dyeing:

One old woman, a neighbour and friend of ours, had been in her young days an expert spinner, and still had both blankets and sheets for which she had spun thread. She often told us about the making of cloth and of the way she used to dye the wool and linen yarn with locally gathered plants. The roots of the *bainnicín* [*Euphorbia hyberna*] gave black, *feileastrom* [*Iris pseudacorus*] gave a grey-blue colour, brier [*Rubus fruticosus*] roots a dark green. Whitethorn [*Crataegus monogyna*] leaves gave a dark blue and alder [*Alnus glutinosa*] a dark green. An expert could get a clear yellow from heather [*Calluna vulgaris*], but less skilful handling gave dull yellows or brown. The blossoms of furze [*Ulex* species] and of *buchalán buí* [*Senecio jacobaea*] also gave various shades of brown and dull yellow. Bright reds and blues were beyond the reach of those who depended entirely upon local dye-plants, but could be got from madder [*Rubia tinctoria*] and indigo [*Indigofera tinctoria*]; these had to be bought from the shop or, at an earlier time, from pedlars, and both these coloured stuffs have been known in Ireland and generally western Europe for many centuries. The most difficult colour of all to get was bright green until modern dyes came on the market, but this did not worry the older generations, indeed, it was commonly held that bright green was an unlucky colour, especially for clothes.

Dyehouses have been excavated in various parts of Ireland, such as by F. J. Bigger in 1880 in Roundstone, Connemara, and by Françoise Henry in 1948, on the Inishkea Islands in north Mayo, where shellfish (*Purpura lapillus* and *Murex* species) were the dye sources used. These shellfish have been used for centuries by the Greeks, Romans and Phoenicians (Mahon, 1983). McGlinchey (2007) mentioned that in the late 1800s in north Donegal where he grew up, a pot of indigo was kept warm beside the fire in every house at the time, to be used in dyeing. J. M. Synge (1979: 15) wrote about visiting the Aran Islands in the early 1900s and described how the women wore petticoats and jackets of the island wool 'stained with madder' to which they usually added a plaid shawl twisted around their chests and tied at the back. The men wore three colours: natural wool colour, indigo and a grey flannel woven from alternate threads dyed with indigo and natural wool.

Natural plant dyes are still used by some traditional weavers working in Ireland today. For example, the well-known Blasket Islands weaver Sue Redican uses the following range of native and wild plants: montbretia (*Crocosmia* ×*crocosmiiflora*) for yellow/greens; *Fuchsia magellanica* for

COLOURS PRODUCED BY PLANTS USED IN DYEING

Black

Alnus glutinosa—Alder (bark)
Empetrum nigrum—Crowberry (berries)
Euphorbia hyberna—Irish spurge
Euphorbia species—Spurges
Filipendula ulmaria—Meadow-sweet (roots)
Iris pseudacorus—Yellow flag (roots)
Lycopus europaeus—Gipsywort
Quercus petraea—Sessile oak; Irish oak (bark and wood chips)
Quercus robur—Pedunculate oak; English oak (bark and wood chips)
Rubus fruticosus agg.—Bramble; Blackberry (roots, stems and leaves)
Rumex obtusifolius—Broad-leaved dock (roots)
Sambucus nigra—Elder (bark)

Blue

Arctostaphylos uva-ursi—Bearberry
Crataegus monogyna—Hawthorn (dark blue) (leaves)
Iris pseudacorus—Yellow flag (roots)
Prunus spinosa—Sloe; Blackthorn (fruits, dyeing linen light blue)
Rubus fruticosus agg.—Bramble; Blackberry (fruits)
Sambucus nigra—Elder (fruits)
Succisa pratensis—Devil's-bit scabious (leaves)
Vaccinium myrtillus—Bilberry
Viola lutea—Mountain pansy

Blue-Green

Iris pseudacorus—Yellow flag
Ligustrum vulgare—Privet (berries)

Brown

Alnus glutinosa—Alder
Betula pendula—Silver birch (bark)
Betula pubescens—Downy birch (bark)
Calluna vulgaris—Heather; Ling
Humulus lupulus—Hop
Juniperus communis—Juniper
Lythrum salicaria—Purple loosestrife
Menyanthes trifoliata—Bogbean
Nymphaea alba—White water-lily
Palmaria palmata—Dulse; Dillisk
Prunus avium—Wild cherry
Quercus petraea—Sessile oak; Irish oak (bark)
Quercus robur—Pedunculate oak; English oak (bark)
Senecio jacobaea—Ragwort (flowers)
Ulex europaeus—Gorse; Whin; Furze
Ulex gallii—Western gorse; Autumn gorse
Vaccinium myrtillus—Bilberry
Veronica beccabunga—Brooklime

Crimson

Drosera species—Sundew

Green

Alnus glutinosa—Alder (catkins)
Digitalis purpurea—Foxglove
Equisetum telmateia—Great horsetail
Euonymus europaeus—Spindle-tree (seeds)
Frangula alnus—Alder buckthorn (fruits)
Iris pseudacorus—Yellow flag
Juncus species—Rush
Ligustrum vulgare—Privet (leaves)
Pteridium aquilinum—Bracken (frond buds)
Rhamnus cathartica—Buckthorn (bark and ripe berries)
Rubus fruticosus agg.—Bramble; Blackberry (dark green)
Rumex acetosa—Sorrel
Sambucus nigra—Elder (leaves)
Urtica dioica—Nettle

Greenish-White

Rumex obtusifolius—Broad-leaved dock (roots)

Grey

Galium verum—Lady's bedstraw
Rubus fruticosus agg.—Bramble; Blackberry (fruits)

Grey-Blue

Iris pseudacorus—Yellow flag

Lime-Green

Fuchsia magellanica—Fuchsia (flowers)

The flowers of *Fuchsia magellanica* produce a lime-green dye.

COLOURS PRODUCED BY PLANTS USED IN DYEING *(continued)*

Magenta

Taraxacum officinale agg.—Dandelion (roots)

Olive-Green

Digitalis purpurea—Foxglove (leaves)
Pteridium aquilinum—Bracken

Pink

Alnus glutinosa—Alder (wood)
Cornus sanguinea—Dogwood

Purple

Drosera species—Sundew
Lythrum salicaria—Purple loosestrife
Quercus petraea—Sessile oak; Irish oak (bark)
Quercus robur—Pedunculate oak; English oak (bark)
Sambucus nigra—Elder
Taraxacum officinale agg.—Dandelion
Vaccinium myrtillus—Bilberry

Purple-Red

Salix species—Willow (bark)

Red

Alnus glutinosa—Alder (bark)
Betula pendula—Silver birch (inner bark)
Betula pubescens—Downy birch (inner bark)
Euonymus europaeus—Spindle-tree (fruits)
Galium boreale—Northern bedstraw (roots)
Galium verum—Lady's bedstraw (stolons and roots)
Lythrum salicaria—Purple loosestrife
Potentilla erecta—Tormentil (roots)
Rubia peregrina—Wild madder (roots)
Rumex acetosa—Sorrel (roots)
Sherardia arvensis—Field madder (roots)
Umbilicus rupestris—Wall pennywort; Navelwort

Rose-Pink

Rubus fruticosus agg.—Bramble; Blackberry (fruits)

Yellow

Agrimonia eupatoria—Agrimony
Alnus glutinosa—Alder (young shoots)
Anthyllis vulneraria—Kidney vetch
Betula pubescens—Downy birch (leaves)
Blackstonia perfoliata—Yellow-wort
Calluna vulgaris—Heather; Ling
Caltha palustris—Marsh-marigold; Kingcup
Cornus sanguinea—Dogwood
Cytisus scoparius—Broom
Dipsacus fullonum—Teasel
Drosera species—Sundew
Drosera rotundifolia—Round-leaved sundew
Equisetum species—Horsetails
Erica tetralix—Cross-leaved heath
Euonymus europaeus—Spindle-tree (seeds)
Filipendula ulmaria—Meadow-sweet (pale yellow)
Frangula alnus—Alder buckthorn (fresh inner bark)
Fraxinus excelsior—Ash (bark)
Hippuris vulgaris—Mare's-tail
Hypericum species—St John's-wort
Ligustrum vulgare—Privet (leaves)
Lonicera periclymenum—Honeysuckle; Woodbine
Malus sylvestris—Crab-apple (bark)
Myrica gale—Bog myrtle (branch tips)
Narthecium ossifragum—Bog asphodel
Persicaria hydropiper (*Polygonum hydropiper*)—Water pepper
Persicaria maculosa (*Polygonum persicaria*)—Redshank or Spotted persicaria
Potentilla erecta—Tormentil (roots)
Pteridium aquilinum—Bracken (roots and young tops)
Quercus petraea—Sessile oak; Irish oak (bark)
Quercus robur—Pedunculate oak; English oak (bark)
Reseda luteola—Yellow-weed
Rhamnus cathartica—Buckthorn (unripe fruits and bark)
Rubus fruticosus agg.—Bramble; Blackberry (roots)
Rumex obtusifolius—Broad-leaved dock
Senecio aquaticus—Marsh ragwort
Stachys palustris—Marsh woundwort
Thalictrum flavum—Common meadow-rue
Ulex europaeus—Gorse; Whin; Furze (flowers)
Ulex gallii—Western gorse; Autumn gorse (flowers)
Umbilicus rupestris—Wall pennywort; Navelwort
Urtica dioica—Nettle

Yellow-Green

Crocosmia ×crocosmiiflora—Montbretia
Pteridium aquilinum—Bracken

Unspecified colours

Corylus avellana—Hazel
Sherardia arvensis—Field madder
Urtica dioica—Nettle
Vaccinium myrtillus—Bilberry; Fraughan

lime green; onion skins for a strong golden yellow; horsetails (*Equisetum* species) for very bright, strong yellows; honeysuckle or woodbine (*Lonicera periclymenum*) leaves, nettles (*Urtica dioica*) or dock (*Rumex* species) leaves for yellow; bracken (*Pteridium aquilinum*) for olive greens/dull yellows; foxglove (*Digitalis purpurea*) leaves for olive green; lichens such as 'Sea Ivory' for pale peach colour; *Parmelia* species for rich deep brown and *Xanthoria parietina* for pale brown (Redican, pers. comm., 2010).

Dinneen (1927) gave the Irish word *ruaim* (*-uama*; *-uamanna*) for a wool-dye made by boiling briar or thorn roots, dock roots, sorrel, alder bark, etc., that was used as the first dye or base colour. Ó Dónaill (1977) gave the Irish word *corcair* for '(lichen producing) kind of purple dye'. In Irish, *corcraigh* means 'purple'.

Peat: An Important Plant-Derived Resource

Peat (or 'turf')[11] has been one of Ireland's most important plant-derived natural resources for several thousand years, probably almost since the earliest human colonisation of Ireland. Largely composed of the preserved remains of various species of *Sphagnum* mixed with other bogland species, such as sedges and some grasses, peat has been used for a wide variety of purposes and was important as a free or cheap fuel source for many of the poorer people in rural parts of the country. Peat has been cut by hand for fuel (with a special turf-spade called a 'slane' or 'shlawn'; *sleán* or *sleaghán*[12] in Irish). Most families who cut their own turf will have a slane that may sometimes have been used for several generations. Today in Ireland, slanes are not easy to find for sale. When I tried to buy one in a hardware shop in Westport, Co. Mayo, the response I got was 'try an antique shop'! In Dingle, Co. Kerry, the best local hardware supplier told me (in 2010) that he had not sold one for about five years. I eventually found one for sale in Letterfrack, Co. Galway, that had been made in East Galway.

Patrick Campbell (1974) described the harvesting of peat in his memoir of growing up in Co. Donegal. Springtime was the time to cut the turf. Farmers would hire a team of turf-cutters. The turf-cutter reputed to be the fastest went to work on the first 'floor' of the turf bank and he would try to keep ahead of the men cutting the second 'floor'. When he was tired, his place would be taken by another, and so on so that the fastest turf-cutting pace possible could be maintained throughout the day. Campbell said that in Donegal the different turf-layers were called 'floors', but elsewhere in Ireland the term 'spit' was used.

Once cut, the sods of peat were laid out individually on the bog surface to dry and were later turned and dried further in small, upright stacks (called *gruagán*[13]), sometimes with a sod of turf placed across the top. Drying could take three to four weeks or even longer depending on the weather. When fully dry, the sods were gathered together in large stacks in the bog before being transported home in a cart or creel. One method of transporting the turf home was on a simple wheel-less slide made with two shafts of holly (*Ilex aquifolium*) or birch (*Betula* species), particularly in the south-west, pulled by a Kerry Bog Pony (Bell & Watson, 2008). In some areas where the peat cutting could not be adequately drained, and the peat was too wet, it was extracted like mud and then moulded before being dried. There are records of 'mud turf' being extracted and used in several parts of Ireland, such as Cos. Cavan, Leitrim, Down and Sligo, and peat was still being extracted this way in Ardee Bog in Co. Louth until the 1980s (Rotherham, 2009).

Peat became particularly important as a fuel after the widespread clearance of Ireland's forests, when timber for firewood became much scarcer. Another option would have been dried cow dung, the smell of which is sweet when burnt and not as unpleasant as one might

11. De Bhaldraithe (1959) gave the Irish words *móin* or *móina* for 'peat' (or 'turf'); *fód móna*, 'a sod of peat' (or turf); *portach móna*, 'a peat bog' and *portach* or *criathrach*, 'a bog'.

12. Dinneen (1927).

13. Dinneen (1927).

imagine. However, dung was regarded as a good fertilizer for agricultural land, so to burn it was considered poor husbandry, and it would have been used as fuel only when peat or timber was scarce or unavailable. Up to the present day, many small farms, particularly in western districts and the midlands, retain special rights to cut peat turf, called turbary rights. According to Rotherham (2009), from AD 1500 up to the 19th century, one household would consume around 8000 turves (sods of cut peat) per annum for fuel and other purposes, and 6 tonnes of peat was equivalent to 1 tonne of coal.

Industrially produced milled peat (peat in crumb or powder form) has also been an important fuel in Ireland and has been used in several power stations up to the present day for electricity production. For example, between 1906 and 1922, the Trans-Atlantic wireless station built by Guglielmo Marconi in Clifden, Connemara, was powered by turf-fuelled steam engines that used 8000 tonnes of hand-cut peat each year (Rotherham, 2009). The station was situated several miles south of the town of Clifden and was an important long-wave wireless telegraphic link between Europe and North America, from when it started operating in October 1907 until it was damaged during the Irish Civil War in 1922. At the height of its operations it employed up to 400 people. Marconi (1874–1937) was born near Bologna, Italy, the second son of Giuseppe Marconi, an Italian landowner, and his Irish wife, Annie Jameson, granddaughter of the founder of the Jameson whiskey distillery.

Peat gas, used for heating and lighting, was produced in various parts of Europe, including Ireland. Although never very commercially viable, it was used to light several small towns at one time, including Mullingar and Tuam in the 1850s, and again during 'The Emergency' during World War II. The Hamilton Robb Linen Factory in Portadown, Co. Down, was also powered by peat gas and used 3500 tonnes of peat each year (Rotherham, 2009).

Peat was also formerly used to make peat charcoal, and in iron smelting and the manufacture of gunpowder. Peat charcoal burns with greater intensity and heat than pure peat and so has been used for iron smelting in Ireland. It is less bulky too than the peat itself. In the mid-19th century, peat charcoal was used at Creevelea, Co. Leitrim, for iron smelting. To make peat charcoal, three or four cart loads of turf were stacked in long low heaps and covered with sods or clay. Then they were lit and allowed to smoulder for a day and a night, or longer, and kept airtight, otherwise the turf would burn away to ash (McGlinchey, 2007). Peat charcoal was a popular intense-burning fuel for blacksmiths in Ireland before coal became widely available. Milled peat is also still used to produce briquettes, a popular domestic fuel. Briquettes are made from milled peat compressed at high temperatures and remain an extremely popular and widely used domestic fuel source in Ireland today. They are manufactured by Bord na Móna (formerly the Turf Development Board) at two locations in Ireland, Littleton, Co. Tipperary, and Derrinlough, Co. Offaly. The peat is harvested through mechanised means from the surface of the bog as milled peat from the Bord na Móna–owned bogs and then transported by 3-foot wide, narrow-gauge railways across the bogs to the factory. The milled peat is then dried to a 10% moisture content. When

Turf and slane.

The peat harvest of blanket bog on mountains of the Dingle Peninsula, Co. Kerry.

cool, it is compressed and extruded through the briquette-making moulds before being baled and sold.

Milled peat production by Bord na Móna in Ireland is still an important industry, despite declining amounts of available peat bog, especially in the Midlands. Bord na Móna was set up in 1946 as a result of the Turf Development Act, taking over from an earlier company, the Turf Development Board Limited, established in 1933. In the year 2008–2009 the Bord na Móna Annual Report noted that almost 3 million tonnes of milled peat was produced by the company during the year, as well as 1,701,000 cubic metres of peat for horticulture, and 217,000 tonnes of peat briquettes. In the same year, sales of milled peat through Bord na Móna amounted to 3,054,000 tonnes to electricity-generating power stations and 882,000 tonnes to various Bord na Móna factories for horticultural peat and briquette manufacturing.

The widespread use of peat for electricity peaked in the mid-1900s. A sod peat-fired station was commissioned at Portarlington, Co. Laois, in 1949, followed by another similar sod-peat fired station in Allenwood, Co. Kildare, in 1951 (that station was closed in 1994). The first milled peat station was commissioned at Ferbane, Co. Offaly, in 1957. According to the Bord na Móna Annual Report and Sustainability Report (2009), the contribution of peat to the electricity sector peaked in the early 1960s, when it provided just under 40% of the total power generated. By the early 1980s the contribution of peat to Ireland's growing electricity requirements had declined significantly. Currently, peat produces around 10% of the Republic of Ireland's electricity requirements. There are six power stations generating electricity from peat today, including in West Offaly at Shannonbridge, Co. Offaly, Edenderry, Co. Offaly, and Lough Ree, Lanesborough, Co. Longford.

Other uses for milled peat have included packing material for vegetables, such as potatoes and tomatoes, absorbent bedding for livestock and even as a food additive for animals. It was believed that peat mixed into animal food would slow down the passage of food through the animals' stomach, and so more nutrients would be absorbed from the food. Peat fibre was used in the 18th and 19th century for making ropes, clothing, cardboard and rough paper. The economic plant collections at the National Botanic Gardens include examples of peat fibre cloth, string made from peat fibre as well as other peat-derived products, such as peat soap and peat oil. Peat fibres used are generally those derived from sedges, grasses and other bogland plants, rather than from *Sphagnum*.

Around 1835, peat paper was made in Ireland, and in 1903 there was a peat paper factory in Celbridge, Co. Kildare (the Callender Paper Company), that manufactured brown wrapping paper and card from moss peat fibre and waste paper pulp (Rotherham, 2009).

Landing turf from a Galway hooker on the Aran Islands.

PEAT PRODUCTION BY BORD NA MÓNA IN IRELAND

Production 2008–2009	
Milled peat—tonnes	2,910,000
Horticulture—cubic metres	1,701,000
Briquettes—tonnes	217,000
Sales	
Milled peat to power stations—tonnes	3054
Milled peat to Bord na Móna factories—tonnes	882
Total milled peat	3936
Horticulture—cubic metres	1724
Briquettes—tonnes	245
Value of Sales (€'000)	
Milled peat to power stations	94,746
Electricity	56,636
Horticulture	47,567
Briquettes	42,894

Source: Bord na Móna Annual Report and Sustainability Report (2008–2009).

The peat was brought to the factory in Celbridge by cart from a bog several miles away. At the factory it was placed in a large, spherical digester with a solution of diluted alkali. The fibre was then transferred to a large machine where it was scarified and beaten to convert it to pulp. The unbleached pulp was then passed through a series of rollers weighing about 150 tonnes to emerge as paper. The paper was then glazed and polished by another series of rollers before being cut into various sizes and tied in bundles by the girls working in the factory. Power to drive the paper mill was provided by an hydro-electric turbine driven by the water from the river Liffey nearby. The factory ran into financial trouble by late 1904, and went out of business with financial liabilities in December 1905 because it could not compete with more economical wood- and straw-pulp papers (Ryan, 1907). Postcards were made from peat paper too and were popular with tourists and to send to Irish emigrants around the world.

Compressed milled peat has been used to make ornaments, mainly for the tourist trade. Such ornaments are hard and relatively robust, somewhat resembling the rarer and more expensive, carved bog oak ornaments. Several companies in Ireland, such as Owencraft in Omagh, Northern Ireland, currently make crosses, paperweights, figurines, animals, models of Irish saints and other souvenirs from compressed peat. Another company, Island Turf Crafts, in Coalisland, Co. Tyrone, makes Celtic crosses, harps, clocks, jewellery, statues, plaques, religious objects, picture frames and trophies from compressed peat.

Moss peat, sometimes sold as peat moss (made from milled peat), is an important product for horticulture in Ireland, used in composts and as a soil additive and conditioner. As mentioned earlier, on the farm, peat was also used as litter for animals, and once soiled, it could be spread on the land as a fertilizer. In England it was widely used in the early part of the 20th century by the military as horse bedding (Rotherham, 2009).

Peat soot has been used as a dye and gives a good shade of brown to wool (Mairet, 1916). The wool is boiled for a couple of hours with the peat soot, then carefully washed with several changes of water.

Peat smoke has had important economic uses too and has been used in Ireland to flavour whiskey and smoke salmon. It was used to fuel the kilns in whiskey distilleries and also to dry the barley used, which gives the whiskey its traditional smoky flavour. Peat

Peat postcard. Text on the postcard reads: 'This Card is made from . . . Peat taken from the Bog of Allen'.

would also have been used in the production of illegal 'poteen', providing the fuel for illegal stills. Peat was apparently even used as a tobacco substitute, and Dinneen (1927) gave the Irish name *tobac móna* for 'a brownish peat' that was used like tobacco. I have even seen a tourist souvenir peat incense-burner on sale, where a small piece of peat is burnt on a tray to give the traditional and much-loved aroma (what Rotherham, 2009, calls 'peat reek') of a traditional turf fire.

Houses made from cut peat blocks were constructed in some regions of Ireland (Danaher, 1964a). Turf houses would have been a frequent feature of the practice of booleying, an Irish form of transhumance, where in some parts of Ireland during the summer months, families would move from the lowlands to the mountains to avail of summer grazing for their stock. There they lived in simple houses called 'booley-houses' or 'boliehouses'. Evans (1957) writes about the tradition of booleying in Cos. Antrim and Derry, recording that the Ordnance Survey letters of the 1830s mention that booley-houses were built of sods upon a foundation of earth and stones, roofed with bog timbers and thatched with heather, probably mainly *Calluna vulgaris*. The ruined stone houses that we see so frequently scattered throughout Ireland would only have made up a proportion of the dwellings in which the population lived in the past; there is now no trace left of the turf-built houses many called home. Although roughly made and primitive in appearance, apparently these turf houses were warm and dry, easy to build and maintain and could be constructed from materials available at no cost.

In his personal account of life and folklore in Co. Donegal in the 1800s, McGlinchey (2007: 15) wrote that he does not remember sod houses during his lifetime (1861–1954):

> [Houses] were all built of stone and plastered with blue clay and lime. The people never bothered much about whitewashing. There wasn't much order on the houses then. The *cál leannógach* (algae, wall fungus) would be growing on the walls two or three feet up, nearly as far as the window-sill. The houses were roofed with bog fir [*Pinus sylvestris*] and laths with scraths (sods) put on top of that with the clay side down. Then the houses were thatched with rushes or straw. The outhouses were often thatched with fine heather or bent from the shore [probably marram, *Ammophila arenaria*]. The thatch was tied down with home-made súgáns (straw ropes) attached to *bacáns* (hooks) stuck along the top of the walls below the eaves. The chimney was just a rod creel [basket] with no bottom.

The top sods of the bog, called 'scraws', were also used on the roofs of many thatched cottages, providing an insulating layer below the thatch onto which the thatch would be laid and secured. In some parts of Ireland, simple houses (cabins) were built of sods, both from the top of the bog and from other habitats. According to Evans (1957), the 1841 Census of Ireland revealed that at that time almost half of the rural population was living in such one-roomed mud cabins. O'Hare (1993) described how these houses could be erected in a few days. The walls would be built the first day to about 7 ft high, the next day the roof timbers and rafters were put on, and on the final day the roof was thatched. Sometimes, sods were used for walls of houses within a framework of upright timbers.

Simple furniture was also made from peat (Kinmouth, 1993), which would have been cut into the required shapes and dried.

Soot encrusted roof scraws were used for medicinal purposes too. Wood-Martin (1902: 36) described how fire was made by rubbing sticks together by the peasantry in Ireland as a form of protection or prevention of disease in cattle:

> ...when a swelling of the head or disease amongst cattle, called 'big-head,' appeared, every fire was extinguished in the townland on which it had broken out. The inhabitants then assembled at the affected farm to kindle what was called a 'need-fire,' which was done as follows: - Two men commenced to rub two sticks together till the friction produced a flame. It was hard work, each man rubbing in turn. When the sticks had ignited, they collected dry 'scraws' (sods) covered with soot from the roofs of the dwelling houses, to produce a great smoke. The affected cattle then had pieces of wood inserted in their mouths to keep them open, and the head was held over the smoke till water ran plentifully from mouth and nostrils, was completed. Every fire that had been extinguished was then re-kindled from the 'need-fire'.

FIG. 6.

Lighting the "Need-Fire." A Scene in the Nineteenth Century.

Lighting the need-fire. From Wakeman, William F. 1898. *Archaeologica Hibernica. A Hand-Book of Irish Antiquities*. Hodges, Figgis and Co., Dublin.

Peat bogs were also used to store and preserve food in the past in Ireland. Ancient butter buried in bogs and forgotten is still occasionally found today. In the past, butter was preserved through the addition of about 5% salt, which had to be dissolved before the butter was eaten. When salt was scarce, it could instead be preserved in the antiseptic conditions of the bog (Feehan & O'Donovan, 1996). Butter was buried in bogs up until the end of the 17th century, placed in baskets or wooden casks, or wrapped in cloth or skin. Preserved artefacts such as boats, weapons, jewellery and even bodies are also occasionally found in Ireland's bogs.

In parts of Europe, peat was used in health spas and baths. In Britain this practice continued up until the 1950s, and it was certainly used in centres such as Buxton and Harrowgate in England and Strathpeffer in Scotland (Rotherham, 2009). The patient was immersed in hot peat and could have electric currents passed through their body, or instead, enjoy a hot rectal peat douche or enema. A peat bath is said to increase blood circulation in the skin and have an anti-inflammatory effect. At least one company in Ireland today produces peat-based skincare products. Óg Ireland, based in Tullamore, Co. Offaly, produces face and foot masks and other cosmetic products containing peat.

By the early 20th century, peat was one of the most vital and valuable natural resources in Ireland. As an indigenous energy source its use provided a cheap and easily accessible source of fuel, not only for domestic usage, but also for electricity generation, while at the same time providing income and employment for many people in rural Ireland. Milled peat became an important soil conditioner and resource for the horticultural industry. Cut-over bogs (i.e., those where there is no peat left to harvest) have become useful natural habitats (especially wetlands) supporting nature conservation and, in some cases, used as new farmland. It is not surprising that the governmental authorities both before and since the establishment of the State have been keen to explore, promote and support the use of peat

INDUSTRIAL-SCALE USES OF PEAT AT THE BEGINNING OF THE 20TH CENTURY

Peat litter

Milled peat was used as livestock bedding and for manure (when mixed with animal dung), soil conditioning, etc.

Peat paper

Peat was used to make coarse grades of paper, such as brown wrapping paper. The peat occurring near the surface of the bog was best for making paper because it included a higher percentage of well-preserved fibres. Extracting the fibres from the general peat pulp was the main challenge in making peat paper. Various processes involving washing the peat and soaking, bleaching or washing the fibres in alkaline solutions were used to extract the fibres.

Peat-wool for clothes, carpets, antiseptic bandages, etc.

Peat fibres were used as a substitute for wool. Ryan (1907) suggested that the best type of peat to use for making a wool substitute was peat composed of *Sphagnum cuspidatum* or 'wool grasses' (*Eriophorum*). *Eriophorum* species are generally known as bog-cotton today. The best results in peat-wool making were obtained from the 'layer of wool-grass peat which usually rests on the sand of mountain-bogs under the layer of peat fuel' (Ryan 1908: 415). To manufacture peat-wool, the peat was well-washed and broken up or scarified by a rotating drum, whose surface was covered with wire needles. In some processes, such as the following, called 'Geige's Process', invented in 1898, it was washed, soaked or heated in weak solutions of alkali or acid to clean the fibres and to remove starch present. It was then fermented with yeast at a temperature of 30°C to remove any traces of glucose left. After washing, the fibres were washed in hot petroleum ether to extract fats and then finally washed in dilute alkali to remove the tannic acid, then bleached and dried. The turf-wool obtained was soft, woolly, strong and flexible and easily spun, bleached and dyed and made into garments that are cool to wear in summer and warm in winter. Ryan (1907) described a turf-wool factory in Düsseldorf, Germany, using Geige's Process, and another in Oldenburg, also in Germany. However, I have not found records of turf-wool being produced in Ireland.

Roofing material for houses

Insulating material for ice

Peat for tanning

A thick mash produced by mixing peat moss with a solution of alum and salt was useful in tanning.

Peat dye

A brown dye was produced as a by-product of the peat-paper production process. The waste alkaline liquid was concentrated and then treated with acid to precipitate the dye, then filtered and dried. When mixed with a sizing agent and water it could be used as paint.

Preservation of fruit, vegetables and manures (such as kainite); and packing of eggs, fish, meat, etc.

Filtration of oils

Manufacture of peat-molassine meal

Peat fibre mixed with molasses was used in food for cattle making it easier to digest than pure molasses.

Preparation of alcohol by fermentation of the previously hydrolysed cellulose

Crude, undried, fibrous peat was washed with water and diluted sulphuric acid, then heated in a closed boiler or digester to a temperature of 120°C for five hours. When cold, the mixture was combined with quicklime or powdered chalk to neutralise it, then concentrated by evaporating off a proportion of the water and filtered. Yeast was then added and the mixture left to ferment at a temperature of 30°C. The alcohol was then extracted by distillation. I cannot find any evidence that peat alcohol was ever successfully produced commercially in Ireland.

Source: Ryan (1907, 1908).

for many purposes. So successful have they been that today urgent efforts must be made to preserve and safeguard the best surviving intact bogs and peatland habitats, which are such an important part of Ireland's traditional culture and natural heritage.

Reports upon the Irish Peat Industries: Parts I and II, written in 1907 and 1908, describe in detail the various uses and processes involved in peat use at the large or industrial scale in Ireland and elsewhere (Ryan, 1907, 1908).

Plants and Children's Games

A wide variety of plants have been used in Ireland for children's games. When I was a child growing up in Cashel, Co. Tipperary, Limerick and on the Dingle Peninsula in Co. Kerry, we would often be told, 'get out of the house and go and amuse yourselves'. The plants we found growing around us were used as the raw materials for our play. Bows and arrows for our games of 'cowboys and Indians' were a common diversion (usually made from bamboos, which we were fortunate to have growing in a thicket at the bottom of the garden), and so were the tree houses we constructed (several were so precarious that they were quickly condemned by the adults when younger members of the family fell through the floor). Fencing with sharp sticks was another pastime we carried out with complete disregard for the possibility of losing an eye or two. A popular game involved climbing to the top of a tall, spreading *Chamaecyparis lawsoniana* and sliding down its canopy that spread like wide petticoats below. That, too, was banned when one small boy had to be taken to the hospital! In autumn, 'conkers' was another popular game, and the fruits of the horse-chestnut (*Aesculus hippocastanum*) were much sought after. In lazy, sunny summer afternoons, making daisy chains while chatting to friends was an ideal way to pass the time. Watching television was only possible in the evenings and computer games had not been invented. Today, the greater supervision of children, often for reasons of their safety, has, in part, resulted in a decline in the simple, nature-based pastimes that have become cherished memories of the childhoods of older generations in Ireland.

Danaher (1964b: 76) recorded a range of childhood games that were popular in the past in Ireland using flowers and various plant parts. He mentions 'making daisy chains [and] blowing dandelion clocks, catching bees in fairy thimbles [*Digitalis purpurea*], making boats from *feileastrom* [*Iris pseudacorus*] blades and plaiting bands and butterfly cages from rushes; little boys made whips and *bastúns* of the long soft rushes which grew in the marshy places'.

As a very young child in Cashel, I remember well the crazes that would occur from time to time for particular games, notably for hula hoops (at that time made from wood, but later from plastic) as well as for wooden spinning tops. I found the following account on the use of spinning tops from Ballon, Co. Carlow, which is very similar to my own memories of them:

> Spinning tops were a common pastime during my childhood. Children would come to the forge with cone-shaped blocks of wood into which my father would hammer a piece of spiked iron. The children would then bind the top with cord or twine and grasping the end of the cord fling the top spinning onto a flat surface. If you had the 'knack' of expert pitching your top could out-spin all others and be the last to fall, or it could strike and knock the others, cracking some of them into pieces. You were then declared the winner. Boxwood was the toughest of woods and if you were lucky enough to get it, your top could demolish all comers. Competitions were held and reputations made and lost and my father enjoyed his part in it all (see <http://ballonvillage.com>).

McGlinchey (2007) also included a chapter on children's pastimes in his book on life in Co. Donegal from the late 1800s up to the mid-1990s and detailed some of the very inventive uses for plants when nature is the playground:

> A great pastime we had, too, when herding, was plaiting rushes into the shape of a whip-handle called a bugaboo, or making rush hats. These were shaped like a dunce's cap and went up into a point. We made leggings, too, the same way. If we were herding about the woods we made bows and arrows out of sally [willow] rods and shot at birds, mostly crows and magpies. Sometimes we put up a cockshot, a stone or a turf, but we hardly ever hit anything. In the spring when the sap would be rising in the bushes every young fellow made a whistle or two for himself out of a branch of the rowan tree [*Sorbus aucuparia*]. We cut a straight piece without knots and tapped the bark round and round with the handle of a knife or stick until the bark got loose and came off in a piece. Then we cut a bit for a mouthpiece and stopped up the other end. Some would put a pea inside the whistle. You could hear it half-a-mile away. They soon got broken up in our pocket but it wasn't hard to make another (p. 124).

Aesculus hippocastanum seed 'conkers'.

A wooden spinning top from the 1940s.

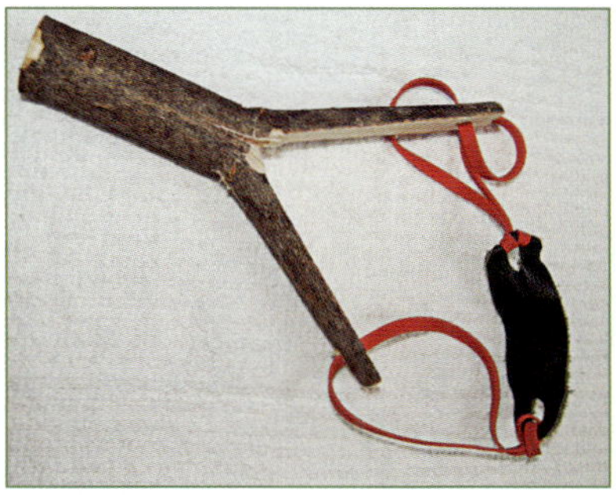

A wooden slingshot was a popular childrens' toy in Ireland.

McGlinchey (2007: 128) also wrote that 'Sitting about the fire at night the children used to light fir-splits [Scots or bog pine, *Pinus sylvestris*] and put them round and round like a wheel to see who could keep the red point the longest. They had a rhyme to say called *Roithlean rotha* but I forget it'.

The collections of the National Museum of Ireland (Museum of Country Life) in Castlebar, Co. Mayo, contain a number of plant-based children's play objects: bull-roarers from Cape Clear Island in Co. Cork (see *Fraxinus excelsior*, ash) (Ref. Nos: F1948.8 and F1934.17); a wooden whistle from Naul, Co. Dublin (Ref. No: F1948.226); a toy gun from Keenagh Beg, Co. Mayo (Ref. No: F1941.281); a buzzer from Dingle, Co. Kerry (Ref. No: F1942.60); and a spinning top from Ardea, Co. Kerry (Ref: No: 1950.164).

Bull-roarers and buzzers were spun in the air with string to make a loud noise. The Museum's collections also include a skull cracker made from bone, from Lissarulla, Co. Donegal (Ref. No: F1951.14), which is a piece of bone bored with a hole through which a piece of string was threaded. It was swung around and used for striking one's friends on the head! Larger versions were used as adult weapons in the past. Other popular toys made by children in the past in Ireland were paper hats, paper planes and gliders and variously folded paper toys (origami).

Various timbers were also used to make dolls. Most homemade dolls were usually made from cloth though. McGlinchey (2007) wrote that a handy carpenter could make a wooden man with jointed arms and legs. These could be danced on a board when you held him over it and hit the board. The dolls could also be held with a stick in their backs with the feet dangling over the board below. In Britain these were called 'jig dolls'. They were generally made for entertainment at home or by street buskers, and not usually as children's toys. A famous Irish song, 'Whiskey on a Sunday'[14], refers to these dancing dolls:

> He sits on the corner of Beggars Bush
> Astride of an old packing case
> And the dolls at the end

14. This song seems to have been written by the Liverpool poet Glyn Hughes, who died in 1972.

of the plank were dancing
As he crooned with a smile on his face

Come day go day
Wishing in me heart it was Sunday
Drinking buttermilk all the week
Whiskey on a Sunday

His tired old hands from the wooden beam
And the puppets they danced up and down
A far better show than you ever would see
In the fanciest theatre in town

But in 1902 old Seth Davey [or Step Daly] died
His song it was heard no more
The three dancing dolls
in the dustbin were thrown
And the plank went to mend the backdoor

But on some stormy night
if you're passing that way
With the wind blowing up from the sea
You can still hear the song
of old Seth Davey
As he croons to his dancing dolls three

Much of our knowledge about children's games that are no longer in practice comes from personal accounts and memories. Eibhlís Ní Shúilleabháin described an interesting game for young people from the Blasket Islands. She was a young woman who lived on the now uninhabited Great Blasket Island in Co. Kerry and carried out a correspondence with an Englishman, George Chambers, over a period of 20 years. Her letters describe much about the life of people on that island at that time. On 29 October 1931 she wrote of a Halloween game for young people that involved beans: 'Roasting beans in the fire, two together at a time and we imagine them two beans to be a boy and a girl. After taking them from the fire we throw them into a cupful of water and if they come near each, ha há dé, that is a good sign you know yourself and if they don't it doesn't matter which, we'll be laughing at the two if they are in the house' (Ní Shúilleabháin, 1988: 58).

Halloween (or *Oíche Shamhna*) was a time when other indoor plant-based games were commonly played too, most notably, bobbing for apples (*Malus domestica*) (trying to take a bite from an apple floating in a basin of water, with one's hands behind your back) or trying to take a bite from an apple hanging from a string.

Another common children's game with apples was to twist the stalk of a fruit and with each twist of the stalk, repeat a letter of the alphabet, A, B, C, etc., until the stalk broke off from the apple. The letter reached indicated the initial letter of the Christian name that one was supposed to be likely to marry. Then, the broken end of the stalk was poked repeatedly into the apple, counting through the alphabet as before. When the skin of the apple was broken this indicated the initial letter of the surname of the person to be married. Cheating was frequent, to ensure that the desired name could be achieved. I'm not sure how widespread this game was in Ireland, but I certainly remember it in Limerick and heard of it too in Co. Donegal.

The following is a list of the main wild plant species that I have been able to locate as having been used for childhood games in Ireland. Details of each can be found in the Systematic List.

WILD PLANTS USED FOR CHILDREN'S GAMES

Acer pseudoplatanus (Sycamore)*
'Helicopters'
Aesculus hippocastanum (Horse-chestnut)*
'Conkers'
Angelica sylvestris (Wild angelica)
'Scouts', 'scoots', or 'squirts', as 'pea-shooters'
Arctium minus (Burdock)
Sticking in the hair or on clothes
Bellis perennis (Daisy)
Daisy chains[1]
'She loves me, she loves me not'
Brassica oleracea (Cabbage)*
'Kicking cabbages'
Centaurea nigra (Common knapweed; Blackheads; Stickybacks)
Sticking in the hair
Chrysanthemum leucanthemum (*Leucanthemum vulgare*) (Ox-eye daisy)
'She loves me, she loves me not'
Crataegus monogyna (Hawthorn)
Berries (haws) used as ammunition for pea-shooters
Young leaves eaten as 'Bread and Cheese'
Crocosmia ×crocosmiiflora (Montbretia)*
Whistles (from the leaves)
Digitalis purpurea[2] (Foxglove)
'Cracks', popping unopened flower buds
Finger gloves
Trapping bees in the flowers ('Fairy thimbles')
Galium aparine (Robin-run-the-hedge; Cleavers; Goosegrass)
Sticking in the hair or on clothes
Grasses
Grass whistles
Hedera helix (Ivy)
Berries used as ammunition for pea-shooters
Heracleum mantegazzianum (Giant hogweed)*[3]
Pea-shooters
Heracleum sphondylium (Hogweed; Cow-parsnip)
Pea-shooters
Hordeum murinum (Wall barley)*
'Darts'
Ilex aquifolium (Holly)
'Building the house'
'She loves me, she loves me not'
Iris pseudacorus (Yellow flag)
Toy boats
Juncus species (Rushes)
Plaited and woven hats, leggings, whip-handles (bugaboos), baskets and various toys
Lonicera periclymenum (Honeysuckle; Woodbine)
Sucking the nectar from flowers
Papaver species
'Poppy dolls'
Pinus sylvestris (Scots pine; Bog pine)
Burning 'fir-splits' around the fire (see above)
Plantago lanceolata (Ribwort plantain)
'Cocks', 'Fighting cocks' or 'Soldiers'
Primula veris (Cowslip)
'Cowslip balls'
Ranunculus species (Buttercups)
'Loves butter'
Rosa species (Rose)
Itching powder from the seed hairs
Salix species (Willows)
Bows and arrows
Sambucus nigra (Elder)
Pea-shooters, pop-guns and water-guns
Solanum tuberosum (Potato)*
Ammunition for a spud gun
Sorbus aucuparia (Rowan or Mountain ash)
Whistles
Stellaria holostea (Greater stitchwort)
Pop the seedpods
Succisa pratensis (Devil's-bit scabious)
'Boomerangs'
Twisting stalks
Taraxacum officinale agg. (Dandelion)
'Dandelion clock'
'She loves me, she loves me not'
Trifolium pratense (Red clover)
Sucking the nectar from flowers
Urtica dioica (Nettle)
'Nettlemas Day stinging'

*Non-natives.
1. In Irish, slabhra nóiníní (Ó Dónaill, 1977).
2. If consumed, *Digitalis purpurea* is quite poisonous.

Prunus spinosa (blackthorn). Watercolour by Lydia Shackleton. Courtesy of the National Botanic Gardens, Glasnevin.

NATURE'S LARDER:
WILD PLANTS USED AS FOOD

THE STAPLE DIET of the early Irish, before the introduction of the potato, consisted of bread and milk products (Kelly, 1997). From the earliest times until the 17th century, oats, barley, wheat and rye (in that order of importance) were the mainstays of the everyday food (Lucas, 1960). Important foods were oaten cakes, cheese, curds, milk, cattle blood and eggs, and in the richer areas and households, beef, pork and lamb (Hobhouse, 1986). A Frenchman's account from 1644 (translated by Crofton Croker and published in the *Irish Penny Journal* of 1840) provides an interesting and useful picture of Irish life nearly 400 years ago, before the growth in dominance of the potato as a food source:

> The Irish gentlemen eat a great deal of meat and butter, and but little bread. They drink milk, and beer, into which they put laurel leaves, and eat bread baked in the English manner. The poor grind barley and peas between two stones, and make it into bread, which they cook upon a small iron table heated on a tripod; they put into it some oats, and this bread, which in the form of cakes they call haraan, they eat with great draughts of buttermilk. Their beer is very good, and the eau de vie, which they call *brandovin* [brandy] excellent. The butter, the beef, and the mutton, are better than in England (Anonymous, 1840: 5).

In his book *Seeds of Change: Five Plants That Transformed Mankind*, Hobhouse (1986) pointed out that the damp climate of Ireland was not very suitable for the cultivation of wheat, and the growing season was too short for adequate ripening of crops, therefore bread made from wheat was not an early staple food. Wheat crops had to be harvested before they were ripe and were burnt; scorching the immature cereal was the only way to make it edible.

Another early traveller in Ireland in 1670, referred to by Wood-Martin (1902: 98), described the meagre Irish food: 'The Irish feed much upon herbs, watercresses, shamrocks, mushrooms, and roots. They also take beef broth, and flesh, sometimes raw, from which they have pressed out the blood. They do not care much for bread; but they give the corn to their horses, of whom they are very careful'. Wood-Martin (1902: 99) reminds us that grain from many plants can be used as food and ground into a kind of flour, and he said, 'for most grasses produce very small edible kernels, as do also many trees and bushes, whilst nuts, acorns, dried roots of plants, and fruits, can all be reduced to a rough species of flour'.

Lucas (1960: 31) said that quite a number of plants, 'some wild, some, apparently cultivated, were used as food from earliest times'. Of the wild plants, one which was very frequently collected and eaten was watercress (*biolar*) (*Nasturtium* species), as well as sorrel (*samhadh*) (*Rumex* species), nettles (*Urtica dioica*) and wild garlic (*creamh*) (*Allium ursinum*). Of the wild fruits collected, Lucas (1960: 37) said that in addition to hazelnuts (*Corylus avellana*), which are mentioned so often in early texts as important food, the list of fruits available was quite limited and 'the sloe, wild cherry, raspberry, blackberry, strawberry, rowan, crabapple, elderberry, whortleberry and cranberry virtually exhaust the list'.

In the 1800s most people living in the countryside subsisted on home-produced food. McGlinchey (2007) described how in Co. Donegal in the 1800s, cereal crops were grown and threshed, and from that crop, every house got a *mealdar*, or a measure of meal, which would last them for a year. For breakfast, porridge made from oats (*brachán*) and milk, or else oaten bread, was eaten. In the middle of the day, at 'dinner-time', potatoes were eaten with salt and buttermilk, or else fish, such as herrings, or rabbit. Less usual foods (i.e., other than staple cereals and potatoes) were called 'kitchen'. In the evening, oaten bread and milk was again eaten and a pot of *brachán* for supper. *Sabhans* was also taken, made from the dust of the oaten meal. It was boiled and became a solid, very nutritious, pudding-like food, similar to that obtained from the seaweed, carrageen moss. Bread was made from oat flour and wheat flour. 'Boxty' (in Irish, *steaimpí*, Ó Dónaill, 1977), or potato bread, was made by mixing grated raw potatoes with flour. Other foods included pig meat, butter and eggs from hens, geese and ducks. In former times, too, a food was also made from the blood of cattle. The cattle were bled in summertime from a vein in their necks, and the blood obtained was then boiled with oat meal to make a nutritious food, very similar to the present-day popular black pudding. MacConghail (1987: 47) described the everyday (plant) food of the Blasket Islanders in Co. Kerry as 'harvested potatoes, cabbage, turnips, parsnips, corn and wheat and the hay to provide for their animals in the winter', which was probably typical of most rural communities in Ireland in the 1800s and 1900s.

If we go back to Ireland's earliest human history to the times of the first hunter–gatherers, wild plant foods would undoubtedly have formed an important part of their diets. We really still know very little about what wild foods were gathered, how they were prepared, stored, cooked and eaten. However, wild collected nuts (hazel, in particular), berries and other fruits, roots and shoot vegetables would have been gathered and eaten, at least at particular times of the year. Such information is rarely part of the evidence that can be obtained from archaeological excavations, and of course there is no written record. Some hints may be obtained from studying the ways in which hunter–gatherer societies and cultures in other parts of the world use or formerly used the same or similar plants (such as amongst the Native Americans, or American Indians as they are often called). Cultural groups didn't always evolve and employ the same methods, or even use the same plants for the same purposes, however, so we should be careful in cross-applying what we know of traditional practices in, for example, northern Asia or America, to ancient Ireland. In any case, some of the most important plants used in these regions did not occur in Ireland (such as the three most famous Indian staple foods—corn [maize], squash and beans). In Ireland, winter diets were probably largely made up of meat

A dinner of potatoes served from a willow skib in an Irish cottage, ca. early 1900s.

Taraxacum officinale agg. Dandelion flowers can be used to make syrup (left); the flowers are boiled to extract flavour (middle). The roots may be roasted and ground for coffee (right).

and fish, but some foods may have been stored during times of plenty for use later in the year. In North America the first settlers in Virginia and the east coast of what is now the United States found underground storage pits lined and covered with tree bark, in which surplus food was placed against the winter (Kavasch, 2005). Underground chambers and tunnels are regularly found associated with ancient settlement in Ireland. These *souterraines* may have been used to store food, kept cool and out of the way of marauding animals. Once farming became established in Ireland, the diet would have evolved to include a higher proportion of plant-derived foods (from cereal crops) and less hunted meat, but some wild gathered plant foods probably still remained important.

CATEGORIES OF FOOD PLANTS

- Asparagus-like shoot vegetables
- Candied sweets
- Coffee substitutes
- Cooked vegetables (edible parts other than green leaves and roots)
- Drinks (non-alcoholic, non-herbal tea, cold and hot)
- Edible fruit and fruit leathers
- Flour, cereals, meal, edible seeds and starch sources
- Fritters
- Green leaf vegetables
- Herbal teas
- Herbs and flavourings
- Jams and jellies
- Nuts
- Oils
- Pickles
- Roots, rhizomes and tubers
- Salads and as a raw vegetable
- Seasoning and condiments
- Soups
- Syrups and sugars
- Vinegars
- Wines, beers and liqueurs

Collecting and Cooking Wild Foods—Some Suggestions and Tips

In this book only very simple guidance is given on the collection and cooking of edible wild foods. The principles in using wild foods are the same as those in any general cookery, and users can be guided by instructions in favourite recipe books. However, I have provided a few tips below related to my experience in collecting, preparing, cooking and eating wild foods. My main advice is to experiment and use what you like. Some wild plants are quite simply not to my taste.

After you collect wild plant materials make sure you wash them well and dry them by dabbing with a dry, clean cloth. Store the material in the refrigerator until you are ready to use it, but don't keep it too long—enjoy it while it is still fresh. If you steam the vegetables you will preserve more of the vitamins they contain. However, some species require to be boiled to reduce their bitterness and destroy or remove any of the toxins they may contain. Don't leave vegetables soaking in water, which will reduce their vitamin content. If vegetables are too tough, they can be softened with a little addition of baking soda to the water in which they are boiled, but remember that will greatly reduce their vitamin content.

Asparagus-type vegetables are best collected in the spring when there is fresh, soft, new growth. Remember that when shoots first emerge, it may not be possible to identify the plant with certainty, so make sure you know what you are collecting and eating.

Instructions on how to prepare candied plant material is given under the entry for *Angelica sylvestris*. Older recipe books may also give you good instructions or tips to follow. Some species, such as hazelnuts, can also be preserved in sugar syrup. Leave them in the syrup for sufficient time to ensure that they are completely saturated with the sugar solution before using them. This may take some weeks before they are as rich and delicious as they can be!

Plant material used as a coffee substitute will need to be roasted and ground. Roast the material (e.g., roots) in a low oven at a temperature of between 80°C and 120°C (175°F and 250°F). When it is fully roasted (brown throughout), it can be ground down and used. Roasted coffee substitutes should be stored in the deep freeze if you don't intend to use them immediately.

Fruits can be dried for storage, as well as used fresh of course. To dry fruits, use an oven or leave them spread out on a tray on a warm sunny windowsill. The making of fruit leathers is described under *Prunus spinosa*.

Fritters are made by dipping edible flowers into batter and then frying them. To make batter, sift 125 g (4 oz) of flour with a little salt into a bowl, add in a few tablespoons of oil, some water and one egg white. Whisk all together until the mixture has the consistency of thick cream. A pancake mixture of flour, egg and milk can also be used to make fritters.

Most good recipe books provide instructions on how to make jams and jellies. Some fruits are high in pectin, which helps the jam or jelly to set. Pectin is available commercially and can be added to the mixture. Alternatively, fruit can be mixed with another pectin-rich fruit (such as crab apples). I often use commercial 'jam sugar' with pectin already mixed in. Remember that when bottling your jams and jellies to sterilise the jars (and lids) that you use well in the oven or in boiling water. If you do, the contents will last much longer. Also, remember to fill the jars close to the top. I have successfully stored (in a cool cupboard) some jams and jelly that have been fine to use even a couple of years later. The best way to make good jams and jellies is to experiment. Remember, if the mixture does not set no matter what you try, you can always pretend that you meant to make syrup all along (and use it as a topping for ice cream or pancakes).

Extracting vegetable oils from wild plants is not easy and somewhat laborious. To extract oil from hazelnuts for instance, shell the nuts then crush them. Place the broken nuts into a pan of water and bring to the boil. Skim off the oil when it rises to the surface. Store in a sealed container in a refrigerator and use it within a couple of weeks before it goes rancid. Oil from hazelnuts is best used in salads, as it breaks down under the high temperatures used for frying.

Flour can be made from a variety of different parts of wild plants, including pollen, the inner bark of some trees, seeds, starch from roots and rhizomes. To make flour from seeds, the grain must be dried and then winnowed to remove the husks. When the grain is very dry, it is rubbed between the hands to separate the husks from the kernels. The kernels can be cleaned by simply gently blowing away the husks. When all the kernels remain, they can be ground down to make flour. Sifting after grinding helps to remove any fibres that may be present in the flour. The process of extracting flour from roots or rhizomes is described under *Typha latifolia*. Small seeds are very difficult to winnow and make into flour, and so instead they can be ground down husks and all, boiled and made into a porridge-type meal.

In using wild plants in salads, make sure that you select the freshest and most tender leaves you can. Discard any leaves that are damaged or discoloured. If a particular wild plant has a very strong flavour, you can mix with other leaves that are milder. Remember to gather salad plants from habitats that you are sure are clean and unpolluted (e.g., be cautious of plants, such as water-cress, that sometimes grow in wet drains that could be fed with sewage run-off or be infected with parasitic flatworms such as liver fluke).

Edible Plants

A list is given here of the main species found in Ireland that are known to be edible and have been used for food either in Ireland or elsewhere in the world. I have divided them into broad categories, such as plants used as leaf vegetables, edible roots, herbal teas, fruits, salad plants and coffee substitutes. Users of the list should treat them as look-up lists for the main entries for each species in the Systematic List. This is important because these entries often indicate which part of the plant is edible and how it may be used. For example, some plants may have edible roots, but poisonous fruits or other parts and so referring to the main entries for each is essential.

Please remember that the lists are not intended to be fool-proof guides to what can be eaten and what is not edible amongst the flora. The lists simply record species that are reported to have been eaten in the past in Ireland or elsewhere. However, do remember that there are relatively few seriously poisonous plants in the Irish flora and not enough to deter most users from sampling and enjoying wild plants as food. Despite the fact that some of the plants are listed, it does not necessarily mean that they taste good!

Vaccinium myrtillus (bilberry). Watercolour by Lydia Shackleton. Courtesy of the National Botanic Gardens, Glasnevin.

EDIBLE WILD IRISH PLANTS

(see Systematic List for additional information)

Asparagus-like shoot vegetables

Acorus calamus—Sweet-flag
Arctium lappa—Great burdock
Arctium minus—Burdock
Asparagus officinalis subsp. *prostratus*—Wild asparagus
Cirsium arvense—Creeping thistle
Cirsium palustre—Marsh thistle
Cirsium vulgare—Common thistle
Crambe maritima—Seakale
Epilobium angustifolium—Rose-bay willow-herb; Fireweed
Equisetum arvense—Field horsetail
Equisetum palustre—Marsh horsetail
Fallopia japonica (*Polygonum cuspidatum; Reynoutria japonica*)—Japanese knotweed
Heracleum sphondylium—Hogweed; Cow-parsnip
Salicornia species—Glasswort
Smyrnium olusatrum—Alexanders
Symphytum officinale—Comfrey
Typha latifolia—Bulrush; Reed mace

Candied sweets

Acorus calamus—Sweet-flag
Althaea officinalis—Marsh-mallow
Angelica sylvestris—Wild angelica
Arctium lappa—Great burdock
Arctium minus—Burdock
Berberis vulgaris—Barberry
Inula helenium—Elecampane
Viola odorata—Sweet violet

Coffee substitutes

Cichorium intybus—Chicory
Cytisus scoparius—Broom (toxic at over-consumption)
Fagus sylvatica—Beech
Fragaria vesca—Wild strawberry
Galium aparine—Robin-run-the-hedge; Cleavers; Goosegrass
Quercus petraea—Sessile oak; Irish oak
Quercus robur—Pedunculate oak; English oak
Silybum marianum—Milk thistle
Taraxacum officinale agg.—Dandelion

Cooked vegetables (of which parts are used other than just for their green leaves)

Alisma plantago-aquatica—Water-plantain
Angelica sylvestris—Wild angelica
Arctium lappa—Great burdock
Arctium minus—Burdock
Crambe maritima—Seakale
Daucus carota—Carrot
Equisetum sylvaticum—Wood horsetail (rhizome buds only)
Erodium cicutarium—Common stork's-bill
Fallopia japonica (*Polygonum cuspidatum; Reynoutria japonica*)—Japanese knotweed
Filipendula ulmaria—Meadow-sweet
Galium aparine—Robin-run-the-hedge; Cleavers; Goosegrass
Geum urbanum—Wood avens
Glaux maritima—Sea-milkwort
Halimione portulacoides—Sea purslane
Honckenya peploides—Sea sandwort
Lathyrus japonicus subsp. *maritimus*—Sea pea
Malva neglecta—Dwarf mallow
Malva sylvestris—Common mallow
Nymphaea alba—White water-lily
Silybum marianum—Milk thistle
Tragopogon pratensis—Goat's-beard
Trifolium dubium—Yellow clover; Shamrock
Trifolium repens—White clover
Typha latifolia—Bulrush; Reed mace
Ulmus glabra—Wych elm; Irish elm
Vicia sativa—Common vetch

Drinks and cordials (non-alcoholic, non-herbal tea drinks, cold and hot)

Adiantum capillus-veneris—Maidenhair fern
Arctium lappa—Great burdock
Arctium minus—Burdock
Berberis vulgaris—Barberry
Filipendula ulmaria—Meadow-sweet
Geum rivale—Water avens
Hippophae rhamnoides—Sea buckthorn
Mentha aquatica—Water mint
Ribes nigrum—Black currant
Rubus chamaemorus—Cloudberry
Rumex acetosella—Sheep's sorrel
Sambucus nigra—Elder
Taraxacum officinale agg.—Dandelion
Vaccinium myrtillus—Bilberry; Fraughan
Vaccinium vitis-idaea—Cowberry

Edible fruits and fruit leathers

Arbutus unedo—Strawberry-tree
Arctostaphylos uva-ursi—Bearberry

EDIBLE WILD IRISH PLANTS *(continued)*

(see Systematic List for additional information)

Edible fruits and fruit leathers *(continued)*

- *Berberis vulgaris*—Barberry
- *Crataegus monogyna*—Hawthorn
- *Empetrum nigrum*—Crowberry
- *Ficus carica*—Fig
- *Fragaria vesca*—Wild strawberry
- *Fuchsia magellanica*—Fuchsia
- *Hippophae rhamnoides*—Sea buckthorn
- *Juniperus communis*—Juniper
- *Malus sylvestris*—Crab-apple
- *Prunus avium*—Wild cherry
- *Prunus domestica*—Bullace; Wild plum
- *Prunus padus*—Bird cherry
- *Prunus spinosa*—Sloe; Blackthorn
- *Pyrola minor*—Common wintergreen
- *Ribes nigrum*—Black currant
- *Ribes rubrum*—Red currant and White currant
- *Ribes uva-crispa*—Gooseberry
- *Rosa* species—Rose
- *Rubus caesius*—Dewberry
- *Rubus chamaemorus*—Cloudberry
- *Rubus fruticosus* agg.—Bramble; Blackberry
- *Rubus idaeus*—Wild raspberry
- *Rubus saxatilis*—Stone bramble
- *Sambucus nigra*—Elder (cooked only)
- *Sorbus aucuparia*—Mountain ash; Rowan
- *Sorbus devoniensis*—Devon whitebeam
- *Sorbus hibernica*—Irish whitebeam
- *Vaccinium myrtillus*—Bilberry; Fraughan
- *Vaccinium oxycoccos*—Cranberry
- *Vaccinium vitis-idaea*—Cowberry
- *Viburnum opulus*—Guelder rose (cooked only)

Flour, cereals, meal, edible seeds and starch sources

- *Arum maculatum*—Lords and Ladies; Cuckoo-pint
- *Betula pendula*—Silver birch (bark)
- *Betula pubescens*—Downy birch (bark)
- *Capsella bursa-pastoris*—Shepherd's purse
- *Castanea sativa*—Spanish chestnut, Sweet chestnut
- *Chenopodium album*—Fat hen
- *Cladium mariscus*—Saw sedge
- *Corylus avellana*—Hazel
- *Crataegus monogyna*—Hawthorn
- *Dactylorhiza* species—Spotted- and Marsh-orchids
- *Deschampsia cespitosa*—Tufted hair-grass
- *Fallopia convolvulus*—Black bindweed
- *Festuca arundinacea*—Tall fescue
- *Hippophae rhamnoides*—Sea buckthorn
- *Lathyrus montanus*—Bitter vetch
- *Lathyrus pratensis*—Meadow vetchling
- *Malva sylvestris*—Common mallow
- *Medicago lupulina*—Black medick
- *Medicago sativa* subsp. *sativa*—Lucerne
- *Menyanthes trifoliata*—Bogbean
- *Nuphar lutea*—Yellow water-lily
- *Nymphaea alba*—White water-lily
- *Papaver rhoeas*—Corn poppy; Common poppy
- *Papaver somniferum*—Opium poppy
- *Persicaria hydropiper*—Water pepper
- *Persicaria lapathifolia*—Pale persicaria
- *Phragmites australis* (*Phragmites communis*)—Reed
- *Pinus sylvestris*—Scots pine (pollen)
- *Plantago coronopus*—Buck's-horn plantain
- *Plantago lanceolata*—Ribwort plantain
- *Plantago major*—Greater plantain
- *Plantago maritima*—Sea plantain
- *Polygonum arenastrum*—Equal-leaved knotgrass
- *Polygonum aviculare*—Knotgrass
- *Polygonum maritimum*—Sea knotgrass
- *Polygonum oxyspermum* subsp. *raii*—Ray's knotgrass
- *Potentilla anserina*—Silverweed
- *Pteridium aquilinum*—Bracken
- *Quercus petraea*—Sessile oak; Irish oak
- *Quercus robur*—Pedunculate oak; English oak
- *Raphanus raphanistrum*—Wild radish
- *Schoenoplectus lacustris*—Common club-rush
- *Sinapis arvensis*—Charlock
- *Sorbus aucuparia*—Mountain ash; Rowan
- *Stellaria media*—Chickweed
- *Spergula arvensis*—Corn spurrey
- *Trifolium repens*—White clover
- *Typha latifolia*—Bulrush; Reed mace
- *Ulmus glabra*—Wych elm; Irish elm

Fritters

- *Sambucus nigra*—Elder (flowers)
- *Symphytum officinale*—Comfrey
- *Taraxacum officinale* agg.—Dandelion

Green leafy vegetables

- *Achillea millefolium*—Yarrow
- *Aegopodium podagraria*—Goutweed; Ground-elder; Bishop's-weed
- *Allium ursinum*—Wild garlic; Ramsons
- *Althaea officinalis*—Marsh-mallow

EDIBLE WILD IRISH PLANTS *(continued)*

(see Systematic List for additional information)

Green leafy vegetables *(continued)*

Aquilegia vulgaris—Columbine (leaves only) (Caution!)
Arctium lappa—Great burdock
Arctium minus—Burdock
Atriplex patula—Common orache
Atriplex prostrata (*Atriplex hastata*)—Spear-leaved orache
Barbarea vulgaris—Winter-cress; Yellow rocket
Bellis perennis—Daisy
Beta vulgaris subsp. *maritima*—Sea beet
Caltha palustris—Marsh-marigold; Kingcup (Caution!)
Capsella bursa-pastoris—Shepherd's purse
Centranthus ruber—Red valerian
Chenopodium album—Fat hen
Chenopodium bonus-henricus—Good King Henry
Chenopodium rubrum—Red goosefoot
Cichorium intybus—Chicory
Crambe maritima—Seakale
Crithmum maritimum—Rock samphire
Epilobium angustifolium—Rose-bay willow-herb; Fireweed
Filipendula ulmaria—Meadow-sweet
Galinsoga species—Gallant-soldier
Galium aparine—Robin-run-the-hedge; Cleavers; Goosegrass
Hypochaeris radicata—Cat's-ear
Lapsana communis—Nipplewort
Leontodon hispidus—Rough hawkbit
Melilotus altissimus—Tall melilot
Melilotus officinalis—Ribbed melilot
Mertensia maritima—Oysterplant
Moehringia trinervia—Three-nerved sandwort
Nasturtium microphyllum—Narrow-fruited water-cress
Nasturtium officinale—Water-cress
Oenothera biennis—Common evening-primrose
Oenothera erythrosepala—Large-flowered evening-primrose
Papaver rhoeas—Corn poppy; Common poppy
Persicaria lapathifolia—Pale persicaria
Persicaria maculosa—Redshank or Spotted persicaria
Pinus sylvestris—Scots pine
Plantago coronopus—Buck's-horn plantain
Plantago lanceolata—Ribwort plantain
Plantago major—Greater plantain
Plantago maritima—Sea plantain
Primula vulgaris (*Primula acaulis*)—Primrose
Rhodiola rosea (*Sedum rosea*)—Roseroot
Rumex acetosa—Sorrel
Rumex acetosella—Sheep's sorrel
Rumex crispus—Curled dock

Leaves of *Rumex acetosa* (sorrel).

Sanguisorba minor—Salad burnet
Sanguisorba officinalis—Great burnet
Sedum telephium—Orpine
Smyrnium olusatrum—Alexanders
Silene vulgaris—Bladder campion
Silybum marianum—Milk thistle
Sinapis arvensis—Charlock
Sisymbrium officinale—Hedge mustard
Sonchus arvensis—Perennial sow-thistle
Sonchus asper—Spiny sow-thistle
Sonchus oleraceus—Smooth sow-thistle
Stellaria graminea—Lesser stitchwort
Stellaria media—Chickweed
Suaeda maritima—Annual sea-blite
Symphytum officinale—Comfrey
Taraxacum officinale agg.—Dandelion
Thlaspi arvense—Field penny-cress
Tragopogon pratensis—Goat's-beard
Ulmus glabra—Wych elm; Irish elm
Urtica dioica—Nettle
Urtica urens—Small nettle
Veronica beccabunga—Brooklime
Vicia sativa—Common vetch
Vicia sepium—Bush vetch

Herbal teas

Achillea millefolium—Yarrow
Agrimonia eupatoria—Agrimony

EDIBLE WILD IRISH PLANTS *(continued)*

(see Systematic List for additional information)

Herbal teas *(continued)*

Agrimonia procera—Fragrant agrimony
Artemisia vulgaris—Mugwort
Calluna vulgaris—Heather; Ling
Castanea sativa—Spanish chestnut; Sweet chestnut
Chamaemelum nobile (*Anthemis nobilis*)—Chamomile
Chrysanthemum leucanthemum (*Leucanthemum vulgare*)—Ox-eye daisy
Clinopodium ascendens—Calamint
Conopodium majus—Pignut
Epilobium angustifolium—Rose-bay willow-herb; Fireweed
Erica cinerea—Heather; Bell heather
Filipendula ulmaria—Meadow-sweet
Foeniculum vulgare—Fennel
Fragaria vesca—Wild strawberry
Glechoma hederacea—Ground ivy
Humulus lupulus—Hop
Juniperus communis—Juniper
Malva sylvestris—Common mallow
Matricaria discoidea—Pineapple-weed
Medicago sativa subsp. *sativa*—Lucerne
Mentha aquatica—Water mint
Origanum vulgare—Marjoram; Oregano
Petroselinum crispum—Parsley
Pinus sylvestris—Scots pine
Potentilla fruticosa—Shrubby cinquefoil
Primula veris—Cowslip
Prunus domestica—Bullace; Wild plum
Prunus padus—Bird cherry
Prunus spinosa—Sloe; Blackthorn
Pyrola minor—Common wintergreen
Ribes nigrum—Black currant
Rosa species—Rose
Rubus caesius—Dewberry
Rubus fruticosus agg.—Bramble; Blackberry
Rubus idaeus—Wild raspberry
Saxifraga oppositifolia—Purple saxifrage
Symphytum officinale—Comfrey
Taraxacum officinale agg.—Dandelion
Thymus polytrichus (*Thymus drucei; Thymus praecox; Thymus serpyllum*)—Wild thyme
Ulex europaeus—Gorse; Whin; Furze
Ulex gallii—Western gorse; Mountain gorse; Autumn gorse
Urtica dioica—Nettle
Urtica urens—Small nettle
Valeriana officinalis—Wild valerian
Verbascum thapsus—Mullein
Veronica chamaedrys—Germander speedwell

Herbs and flavourings

Alliaria petiolata—Garlic mustard; Hedge garlic
Allium schoenoprasum—Chives
Allium ursinum—Wild garlic; Ramsons

***Allium ursinum* (wild garlic).**

Angelica sylvestris—Wild angelica
Anthriscus sylvestris—Cow parsley
Artemisia absinthium—Wormwood
Artemisia vulgaris—Mugwort
Berberis vulgaris—Barberry
Filipendula ulmaria—Meadow-sweet
Foeniculum vulgare—Fennel
Geum urbanum—Wood avens
Ligusticum scoticum—Lovage
Melilotus officinalis—Ribbed melilot
Mentha aquatica—Water mint
Myrica gale—Bog myrtle
Myrrhis odorata—Sweet cicely
Origanum vulgare—Marjoram; Oregano
Petroselinum crispum—Parsley
Salvia horminioides—Wild clary; Wild sage
Tanacetum vulgare—Tansy
Thymus polytrichus (*Thymus drucei*; *Thymus praecox*; *Thymus serpyllum*)—Wild thyme

Jams and jellies

Arbutus unedo—Strawberry-tree
Berberis vulgaris—Barberry
Crataegus monogyna—Hawthorn
Empetrum nigrum—Crowberry
Eryngium maritimum—Sea holly (roots)

EDIBLE WILD IRISH PLANTS *(continued)*

(see Systematic List for additional information)

Jams and jellies *(continued)*

- *Fallopia japonica* (*Polygonum cuspidatum; Reynoutria japonica*)—Japanese knotweed
- *Ficus carica*—Fig
- *Fragaria vesca*—Wild strawberry
- *Hippophae rhamnoides*—Sea buckthorn
- *Malus sylvestris*—Crab-apple
- *Prunus avium*—Wild cherry
- *Prunus domestica*—Bullace; Wild plum
- *Prunus spinosa*—Sloe; Blackthorn
- *Ribes nigrum*—Black currant
- *Ribes rubrum*—Red currant and White currant
- *Ribes uva-crispa*—Gooseberry
- *Rosa* species—Rose
- *Rubus caesius*—Dewberry
- *Rubus chamaemorus*—Cloudberry
- *Rubus fruticosus* agg.—Bramble; Blackberry
- *Rubus idaeus*—Wild raspberry
- *Rubus saxatilis*—Stone bramble
- *Sambucus nigra*—Elder
- *Sorbus aucuparia*—Mountain ash; Rowan
- *Sorbus devoniensis* –Devon whitebeam
- *Sorbus hibernica*—Irish whitebeam
- *Tussilago farfara*—Colt's foot
- *Vaccinium myrtillus*—Bilberry; Fraughan
- *Vaccinium oxycoccus*—Cranberry
- *Vaccinium vitis-idaea*—Cowberry
- *Viburnum opulus*—Guelder rose

Nuts

- *Corylus avellana*—Hazel
- *Fagus sylvatica*—Beech

Oils

- *Brassica napus*—Swede; Oil-seed rape
- *Cornus sanguinea*—Dogwood
- *Corylus avellana*—Hazel
- *Fagus sylvatica*—Beech
- *Origanum vulgare*—Marjoram; Oregano

Pickles

- *Aphanes arvensis*—Parsley-piert
- *Aphanes australis*—Slender parsley-piert
- *Bellis perennis*—Daisy
- *Caltha palustris*—Marsh-marigold; Kingcup
- *Crambe maritima*—Seakale
- *Crithmum maritimum*—Rock samphire
- *Cytisus scoparius*—Broom
- *Fraxinus excelsior*—Ash
- *Glaux maritima*—Sea-milkwort
- *Halimione portulacoides*—Sea purslane
- *Rhodiola rosea* (*Sedum rosea*)—Roseroot
- *Salicornia* species—Glasswort
- *Stachys palustris*—Marsh woundwort
- *Typha latifolia*—Bulrush; Reed mace

Roots, rhizomes and tubers

- *Alisma plantago-aquatica*—Water-plantain
- *Althaea officinalis*—Marsh-mallow
- *Anacamptis morio* (*Orchis morio*)—Green-winged orchid
- *Armoracia rusticana*—Horse-radish
- *Bolboschoenus maritimus* (*Scirpus maritimus*)—Sea club-rush
- *Brassica napus*—Swede; Oil-seed rape
- *Brassica rapa*—Wild turnip
- *Butomus umbellatus*—Flowering rush
- *Centranthus ruber*—Red valerian
- *Cirsium vulgare*—Common thistle
- *Conopodium majus*—Pignut
- *Daucus carota*—Carrot
- *Elytrigia repens* (*Elymus repens*)—Scutch; Couch-grass
- *Eryngium maritimum*—Sea holly
- *Filipendula vulgaris*—Dropwort
- *Foeniculum vulgare*—Fennel
- *Glaux maritima*—Sea-milkwort
- *Lathyrus montanus*—Bitter vetch
- *Menyanthes trifoliata*—Bogbean
- *Myrrhis odorata*—Sweet cicely
- *Nuphar lutea*—Yellow water-lily
- *Nymphaea alba*—White water-lily
- *Oenothera biennis*—Common evening-primrose
- *Oenothera erythrosepala*—Large-flowered evening-primrose
- *Ononis repens*—Common restharrow
- *Orchis mascula*—Early purple orchid
- *Pastinaca sativa*—Wild parsnip
- *Persicaria vivipara*—Alpine bistort
- *Phragmites australis* (*Phragmites communis*)—Reed
- *Potamogeton natans*—Broad-leaved pondweed
- *Potentilla anserina*—Silverweed
- *Pteridium aquilinum*—Bracken
- *Sagittaria sagittifolia*—Arrowhead
- *Schoenoplectus lacustris*—Common club-rush
- *Sedum telephium*—Orpine
- *Smyrnium olusatrum*—Alexanders

EDIBLE WILD IRISH PLANTS *(continued)*

(see Systematic List for additional information)

Roots, rhizomes and tuber *(continued)*

- *Solanum tuberosum*—Potato
- *Stachys palustris*—Marsh woundwort
- *Taraxacum officinale* agg.—Dandelion
- *Tragopogon pratensis*—Goat's-beard
- *Typha latifolia*—Bulrush; Reed mace

Salads and as a raw vegetable

- *Acorus calamus*—Sweet-flag
- *Aegopodium podagraria*—Goutweed; Ground-elder; Bishop's-weed
- *Alliaria petiolata*—Garlic mustard; Hedge garlic
- *Allium schoenoprasum*—Chives
- *Allium ursinum*—Wild garlic; Ramsons
- *Althaea officinalis*—Marsh-mallow
- *Angelica sylvestris*—Wild angelica
- *Anthriscus sylvestris*—Cow parsley
- *Aphanes arvensis*—Parsley-piert
- *Aphanes australis*—Slender parsley-piert
- *Aquilegia vulgaris*—Columbine (leaves only)
- *Armoracia rusticana*—Horse-radish
- *Barbarea vulgaris*—Winter-cress; Yellow rocket
- *Bellis perennis*—Daisy
- *Beta vulgaris* subsp. *maritima*—Sea beet
- *Cakile maritima*—Sea rocket
- *Capsella bursa-pastoris*—Shepherd's purse
- *Cardamine flexuosa*—Wood bittercress
- *Cardamine hirsuta*—Hairy bittercress
- *Cardamine pratensis*—Cuckoo-flower; Lady's smock
- *Centranthus ruber*—Red valerian
- *Chenopodium album*—Fat hen
- *Chenopodium bonus-henricus*—Good King Henry
- *Chrysanthemum leucanthemum* (*Leucanthemum vulgare*)—Ox-eye daisy
- *Chrysosplenium oppositifolium*—Golden saxifrage
- *Cichorium intybus*—Chicory
- *Corylus avellana*—Hazel
- *Crataegus monogyna*—Hawthorn
- *Crithmum maritimum*—Rock samphire
- *Cymbalaria muralis*—Ivy-leaved toadflax
- *Cytisus scoparius*—Broom (flowers)
- *Erodium cicutarium*—Common stork's-bill
- *Fagus sylvatica*—Beech
- *Foeniculum vulgare*—Fennel
- *Galium aparine*—Robin-run-the-hedge; Cleavers; Goosegrass
- *Glaux maritima*—Sea-milkwort
- *Halimione portulacoides* (*Atriplex portulacoides*)—Sea purslane
- *Honckenya peploides*—Sea sandwort
- *Hypochaeris radicata*—Cat's-ear
- *Lamium hybridum*—Cut-leaved dead-nettle
- *Lamium purpureum*—Red dead-nettle; Archangel
- *Lapsana communis*—Nipplewort
- *Ligusticum scoticum*—Lovage
- *Lythrum portula* (*Peplis portula*)—Water purslane
- *Malva neglecta*—Dwarf mallow
- *Malva sylvestris*—Common mallow
- *Medicago sativa* subsp. *sativa*—Lucerne
- *Melilotus altissimus*—Tall melilot
- *Melilotus officinalis*—Ribbed melilot
- *Mentha aquatica*—Water mint
- *Mertensia maritima*—Oysterplant
- *Myrrhis odorata*—Sweet cicely
- *Nasturtium microphyllum*—Narrow-fruited water-cress
- *Nasturtium officinale*—Water-cress
- *Origanum vulgare*—Marjoram; Oregano
- *Oxalis acetosella*—Wood sorrel
- *Papaver rhoeas*—Corn poppy; Common poppy
- *Persicaria lapathifolia*—Pale persicaria
- *Persicaria maculosa*—Redshank or Spotted persicaria
- *Plantago coronopus*—Buck's-horn plantain
- *Plantago lanceolata*—Ribwort plantain
- *Plantago major*—Greater plantain
- *Plantago maritima*—Sea plantain
- *Primula veris*—Cowslip
- *Primula vulgaris* (*Primula acaulis*)—Primrose
- *Pyrola minor*—Common wintergreen
- *Raphanus raphanistrum*—Wild radish
- *Raphanus raphanistrum* subsp. *maritimus*—Sea radish
- *Rhodiola rosea* (*Sedum rosea*)—Roseroot
- *Rumex acetosa*—Sorrel
- *Rumex acetosella*—Sheep's sorrel
- *Salicornia* species—Glasswort
- *Sanguisorba minor*—Salad burnet
- *Sanguisorba officinalis*—Great burnet
- *Saxifraga oppositifolia*—Purple saxifrage (flowers)
- *Smyrnium olusatrum*—Alexanders
- *Silybum marianum*—Milk thistle
- *Sonchus arvensis*—Perennial sow-thistle
- *Sonchus asper*—Spiny sow-thistle
- *Sonchus oleraceus*—Smooth sow-thistle

EDIBLE WILD IRISH PLANTS *(continued)*

(see Systematic List for additional information)

Salads and as a raw vegetable *(continued)*

Spergularia media—Greater sea-spurrey
Stellaria graminea—Lesser stitchwort
Stellaria media—Chickweed
Suaeda maritima—Annual sea-blite
Taraxacum officinale agg.—Dandelion
Thlaspi arvense—Field penny-cress
Tragopogon pratensis—Goat's-beard
Trifolium pratense—Red clover
Trifolium repens—White clover
Ulmus glabra—Wych elm; Irish elm
Umbilicus rupestris—Wall pennywort; Navelwort
Valerianella locusta—Lamb's lettuce; Corn salad
Veronica beccabunga—Brooklime
Vicia sativa—Common vetch
Vicia sepium—Bush vetch

Seasoning and condiments

Alliaria petiolata—Garlic mustard; Hedge garlic
Armoracia rusticana—Horse-radish
Capsella bursa-pastoris—Shepherd's purse
Juniperus communis—Juniper
Sedum acre—Stonecrop; Wall pepper (leaves)
Smyrnium olusatrum—Alexanders
Thlaspi arvense—Field penny-cress
Tussilago farfara—Colt's foot

Soups

Aegopodium podagraria—Goutweed; Ground-elder; Bishop's-weed
Alliaria petiolata—Garlic mustard; Hedge garlic
Allium schoenoprasum—Chives
Allium ursinum—Wild garlic; Ramsons
Atriplex patula—Common orache
Atriplex prostrata (*Atriplex hastata*)—Spear-leaved orache
Beta vulgaris subsp. *maritima*—Sea beet
Capsella bursa-pastoris—Shepherd's purse
Centranthus ruber—Red valerian
Chenopodium album—Fat hen
Filipendula ulmaria—Meadow-sweet
Galinsoga species—Gallant-soldier
Galium aparine—Robin-run-the-hedge; Cleavers; Goosegrass
Geum urbanum—Wood avens
Glechoma hederacea—Ground ivy
Ligusticum scoticum—Lovage
Malva sylvestris—Common mallow
Melilotus officinalis—Ribbed melilot
Myrrhis odorata—Sweet cicely
Nasturtium microphyllum—Narrow-fruited water-cress
Nasturtium officinale—Water-cress
Origanum vulgare—Marjoram; Oregano
Potentilla anserina—Silverweed
Prunus avium—Wild cherry
Rosa species—Rose
Rumex acetosa—Sorrel
Rumex acetosella—Sheep's sorrel
Salicornia species—Glasswort
Sambucus nigra—Elder
Smyrnium olusatrum—Alexanders
Sonchus arvensis—Perennial sow-thistle
Sonchus asper—Spiny sow-thistle
Sonchus oleraceus—Smooth sow-thistle
Stachys sylvatica—Hedge woundwort
Stellaria media—Chickweed
Symphytum officinale—Comfrey
Typha latifolia—Bulrush; Reed mace
Ulmus glabra—Wych elm; Irish elm
Urtica dioica—Nettle
Urtica urens—Small nettle
Vaccinium myrtillus—Bilberry; Fraughan

Syrups and sugars

Acer pseudoplatanus—Sycamore
Adiantum capillus-veneris—Maidenhair fern
Betula pubescens—Downy birch
Betula pendula—Silver birch
Fraxinus excelsior—Ash
Primula veris—Cowslip
Rosa species—Rose
Rubus fruticosus agg.—Bramble; Blackberry
Sambucus nigra—Elder
Taraxacum officinale—Dandelion
Vaccinium myrtillus—Bilberry; Fraughan

Vinegars

Malus sylvestris—Crab-apple
Prunus spinosa—Sloe; Blackthorn
Rubus idaeus—Wild raspberry

Wines, beers and liqueurs

Arbutus unedo—Strawberry-tree
Arctium lappa—Great burdock
Arctium minus—Burdock

EDIBLE WILD IRISH PLANTS *(continued)*

(see Systematic List for additional information)

Wines, beers and liqueurs *(continued)*

- *Arbutus unedo*—Strawberry-tree
- *Arctium lappa*—Great burdock
- *Arctium minus*—Burdock
- *Artemisia maritima* (*Seriphidium maritimum*)—Sea wormwood
- *Betula pubescens*—Downy birch
- *Betula pendula*—Silver birch
- *Calluna vulgaris*—Heather; Ling

Calluna vulgaris **and** ***Erica cinerea.***

- *Crataegus monogyna*—Hawthorn
- *Cytisus scoparius*—Broom (flowers)
- *Daucus carota*—Carrot
- *Fagus sylvatica*—Beech
- *Filipendula ulmaria*—Meadow-sweet
- *Fraxinus excelsior*—Ash
- *Malus sylvestris*—Crab-apple
- *Menyanthes trifoliata*—Bogbean
- *Myrica gale*—Bog myrtle
- *Primula veris*—Cowslip
- *Primula vulgaris* (*Primula acaulis*)—Primrose
- *Prunus avium*—Wild cherry
- *Prunus domestica*—Bullace; Wild plum
- *Prunus padus*—Bird cherry
- *Prunus spinosa*—Sloe; Blackthorn
- *Quercus petraea*—Sessile oak; Irish oak
- *Quercus robur*—Pedunculate oak; English oak
- *Rosa* species—Rose
- *Rubus fruticosus* agg.—Bramble; Blackberry
- *Rubus idaeus*—Wild raspberry
- *Salvia horminioides*—Wild clary; Wild sage
- *Sambucus nigra*—Elder
- *Smyrnium olusatrum*—Alexanders
- *Sorbus aria*—Common whitebeam
- *Sorbus aucuparia*—Mountain ash; Rowan
- *Teucrium scorodonia*—Wood sage
- *Trifolium pratense*—Red clover
- *Tussilago farfara*—Colt's foot
- *Ulex europaeus*—Gorse; Whin; Furze
- *Ulex gallii*—Western gorse; Mountain gorse; Autumn gorse
- *Urtica dioica*—Nettle
- *Urtica urens*—Small nettle
- *Vaccinium myrtillus*—Bilberry; Fraughan
- *Vaccinium vitis-idaea*—Cowberry
- *Viburnum opulus*—Guelder rose

Guidelines for Eating Wild Plants

If using wild plants for food in Ireland the following guidelines should be applied:

1. Learn to identify plants correctly so that you can distinguish between the edible and the poisonous.
2. Acquire and use a good illustrated field guide and only use plants that you are certain about their identify.
3. Remember that some edible plants can closely resemble other poisonous plants so make be sure of your identifications before use. If you have **any** doubts do not use a particular plant.
4. Teach children not to put plants in their mouths. Explain what plants are poisonous and why they should not be eaten (e.g., fruits of the yew tree). Equally, point out plants that are easily identified and can be eaten without fear (e.g., blackberries).
5. Don't collect and eat plants that may have been sprayed recently with pesticides and herbicides.
6. Don't collect and eat plants that are growing in habitats that may potentially be polluted or contain toxic wastes (such as on busy roadsides or where the plant may have been fouled by dogs or by sewage).
7. Remember that susceptibility to different chemical compounds and allergies can vary from person to person. A plant that is harmless to some may be toxic to others. If you do experiment with using wild plants for food, proceed with caution, trying out new species in moderation.
8. Remember that just because an animal or bird may eat a particular plant (such as its fruits) it does not necessarily indicate that it is edible for us too.
9. Remember to wash well the plants you intend to eat.
10. Some plants contain edible parts and other parts that are either indigestible or even poisonous (the well-known case of rhubarb leaves, which are poisonous, illustrates this point).

Conservation of Wild Plants

Conservation of wild plants should be the number one concern when collecting plants or fungi. When taking plants from the wild you should follow these guidelines:

1. Never take more than a small proportion of the plants of a particular species that are present at any site.
2. Make sure that the species you are harvesting is not legally protected in Ireland or on the national endangered plant list (see <http:botanicgardens.ie>).
3. Don't pick more plant material than you will use.
4. Always leave enough material for the plant and population to reproduce and grow undiminished the next year.
5. When you collect plant material leave the roots (unless it is the root you need to collect), so that it can resprout in the future.
6. If you plan to collect the same plant from the same location the following year it is a good idea to keep a record of the abundance or number of plants present and if it seems to be declining then leave the population to recover.
7. Ask permission from the landowner before collecting plants on private land and don't collect plants in nature reserves, national parks and other protected or vulnerable sites.
8. Be careful not to damage or disturb the habitat in which the plant is growing that you are collecting.

POISONOUS WILD PLANTS IN IRELAND

Aconitum napellus—Monk's-hood (lethally toxic; historic uses as weapon and intentional poisons across Europe and Asia)
Aesculus hippocastanum—Horse-chestnut
Aethusa cynapium—Fool's parsley
Agrostemma githago -Corncockle
Anagallis arvensis—Scarlet pimpernel
Anemone nemorosa—Wood anemone
Apium nodiflorum—Fool's water-cress
Artemisia absinthium—Wormwood
Arum maculatum—Lords and Ladies; Cuckoo-pint (toxic, but the intense stinging irritation of the oxalate crystals deter consumption)
Atropa belladonna—Deadly nightshade
Caltha palustris—Marsh-marigold; Kingcup (if eaten raw)
Chelidonium majus—Greater celandine
Cicuta virosa—Water hemlock; Cowbane
Clematis vitalba—Traveller's joy; Wild clematis
Colchicum autumnale—Autumn crocus
Conium maculatum—Hemlock (lethally toxic if ingested in any quantity; neurotoxins cause death by respiratory paralysis and collapse)
Convallaria majalis—Lily-of-the-valley
Cytisus scoparius—Broom
Datura stramonium—Thorn-apple (lethally toxic; tropane alkaloids such as atropine and scopolamine)
Digitalis purpurea—Foxglove
Equisetum species—Horsetails
Euonymus europaeus—Spindle-tree
Euphorbia species—Spurges
Frangula alnus—Alder buckthorn
Hedera helix—Ivy
Holcus lanatus—Yorkshire fog
Hyoscyamus niger—Henbane (lethally toxic; tropane alkaloids such as hyoscyamine and scopolamine)
Hyacinthoides non-scripta—Bluebell
Iris pseudacorus—Yellow flag
Leucojum aestivum—Summer snowflake
Ligustrum vulgare—Privet
Lolium temulentum—Darnel
Mercurialis annua—Annual dog's mercury
Mercurialis perennis—Perennial dog's mercury
Oenanthe aquatica—Water-dropwort
Oenanthe crocata—Hemlock water-dropwort
Oenanthe fistulosa—Water-dropwort
Papaver somniferum—Opium poppy
Ranunculus species—Buttercups; Crowfoots
Rhamnus cathartica—Buckthorn
Rhododendron ponticum—Rhododendron
Saponaria officinalis—Soapwort
Scilla verna—Spring squill

***Scilla verna* (spring squill). Watercolour by Lydia Shackleton. Courtesy of the National Botanic Gardens, Glasnevin.**

Senecio jacobaea—Ragwort
Sium latifolium—Greater water-parsnip
Solanum dulcamara—Bittersweet
Taxus baccata—Yew (all plant parts except the attractive red aril around the poisonous seed are lethally toxic)
Viscum album—Mistletoe

Treatment in the Case of Poisoning

If you do eat a plant you think may be poisonous or suspect that someone else may have eaten a poisonous plant it is essential to act quickly. Call your local doctor. Make sure you do the following:

> Know the name of the plant eaten, or
> Have a sample of it available to check the identity, or
> Be able to describe it well.

Be sure you can tell the doctor what parts and how much of the plant was eaten. You should also pass on to the doctor information on when the plant was eaten, as well as information on the patient, giving his or her age, weight and any information you have on his or her general state of health or known allergies or illnesses. If a doctor is not available, try to get the patient to drink lots of water and induce vomiting before rushing them to the nearest hospital or health centre. Vomiting can be induced by drinking soapy water, or diluted coffee grounds, or by triggering a gagging mechanism by putting fingers down the throat. If the patient is having convulsions or is unconscious do not try to induce vomiting because this may cause the patient to choke.

Cultivated Cereal Crops

The main cultivated cereal species found in Ireland are wheat, barley, oats and rye. All are annual grass species and occur occasionally in the wild as 'escapees' from cultivation, usually derived from spilled seeds (Reynolds, 2002). All four crops have been cultivated in Ireland for centuries, and wheat and barley in particular are known from the earliest times in Ireland, although Lucas (1960: 9) suggested that rye was of little importance in ancient Ireland and that there is no record of oats 'from the prehistoric period'. Before the introduction of the potato, cereals formed the main staple foodstuffs for Ireland's human population, providing a large proportion of the calories and protein needed in the human diet. Cereal-derived products, mainly flour, were used to make bread, porridge and gruel, as well as to brew drinks such as beer.

The English author Dr Samuel Johnson in *A Journey to the Western Islands of Scotland* (1775: 122) described bread and cakes eaten that were made from barley and oats, which would not have been much different from those eaten very widely in Ireland in pre-potato days.

> Their native bread is made of oats, or barley. Of oatmeal they spread very thin cakes, coarse and hard, to which unaccustomed palates are not easily reconciled. The barley cakes are thicker and softer; I began to eat them without unwillingness; the blackness of their colour raises some dislike, but the taste is not disagreeable. In some houses there is wheat flour, with which we were sure to be treated, if we stayed long enough to have it kneaded and baked. As neither yeast nor leaven are used among them, their bread of every kind is unfermented. They make only cakes and never mould a loaf.

All the cereal crop species have been bred and selected over thousands of years to enhance their properties and yields for agriculture and as food. No cereal species are native to Ireland but have been introduced to this island many times from the earliest farmers, down through the ages right up to the present when new varieties are constantly developed, introduced and available for agriculture.

The history of human use of wheat and barley goes back more than 10,000 years when they were first grown as crops in the Middle East (the Fertile Crescent) in what are today parts of Iraq and Iran. Kelly (1997) outlined much valuable information on the history of cultivation of cereals in early Ireland. He listed seven cereal grains given, in order of importance, in an 8th-century law-text, *Bretha Déin Chécht: cruithneacht* (bread-wheat), *secal* (or *seagal*) (rye), *suillech* (spelt wheat?), *ibdach* (two-rowed barley), *rúadán* (emmer wheat?), *éornae* (six-rowed barley) and *corcae* (common oat). Analysis of dental remains from archaeological sites in Ireland indicates that stone-ground cereals were an important part of the diet for early inhabitants in these islands (fragments of stone in the ground flour would have caused significant wear on their teeth).

A windmill near Blennerville, Co. Kerry.

Wheat

The taxonomy of wheat (in Irish, *cruithneacht* or *tuireann*) is complex and more than 20 different species and many more subspecies can be included under the name 'wheat'. Some wild species of wheat are known, while others are only found as cultivated crops. *Triticum aestivum* (bread wheat) is the most popular and widely grown species, one that is of complex genetic origin and containing the genes of a number of related wheat species as well as those of a wild grass, *Aegilops squarrosa*. Wheat is not ideally suited to the Irish climate, which is rather too cool and damp to suit it best. Nevertheless, we know that the cultivation of wheat in Ireland goes back to ancient times. Joyce (1875) quotes evidence of early wheat cultivation in Ireland from a poem included in the 12th-century chronicles of medieval Irish history, the *Annals of Tighernach*, the *Annals of Ulster* and the *Annals of the Four Masters*: 'O mill, what hast thou ground? Precious thy wheat! It is not oats that thou hast ground, but the offspring of Cearbhall. The grain which the mill has ground is not oats but blood-red wheat[15]; With the scions of the great tree (Cearbhall) Maeloran's mill was fed' (as quoted in Joyce, 1875: 319). Today, wheat is primarily grown in Ireland for animal feed, but is also used for seed and for milling as bread and biscuit flour. Much of the flour used in the baking industry today is imported from drier climates than Ireland which are better suited to the production of hard wheats. Wheat straw is used for bedding for animals, as animal feed and in mushroom composts.

CULTIVATED CEREAL CROPS

Avena sativa—Oat
Avena strigosa—Bristol oat; Small oat
Hordeum distichon—Two-rowed barley
Hordeum vulgare—Six-rowed barley; Four-rowed barley
Secale cereale—Rye
Triticum aestivum—Bread wheat
Triticum dicoccum—Emmer wheat
Triticum spelta—Spelt wheat
Triticum turgidum—Rivet wheat
Zea mays—Maize; Corn

Barley

Barley (in Irish, *eorna*) is a tougher cereal than wheat and can be grown in habitats where many other cereals cannot survive. Barley is the fourth most important cereal crop in the world after wheat, maize and rice. Like wheat, it was

15. According to Joyce (1875), early Irish wheat was distinguished by its red colour. The Irish word for wheat, *cruithneacht*, is derived from *cruith*, meaning 'blood-coloured' or 'red'.

domesticated in the Fertile Crescent about 10,000 years ago, probably in what is present-day Jordan and Israel. Remains of barley (*Hordeum vulgare*) grains found at archaeological sites in the Fertile Crescent indicate that it was domesticated there from its wild relative *Hordeum spontaneum* (Badr et al., 2000). Today, it is used for food, brewing and as animal feed. Barley straw is also used for animal bedding. About 10% of Ireland's barley production (+90,000 tonnes) is used in the brewing of Guinness stout, and in ancient Ireland, most beer was brewed from malted barley too (although some was made from wheat) (Kelly, 1997). Both two-rowed and six-rowed barley[16] are known from early archaeological excavations in early Christian Ireland (Kelly, 1997). Malted barley (or potatoes) is traditionally used in Ireland for the making of poitín (More information on poitín is given under the entry for *Solanum tuberosum*, potato).

Rye

Rye (*Secale cereale*) (in Irish, *seagal*) is a hardy cereal crop and more tolerant of frost and drought than wheat. It is the most winter hardy of all cereals, and is frequently grown under conditions where other cereals fail. Unlike wheat and barley, which had their origins in the Middle East, wild rye is indigenous to Anatolia and was domesticated there in the early Neolithic period. It migrated into Central Europe as a weed of other cultivated cereal crops and, as the Bronze and Iron Ages progressed, it gradually changed from being a weed to a cultivated crop in its own right. As cereal crops were grown on poorer soils, the weedy rye was able to take a competitive advantage of these conditions and gradually replace the cultivated crop (Behre, 1992). Kelly (1997) mentioned that there is evidence of rye in Ireland from a Bronze Age site at Carrowmore, Co. Sligo, even though its Irish name (*secal* or *seagal*) is derived from post-Classical Latin *secale* when new strains of the crop may have entered the country as Roman influences spread throughout Europe. Rye is the tallest cereal and thus provides the longest straw, making it excellent for thatching and rope-making. Some of the older varieties can grow up to 1.5 m (5 ft) tall.

Oats

Oats (*Avena sativa*) (in Irish, *coirce*) have also been grown in Ireland for centuries, not only for human food (for example, oatmeal to make porridge, rolled oats and as oat flour), but also for animal feed. Oat chaff was used as bedding too. Rutty (1772: 103) noted that oats were preferable to straw because the chaff is light and dry and 'not so easily compressible, and much cooler than flocks or feathers'. Oats are easier to cultivate than wheat and grow well on poorer soils and in harsher climates. Like rye, oats probably also arose as weedy contaminants of wheat and barley crops, which were gradually domesticated through selection, and they achieved a competitive advantage over wheat in cooler wetter areas as these crops spread from the Fertile Crescent into Europe. *Avena sativa* is the most widely cultivated species, although *A. strigosa* was formerly cultivated in Ireland, particularly along the western coastal regions of the country. *Avena strigosa* is very tolerant of light soils and high winds so it was suited for cultivation in exposed regions where other crops may not have performed well (it is still cultivated as a rare crop in parts of the Hebrides in Scotland for human consumption, fodder for animals and for straw used in thatching and for furniture). In Ireland, *A. strigosa* is now found only as a rare arable weed (Reynolds, 2002); however, both species of *Avena* are known from early Christian sites in Ireland (Kelly, 1997). Up to the arrival of the potato as the main staple food crop for the majority of poorer people in Ireland, oats were an important staple

16. Wild barley has only a single fertile spikelet but through domestication a number of mutations have been selected with two or six fertile rows (known as 'two-rowed barley' and 'six-rowed barley').

food. A quern or hand-mill made from stone, primarily used for grinding oats, was a common tool in houses of country people in Ireland (Joyce, 1875).

Maize

Maize[17] (*Zea mays*) is another cereal grass, with its origins and domestication in the Americas. It is also grown as a crop in Ireland. However, it does not have a long history of cultivation in Ireland and has only been grown commercially in the country since the 1970s, and more successfully, following the development of lower-temperature-tolerant varieties.

In 2009, 83,000 ha were cultivated with wheat in the Republic of Ireland; 185,900 ha with barley and 20,100 ha with oats (Central Statistics Office of Ireland; see <http:cso.ie>).

Straw

Straw (in Irish, *tuí*) has been an extremely important plant-based resource in Ireland for centuries and used on the island for a wide variety of purposes. It is produced as a by-product of cereal production and represents the dried stalks of the cereal plants after the grain has been removed. In Ireland, straw has been produced from four cereal crops—barley, oats, rye and wheat—all of them at one time or another have been important agricultural crops.

Straw from different cereal crops has different characteristics and so is used for particular purposes. Rye straw is long and is favoured for thatching and rope-making. Wheat, oats and barley usually provide a shorter straw. Thatchers need a length of 700–900 mm (28–35 in.) or more (Staniforth, 2008). Barley straw is soft and supple, whereas wheat straw is often stiff and hard. For that reason, barley straw is suitable for stuffing fabrics and wheat straw is good for thatching (see Straw discussion beginning on p. 39).

Coffee, Tea and Other Beverages

If we were to travel back in time in Ireland several centuries ago to see what beverages were commonly drunk at that time, the most significant difference from today would be the absence of tea and coffee. Milk, water and ale would have been the everyday drinks. Infusions made by adding hot water to wild herbs would have also been drunk but generally, if not exclusively, for medicinal purposes.

Coffee (*Coffea arabica*) was first used as a drink in the Arabian world in the 16th century and gradually spread throughout Europe. It is made from the roasted beans (seeds) of an evergreen tropical shrub, which are ground after roasting and infused in hot water. It probably first arrived in Ireland in the late 1600s, the first coffee house having opened in England in the 1650s. Today it is one of Ireland's most popular and ubiquitous beverages and about 3.5 kilos of coffee are consumed per person per annum (pppa). Although this is small compared to that of Finland (12.0 kilos pppa), it is larger consumption than that of the United Kingdom which has a pppa of 2.8 kilos (Coffee consumption, World Resources Institute, <http://wri.org>).

Irish Coffee is a recent invention, being made from coffee, sweetened with brown sugar, with a good measure of Irish whiskey and topped off with whipped cream. Irish Coffee was invented by Joe Sheridan in Foynes, Co. Limerick, in 1942 to warm passengers coming off trans-Atlantic flights into that seaplane port.

Tea (*Camellia sinensis*) is today Ireland's most popular and widely consumed beverage. Its origins are Asian, and it has been grown in China and other countries in Asia from the

17. Irish words for maize are *min bhuí* and *arbhar indiach* (De Bhaldraithe, 1959).

earliest of times. The earliest reference to tea drinking goes back to more than 2500 years in China. It arrived in Britain and Ireland in the early 1800s as an expensive and luxurious beverage for the upper and middle classes and later spread to become a common and popular drink for the rural and lower class in the mid-1800s. Wooden tea caddies from the Georgian period were typically protected with a lock so that the tea they contained could be dispensed as needed and used sparingly and so that there was no danger that the tea they contained would be stolen by the servants. The tea that was available for most people in Ireland was not the best quality and therefore tended to be brewed strong and to be served with milk and was often sweetened with sugar. The tradition of drinking tea strong in Ireland survives to the present day. The Republic of Ireland has one of the largest per capita tea consumptions in the world, with an average of four cups per person per day consumed.

Over the last few decades caffeine-free tea-like infusions of herbs have become popular drinks in Ireland, such as peppermint and spearmint teas, chamomile tea, lemon and ginger tea, raspberry tea and rooibos (South African) tea. There are some Irish producers of such teas but most are made from imported ingredients, and few if any are derived from wild collected Irish plants. In the past many infusions made from wild plants were in frequent usage in Ireland, and in this study I have recorded 47 different species used at one time or another as herbal teas (see Edible Wild Irish Plants).

Beer and Wine and Other Alcoholic Drinks

Beer

In 1903 Patrick Joyce wrote, 'In old times people were quite as fond of intoxicating drink at dinner and banquets as they are now' (p. 114). Ancient Irish literature is rich in references to banquets where drink, and large quantities of it, was an important part of the proceedings (Danaher, 1964a: 45). Danaher mentioned an old tale of a banquet in Ulster where 'hundreds of barrels of ale were drunk, the effect of which was that the Ulstermen found themselves disturbing the peace of Munster before they recovered their senses'. Joyce (1903) pointed out that Dioscorides, writing in the 1st century, said that the Britons and Hiberi, or Irish, instead of wine use a liquor made of barley. Kelly (1997) also pointed out the importance of beer in ancient Ireland and mentioned that a proper king was expected to drink beer with his household on Sundays and that it was used as an expression of hospitality. He wrote, 'In an Old Irish mythological text the reputation of a king is dismissed in disparaging terms because his guests did not leave his house 'with the smell of beer on their breath' (p. 332). People of lower rank also brewed and drank beer frequently, and it must have been an important part of the early Irish diet, as it was sometimes safer than water, rich in nutrients and available at times of the year when other foods and drinks (such as milk) were not common. Monks drank beer in their monasteries, and it is said that even St Patrick had his own brewer, a priest named Mescan (Kelly, 1997).

The history of beer goes back thousands of years and has been documented in the early civilizations of the valleys of the rivers Tigris and Euphrates (Louw, 1983). Beer breeding also spread into the Egyptian and Roman civilizations as well as westwards into Europe well before the Middle Ages. Ale was the earliest brew, using malted barley, as hops were neither native to nor grown in Ireland until imported into the country from the 18th century onwards. Kelly (1997) agreed that barley was the main ingredient used in beer making, but he also mentioned one law text where beer from wheat was referred to.

Although Ireland is better known for stout (Guinness, Murphy's and Beamish in particular), according to Wikipedia (accessed 4 September 2011), 63% of the beer sold in the country is lager. In 2006, stout made up 32% of the market, with ale the remaining 5%. Brewing in Ireland has a long history. In the early 19th century there were several hundred breweries in the country; now there are only about a dozen.

See Edible Wild Irish Plants on page 66 for those plants that have been used in making wines, beers and liqueurs in Ireland.

Wine

The history of wine (made from grapes) in Ireland too goes back to ancient times, although it seems that wine was always imported and vines were not successfully grown. Gerald of Wales (Geraldus Cambrensis) (ca. 1185) said that Ireland never had vineyards (O'Meara, 1982). Danaher (1964a) wrote that even in early times there was trade with Gaul for the import of wine, and furs and hides were exported. Lucas (1960) suggested that, as an import, wine was probably always expensive and therefore a luxury. A very early date for a mention of wine in Ireland is AD 433 when St Patrick went to Tara on Easter Sunday and found kings, princes and druids feasting and drinking wine in the Palace with King Laegaire, according to Muirchu writing in the 7th century (Joyce, 1903). A further early record is when Domnall, king of Ireland from AD 627 to 641, in preparation for a banquet of Dun-nan-gedh, provided three kinds of drink: wine, mead and ale. Also, in Cormac's Glossary, a 9th-century manuscript providing meanings for about 1400 Irish words, wine is mentioned several times (O'Donovan, 1868).

French and Spanish wines were popular in early times and still are today. It is probably no coincidence that some of the Wild Geese (members of the Irish Jacobite army leaving Ireland in 1691 after the Treaty of Limerick at the end of the Williamite War) settled in the Bordeaux region. There are still Irish names associated with some of the most famous wines of that region, such as Châteaux Kirwan, Lynch-Bages, Clarke, Léoville-Barton, Phélan Ségur and so on, indicating a long history of Irish connections in Europe along the Atlantic coasts.

In historic times in Ireland, hot alcoholic drinks were popular, including mulled ale and porter, punch, hot spiced wine, possets, buttered whiskey or rum and *scáiltín* (a mixture of hot milk, butter, whiskey and sugar) (Danaher, 1964a). Today hot alcoholic drinks are still enjoyed, particularly during winter. Mulled wine and mince pies are a now a traditional part of Christmas festivities, and the merits of a hot whiskey or hot port (whiskey and port to which are added hot water, lemon, sugar and sometimes cloves) are popular after a cold winter walk or as a winter cure for colds or flu.

Wine is one of the most popular alcoholic drinks today in Ireland. The Irish wine market was worth €1.65 billion in 2009 and had grown from an 8% share of the overall alcoholic beverage market in Ireland in 1994 to 22% in 2007, with approximately 1,451,000 wine drinkers (Geraghty, 2009). Although the European Union has designated Ireland as a wine-producing region, there are still very few wineries in Ireland, most of them in Co. Cork and few over 5 acres in extent. These include the Blackwater Valley Vineyard, which is located in Mallow and has 5 acres of vines. The Longueville House vineyard is also in Mallow and has just over 1 acre of vines. It produces a light white wine called 'Coisreal Longueville'. In Kinsale is the Thomas Walk Vineyard. Another producer is David Llewellyn, who has produced several varieties of red and white wine, Lusca wine, in Lusk, Co. Dublin, which he has sold at farmers' markets. Wine production, even at these few southern wineries, is only possible after a good summer, which does not come every year. As the effects of climate change impact Irish weather over the coming decades, perhaps Ireland will one day become an increasingly important wine producer.

Country wines made by home enthusiasts from fruits and other plant materials include blackberry, elderberry, cowslip, crab apple, dandelion, bilberry, hawthorn flower and sloe wines. Some of these occasionally appear for sale in markets and fairs but are generally small-scale operations and largely produced for home consumption.

I have been singularly unsuccessful in my efforts to make beer and wines from wild plants. This is mainly because I am impatient and generally don't follow the precise instructions that are given in books on wine and beer making. If you want to make them, my advice is to get a good book on the subject, give yourself plenty of time and follow the instructions to the letter, until you are confident that you know what you are doing. If you are successful you can always invite me around to sample your efforts!

Cider

Cider is a popular drink today in Ireland, both imported and made by the famous Bulmers factory in Clonmel, Co. Tipperary. A popular non-alcoholic version was also originally made by Bulmers called 'Cidona'. It is now part of a different company, Britvic Ireland. Homemade cider can be made from home-grown apples and crab-apples too.

Mead

Mead was also a beverage made and well known in ancient Ireland, although according to Lucas (1960: 40), 'seems, in reality, to have been of little importance'. Mead is made by fermenting honey with water to produce a mildly intoxicating drink. It may well be one of the most ancient alcoholic drinks of all, pre-dating agriculture. Honey was the only source of sweetener available in Ireland until methods to manufacture sugar were invented. In ancient Ireland, mead was regarded as a more prestigious drink than beer, and was described as 'the relish of noble stock' (Kelly, 1997: 113). St Ailbe, a contemporary of St Patrick, said that when the monks sit down to dine they should be served on clean dishes, herbs or roots, washed in water, apples and 'mead from the hive to the depth of a thumb' (Joyce, 1903: 120). Joyce also wrote in the same book that in the story of the Children of Lir (children magically turned into swans), one of them, Finola, talking of their former happy life, spoke of drinking hazel-mead (*miodh cuill*), presumably mead flavoured with hazelnuts. Honey was also added to beer in ancient Ireland and is mentioned quite frequently in ancient law texts, and another drink produced from malt and honey, called 'bragget', was also made (Kelly, 1997). Bees were widely kept for honey production in ancient Ireland.

Whiskey

Danaher (1964a) noted that whiskey is of comparatively recent origin, but it is not possible to say when distilling was introduced to Ireland. He said that the first mention of whiskey comes from 1405 when one Richard Mac Raghnaill from Co. Leitrim is recorded as having died from drinking uisce beatha (*uisce beatha* in Irish is the 'water of life'; *uisce* has been anglicised as 'whiskey'). According to Kelly (1997), there is no evidence that distilling occurred in Ireland earlier than the 13th century (which means that the legend that St Patrick invented whiskey cannot be true—St Patrick is said to have lived sometime in the 5th century). Danaher (1964a: 48) also wrote that during the Middle Ages Irish whiskey was famous and highly esteemed; quoting an early English visitor to Ireland thus: 'It is a very wholesome drink, and natural to digest the crudities of the Irish feeding. You may drink a naggin without offence'. Whiskey was also flavoured with the addition of various herbs, seeds and fruits (something that is rarely done today). Danaher (1964a: 48) goes on to say that 'This custom of adding herbs or flavourings to whiskey is now almost completely dead in Ireland, but in other countries people still buy quantities of the local spirits ... adding to each bottle some substance which improves the flavour or the colour or gives it a curative quality. Some of the things added to the spirits are blackberries, whortleberries, cherries, caraway seeds, cowslips, sloes, black-currants, aniseed, elder-berries, elder flowers, chamomile, lily of the valley, wormwood, tansy, yarrow, thyme, rosemary, angelica and lavender flowers, and there are dozens of others'.

Whiskey in Ireland is of course a popular part of Irish culture today and one of its best known products abroad. An example comes from my father, Robert Wyse Jackson (1908–1976), who was formerly Bishop of Limerick and won an international Limerick competition with a verse on whiskey:

The traditional poets of Maigue[18]
Knew nothing of White Horse or Haig,
But *uiscebah* hot
Distilled in a pot
Kept them merry, poetic and vague.

(Wyse Jackson & McDonnell, 2007: 357)

The most important regulations defining what is 'Irish whiskey' today are laid out in the Irish Whiskey Act of 1980. This Act says that Irish whiskey must be distilled and aged in the Republic of Ireland or in Northern Ireland. It must be distilled to an alcohol by volume level of less than 94.8% from a yeast-fermented mash of cereal grains (saccharified by the diastase of malt contained therein, with or without other natural diastases) in such a way that the distillate has an aroma and flavour derived from the materials used. The whiskey must also be aged for at least three years in wooden casks and it is made up of more than one spirit source then it must be described as 'blended'. Ireland has relatively few distilleries (by comparison to Scotland) and most of them produce a number of different whiskeys. Some of the best known brands are Jameson, Paddy, Powers, Midleton, Bushmills (of various ages: 10-, 12-, 16- and 21-year-old single malts), Black Bush, Connemara, Knappogue Castle, Michael Collins, Tullamore Dew, Locke's and Kilbeggan.

A new distillery in Dingle, Co. Kerry, is now preparing gin using local plants as the 'botanical' flavors, including rowan berries, bog myrtle, heather, hawthorn, angelica and *Fuchsia magellanica.*

The making of poteen, another Irish distilled spirit, is described in the Systematic List under *Solanum tuberosum* (potato).

Plants Smoked and Plants Used as Tobacco Substitutes

Cultivated tobacco (*Nicotiana tabacum*) is one of the most widely smoked plants on earth. Although it is predominantly smoked for pleasure, or because smokers are addicted to it, in some cultures it has been smoked for medicinal and ritual purposes, sometimes going back thousands of years (Pennacchio et al., 2010). It is native to tropical and subtropical Americas and was widely used by Native Americans. Sir Walter Raleigh (ca. 1554–1618) is often said to have brought tobacco to Ireland. Danaher (1964a: 57) wrote that 'Sir Walter Raleigh sat at ease in his garden in Youghal [Co. Cork], smoking his pipe and dreaming of further piratical raids, when his new servant boy, terrified at the cloud of smoke, "quenched" him with a big pail of water'. By 1698 the use of tobacco was certainly widespread in Ireland. John Dunton in his travels in Ireland mentioned its use several times (Dunton, 1698). Smoking tobacco has been a favoured pastime in Ireland for several centuries, since the 16th century, and it was only from the 1950s onwards

18. The Maigue region is named after a river in Co. Limerick, where this verse form is said to have been invented.

Bog oak pipes. Courtesy of the National Botanic Gardens, Glasnevin.

that the health dangers associated with smoking began to be stressed. The common way of smoking tobacco in Ireland up to the last century was with a pipe, first by men and later also by women. Tobacco was sold in the form of 'twist' where the tobacco was twisted into a thin rope and usually sold by length (Danaher, 1964a). The economic botany collections at the National Botanic Gardens in Glasnevin include several examples of 'twist' ships tobacco, that would have been issued to sailors before a voyage, the length of the tobacco based likely on the duration of the journey. This tobacco was often chewed, as lighting it on board ship, surrounded by lots of flammable materials, would have been dangerous. Cigar smoking and taking snuff was also popular amongst men. The mass-produced cigarette was a later invention, which appeared during the latter half of the 19th century. Cigarettes made tobacco cheap and accessible to a mass market throughout the world. At first, cigarettes were mainly smoked by men, but in the 1920s women increasingly smoked them. Women's use of cigarettes really expanded, though, after World War II.

In Ireland today the smoking of tobacco is very actively discouraged. For example, on 29 March 2004, under provisions in the Public Health (Tobacco) Acts 2002 and 2004, enclosed workplaces became smoke-free in the Republic of Ireland. Further, on 1 July 2009, new legislation came into place in Ireland prohibiting all point-of-sale advertising in shops and all other retail outlets and made it illegal to store tobacco within the sight of customers. Nevertheless, very large numbers of people, young and old in Ireland, smoke tobacco. The Department of Health/An Roinn Sláinte (2011) pointed out that 7000 people die from smoking-related diseases in Ireland every year, and that 90% of lung cancers are caused by smoking. Furthermore, a recent study from the Department of Health found that the number of smokers in Ireland had increased from 27% in 2002 to 29% in 2007.

A small number of native plant species have been smoked in Ireland as tobacco substitutes. Probably the most popular species used was *Tussilago farfara* (colt's foot, or in Irish, *sponc*). Sharkey (2000) noted that colt's foot was used as a substitute for expensive tobacco when it was first introduced, mixed with wild rose petals and other aromatic herbs. A recipe for a herbal smoking mixture is given by Howes (1974) as colt's foot (50%), clover flowers (30%), lavender flowers (10%) and rose petals (10%).

Peat fibres were also smoked as a tobacco substitute. Dinneen (1927: 1217) gave the Irish name *tobac móna* to mean 'a brownish peat used as a tobacco substitute'. Pat O'Shea from Killorglin, Co. Kerry (in King, 2009), wrote that during the years of World War II in Ireland, 'The Emergency', when many essentials were rationed, people smoked tobacco *móna* instead. He said that the tobacco *móna* was found at the bottom of the bog and was impossible to cut with a *sleán*. He also reported that people also smoked dandelion (*caisearbhán*) roots.

Cannabis sativa (hemp, marijuana), a non-native plant, is illegal to grow, sell or possess in Ireland but, nevertheless, is smoked widely by young and middle-aged people of all social standings as a recreational drug. It is native to Asia but very widely cultivated throughout the world. It is known by a wide variety of popular and slang names (those used in Ireland have included the following: 'pot', 'blow', 'grass', 'ganga', 'hash', 'hashish', 'skunk', 'weed' and probably dozens more). Although much is imported into Ireland, some is grown secretly by people at home for their own use. Both shredded leaf material and extracted *Cannabis*

PLANTS SMOKED OR USED AS TOBACCO SUBSTITUTES

Achillea millefolium—Yarrow
Arctostaphylos uva-ursi—Bearberry (in North America)
Artemisia vulgaris—Mugwort
Cannabis sativa—Hemp; Marijuana
Mentha aquatica—Water mint
Mentha arvensis—Corn mint; Field mint
Myrica gale—Bog myrtle; Sweet-willow
Prunus spinosa—Sloe; Blackthorn
Rosa species—Rose (petals)
Stachys palustris—Marsh woundwort
Taraxacum officinale agg.—Dandelion (leaves and roots)
Tussilago farfara—Colt's foot

resin is available, usually mixed with tobacco and smoked in a home-made cigarette (a 'joint') or in a pipe.

The Misuse of Drugs Act 1977, the Misuse of Drugs Act 1984 and the Criminal Justice (Psychoactive Substances) Act 2010 are the acts of the Oireachtas in the Republic of Ireland regulating drugs. These acts define the penalties for unlawful production, possession and supply of drugs.

The National Documentation Centre on Drug Use in Ireland has said that the proportion of adults who reported using *Cannabis* at some point in their lives (ever used) increased from 17% in 2002–2003 to 22% in 2006–2007. The proportion of young adults who reported using cannabis in their lifetime also increased, from 24% in 2002–2003 to 29% in 2006–2007. The proportion of adults who reported using *Cannabis* in the last year (recent use) increased from 5% in 2002–2003 to 6% in 2006–2007. The proportion of young adults who reported using *Cannabis* increased from 9% in 2002–2003 to 10% in 2006–2007. The proportion of adults who reported using *Cannabis* in the last month (current use) in 2006–2007 was 2.6%. The majority (60%) of *Cannabis* users reported using a form of *Cannabis* resin. Almost two in five (38%) reported using a form of herbal *Cannabis*. In a 2006–2007 survey carried out by the National Documentation Centre on Drug Use (National Advisory Committee on Drugs and Public Health Information Research Branch, 2008), men were more frequent users than women. Almost all (99%) smoked it, as a joint (93%), in a pipe (4%) or in a bong (3%); less than 1% ate the drug.

The psychoactive effects of *Cannabis* are known to induce a state of relaxation, and to a lesser degree, euphoria. Secondary psychoactive effects are reported to include a facility for philosophical thinking and introspection, as well as, in some cases, anxiety and paranoia. A further psychoactive effect of the drug can include an increase in heart rate and hunger.

Geranium robertianum (herb-Robert). Watercolour by Lydia Shackleton. Courtesy of the National Botanic Gardens, Glasnevin.

NATURE'S MEDICINE CHEST: MEDICINAL PLANTS IN IRELAND

THROUGHOUT HISTORY, plants have been the most important source of medicine for human health. Many modern pharmaceuticals are derived from leaves, bark, roots, stems, seeds, fruits and flowers, either extracted from the plants themselves, or artificially synthesized to copy plant chemical compounds. Medicinal plants form an important part of the world's economy with millions used annually. Western pharmaceuticals are often expensive, inaccessible or unsuitable in many countries, particularly in rural regions, and in the 1990s the World Health Organisation (WHO) estimated that 80% of the world's population relied on plants for their primary healthcare (WHO, IUCN, WWF, 1993).

In China, for example, traditional medicine is largely based on more than 5000 plant species, many of which are collected in the wild (Xiao, 1991), and in the 1990s, 40% of patients in urban areas and 90% in rural areas used plants from traditional medicine (He & Cheng, 1991). Likewise, in India, systems of traditional medicine—Ayurveda, Unani and Siddha—operate alongside modern healthcare. There are more than 400,000 registered traditional medical practitioners in India, a number comparable with the number of registered doctors (Akerele et al., 1991).

In industrialised countries the medicinal use of unprocessed plants has declined. Nevertheless, more than 7000 different plant compounds are used in pharmaceutical medicine as heart drugs, laxatives, anti-cancer agents, hormones, contraceptives, diuretics, antibiotics, decongestants, analgesics, anaesthetics, ulcer treatments and anti-parasitic compounds. Well-known examples of plant-derived drugs include the heart drug digoxin, derived from *Digitalis purpurea* (foxglove), and aspirin, which is probably the world's most widely used medicine. Aspirin was first extracted from the leaves and bark of *Salix alba* (willow tree), and from the herb *Filipendula ulmaria* (meadow-sweet). Worldwide, extensive pharmaceutical screening programmes have been carried out to identify promising chemical compounds to provide new treatments for human diseases.

It should be remembered that for the majority of human history in Ireland, plants provided most of the resources needed for human healthcare and for medical treatment. Modern pharmaceutical-based medicine is really no more than a few centuries old, and until the 20th century, plant treatments very often provided the only effective means to cure common human ailments and conditions. It is recorded, for example, that in 1841 there were only 39 infirmaries operating in the whole of Ireland, serving a population of more than 8 million people (Woodham-Smith, 1962), and until the 19th century, there were few qualified doctors in rural Ireland (O'Regan, 1997). Traditional medicines from plants would have been then the primary resource for healthcare and used by the majority of the Irish population. Indeed, medicinal plants have been used in Ireland to treat such a remarkable range of ailments, conditions, diseases and illnesses that the list reads like a lexicon of human healthcare. The following are only some of the health problems that have been treated with wild plants in Ireland at one time or another: ague, back aches, blood cleansing, blood clotting, boils, breast lumps, bruises, burns, catarrh, chest ailments, cleaning teeth, cleansing the skin, colds, constipation, coughs, dislocations, dog bites, dysentery, eczema, epilepsy, excessive tiredness, eye ailments, fever, flatulence, loss of appetite, gall stones, hair loss, hangovers, headaches, head lice, healing bones, healing wounds and cuts, heartburn, haemorrhoids, indigestion, inflammations, influenza, insect stings, intestinal pains, jaundice, kidney stones, liver ailments, measles, neuralgia, nose bleeds, removing warts, rheumatism, rickets, ringworm, scalds, sciatica, smallpox, sore throats, spleen and heart ailments, sprains, stomach pains, swellings, swollen glands, tape worms and other intestinal worms, toothaches, vomiting and whooping cough. They have also been used as abortifacts, cancer treatments, counter-irritants, diuretics, emetics, emmenagogues, laxatives, male and female contraceptives and skin tonics.

We can be thankful, however, that much information has survived in written and published literature. Early Irish tales and legends include mentions of plants being used for medicinal purposes. An important medieval manuscript (ca. 1160), *The Book of Leinster*, mentions physicians in ancient Ireland and names Finghin Faithliagh, physician to Conchubhar MacNeasa, king of Ulster in AD 33 (Moloney, 1919). This composite work of Irish literature, mythology and genealogy includes the legend *Táin Bó Cúailnge*, or the *Cattle Raid of Cooley*. In *Táin Bó Cúailnge*, the wounded hero of the tale, Cuchulainn, is placed in a stream strewn with plants and herbs to clean his wounds: 'dry wisps and pads of grass and moss were placed against his wounds. And supple willow wands were arched above him to sustain his brath [cloak] … he might not endure it if his raiment should reach his skin' (as quoted in Synnott, 1979: 38).

Synnott (1979) has suggested that the Normans may have been key players in the introduction of particular medicinal plant practices to Ireland, and some herbs and herbal uses may also have been introduced by Irish monks returning from continental Europe during the early Christian period. Synnott (1979: 39) makes the keen observation that 'the Normans probably were to Irish cooking and medicine what the Romans were to the British,' and, indeed, the Normans, like the Romans, introduced many plants and plant uses to the lands they occupied.

Early important works on plants used in medicine in Ireland include Caleb Threlkeld's *Synopsis Stirpium Hibernicarum*, published in Dublin in 1726. Threlkeld was born in England in 1676 and came to Ireland in 1712 and practiced medicine and studied botany in Dublin. His 1726 work is regarded as the first book on wild Irish plants and includes information on their distribution and uses, mainly for medicine. Threlkeld died in 1728. Another work is by John K'Eogh (or Keogh), *Botanalogia Universalis Hibernica*, published in Cork in 1735, which is a classic early account of the herbal use of plants for Ireland. K'Eogh was born in about 1681, probably in Strokestown, Roscommon. He was educated at Trinity College Dublin, graduated with a Bachelor of Arts in 1711 and was ordained as a priest in the Church of Ireland. At one time he was chaplain to James King, Lord Kingston and later lived in Mitchelstown, Co. Cork. He died in 1754.

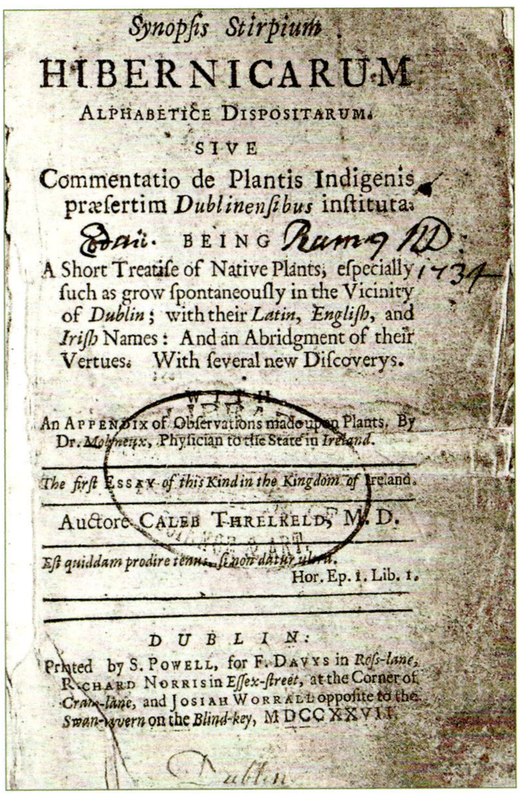

Synopsis Stirpium
HIBERNICARUM
Alphabeticè Dispositarum.
SIVE
Commentatio de Plantis Indigenis præsertim *Dublinensibus* instituta.
BEING
A Short Treatise of Native Plants, especially such as grow spontaneously in the Vicinity of *Dublin*; with their *Latin*, *English*, and *Irish* Names: And an Abridgment of their Vertues. With several new Discoverys.
WITH
An Appendix of Observations made upon Plants, By Dr. *Molyneux*, Physician to the State in *Ireland*.
The first Essay of this Kind in the Kingdom of Ireland.
Auctore Caleb Threlkeld, M.D.
Est quiddam prodire tenus, si non datur ultra.
Hor. Ep. 1. Lib. 1.
DUBLIN:
Printed by S. Powell, for F. Davys in *Ross-lane*, Richard Norris in *Essex-street*, at the Corner of *Crane-lane*, and Josiah Worrall opposite to the *Swan-Tavern* on the *Blind-key*, MDCCXXVII.

Threlkeld's *Synopsis Stirpium Hibernicarum* from 1726.

Threlkeld and K'Eogh include many names of plants in Irish in their works, which suggests that knowledge of plants and their uses was well known in Ireland if such a wide diversity of species could be named in the vernacular language. Both Threlkeld's and K'Eogh's books owe a lot to the work and writings of early British botanists and herbalists, such as John Gerard (1545–1611 or 1612), John Ray (1627–1705), John Parkinson (1587–1650) and Nicholas Culpeper (1616–1654). Prior to the 18th century the study of botany was intrinsically linked to medical studies and teaching. Indeed, most of the first botanic gardens in Europe were started as 'physic gardens', or university gardens for the cultivation, study and teaching of medicine, in such institutions as the Universities of Pisa (established in 1543), Padua (1545), Florence (1545) and Bologna (1547) in Italy; Zurich in Switzerland (1560); and Leiden in The Netherlands (1577). In these islands, the earliest botanic garden, the Oxford Botanic Garden, was established for similar purposes in 1621. In Ireland, the oldest botanic garden was also created as a Physic Garden at Trinity College Dublin in 1687.

The use of plants for medicine by early European herbalists did not follow any system based on the chemical constituents of the plants. Instead, they followed a system that came to be formalised in the 17th century as the 'doctrine of signatures', in which a particular plant shape or colour suggested a use for the plant in the treatment of a particular ailment. For example, yellow flowers were used to treat jaundice, a condition characterised by a yellowing of the skin and the eyes; plants with kidney-shaped leaves were used to treat renal (kidney) ailments, and so on. Their belief was that God had marked each plant with a 'clue' as to their usage and purpose. Such associations were often, then, the basis for the use of a plant for a particular treatment, some of which proved to be effective. The ways in which particular medicinal treatments based on each plant made their way into traditional medicine at local levels is generally hard to determine. Without doubt, some plant uses were derived from widespread ancient European traditions, including Greek and Roman practices, but other remedies were probably based on local discoveries and practices and handed down within local communities. This may help to explain the great variations in usage of plants for medicinal purposes in different parts of the country. The extent to which traditional use of plants followed the doctrine of signatures in Ireland is hard to determine. I am sure that in Ireland, much medicinal usage of plants predates this system by many centuries, and I suspect that it was later applied to many well-established cures and treatments during the Middle Ages.

The belief that there is a herb for every ailment is supported by early herbals and Old Irish proverbs, such as, '*Tá luibh ar gach leigheas*', and there is not a herb without a cure, '*Níl luibh gan leigheas*' (O'Regan, 1997). In the classic 1735 account of herbal plants in Ireland, *Botanalogia Universalis Hibernia*, or *A General Irish Herbal*, John K'Eogh wrote in his preface 'there is not a herb, shrub or tree in nature but is serviceable to Man, either for food or medicine, or for both, for the omniscient and omnipotent God created nothing in vain'. Similarly, the Gaelic scholar John MacKenzie (1806–1846) noted, 'The belief was common among the Caledonians [in Scotland] that for all the diseases to which mankind is liable there grows a herb some-where, and not far from the locality where the particular disease prevails, the proper application of which would cure it' (as quoted in Cameron, 1883: 10).

In the latter part of the 19th century, Lady Wilde (Jane Francesca Agnes) brought to light many aspects of Irish culture, history and folklore, which she actively collected and published, including medicinal uses and 'cures' derived from plants (many of which are noted in the Systematic List). In 1851 she married Sir William Wilde, who himself was an author of significant works on medicine, and their son Oscar Wilde, became the well-known Irish writer and poet. In *Ancient Legends, Mystic Charms and Superstitions of Ireland* (1887), Lady Wilde noted that 'Irish physicians were much celebrated for their learning, and numerous Irish medical manuscripts are in existence, both in Ireland and England, and are also scattered through the public libraries of the continent. They are chiefly written in Latin, with a commentary in Irish, and show a thorough knowledge on the part of the writers of the works of Hippocrates, Galen, Aristotle, and others as celebrated'. Wilde (1887: 255) said there were seven herbs of great medicinal value in ancient Ireland:

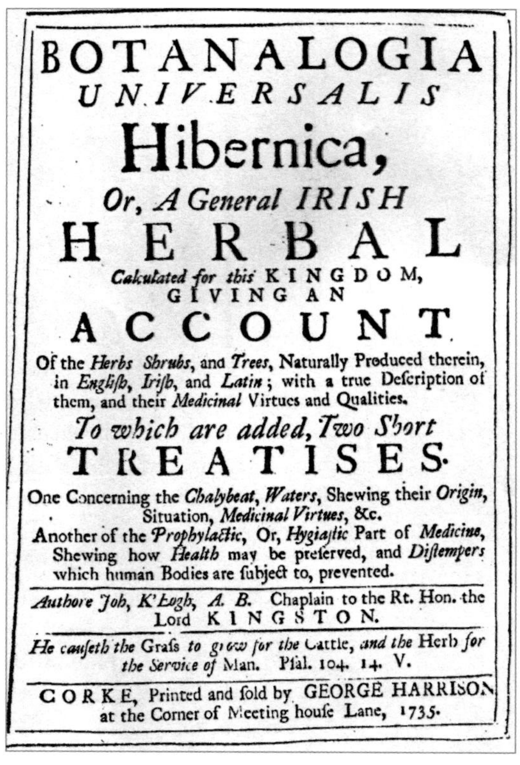
BOTANALOGIA
UNIVERSALIS
Hibernica,
Or, A General IRISH
HERBAL
Calculated for this KINGDOM,
GIVING AN
ACCOUNT
Of the Herbs Shrubs, and Trees, Naturally Produced therein, in English, Irish, and Latin; with a true Description of them, and their Medicinal Virtues and Qualities.
To which are added, Two Short
TREATISES.
One Concerning the Chalybeat, Waters, Shewing their Origin, Situation, Medicinal Virtues, &c.
Another of the Prophylactic, Or, Hygiastic Part of Medicine, Shewing how Health may be preserved, and Distempers which human Bodies are subject to, prevented.
Authore Joh, K'Eogh, A. B. Chaplain to the Rt. Hon. the Lord KINGSTON.
He causeth the Grass to grow for the Cattle, and the Herb for the Service of Man. Psal. 104. 14. V.
CORKE, Printed and sold by GEORGE HARRISON at the Corner of Meeting house Lane, 1735.

K'Eogh's *Botanalogia Universalis Hibernica* from 1735.

> [There are] seven herbs of great value and power; they are ground ivy, vervain, eyebright, groundsel, foxglove, the bark of the elder-tree, and the young shoots of the hawthorn. Nine balls of these mixed together may be taken, and afterwards a potion made of bog-water and salt, boiled in a vessel, with a piece of money and an elf-stone. The elf-stone is generally found near a rath; it has great virtues, but being once lifted up by the spade it must never again touch the earth, or all its virtue is gone.

The view that plants provided a cure for all ailments seemed to have survived into the 20th century in Ireland. McGlinchey's (2007: 83) autobiographical observations about rural life in Co. Donegal at the turn of the 20th century indicate an Ireland where traditional cures were present in the modern practice of medicine: 'In olden times there were few ailments in man or beast that the old people hadn't a cure for. They worked greatly with herbs and old Dr McEleney of Tullagh said that there was a cure of some kind in every plant that grows if people only knew it'. However, McGlinchey goes on to say that 'Some of the cures they went in for were only pisreog [superstitious] cures that did neither good nor harm. People only laugh at them now'.

Throughout Irish mythology and folklore, it is clear that the importance of herbs for healing went far beyond simply their medicinal value, and they were often believed to hold magical properties, and rituals were required in their usage if they were to be effective as cures. For example, O'Regan (1997: 5) reported the story of a postman in Co. Wicklow named Daniel Lowe, who in the late 1900s had a cure for stopping haemorrhages by using a rhyme in the form of a short prayer:

> One day a woman ran out of a house to tell him that there was an old lady dying inside. This lady had the cure for stopping bleeding and wished to pass it on to someone before she died. This had to be done from a male to a female or vice versa, which explained the urgency. Mr Lowe was given the 'cure' with the accompanying words by placing his hand above the injury and repeating the prayer. He would accept no money for his work. Several times he was called to Baggot Street Hospital, Dublin and St Columcille's Hospital, Loughlinstown, where he was asked to join the staff. He refused but continued to travel to the hospital for urgent calls. At the time of his own death, he was so ill with cancer that the cure was not passed on to anyone else.

Bridget Dirrane, an Irish centenarian, wrote in her memoir *A Woman of Aran* (1997: 16) that 'In days gone by people believed in the cures handed down from generation to generation. So it was in the Aran of my youth'. Even though there were doctors present in Inishmore where she grew up as a child in the 1890s, local people still used plant-based medicinal cures extensively. Even today, people can remember old people, perhaps a grandmother, who was known to have particular cures mainly based on plants. The traditional knowledge of the use of plants for medicine was passed down within families, usually in the female line, as part of the oral tradition.

We do not know to what extent information on the value of specific plants remedies was kept secret in Ireland and only passed to others under particular circumstances. In some societies such information is passed down through generations very deliberately, sometimes following the female or male line. In others, it is passed from an 'elder' to his or her 'apprentice', who may or may not be a relative. In most traditional societies medicinal plant knowledge was regarded as an important cultural asset and safeguarded. In some parts of the world it is only now being written down as people realise that it will otherwise be lost forever. In Ireland, which has such a strong oral tradition of handing down stories and songs to the next generations, medicinal plant knowledge would have been similarly passed on. The knowledge that remains as part of an oral tradition in

Ireland today survives as only a shadowy hint and fleeting reminder of a rich plant lore that has been lost.

Within most communities in Ireland there were particular people who were regarded as 'healers', both women and men who were much admired for their traditional medicinal knowledge. In earlier times, women healers were sometimes considered to be witches, or 'fairy-women', practicing sorcery or magic. If looked on more favourably, they were considered 'wise women' or 'herb-women' (in Irish, *bean na luibheanna*). Men were called 'herb doctors', 'medicine men', 'fairy doctors' or *sheeogues*. Several famous Irish witches can be remembered, such as the celebrated (notorious?) Dame Alice Kyteler from Kilkenny who was accused of witchcraft in 1324. According to Wood-Martin (1902) every district, and nearly every village, had such a person whose skills and knowledge were sought after by local people. However, Wood-Martin also noted that they did not ply their trades solely through the use of medicinal herbs but that they also relied on a variety of charms and rituals to affect their cures. For example, many such healers would have a bottle containing a decoction of various herbs that was dispensed as a 'cure-all'. Likewise, a wise woman's house would have certain symbols associated with it: a horseshoe nailed over the door for luck, a bottle of holy water and a bunch of fairy flax (*Linum catharticum*) hanging near the door, and withered yew (*Taxus baccata*) branches hung over the door, beds and cattle in the byre. She also often had a four-leaved clover sewn onto her clothes.

The plants a wise woman might use were varied:

> In the garden grow house-leeks—a specific for sore eyes—tansy, rose noble, Solomon's Seal, bugloss, bogbane, and numerous other herbs, each for some medicinal purpose, and various charms for toothache, headache, for removing warts and taking motes out of the eye. Many condiments are kept in stock. Seal oil for sprains and rheumatism, the tongue of a fox for a poultice to extract thorns, needles, &c., dandelion for liver complaints, comfrey as a styptic, samphire boiled in milk for heartburn; its leaves are also employed in urinary diseases… . A wise-woman would also have belladonna, heartsease, ground ivy, or mountain sage for palpitations or for coughs, bog bark or parsley boiled in milk for gravel, nettles with ginger, for wind in the stomach, horehound as an expectorant, mullein as a cough mixture, ivy leaves for a scald head, and furze tops, broom and carrageen moss combined for a cough mixture; so there was within easy reach a good pharmacopeia (Wood-Martin, 1902: 177).

The best known of these wise women was Biddy Early who lived from 1798 until 1874 in Co. Clare. She learned her skills with herbs from her mother and treated family and friends in her own area. As her reputation grew, others from further afield would come for her treatments. She didn't accept payment but instead would be given gifts of food and alcohol. Her cures involved the use of plants such as nettles, cabbage, water-cress and flax, and she had a famous dark bottle that she used as a talisman (O'Regan, 1997). Her life is documented in books by Edmund Lenihan (1987) and Meda Ryan (1978: 40), which includes a ballad composed during her lifetime:

> In '41 when her first born son
> Played a fairy game of hurley
> I tell you true that the Bottle o' Blue
> Was given to Biddy Early.

She was widely regarded as a witch and was even brought to court for witchcraft (although acquitted for lack of evidence). Her house at Dromore, just south of Feakle,

The overgrown former home of wise woman Biddy Early.

Co. Clare, survives today as an overgrown ruin. A rich mythology has grown up around Biddy Early and her exploits that has been well documented (e.g., by Lenihan, 1987 and Ryan, 1978). I even found that in 2011 someone was reputedly selling her famous 'blue bottle' on the website, eBay.

In August 2011, I visited Biddy Early's house to see whether any important medicinal plants survived growing around her house. Amongst the dark and rather spooky woodlands surrounding her ruined, now roofless, cottage were at least 27 different plant species, including large old trees of *Salix cinerea, Prunus domestica* and *Crataegus monogyna*. Of the 27 species I noted, 22 of them have recorded medicinal uses in Ireland. They included *Hypericum tetrapterum, Sambucus nigra, Lonicera periclymenum, Sanicula europaea, Filipendula ulmaria, Arum maculatum, Geum urbanum* and *Angelica sylvestris*. There were obviously plenty of plants present that could be used by Biddy Early to make health-giving elixirs! Her memory is obviously still revered in Ireland. At her cottage tributes had been left, including coins, crosses, beads, coloured ribbons and cloths, a Guinness bottle, a Bulmers cider can, fresh fruits, religious pictures, a blue liqueur bottle, a silver ring and other jewellery, ornaments, paper notes and drawings and other objects.

There were other similar wise women in various parts of Ireland. My mother told me of visiting the cottage of an old lady in Annestown, Co. Waterford, when she was a young girl, sometime in the early 1930s. The cottage had bunches of wild herbs strung around the walls to dry. The old lady, whose name she remembers as Moll, was widely known as a healer. Erin Kraus has written extensively about another wise woman in Kildare, Moll Anthony (Kraus, 2011). Kraus points out that traditional healers in Ireland could be categorised into those who have cures for particular ailments, and those who were thought to have obtained their powers from the fairies. Moll Anthony and Biddy Early fall into the latter category but would also have used medicinal plants for their healing tasks.

Our knowledge on the application of medicinal plants in early Ireland is poorly known. For example, I have not located information on whether there were seasonal practices related to the gathering of medicinal plants, or whether they were gathered throughout the year or during particular seasons. I also don't know whether medicinal plants were collected, dried and stored for later use, or by and large, always used fresh. In ancient Ireland, the cultivation of medicinal plant gardens may have been a practice particularly associated with monastic centres. The gardener (*lubgortóir*) is listed amongst the seven officers of the church, and there is no mention of gardeners being amongst the servants normally employed by a king or lord (Kelly, 1997). Such practices vary greatly amongst traditional medicine systems in other regions of the world. For example, amongst the Chippewas of North America, plants were usually gathered for medicinal remedies during the late summer (Densmore, 1974). One must remember, too, in studying early herbals, where the medicinal uses and qualities of a particular species are described, we cannot be certain that such plants were ever used for these purposes mentioned in Ireland. Different authors copied earlier works, and it is generally difficult, and sometimes impossible, to determine whether there was genuine traditional usage of some species in Ireland or not.

The use of medicinal herbs was not confined to rural Ireland. Even in the cities, medicinal treatments based on plants (both wild collected and cultivated garden plants) were widespread. Synnott (1979: 37) noted that in 1898 the Irish naturalist Nathaniel Colgan observed that 'Herb shops were still to be seen in Dublin not many years since amongst other places in Mark's Abbey where Threlkeld himself lived'. O'Regan (1997) mentioned that in Cork in the early 1700s there were herb markets also in Cork city where medicinal plants were sold. Market stalls for the sale of medicinal plants are still to be found in many countries, primarily in the developing world, and where the direct use of wild-collected plants in healthcare continues. I have seen such markets in countries including Mauritius, South Africa, Brazil, Vietnam, China and Haiti, and they may not be very different from the ways in which medicinal plants were collected, sold and used in cities and towns in Ireland up to the beginning of the 18th century.

Before pharmaceutical drugs were so readily available, the use of herbs for medicine often involved complex combinations of various plant species and special preparations to treat particular ailments. Sometimes, these recipes were passed down from generation to generation as valuable family lore. One such botanical preparation from the 19th century or early 20th century involved parts from nine different plants to make a poultice to cure the 'rose' (erysipelas, a streptococcal skin infection with lymphatic involvement): 'the root of the stinking roger, or rose noble, the daisy plant roots and leaves, fern leaves, docken leaves, primrose leaves and woodbine leaves, and pennywort leaves. These were all ground together and heated in a pot and then the poultice was ready' (McGlinchey, 2007: 83). Lady Wilde (1890: 37) gave another 'recipe' involving several plants as a treatment for 'the Liver Complaint':

> The leaves of plantain, wild sage, the shamrock and dock-leaf, with valerian and the flower of the daisy, are to be plucked by the person before sunrise, and fasting, on Mondays and Wednesdays, while Hail Mary is said, and the Paternoster; all these are to be boiled and strained, and the herbs afterwards to be carefully burned. A glassful of the liquor to be taken twice a day.

Another recipe aimed to treat 'the decline' (any disease that gradually weakens or wastes the body):

> Seven stalks of sidhe lint or fairy lint [*Linum catharticum*], seven stalks of the *dubhchosach* [*Adiantum capillus-veneris*], seven more of the *Bata cogaidh* (black knapweed) [*Centaurea nigra*]. These were pounded together and mixed with seven meadars, or noggins, of water lifted from a place where three waters meet. All were heated in a cam, that's a piece of an old pot, and two spoonfuls given to the *breoiteachán* [patient] on his fast every morning for seven weeks (McGlinchey, 2007: p. 87).

The number seven is generally said to be a magical number, strongly associated with mythology and various religions, including Christianity, Judaism, Hinduism and Islam. (In Christianity, for example there are seven days of Creation; seven pillars of wisdom, the seven last words of Jesus of the Cross, etc.) One of my mother's often repeated and unexplained phrases, probably of Irish origin, is 'Glory be to the seven churches', perhaps ultimately derived from the Seven Churches of Asia mentioned in the Book of Revelation, the last book of the Christian New Testament.

By the 20th century in Ireland, with the increased availability and use of pharmaceutical drugs, wild plants were used for medicinal purposes only to a very limited extent. Keller (1918: 151) wrote in a pamphlet published by the Department of Agriculture and Technical Instruction for Ireland on the cultivation and collection of medicinal plants that 'at the outbreak of the present war [World War I] the old-fashioned herbalist had almost disappeared and many of his remedies, as well as his methods of prescribing them, had fallen into disuse....' Nevertheless, Keller points out that during the war, concern rose for securing the supply of certain wild-plant–derived drugs with the majority of plant material imported into Britain and Ireland from Austro-Hungary and Germany. Such concern stimulated the growth of the Association, established by the United Irishwomen in 1916, to promote the use of herbal medicine in Ireland. By the end of 1917 the Association was organised into 19 districts or counties and had 650 members. Shortly after its inception, it was organised as a series of Provincial Herb Associations, linked by means of an Irish Herb Council. According to Keller (1918) the goal of forming the Association was not only to dispel the immediate shortage of medicinal herbs but also to build an industry that could survive the war. For example, in 1916, the total quantity of herbs produced by members of the Association was approximately 10 cwt (a hundredweight or centum weight[19]), but by 1917 the total had grown to 5 tonnes, and it was mainly wild collected herbs, which were prepared in five drying sheds throughout the country. Forty-four species required by the drug trade were evaluated for potential commercial exploitation, but only 13 were found to occur in the wild in sufficient quantity to justify their collection (Keller, 1918: 152). I have not found details of which species were considered, but as a result of the evaluation, only three species were regarded as an economic proposition: dandelion (*Taraxacum officinale*), foxglove (*Digitalis purpurea*) and couch-grass (*Elymus repens* [*Elytrigia repens*]). As a result of the War the price of dandelion root had risen three-fold, from 4d (pence) to 1s (shilling, or 12p) per pound. The price of prepared rhizomes (underground stems similar to roots) of couch-grass had also risen to as much as 35s for 1 cwt, and dried foxglove leaves sold for 1s 3d per pound. Serious consideration was given to establishing each of these three species as an economically viable cultivated crop, although I am unsure to what extent this actually happened. Keller (1918) recommended wild-growing pharmacopoeial herbs and noted experiments had shown that belladonna, henbane, digitalis, colchicum, stramonium, valerian and aconite were optimum plants for cultivation in Ireland. Other sweet and medicinal herbs also flourished but were not as important and did not command such remunerative prices. Keller (1918: 151) also provided a useful catch-all insight into how medicinal plants were applied: 'In most cases the patient was directed, with remarkable uniformity, to steep one ounce of the dried herb in one pint of boiling water, and of this infusion he was ordered to drink a wineglass full three times daily until cured'.

The following is a list of the main known and documented uses of different native plants for medicinal purposes. It is not necessarily comprehensive nor have I made any attempt to say whether the use of a particular plant is effective in the treatment of the ailment or condition outlined. The efficacy of each plant is beyond the scope of this work; however, it is clear that in some cases there is a genuine pharmacological basis to the use of certain species, whereas other species are probably unlikely to have any effect whatsoever. In the Systematic List of this book, the use of each plant for particular medicinal purposes is outlined. However, it should be remembered that the information I have provided is not exhaustive, and locally in different parts of Ireland many more plants may have been used as part of particular treatments. In most cases I have provided the source of the information. I have not repeated the sometimes numerous references to the use of plant species for a particular ailment, which in some instances are copied from each other in any case.

19. The hundredweight in Ireland and Great Britain is defined as 112 lb, which is equal to 50.8 kg. There are 20 hundredweights in a tonne.

SUMMARY OF MEDICINAL PLANT USES

Abortions (to induce)
- *Umbilicus rupestris*—Wall pennywort; Navelwort

Aches and pains (general)
- *Persicaria hydropiper* (*Polygonum hydropiper*)—Water pepper
- *Salix* species—Willow
- *Vaccinium myrtillus*—Bilberry; Fraughan
- *Valeriana officinalis*—Wild valerian
- *Verbascum thapsus*—Mullein

Anaemia
- *Nasturtium microphyllum* (*Rorippa microphylla*)—Narrow-fruited water-cress
- *Nasturtium officinale* (*Rorippa nasturtium-aquaticum*)—Water-cress
- *Rumex obtusifolius*—Broad-leaved dock
- *Taraxacum officinale* agg.—Dandelion

Angina
- *Trifolium pratense*—Red clover

Anti-inflammatory uses
- *Agrimonia eupatoria*—Agrimony
- *Chamaemelum nobile* (*Anthemis nobilis*)—Chamomile
- *Fraxinus excelsior*—Ash
- *Galium verum*—Lady's bedstraw
- *Hedera helix*—Ivy
- *Lamium purpureum*—Red dead-nettle; Archangel
- *Lamium hybridum*—Cut-leaved dead-nettle
- *Ophioglossum vulgatum*—Adder's-tongue
- *Persicaria maculosa* (*Polygonum persicaria*)—Redshank; Spotted persicaria
- *Potentilla erecta*—Tormentil
- *Stellaria media*—Chickweed
- *Ulmus glabra*—Wych elm; Irish elm

Anti-spasmodic
- *Conium maculatum*—Hemlock (lethally toxic if ingested)
- *Thymus polytrichus* (*Thymus drucei*; *Thymus praecox*; *Thymus serpyllum*)—Wild thyme
- *Tussilago farfara*—Colt's foot

Anus, cracked
- *Potentilla erecta*—Tormentil

Appetite, loss of
- *Artemisia maritima* (*Seriphidium maritimum*)—Sea wormwood
- *Centaurium erythraea*—Common centaury
- *Menyanthes trifoliata*—Bogbean

Arthritis
- *Apium graveolens*—Wild celery
- *Conium maculatum*—Hemlock
- *Hyoscyamus niger*—Henbane
- *Persicaria hydropiper* (*Polygonum hydropiper*)—Water pepper

Asthma
- *Adiantum capillus-veneris*—Maidenhair fern
- *Allium ursinum*—Wild garlic; Ramsons
- *Althaea officinalis*—Marsh-mallow
- *Asplenium ruta-muraria*—Wall-rue
- *Centaurea nigra*—Common knapweed; Blackheads
- *Chrysanthemum leucanthemum* (*Leucanthemum vulgare*)—Ox-eye daisy
- *Drosera anglica*—Great sundew
- *Drosera intermedia*—Oblong-leaved sundew
- *Drosera rotundifolia*—Round-leaved sundew
- *Erica cinerea*—Heather

Trifolium pratense (red clover). The flower colour in this image appears to have faded from pink to white. Water-colour by Lydia Shackleton. Courtesy of the National Botanic Gardens, Glasnevin.

SUMMARY OF MEDICINAL PLANT USES *(continued)*

Asthma *(continued)*
Glechoma hederacea—Ground ivy
Lavatera arborea—Tree mallow
Rumex hydrolapathum—Water dock
'Seaweed'
Sorbus aucuparia—Mountain ash; Rowan
Taraxacum officinale—Dandelion
Tussilago farfara—Colt's foot
Ulex europaeus—Gorse
Vaccinium myrtillus—Bilberry; Fraughan
Verbascum thapsus—Mullein

Backache
Althaea officinalis—Marsh-mallow
Geranium robertianum—Herb-Robert
Parietaria judaica (*Parietaria diffusa*)—Wall pellitory
Petroselinum crispum—Parsley
Prunella vulgaris—Selfheal

'Bad Humour'
Silene uniflora (*Silene vulgaris* subsp. *maritima*)—Sea campion

Baldness
Centaurium erythraea—Common centaury
Salix species—Willow

Bed-wetting
Hypericum perforatum—Perforate St John's-wort

Bladder problems
Agrimonia eupatoria—Agrimony
Umbilicus rupestris—Wall pennywort; Navelwort (bladder stones)
Vaccinium myrtillus—Bilberry; Fraughan (bladder stones)

Bleeding (to staunch external bleeding; astringents)
Achillea millefolium—Yarrow
Adiantum capillus-veneris—Maidenhair fern
Allium babingtonii (*Allium ampeloprasum* var. *babingtonii*)—Wild leek
Antennaria dioica—Mountain everlasting; Cat's foot
Capsella bursa-pastoris—Shepherd's purse
Geranium robertianum—Herb-Robert (for bleeding piles)
Heracleum sphondylium—Hogweed; Cow-parsnip
Hippuris vulgaris—Mare's-tail
Lythrum salicaria—Purple loosestrife
Nasturtium microphyllum (*Rorippa microphylla*)—Narrow-fruited water-cress
Nasturtium officinale—Water-cress
Persicaria lapathifolia (*Polygonum lapathifolium*)—Pale persicaria
Potentilla anserina—Silverweed
Potentilla erecta—Tormentil
Prunella vulgaris—Selfheal
Prunus spinosa—Sloe; Blackthorn
Rubus fruticosus agg.—Bramble; Blackberry
Rumex acetosa—Sorrel
Rumex hydrolapathum—Water dock
Rumex obtusifolius—Broad-leaved dock
Sanicula europaea—Wood sanicle; Sanicle
Silene uniflora (*Silene vulgaris* subsp. *maritima*)—Sea campion
Sisymbrium officinale—Hedge mustard
Ulmus glabra—Wych elm; Irish elm
Veronica officinalis—Heath speedwell

Bleeding, internal
Alchemilla vulgaris—Lady's-mantle
Capsella bursa-pastoris—Shepherd's purse
Lamium purpureum—Red dead-nettle; Archangel
Lamium hybridum—Cut-leaved dead-nettle
Sanicula europaea—Wood sanicle; Sanicle

Alchemilla vulgaris (Lady's mantle).

Blisters
Allium ursinum—Wild garlic; Ramsons
Glechoma hederacea—Ground ivy
Ranunculus flammula—Lesser spearwort (to cause blisters and draw out disease)
Ranunculus repens—Meadow buttercup (to cause blisters and draw out disease)
Senecio vulgaris—Groundsel

Blood loss or other blood disorders
Ballota nigra—Black horehound

Blood poisoning
Malva sylvestris—Common mallow

SUMMARY OF MEDICINAL PLANT USES *(continued)*

Blood, blood clots (prevention or cure)

Allium ursinum—Wild garlic; Ramsons
Hypericum elodes—Marsh St John's-wort
Hypericum hirsutum—Trailing St John's-wort
Hypericum humifusum—Hairy St John's-wort
Hypericum maculatum—Imperforate St John's-wort
Hypericum perforatum—Perforate St John's-wort
Hypericum pulchrum—Slender St John's-wort
Hypericum tetrapterum—Square-stalked St John's-wort

Blood (to purify or cleanse)

Achillea millifolium—Yarrow (as a tonic for weak blood)
Allium ursinum—Wild garlic; Ramsons
Arctium minus—Burdock
Cochlearia officinalis—Common scurvygrass
Conopodium majus—Pignut
Daucus carota—Carrot
Geranium robertianum—Herb-Robert
Menyanthes trifoliata—Bogbean
Oxalis acetosella—Wood sorrel
Plantago lanceolata—Ribwort plantain (for 'weak blood')
Rumex acetosa—Sorrel
Rumex hydrolapathum—Water dock
Senecio vulgaris—Groundsel
Sorbus aucuparia—Mountain ash; Rowan
Urtica dioica—Nettle

Blood circulation

Achillea millifolium—Yarrow

Blood pressure, hypertension

Apium graveolens—Wild celery

Boils, abscesses and carbuncles

Apium graveolens—Wild celery
Arctium minus—Burdock
Chamaemelum nobile (*Anthemis nobilis*)—Chamomile
Chrysanthemum leucanthemum (*Leucanthemum vulgare*)—Ox-eye daisy
Conium maculatum—Hemlock
Corylus avellana—Hazel
Cytisus scoparius—Broom
Hedera helix—Ivy
Hypericum androsaemum—Tutsan
Menyanthes trifoliata—Bogbean
Oxalis acetosella—Wood sorrel
Parietaria judaica (*Parietaria diffusa*)—Wall pellitory
Plantago coronopus—Buck's-horn plantain
Plantago lanceolata—Ribwort plantain
Plantago major—Greater plantain
Plantago maritima—Sea plantain
Senecio jacobaea—Ragwort
Senecio vulgaris—Groundsel
Stellaria media—Chickweed
Symphytum officinale—Comfrey
Verbascum thapsus—Mullein

Bones, broken

Agrimonia eupatoria—Agrimony
Althaea officinalis—Marsh-mallow
Bellis perennis—Daisy
Filipendula ulmaria—Meadow-sweet
Symphytum officinale—Comfrey
Taraxacum officinale agg.—Dandelion

Brain diseases

Cuscuta epithymum—Dodder

Breasts (to reduce swelling of)

Hyoscyamus niger—Henbane
Sisymbrium officinale—Hedge mustard

Breasts, soreness

Veronica species—Speedwell

Bright's disease (nephritis, kidney)

Pteridium aquilinum—Bracken

Bronchitis

Armoracia rusticana—Horse-radish
Glechoma hederacea—Ground ivy
Hedera helix—Ivy
Nasturtium microphyllum (*Rorippa microphylla*)—Narrow-fruited water-cress
Nasturtium officinale (*Rorippa nasturtium-aquaticum*)—Water-cress
Rumex obtusifolius—Broad-leaved dock
Sambucus nigra—Elder
Verbascum thapsus—Mullein

Bruises and sores

Ajuga reptans—Bugle
Allium ursinum—Wild garlic; Ramsons
Alnus glutinosa—Alder
Anemone nemorosa—Wood anemone
Arctium lappa—Great burdock
Calendula officinalis—Garden marigold; Pot marigold
Conium maculatum—Hemlock
Foeniculum vulgare—Fennel

SUMMARY OF MEDICINAL PLANT USES *(continued)*

Bruises and sores *(continued)*

- *Glechoma hederacea*—Ground ivy
- *Lapsana communis*—Nipplewort
- *Ophioglossum vulgatum*—Adder's-tongue
- *Osmunda regalis*—Royal fern
- *Pilosella officinarum* (*Hieracium pilosella*)—Mouse-ear hawkweed
- *Plantago coronopus*—Buck's-horn plantain
- *Plantago lanceolata*—Ribwort plantain
- *Plantago major*—Greater plantain
- *Plantago maritima*—Sea plantain
- *Ranunculus flammula*—Lesser spearwort
- *Rumex acetosa*—Sorrel
- *Rumex crispus*—Curled dock
- *Rumex obtusifolius*—Broad-leaved dock
- *Succisa pratensis*—Devil's-bit scabious
- *Taraxacum officinale* agg.—Dandelion
- *Urtica dioica*—Nettle
- *Verbascum thapsus*—Mullein

Bubonic plague

- *Ranunculus flammula*—Lesser spearwort

Burns and scalds

- *Alisma plantago-aquatica*—Water-plantain
- *Alchemilla vulgaris*—Lady's-mantle
- *Arctium minus*—Burdock
- *Bellis perennis*—Daisy
- *Conium maculatum*—Hemlock
- *Crataegus monogyna*—Hawthorn
- *Fraxinus excelsior*—Ash
- *Galium aparine*—Robin-run-the-hedge; Cleavers; Goosegrass
- *Hedera helix*—Ivy
- *Ilex aquifolium*—Holly
- *Iris pseudacorus*—Yellow flag
- *Lapsana communis*—Nipplewort
- *Ophioglossum vulgatum*—Adder's-tongue
- *Pilosella officinarum* (*Hieracium pilosella*)—Mouse-ear hawkweed
- *Plantago coronopus*—Buck's-horn plantain
- *Plantago lanceolata*—Ribwort plantain
- *Plantago major*—Greater plantain
- *Plantago maritima*—Sea plantain
- *Potamogeton natans*—Broad-leaved pondweed
- *Potentilla erecta*—Tormentil
- *Primula vulgaris* (*Primula acaulis*)—Primrose
- *Pteridium aquilinum*—Bracken
- *Rubus fruticosus* agg.—Bramble; Blackberry
- *Rumex acetosa*—Sorrel
- *Sambucus nigra*—Elder
- *Scrophularia nodosa*—Figwort
- *Sonchus asper*—Spiny sow-thistle
- *Sonchus oleraceus*—Smooth sow-thistle
- *Stellaria media*—Chickweed
- *Ulmus glabra*—Wych elm; Irish elm

Cancer, tumours

- *Cynoglossum officinale*—Hound's-tongue
- *Digitalis purpurea*—Foxglove
- *Frangula alnus*—Alder buckthorn
- *Galium aparine*—Robin-run-the-hedge; Cleavers; Goosegrass
- *Galium odoratum*—Woodruff
- *Malus sylvestris*—Crab-apple
- *Oenanthe crocata*—Hemlock water-dropwort
- *Oxalis acetosella*—Wood sorrel
- *Rumex acetosa*—Sorrel
- *Scrophularia auriculata*—Water figwort; Water betony (cancerous ulcers)
- *Senecio jacobaea*—Ragwort
- *Taraxacum officinale* agg.—Dandelion (external cancers)
- *Taxus baccata*—Yew (modern pharmaceutical use only; toxic alkaloids are extremely poisonous and occur in wood, needles and seeds)
- *Trifolium pratense*—Red clover
- *Viola odorata*—Sweet violet

Cellulite

- *Taraxacum officinale* agg.—Dandelion

Chest and lung ailments

- *Allium ursinum*—Wild garlic; Ramsons
- *Althaea officinalis*—Marsh-mallow
- *Anemone nemorosa*—Wood anemone

Galium aparine **(robin-run-the-hedge; cleavers; goosegrass) with fruits.**

SUMMARY OF MEDICINAL PLANT USES *(continued)*

Chest and lung ailments *(continued)*

Angelica sylvestris—Wild angelica
Lavatera arborea—Tree mallow
Nasturtium microphyllum—Narrow-fruited water-cress
Nasturtium officinale—Water-cress
Parietaria judaica (*Parietaria diffusa*)—Wall pellitory
Scrophularia nodosa—Figwort
Sisymbrium officinale—Hedge mustard
Veronica species—Speedwell

Chickenpox

Juncus inflexus—Hard rush (ointment for pustules)

Chilblains

Bellis perennis—Daisy
Malus domestica—Apple
Quercus petraea—Sessile oak; Irish oak
Quercus robur—Pedunculate oak; English oak
Solanum tuberosum—Potato

Childbirth (pain during)

Rubus idaeus—Wild raspberry

Childbirth (to expel an expired foetus)

Salsola kali—Saltwort

Circulation of blood

Agrimonia eupatoria—Agrimony

Colds and chills

Achillea millefolium—Yarrow
Achillea ptarmica—Sneezewort
Agrimonia eupatoria—Agrimony
Allium ursinum—Wild garlic; Ramsons
Althaea officinalis—Marsh-mallow
Arctium minus—Burdock
Bellis perennis—Daisy
Chrysanthemum leucanthemum (*Leucanthemum vulgare*)—Ox-eye daisy
Corylus avellana—Hazel (buds mixed with dandelions, chickweed, wood sorrel and oatmeal)
Cytisus scoparius—Broom
Geum urbanum—Wood avens
Hedera helix—Ivy
Lavatera arborea—Tree mallow
Mentha aquatica—Water mint
Menyanthes trifoliata—Bogbean
Nasturtium microphyllum—Narrow-fruited water-cress
Nasturtium officinale—Water-cress
Oxalis acetosella—Wood sorrel
Rubus fruticosus agg.—Bramble; Blackberry
Rumex obtusifolius—Broad-leaved dock
Sambucus nigra—Elder
Senecio jacobaea—Ragwort

Colic

Chamaemelum nobile (*Anthemis nobilis*)—Chamomile
Chrysosplenium oppositifolium—Golden saxifrage
Daucus carota—Carrot
Ilex aquifolium—Holly
Juniperus communis—Juniper
Ranunculus flammula—Lesser spearwort
Stellaria media—Chickweed
Teucrium scorodonia—Wood sage

Convulsions

Arctium minus—Burdock
Asplenium ruta-muraria—Wall-rue
Sinapis arvensis—Charlock

Coughs and sore throats

Adiantum capillus-veneris—Maidenhair fern
Agrimonia eupatoria—Agrimony
Allium ursinum—Wild garlic; Ramsons
Alnus glutinosa—Alder
Althaea officinalis—Marsh-mallow
Bellis perennis—Daisy
Blechnum spicant—Hard fern
Calluna vulgaris—Heather; Ling
Centaurium erythraea—Common centaury
Chamaemelum nobile (*Anthemis nobilis*)—Chamomile
Chrysanthemum leucanthemum (*Leucanthemum vulgare*)—Ox-eye daisy
Corylus avellana—Hazel (buds mixed with dandelions, chickweed, wood sorrel and oatmeal)
Cynoglossum officinale—Hound's-tongue
Cytisus scoparius—Broom
Drosera anglica—Great sundew
Drosera intermedia—Oblong-leaved sundew
Drosera rotundifolia—Round-leaved sundew
Erica cinerea—Heather
Filipendula ulmaria—Meadow-sweet
Geranium robertianum—Herb-Robert
Hedera helix—Ivy
Hyacinthoides non-scripta—Bluebell
Hyoscyamus niger—Henbane
Iris pseudacorus—Yellow flag
Juniperus communis—Juniper

SUMMARY OF MEDICINAL PLANT USES *(continued)*

Coughs and sore throats *(continued)*

Lavatera arborea—Tree mallow
Ligustrum vulgare—Privet
Lycopus europaeus—Gipsywort
Lysimachia nummularia—Creeping-Jenny
Malus sylvestris—Crab-apple
Malva sylvestris—Common mallow
Mentha aquatica—Water mint
Mentha pulegium—Penny-royal
Menyanthes trifoliata—Bogbean
Myrica gale—Bog myrtle
Nasturtium microphyllum—Narrow-fruited water-cress
Nasturtium officinale—Water-cress
Oxalis acetosella—Wood sorrel
Parietaria judaica (*Parietaria diffusa*)—Wall pellitory
Pilosella officinarum (*Hieracium pilosella*)—Mouse-ear hawkweed
Plantago coronopus—Buck's-horn plantain
Plantago lanceolata—Ribwort plantain
Plantago major—Greater plantain
Plantago maritima—Sea plantain
Polygala serpyllifolia—Heath milkwort
Polygala vulgaris—Common milkwort
Potentilla anserina—Silverweed
Potentilla erecta—Tormentil

Potentilla anserina **(silverweed).**

Rubus fruticosus agg.—Bramble; Blackberry
Rubus idaeus—Wild raspberry
Rumex acetosa—Sorrel
Rumex obtusifolius—Broad-leaved dock
Sambucus nigra—Elder
Scrophularia auriculata—Water figwort; Water betony
Scilla verna—Spring squill
Senecio jacobaea—Ragwort
Sinapis arvensis—Charlock
Sisymbrium officinale—Hedge mustard
Sorbus aucuparia—Mountain ash; Rowan
Stellaria media—Chickweed
Succisa pratensis—Devil's-bit scabious
Taraxacum officinale agg.—Dandelion
Teucrium scorodonia—Wood sage
Trifolium pratense—Red clover
Trifolium repens—White clover
Ulex europaeus—Gorse; Whin; Furze
Ulex gallii—Western gorse
Urtica dioica—Nettle
Verbascum thapsus—Mullein
Veronica species—Speedwell
Veronica beccabunga—Brooklime

Cramps (menstrual, muscle or stomach)

Chamaemelum nobile (*Anthemis nobilis*)—Chamomile
Origanum vulgare—Marjoram; Oregano
Sinapis arvensis—Charlock
Verbascum thapsus—Mullein

Cuts and wounds

Achillea millefolium—Yarrow
Agrimonia eupatoria—Agrimony
Ajuga reptans—Bugle
Alchemilla vulgaris—Lady's-mantle
Anemone nemorosa—Wood anemone
Anthyllis vulneraria—Kidney vetch
Arctium minus—Burdock
Calendula officinalis—Garden marigold; Pot marigold
Chrysanthemum leucanthemum (*Leucanthemum vulgare*)—Ox-eye daisy
Corylus avellana—Hazel
Daucus carota—Carrot
Galium verum—Lady's bedstraw
Hedera helix—Ivy
Hydrocharis morsus-ranae—Frogbit
Hypericum androsaemum—Tutsan
Hypericum perforatum—Perforate St John's-wort
Iris pseudacorus—Yellow flag
Juniperus communis—Juniper
Lamium hybridum—Cut-leaved dead-nettle
Lamium purpureum—Red dead-nettle; Archangel
Lapsana communis—Nipplewort
Malva sylvestris—Common mallow
Nasturtium microphyllum—Narrow-fruited water-cress
Nasturtium officinale—Water-cress
Ophioglossum vulgatum—Adder's-tongue
Osmunda regalis—Royal fern

SUMMARY OF MEDICINAL PLANT USES *(continued)*

Cuts and wounds *(continued)*

Persicaria maculosa (*Polygonum persicaria*)—Redshank; Spotted persicaria
Phleum pratense—Timothy
Plantago coronopus—Buck's-horn plantain
Plantago lanceolata—Ribwort plantain
Plantago major—Greater plantain
Plantago maritima—Sea plantain
Primula vulgaris (*Primula acaulis*)—Primrose
Rubus fruticosus agg.—Bramble; Blackberry
Scrophularia auriculata—Water figwort; Water betony (to clean wounds)
Senecio aquaticus—Marsh ragwort
Senecio jacobaea—Ragwort
Senecio vulgaris—Groundsel (to draw pus from)
Solanum dulcamara—Bittersweet
Sphagnum species—Sphagnum moss
Stachys palustris—Marsh woundwort
Stachys sylvatica—Hedge woundwort
Symphytum officinale—Comfrey
Taraxacum officinale agg.—Dandelion
Valeriana officinalis—Wild valerian

Cystitis

Equisetum arvense—Field horsetail

Deafness

Mentha aquatica—Water mint
Primula veris—Cowslip

Depression and melancholia

Cuscuta epithymum—Dodder
Nasturtium microphyllum—Narrow-fruited water-cress
Nasturtium officinale—Water-cress
Salsola kali—Saltwort
Solidago virgaurea—Goldenrod

Diabetes

Geranium robertianum—Herb-Robert
Taraxacum officinale agg.—Dandelion

Diarrhoea

Adiantum capillus-veneris—Maidenhair fern
Capsella bursa-pastoris—Shepherd's purse
Equisetum arvense—Field horsetail
Euphorbia helioscopia—Sun spurge
Euphorbia paralias—Sea spurge
Euphorbia peplus—Petty spurge
Filipendula ulmaria—Meadow-sweet
Hypericum elodes—Marsh St John's-wort
Hypericum hirsutum—Trailing St John's-wort
Hypericum humifusum—Hairy St John's-wort
Hypericum maculatum—Imperforate St John's-wort
Hypericum perforatum—Perforate St John's-wort
Hypericum pulchrum—Slender St John's-wort
Hypericum tetrapterum—Square-stalked St John's-wort
Limonium humile—Lax-flowered sea lavender
Orchis mascula—Early purple orchid
Oxalis acetosella—Wood sorrel
Potentilla anserina—Silverweed
Potentilla erecta—Tormentil
Quercus petraea—Sessile oak; Irish oak
Quercus robur—Pedunculate oak; English oak
Ranunculus flammula—Lesser spearwort
Rubus fruticosus agg.—Bramble; Blackberry
Scrophularia auriculata—Water figwort; Water betony
Vaccinium myrtillus—Bilberry; Fraughan
Verbascum thapsus—Mullein

Diuretics

Adiantum capillus-veneris—Maidenhair fern
Alchemilla vulgaris—Lady's-mantle
Alliaria petiolata—Garlic mustard; Hedge garlic
Angelica sylvestris—Wild angelica
Arctium minus—Burdock
Capsella bursa-pastoris—Shepherd's purse
Cytisus scoparius—Broom
Daucus carota—Carrot
Digitalis purpurea—Foxglove
Glechoma hederacea—Ground ivy
Hypericum androsaemum—Tutsan
Juniperus communis—Juniper
Parietaria judaica (*Parietaria diffusa*)—Wall pellitory
Populus tremula—Aspen
Rubus fruticosus agg.—Bramble; Blackberry
Rumex acetosella—Sheep's sorrel
Salsola kali—Saltwort
Sedum telephium—Orpine
Senecio vulgaris—Groundsel
Taraxacum officinale agg.—Dandelion
Vaccinium myrtillus—Bilberry; Fraughan
Veronica beccabunga—Brooklime

Dog bites (poultices)

Adiantum capillus-veneris—Maidenhair fern
Stachys palustris—Marsh woundwort
Stachys sylvatica—Hedge woundwort

Dropsy[1]

Adiantum capillus-veneris—Maidenhair fern

SUMMARY OF MEDICINAL PLANT USES *(continued)*

Dropsy *(continued)*

- *Alliaria petiolata*—Garlic mustard; Hedge garlic
- *Arctium minus*—Burdock
- *Artemisia maritima* (*Seriphidium maritimum*)—Sea wormwood
- *Betula* species—Birch
- *Cytisus scoparius*—Broom
- *Digitalis purpurea*—Foxglove
- *Filipendula ulmaria*—Meadow-sweet
- *Fraxinus excelsior*—Ash
- *Menyanthes trifoliata*—Bogbean
- *Primula veris*—Cowslip
- *Urtica dioica*—Nettle

Drunkenness

- *Artemisia vulgaris*—Mugwort

Dysentery

- *Adiantum capillus-veneris*—Maidenhair fern
- *Capsella bursa-pastoris*—Shepherd's purse
- *Limonium humile*—Lax-flowered sea lavender
- *Lonicera periclymenum*—Honeysuckle, Woodbine
- *Lysimachia nummularia*—Creeping-Jenny
- *Lythrum salicaria*—Purple loosestrife
- *Pulicaria dysenterica*—Fleabane

Ear ache

- *Daucus carota*—Carrot
- *Fraxinus excelsior*—Ash
- *Hedera helix*—Ivy
- *Persicaria maculosa* (*Polygonum persicaria*)—Redshank or Spotted persicaria
- *Populus tremula*—Aspen
- *Tussilago farfara*—Colt's foot

Emetic (to promote vomiting)

- *Stachys officinalis*—Betony

Epilepsy

- *Anagallis arvensis*—Scarlet pimpernel
- *Angelica sylvestris*—Wild angelica
- *Artemisia absinthium*—Wormwood
- *Artemisia vulgaris*—Mugwort
- *Cardamine pratensis*—Cuckoo-flower; Lady's smock
- *Conium maculatum*—Hemlock
- *Daucus carota*—Carrot
- *Digitalis purpurea*—Foxglove
- *Glechoma hederacea*—Ground ivy
- *Lonicera periclymenum*—Honeysuckle; Woodbine
- *Melilotus officinalis*—Ribbed melilot
- *Potentilla anserina*—Silverweed
- *Sambucus nigra*—Elder
- *Thalictrum flavum*—Common meadow-rue

Erysipelas (St Anthony's fire, sometimes also called 'nettle-rash')

- *Bellis perennis*—Daisy
- *Cytisus scoparius*—Broom
- *Hedera helix*—Ivy
- *Ophioglossum vulgatum*—Adder's-tongue
- *Ranunculus species*—Buttercups; Crowfoots
- *Ranunculus acris*—Meadow buttercup
- *Ranunculus repens*—Creeping buttercup

Eye ailments

- *Allium ursinum*—Wild garlic; Ramsons
- *Althaea officinalis*—Marsh-mallow
- *Anagallis arvensis*—Scarlet pimpernel
- *Anemone nemorosa*—Wood anemone
- *Arrhenatherum elatius*—False oat
- *Bellis perennis*—Daisy
- *Centaurea cyanus*—Cornflower
- *Centaurium erythraea*—Common centaury
- *Chrysanthemum leucanthemum* (*Leucanthemum vulgare*)—Ox-eye daisy
- *Daucus carota*—Carrot
- *Euphrasia officinalis* agg.—Eyebright
- *Geranium robertianum*—Herb-Robert
- *Hedera helix*—Ivy
- *Nasturtium microphyllum*—Narrow-fruited water-cress
- *Nasturtium officinale*—Water-cress
- *Plantago coronopus*—Buck's-horn plantain
- *Plantago lanceolata*—Ribwort plantain
- *Plantago major*—Greater plantain
- *Plantago maritima*—Sea plantain
- *Potentilla anserina*—Silverweed
- *Sorbus aucuparia*—Mountain ash; Rowan
- *Succisa pratensis*—Devil's-bit scabious
- *Taraxacum officinale* agg.—Dandelion (for styes)
- *Umbilicus rupestris*—Wall pennywort; Navelwort
- *Vaccinium myrtillus*—Bilberry; Fraughan
- *Valeriana officinalis*—Wild valerian

Feet, sore (corns, bunions and verrucas)

- *Allium ursinum*—Wild garlic; Ramsons
- *Fraxinus excelsior*—Ash
- *Hedera helix*—Ivy
- *Plantago coronopus*—Buck's-horn plantain

SUMMARY OF MEDICINAL PLANT USES *(continued)*

Feet, sore *(continued)*

Plantago lanceolata—Ribwort plantain
Plantago major—Greater plantain
Plantago maritima—Sea plantain
Potentilla anserina—Silverweed
Quercus petraea—Sessile oak; Irish oak (to reduce foot odour)
Quercus robur—Pedunculate oak; English oak (to reduce foot odour)
Rumex obtusifolius—Broad-leaved dock
Solanum tuberosum—Potato
Umbilicus rupestris—Wall pennywort; Navelwort

Felons (ingrown nails, toes or fingers)

Allium ursinum—Wild garlic; Ramsons

Fevers

Aconitum napellus—Monk's-hood (tincture of aconite)[2]
Artemisia maritima (*Seriphidium maritimum*)—Sea wormwood
Centaurium erythraea—Common centaury
Fraxinus excelsior—Ash
Parietaria judaica (*Parietaria diffusa*)—Wall pellitory
Polypodium australe—Southern polypody
Polypodium interjectum—Intermediate polypody
Polypodium vulgare—Common polypody
Potentilla reptans—Creeping cinquefoil
Prunella vulgaris—Selfheal
Rumex acetosa—Sorrel
Rumex acetosella—Sheep's sorrel
Vaccinium myrtillus—Bilberry; Fraughan

Flatulence

Apium graveolens—Wild celery
Arctium minus—Burdock
Centaurium erythraea—Common centaury
Chamaemelum nobile (*Anthemis nobilis*)—Chamomile
Chrysosplenium oppositifolium—Golden saxifrage
Eryngium maritimum—Sea holly
Juniperus communis—Juniper
Ligusticum scoticum—Lovage
Solidago virgaurea—Goldenrod
Tanacetum vulgare (*Chrysanthemum vulgare*)—Tansy
Thymus polytrichus (*Thymus drucei*; *Thymus praecox*; *Thymus serpyllum*)—Wild thyme

Gall bladder complaints, gallstones

Frangula alnus—Alder buckthorn
Geranium robertianum—Herb-Robert
Lysimachia nemorum—Yellow pimpernel
Taraxacum officinale agg.—Dandelion

Gangrene

Alliaria petiolata—Garlic mustard; Hedge garlic
Hypericum elodes—Marsh St John's-wort
Hypericum hirsutum—Trailing St John's-wort
Hypericum humifusum—Hairy St John's-wort
Hypericum maculatum—Imperforate St John's-wort
Hypericum perforatum—Perforate St John's-wort
Hypericum pulchrum—Slender St John's-wort
Hypericum tetrapterum—Square-stalked St John's-wort

Glands, swollen

Ranunculus acris—Meadow buttercup
Ranunculus bulbosus—Bulbous buttercup

Gluttony (see also Obesity)

Potentilla erecta—Tormentil

Gout

Arum maculatum[3]—Lords and Ladies; Cuckoo-pint
Asparagus officinalis subsp. *prostratus*—Wild asparagus
Colchicum autumnale—Autumn crocus
Conium maculatum—Hemlock
Cytisus scoparius—Broom
Daucus carota—Carrot
Fraxinus excelsior—Ash
Hyoscyamus niger—Henbane
Persicaria hydropiper (*Polygonum hydropiper*)—Water pepper
Plantago coronopus—Buck's-horn plantain
Plantago lanceolata—Ribwort plantain
Plantago major—Greater plantain
Plantago maritima—Sea plantain

Gravel (concretions in the bile or urinary tracts)

Centaurium erythraea—Common centaury
Galium aparine—Robin-run-the-hedge; Cleavers; Goosegrass
Hypericum species—St John's-Wort
Petroselinum crispum—Parsley
Senecio vulgaris—Groundsel
Verbena officinalis—Vervain
Veronica beccabunga—Brooklime

Gripes

Anagallis arvensis—Scarlet pimpernel
Ilex aquifolium—Holly
Teucrium scorodonia—Wood sage

Growing pains in children

Centaurium erythraea—Common centaury

SUMMARY OF MEDICINAL PLANT USES *(continued)*

Gums, sore

Potentilla erecta—Tormentil

Halitosis (bad breath)

Cochlearia officinalis—Common scurvygrass
Lonicera periclymenum—Honeysukle; Woodbine

Hay fever

Trifolium pratense—Red clover

Headaches

Achillea millefolium—Yarrow
Achillea ptarmica—Sneezewort
Alchemilla vulgaris—Lady's-mantle
Bellis perennis—Daisy
Hedera helix—Ivy
Lamium purpureum—Red dead-nettle; Archangel
Lamium hybridum—Cut-leaved dead-nettle
Melilotus officinalis—Ribbed melilot
Mentha aquatica—Water mint
Nasturtium microphyllum—Narrow-fruited water-cress
Nasturtium officinale—Water-cress
Salix repens—Creeping willow
Taraxacum officinale agg.—Dandelion
Umbilicus rupestris—Wall pennywort; Navelwort
Viola odorata—Sweet violet

Headlice

Centaurium erythraea—Common centaury
Euonymus europaeus—Spindle-tree

Heart problems

Adiantum capillus-veneris—Maidenhair fern
Angelica sylvestris—Wild angelica
Caltha palustris—Marsh-marigold; Kingcup
Crataegus monogyna—Hawthorn
Digitalis purpurea—Foxglove (highly toxic except at narrow therapeutic range)
Erica cinerea—Heather
Menyanthes trifoliata—Bogbean
Nasturtium microphyllum—Narrow-fruited water-cress
Nasturtium officinale—Water-cress
Oxalis acetosella—Wood sorrel
Potentilla anserina—Silverweed
Prunella vulgaris—Selfheal
Rumex obtusifolius—Broad-leaved dock
Solidago virgaurea—Goldenrod
Taraxacum officinale agg.—Dandelion
Teucrium scorodonia—Wood sage
Ulex europaeus—Gorse; Whin; Furze
Ulex gallii—Western gorse

Heartburn

Calluna vulgaris—Heather; Ling
Cytisus scoparius—Broom
Erica cinerea—Heather
Fraxinus excelsior—Ash
Ranunculus acris—Meadow buttercup
Ranunculus bulbosus—Bulbous buttercup
Trifolium pratense—Red clover
Ulex europaeus—Gorse; Whin; Furze
Ulex gallii—Western gorse

Hemorrhoids (piles)

Agrimonia eupatoria—Agrimony
Alnus glutinosa—Alder
Ficaria verna (*Ranunculus ficaria*)—Lesser celandine
Parietaria judaica (*Parietaria diffusa*)—Wall pellitory
Potentilla anserina—Silverweed
Potentilla erecta—Tormentil
Primula vulgaris (*Primula acaulis*)—Primrose
Prunella vulgaris—Selfheal
Ranunculus species—Buttercups; Crowfoots
Sambucus nigra—Elder
Sanicula europaea—Wood sanicle; Sanicle (staunch bleeding)
Scrophularia auriculata—Water figwort; Water betony
Scrophularia nodosa—Figwort
Verbascum thapsus—Mullein

Herpes infections

Lonicera periclymenum—Honeysuckle; Woodbine

Hernias

Sambucus nigra—Elder

Sambucus nigra **berries (elderberries).**

SUMMARY OF MEDICINAL PLANT USES *(continued)*

Hiccups

Ulex europaeus—Gorse; Whin; Furze
Ulex gallii—Western gorse

Hives of the bowel

Senecio jacobaea—Ragwort

Hypertension (high blood pressure)

Crataegus monogyna—Hawthorn

Impetigo

Galium aparine—Robin-run-the-hedge; Cleavers; Goosegrass

Impotence

Eryngium maritimum—Sea holly

Incontinence

Equisetum arvense—Field horsetail
Sisymbrium officinale—Hedge mustard

Influenza

Rubus fruticosus agg.—Bramble; Blackberry
Sambucus nigra—Elder

Injuries, general or internal

Teucrium scorodonia—Wood sage

Jaundice

Adiantum capillus-veneris—Maidenhair fern
Agrimonia eupatoria—Agrimony
Arctium minus—Burdock
Artemisia maritima (*Seriphidium maritimum*)—Sea wormwood
Arum maculatum—Lords and Ladies; Cuckoo-pint (toxic, calcium oxalate crystals)
Berberis vulgaris—Barberry
Centaurea nigra—Common knapweed; Blackheads
Chamaemelum nobile (*Anthemis nobilis*)—Chamomile
Chelidonium majus—Greater celandine
Chrysanthemum leucanthemum (*Leucanthemum vulgare*)—Ox-eye daisy
Colchicum autumnale—Autumn crocus
Cytisus scoparius—Broom
Drosera anglica—Great sundew
Drosera intermedia—Oblong-leaved sundew
Drosera rotundifolia—Round-leaved sundew
Filipendula ulmaria—Meadow-sweet
Frangula alnus—Alder buckthorn
Galium odoratum—Woodruff
Hedera helix—Ivy
Iris pseudacorus—Yellow flag
Juncus inflexus—Hard rush
Linum catharticum—Fairy flax
Lonicera periclymenum—Honeysuckle; Woodbine
Mentha aquatica—Water mint
Plantago coronopus—Buck's-horn plantain
Plantago lanceolata—Ribwort plantain
Plantago major—Greater plantain
Plantago maritima—Sea plantain
Primula vulgaris (*Primula acaulis*)—Primrose
Ranunculus acris—Meadow buttercup
Ranunculus bulbosus—Bulbous buttercup
Rumex acetosa—Sorrel
Rumex obtusifolius—Broad-leaved dock
Senecio jacobaea—Ragwort
Sinapis arvensis—Charlock
Stellaria media—Chickweed
Taraxacum officinale agg.—Dandelion
Ulex europaeus—Gorse; Whin; Furze
Ulex gallii—Western gorse
Ulmus glabra—Wych elm; Irish elm
Umbilicus rupestris—Wall pennywort; Navelwort
Urtica dioica—Nettle
Vaccinium myrtillus—Bilberry; Fraughan
Veronica beccabunga—Brooklime
Veronica chamaedrys—Germander speedwell

Kidney and urinary complaints

Achillea millefolium—Yarrow
Agrimonia eupatoria—Agrimony
Allium ursinum—Wild garlic; Ramsons
Anthriscus sylvestris—Cow parsley
Aphanes arvensis—Parsley-piert
Arctium minus—Burdock
Arctostaphylos uva-ursi—Bearberry
Asparagus officinalis subsp. *prostratus*—Wild asparagus
Calystegia sepium—Hedge bindweed
Capsella bursa-pastoris—Shepherd's purse
Chrysosplenium oppositifolium—Golden saxifrage
Cuscuta epithymum—Dodder
Cytisus scoparius—Broom
Elytrigia repens (*Elymus repens*)—Scutch; Couch-grass
Filipendula ulmaria—Meadow-sweet
Galium aparine—Robin-run-the-hedge; Cleavers; Goosegrass
Geranium robertianum—Herb-Robert
Geranium sanguineum—Bloody crane's-bill
Geum urbanum—Wood avens

SUMMARY OF MEDICINAL PLANT USES *(continued)*

Kidney and urinary complaints *(continued)*

Glechoma hederacea—Ground ivy
Hedera helix—Ivy
Linum catharticum—Fairy flax
Malva sylvestris—Common mallow
Myrica gale—Bog myrtle
Nasturtium microphyllum—Narrow-fruited water-cress
Nasturtium officinale—Water-cress
Ononis repens—Common restharrow
Parietaria judaica (*Parietaria diffusa*)—Wall pellitory
Persicaria hydropiper (*Polygonum hydropiper*)—Water pepper
Persicaria maculosa (*Polygonum persicaria*)—Redshank or Spotted persicaria
Petroselinum crispum—Parsley
Pilosella officinarum (*Hieracium pilosella*)—Mouse-ear hawkweed
Ranunculus acris—Meadow buttercup
Ranunculus bulbosus—Bulbous buttercup
Rubus fruticosus agg.—Bramble; Blackberry
Rumex acetosa—Sorrel
Sambucus nigra—Elder
Scleranthus annuus—Annual knawel
Scrophularia nodosa—Figwort
Sedum acre—Stonecrop; Wall pepper
Silene uniflora (*Silene vulgaris* subsp. *maritima*)—Sea campion
Solidago virgaurea—Goldenrod
Symphytum officinale—Comfrey
Umbilicus rupestris—Wall pennywort; Navelwort
Vaccinium myrtillus—Bilberry; Fraughan
Verbascum thapsus—Mullein
Veronica beccabunga—Brooklime

Kidney stones

Adiantum capillus-veneris—Maidenhair fern
Cardamine pratensis—Cuckoo-flower; Lady's smock
Centaurium erythraea—Common centaury
Fraxinus excelsior—Ash
Galium aparine—Robin-run-the-hedge; Cleavers; Goosegrass
Persicaria maculosa (*Polygonum persicaria*)—Redshank or Spotted persicaria
Rubus fruticosus agg.—Bramble; Blackberry
Silene uniflora (*Silene vulgaris* subsp. *maritima*)—Sea campion
Taraxacum officinale agg.—Dandelion

Labour pains during childbirth

Nasturtium microphyllum—Narrow-fruited water-cress
Nasturtium officinale—Water-cress
Veronica beccabunga—Brooklime

Leprosy

Salsola kali—Saltwort
Succisa pratensis—Devil's-bit scabious

Liver problems (including hepatitis and cirrhosis)

Agrimonia eupatoria—Agrimony
Athyrium filix-femina—Lady-fern
Bellis perennis—Daisy
Berberis vulgaris—Barberry
Bidens tripartita—Trifid bur-marigold
Centaurea nigra—Common knapweed; Blackheads
Centaurium erythraea—Common centaury
Dryopteris affinis—Scaly male-fern
Euonymus europaeus—Spindle-tree
Frangula alnus—Alder buckthorn
Fraxinus excelsior—Ash
Geranium robertianum—Herb-Robert
Menyanthes trifoliata—Bogbean
Ophioglossum vulgatum—Adder's-tongue
Osmunda regalis—Royal fern
Plantago coronopus—Buck's-horn plantain
Plantago lanceolata—Ribwort plantain
Plantago major—Greater plantain
Plantago maritima—Sea plantain
Potentilla erecta—Tormentil
Rumex crispus—Curled dock
Rumex obtusifolius—Broad-leaved dock
Scrophularia nodosa—Figwort
Succisa pratensis—Devil's-bit scabious
Taraxacum officinale agg.—Dandelion
Trifolium repens—White clover
Vaccinium myrtillus—Bilberry; Fraughan
Valeriana officinalis—Wild valerian
Verbascum thapsus—Mullein
Veronica beccabunga—Brooklime

Lumbago (lower back pain)

Trifolium dubium—Yellow clover; Shamrock

Lymphatic disorders

Vaccinium myrtillus—Bilberry; Fraughan

Madness

Anagallis arvensis—Scarlet pimpernel
Angelica sylvestris—Wild angelica

SUMMARY OF MEDICINAL PLANT USES *(continued)*

Measles

Allium ursinum—Wild garlic; Ramsons
Euphrasia officinalis agg.—Eyebright
Lamium purpureum—Red dead-nettle; Archangel
Lamium hybridum—Cut-leaved dead-nettle
Mentha aquatica—Water mint
Myrica gale—Bog myrtle
Potentilla erecta—Tormentil
Senecio jacobaea—Ragwort
Urtica dioica—Nettle
Vaccinium myrtillus—Bilberry; Fraughan

Migraine

Achillea millefolium—Yarrow

Mouth sores

Alisma plantago-aquatica—Water-plantain

Mouth ulcers and inflammations

Cochlearia officinalis—Common scurvygrass
Lonicera periclymenum—Honeysuckle; Woodbine
Rumex hydrolapathum—Water dock
Senecio jacobaea—Ragwort

Mumps

Allium ursinum—Wild garlic; Ramsons
Iris pseudacorus—Yellow flag
Ranunculus acris—Meadow buttercup
Ranunculus bulbosus—Bulbous buttercup
Stellaria media—Chickweed

Narcolepsy (excessive sleepiness, sleep disorder)

Artemisia vulgaris—Mugwort

Nasal polyps

Polypodium australe—Southern polypody
Polypodium interjectum—Intermediate polypody
Polypodium vulgare—Common polypody

Neck (stiffness)

Ilex aquifolium—Holly

Nervous conditions

Arctium minus—Burdock
Cheiranthus cheiri—Wallflower
Primula vulgaris (*Primula acaulis*)—Primrose
Taraxacum officinale agg.—Dandelion
Thymus polytrichus (*Thymus drucei*; *Thymus praecox*; *Thymus serpyllum*)—Wild thyme
Valeriana officinalis—Wild valerian

Neuralgia

Cytisus scoparius—Broom
Quercus petraea—Sessile oak; Irish oak
Quercus robur—Pedunculate oak; English oak

Nipples, cracked and sore

Potentilla erecta—Tormentil

Nosebleeds

Achillea millefolium—Yarrow
Alchemilla vulgaris—Lady's-mantle
Galium verum—Lady's bedstraw
Senecio vulgaris—Groundsel

Nursing mothers, milk flow

Polygala serpyllifolia—Heath milkwort
Polygala vulgaris—Common milkwort

Obesity

Fraxinus excelsior—Ash
Taraxacum officinale agg.—Dandelion

Pain relief

Cheiranthus cheiri—Wallflower
Conium maculatum—Hemlock

Palpitations of the heart

Teucrium scorodonia—Wood sage

Palsy

Primula veris—Cowslip
Thymus polytrichus (*Thymus drucei*; *Thymus praecox*; *Thymus serpyllum*)—Wild thyme

Period pains (menstrual)

Rubus idaeus—Wild raspberry

Phlegm

Corylus avellana—Hazel (buds mixed with dandelions, chickweed, wood sorrel and oatmeal)

Pleurisy

Bellis perennis—Daisy

Poison antidote

Sisymbrium officinale—Hedge mustard

Purges (and laxatives)[4]

Agrimonia eupatoria—Agrimony
Arctium minus—Burdock
Atriplex patula—Common orache
Calystegia sepium—Hedge bindweed
Euonymus europaeus—Spindle-tree
Euphorbia helioscopia—Sun spurge
Euphorbia hyberna—Irish spurge
Euphorbia paralias—Sea spurge
Euphorbia peplus—Petty spurge

SUMMARY OF MEDICINAL PLANT USES *(continued)*

Purges *(continued)*

- *Frangula alnus*—Alder buckthorn
- *Galium aparine*—Robin-run-the-hedge; Cleavers; Goosegrass
- *Iris pseudacorus*—Yellow flag
- *Linum catharticum*—Fairy flax
- *Lonicera periclymenum*—Honeysuckle; Woodbine
- *Menyanthes trifoliata*—Bogbean
- *Parietaria judaica* (*Parietaria diffusa*)—Wall pellitory
- *Rhamnus cathartica*—Buckthorn
- *Rumex crispus*—Curled dock
- *Sambucus nigra*—Elder
- *Scorzoneroides autumnalis* (*Leontodon autumnalis*)—Autumn hawkbit
- *Senecio vulgaris*—Groundsel (for children in particular)
- *Teucrium scorodonia*—Wood sage
- *Urtica dioica*—Nettle
- *Verbascum thapsus*—Mullein
- *Viola lutea*—Mountain pansy

Quinsy (inflammation of the tonsils and throat)

- *Asperula cynanchica*—Squinancywort

Rabies

- *Anagallis arvensis*—Scarlet pimpernel

Respiratory problems (including lung complaints)

- *Allium babingtonii* (*Allium ampeloprasum* var. *babingtonii*)—Wild leek
- *Arctium minus*—Burdock
- *Drosera anglica*—Great sundew
- *Drosera intermedia*—Oblong-leaved sundew
- *Drosera rotundifolia*—Round-leaved sundew
- *Euphrasia officinalis* agg.—Eyebright
- *Hedera helix*—Ivy
- *Prunella vulgaris*—Selfheal
- *Sambucus nigra*—Elder
- *Taraxacum officinale* agg.—Dandelion
- *Thymus polytrichus* (*Thymus drucei*; *Thymus praecox*; *Thymus serpyllum*)—Wild thyme
- *Tussilago farfara*—Colt's foot

Rheumatism

- *Achillea millefolium*—Yarrow
- *Agrimonia eupatoria*—Agrimony
- *Allium ursinum*—Wild garlic; Ramsons
- *Althaea officinalis*—Marsh-mallow
- *Anagallis arvensis*—Scarlet pimpernel
- *Apium graveolens*—Wild celery
- *Arctium minus*—Burdock
- *Asparagus officinalis* subsp. *prostratus*—Wild asparagus
- *Centaurea nigra*—Common knapweed; Blackheads
- *Centaurium erythraea*—Common centaury
- *Conium maculatum*—Hemlock
- *Cytisus scoparius*—Broom
- *Erica cinerea*—Heather
- *Fraxinus excelsior*—Ash
- *Geranium robertianum*—Herb-Robert
- *Malva sylvestris*—Common mallow
- *Menyanthes trifoliata*—Bogbean
- *Nasturtium microphyllum*—Narrow-fruited water-cress
- *Nasturtium officinale*—Water-cress
- *Oenanthe fistulosa*—Water-dropwort
- *Origanum vulgare*—Marjoram
- *Osmunda regalis*—Royal fern
- *Polypodium australe*—Southern polypody
- *Polypodium interjectum*—Intermediate polypody
- *Polypodium vulgare*—Common polypody
- *Rumex crispus*—Curled dock
- *Rumex hydrolapathum*—Water dock
- *Rumex obtusifolius*—Broad-leaved dock
- *Senecio jacobaea*—Ragwort
- *Sorbus aucuparia*—Mountain ash; Rowan
- *Stellaria graminea*—Lesser stitchwort
- *Stellaria media*—Chickweed
- *Symphytum officinale*—Comfrey
- *Taraxacum officinale* agg.—Dandelion
- *Teucrium scorodonia*—Wood sage
- *Urtica dioica*—Nettle
- *Verbascum thapsus*—Mullein

Rheumatoid arthritis

- *Equisetum arvense*—Field horsetail

Rickets

- *Asplenium ruta-muraria*—Wall-rue
- *Capsella bursa-pastoris*—Shepherd's purse
- *Osmunda regalis*—Royal fern
- *Dryopteris affinis*—Scaly male-fern
- *Juncus inflexus*—Hard rush
- *Pteridium aquilinum*—Bracken

Ringworm (fungal infections)

- *Arctium lappa*—Great burdock (or *A. minus*?)[5]
- *Bellis perennis*—Daisy
- *Fraxinus excelsior*—Ash
- *Geranium robertianum*—Herb-Robert

SUMMARY OF MEDICINAL PLANT USES *(continued)*

Ringworm *(continued)*

Juncus inflexus—Hard rush
Quercus petraea—Sessile oak; Irish oak
Quercus robur—Pedunculate oak; English oak
Rumex obtusifolius—Broad-leaved dock
Salix species—Willow
Sambucus nigra—Elder
Scrophularia auriculata—Water figwort; Water betony
Scrophularia nodosa—Figwort
Taxus baccata—Yew
Ulex europaeus—Gorse; Whin; Furze
Ulex gallii—Western gorse

Sciatica

Allium ursinum—Wild garlic; Ramsons
Hypericum elodes—Marsh St John's-wort
Hypericum hirsutum—Trailing St John's-wort
Hypericum humifusum—Hairy St John's-wort
Hypericum maculatum—Imperforate St John's-wort
Hypericum perforatum—Perforate St John's-wort
Hypericum pulchrum—Slender St John's-wort
Hypericum tetrapterum—Square-stalked St John's-wort
Osmunda regalis—Royal fern
Populus tremula—Aspen
Sambucus nigra—Elder
Senecio jacobaea—Ragwort
Sisymbrium officinale—Hedge mustard

Scrofula (tuberculous infection of lymph nodes in neck)

Arctium lappa—Great burdock
Arctium minus—Burdock
Bellis perennis—Daisy
Chrysanthemum leucanthemum (*Leucanthemum vulgare*)—Ox-eye daisy
Conium maculatum—Hemlock
Filipendula ulmaria—Meadow-sweet
Lamium purpureum—Red dead-nettle; Archangel
Lamium hybridum—Cut-leaved dead-nettle
Menyanthes trifoliata—Bogbean
Nasturtium microphyllum—Narrow-fruited water-cress
Nasturtium officinale—Water-cress
Scrophularia nodosa—Figwort
Verbena officinalis—Vervain

Scurvy

Agrimonia eupatoria—Agrimony
Betula species—Birch
Cardamine pratensis—Cuckoo-flower; Lady's smock
Centaurea nigra—Common knapweed; Blackheads
Cochlearia officinalis—Common scurvygrass
Nasturtium microphyllum—Narrow-fruited water-cress
Nasturtium officinale—Water-cress
Rubus saxatilis—Stone bramble
Rumex acetosella—Sheep's sorrel
Rumex hydrolapathum—Water dock
Vaccinium myrtillus—Bilberry; Fraughan
Veronica beccabunga—Brooklime

Sedatives

Ligusticum scoticum—Lovage
Solanum dulcamara—Bittersweet
Thymus polytrichus (*Thymus drucei*; *Thymus praecox*; *Thymus serpyllum*)—Wild thyme
Valeriana officinalis—Wild valerian

Shingles, 'Wildfire'

Helianthemum oelandicum subsp. *piloselloides* (*Helianthemum canum*)—Hoary rock-rose
Juncus inflexus—Hard rush
Tuberaria guttata—Spotted rock-rose

Skin complaints (including rashes, eczema, pimples, dermatitis, blackheads, itches and cleaning scabs)

Agrimonia eupatoria—Agrimony
Apium graveolens—Wild celery
Arctium lappa—Great burdock
Arctium minus[5]—Burdock
Arum maculatum—Lords and Ladies; Cuckoo-pint
Betula pubescens—Downy birch
Betula pendula—Silver birch
Carlina vulgaris—Carline thistle
Cochlearia officinalis—Common scurvygrass
Conopodium majus—Pignut
Cuscuta epithymum—Dodder
Cynoglossum officinale—Hound's-tongue
Cytisus scoparius—Broom
Geranium robertianum—Herb-Robert
Glechoma hederacea—Ground ivy
Hedera helix—Ivy
Hydrocharis morsus-ranae—Frogbit
Menyanthes trifoliata—Bogbean
Nasturtium microphyllum—Narrow-fruited water-cress
Nasturtium officinale—Water-cress
Ononis repens—Common restharrow (for chapped hands)

SUMMARY OF MEDICINAL PLANT USES *(continued)*

Skin complaints *(continued)*

Parietaria judaica (*Parietaria diffusa*)—Wall pellitory
Plantago coronopus—Buck's-horn plantain
Plantago lanceolata—Ribwort plantain
Plantago major—Greater plantain
Plantago maritima—Sea plantain
Potentilla erecta—Tormentil (for chapped hands)
Primula veris—Cowslip (for spots and wrinkles)
Primula vulgaris (*Primula acaulis*)—Primrose
Prunella vulgaris—Selfheal
Quercus petraea—Sessile oak; Irish oak
Quercus robur—Pedunculate oak; English oak
Ranunculus acris—Meadow buttercup
Rumex crispus—Curled dock
Rumex obtusifolius—Broad-leaved dock
Salsola kali—Saltwort
Sambucus nigra—Elder
Scrophularia nodosa—Figwort
Senecio vulgaris—Groundsel (for chapped hands)
Stellaria media—Chickweed
Succisa pratensis—Devil's-bit scabious
Symphytum officinale—Comfrey
Tussilago farfara—Colt's foot
Ulex europaeus—Gorse; Whin; Furze
Ulex gallii—Western gorse
Urtica dioica—Nettle

Sleeping aids, insomnia

Chamaemelum nobile (*Anthemis nobilis*)—Chamomile
Potentilla erecta—Tormentil (with *Lysimachia nemorum*)
Primula veris—Cowslip
Primula vulgaris (*Primula acaulis*)—Primrose
Thymus polytrichus (*Thymus drucei*; *Thymus praecox*; *Thymus serpyllum*)—Wild thyme

Urtica dioica (nettle).

Spleen complaints

Athyrium filix-femina—Lady-fern
Bidens tripartita—Trifid bur-marigold
Dryopteris affinis—Scaly male-fern
Osmunda regalis—Royal fern
Polypodium australe—Southern polypody
Polypodium interjectum—Intermediate polypody
Polypodium vulgare—Common polypody
Veronica beccabunga—Brooklime (swellings)

Splinters and thorns, to draw from the skin

Agrimonia eupatoria—Agrimony
Hedera helix—Ivy
Stellaria media—Chickweed

Sprains

Althaea officinalis—Marsh-mallow
Calendula officinalis—Garden marigold; Pot marigold
Conium maculatum—Hemlock
Cytisus scoparius—Broom
Foeniculum vulgare—Fennel
Osmunda regalis—Royal fern
Quercus petraea—Sessile oak; Irish oak
Quercus robur—Pedunculate oak; English oak
Scrophularia auriculata—Water figwort; Water betony
Scrophularia nodosa—Figwort
Senecio jacobaea—Ragwort
Solanum tuberosum—Potato
Stellaria media—Chickweed
Taraxacum officinale agg.—Dandelion
Teucrium scorodonia—Wood sage

Sterility, aid to conception

Orchis mascula—Early purple orchid

Stimulants

Agrimonia eupatoria—Agrimony

Stings, insect

Silene uniflora (*Silene vulgaris* subsp. *maritima*)—Sea campion
Trifolium pratense—Red clover (for bee stings)
Umbilicus rupestris—Wall pennywort; Navelwort (for bee stings)
Verbascum thapsus—Mullein (for bee stings)

Stings, nettle

Mentha aquatica—Water mint
Rumex obtusifolius—Broad-leaved dock
Senecio jacobaea—Ragwort

SUMMARY OF MEDICINAL PLANT USES *(continued)*

'Stitches' (sharp muscle pains, pains in the side)

Ranunculus species—Buttercups; Crowfoots
Ranunculus repens—Creeping buttercup
Stellaria graminea—Lesser stitchwort
Stellaria holostea—Greater stitchwort

Stomach upsets, aches and indigestion

Agrimonia eupatoria—Agrimony
Allium ursinum—Wild garlic; Ramsons
Apium graveolens—Wild celery
Artemisia maritima (*Seriphidium maritimum*)—Sea wormwood
Artemisia vulgaris—Mugwort
Bellis perennis—Daisy
Bidens tripartita—Trifid bur-marigold
Cardamine pratensis—Cuckoo-flower; Lady's smock
Centaurium erythraea—Common centaury
Chamaemelum nobile (*Anthemis nobilis*)—Chamomile
Erica cinerea—Heather
Fumaria species—Fumitory
Galium aparine—Robin-run-the-hedge; Cleavers; Goosegrass
Geranium robertianum—Herb-Robert
Lathyrus montanus (*Lathyrus linifolius*)—Bitter vetch
Mentha aquatica—Water mint
Menyanthes trifoliata—Bogbean
Nasturtium microphyllum—Narrow-fruited water-cress
Nasturtium officinale—Water-cress
Origanum vulgare—Marjoram
Parietaria judaica (*Parietaria diffusa*)—Wall pellitory
Polypodium vulgare—Common polypody
Primula vulgaris (*Primula acaulis*)—Primrose
Rubus fruticosus agg.—Bramble; Blackberry
Scrophularia nodosa—Figwort
Solidago virgaurea—Goldenrod
Taraxacum officinale agg.—Dandelion
Teucrium scorodonia—Wood sage
Thymus polytrichus (*Thymus drucei*; *Thymus praecox*; *Thymus serpyllum*)—Wild thyme

St Vitus dance (Sydenham chorea)

Cardamine pratensis—Cuckoo-flower; Lady's smock

Styes, eye

Ribes uva-crispa—Gooseberry

Sunburn

Drosera anglica—Great sundew
Drosera intermedia—Oblong-leaved sundew
Drosera rotundifolia—Round-leaved sundew
Potentilla anserina—Silverweed
Stellaria media—Chickweed

Swellings (often poultices to reduce inflammations, diverse causes)

Althaea officinalis—Marsh-mallow
Arctium minus—Burdock
Bellis perennis—Daisy
Chamaemelum nobile (*Anthemis nobilis*)—Chamomile
Cytisus scoparius—Broom
Digitalis purpurea—Foxglove
Galium aparine—Robin-run-the-hedge; Cleavers; Goosegrass
Lavatera arborea—Tree mallow
Nasturtium microphyllum—Narrow-fruited water-cress
Nasturtium officinale—Water-cress
Oenanthe crocata—Hemlock water-dropwort
Persicaria hydropiper (*Polygonum hydropiper*)—Water pepper
Plantago coronopus—Buck's-horn plantain
Plantago lanceolata—Ribwort plantain
Plantago major—Greater plantain
Plantago maritima—Sea plantain
Rubus fruticosus agg.—Bramble; Blackberry
Sambucus nigra—Elder
Scrophularia auriculata—Water figwort; Water betony
Sisymbrium officinale—Hedge mustard (of the breasts and testicles)
Stellaria media—Chickweed
Taraxacum officinale agg.—Dandelion
Ulex europaeus—Gorse; Whin; Furze
Ulex gallii—Western gorse
Ulmus glabra—Wych elm; Irish elm
Umbilicus rupestris—Wall pennywort; Navelwort

Teeth, loose teeth

Potentilla erecta—Tormentil

Thrush (fungal infections, mouth and tongue)

Lonicera periclymenum—Honeysuckle; Woodbine
Taraxacum officinale agg.—Dandelion

Tiredness, fatigue

Centaurium erythraea—Common centaury

Tonic, general health

Achillea millefolium—Yarrow
Agrimonia eupatoria—Agrimony
Artemisia maritima (*Seriphidium maritimum*)—Sea wormwood

SUMMARY OF MEDICINAL PLANT USES *(continued)*

Tonic *(continued)*

- *Bidens tripartita*—Trifid bur-marigold
- *Centaurea nigra*—Common knapweed; Blackheads
- *Fragaria vesca*—Wild strawberry
- *Glechoma hederacea*—Ground ivy
- *Menyanthes trifoliata*—Bogbean
- *Nasturtium microphyllum*—Narrow-fruited water-cress
- *Nasturtium officinale*—Water-cress
- *Taraxacum officinale* agg.—Dandelion
- *Ulex europaeus*—Gorse; Whin; Furze
- *Ulex gallii*—Western gorse
- *Urtica dioica*—Nettle

Toothaches

- *Achillea millefolium*—Yarrow
- *Allium ursinum*—Wild garlic; Ramsons
- *Anagallis arvensis*—Scarlet pimpernel
- *Crataegus monogyna*—Hawthorn
- *Cytisus scoparius*—Broom
- *Hyoscyamus niger*—Henbane
- *Iris pseudacorus*—Yellow flag
- *Limonium humile*—Lax-flowered sea lavender
- *Lonicera periclymenum*—Honeysuckle; Woodbine
- *Potentilla anserina*—Silverweed
- *Potentilla erecta*—Tormentil
- *Primula vulgaris* (*Primula acaulis*)—Primrose
- *Quercus petraea*—Sessile oak; Irish oak
- *Quercus robur*—Pedunculate oak; English oak
- *Ranunculus flammula*—Lesser spearwort
- *Senecio vulgaris*—Groundsel
- *Symphytum officinale*—Comfrey
- *Taraxacum officinale* agg.—Dandelion
- *Typha latifolia*—Bulrush; Reed mace

Tuberculosis

- *Allium schoenoprasum*—Chives
- *Allium ursinum*—Wild garlic; Ramsons
- *Arctium minus*—Burdock
- *Artemisia vulgaris*—Mugwort
- *Arum maculatum*—Lords and Ladies; Cuckoo-pint
- *Bellis perennis*—Daisy
- *Centaurea nigra*—Common knapweed; Blackheads
- *Chrysanthemum leucanthemum* (*Leucanthemum vulgare*)—Ox-eye daisy
- *Lonicera periclymenum*—Honeysuckle; Woodbine
- *Orchis mascula*—Early purple orchid
- *Plantago lanceolata*—Ribwort plantain
- *Primula vulgaris* (*Primula acaulis*)—Primrose
- *Ranunculus flammula*—Lesser spearwort
- *Rumex acetosa*—Sorrel
- *Sanicula europaea*—Wood sanicle; Sanicle
- *Scorzoneroides autumnalis* (*Leontodon autumnalis*)—Autumn hawkbit
- *Scrophularia nodosa*—Figwort
- *Solidago virgaurea*—Goldenrod
- *Taraxacum officinale* agg.—Dandelion
- *Teucrium scorodonia*—Wood sage
- *Thymus polytrichus* (*Thymus drucei*; *Thymus praecox*; *Thymus serpyllum*)—Wild thyme
- *Ulex europaeus*—Gorse; Whin; Furze
- *Ulex gallii*—Western gorse
- *Umbilicus rupestris*—Wall pennywort; Navelwort
- *Valeriana officinalis*—Wild valerian
- *Verbascum thapsus*—Mullein

Ulcers, skin and mouth

- *Agrimonia eupatoria*—Agrimony
- *Alnus glutinosa*—Alder
- *Cardamine pratensis*—Cuckoo-flower; Lady's smock
- *Chrysanthemum leucanthemum* (*Leucanthemum vulgare*)—Ox-eye daisy
- *Digitalis purpurea*—Foxglove
- *Galium aparine*—Robin-run-the-hedge; Cleavers; Goosegrass
- *Glechoma hederacea*—Ground ivy
- *Hedera helix*—Ivy
- *Persicaria maculosa* (*Polygonum persicaria*)—Redshank or Spotted persicaria
- *Potentilla anserina*—Silverweed
- *Rubus fruticosus* agg.—Bramble; Blackberry
- *Senecio jacobaea*—Ragwort
- *Silene uniflora* (*Silene vulgaris* subsp. *maritima*)—Sea campion

Varicose veins

- *Corylus avellana*—Hazel

Yellow fever

- *Digitalis purpurea*—Foxglove (highly toxic, except in narrow therapeutic range)

Venereal disease

- *Galium aparine*—Robin-run-the-hedge; Cleavers; Goosegrass

Vein disorders

- *Vaccinium myrtillus*—Bilberry; Fraughan

Water retention

- *Taraxacum officinale* agg.—Dandelion

SUMMARY OF MEDICINAL PLANT USES *(continued)*

Warts

Allium ursinum—Wild garlic; Ramsons
Caltha palustris—Marsh-marigold; Kingcup
Centaurea nigra—Common knapweed; Blackheads
Chelidonium majus—Greater celandine
Daucus carota—Carrot
Euphorbia helioscopia—Sun spurge
Euphorbia hyberna—Irish spurge
Euphorbia paralias—Sea spurge
Euphorbia peplus—Petty spurge
Fraxinus excelsior—Ash
Plantago coronopus—Buck's-horn plantain
Plantago lanceolata—Ribwort plantain
Plantago major—Greater plantain
Plantago maritima—Sea plantain
Polygala serpyllifolia—Heath milkwort
Polygala vulgaris—Common milkwort
Ranunculus acris—Meadow buttercup
Ranunculus bulbosus—Bulbous buttercup
Sambucus nigra—Elder
Sedum acre—Stonecrop; Wall pepper
Senecio jacobaea—Ragwort
Solanum tuberosum—Potato
Sonchus asper—Spiny sow-thistle
Sonchus oleraceus—Smooth sow-thistle
Taraxacum officinale agg.—Dandelion
Umbilicus rupestris—Wall pennywort; Navelwort

Whitlows (herpetic or chronic paronychia, abscesses of the finger tips)

Achillea millefolium—Yarrow
Bellis perennis—Daisy
Hyacinthoides non-scripta—Bluebell
Persicaria maculosa (*Polygonum persicaria*)—Redshank; Spotted persicaria
Pilosella officinarum (*Hieracium pilosella*)—Mouse-ear hawkweed
Plantago lanceolata—Ribwort plantain

Whooping cough

Drosera anglica—Great sundew
Drosera intermedia—Oblong-leaved sundew
Drosera rotundifolia—Round-leaved sundew
Galium aparine—Robin-run-the-hedge; Cleavers; Goosegrass
Hyoscyamus niger—Henbane
Mentha aquatica—Water mint
Mentha pulegium—Penny-royal
Pilosella officinarum (*Hieracium pilosella*)—Mouse-ear hawkweed
Stellaria holostea—Greater stitchwort
Thymus polytrichus (*Thymus drucei*; *Thymus praecox*; *Thymus serpyllum*)—Wild thyme
Veronica serpyllifolia—Thyme-leaved speedwell

Women's ailments (gynaecological problems)

Artemisia vulgaris—Mugwort
Calendula officinalis—Garden marigold; Pot marigold
Capsella bursa-pastoris—Shepherd's purse
Halimione portulacoides—Sea purslane
Lamium purpureum—Red dead-nettle; Archangel
Lamium hybridum—Cut-leaved dead-nettle
Limonium humile—Lax-flowered sea lavender
Mentha aquatica—Water mint
Nasturtium microphyllum—Narrow-fruited water-cress
Nasturtium officinale—Water-cress
Parietaria judaica (*Parietaria diffusa*)—Wall pellitory

Women's fertility

Geranium robertianum—Herb-Robert

Worms (parasitic pinworms, roundworms, hookworms, tapeworms)

Allium babingtonii (*Allium ampeloprasum* var. *babingtonii*)—Wild leek
Allium ursinum—Wild garlic; Ramsons
Artemisia maritima (*Seriphidium maritimum*)—Sea wormwood
Arum maculatum—Lords and Ladies; Cuckoo-pint (toxic, calcium oxalate crystals)
Centaurium erythraea—Common centaury
Dryopteris affinis—Scaly male-fern (tapeworm)
Dryopteris filix-mas—Male-fern
Eryngium maritimum—Sea holly
Fraxinus excelsior—Ash
Hypericum elodes—Marsh St John's-wort
Hypericum hirsutum—Trailing St John's-wort
Hypericum humifusum—Hairy St John's-wort
Hypericum maculatum—Imperforate St John's-wort
Hypericum perforatum—Perforate St John's-wort
Hypericum pulchrum—Slender St John's-wort
Hypericum tetrapterum—Square-stalked St John's-wort
Nasturtium microphyllum—Narrow-fruited water-cress
Nasturtium officinale—Water-cress
Plantago lanceolata—Ribwort plantain
Prunella vulgaris—Selfheal
Prunus spinosa—Sloe; Blackthorn

SUMMARY OF MEDICINAL PLANT USES *(continued)*

Worms *(continued)*

Pteridium aquilinum—Bracken
Quercus petraea—Sessile oak; Irish oak (pin-worms)
Quercus robur—Pedunculate oak; English oak (pin-worms)
Sedum acre—Stonecrop; Wall pepper
Sisymbrium officinale—Hedge mustard (as a general vermifuge for children)
Sorbus aucuparia—Mountain ash; Rowan
Teucrium scordium—Water germander
Ulex europaeus—Gorse; Whin; Furze
Ulex gallii—Western gorse
Umbilicus rupestris—Wall pennywort; Navelwort
Urtica dioica—Nettle

1. Lady Wilde (1890: 22) said that dropsy was not prevalent in Ireland in early times but much more common when potatoes became a common part of the diet and with the 'copious amount of whisky drank in the towns'.
2. Monk's-hood is very poisonous. Nevertheless, Oliver Phelps Brown (1885: 123–124) noted that roots of monk's-hood were used for medicinal tinctures, with 'its action is more especially displayed in the highest grades of fever and inflammation'. Brown also observed, 'in the hands of the intelligent physician is of great service . . . as a sedative and anodyne'.
3. Oxalate crystals found in the berries and vegetation cause intense tingling, irritation and swelling of skin and anything that comes in contact. Allergic reactions may be severe and life-threatening.
4. All of these plants used in Ireland as laxatives, to treat constipation, seem to have been administered orally. I have not found evidence of any constipation treatments given anally, as enemas or in other ways, although such treatments are known elsewhere in the world in traditional medicine. For example, a mixture of water and ground bark roots of the *Gymnocladus dioicus*, the Kentucky coffee-tree, was used by Native American tribes in the Missouri River region, administered by means of a home-made syringe made from an animal bladder tied with animal sinews onto a hollow bird's bone (Gilmore, 1991).
5. Many early records of *Arctium lappa* in Ireland probably refer to *A. minus*, which was formerly included in that species.

Wild Plants Used in Veterinary Medicine

Alongside extensive use of medicine based on plants in many traditional societies throughout the world, complex and sophisticated traditions on the use of plants for veterinary medicine also exists. For example, ancient veterinary texts from India, Egypt and China can be dated back to BC 5000 in India (Jain, 1999). In his dictionary of ethnoveterinary plants in India (1999), Jain documents more than 800 plant species used in animal healthcare in India for a great range of ailments and conditions. In North America more than 250 plant species have been documented as being used by native North Americans for the treatment of animals (Gehring & Kindscher, 2010). In Europe, there is a similar tradition, stretching back many centuries, and for the Netherlands, van Asseldonk (2010) recorded 168 folk remedies for animals, involving 68 plant species. Some of the species used for veterinary purposes in the Netherlands include common Irish wild or cultivated plants, such as *Allium sativum* (garlic), *Mentha* species (mint), *Taraxacum officinale* (dandelion) and *Urtica dioica* (nettle).

Similarly, in Ireland a rich tradition of the use of wild plants for veterinary purposes has been recorded. Not surprisingly, the plants used are generally widespread or common species. In particular, they were used to treat farm animals. Cattle, horses, sheep and poultry have the most herbal remedies and treatments recorded for them. I am somewhat surprised by how few herbal treatments for pigs I have been able to document from Ireland. There are also relatively few recorded treatments for domestic animals such as dogs and cats that I have been able to find, and none for smaller pets such as guinea pigs, rabbits, hamsters and cage birds, other than *Stellaria media* (chickweed), which was often given to canaries and other birds as a general tonic.

Allen and Hatfield (2004) presented 105 genera of vascular plants that were used for the treatment of animals in Britain and Ireland. I have included many of the results of their work from Ireland in my survey. In total, I noted at least 78 species that have been used for veterinary treatments in Ireland. Of those, 48 have been used to treat cattle; 30

to treat horses; eight to treat sheep; 12 to treat poultry and eight to treat dogs. Only five species are recorded as having been used to treat pigs. Some species have been used for veterinary purposes for unspecified animals and some to treat unspecified ailments. I suspect that I have barely scratched the surface of this fascinating subject, and much work remains to be undertaken in documenting ethnoveterinary medicine in Ireland. Of course, it is not surprising that a rich lore of traditional practices in relation to herbal veterinary medicine would have grown up in Ireland. Farm animals were, and to some extent still are, valuable, and losing a cow or horse could have come as a devastating blow for a poor family. I suspect that amongst the farming community in Ireland there is probably still much extant information to be collected on this subject. Traditions involving herbal usage with animals may be passed down from generation to generation on the farm, and farmers might obviously wish to avoid the cost of calling in a veterinary surgeon for routine care, especially if they have an effective traditional remedy to hand.

Stories in folklore often imply that animals instinctively search out and eat wild plants to treat their own ailments, and it has been suggested that traditional herbalists around the world have observed the plants that sick wild animals consumed in an effort to discover new remedies (Engel, 2007). In an Irish context, we have no evidence to support this suggestion, but anecdotal evidence seems to suggest that animals do indeed treat themselves with herbs (e.g., a dog with a stomach-upset commonly eats grass). This is a potentially fruitful topic for further research in Ireland.

WILD PLANTS USED FOR VETERINARY PURPOSES[1]

General Use in Animals

- *Achillea millifolium*—Yarrow
- *Alchemilla* species—Lady's mantle
- *Allium ursinum*—Wild garlic; Ramsons
- *Anemone nemorosa*—Wood anemone
- *Arrhenatherum elatius*—False oat
- *Artemisia vulgaris*—Mugwort
- *Berberis vulgaris*—Barberry
- *Centaurea nigra*—Common knapweed; Blackheads
- *Chelidonium majus*—Greater celandine
- *Cirsium* species—Thistle
- *Conium maculatum*—Hemlock (liver fluke, sheep scab, cuts, swollen feet, sore breasts, farcy, swellings)
- *Crataegus monogyna*—Hawthorn
- *Cytisus scoparius*—Broom ('worms or bots'[2], gravel)
- *Daucus carota*—Carrot; Wild carrot
- *Digitalis purpurea*—Foxglove
- *Dryopteris filix-mas*—Male-fern (red-water fever[3], liver fluke, kidney failure, white scour)
- *Equisetum* species—Horsetails
- *Euphorbia hyberna*—Irish spurge
- *Euphorbia peplus*—Petty spurge (blackleg)
- *Euphrasia officinalis* agg.—Eyebright (red-water fever)
- *Filipendula ulmaria*—Meadow-sweet; Bridewort (poultice for broken bones)
- *Fragaria vesca*—Wild strawberry (red-water fever)
- *Galium verum*—Lady's bedstraw
- *Gentianella campestris*—Field gentian
- *Geranium robertianum*—Herb-Robert
- *Glechoma hederacea*—Ground ivy
- *Hedera helix*—Ivy (mastitis and digestive problems)
- *Heracleum sphondylium*—Hogweed; Cow-parsnip
- *Hypericum* species—St John's-wort
- *Hypericum elodes*—Marsh St John's-wort
- *Hypericum pulchrum*—Slender St John's-wort
- *Inula helenium*—Elecampane
- *Juniperus communis*—Juniper
- *Lemna* species—Duckweed (diarrhoea)
- *Linum catharticum*—Fairy flax (to treat constipation)
- *Lonicera periclymenum*—Honeysuckle; Woodbine
- *Malva sylvestris*—Common mallow
- *Mentha* species—Mint
- *Menyanthes trifoliata*—Bogbean; Bog hop; Marsh trefoil
- *Myrica gale*—Bog myrtle; Sweet-willow
- *Nasturtium officinale*—Water-cress
- *Oxalis acetosella*—Wood sorrel
- *Persicaria maculosa* (*Polygonum persicaria*)—Redshank; Spotted persicaria
- *Pinguicula vulgaris*—Common butterwort
- *Plantago lanceolata*—Ribwort plantain
- *Polygonum aviculare*—Knotgrass
- *Potentilla erecta*—Tormentil (red-water fever, white scour)

WILD PLANTS USED FOR VETERINARY PURPOSES *(continued)*

General Use in Animals *(continued)*

Primula vulgaris (*Primula acaulis*)—Primrose
Prunus spinosa—Sloe; Blackthorn (diarrhoea)
Pteridium aquilinum—Bracken
Puccinellia maritima (*Poa maritima*)—Common saltmarsh-grass
Quercus petraea—Sessile oak; Irish oak
Quercus robur—Pedunculate oak; English oak
Ranunculus species—Buttercup (blackleg, red-water fever)
Rhamnus cathartica—Buckthorn
Rubus fruticosus—Bramble; Blackberry
Rumex crispus—Curled dock
Salix alba—White willow
Sambucus ebulus—Dwarf elder
Sambucus nigra—Elder
Sanicula europaea—Wood sanicle; Sanicle
Saxifraga spathularis—St Patrick's-cabbage (possibly used to treat dysentery)
Scrophularia nodosa—Figwort (red-water fever, combined with *Stellaria media*)
Sedum acre—Stonecrop; Wall pepper (worms)
Sempervivum tectorum—House-leek
Senecio aquaticus—Marsh ragwort; Marsh ragweed (wounds, red-water fever)
Senecio jacobaea—Ragwort; Ragweed (wounds, red-water fever)
Senecio vulgaris—Groundsel (red-water fever, blackleg)
Sorbus aucuparia—Mountain ash; Rowan; Quicken tree ('strangles')
Stellaria media—Chickweed (blackleg, red-water fever, combined with *Scrophularia nodosa*)
Symphytum officinale—Comfrey (sprains and broken legs, scour, sore feet)
Tamus communis—Black bryony
Tanacetum vulgare (*Chrysanthemum vulgare*)—Tansy (red-water fever)
Taraxacum officinale agg.—Dandelion (cow-pox)
Trifolium pratense—Red clover
Ulex europaeus—Gorse; Whin; Furze (worms)
Ulex gallii—Western gorse; Mountain gorse; Autumn gorse (worms)
Umbilicus rupestris—Wall pennywort; Navelwort
Urtica dioica—Nettle (red-water fever)
Verbascum thapsus—Mullein (sores)
Lycoperdaceae—Puffballs (basidomycetes or mushroom fungus)

Cattle

Alchemilla species—Lady's mantle (unspecified)
Allium sativum—Garlic (unspecified)
Allium ursinum—Wild garlic; Ramsons (hoose[4], coughs, worms, blackleg, ringworm in calves)
Berberis vulgaris—Barberry (jaundice)
Chelidonium majus—Greater celandine (warts)
Cirsium species—Thistle (unspecified)
Conium maculatum—Hemlock (poultice for swollen udders after milking)
Crataegus monogyna—Hawthorn (diarrhoea in bullocks)
Daucus carota—Carrot; Wild carrot (staggers)
Digitalis purpurea—Foxglove (swollen udders)
Equisetum species—Horsetails (hoose)
Euphorbia hyberna—Irish spurge (purgative)
Filipendula ulmaria—Meadow-sweet; Bridewort (diarrhoea in calves)
Gentianella campestris—Field gentian (*crúbain*[5])
Geranium robertianum—Herb-Robert ('red-water'—blood in the urine, worms)
Glechoma hederacea—Ground ivy (kidney problems)
Hedera helix—Ivy (eye problems, warts, expelling afterbirth, mastitis)
Heracleum sphondylium—Hogweed; Cow-parsnip (rheumatism)
Hypericum species—St John's-wort ('hard udder')
Hypericum elodes—Marsh St John's-wort (diarrhoea)
Hypericum pulchrum—Slender St John's-wort (red-water fever)
Inula helenium—Elecampane (unspecified)
Malva sylvestris—Common mallow (sore or swollen udders, scabs and sores, swellings under the jaw)
Menyanthes trifoliata—Bogbean; Bog hop; Marsh trefoil ('darn', expel afterbirth, 'pine', a trace-element deficiency)
Myrica gale—Bog myrtle; Sweet-willow (liver fluke in calves, worms)
Nasturtium officinale—Water-cress (heart ailments)
Pinguicula vulgaris—Common butterwort (swellings and 'clefts' in udders)
Plantago lanceolata—Ribwort plantain (red murrain, 'red milk', udder cures, bites)
Potentilla erecta—Tormentil (diarrhoea)
Pteridium aquilinum—Bracken (cough)
Puccinellia maritima (*Poa maritima*)—Common saltmarsh-grass (purgative)

WILD PLANTS USED FOR VETERINARY PURPOSES *(continued)*

Cattle *(continued)*

Quercus petraea—Sessile oak; Irish oak (diarrhoea)
Quercus robur—Pedunculate oak; English oak (diarrhoea)
Rhamnus cathartica—Buckthorn ('dry murrain')
Rubus fruticosus—Bramble; Blackberry (diarrhoea and scour)
Rumex crispus—Curled dock (coughs)
Sambucus ebulus—Dwarf elder (ulcerated udders and teats)
Saxifraga spathularis—St Patrick's-cabbage (possibly used to treat dysentery in calves and young cattle)
Scrophularia nodosa—Figwort (expel afterbirth, 'blasts (swellings) on cows' nipples)
Sempervivum tectorum—House-leek (diarrhoea, expel afterbirth)
Symphytum officinale—Comfrey (swollen udders, sore feet, diarrhoea)
Tamus communis—Black bryony ('wasting cows')
Taraxacum officinale agg.—Dandelion—(liver fluke in calves, blackleg[6], cow-pox and diarrhoea in calves)
Trifolium pratense—Red clover (sore tongue)
Ulex europaeus—Gorse; Whin; Furze (coughs)
Ulex gallii—Western gorse; Mountain gorse; Autumn gorse (coughs)
Umbilicus rupestris—Wall pennywort; Navelwort (expel afterbirth)
Verbascum thapsus—Mullein (pulmonary diseases)
Lycoperdaceae—Puffballs (bleeding when polled)

Sheep

Anemone nemorosa—Wood anemone (sheep scab)[7]
Conium maculatum—Hemlock (liver fluke, sheep scab)
Digitalis purpurea—Foxglove (staggers)
Hedera helix—Ivy (eye problems, pains)
Inula helenium—Elecampane (sheep scab)
Myrica gale—Bog myrtle; Sweet-willow (eye ailments)
Oxalis acetosella—Wood sorrel (ticks, lice, mange)
Sambucus nigra—Elder (rot, blackleg)

Horses

Allium ursinum—Wild garlic; Ramsons (farcy[8])
Arrhenatherum elatius—False oat (white scale)
Artemisia vulgaris—Mugwort (gangrene)
Centaurea nigra—Common knapweed; Blackheads ('strangles'[9])
Cirsium species—Thistle (staunch bleeding)
Conium maculatum—Hemlock (dislocated joints, sprains, farcy, sore breast (combined with *Malva sylvestris*)
Cytisus scoparius—Broom (gravel[10], broken wind)
Daucus carota—Carrot; Wild carrot (unspecified)
Dryopteris filix-mas—Male-fern (kidney disease)
Euphorbia hyberna—Irish spurge (purgative)
Geranium robertianum—Herb-Robert (worms)
Glechoma hederacea—Ground ivy (eye problems, worms and other parasites)
Inula helenium—Elecampane (improve appetite, skin conditions, farcy)
Juniperus communis—Juniper (abortifacient and contraceptive)
Lonicera periclymenum—Honeysuckle; Woodbine (farcy)
Malva sylvestris—Common mallow (sore breast—combined with *Conium maculatum*)
Myrica gale—Bog myrtle; Sweet-willow (red-water fever, collar sores, gravel)
Potentilla erecta—Tormentil (diarrhoea)
Primula vulgaris (*Primula acaulis*)—Primrose (coughs)
Pteridium aquilinum—Bracken (fern used for kidney disease)
Quercus petraea—Sessile oak; Irish oak (harness sores)
Quercus robur—Pedunculate oak; English oak (harness sores)
Rumex crispus—Curled dock (farcy)
Rumex species—Docks (coughs, farcy)
Scrophularia nodosa—Figwort (farcy)
Symphytum officinale—Comfrey (swellings)
Tanacetum vulgare (*Chrysanthemum vulgare*)—Tansy (worms)
Taraxacum officinale agg.—Dandelion (pain)
Ulex species—Gorse; Furze; Whin (worms)
Ulex europaeus—Gorse; Whin; Furze (swellings and pains, bots)
Ulex gallii—Western gorse; Mountain gorse; Autumn gorse (swellings and pains, bots[11])
Umbilicus rupestris—Pennywort; Navelwort ('scruff')
Lycoperdaceae (Agaricales)—Puffballs (collar and saddle sores)

WILD PLANTS USED FOR VETERINARY PURPOSES *(continued)*

Pigs

Mentha species—Mint (fits)
Polygonum aviculare—Knotgrass (swine fever)
Sambucus nigra—Elder ('to dose pigs')
Stellaria media—Chickweed (convulsions)
Symphytum officinale—Comfrey (swine fever)

Dogs

Allium ursinum—Wild garlic; Ramsons
Arrhenatherum elatius—False oat (white scale[12])
Cytisus scoparius—Broom (fits)
Galium verum—Lady's bedstraw (fits)
Geranium robertianum—Herb-Robert (constipation, diarrhoea and tuberculosis)
Mentha species—Mint (fits)
Myrica gale—Bog myrtle; Sweet-willow (fits)
Plantago lanceolata—Ribwort plantain (madness)

Cats

Euphrasia officinalis agg.—Eyebright (weak kittens with mucus-blocked eyes and noses)

Poultry

Allium ursinum—Wild garlic; Ramsons (pip[13])
Rubus fruticosus—Bramble; Blackberry (pip)
Salix alba—White willow (pip)
Sanicula europaea—Wood sanicle; Sanicle (gapes[14] in young pheasants)
Senecio aquaticus—Marsh ragwort; Marsh ragweed (gapes)
Senecio jacobaea—Ragwort; Ragweed (gapes)
Senecio vulgaris—Groundsel (pip)
Stellaria media—Chickweed (pip, gapes)
Symphytum officinale—Comfrey (plasters for broken limbs)
Tanacetum vulgare (*Chrysanthemum vulgare*)—Tansy (pip)
Taraxacum officinale agg.—Dandelion (pip, 'pine' [a trace-element deficiency] in turkeys)
Urtica dioica—Nettle (pip)

Cagebirds

Stellaria media—Chickweed (moulting)

---

1. Veterinary information is taken from Merck (2010) and other sources.
2. Allen and Hatfield (2004: 351).
3. Water fever is also called 'red murrain and 'blood murrain' and is a zoonotic, tick-borne disease of cattle and sheep, and also humans. Babesiosis is caused by the parasitic protozoan *Babesia divergens*.
4. Hoose is a bronchial infestation by nematodes (*Dictyocaulus, Metastrongylus, Protostrongylus*) and seen in cattle, sheep, goats and pigs.
5. *Crúbain* is a warm-season condition characterised by the crouched or cringing position of the animal.
6. Blackleg is a rapidly fatal disease, usually seen in summer and autumn, in cattle and sheep. It is caused by *Clostridium*, anaerobic bacteria also associated with botulism and tetanus. Blackleg is characterised by emphysematous swellings, more evident in heavier musculature such as the legs.
7. Sheep scab refers to psoroptic mange caused by the skin mite *Psorergates ovis*. The intense itching and lesions damage the wool and value of the sheep. Dipping or spraying with 0.2% malathion solution has eradicated sheep scab in many parts of the world today (Merck, 2010).
8. Farcy is the coetaneous form of glanders, a serious bacterial infection (*Burkholderia mallei*) of the skin and respiratory tract in horses. It is zoonotically transmissible to dogs, cats, goats, and humans, causing painful ulcers and superficial lymphatic involvement and swelling (Merck, 2010).
9. Strangles is a serious upper respiratory infection in horses (equine distemper), caused by *Streptococcus equi*.
10. Gravel refers to suppurative pododermatitis of the hoof. This bacterial infection causes lameness that would occur after a farrier has trimmed and shod a horse's hooves. Nails that secure the horseshoe introduce the bacteria into more sensitive areas of the hoof.
11. There are three species of bot flies that infest horses: the common bot *Gasterophilus intestinalis*; the throat bot *G. nasalis*; and the nose bot *G. haemorrhoidalis*. These hairy-bodied flies are pesky and active in warm weather. The adults lack mouth parts to bite, but lay eggs on horse hair and skin. These hatch and the larvae migrate through the mouth and nose into the gastrointestinal tract. These feed and pupate, only to pass out in dung and begin the life cycle again.
12. White scale refers to exfoliating dermatoses that may lead to alopecia in dogs. This skin disorder is more often seen in sporting breeds such as the cocker spaniel, English Springer spaniel and the Irish setter, but also occurs in other breeds such as the West Highland terrier and the retrievers.
13. Pip generally refers to respiratory disease in birds. A thick mucous discharge crusts in the mouth and throat of the affected bird.
14. Gapes refers to laboured breathing seen in poultry. The red or gape worm *Syngamus trachea* is a parasitic nematode that infests the trachea, bronchi and bronchioles of the lungs. Gapes is a serious threat in rearing pheasants, but young poults and chicks of all poultry are suspectible. Adult birds rarely gape. These nematodes are found in soil, in earthworms, slugs and snails, and are then consumed by the birds.

Aphrodisiacs and Love Potions in the Irish Flora

An aphrodisiac is any substance that increases sexual desire. These substances are usually foods, drinks or scents. They obtain their name from Aphrodite, the Greek goddess of love and sexuality. Throughout history, many substances are reputed to have been of value as aphrodisiacs, both to increase male and female sexual desire and to increase human ability to make love. A number of supposed aphrodisiacs are also recognised as being fatally poisonous, such as *Atropa belladonna* (deadly nightshade) and *Conium maculatum* (hemlock), or hallucinogenic, such as the non-native *Cannabis sativa* (marijuana). Experimentation with any of these plants is not recommended.

The folklore in many countries includes stories of spells and potions designed to promote love and carnal desire and erotic pleasure (Rätsch, 1997). Ireland is no exception, and throughout history a range of wild plant species have been reputed to act as aphrodisiacs and love potions. Lady Wilde (1890: 32) wrote that 'Philters [a potion thought to arouse sexual love, particularly directed towards a specific person], love powders, and charms to procure affection were frequently used in Ireland, and the belief in them. The bardic legends have frequent allusions to love charms'. Sadly, in most, if not all, cases involving the use of native Irish plants as aphrodisiacs, there is probably no physiological basis to their use as love potions and they are probably no more effective than, for example, giving a large bunch of red roses to a loved one, or chocolate, strawberries and champagne. Foods that had a reputation for being aphrodisiacs were often unusual or exotic ones, such as artichokes and asparagus, whose reputations faded as these foods became more common.

A small number of species (i.e., *Euphorbia helioscopia* and *Urtica dioica*) are reputed to have been used externally to enhance sexual pleasure, but I have not found any direct evidence of their use in Ireland for such purposes. Alcohol is sometimes regarded as an aphrodisiac, and may decrease sexual inhibitions, if nothing else. Some aromatic oils

PLANTS REPUTED TO BE APHRODISIACS OR LOVE POTIONS

Achillea millefolium—Yarrow
Anacamptis morio (*Orchis morio*)—Green-winged orchid
Apium graveolens—Wild celery
Asparagus officinalis—Asparagus
Atropa belladonna—Deadly nightshade
Cannabis sativa—Hemp, Marijuana
Conium maculatum—Hemlock
Daucus carota—Carrot; Wild carrot
Drosera rotundifolia—Round-leaved sundew (love charm)
Euphorbia helioscopia—Sun spurge (external use)
Ligusticum scoticum—Lovage; Scottish lovage
Orchis mascula—Early purple orchid
Porphyra umbilicalis—Laver
Thymus polytrichus (*Thymus drucei; Thymus praecox; Thymus serpyllum*)—Wild thyme
Urtica dioica—Nettle (external use)
Viscum album—Mistletoe (love charm)

Daucus carota (wild carrot).

derived from plants that are known in the wild and in cultivation in Ireland have been used for erotic massage purposes.

For obvious reasons, little direct evidence of the use of aphrodisiacs in Ireland is available and no recent references. However, in addition to those species listed below, the following are those species found in the wild or cultivated in Ireland that are said to be effective aphrodisiacs according to Rätsch (1997): *Acorus calamus* (sweet-flag), *Allium sativum* (garlic), *Armoracia rusticana* (horse-radish), *Artemisia absinthium* (wormwood), *Brassica nigra* (mustard), *Datura stramonium* (thorn-apple), *Hyoscyamus niger* (henbane), *Papaver somniferum* (opium poppy) and *Petroselinum crispum* (parsley).

A few native Irish fungi have also been regarded as having aphrodisiac qualities in other countries, such as *Amanita muscaria* (fly agaric) (further described in the Systematic List), as well as *Phallus impudicus* (stinkhorn), which gives out a strong smell of rotting meat detectable for several hundred yards in a woodland. Rätsch (1997) said that it caused people in the Middle Ages to think of the vapours from their own genitals, but that any recipes to prepare love drinks from it have been lost. A number of species of 'magic mushroom', such as the liberty cap mushroom (*Psilocybe semilanceata*), have been collected, sold or eaten in Ireland and cause hallucinogenic effects. The sale of 'magic mushrooms' in Ireland is illegal.

In general, the best aphrodisiacs seem to be the ones where those using them believe them to be effective. There are no known foods that have substantiated proof that they physiologically enhance sexual performance or desire.

Cosmetics and Skin Care Products from Wild Irish Plants

The use of natural plant-based products in cosmetics has grown in popularity in Ireland particularly since the 1980s. In part, this has been boosted by the success of such international enterprises as The Body Shop (opened in 1976), which produces beauty products, as well as greater affluence and a growing interest in 'natural' products. Artisan and other producers of cosmetics have now developed a wide range of cosmetics and skin care products in Ireland. In general, most producers use imported materials or ingredients derived from cultivated non-native plants. However, a significant number of wild plants are also harvested and used for their medicinal and soothing properties and natural fragrances, such as roses, chickweed, nettles, comfrey and seaweeds.

Homemade hand cream is straightforward to prepare. One recipe is as follows: ¾ cup of almond oil; ¼ cup of coconut oil; 1 teaspoon of lanolin and ½ ounce of grated beeswax. These are mixed and melted together over a low heat then allowed to cool before ½ cup of rosewater, ½ cup of *Aloe vera* gel and a few drops of rose oil are added. Finally, these are all whipped together until thick, and then stored sealed in a jar before use. The recipe can be varied to change the consistency of the hand cream according to individual preferences, and different aromatic and scented ingredients can also be added.

One of these producers of skincare products, Harebell Herbs, makes organic skincare products in Ventry on the Dingle Peninsula in West Kerry. Their products include creams, moisturisers and oils. Some of the ingredients used by Harebell Herbs include oils, plant extracts and other plant products (Rosemarie Lieb, pers. comm., 2010). Rí Na Mara, a company making natural health and beauty seaweed products, uses a variety of wild-collected Irish

SOME NATURAL COSMETICS COMPANIES IN IRELAND

Sally Miller Natural Cosmetics, West Cork
Harebell Herbs, Ventry, Co. Kerry
Flourish Organics, Killarney, Co. Kerry
Rí Na Mara, Irish Seaweed Cosmetics, Spiddal, Co. Galway
Natures.ie, Natural Cosmetics, Dublin
Enniskerry Essentials, Enniskerry, Co. Wicklow

seaweeds, including *Laminaria digitata* (oarweed), *Fucus serratus* (serrated wrack), *Ascophyllum nodosum* (knotted wrack) and *Chondrus crispus* (carrageen).

In 2010, perfume, cologne and other fragrance producers in Ireland using plant extracts included:

- The Burren Perfumery, Carran, Co. Clare, which produces a range of products called Burren Botanicals
- Fragrances of Ireland, Kilmacanogue, Co. Wicklow, which makes 'Inis' products
- Bog Standard, Belfast, whose products include soaps and hand care items, including one called 'Irish Hedgerow Soap'
- L'aura Ireland Ltd from Dublin, which makes a perfume called 'Irish Rain'.

An assortment of cosmetics made by Harebell Herbs.

Seaweed soap.

INGREDIENTS USED IN COSMETICS BY HAREBELL HERBS

Oils

Plant oils
Almond oil
Coconut oil
Wheat germ oil
Rose oil
Sunflower oil
Evening primrose seed oil (*Oenothera* species)
Lavender essential oil
Rosemary essential oil
Jasmine oil
Grape seed oil

Plants and plant extracts

Rose water (*Rosa* species)
Rose buds (*Rosa* species)
Borage (*Borago officinalis*)
Witch hazel (*Hamamelis* species)
Comfrey (*Symphytum officinale*)
Chickweed* (*Stellaria media*)
Feverfew (*Tanacetum parthenium*)
Elderflower* (*Sambucus nigra*)
Frankincense (*Boswellia* species)
Lemon grass (*Cymbopogon* species)
Lime (*Citrus aurantifolia*)
St John's-wort flowers* (*Hypericum* species)
Rosemary (*Rosmarinus officinalis*)
Lavender (*Lavandula* species) (in Irish, *lus liath*, de Vere Coneys, 1849)
Geranium (*Pelargonium* species)
Bergamot (*Monarda didyma*, Lamiaceae)
Lemon balm or Melissa (*Melissa officinalis*)
Chamomile (*Chamaemelum nobile* [*Anthemis nobilis*])
Marigold (*Calendula officinalis*)
Carrageen* (*Chondrus crispus*)

Other products

Borax**
Beeswax

* Native Irish plants.
** Borax (sodium borate) is a white powder that is used in many cosmetics, detergents and ceramic glazes and sometimes as a food preservative.

Primula vulgaris (primrose; formerly called *Primula acaulis*). Watercolour by Lydia Shackleton. Courtesy of the National Botanic Gardens, Glasnevin.

HORTICULTURAL USES FOR PLANTS

THE CULTIVATION OF PLANTS for ornament is a practice that probably goes back only a few hundred years in Ireland. Early Irish texts contain few if any references to ornamental plantings or decorative gardens (Kelly, 1997). In ancient Ireland the cultivation of plants in gardens was probably more or less entirely for culinary and medicinal purposes. However, gardening is a hugely popular activity in Ireland today, and a wide range of ornamental plant species, hybrids and varieties are available for Irish gardens, mainly species that are native to other countries around the world and not to Ireland. The collections of the National Botanic Gardens in Glasnevin, for example, include more than 17,000 different species, hybrids and cultivated forms. It is notable that very few home gardeners in Ireland are currently cultivating many of our own native flora or are sufficiently aware how many deserve to be more widely grown. There is increasing interest in Ireland in growing 'wild flowers', which often means that well-meaning gardeners will invest in packets of mixed wild flower seeds that are likely made up of mixtures of native and non-native species, sometimes sourced from imported stock such as bluebells, poppies, foxgloves, corncockle, cornflower, marigolds, chamomile and ox-eye daisies. In contrast, I suggest that a significant number of native plants have considerable garden potential and are worthy of cultivation as ornamentals. Some are already grown in Irish gardens and are listed as such in the Systematic List, but others have been neglected as far as I am aware. In many countries, notably Australia, South Africa and parts of the United States, there is a growing movement to promote the cultivation of native plants in gardens. Often, this is because they are particularly well suited to local climatic conditions and don't require expensive irrigation using scarce water resources. However, in the Irish context I am not suggesting that the cultivation of native plants is preferable to growing plants from different parts of the world. Instead, it merely provides a fascinating extra dimension for Irish gardening and further contributes to plant conservation and raises public awareness of the value of native plant diversity.

Plants for the Rock Garden

There are several obvious native Irish plants suitable for the rock garden, including the common *Armeria maritima* (thrift or sea pink), *Juniperus communis* (juniper), *Dryas octopetala* (mountain avens), *Primula veris* (cowslip) and Saxifrages, such as *Saxifraga rosacea* and its Irish endemic subspecies from Co. Donegal, *S. rosacea* subsp. *hartii* (or *S. hartii*). The beautiful, little bulbous plant of the east coast of Ireland, *Scilla verna* (spring squill), should be ideal to naturalise on a rockery. Another native bulb, *Colchicum autumnale* (autumn crocus), can be grown too. It is found only in meadows in the Nore river valley in south-east Ireland and is now threatened by extinction.

Amongst plants suitable for the rock garden is one of Ireland's most beautiful alpines, *Silene acaulis*, a small cushion-forming tufted plant with pink flowers, which is found only rarely in the wild in Ireland in the mountains of the north-west in Cos. Mayo, Sligo, Leitrim, Donegal and Derry. *Meconopsis cambrica* (Welsh poppy) is another plant grown in many rock gardens. Although it can become quite weedy and even invasive at times, it is an attractive plant appropriate for wilder cottage gardens. In the wild in Ireland the Welsh poppy is quite a rare native and found only in rocky places in some mountain ranges. However, it has naturalised in the wild more frequently as a fugitive from gardens.

Primula veris (cowslip).

Several native ferns, such as *Asplenium scolopendrium* (hart's-tongue fern), *Adiantum capillus-veneris* (maidenhair fern) and *Asplenium ceterach* (*Ceterach officinarum*) (rustyback fern), are all excellent choices for rockeries. All will thrive if planted in rocky crevices, even in vertical rock faces, just like they often grow in the wild. An attractive plant to naturalise in similar situations is the white-flowered variety of the common herb-Robert, *Geranium robertianum*. Its pure white flowers contrast well with the red stems and foliage so typical of this species. Be warned, though, that it is seriously invasive and may run rampant in your garden. A close relative is the bloody crane's-bill, *G. sanguineum*, a hairy, large-flowered, red-purple in colour, perennial geranium with deeply dissected leaves. It is found commonly on dry limestone banks and rocky places in the west, particularly in the Burren region of Co. Clare. In the rock garden it grows well in clumps and rarely becomes over-vigorous or invasive and is an extremely useful ground-cover plant, flowering freely in July and August.

An Irish plant that deserves to be cultivated as a garden plant is a small crucifer *Cochlearia officinalis* subsp. *scotica*. This tiny tufted plant forms a tight rosette of dark green fleshy leaves and small white flowers typical for crucifers that have four paddle-shaped petals and attractive yellow-green anthers in the centre. In the wild it grows on exposed rocky and gravel shores on the Atlantic coasts of Ireland, occurring as smaller and more distinctive forms the more exposed the situation. It is a variant of the more frequently occurring common scurvygrass (*Cochlearia officinalis*), so called because its leaves were once a source of vitamin C and eaten by northern voyagers to ward off the deadly scurvy brought on by vitamin C deficiency.

A plant that I have tried and failed to grow is *Simethis mattiazii* (*Simethis planifolia*), the Kerry lily. In the wild in Ireland it is recorded only from coastal heaths in south Kerry and west Cork. It is deeply rooted and slow to establish but could probably be raised from seed. Its white flowers, tinged with purple on the outside, are a delicate and beautiful sight during the few weeks in summer when it is in flower. A similarly rare native is *Astragalus danicus*, found only on the Aran Islands. It is straightforward to cultivate and will reward the grower in early June with attractive purple-violet flowers borne in short round heads extending from its divided leguminous silky, grey leaves. Less easy to grow well is *Otanthus maritimus* (cottonweed), a woolly, white perennial that is almost extinct in Britain and Ireland. Formerly recorded in a few coastal localities in Cos. Kerry, Waterford, Wicklow and Wexford, it now persists only at a single site in the latter county. It grows well from cuttings, but in the garden setting, I've failed to establish the large, healthy colonies in which it looks its best when it is thriving in the wild. It would seem to be somewhat frost sensitive.

Other native plants worth trying in the rock garden or raised beds include *Trifolium occidentale* (western clover), *Anthyllis vulneraria* (kidney vetch), various *Alchemilla* species (such as *A. xanthochlora, A. filicaulis* and *A. glabra*), *Rhodiola rosea* (roseroot), *Antennaria dioica* (cat's foot; mountain everlasting) and *Achillea ptarmica* (sneezewort). *Wahlenbergia hederacea* (ivy-leaved bell-flower) is a very classy but tiny pale blue-flowered, creeping perennial found only occasionally on river-banks and damp grassy places, mainly in the south and west of Ireland. It would be a fascinating plant to try naturalising in a damp lawn.

Cold Greenhouse Plants

A range of smaller native species are very well suited to cultivation as cold greenhouse or cold frame subjects. If grown in pots or shallow pans they can be easily cared for and don't always require the soil conditions from which they are derived in the wild. The famous brilliant blue-flowered *Gentiana verna* (spring gentian) of Burren fame[20], can be grown in such conditions, but I have never managed to maintain plants in cultivation for long. *Campanula rotundifolia* (harebell) is similarly worth trying in a pan too. Another useful subject is *Samolus valerandi* (brookweed), a little member of the Primulaceae with small white flowers. It is a plant of muddy ditches and sea and lake-shores. Rather overlooked as a garden plant, I have grown it successfully in a cold greenhouse in a shallow pan, where it thrived on ordinary garden soil and flowered well for me in midsummer. Another interesting and unusual plant to grow in similar conditions is the sea lavender, *Limonium*, particularly *L. binervosum* and its close relatives. There are several rare *Limonium* species found in the west and south-west of Ireland found on gravel and rocky shores that might be tried. Another plant I have always admired in the wild but never yet attempted to grow is *Blackstonia perfoliata* (yellow-wort). It is an erect annual plant with yellow flowers and blue-green leaves found on sand-hills and gravel banks. I suspect that it too would grow well from seed and could be managed in the cold greenhouse.

A native plant that can be grown as a curiosity in the alpine house is *Sibthorpia europaea*, the so-called Cornish moneywort. It is native only to a small region of Ireland, in the Dingle Peninsula, Co. Kerry. It has delicate, long prostrate stems and round leaves, with tiny flowers. When confined in a pot, the plant will quickly fill it and then begin to hang down the sides like delicate curtains. It is said to be slightly frost sensitive. The blue-eyed grass, *Sisyrinchium bermudiana*, also grows well in cultivation and is suited to pot culture. It is found mainly in moist meadows in the west and in Kerry. *Ophioglossum vulgatum* (adder's tongue) is a small fern with an underground stem that pushes up its single olive-green leaf from below ground, from which it then develops its fertile spore-bearing spike. The adder's tongue fern grows extremely well in cultivation and will thrive in a pot, or when confined in a soil pocket, will grow between boulders in the rockery.

Campanula rotundifolia (harebell).

Woodlands and Hedgerows

In Ireland the natural 'climax' vegetation over much of the landscape would be woodlands, had they not been largely removed for agriculture and settlement. It is not surprising, therefore, that a diverse range of species occur in the tree, shrub and herb layers of our woodland ecosystems. Of course a wide range of native tree species are already widely grown and find appropriate homes in most Irish gardens. These include *Ilex aquifolium* (holly), *Sorbus aucuparia* (rowan or mountain ash), *Sambucus nigra* (elder), *Crataegus monogyna* (hawthorn), *Betula pendula* and *B. pubescens* (birch), *Alnus glutinosa* (alder), *Corylus avellana* (hazel), *Taxus baccata* (yew) and the two native oaks *Quercus petraea* and *Q. robur*. Less common are *Euonymus europaeus* (spindle-tree), *Malus sylvestris* (crab apple), *Cornus sanguinea*

Crataegus monogyna (hawthorn).

20. The Burren is an area of approximately 250 sq. km. in north Co. Clare of karst limestone topography, one of the largest karst landscapes in Europe. It is famed for the richness and diversity of its flora, which includes species that are also found in Arctic, Alpine and Mediterranean regions.

(dogwood) and *Viburnus opulus* (guelder-rose), several species of willow (*Salix* species) and the cherry species *Prunus padus* (bird cherry), *P. avium* (wild cherry) and *P. spinosa* (sloe or blackthorn). Most of these occur in gardens that have incorporated previously wild hedgerows or field margins.

Several other native trees and shrubs with considerable garden merit are little grown, however. These include the small, thorny tree *Rhamnus cathartica* (buckthorn) and *Frangula alnus* (alder buckthorn). The magnificent and much loved *Ulmus glabra*, our native wych elm, is sadly diminished through Dutch elm disease and therefore probably of limited garden potential. *Populus tremula* (aspen) is also native and well worth growing. I think that it is one of the only Irish plants one can identify by sound![21] The rustling of its leaves in the wind is a delightful addition to any garden. Its only drawback as a garden plant is its tendency to produce copious suckers. However, if grown in the centre of a lawn these can be easily kept in check by the lawnmower.

A useful, smaller native shrub for open (unshaded) locations in the garden is *Potentilla fruticosa* (shrubby cinquefoil), a common garden plant. In cultivation it is represented by a numerous range of cultivars with different growth forms, sizes and flower colours. As a native plant in Ireland, it is only found in north Clare and south-east Mayo, in rocky places that are seasonally flooded. *Cytisus scoparius* (broom) can be extremely attractive in the garden, although it is a relatively short-lived shrub. Rather more interesting for the garden are the prostrate forms of this species (subsp. *maritimus*) that occur in a few places on coastal cliffs in Cos. Cork, Dublin and Kerry. Amongst the wild roses, few are easy to 'tame' as garden plants, and most will not flower sufficiently freely enough to warrant inclusion in smaller gardens. However, one exception might be the smaller white or pink-flowered *Rosa pimpinellifolia* (Scotch or Burnet rose). It is quite widespread in the wild on sand dunes, rocky heaths and limestone pavements that are especially close to the sea. *Artemisia maritima* (*Seriphidium maritimum*) (sea wormwood) is a rare native shrub that occurs on several muddy and rocky shores in the east and west. In the garden it can also be usefully grown.

In shady places under trees, many native spring flowers of Irish woodlands can be easily naturalised, including *Ficaria verna* (*Ranunculus ficaria*) (lesser celandine), *Anemone nemorosa* (wood anemone), *Oxalis acetosella* (wood sorrel), *Primula vulgaris* (primrose) and the common bluebell, *Hyacinthoides non-scripta*. Worth growing, too, on damp, shady banks are the wild saxifrages, *Saxifraga spathularis* (London pride or St Patrick's-cabbage) and *S. hirsuta*. I particularly like the hairy kidney-shaped leaves and the delicate hairy stems of *S. hirsuta*. However, *S. hirsuta* is less frequently seen in gardens or in the wild than its hybrid with *S. spathularis* (*S.* ×*polita*). Another saxifrage, *S. granulata* (meadow saxifrage), is native to sand-hills and pastures near the east coast but is now very rare and endangered in Ireland. The meadow saxifrage is quite a robust hairy perennial, growing up to about 50 cm tall with small white flowers which come up year after year.

As groundcover plants in informal woodland garden settings, *Luzula sylvatica* (woodland wood-rush), *Hypericum androsaemum* (tutsan), *Lysimachia vulgaris* (yellow loosestrife), *Arum maculatum* (lords and ladies or cuckoo-pint) and *Carex pendula* may be grown. *Carex pendula*, as its name suggests, is an attractive, robust sedge, a tufted plant with pendulous flowers borne on long stems. Another beautiful but very rare, little woodland plant in Ireland with red-purple flowers is *Stachys officinalis* (betony). Native populations are protected by law in Ireland, but plants are freely available within the nursery trade.

Lathraea squamaria (toothwort) is an extraordinary perennial member of the Orobanchaceae. It grows as a parasite on the roots of various native trees such as hazel. It

21. The only other candidate I can think of is *Heracleum sphondylium*, sometimes called 'The Singer', from the noise made when the wind blows through its seed heads.

is entirely without green colour and survives purely on what it can gain from its host. It can be established in the garden, and there are small populations growing well on several *Corylus* (hazel) trees at the Trinity College Botanic Gardens in Dartry, which were introduced by Professor D. A. Webb in the 1960s. This plant also grows well in the National Botanic Gardens in Glasnevin.

Herb and Vegetable Gardens

Native plants for the vegetable garden are quite limited. Nevertheless, the real aristocrat of vegetables, *Crambe maritima* (seakale), is a native plant that sadly has become rare both in the wild and in cultivation. *Crithmum maritimum* (rock samphire) is another coastal plant, commonly found on exposed rocks, that is collected and pickled in some European countries. *Beta vulgaris* subsp. *maritima* (sea beet) can be grown as a vegetable and treated as spinach. *Asparagus officinalis* is also native, but found only rarely in some eastern coastal sand dune systems where it occurs as a prostrate variety, unsuitable for cultivation as a vegetable.

Crambe maritima (seakale).

Angelica sylvestris (wild angelica) can be used as a culinary herb in the same way as its commonly grown garden relative, *A. archangelica*. *Angelica sylvestris* can be used in salads, cooked as a vegetable, chopped and added to stewed rhubarb. Its seeds are useful aromatics for culinary purposes, and its leaf stalks and stems can be candied.

The common wild *Daucus carota* (wild carrot) is a relative of the ordinary garden carrot. It is a widespread plant of dry banks and pastures, particularly near the sea and on limestone soils. If you have ever tried digging a wild carrot for its root, it takes some imagination to believe that it is this paler form from which the large, orange root vegetable was derived. Cleaning and peeling the roots of the wild carrot as a vegetable is not a productive occupation. Care must be taken not to confuse the wild carrot with the similar but fatally deadly hemlock (*Conium maculatum*) that may grow nearby.

For the herb garden, however, the Irish native flora provides a range of useful species such as *Ligusticum scoticum* (lovage), *Origanum vulgare* (marjoram), *Thymus polytrichus* (*T. drucei*, *T. praecox*, *T. serpyllum*) (wild thyme), *Calamintha sylvatica* (calamint), *Symphytum officinale* (comfrey) and *Rumex acetosa* (sorrel). Native wild mint species with deliciously aromatic foliage such as *Mentha arvensis* (corn mint) and *M. pulegium* (penny-royal) might also be grown. Penny-royal was formerly more widespread in Ireland, and occurred in mainly damp sandy habitats, particularly in Co. Kerry. It has declined dramatically over the last century and is now both rare and endangered in Ireland and Britain.

Bog and Water Gardens

Water gardens are well served by having several excellent aquatic and marginal aquatic plants native to Ireland that can be grown easily and effectively. These include *Caltha palustris* (marsh-marigold or kingcup), *Butomus umbellatus* (flowering rush), *Osmunda regalis* (royal fern), *Mentha aquatica* (water mint) and *Iris pseudacorus* (yellow flag iris). Other species worth evaluating might include *Comarum palustre* (*Potentilla palustris*) (marsh cinquefoil), *Menyanthes trifoliata* (bogbean) and *Lychnis flos-cuculi* (ragged robin), although I have not yet tried to grow any of these. *Teucrium scordium* is a

Mentha aquatica (water mint).

J. T. Mackay, after whom *Erica mackayana*, Mackay's heath, is named.

low-growing, hairy perennial found most often in Ireland on some lake-shores and along the river Shannon. With its delicate pink-mauve spikes of flowers, *T. scordium* is worth cultivating when confined to a pot and grown in semi-waterlogged conditions.

Growing in the ponds and lakes scattered throughout Ireland are *Nuphar lutea* (yellow water-lily) and *Nymphaea alba* (white water-lily), both of which should be standard for any larger garden pond. In the bog garden I have tried and failed to grow several species of butterwort (*Pinguicula vulgaris* and *P. grandiflora*) successfully. The latter was considered by Reginald Scully, author of the *Flora of County Kerry*, as Ireland's most beautiful wild plant (Scully, 1916). *Pinguicula grandiflora* has large violet flowers borne on stems 15 cm tall that arise from the sticky, insect-trapping, yellow-green leaves. It is found mainly in Cos. Kerry and Cork, in south-west Ireland. *Pinguicula vulgaris* has smaller flowers and is considerably more widespread in the bogs and marshes of Ireland. I suspect that both species are only suited to the determined specialist carnivorous plant grower (although I recollect that *P. grandiflora* can be grown well in a hollowed-out crevice of a boulder, kept moist in summer but not too sodden in winter).

In addition to *Arbutus unedo* (strawberry-tree), a range of native plants in the Ericaceae family are suited to cultivation such as in bog gardens on peat and in acid soils. *Vaccinium oxycoccos* (cranberry) is a neat little plant with red flowers. It can be grown in a peaty border but is probably best maintained in a pot or pan standing in water. In Ireland it is a relatively rare plant of the midland raised bogs[22]. *Arctostaphylos uva-ursi* (bearberry), a close relative, will grow in similar situations in the garden. Bearberry is found mainly on inaccessible moors and amongst rocks along the north and western coasts. However, of the Ericaceae, several species of *Erica* are more easily grown. These include the common *E. cinerea* (heather), as well as its rare native relatives, *E. ciliaris* (Dorset heath), known as a wild plant in only one location in Co. Galway; *E. erigena* (Mediterranean heath), Ireland's largest heath, which grows to be a robust shrub up to 150 cm tall and *E. mackayana* (Mackay's heath), which is found in several locations restricted along the western coast of Ireland. However, my favourite of the heaths is the large-flowered *Daboecia cantabrica* (St Dabeoc's heath), with flowers pinkish purple in colour. This heath is quite frequent in peaty habitats in Cos. Galway and Mayo and thrives in similar soils in the garden. A range of cultivated varieties selected from these Irish species are described by Charles Nelson in *A Heritage of Beauty: The Garden Plants of Ireland—An Illustrated Encyclopaedia* (Nelson, 2000b) (there is even a double-flowered variety named *D. cantabrica* 'Charles Nelson', found in 1978 near

22. Raised bogs are dome-shaped bogs that developed since post-glacial times in former lake basins and depressions or hollows. They are particularly prevalent in Ireland in the midlands. Most raised bogs in Ireland have been extensively cut-over for peat. By contrast, in mountainous regions with high rainfall, blanket bogs developed.

Carna, Connemara, Co. Galway), but the 'unimproved' wild forms are still excellent garden plants and generally easy to propagate and cultivate.

In Ireland there are eight or nine species of horsetails (*Equisetum* species). However, most are too vigorous and unruly for the garden and should be avoided. The rare *E.* ×*moorei*, known from several sites on the coasts of Cos. Wicklow and Wexford, from south-east Ireland, was introduced to the native plant area of the National Botanic Gardens some years ago and is probably now unlikely ever to be eradicated from the garden.

Herbaceous Borders

The development of a native herbaceous border would probably be one of the best ways to use a range of Irish herbaceous perennial plants. This is one of the forthcoming projects that the National Botanic Gardens plans to undertake, creating a traditional herbaceous border entirely from Irish stock. A wide range of native species could be used, including *Aquilegia vulgaris* (columbine), *Digitalis purpurea* (foxglove), *Althaea officinalis* (marsh-mallow), *Lavatera arborea* (tree mallow), *Eupatorium cannabinum* (hemp agrimony), *Campanula trachelium* (nettle-leaved bellflower), *Verbascum thapsus* (mullein) and *Trollius europaeus* (globeflower). The latter is found now only rarely in the wild, around several lakes, mainly in the north-west, but is well known as an attractive and useful perennial border plant. Others worth evaluating are *Centaurea scabiosa* (greater knapweed), *Valeriana officinalis* (wild valerian), *Knautia arvensis* (field scabious) and *Pulicaria dysenterica* (fleabane). The handsome *Dipsacus fullonum* (teasel) and *Solidago virgaurea* (goldenrod) are also excellent garden plants. Several non-native relatives of the goldenrod, notably *S. canadensis*, are already frequently grown in gardens. In Ireland, *S. virgaurea* is quite variable, and several attractive, small forms can be found that retain their size in cultivation and could be used well in a rockery or the front of a herbaceous border.

Another beautiful native herbaceous plant for cultivation is *Sanguisorba officinalis* (burnet). This plant is best known from neutral grasslands along lake-shores of Lough Conn, in Co. Mayo. It is an extremely handsome herb that grows to about 100 cm tall with dark red flowers. *Lythrum salicaria* (purple loosestrife) might also grow well in the garden, although in several parts of the world, and most notably in North America, purple loosestrife is an extremely invasive and noxious weed.

Several native seashore plants would do well too, including the crucifer *Matthiola sinuata* (sea stock), which was formerly recorded in Cos. Clare, Wexford, Kerry and Galway but now thought to be completely extinct in the wild in Ireland. It grows as a slightly unruly, tall biennial but provides an attractive sight in the garden with its bluish-green foliage and lilac flowers. Other seashore plants to consider include another crucifer, *Cakile maritima* (sea rocket), *Eryngium maritimum* (sea holly), the saltmarsh plant, *Aster tripolium* (sea aster) and the rare *Lathyrus japonicus* subsp. *maritimus* (sea pea). *Mertensia maritima* (oysterplant) is one of the most critically endangered Irish plants confined to shingle shores on the northern coasts. It is a prostrate perennial, blue-green in colour, with blue to pink flowers. I've seen this grown well in several gardens in Ireland and elsewhere, when kept in a sunny, dry position and protected from slugs.

For many years I have grown the rare native *Filipendula vulgaris* (meadow-sweet) as a border plant. It is smaller than its relative, *F. ulmaria*, a common plant of Irish hedgerows, and it has attractive, more finely divided foliage and larger white flowers tinged with red. In the wild in Ireland, it is found only in a small area of north Clare and south Galway, growing in rocky limestone pastures.

Another Irish rarity is *Inula salicina* (Irish fleabane), which is now restricted to a single site on Lough Derg, its last surviving wild locality in Ireland or Britain. It grows well in the garden. Amongst the native thistles, probably the only one worth growing is the *Carlina vulgaris* (Carline thistle), which is more compact, tidier and, dare I suggest, more attractive than its more rank and commoner relatives *Cirsium vulgare* (spear thistle) and *C. arvense* (creeping thistle), which are common pasture and garden weeds. *Scrophularia nodosa* (figwort) and its rarer relatives *S. auriculata* and *S. umbrosa* will grow as robust and tall herbaceous perennials. Their reddish or greeny-brown flowers are small and not very showy, so each is of marginal garden value.

Finally, the Irish spurge, *Euphorbia hyberna*, well known from Cos. Kerry and Cork, is an extremely useful garden plant that is easy to maintain in cultivation. Once established, it will form a robust clump with unbranched leafy stems and bright yellow flowers.

PLANTS AS SYMBOLS

Hurleys.

Plants have been used widely around the world as symbols. A few examples include the Tudor rose in Britain; the bamboo in China, signifying strength, grace and longevity; the jasmine in Hinduism, symbolizing love; and the chrysanthemum in Japan, symbolizing longevity. In Scotland, the thistle is a national symbol[23], and in Wales, the leek[24]. In Ireland, too, plants have been viewed as national symbols, such as the lucky white heather, the shamrock and the Easter lily. Some are native species, while others are introduced garden flowers. Examples of plants that are or have been used as symbols in Ireland are given below.

National Tree, the Irish Oak

The Irish oak, *Quercus petraea*, is widely regarded as the national tree of Ireland, although as far as I can find, it has no particular official status as such. I have been told that during the 1980s the suggestion was made that a more appropriate choice of national tree for Ireland would be the ash (*Fraxinus excelsior*), because it is common in woodlands and hedgerows and its timber has been used for many purposes. It has also been used in the making of hurleys for Ireland's national sport and has a rich folklore associated with it. However, the story that was related to me (I cannot say whether the story is strictly true or not) was that the Taoiseach at that time, Charles J. Haughey[25], would have none of this suggestion and insisted that it should be the Irish oak. *Quercus petraea* is the national tree of Wales, where it is appropriately called the 'Welsh oak'.

Shamrock

The three leaflets that make up a clover leaf of various *Trifolium* species, called 'shamrock', is undoubtedly the best known plant symbol of Ireland and is widely used both

23. The origin of the thistle as the symbol of Scotland is said to have come from an incident where the Scottish army was camped one night on the banks of the river Tay. The Norsemen attempted to come over a dyke at night to attack them, but while doing so one bare-footed Norseman stood on a thistle. His cry of pain roused the Scottish army who then went on to defeat their Norsemen attackers (Clyne, 1989).

24. The leek is said to have arisen as the Welsh national symbol when St David, patron saint of Wales, ordered his army to wear leeks on their helmets before doing battle with the invading pagan Saxons.

25. Charles J. Haughey served as Taoiseach (Prime Minister of Ireland) for three terms, from December 1979 to June 1981, March 1982 to December 1982, and March 1987 to February 1992.

The shamrock, a clover, is internationally recognised as a symbol of Ireland. Four-leaved clovers (i.e., leaves divided into four leaflets, rather than the usual three) are regarded as being lucky. Such objects are widely available in the United States and are used as decorations for St Patrick's Day, such as in the Missouri Botanical Garden conservation float at the St Patrick's Day Parade, 2011 (above left). A wooden four-leaved clover made in China for the U.S. market, dated 1999, marked 'Erin go Braugh!', an old Irish blessing (above right). Erin go braugh (also Erin go bragh) is a phonetic English version of *Éirinn go brách*, which in Irish means 'Ireland for ever'. The phrase is rarely used in Ireland.

in Ireland and throughout the world wherever anything Irish is displayed. While not an official symbol, it is still a powerful image in all parts of the island. Shamrocks are said to be lucky and so have been used as a badge for sports teams, state organisations and troops serving abroad from Ireland. The national airline, Aer Lingus, uses a shamrock as its logo. The leaf is said to symbolise a story told of St Patrick's use of the three-leaved shamrock to explain the Holy Trinity (God the Father, God the Son and God the Holy Ghost).

The shamrock is worn widely on St Patrick's Day, 17 March. Much mythology about the shamrock has grown up in Ireland and elsewhere and is very well documented by Charles Nelson in his comprehensive book, *Shamrock* (1991). Nevertheless, one can find Irish people who believe that shamrock will not grow in any country other than Ireland and others cannot agree what plant species should be regarded as the one true shamrock. Today, *Trifolium dubium* is generally the species that is grown and worn on St Patrick's Day. This is the plant species that is put on display at the National Botanic Gardens in Glasnevin each year, representing the 'real' shamrock.

'The Wearing of the Green'.

The Emerald Isle

Green is the colour most often associated with Ireland. The green pigment found in most plants (chlorophyll) is predominant in the Irish landscape, and allusions to it crop up in so many aspects of Irish culture and folklore. Traditional songs such as 'The Forty Shades of Green', 'The Green Glens of Antrim', 'The Green Hills of Clare', 'Four Green Fields', 'Free and Green' and 'Land of Green' all refer to Ireland's verdant green colour of fields, hedgerows, woods and mountainsides. There has also been a strong connection between the colour green and Ireland's nationalist movements over the centuries. The tricolour flag of Ireland of green, white and orange was introduced by Thomas Francis Meagher in 1848. The colour green represents the Irish people, the orange the English supporters of King William of Orange, and the white, peace.

One of the best known songs referring to the symbolism of green is 'The Wearing of the Green', which refers back to the rebellion of 1798. One version of the lyrics is given below by Dion Boucicault, which is included in his 1865 play, *Arrah Na Pogue* (Herder, 1998).

O Paddy dear, and did ye hear the news that's goin' round?
The shamrock is by law forbid to grow on Irish ground!
No more Saint Patrick's Day we'll keep, his colour can't be seen
For there's a cruel law ag'in the Wearin' o' the Green.'
I met with Napper Tandy, and he took me by the hand
And he said, 'How's poor old Ireland, and how does she stand?'
'She's the most distressful country that ever yet was seen
For they're hanging men and women there for the Wearin' o' the Green.'

'So if the colour we must wear be England's cruel red
Let it remind us of the blood that Irishmen have shed
And pull the shamrock from your hat, and throw it on the sod
But never fear, 'twill take root there, though underfoot 'tis trod.

When laws can stop the blades of grass from growin' as they grow
And when the leaves in summer-time their colour dare not show
Then I will change the colour too I wear in my caubeen
But till that day, please God, I'll stick to the Wearin' o' the Green.

County Flowers

Apart from in Northern Ireland, Ireland does not generally have any official designation of so-called county flowers. In the United Kingdom every county has a flower that is the symbol of that county, and the practice is common in the United States and elsewhere. However, apart from in Co. Offaly, which is said to have *Andromeda polifolia* (bog rosemary) as its symbol, I have not been able to find any other scheme in the Republic that has designated county flowers.

In Northern Ireland, the county (and in one case a city) flowers are as follows (all are native species):

Antrim—*Campanula rotundifolia* (Harebell)
Armagh—*Cicuta virosa* (Cowbane)
Belfast—*Ulex europaeus* (Gorse)
Derry—*Saxifraga oppositifolia* (Purple saxifrage)
Down—*Scilla verna* (Spring squill)
Fermanagh—*Trollius europaeus* (Globeflower)
Tyrone—*Andromeda polifolia* (Bog rosemary)

Orange Lilies

In Northern Ireland, the orange lily (*Lilium bulbiferum*) is a traditional symbol of loyalist support for the Union of the United Kingdom of Great Britain and Northern Ireland. Over the past several decades on 12 July, orange lilies have been worn or displayed by marching loyalists, particularly members of the Orange Order. It is not clear to me how long the orange lily has been a symbol of the Orange Order in Northern

Ireland, but I have found references to them being worn during marches in Northern Ireland on 12 July going back at least fifty years. To give its correct title, the Orange Institution (which is better known as the Orange Order) is a Protestant fraternal organisation that is particularly centred in Northern Ireland but that also has lodges in Scotland, England, the United States and some countries of the Commonwealth. It was established in 1796, in Co. Armagh, as a tribute to the Protestant King William of Orange, who had defeated the Roman Catholic King James II at the Battle of the Boyne in 1690. The Orange Order traditionally has been a strong supporter of unionism. The Orange Order adopted a new logo in 2007, which was designed by Mark Thompson in Co. Down to include a stylised orange lily. The intention of the new logo was to bring together the principles of reformation, using a combination of Martin Luther's rose, the orange lily and the purple star, the traditional symbol of the Orange Order. The purple star was a symbol used by the armies of King William of Orange. *Lilium bulbiferum*, alternatively known as the 'Orange Lily' or 'Fire Lily', is not native to Ireland but is found in the wild in Central Europe (Tutin et al., 1980).

Easter Lilies

Easter lilies are a traditional Irish republican symbol. During the early part of the 20th century, churches in Ireland are said to have been decorated with Easter lilies, and therefore this bloom became closely associated with the Easter Rising in 1916. The badge of an Easter lily has been worn widely in Ireland at Easter to commemorate those who died (on the Irish side) during the Rising (both those who were killed during the combat and the leaders who were subsequently executed). According to Hanley (2002: 49), the Easter lily was introduced in 1926 by Cumann na mBan (the Irish Republican women's association), who sold the badge outside church gates on Easter Sunday to raise funds for the Prisoners' Dependents Fund. During the early years they were widely worn at commemorative events by all of the nationalist and republican parties, but as time went by, they became more closely associated with Sinn Féin and the Irish Republican Army (I.R.A.). In 1935 the main Republican political party, Fianna Fáil, decided to cease selling and using the Easter lily symbol because of its association with the I.R.A., but many of its members continued to wear it nevertheless. In the earliest years the badge was sold and worn with a pin attached. However, in 1967 a version with a self-adhesive back was introduced. In 1969 and 1970 the I.R.A. split into the 'Official IRA' and the 'Provisional IRA'. The new breakaway 'Provisionals' returned to selling and wearing the older badges with pins, whereas the 'Officials' continued to use the self-adhesive versions. This led to the two organisations being given pejorative nicknames of 'Stickies' or 'Pinnies', or 'Pinheads'. While the name 'Stickies' lasted for many years, the other names died out quickly. In recent years the symbol was also used to commemorate those who had died during the more recent I.R.A. active campaigns. Metal (rather than paper) versions have also been made and worn since the 1990s, and not just at Easter time.

The Easter lily, *Zantedeschia aethiopica*.

In 2001 the use of Easter lilies as a symbol of Irish republican commemoration gave rise to a serious political row in Northern Ireland when the Northern Ireland Assembly Commission voted in early April to allow Easter lilies to be displayed in Parliament Buildings at Stormont. The Sinn Féin members of the Assembly welcomed the move, but representatives of the Democratic Unionist Party (D.U.P.) strongly objected. A Sinn Féin member of the Assembly had put forward a motion calling for the Easter lilies to be sold at Stormont on behalf of the National Graves Association, which looks after the republican plot at Milltown Cemetery in Belfast. The compromise proposed was that the flowers would be displayed in the central hall of Stormont. The D.U.P. tried to overturn this compromise by calling an emergency session of the Assembly for a debate during its Easter recess. Despite winning their motion by 48 to 36 votes, it did not gain the cross-community support necessary for it to be passed in the power-sharing institution. The following year, in 2002, Easter lilies were again displayed at Stormont when a further compromise was reached that would allow white lilies to be displayed at Easter, orange lilies in July and poppies in November.

The Easter lily is not actually a lily, but instead is the unrelated, *Zantedeschia aethiopica*, a member of the Araceae family and sometimes also called the Arum lily, and in Ireland, the 'Altar lily'. It is native to southern Africa and has been cultivated for over a century in Britain and Ireland. Because it flowers early in the year, it became useful for the Easter floral trade.

Rose

The rose is a universal symbol of beauty and has been widely used in Ireland when referring to the beauty of women. The Rose of Tralee Festival, held annually in Tralee, Co. Kerry, includes a contest where a beautiful young woman of Irish descent is crowned as the Rose of Tralee. The Festival has been held annually since 1959 and was inspired by a 19th century ballad 'The Rose of Tralee'. Each year, a young woman of Irish descent is selected through an international competition to become that year's Rose of Tralee. Some lines from the song are:

> She was lovely and fair as the rose of the summer,
> Yet 'twas not her beauty alone that won me;
> Oh no, 'twas the truth in her eyes ever dawning,
> that made me love Mary, the Rose of Tralee.

Red roses are widely used as a symbol of eternal love and ideal beauty. On St Valentine's Day in Ireland, tens of thousands of red roses are bought by men to give to their favourite loved ones. The symbolism of roses for love can be found in many Irish traditional songs, such as 'Banks of the Roses', 'Red Is the Rose', 'The Rose of Mooncoin' and 'My Wild Irish Rose':

> My wild Irish rose,
> The sweetest flower that grows.
> You may search everywhere,
> But none can compare
> To my wild Irish rose.

The rose has also been used as a symbol of Ireland in literature, beautiful but often with thorns. The famous Irish poem, 'Dark Rosaleen', by the Dublin poet James Clarence Mangan (1803–1849), sees Ireland as a rose. It is a patriotic poem disguised as a love song, in which the name 'Dark Rosaleen' is used as a poetic symbol of Ireland.

> O MY Dark Rosaleen
>   Do not sigh, do not weep!
> The priests are on the ocean green,

They march along the deep.
There's wine from the royal Pope,
Upon the ocean green;
And Spanish ale shall give you hope,
My Dark Rosaleen!
My own Rosaleen!
Shall glad your heart, shall give you hope,
Shall give you health, and help, and hope,
My Dark Rosaleen'

(James Clarence Mangan, 'Dark Rosaleen',
from Quiller-Couch, 1919: 664)

Another example is a well-known Irish nationalist poem 'The Rose Tree' by William Butler Yeats (1865–1939), representing an imagined conversation between the two leaders of the 1916 Easter Rising, Padraig Pearse and James Connolly.

'O words are lightly spoken',
Said Pearse to Connolly,
'Maybe a breath of politic words,
Has withered our Rose Tree;
Or maybe but a wind that blows
Across the bitter sea'

(William Butler Yeats, 'The Rose Tree',
from Yeats, 1949: 67)

Ivy

The ivy leaf became a symbol for supporters of Charles Stewart Parnell after his death in 1891. Parnell (1846–1891) was a great Irish political leader, founder of the Irish Parliamentary Party and a campaigner for land reform. The ivy leaf was worn on the lapel of his supporters on the first Sunday after the anniversary of his death on 6 October, called 'Ivy Day'. The symbol originated when a Cork woman sent a wreath to his funeral made from ivy as it was all she could afford. In 1914, James Joyce published a short story called 'Ivy Day in the Committee Room' in his book *Dubliners*. Joyce (1922: 106) also mentions Ivy Day in *Ulysses*: 'Even Parnell. Ivy day dying out'.

Red Poppies

The poppy (*Papaver* species) is used as a symbol for Remembrance Day (11 November), which commemorates those who died in the British Armed Forces, particularly during World Wars I and II. The symbol arose after the devastation of the European battlefields during World War I, when magnificent displays of blood-red poppies sprang up the following year, colonizing the bare ground. Millions of poppy replicas have been sold by the Royal British Legion, a British charity established in 1921, to raise funds to help those who served in the British Armed Forces and their dependants. These poppies are worn, particularly in Britain, from late October to mid-November, but

The poppy is worn for Remembrance Day, on 11 November, in Ireland.

in Ireland, they are generally seen only on Remembrance Sunday. Remembrance Day poppies have been made since the early 1920s and have been sold in Ireland for many years. Despite the fact that many Irish died during World Wars I and II in Europe, the wearing of poppies in Ireland has not always been welcomed, and it has been seen by some as an unwanted symbol of Britishness.

Green carnation.

Green Carnation

The green carnation is an interesting plant symbol associated with the great Irish playwright, writer and poet, Oscar Wilde (1854–1900). Wilde was asked what the green carnation represented, but responded, 'Nothing whatever, but that is just what nobody will guess' (McKenna, 2005: 169). The English playwright Noël Coward (1899–1973) helped to make the green carnation even better known when he included the song 'Green Carnation' in his musical 'Bitter Sweet' in 1929:

> Pretty boys, witty boys, you may sneer
> At our disintegration,
> Haughty boys, naughty boys, dear, dear, dear!
> Swooning with affectation. . .
> Faded boys, jaded boys, come what may,
> And as we are the reason for the 'Nineties' being gay,
> We all wear a green carnation (Coward, 1973: 83).

Green carnations have also been used elsewhere as a more general symbol of Ireland. For example, I have seen a bunch of green carnations presented to the wife of the Grand Marshall of the St Patrick's Day parade during the parade in St Louis, Missouri, in the United States. In reality, the green carnation is not a natural phenomenon, or even a cultivated variety, but rather, a white carnation that has been dyed with green ink.

Daffodils in bloom.

Daffodils

The wearing of daffodils (*Narcissus* species) on a particular day during the spring has been promoted by various charitable organizations around the world to raise funds, particularly for cancer research. Since 1988 an annual daffodil day has been held in late March by the Irish Cancer Society as a 'flag day', when small replica yellow daffodils are sold. Each daffodil day has successfully raised upwards of four or five million Euros for this charitable cause in Ireland. Daffodil day is not held on the same day every year, rather a different day towards the end of March is chosen each year. Colgan (1904) gave two additional local Dublin common names for the daffodil: daffydowndilly and daffy, both of which are still occasionally heard today.

Irish Wild Plants Featured on Irish Postage Stamps

Before 1978 no Irish postage stamps depicted any Irish wild plants with the exception of the decorative sprigs of shamrock as background on some definitive stamps issued in 1922. A shamrock (*Trifolium*) leaf was also included on a special St Patrick's Day stamp issued in 2007.

Since 1978, a total of 49 different Irish stamps (Table 9) have depicted 46 different plant species and varieties of Irish wild plants, as well as three fungi (on a set of three commemorative stamps issued in 2008). Three species (*Gentiana verna*, spring gentian; *Arbutus unedo*, strawberry-tree and *Pinguicula grandiflora*, large-flowered butterwort) have been depicted twice. During the years 2004–2008 a set of definitive stamps (the sixth definitive stamp series) was issued depicting Irish plants. These stamps were commissioned by the Irish Post Office organization, An Post, from Irish artist Susan Sex. Themes of some of the commemorative issues included Irish trees (two separate issues), Irish ferns, Irish orchids and endangered Irish plant species.

Table 9. Irish wild plants featured on Irish postage stamps (1978–2010).

Species	Common name	Date of Issue	Denomination
Gentiana verna	Spring gentian	1978	8p
Arbutus unedo	Strawberry-tree	1978	10p
Pinguicula grandiflora	Large-flowered butterwort	1978	11p
Daboecia cantabrica	St Daboec's heath	1978	17p
Sorbus hibernica	Irish whitebeam	1984	22p
Taxus baccata 'Fastigiata'	Irish yew	1984	26p
Salix hibernica	Irish willow	1984	29p
Betula pubescens	Birch	1984	44p
Phyllitis scolopendrium[1]	Hart's-tongue fern	1986	24p
Ceterach officinarum[2]	Rustyback fern	1986	28p
Trichomanes speciosum	Killarney fern	1986	46p
Otanthus maritimus	Cottonweed	1988	24p
Saxifraga hartii	Hart's saxifrage	1988	28p
Astragalus danicus	Purple milk-vetch	1988	46p
Ophrys apifera	Bee orchid	1993	28p
Dactylorhiza fuchsii var. *okellyi*	O'Kelly's orchid	1993	32p
Epipactis atrorubens	Dark-red helleborine	1993	38p
Spiranthes romanzoffiana	Irish lady's-tresses	1993	52p
Geranium sanguineum	Bloody crane's-bill	2004–2008	1c
Neotinea maculata	Irish orchid	2004–2008	2c
Iris pseudacorus	Yellow flag	2004–2008	3c
Viola riviniana	Common dog-violet	2004–2008	4c
Taraxacum officinale	Dandelion	2004–2008	5c

Table 9. Irish wild plants featured on Irish postage stamps (1978–2010). *(continued)*

Species	Common name	Date of Issue	Denomination
Ophrys apifera	Fly orchid	2004–2008	7c
Dryas octopetala	Mountain avens	2004–2008	10c
Ulex gallii	Autumn gorse	2004–2008	12c
Armeria maritima	Thrift	2004–2008	20c
Centaurea nigra	Common knapweed	2004–2008	25c
Primula vulgaris	Primrose	2004–2008	48c
Bellis perennis	Daisy	2004–2008	48c
Sedum acre	Biting stonecrop	2004–2008	50c
Pinguicula grandiflora	Large-flowered butterwort	2004–2008	55c and N[3]
Crataegus monogyna	Hawthorn	2004–2008	60c
Hyacinthoides non-scripta	Bluebell	2004–2008	65c
Umbilicus rupestris	Wall pennywort	2004–2008	75c
Schoenus nigricans	Black bog-rush	2004–2008	78c
Aster tripolium	Sea aster	2004–2008	82c
Echium vulgare	Viper's-bugloss	2004–2008	90c
Lythrum salicaria	Purple loosestrife	2004–2008	95c
Digitalis purpurea	Foxglove	2004–2008	1 Euro
Arum maculatum	Lords and ladies	2004–2008	2 Euro
Rosa canina	Dog-rose	2004–2008	5 Euro
Gentiana verna	Spring gentian	2004–2008	10 Euro
Quercus petraea	Sessile oak	2006	48c
Taxus baccata	Yew	2006	60c
Fraxinus excelsior	Ash	2006	75c
Arbutus unedo	Strawberry-tree	2006	1 Euro
Sisyrinchium bermudiana	Blue-eyed-grass	2007	N[3]
Glaucium flavum	Yellow horned-poppy	2008	N[3]
Leccinum versipelle	Orange birch bolete	2008	55c
Macrolepiota procera	Parasol	2008	55c
Hygrocybe calyptriformis	Pink waxcap	2008	82c

1. The accepted name of this fern is now *Asplenium scolopendrium*.
2. The accepted name for this fern is now *Asplenium ceterach*.
3. N represents the national letter rate, for items up to 100 g in weight.

TROPICAL DRIFT SEEDS AND FRUITS

A range of tropical drift seeds and fruits have been gathered on the Irish coasts probably for thousands of years. These have been carried in ocean currents from tropical Central and South America and probably mainly from the West Indies. Charles Nelson has researched and published extensively on the subject (e.g., Nelson, 1978, 1982, 2000a), and much of the information presented here is derived from his work.

Kelly (1997) noted that non-native nuts were recorded in early Irish law texts and were called *cnó finn* ('beautiful nut') and listed amongst valuables that could be given as pledges. Kelly suggested that these may be a reference to various tropical plants (especially *Entada gigas*; see below), some of which may have occurred as drift seeds. He also mentioned that these seeds have traditionally been prized in Ireland as charms because of their attractive appearance.

Milliken and Bridgewater (2004) observed that in the Outer Hebrides of Scotland, drift seeds were frequently gathered, used and treasured by local people. Some were mounted in silver to become family heirlooms and some were even consecrated by local clergy. Milliken and Bridgewater (2004: 170) quoted a 19th-century account by Alexander Carmichael that in the Hebrides, 'every nurse [midwife] has one which she places in the hand of the woman to increase her faith and distract her attention'. In Scotland, the seeds were worn also as charms around the necks of children, protecting them from evil, witchcraft and ill luck and by young women, protecting their virtue. They were also used in Scotland to make children's toys, matchboxes and snuff-boxes and for some medicinal purposes. In Ireland little if any of the rich folklore associated with drift seeds in other parts of north-west Europe has survived

Necklace of seeds from Cuba that are amongst the most frequent drift seeds recorded in Ireland: *Caesalpinia bonduc* (nickar nuts, grey) and *Entada gigas* (sea beans, brown).

DRIFT SEED AND FRUIT SPECIES RECORDED ON IRISH BEACHES

Astrocaryum species (Palmae)—Starnut palm

These have recently been gathered on several beaches in Co. Galway, including Inishbofin. These palms are native to tropical America and the West Indies (Nelson, 2000a).

Barringtonia asiatica (Lecythidaceae)—Box fruit

There is one record of this species occurring in Ireland, on the coast of Co. Mayo (Nelson, 2000a). This is also known as Fish poison tree for the toxic saponins (alkaloids) present in all parts of the plant. The ground seeds have been used to stun and collect fish, when they float to the surface. However, such usage is not recorded in Ireland.

Caesalpinia bonduc (Leguminosae)—Nickar nut; Grey nickar; Molucca bean

Less than a dozen collections of this species are known from Irish beaches. It is a widely distributed shrub in the tropics. Although Nickar nut is native to south-east Asia, it is widely grown and occurs in the Americas and the West Indies. This was one of the first drift seeds recorded in Ireland, and Sir Hans Sloane mentioned it in 1725 in his *Natural History of Jamaica* as having been 'cast ashore by the sea on the north west coasts of Ireland and Scotland' (p. 41).

Canavalia nitida (Leguminosae)—There is no known common name for the drift seeds

This species is native to the West Indies and recorded on only one occasion in Ireland, on Ballinesker beach in Co. Wexford (Nelson, 1982). This is unusual as most drift seeds are recorded on coasts in the west and north.

Cocos nucifera (Palmae)—Coconut

Various records of coconuts found on western and northern beaches are known in Ireland. Most have still been enclosed in their husks, which would help them to float the thousands of miles from the Americas. In some cases they may occur as refuse thrown or washed overboard from ocean-going ships. However, Nelson (1978) suggested that several specimens found appeared to have floated for many months as they were colonised by marine molluscs.

Dioclea reflexa (Leguminosae)—Sea-purse

This is a woody vine with a pan-tropical distribution (Nelson, 1978). It has been recorded on several occasions on south-western and north-western Irish beaches.

Entada gigas (Leguminosae)—Sea-heart; Sea bean (in Irish, *sliogán bóileid*)

This the most widely recorded and best known drift seed in Ireland. Seeds of this species have been known for centuries in Ireland, and there are recorded ethnobotanical uses and folklore associated with them. The fact that this drift seed has an Irish name is significant in this regard. Nelson (1978) mentioned that these seeds have been used to make snuff-boxes, vesta boxes (matchboxes) and used as play-objects for children (noted by Nathaniel Colgan in 1919). Nelson (2000a) also mentioned their use as key-minders, door knockers and as a substitute for padlocks, and noted that there are superstitions associated with them in western Ireland.

Merremia discoidesperma (Convolvulaceae)—Virgin Mary's bean

This is a rare drift seed known from a handful of records of findings on Irish beaches (Nelson, 1978). This tropical vine, or liana, a member of the Morning Glory family, is native to tropical Central America as well as the islands of Cuba and Hispaniola (Haiti and the Dominican Republic). The seeds are marked with a cross, giving rise to their vernacular name. In Central America the seeds are used as a cure for haemorrhoids (in Mexico) and as an antidote for snakebite (in Nicaragua). In Ireland, I have not come across any records or their use for medicinal purposes, and Ireland has no snakes. (Legend has it that St Patrick was bothered by snakes while undertaking a 40-day fast on the top of a hill in Ireland. As a consequence he banished all snakes from the island. In reality, there is no evidence that Ireland ever had snakes, having become separated from continental Europe early in the post-glacial period before snakes might have had an opportunity to colonise the country.)

DRIFT SEED AND FRUIT SPECIES RECORDED ON IRISH BEACHES *(continued)*

Mucuna sloanei (Leguminosae)—Sea bean; Horse-eye bean

This is a woody pan-tropical vine. Nelson (1978) reported 33 records of this drift seed occurring on Irish beaches. Other species of *Mucuna* may also have been recorded (i.e., *M. fawcettii* and *M. urens*).

Sacoglottis amazonica (Humiriaceae)—There is no known common name for the drift seeds

This is a forest tree native to the estuaries of the Amazon and Orinoco rivers in South America. Nelson (1978) listed only one record of this species occurring on an Irish beach, in about 1890.

today, and Nathaniel Colgan (1919) pointed out that even in his time, little folklore on this subject had been written down.

Various experiments to germinate and grow drift seeds have been undertaken in Ireland, and elsewhere in these islands, with mixed success (Nelson et al., 1990), and no records of any of these tropical species becoming established in the wild are known. The same authors recorded that on the small island of Inishbofin, Co. Donegal, a craftsman, Neal MacGregor has recently revived an old craft, using locally gathered *Entada gigas* to make snuff-boxes, chased with silver.

Nelson (2000a) recorded a variety of uses for drift seeds in north-west Europe. Large sea beans have been used as teething objects for infants, as they are unbreakable and impossible for a baby to swallow. Some can be polished and drilled and strung to make jewellery such as necklaces and bracelets. Drift seeds have been also been used as gaming pieces and counters, for solitaire, checkers or draughts.

The small seeds of a native temperate plant, *Lathyrus japonicus* subsp. *maritimus* (Fabaceae [Leguminosae]), are also found on Irish beaches. This plant, the sea pea, is also native to North America. Although the growing plant is very rare in Ireland, it has been found on the coasts of Kerry, Cork, Wexford and Mayo counties. Nelson (2000a) and other authors suggested that these plants or colonies may be derived originally from drift seeds.

A necklace of *Mucuna* seeds from Belize.

INTRODUCED (NON-NATIVE) PLANTS

Thousands of species from other temperate parts of the world are grown in the gardens of Ireland, from Europe, China and other parts of Asia, North and South America, South Africa and Australasia. Many species have escaped from cultivation, and there are now as many non-native plants established in the wild in Ireland as there are natives. *A Catalogue of Alien Plants in Ireland* (Reynolds, 2002) lists approximately 920 plant taxa (species, subspecies, varieties, forms and hybrids) that have been recorded on at least one occasion in the wild in Ireland. Of those, more than half are casuals and may not persist for more than a few years, and more than a quarter have been recorded only once or at one locality. The total continues to rise as more plants escape from cultivation or arrive as weeds in other ways.

In her comprehensive review of alien plants in Ireland, Reynolds (2002) reported that about 70% of the total recorded in the wild in Ireland originated as cultivated plants. Many of these plants were first introduced to Ireland because of their useful value as food sources, culinary herbs, medicines and ornamentals and for a wide variety of other purposes. Today the uses of many of these plants have been forgotten by most people, but the plants for the most part linger on in the wild in Ireland.

In recent years changes in the weather and temperatures in Ireland, perhaps as a result of climate change, have allowed some non-native plants to become better established and to spread. An example is *Gunnera tinctoria* (sometimes called giant rhubarb), which was for many years grown only in large gardens but is now well established and spreading in boggy habitats in parts of the west of Ireland, particularly in Connemara in west Galway and west Mayo. As climate changes continue, we can expect conditions to favour species from warmer climates, such as Mediterranean countries, if the predicted warmer, wetter winters and drier summers occur.

Some non-native plants were introduced to Ireland in the earliest of times, perhaps deliberately for food, medicinal or other purposes. Others were introduced inadvertently, as weeds of crops and by other means. These early introductions are termed 'archaeophytes'. Archaeophytes are defined as species that arrived in Ireland prior to 1500, but were not deemed as native species before that date. Many of these plants were common weeds up to the end of the 19th century but gradually became rarer as herbicides and better seed-cleaning methods were introduced in the 20th century. Some examples of these are cornflower (*Centaurea cyanus*) and corncockle[26] (*Agrostemma githago*), which were formerly common in Ireland but are extremely rare, and are now almost extinct. 'Neophytes' is the term used to define introduced species that arrived in Ireland after 1500, and are therefore non-natives. Due to the lack of historic records or other evidence it is often difficult to determine whether a species is a native, archaeophyte or neophyte. In the Systematic List, I have listed species on a continuum from definitely native to certainly introduced, as follows:

- Native
- Probably Native
- Possibly Introduced
- Probably Introduced
- Introduced

---

26. Irish words for 'corncockle' are *cogal* or *cogail* (de Vere Coneys, 1849; Dinneen, 1927); *luibh laoibheach* (Dinneen, 1927); *diasa fia* (Ó Dónaill, 1977) or *roithléith/roidhléith* (Dinneen, 1927).

Possible archaeophytes are also highlighted. In some instances there is uncertainty as to whether taxa are actually native or not. In other instances, the date of introduction is unknown and may have been in the country for many centuries.

The Systematic List of Wild Plants also includes information on the use of many of the non-native plants that were formerly used in Ireland for a variety of important purposes and some that are still used today. The section does not include the large number of plants that may occur in the wild as casuals. 'Casuals' is the term used to describe plants that only rarely occur in the wild but do not persist for longer than a few years and do not usually become established. Many of these are either recent imports growing from seed contaminating various imported cargos or are ornamental plants discarded from gardens.

In addition, the Systematic List does not include all plant species and varieties that were deliberately grown as agricultural or horticultural crops, for food and other purposes, such as wheat, barley, flax, oats, oil-seed rape, rye, sugar beet, rhubarb, cabbages, onions, carrots, peas and beans and many other cultivated fruit, cereal, vegetable and fibre plants that were formerly or sometimes today regular and important components of the agricultural and rural economy of Ireland. In some cases such plants have been included when ethnobotanical usage of the plant is known in Ireland for purposes other than their primary use for food. In recent years some new biomass crops have also been introduced, such as elephant grass (*Miscanthus* species) and some willows, but these have not been included either. Although an extremely important part of Ireland's agricultural economy, all these plants generally fall beyond the scope of this book.

Bellis perennis (daisy). Watercolour by Lydia Shackleton. Courtesy of the National Botanic Gardens, Glasnevin.

IRISH AND OTHER VERNACULAR NAMES

The usage and application of common names both in English and Irish are somewhat controversial. Many species have had English common names created for them over the last few decades, which have never been in widespread vernacular usage. Many of these are merely modern translations of existing scientific (Latin) or English names. I have not attempted to determine or document whether a particular name had any traditional usage. Further complicating the issue is the fact that names given to plants varied greatly over time and in different parts of the country. My knowledge of the Irish language is inadequate to do justice to the task of outlining the origins, roots and meanings of Irish names of plants, and so I leave this work to others, although I have mentioned some meanings interpreted by others when the name seems to point to a former use of a particular plant. However, when I have found the names of plants in Irish in a wide variety of publications, and if I can be relatively sure that they relate to a specific species or genus or where the name has been applied to a particular species by a previous author, I have listed them in the Systematic List. In part, I have documented these names and their sources to provide an easy reference for others who will follow in ethnobotanical studies of the Irish flora and also because no other previous work on the Irish flora has included a wide survey of Irish plant names and their variants documented in major dictionaries.

For the English names, the first name listed is generally the 'official' name given in the *Census Catalogue* (Scannell & Synnott, 1987), which documents both English and Irish names of plant species. I have also included those English names that are (or were) probably used in the vernacular language in Ireland as well as additional names given in the Irish botanical literature. Some of the names were (or are) in local or limited usage only. I have also given many of the common vernacular names recorded by Colgan (1904), which he collected in the Dublin region and listed in his *Flora of the County Dublin* and others that he recorded in Co. Mayo (Colgan, 1914). Henry Chichester Hart included a valuable appendix of English plant names (sometimes these are anglicised forms of Irish names) in his *Flora of Donegal* (1898) and I have listed many of these as well.

Henry Chichester Hart (left) and Nathaniel Colgan (right) produced works important to the documentation of Irish native plants and their vernacular names at the turn of the 20th century.

Additional English names for plants in Ireland are provided in several other works, including Geoffrey Grigson (1958), *The Englishman's Flora*; Niall MacCoitir (2003), *Irish Trees: Myths, Legends & Folklore*; and Niall MacCoitir (2006), *Irish Wild Plants: Myths, Legends & Folklore*. However, I have not included many of these names in this work, because the source of these names is often not clear. I also have not referred to the multitude of local English names that have been applied to plants in different regions of England and elsewhere, most of which have probably been little used in Ireland. It is not always easy to be certain whether particular common names are in actual usage today in Ireland. Further work on this aspect of ethnobotany in Ireland would be useful and rewarding, as would an investigation of what common names are used locally in different parts of the island today.

In the case of Irish names, the first name given in the Systematic List for each species is that used in the *Census Catalogue*, the 'official' Irish name (Scannell & Synnott, 1987). When the same name is given in other works I have not noted these additional references. Other names in Irish recorded in the published literature (that I have been able to find) are as given in the following publications (many of which are minor variants, sometimes used or formerly used, in different parts of Ireland):

- Anonymous (1978). *Ainmneacha Plandaí agus Ainmhithe* (*Flora and Fauna Nomenclature*). An Roinn Oideachais, Oifig an tSoláthair, Baile Átha Cliath.
- Cameron, John (1883). *The Gaelic Names of Plants*. William Blackwood and Sons. Edinburgh and London. (This work lists the Scottish Gaelic names of plants but also gives many Irish and Welsh names too. Many of the Irish and Scottish Gaelic names are very similar.)
- Colgan, Nathaniel (1911). Gaelic Plant and Animal Names, and Associated Folk-Lore. *Proceedings of the Royal Irish Academy. Section B: Biological, Geological, and Chemical Science, 31, A Biological Survey of Clare Island in the County of Mayo, Ireland and of the Adjoining District* (*Sections 1–3*) *1911–1915*, pp. 4.1–3.30.
- De Bhaldraithe, Tomás (Ed.) (1959). *English—Irish Dictionary*. Oifig an tSoláthair, Baile Átha Cliath.
- de Vere Coneys, Thomas (1849). *Focloir Gaoidhilge-Sacs Bearla, or An Irish-English Dictionary, intended for the use of students and teachers of Irish*. Hodges and Smith, for the Irish Society, Dublin.
- Dinneen, Patrick S. (1927). *Foclóir Gaedhilge agus Béarla: An Irish-English Dictionary*. Educational Company of Ireland for the Irish Texts Society, Dublin.
- Kelly, Fergus (1997). *Early Irish Farming*, School of Celtic Studies, Dublin Institute for Advanced Studies, Dublin.
- Ó Criomhthain, Tomás (MSS, n.d., ca. 1908). A manuscript list of Irish plant names compiled for the Norwegian scholar, Carl Marstrander, by the Islandman, Tomás Ó Criomhthain, in the early 1900s, from Kerry's Great Blasket Island. Located in Oslo University Library, Oslo.
- Ó Dónaill, Niall (1977). *Foclóir Gaeilge-Béarla*. Oifig an tSoláthair, Baile Átha Cliath.
- Moloney, Michael F. (1919). *Irish Ethno-botany and the Evolution of Medicine in Ireland*. M. H. Gill & Son, Ltd, Dublin.
- Tóibín, Seán (1965). *Blátha an Bhóithrín*. Oifig An tSoláthair, Baile Átha Cliath.
- Ua Maoileoin, Pádraig (2001). *Dúlra Duibhneach* (Ed. 2), Binn Éadair, Baile Átha Cliath.
- Uí Chanainn, Eibhlín (1962). *Foclóir an Mhic Léinn: Scholar's Irish-English Dictionary*. An Press Náisiúnta, Baile Átha Cliath.
- Uí Chonchubhair, Máirín (1995). *Flóra Chorca Dhuibhne: Aspects of the Flora of Corca Dhuibhne*. Oidhreacht Chorca Dhuibhne, Baile an Fheirtéaraigh, Co. Kerry.

- Wade, Walter (1808a). *Sketch of Lectures on Meadow and Pasture Grasses Delivered in the Dublin Society's Botanical Garden, Glasnevin*. Graisberry and Campbell, Dublin. (My copy of this publication is bound into a volume of Wade's *Plantae Rariores in Hibernia Inventae*, 1804. A contemporary note stuck into my copy largely attributes the Irish names in this article to John White, a gardener at the Botanic Gardens).
- Wade, Walter (1808b). *Sketch of Lectures on Artificial or Sown Grasses as Lucern, Saint-Foin, Clovers, Trefoils, Vetches, &c. &c*. Graisberry and Campbell, Dublin. (In my copy bound with Wade, 1804).
- White, John (1803). *An Essay on the Indigenous Grasses of Ireland*. Graisberry and Campbell, Dublin.

Early Irish plant names are given in such important publications as Threlkeld's *Synopsis Stirpium Hibernicarum* (1726) and K'Eogh's *Botanalogia Universalis Hibernica* (1735). I have listed these separately (see Appendix 1). The similarities between these works in relation to Irish names and the polynomial Latin names applied to each species are significant and suggest that K'Eogh had access to Threlkeld's work or the same sources used by Threlkeld while preparing his own book. Threlkeld (1726: unnumbered [17–18]) said that 'As to the Irish names I copied them from a Manuscript; which bears great Authority with me, and seems to be written sometime before the civil Wars in 1641 and probably by that Revd. Irish Divine, Mr. Heaton'. Williams (1987) has pointed out that K'Eogh even copied some of the errors in Threlkeld's work. In the Preface to his book, K'Eogh wrote: 'I take it to be much better to write the Irish names of the herbs in Roman characters, as they are pronounced, not as they are generally spelt. For one, that does not understand to read the Irish language (as there are a hundred to one that do not) could never make sense of them, there being so many letters which have no sound at all'. K'Eogh's Irish names of plants are therefore presented phonetically, both in his work and in the list I have compiled from his book. Cameron (1883: 118) was unimpressed, though, by K'Eogh's work, saying that it was 'of no value save for the names, for they contain no information except the supposed medicinal virtues of the plants … given in them' and 'a large number of K'Eogh's prescriptions are compounds of the most disgusting ingredients'.

Earlier works in the Irish language also give Irish names. One example is work by Richard Plunkett (Risdeard Pluincéad), a Franciscan living in the Friary of Trim, Co. Meath, who finished the first great bilingual dictionary in Latin and Irish in 1662. This was never published but is held in Marsh's Library in Dublin. Subsequent authors made use of this work, such as Edward Lhuyd (1660–1709), a Welsh linguist and naturalist who made extensive use of Plunkett's dictionary, published in 1707, *Archaeologia Britannica: Giving Some Account Additional to What Has Been Hitherto Publish'd, of the Languages, Histories and Customs of the Original Inhabitants of Great Britain*. Lhuyd's work contains lists of words from various Celtic languages. I have not reviewed these works to extract Irish plant names.

During the 18th century several major English–Irish dictionaries and Irish–English dictionaries were compiled:

- In 1732, the first comprehensive English–Irish dictionary was published and compiled by Conchubhar Ó Beaglaoich, a priest and tutor in Paris, assisted by Aodh Buí Mac Cruitín, a scholar from Co. Clare.
- In 1739 Tadhg Ó Neachtain published an Irish–English dictionary.
- In the 1760s a large manuscript was completed by Mr Crab, a school master from Ringsend in Dublin of an English–Irish dictionary that remained in manuscript form and was never published.
- In 1768, in Paris a similar Irish–English work was published by Seán Ó Briain, Bishop of Cloyne and Ross.

In the 19th century other English–Irish and Irish–English dictionaries were published that included plant names in Irish:

- Tadhg Ó Coiniallláin's English–Irish dictionary (1814).
- Edward O'Reilly's Irish–English dictionary, first published in 1817 and republished in subsequent editions in 1821 and 1864.
- Robert MacAdam's English–Irish Dictionary (unpublished but completed ca. 1850).

SOME YOLA WORDS RELATED TO PLANTS AND PLANT PRODUCTS[1]

Baarich, Barish	Barley
Baulkès	Timber (unsawn)
Beanès	Beans
Bidaades	Potatoes
Boouchelawn	Ragwort (*Senecio jacobaea*)
Brimel, Brimelès (pl.)	Briar
Bremel, Brimelès (pl.)	Bramble
Busheen	Bushes (growing)
Copprouse	Red poppy (*Papaver rhoeas*)
Correate, Coureate Correatès (pl.)	Carrot
Detch	Thatch, to
Dhernapès	Turnip
Dhrashel	Flail, a
Dhrivès	Turf
Ellena ghou	Elder tree (*Sambucus nigra*)
Fearn	Fern
Goss	Gorse (*Ulex* species)
Greve	Grove of trees
Harnothes	Pignuts
Hey, Hye	Garden, field or enclosure
Hye	Hay
Kaayle	Cabbage
Koorn	Corn
Kraanberry	Gooseberry (*Ribes uva-crispa*)
Levers	Flag iris (*Iris pseudacorus*)
Measkeen	Basket[2]
Meelough, Milagh	Trefoil, Clover (*Trifolium* species)
Neape, Neapès (pl.)	Parsnip
Neape, Neapès (pl.)	Turnip
Piz, Pizzen	Pea
Rusheen	Rushes
Spraay	A twisted twig for thatching
Strat	Rope of straw or hay
Stre, Strew	Straw
Thrive	Sod of turf
Tren	Trees

1. Poole, Jacob (1867). *A Glossary of the Old Dialect of the English Colony in the Baronies of Forth and Bargy, County of Wexford, Ireland.* J. Russell Smith, London.
2. Flat-bottomed for draining potatoes.

In the 20th century a range of English–Irish dictionaries were published as were several major Irish–English dictionaries. The English–Irish dictionaries were produced by Edmund Fournier d'Albe (1903), Timothy O'Neill Lane (1904, 1918), Lambert McKenna (1935) and Tomás de Bhaldraithe (1959). The Irish–English dictionaries included Father Patrick S. Dinneen's famous work first published in 1904 and a subsequent edition in 1927[27]. There was also *Contributions to a Dictionary of the Irish Language* (1913–1976) published by the Royal Irish Academy, which included both Old and Middle Irish, and Ó Dónaill's *Foclóir Gaeilge-Béarla* (1977). Koch (2006) provided a useful list of both manuscript and published Irish language dictionaries.

A more comprehensive trawl through all of these works would probably add many more Irish plant names than I have listed, but this is a huge task and not one which I have attempted. Nevertheless, many other authors listing Irish names may have been able to benefit from the earlier published and unpublished work of others. For example, Moloney (1919) listed many of the same names as Cameron (1818) but does not reference that work in his *Irish Ethnobotany*, nor does he say which other works he consulted. So we cannot be sure what sources he was able to draw upon. Similarly, Cameron listed the sources of only some of the names he gave in his book on *The Gaelic Names of Plants*, but merely by name (e.g., 'O'Reilly', 'Shaw', 'Fergusson', 'Armstrong', etc.). Many of the names given by Moloney (1919) are also listed in Dinneen (1927), and so on.

In some cases the same Irish name has been applied to several different species. In those instances I have listed all of the species to which it has been applied. In others the same name with minor variations in spelling has also been used in different publications. I list many of these spelling differences.

A handful of plant names are also included in 'Yola'. This is an extinct dialect of Middle English that survived up to the 19th century in parts of Co. Wexford. The names are as recorded in Jacob Poole's glossary, which described the old dialect used in the Baronies of Forth and Bargy and listed some 1700 words, phrases and songs (Dolan & Ó Muirithe, 1979).

Finally, a number of words in Shelta, the Traveller cant, related to plants and plant products are also included, as well as a very few plant names. Many are derived from an on-line dictionary of Shelta (<http://www.travellersrest.org/sheltanocant990418.htm>).

SOME SHELTA WORDS RELATED TO PLANTS AND PLANT PRODUCTS

Brouen	Grain, corn
Galapa	Fern
Glask	Grass
Grayer	Hay
Greenlesk	Flax
Greesk	Straw
Groola	Apple
Gweesh	Hay, straw
Kawb	Cabbage
Kogi	Turnip
Koldni	Bud
Lashool	Flower
Loorp	Flour
Muntches	Tobacco
Noabyri	Bog, swamp
Raagli	Garden
Raib	Hay
Reesbath	Basket
Rooski	Basket
Shee-u	Sow seed
Sheerk	Grass
Shkrawkh	Tree
Sloofa	Wood
Slukhul	Greenwood, copse, grove
Snail	Rush
Sorsh	Resin
Sreedja	Wine
Strumble	Straw
Surgu	Thorn
Taryiu	Rope

Source (2010): <http://www.travellersrest.org/sheltanocant990418.htm>.

27. In a copy of Dinneen's dictionary held in the Library at the National Botanic Gardens in Glasnevin, a hand-written note has been stuck into the front of the book that says 'I have found the names of plants given in Dinneen's dictionary to be unreliable. P. O'Connor, 13/11/1944'. Be that as it may, it is still one of the most comprehensive and important published sources of Irish names of plants.

Acer pseudoplatanus (sycamore). Watercolour by Lydia Shackleton. Courtesy of the National Botanic Gardens, Glasnevin.

AN ETHNOBOTANY OF WILD PLANTS IN IRELAND: A SYSTEMATIC LIST

IRISH WILD PLANTS are listed below in alphabetical order by genus and species within four major groups: flowering plants, conifers, ferns and fern allies, and algae and miscellaneous. Family names used in capital letters are those generally recognized by the Angiosperm Phylogeny Group (APG) III (2009) and are accepted herein. Notable exceptions include the Orobanchaceae (broomrapes) and Veronicaceae (speedwells), which are distinctive families segregated from the Scrophulariaceae s.l. The Valerianaceae (valerians) are recognized separately from the Caprifoliaceae, and the Montiaceae (blinks) are removed from the Portulacaceae. Other family names that have been used in the flora area are listed in parentheses after the name in capital letters. In general, family name usage also follows *New Flora of the British Isles* (Stace, 2010). Species preceded by an asterisk (*) are those for which there are no known uses, past or present, or for which uses have yet to be located and documented.

The first given common English and Irish names are generally those included by Scannell and Synnott (1987) in *Census Catalogue of the Flora of Ireland*. Other English names are included where I have come across their use in Ireland in a wide variety of sources. Generally, the most common source is given first. I have not included the multitude of English vernacular names in some publications from Great Britain; most of these names have never (or rarely) been used in Ireland.

Scientific names for marine algae are as given in Hardy and Guiry (2003).

Plant distribution and typical habitat information are derived from a variety of sources, including the author's personal knowledge, as well as from Clapham et al. (1962), Webb and Scannell (1983), Webb et al. (1996), Doogue et al. (1998) and Preston et al. (2002). Online resources now exist from the Botanical Society of Britain and Ireland (BSBI) and the Biological Records Centre (BRC) and include the *Online Atlas of the British & Irish Isles*, <http://www.brc.ac.uk/plantatlas>, with mapped detail for plant collections found at <http://www.bsbimaps.org.uk/atlas/main.php>. Information from these sites comes mainly from the *New Atlas of the British and Irish Flora* (Preston et al., 2002) and *PLANTATT—Attributes of British and Irish Plants: Status, Size, Life History, Geography and Habitats* (Hill et al., 2004).

Major Categories of Plant Use in Ireland

Plants that are known to be poisonous or with poisonous parts

 Edible or food plants

 Plants used for medicinal purposes

 Plants used as ornamental plants in gardens and for horticulture

 Plants used as a source of dyes

 Plants used for teas and other beverages

 Plants used for fibres, including wicker, matting, ropes and basketry

 Plants used as aphrodisiacs and love potions

 Plants used for timber

 Plants used for children's games

 Plants used for wine and beer making

 Plants used as a tobacco substitute

 Plants used for tanning leather

 Plants used for veterinary purposes

 Plants used for thatching

 Plants used for cosmetics

 Plants used for making musical instruments

FLOWERING PLANTS

Acer pseudoplatanus

- Sycamore (Scannell & Synnott, 1987); Great maple (Harron, 1986)
- *Seiceamóir* (Scannell & Synnott, 1987); *Crann bán* (Webb & Scannell, 1983; Webb et al., 1996); *Crann sicimír* (Dinneen, 1927); *Crann siceamóir* (Dinneen, 1927)

SAPINDACEAE (Aceraceae)—THE MAPLE FAMILY
Introduced

The sycamore is one of Ireland's most common trees, introduced more than four centuries ago. Although it is not a native species, *Acer pseudoplatanus* is nevertheless established and naturalised throughout Ireland as a common element of the wild flora of woodlands and hedgerows and as a weed in gardens. It was grown as early as about 1610 in Derry (Nelson & Walsh, 1993). It self-sows freely in most habitats, grows quickly and can be a real nuisance in many gardens. However, a mature sycamore can grow into a beautiful and stately tree, providing shelter, useful timber and important habitat for native wildlife. Some exceptional sycamore trees can be found in Ireland, such as one specimen growing at Money House, Coolkenna, Co. Wicklow, which measures 19.5 m tall and 5.2 m in girth (Anonymous, 2005). The tallest known sycamore in Ireland is recorded at Shelton Abbey, Arklow, Co. Wicklow, and is 29.5 m in height (Anonymous, 2005). The current girth champion of this species in Ireland is a specimen from Gormanstown, Co. Meath, that has been measured at 8.03 m in girth (and 25.5 m tall) (Anonymous, 2005).

In some parts of Ireland where native woodlands are rare, the sycamore has become an important component of regenerated woodlands, such as in river valleys, along stream banks and railway embankments, and in waste places, where it grows along with ash, alder, willows, birch and other tree species. In some ways it replaces oaks as canopy dominants, which are often not available locally to recolonise land that is no longer used for agriculture. In exposed coastal areas sycamore is sometimes planted as a windbreak because it is very resistent to salt-laden winds. Sycamore and other *Acer* species are popular trees with bee-keepers as important nectar sources (John McLoughlin, pers. comm., 2008).

Sycamore wood is fine grained, quite hard and heavy and light in colour, and has traditionally been used for turnery and to make fine implements, particularly for the kitchen, such as breadboards, rolling pins, platters and the rollers of old-fashioned laundry mangles. In my collection I have a potato masher (or potato beetle), an elegantly turned club-like implement, made from sycamore wood. The best old kitchen tables were made of sycamore and could be regularly scrubbed clean. Sycamore wood was used to make the O'Carolan Harp, which dates from the late 17th or early 18th century, that is now in the National Museum of Ireland (Rimmer, 1969). The neck, forepillar and one-piece soundbox are all of sycamore wood. Because of its interesting decorative patterned grains, sycamore wood has also been used to make the backs of violins. Another use for sycamore was by the linen industry in Ireland to provide the great beams in the engines of beetling

Acer pseudoplatanus.

mills. Beetling is a finishing process for linen that involves pounding the textile with heavy weights (Evans, 1957).

Several *Acer* species are used as a source of sugar syrup made from the tree sap, such as *A. saccharum*, (sugar maple), *A. nigrum* (black maple) and occasionally *A. rubrum* (red maple) and *A. saccharinum* (silver maple). A British native tree, *A. campestre*, the field maple, can also be harvested for its sap. Most *Acer* species contain relatively high levels of sugar in their sap, ranging from 2% to 5% (Ciesla, 2002). During the early spring when the sap is rising, a small hole may be bored into the trunk of the tree, at an upwards slope and to a depth of 5–7.5 cm (2–3 in.). The sap then runs off into containers, after which it may be boiled gently to reduce its volume by as much as 30 to 40 times. During boiling, great care needs to be taken not to burn the syrup. After collecting the sap, the hole bored into the tree must be plugged. I presume that the common sycamore can also be used to produce syrup, but I have not tried it, nor have I heard of it being used in this way in Ireland.

The winged fruits of sycamore have been used in a popular game for children in Ireland. They were thrown in the air and watched as they drop, spinning like helicopters.

In Irish, a maple tree is called *mailp* (Dinneen, 1927; Uí Chanainn, 1962). Earlier, the words *mailp* or *craobh mhalip* referred to *Acer campestre*, or field maple (Cameron, 1883).

Achillea millefolium

- Yarrow (Scannell & Synnott, 1987); Tansy (Colgan, 1904); Lady's fingers (Colgan, 1904); Elfshot cow (Hart, 1898); Milfoil (Praeger, 1938)
- *Athair thalún* (Scannell & Synnott, 1987); *Luibh an tinneas droma* (De Bhaldraithe, 1959); *Luibh na ndaitheacha* (Ó Dónaill, 1977; Anonymous, 1978); *Athair talmhan* (Moloney, 1919; Dinneen, 1927); *Lus na fola* (Moloney, 1919; Dinneen, 1927); *Lus na gcluas* (Moloney, 1919; Dinneen, 1927); *Athair talmhuin* (de Vere Coneys, 1849); *Athairthalmhuin coitchion* (Wade, 1808b)

ASTERACEAE (Compositae)—THE DAISY FAMILY

Native

Achillea millefolium is a common perennial herb of grasslands, roadsides and waste-ground throughout Ireland.

Achillea millefolium.

Yarrow traditionally has been one of the most popular and widely used medicinal herbs in Ireland. Lady Wilde (1887) observed that of all herbs yarrow is the best for cures and potions, and it was even sewn up in clothes as a preventive of disease. Its use to treat arthritis is still quite widely remembered in Ireland today. Throughout Ireland and elsewhere in the world it was formerly widely known and used for its haemostatic properties. Even today, yarrow remains a commonly used herbal remedy in many countries (Shaw, 1998). Allen and Hatfield (2004) noted it was used in Ireland as a poultice or ointment to staunch bleeding from cuts and wounds, scratches and sores. These same authors also noted that in some parts of Ireland it was used to induce nosebleeds and as a cure for migraines, headaches and toothaches (chewing the leaves). Moloney (1919: 29) indicated a similar use to treat congestive headaches, and that when dried and made into a powder, yarrow was used as snuff, to 'draw blood from the nose'. Smith (1815) recorded that in Co. Cork it was used with success in curing agues (sharp fever). In Co. Cavan yarrow leaves ground up with fresh grass and mixed with egg white and left undisturbed for 48 hours was regarded as a good treatment to prevent infections (Ui Sheasnáin, 1982). In Co. Kerry the flowers of yarrow were boiled in water and the liquid drunk as a tonic for weak blood and to improve circulation. The flowers were also recommended as a cure for rheumatism and 'pains in the bones' (O'Sullivan, 1994: 122). According to Egan (1887), a decoction of yarrow was also used to treat kidney diseases. Another use was noted by Logan (1981), in which yarrow leaves were included with grass and other herbs as ingredients in a poultice made to treat a whitlow (an infected tendon sheath on a finger). Yarrow is widespread in Ireland, easily recognized and a popular plant for home herbalists in tonics to treat colds and rheumatism. Yarrow tea can be brewed as a remedy for severe colds (Phillips, 1983). To make yarrow tea, the dried leaves should be steeped in hot water for 10 to 15 minutes. It should be noted that fresh plants can cause dermatitis in sensitive people (van Wyk & Wink, 2004). Elsewhere, yarrow has a long history of medicinal use. For example, the Chippewas of Minnesota and Wisconsin in the United States used yarrow as a diuretic and stimulant tonic, as well as to treat headaches (Densmore, 1974), and the Forest Potawatomi people of North America burned the florets to reduce fevers and revive comatose patients (Pennacchio et al., 2010).

An early beer called 'gruit' was brewed throughout much of continental Europe, but probably not in Ireland. In this beer a variety of herbs were used for flavour, including yarrow (as well as bog myrtle, *Myrica gale*, and marsh rosemary, *Ledum palustre*). *Ledum palustre* is not a native plant in Ireland. In the 1700s the Swedish botanist Carl Linnaeus suggested in his work *Lachesis Naturalis* that drinks made from yarrow caused madness, and other authors have suggested that it considerably enhances the intoxicating effect of beer.

Another use for yarrow was as a tobacco substitute, where the leaves were dried and smoked (Williams, 1993). It was also smoked in pipes to relieve toothaches (Fargher, 1969).

The leaves were used in veterinary remedies in Donegal and pounded with salt and water and given to a cow to treat an ailment called 'elfshot cow' (Hart, 1898). In Co. Cavan it was used in mixtures with other herbs for veterinary purposes (Allen & Hatfield, 2004).

MacMahon (1994) wrote that yarrow was hung in the home on St John's Eve in Ballyheigue, north Kerry. St John's Eve is an important cultural and religious date that is celebrated on 24 June and coincides with the summer solstice. In several cultures, including in Ireland, it was a time when festivals were held, bonfires lit and herbs gathered for their medicinal properties.

In some parts of the world yarrow has been burnt as an insect repellant (Pennacchio, 2010), and there are some commercially available insect repellants whose active compounds, called sesquiterpenes, are derived from yarrow (Milliken & Bridgewater, 2004).

Mabey (1972) suggested that its leaves can be used as a vegetable, removed from the stem, boiled for about 20 minutes, strained and then simmered in butter.

In North America, Navajos used it as an aphrodisiac and drank tea made from it or chewed its leaves one to two hours before intercourse (Rätsch, 1997). In Ireland it had an interesting related folklore use, recorded by Lady Wilde (1887) and H. C. Hart (1898) in Cos. Derry and Donegal. Hart (1898: 391) wrote of the following charm involving yarrow to foretell future loves:

> Young people used to gather yarrow on May Eve and pull ten stalks. At every stalk the charm given below was repeated. The tenth stalk was thrown away over the left shoulder. The other nine were put into the left stocking, and placed under the pillow. No word must be spoken from the pulling of the yarrow till the following morning, and the sleeper will dream of future wife or husband.
>
> Good morrow, good morrow, fair yarrow,

Thrice good morrow to thee;
I hope, before this time to-morrow,
You'll show my true lover to me.

A range of cultivated varieties of yarrow with white or pink flowers are grown as herbaceous perennials in Irish gardens.

Moloney (1919) and O'Regan (1997) listed the alternative Irish names *lus na fola* and *lus na gcluas* (the 'blood herb' and the 'ear herb'), suggesting its former medicinal usage.

Achillea ptarmica

- Sneezewort (Scannell & Synnott, 1987); Wild fire (Hart, 1898); White weed (Praeger, 1938)
- *Lus corráin* (Scannell & Synnott, 1987); *Roibhe* (Moloney, 1919; Dinneen, 1927; Uí Chonchubhair, 1995); *Cruadhlus*[1] (Moloney, 1919; Dinneen, 1927)

ASTERACEAE (Compositae)—THE DAISY FAMILY
Native

Achillea ptarmica is an occasional perennial herb throughout Ireland of damp habitats, such as lake-shores, cultivated ground and waste places. It is found more frequently in the northern half of Ireland than in the south.

The common name sneezewort comes from its former use as a medicinal plant to help clear the head of mucus (when the leaves were dried and made into snuff). According to Uí Chonchubhair (1995), it was also used to treat headaches.

Aconitum napellus

- Monk's-hood (Harron, 1986); Wolfsbane (or Wolf's-bane) (Smith, 1882)
- *Dathabha dubh* (Moloney, 1919; Dinneen, 1927; Ó Dónaill, 1977); *Fuath an mhadra* (Ó Dónaill, 1977)

RANUNCULACEAE—THE BUTTERCUP AND CROWFOOT FAMILY
Introduced

Aconitum napellus is a commonly grown and attractive garden plant that is occasionally found in the wild in Ireland as a garden escape or as a discard (Reynolds, 2002). It is easily recognizable by its attractive racemes of blue to violet hooded flowers. Known since antiquity for its lethality, *A. napellus* and other species have been cultivated from wild populations across temperate Europe. Beyond its historic use as a weapon and cardiac poison, as well as a favorite in literary poisoning, monk's-hood has been used therapeutically as a numbing ointment and potent tincture in homoeopathic medicine. Among its alkaloids are aconitine, mesaconitine and jesaconitine, which affect the

1. Dinneen (1927) also gave this name for white hellebore (presumably *Veratrum album* or *V. viride*, which are non-native plants occasionally grown in Irish gardens).

cardiac and central nervous systems. *Aconitum napellus* is lethally poisonous and definitely should not be consumed in any form. Other monk's-hood species are used in Asian traditional medicine at precise and miniscule concentrations. The similar roots of monk's-hood can be mistaken for horse-radish, so beware. Indeed, the murder mystery novel *Common or Garden Crime*, by the Irish author Sheila Pim, had a plot that was based around the victim being given cold roast beef, together with horse-radish sauce to which grated *Aconitum* root had been added (Pim, 1945). Accidental poisonings from mistaken horse-radish sauce date back to the mid-19th century, and a particular instance was noted from Newry, Cos. Armagh and Down, where a prisoner died in four hours after accidentally receiving 1 drachm of tincture of aconite rather than one of chiretta (Headland, 1856: 19–20). Even modern herbals caution: 'It is wiser not to grow Aconite among kitchen herbs of any sort. The root has occasionally been mistaken for horse-radish, with fatal results—it is, however, shorter, darker and more fibrous—and the leaves have produced similar fatal results. In Ireland a poor woman once sprinkled powdered Aconite root over a dish of greens, and one man was killed and another seriously affected by it.' (Grieve, 1931).

Acorus calamus

- Sweet-flag (Scannell & Synnott, 1987)
- *Feileastram cumhra* (Scannell & Synnott, 1987)

ACORACEAE (Araceae)—THE SWEET-FLAG FAMILY

Introduced

Acorus calamus is an attractive and decorative species that is used as an ornamental plant in Ireland for aquatic gardens, ponds, stream and canal banks and lakes. It occasionally escapes into the wild, particularly in the east, but is nowhere common. It is native to eastern Asia but now widespread in many temperate regions of the world.

Sweet-flag has been widely used as a medicinal plant in other parts of the world. Material of this species has been found, for example, in Tutankhamun's tomb in Egypt. The rhizome has been used to stimulate the appetite and treat indigestion and flatulence. Today, its use in digestive medicines has been discontinued in some countries because of possible harmful effects (van Wyk & Wink, 2004). In North America, the Chippewas used it to treat colds, and other indigenous peoples used it for fevers and coughs (Densmore, 1974). In parts of India it was used as a fumigant for the relief of pain associated with piles (haemorrhoids), and in West Bengal, smoke from burning rhizomes was inhaled by people suffering from epilepsy and hysteria (Pennacchio et al., 2010). The same authors mention the use of this species by the Iroquois of North America, who smoked the roots to relieve toothache; after inhaling the smoke, they sucked it into the hollows of the decayed teeth.

The rhizome of sweet-flag can be candied (see *Angelica sylvestris* also for details of the method). It is boiled in several changes of water for about an hour until tender, then simmered in a rich sugar solution for about 20 minutes before being dried (Peterson, 1977). The inner parts of spring shoots are also edible and can be used in salads, particularly early in the year when they are somewhat more tender.

Acorus calamus also has been used for its fragrance, and Cameron (1883) observed that the leaves were used to strew the floors of houses, churches and monasteries (presumably in Scotland). I have not found records of its use in this way in Ireland, though.

Adoxa moschatellina (moschatel). Watercolour by Lydia Shackleton. Courtesy of the National Botanic Gardens, Glasnevin.

**Adoxa moschatellina*

- Moschatel (Scannell & Synnott, 1987); Town Hall clock (Hackney, 1992); Five-faced bishop (Clapham et al., 1962)
- *Moscadal* (Scannell & Synnott, 1987)

ADOXACEAE—THE MOSCHATEL FAMILY
Introduced

Adoxa moschatellina is a small perennial herbaceous plant with heads of little green flowers and 2- or 3-lobed leaves. It is found only in scattered localities in woodland habitats throughout Ireland and is much more common in Britain and temperate Europe.

*No ethnobotanical uses of this family have been noted in Ireland.

Aegopodium podagraria

- Ground-elder (Scannell & Synnott, 1987); Goutweed (Scannell & Synnott, 1987); Bishop's-weed (Scannell & Synnott, 1987); Elder-plant (Hart, 1898); Farmer's plague (Praeger, 1938)
- *Lus an easpaig* (Scannell & Synnott, 1987); *Lus an ghúta* (Moloney, 1919; Dinneen, 1927; Ó Dónaill, 1977); *Lus an easbuig* (Moloney, 1919); *Lus an easpuig* (Dinneen, 1927); *Gúta* (Ó Dónaill, 1977)

APIACEAE (Umbelliferae)—THE CARROT FAMILY
Introduced (possible archaeophyte)

Aegopodium podagraria is a common weed throughout Ireland of woodlands, gardens and waste places and is often found growing near houses.

This plant was used medicinally in Ireland and boiled leaves were made into a poultice and applied to relieve pain, such as from sciatica (Moloney, 1919). According to Threlkeld (1726), this plant was also formerly used in Ireland to treat gout.

Throughout Europe during the Middle Ages it was cultivated and used as a potherb and spinach-like vegetable. It has a strong flavour when cooked but is said to be quite tasty. The leaves and stems can also be used raw in salads or added to soups (Hilton, 2007). It is best collected and used in spring when the leaves are the most tender.

Aesculus hippocastanum

- Horse-chestnut (Scannell & Synnott, 1987)
- *Crann cnó capaill* (Scannell & Synnott, 1987); *Cran-geanmchno* (de Vere Coneys, 1849)

SAPINDACEAE (Hippocastanaceae)—THE HORSE-CHESTNUT FAMILY
Introduced

Aesculus hippocastanum is a familiar tree throughout Ireland and is commonly planted as an ornamental in large gardens and parkland, or as a street tree. It occasionally self-sows into wild habitats such as hedgerows. The tree was introduced to Western Europe in

1576 from the Balkans where it is native. The tallest horse-chestnut tree known in Ireland is recorded from Johnstown Castle, Johnstown, Co. Wexford, and has been measured at 29 m tall (Anonymous, 2005). The current girth champion of this species in Ireland is a specimen from Lough Key Forest Park that has been measured at 6.29 m in girth (and 22.5 m in height) (Anonymous, 2005).

Aesculus hippocastanum.

Its fruits are large and covered in spines and contain one or two large brown seeds (conkers). These have been used for several generations in Ireland for a childhood game where the conkers are pierced and hung on a string and used to battle other conkers by bashing them together. Each participant takes a turn to try to hit his opponent's conker, and the last one to remain on its string is the winner. When I played with conkers in Limerick in the early 1960s, the number of victories for a particular conker was often recorded so that it became a 'one-er', 'two-er', etc. If a victory was achieved over another conker, which itself had been successful, its total was added to the winning conker so that high numbers of conquests were often recorded for particularly illustrious seeds. Great efforts were made to artificially toughen conkers before use, such as by soaking them in vinegar and other substances, baking them in the oven (or even the microwave), or piercing them and saving them for the following season, generally with only mixed success. Vickery (1995) noted that playing conkers became widespread in Britain and Ireland only early in the 20th century.

The collections of the National Museum of Ireland (Museum of Country Life) in Castlebar, Co. Mayo, include a necklace made of horse-chestnut seeds from South Ulster from the 1940s (Ref. No: F1944.61).

Horse-chestnut seeds and bark have been used medicinally, although probably not in Ireland. According to Maloney (1972), the leaves were used in Ireland as a cure for ague; the leaves were boiled and the water drunk. Extracts of the seeds have been used to treat varicose veins, ulcerations, piles, rheumatism and fevers (van Wyk & Wink, 2004). The seeds of this species are poisonous and contain saponins and glycosides, and so they should definitely not be eaten.

The horse-chestnut has not been having an easy time in Europe in recent years and is subject to three major pests. One is Guignardia leaf blotch (*Guignardia aesculi*), a fungus that causes reddish or dull brown irregular blotches on the leaves. These blotches can often be identified as *G. aesculi*, as they are surrounded by a conspicuous yellow border. The second disease is horse-chestnut bleeding canker, which is a bacterial attack by *Pseudomonas syringae* that results in black tarry exudates on the trunk and larger branches and which may eventually kill the tree. No treatment for this disease seems to be available. The third pest is the horse-chestnut leaf miner (*Cameraria ohridella*). This moth has been spreading throughout Europe over the last decade or so and arrived in the United Kingdom in 2002. It is heading for Ireland gradually. The moth lays its eggs on the leaf, which hatch and then burrow into the leaf and cause much of it to turn brown, in late summer and early autumn. The attack is unsightly and slows the growth of the trees but does not seem to cause its death.

Guignardia leaf blotch.

Horse-chestnut wood is white in colour, sometimes with a slight yellow tinge, light, with a very fine grain, not very durable, and little used for timber. In the past its timber has been used for a variety of purposes, including brush backs, bobbins, general turnery, dairy and kitchen utensils, fruit storage racks and trays, toys, box making and moulders' patterns (Anonymous, 1941). Its value has been the bone-whiteness of its colour and lightness. In Ireland today it is probably little used.

Rutty (1772) noted that the nuts were used for whitening hemp, flax and cloth, and also that starch may be made from them.

**Aethusa cynapium*

- Fool's parsley (Scannell & Synnott, 1987)
- *Peirsil amaide* (Scannell & Synnott, 1987); *Peirsil na n-amadán* (Dinneen, 1927)

APIACEAE (Umbelliferae)—THE CARROT FAMILY

Introduced (possible archaeophyte)

Aethusa cynapium is a common weed throughout Ireland and temperate Europe of waste places. Like so many other plants in this family, *A. cynapium* is quite poisonous but is less deadly than its larger relative, hemlock (*Conium maculatum*), which is associated with its historic use as a poison.

*No ethnobotanical uses of this plant have been noted in Ireland.

Agrimonia species

- Agrimony (Scannell & Synnott, 1987)
- *Marbhdhraighean* (Scannell & Synnott, 1987); *Airgeadán* (Dinneen, 1927; De Bhaldraithe, 1959)

ROSACEAE—THE ROSE FAMILY

Including:

Agrimonia eupatoria

- Agrimony (Scannell & Synnott, 1987)
- *Marbhdhraighean* (Scannell & Synnott, 1987); *Méirín na mágh* (de Vere Coneys, 1849; Moloney, 1919; Dinneen, 1927); *Geirbhileach airgeadán* (Moloney, 1919); *Marbh-dhruidhm* (Cameron, 1883); *Méirínína maighe* (Uí Chonchubhair, 1995); *Mar-dhroighean* (Dinneen, 1927)

Native

Agrimonia procera

- Fragrant agrimony (Scannell & Synnott, 1987)
- *Marbhdhraighean cumhra* (Scannell & Synnott, 1987)

Native

Both Irish species of agrimony are tall, hairy perennial herbs with bright yellow flowers, and are found occasionally throughout Ireland in rough rocky grasslands and roadsides. *Agrimonia procera* usually occurs on acid soils and is rarer than the more widespread *A. eupatoria*.

Agrimonia eupatoria was important in ancient times as a medicinal plant and used as an astringent herb to treat external wounds and stop bleeding. It has been used medicinally throughout much of its original range in Europe and the Near East. *Agrimonia procera*, or fragrant agrimony, has had similar uses in the same regions as well as in Asia Minor and Asia where it also occurs. However, present-day usages are largely confined to treating throat infections, diarrhoea, haemorrhoids, cystitis and urinary incontinence (van Wyk & Wink, 2004). According to Uí Chonchubhair (1995), *A. eupatoria* was highly regarded in ancient times for its medicinal value for cleansing and purging. The same author recorded that a drink made from the flowers was used to treat the common cold, and that the seeds and leaves crushed and mixed with hog's lard were used to draw out splinters from the skin. *Agrimonia eupatoria* also has been used as an anti-inflammatory. In Ireland the flower heads have been used in an infusion to treat skin rashes, stomach upsets, sore throats, rheumatism, haemorrhoids and circulation problems (Reader's Digest Association, 2006). It has also been used as an ingredient in a lotion for wounds (Moloney, 1919). It was formerly taken as a tea that was mildly stimulating and used as a tonic (Allen & Hatfield, 2004) and was used to treat various health complaints, such as colds, liver problems, jaundice, scurvy and ulcers (Threlkeld, 1726; K'Eogh, 1735). Williams (1993) recorded that agrimony was used medicinally in Ireland for broken bones, kidney complaints and weak bladders in boys, and in parts of Co. Galway, it was called 'wild tea' and used instead of tea in times of shortages. Hart (1898: 365) noted that in Donegal, *A. eupatoria* was 'much valued for weak or sore eyes'.

Agrimonia eupatoria also yields a yellow or yellow-grey dye (Mahon, 1983; Williams, 1993).

Agrostis species

- Bent grasses (Scannell & Synnott, 1987); Fiorin grasses
- *Feorainn* (Scannell & Synnott, 1987)

POACEAE—THE GRASS FAMILY

Including:

Agrostis canina

- Velvet bent (Scannell & Synnott, 1987)
- *Feorainn shlim* (Scannell & Synnott, 1987); *Taenfhér donn* (White, 1803)

Native

Agrostis capillaris (*Agrostis pumila*; *Agrostis tenuis*; *Agrostis vulgaris*[2])

- Common bent (Scannell & Synnott, 1987)
- *Feorainn mhín* (Scannell & Synnott, 1987); *Taenfheur coitchion* (Wade, 1808a); *Beinteach* (Ó Dónaill, 1977); *Taenfhér coitchion* (White, 1803); *Taenfhér abhach*[3] (White, 1803)

Native

2. Moloney (1919) listed the following Irish names under the older grass name *Agrostis vulgaris* (fine-leaved bent grass): *bíríneach*, *bi gíneach* and *biníneach*.

3. White (1803) applied this name to *Agrostis pumila*.

**Agrostis gigantea*

- Black bent (Scannell & Synnott, 1987)
- *Feorainn dhubh* (Scannell & Synnott, 1987)

Native (or Possibly Introduced)

Agrostis stolonifera (*Agrostis alba*)

- Creeping bent (Scannell & Synnott, 1987); Fiorin grass (Webb & Scannell, 1983)
- *Feorainn* (Scannell & Synnott, 1987); *Taenfheur ban*[4] (Wade, 1808a); *Taenfheur forin* (Wade, 1808a); *Feorna* (*Feora* or *Feorthainn*) (Ó Dónaill, 1977); *Taenfhér forín* (White, 1803); *Taenfhér bán* (White bent-grass) (White, 1803)

Native

There are four bent grasses recorded in Ireland. *Agrostis canina* is a common perennial grass of moorland, heaths and upland habitats found on peaty or acid soils throughout Ireland. It also occurs sporadically throughout temperate Europe and North America. Another bent grass, *A. capillaris*, is a perennial grass of dry mountain and moorland habitats and rough grasslands and is abundant throughout Ireland. This grass is used in some commercial wildflower seed mixes and included under the name 'brown top bent grass' from such suppliers as Irish Wildflower Growers, Crettyard, Co. Carlow. The tallest of the bent grasses is *A. gigantea*, reaching a metre in height. It is found sporadically in Ireland in waste places and lowland habitats and on cultivated ground. It is seen more commonly in the east in Cos. Wicklow to Waterford and is much rarer in the west. *Agrostis stolonifera* is one of Ireland's most abundant grasses and is found throughout the country in a variety of rough grasslands and waste places, such as damp meadows, wastegrounds, drains, ditches and some rocky seashores. It occurs throughout the Northern hemisphere in similar settings.

Agrostis stolonifera is considered by some authors (e.g., Allen & Hatfield, 2004) to be the 'hungry grass', or grassy patches on the mountainside that cause extreme hunger when stepped on. However, Clyne (1989) associated this superstition with *Briza media*. Logan (1981) said that the condition of hungry grass is likely to occur after a long day in the mountains when rounding up sheep, or more likely, shooting. The sufferer feels faint and weak, and in severe cases, may lose consciousness. Vickery (1995) suggested that it may have been caused by the consumption of alcohol on occasion too. He also proposed that hungry grass couldn't be attributed to any particular species. Lady Wilde (1887: 69) described a similar piece of Irish folklore:

> There is an herb, also, or fairy grass, called the *Faud Shaughran*, or the 'stray sod', and whoever treads the path it grows on is compelled by an irresistible impulse to travel on without stopping, all through the night, delirious and restless, over bog and mountain, through hedges and ditches, till wearied and bruised and cut, his garments torn, his hands bleeding, he finds himself in the morning twenty or thirty miles, perhaps, from his own home. And those who fall under this strange influence have all the time the sensation of flying and are utterly unable to pause or turn back or change their career. There is, however, another herb that can neutralize the effects of the *Faud Shaughran*, but only the initiated can utilize its mystic properties.

Danaher (1964a: 74) said that hungry grass 'cannot be distinguished from other kinds of grass', suggesting that this condition cannot be attributable to any particular plant species

4. Wade (1808a) applied this name to *Agrostis alba*. *Agrostis stolonifera* is generally a synonym of *A. alba* (although sometimes it is also considered a synonym of *A. gigantea*).

but is a purely legendary or magical. Hart (1898: 374) recorded the names *farragotha* or *firgotha* for hungry grass from Derry and Donegal.

Mackay (1836) noted that *Agrostis stolonifera* was used as a hay crop in Connemara on reclaimed bog and as manure for land where potatoes had been planted as a first crop and barley and oats the following year. Danaher (1964a) also suggested that bent grass was used for thatching in parts of Ireland where better materials such as straw and reeds were not so easily available. H. C. Hart (1898: 365) recorded that the 'fibrous roots of bent or some other sand plant [were] used in scouring pans, kitchen tables, etc.' in Fanet, Co. Donegal, under the name *barran a dhu* or *barradhu*. The name 'bent' has been applied to a range of plants, including *Ammophila arenaria* (marram), but is mainly understood for *Agrostis* species today. Dinneen (1927) gave the Irish name *muirleach* (*-ligh*), meaning a 'marsh' or 'puddle', as the name for 'a variety of long marine grass growing in wet locations and used in rope-making'. It is not clear what species this name was applied to, but it may have referred to *Agrostis stolonifera*.

**Aira* species

- Hair-grasses (Scannell & Synnott, 1987)
- *Mionfhéar* (Scannell & Synnott, 1987)

POACEAE—THE GRASS FAMILY

Including:

**Aira caryophyllea*

- Silver hair-grass (Scannell & Synnott, 1987)
- *Mionfhéar geal* (Scannell & Synnott, 1987); *Gruagfhér airgid* (White, 1803)

Native

**Aira praecox*

- Early hair-grass (Scannell & Synnott, 1987)
- *Mionfhéar luath* (Scannell & Synnott, 1987); *Gruagfhér moch* (White, 1803)

Native

Aira caryophyllea is a small annual grass species found in most parts of Ireland except in the midlands. It occurs in a variety of dry habitats, such as on walls and in dry banks and sandy waste places. *Aira praecox* is very similar in appearance to *A. caryophyllea* and is found in dry heaths and sandy waste places across the Irish landscape.

*No ethnobotanical uses of either species of hair-grass have been noted in Ireland.

**Ajuga pyramidalis*

- Pyramidal bugle (Scannell & Synnott, 1987)
- *Glasair bheannach* (Scannell & Synnott, 1987)

LAMIACEAE (Labiatae)—THE DEAD-NETTLE FAMILY

Native

Ajuga pyramidalis is very rare in Ireland and found mainly around Galway Bay and from Rathlin Island in the north, offshore from Ballycastle, Co. Antrim. Given its rarity, it is not surprising that no ethnobotanical uses of this plant have been documented.

Ajuga pyramidalis.

*No ethnobotanical uses of this plant have been noted in Ireland.

Alchemilla alpina.

Ajuga reptans

- Bugle (Scannell & Synnott, 1987); Betony (Hart, 1898; Praeger, 1938); Creeping bugle (Brunker, 1950)
- *Glasair choille* (Scannell & Synnott, 1987); *Meacan dubh fiadhain* (de Vere Coneys, 1849; Moloney, 1919; Dinneen, 1927); *Glasair choilleadh* (Moloney, 1919); *Lasair léana* (Dinneen, 1927)

LAMIACEAE (Labiatae)—THE DEAD-NETTLE FAMILY
Native

Ajuga reptans is a common perennial herb with attractive, bright blue flowers found throughout Ireland in woods, hedges, ditches and damp, shady habitats.

It was formerly used medicinally for healing wounds and in general healing applications (Moloney, 1919). According to Allen and Hatfield (2004) the juice of *Ajuga reptans* was applied to bruises.

Dark-leaved cultivated forms are frequently grown as spring-flowering garden plants and are most useful in shady places.

Alchemilla alpina

- Alpine lady's-mantle (Scannell & Synnott, 1987)
- *Bratóg Mhuire* (Scannell & Synnott, 1987); *Trusgán* (Moloney, 1919)

ROSACEAE—THE ROSE FAMILY
Native

Alchemilla alpina is a very rare perennial of high mountain alpine scree and rocky slopes. It is found as a native only on a few mountain cliffs in Cos. Kerry and Wicklow.

In parts of Europe this plant has been used in traditional medicine to treat menstrual disorders and as a diuretic and antispasmodic (van Wyk & Wink, 2004). However, I have been unable to find any records of its use as a medicinal plant in Ireland.

Alpine lady's-mantle is sometimes grown as a rock garden plant in Irish gardens.

Alchemilla vulgaris agg., s.l.

- Lady's-mantle
- *Dearna Mhuire* (Scannell & Synnott, 1987); *Bratóg Mhuire* (Moloney, 1919; Webb, 1977); *Fallaing Mhuire* (Moloney, 1919; Ó Dónaill, 1977); *Cóta preasach* (de Vere Coneys, 1849; Moloney, 1919; Dinneen, 1927); *Leathach bhuidhe* (de Vere Coneys, 1849; Moloney, 1919; Dinneen, 1927); *Mionán Mhuire* (Moloney, 1919); *Crúba-leomhain* (de Vere Coneys, 1849; Dinneen, 1927); *Leamhach mhaighe* (Dinneen, 1927)

ROSACEAE—THE ROSE FAMILY
Native

Including:

Alchemilla filicaulis (Alchemilla vestita)
- Slender lady's-mantle (Scannell & Synnott, 1987)
- *Dearna Mhuire chaol* (Scannell & Synnott, 1987)

Alchemilla glabra
- Smooth lady's-mantle (Scannell & Synnott, 1987)
- *Dearna Mhuire mhín* (Scannell & Synnott, 1987)

Alchemilla glaucescens (Alchemilla minor)
- Silky lady's-mantle (Scannell & Synnott, 1987)
- *Dearna Mhuire bheag* (Scannell & Synnott, 1987)

Alchemilla xanthochlora
- Intermediate lady's-mantle (Scannell & Synnott, 1987)
- *Dearna Mhuire bhuí* (Scannell & Synnott, 1987)

Alchemilla vulgaris.

Alchemilla vulgaris (lady's-mantle). Watercolour by Lydia Shackleton. Courtesy of the National Botanic Gardens, Glasnevin.

Lady's-mantle species grow in a variety of different grassland and mountain habitats throughout Ireland. *Alchemilla filicaulis* is most frequently found on limestone soils growing in short, dry grassland. *Alchemilla glabra* is rarer and occurs mainly in the north and west on moist acidic upland habitats and beside mountain streams. *Alchemilla glaucescens* is a rare species in Ireland, known only on limestone hills north of Dromahair, Co. Leitrim, and in adjacent Co. Sligo. *Alchemilla xanthochlora* is generally found in unimproved grasslands, particularly in the northern half of the country. These four species are morphological segregates that represent local populations that fall within the larger variation of the more widespread species *A. vulgaris*.

Alchemilla vulgaris occurs across Europe, North America and Asia and accordingly has diverse medicinal uses. Lady's-mantle has been used to treat diarrhoea, gynaecological ailments, and sore throats and as an astringent for sores and septic wounds (van Wyk & Wink, 2004). Harrop (1977) and Ody (1997b) mentioned the use of lady's-mantle to treat premenstrual and menstrual tension, but it is unclear if Irish native species of *Alchemilla* have been used in this way. According to Allen and Hatfield (2004), *A. vulgaris* was formerly used in Irish traditional medicine as a diuretic and to treat cuts, nosebleeds, burns, scalds and sore heads. Threlkeld (1726) noted its use to stop internal bleeding. Lady's-mantle was also used to treat headaches and 'against bad eyes', as well as to cure ailments of cattle (Williams, 1993). Meehan (1906: 207–208) described its use in cattle in Co. Leitrim in north-west Ireland:

> . . . the Elf-Doctor himself steals out softly and alone to gather 'erribs'—*vulgo* 'herbs.' What these 'erribs' are is a secret beyond price, and none dare ask it. Still, for the sake of humanity and cattle, I will venture to disclose it.
>
> It springs up profusely in light moss, or in a track where a last year's 'whin-bush' (as we call 'furze' around here) was burned down. It grows close to the ground with corymbs of yellow-green flowers. Its popular name is Lady's Mantle, (more

> properly, Our Lady's Mantle), and its botanical one *Alchemilla vulgaris*. . . .
>
> As soon after the messenger's return as they can detach themselves from the enchantments of good-fellowship, all attention is once more focussed on the stricken cow, and the cure at last is made, with all due solemnity. It is simple enough. In a pail are put (*a*) the expressed juice of the 'erribs,' (*b*) the coins, copper and silver, and one flint from the elf-pouch, (*c*) a pinch of salt, and (*d*) finally the three-mearne-water, with ordinary spring water *ad lib*.
>
> It is the supreme fact of the cure that the cow takes *three* sups of this concoction. . . .If she refuse it, it has to be bottled down her throat, three several glugs of it, for her health's sake. The last tin-pannikin-full' is reserved and sprinkled along her spine, where the hair is most rebellious, with slight massage treatment and a carrying of some of the dirt out of the cleets of the fore-feet over the back. Finally, the last drops are tossed into her ears. If she keeps on never minding, it is a very bad sign. But if, on the other hand, she rouses up and shakes herself, the 'cure' is already working, and the little festivities are gone back to with a light heart.
>
> Details in the ritual vary slightly with the individual practitioner. . . .

Allen and Hatfield (2004) also mentioned the use of *Alchemilla* species for unspecified veterinary uses in Co. Donegal.

**Alisma lanceolatum*

- Narrow-leaved water-plantain (Scannell & Synnott, 1987)
- *Corrchopóg chaol* (Scannell & Synnott, 1987); *Biorshlánlus* (Moloney, 1919); *Corchopóg beag* (Moloney, 1919)

ALISMATACEAE—THE WATER-PLANTAIN FAMILY
Native

Alisma lanceolatum is a rare plant throughout Ireland of wetland habitats, such as ponds, wet drains and ditches.

*No ethnobotanical uses of this plant have been noted in Ireland.

Alisma plantago-aquatica

- Water-plantain (Scannell & Synnott, 1987); Slanlis (Hart, 1898)
- *Corrchopóg* (Scannell & Synnott, 1987); *Biorshlánlus* (Moloney, 1919; Dinneen, 1927; Ó Dónaill, 1977); *Corchopóg mór* (Moloney, 1919)

Native

Alisma plantago-aquatica is a marshland perennial plant found frequently throughout Ireland in a variety of wet habitats, such as ditches, slow-flowing streams, ponds and drains. It also occurs throughout temperate Europe, northern China and North America.

This plant was used medicinally to treat a sore mouth (Allen & Hatfield, 2004). Its juice was used in Co. Londonderry to stop the spitting of blood (Moore, 1834–5). In Donegal it was used to treat burns (Hart, 1898). According to Mears and Hillman (2007), the corm is edible raw, steamed or roasted but is not very palatable. The leaves and corms are in fact toxic, but heat and drying mitigate this. Native Americans used the dried corm for a variety of ailments, most often gastrointestinal, and tea from its leaves was used by the Iroquois (Moerman, 1998: 56).

Alliaria petiolata

- Garlic mustard (Scannell & Synnott, 1987); Hedge garlic (Harron, 1986); Jack-by-the-hedge (Praeger, 1938)
- *Bóchoinneal* (Scannell & Synnott, 1987); *Gáirleóg choilleadh* (Moloney, 1919; Dinneen, 1927); *Bó chuinneall* (Moloney, 1919); *Gairealach (-aigh) collaim* (de Vere Coneys, 1849); *Garbhlach cuilm (colaim)* (Dinneen, 1927); *Lus an óir*[5] (Ó Dónaill, 1977)

BRASSICACEAE (Cruciferae)—THE CABBAGE AND CRESS FAMILY
Native

Alliaria petiolata is a common wayside herb throughout much of Ireland, although it can be rare in parts of the western half of the country. It is often found nestling on the roadside in a damp spot in the shade of a small woodland or hedge, living up to one of its alternative common names, Jack-by-the-hedge. It is widely distributed across Europe into Central Asia and was introduced in North America in the 19th century. It has become a noxious weed primarily in the eastern United States and invades moist forest floors and flood plains.

Alliaria petiolata.

Hedge garlic or hedge mustard was formerly widely used in the British Isles as a salad or flavouring herb (Phillips, 1983) and tastes mildly of garlic. Mears and Hillman (2007) noted that the seeds can be used to make a fiery mustard, and the leaves can be added to nettle soup. In some parts of Britain and Ireland the leaves were chopped and added to mint sauce to accompany lamb or mutton, and it was also used to flavour fish, particularly herrings (Hilton, 2007). A pesto can also be made from this plant by blending together its leaves with pine nuts, grated parmesan cheese and olive oil. Rutty (1772: 36) said that 'it is used in sauces, and to actuate the colder ingredients in salads'.

Alliaria petiolata was formerly used medicinally in Ireland as a diuretic, and to treat dropsy, 'pestilential distempers' and gangrene (Threlkeld, 1726: unnumbered [7]).

Lucas (1959) mentions that the word *creamh*, sometimes translated as 'garlic', is frequently referred to in early Irish literature as having been used for food. Although it is not always entirely certain what botanical species *creamh* refers to, Lucas (1959) suggested that it may well be *Alliaria petiolata*.

Allium babingtonii (Allium ampeloprasum var. *babingtonii)*

- Wild leek (Webb, 1977); Wild garlic (Hart, 1898)
- *Cainneann* (Scannell & Synnott, 1987); *Gairleog fiáin* (O'Rourke, 2006); *Oinniún fiáin* (O'Rourke, 2006)

AMARYLLIDACEAE (Alliaceae, Liliaceae)—THE ONION FAMILY
Native or Possibly Introduced (possible archaeophyte)

5. Ó Dónaill (1977) defined *lus an óir* as 'hedge-mustard'.

Allium babingtonii.

Allium babingtonii is a rare species of wild leek in Ireland, with scattered distribution along the west coastal regions from Cos. Clare to Donegal. Elsewhere in Ireland (e.g., in the south-east) it is a garden escape or discard. It grows on roadsides and limestone pavement and in waste places near the sea and in the margins of fields. Its nativity is not certain, although Webb and Scannell (1983: 207) said that it is 'at least arguable that it is an endemic species of N.W. Europe which has become to some extent commensal with man.' It also occurs in Cornwall. It is possible that it may have been introduced in early times and gone wild. It is very closely related to the cultivated leek *A. ampeloprasum* and is often regarded just as a variety of this. Webb and Scannell (1983) stated that it has not been cultivated in Ireland within the last 150 years.

Lucas (1959) mentions that the Irish word *cainneann*, which is sometimes translated as 'leek', is frequently referred to in early Irish literature as having been used for food in Ireland. Although it is not entirely certain what botanical species *cainneann* refers to, it may well be *Allium babingtonii* or one of the other wild *Allium* species found in Ireland. Kelly (1997) suggested that *cainneann* (or *cainnenn* in Old Irish) may be translated as 'onion'. Earlier, de Vere Coneys (1849) gave the Irish word *creamh-garaidh* for 'leek'.

In the Aran Islands *Allium babingtonii* is still used occasionally as a culinary herb and was formerly an important medicinal plant there. It was used to stop bleeding from wounds, and to treat various respiratory ailments and against worms (O'Rourke, 2006).

According to Hart (1898), blackleg, a disease of cattle, was cured by using this species. Half a bulbil was inserted into a slit cut into a calf's tail and bound. The treated calf was then regarded as immune to this disease.

Allium schoenoprasum

- Chives (Scannell & Synnott, 1987)
- *Síobhas* (Scannell & Synnott, 1987); *Siabal* (Moloney, 1919); *Saidhbhín* (Moloney, 1919); *Feurán* (Moloney, 1919); *Fearan*[6] (Cameron, 1883; Dinneen, 1927); *Sibhín*[7] (Dinneen, 1927)

AMARYLLIDACEAE (Alliaceae, Liliaceae)—THE ONION FAMILY

Possibly Introduced

Allium schoenoprasum is rare in the wild in Ireland and possibly introduced. However, it may be native around Lough Mask in Co. Mayo; no one is quite certain. In its natural habitats *A. schoenoprasum* grows as tufts in rocky places. Nevertheless, it is a commonly cultivated plant of gardens throughout Ireland and is used in salads and for a variety of culinary purposes.

This plant was formerly used for several medicinal purposes both internally and externally, particularly in the treatment of tuberculous lesions (Moloney, 1919).

6. Uí Chanainn (1962) gave the spelling *féarán*.
7. Dinneen (1927) defined *sibhín* as 'a chive, a leek'.

Allium ursinum

- Wild garlic (Webb, 1977); Ramsons (Scannell & Synnott, 1987); Ramps (Hart, 1898); Common garlic (Brunker, 1950)
- *Creamh* (Scannell & Synnott, 1987); *Cneamh* (de Vere Coneys, 1849; Moloney, 1919); *Gáirleóg Mhuire* (Moloney, 1919); *Gáirleóg fhiadain* (Moloney, 1919; Dinneen, 1927); *Farach-dubh (-duibh)*[8] (de Vere Coneys, 1849); *Féarán* (Dinneen, 1927; Uí Chanainn; 1962); *Gairleog na mBos* (Tóibín, 1965)

AMARYLLIDACEAE (Alliaceae, Liliaceae)—THE ONION FAMILY
Native

Allium ursinum.

Allium ursinum is a bulbous plant of damp woods, hedges and shady places. It is common throughout Ireland, although in some regions it is quite scarce. This plant is remarkable for sometimes forming almost pure colonies under a canopy of trees. It is also distributed throughout Great Britain and temperate Europe.

Wild garlic was an extremely important wild food in ancient Ireland and was much used as a medicinal plant throughout Irish history (Kelly, 1997). Both the leaves and the bulbs can be eaten raw or cooked. The leaves are mildly flavoured of garlic. The bulbs of *Allium ursinum* taste stronger. Chopped leaves add interest to salads or can be added to flavour other foods, such as stews, sauces, soup or soft cheeses and cottage cheese. The leaves can also be made into a puree with nuts, mustard leaves, olive oil and lemon juice to make a pesto that can be used with pasta or added as flavouring to stews, burgers and other meats. The flowers can also be added to salads. Rutty (1772) said that the young leaves were sometimes eaten with bread and butter. Houston and Milne (2008) suggested baking or grilling lamb or fish wrapped in wild garlic leaves, or making wild garlic bread—chopped leaves are mixed with soft butter and liberally spread over split French bread; the mixture is then wrapped in foil and placed in a moderately hot oven to melt the butter. Williams (1993) reported that garlic was used to flavour butter in the 19th century in Ireland; butter was wrapped in garlic and then buried in the bogs to flavour and preserve it (see also *Sphagnum* for more about bog butter).

Allen and Hatfield (2004) suggested that *Allium ursinum* was used medicinally, both for its curative and preventative properties, and a traditional practice of walking around with ramsons inside one's shoes or socks existed, at least in Cos. Clare and Galway, to cure a cold or cough. It was often eaten raw or boiled, usually in milk, and the liquid rubbed into the skin, but its medicinal properties are most effective if eaten. These same authors note that it was used to cure a variety of ailments: toothache, intestinal worms, warts and corns, measles, rheumatism, inflamed sores on the fingers (called 'felons'), sore eyes, chest and lung infections, sore throats, coughs and colds, asthma, indigestion and stomach aches, and to purify the blood and in the prevention of blood clots, and in Galway it was used in a poultice to treat swellings from mumps on the neck. Egan (1887) mentioned that garlic (presumably *A. ursinum*) was used in traditional medicine in Ireland to treat severe rheumatism and sciatica. The leaves were boiled in water to make a strong bath for the whole body. A barrel was sometimes used to bathe in. Threlkeld (1726) also noted it was used to treat blisters and kidney problems. Wild garlic was also widely used in Ireland as a much-valued treatment for tuberculosis (Williams, 1993).

Allium ursinum was used as a veterinary remedy in various parts of Ireland to treat coughs

8. De Vere Coneys (1849) also gave this Irish name for *Allium carinatum* and 'wild garlick'.

in horses and cattle; hoose[9], ringworm in calves, farcy, blackleg and worms (Allen & Hatfield, 2004). O'Regan (1997) mentioned that until recently in Co. Westmeath wild garlic was encouraged to grow in fields where cattle were grazed because of its healing effects. Doherty (2001) mentioned records of cloves of garlic being used in Ireland to treat cattle diseases, such as 'recumbency in post-parturient cows'. A clove of garlic was inserted under the skin in the tail region, often during bad winters when animals went down. In other cases a mixture of garlic, soot and goose dung was applied and bandaged in position.

Cultivated garlic is *Allium sativum*; in Irish, cultivated garlic is *gáirleóg*, flower of garlic is *lurachán*, and a clove of garlic is *ionga gháirleóige* (Moloney, 1919). Several other species of *Allium* occur in the wild in Ireland as introductions, mainly as escapes from gardens. *Allium triquetrum* (in Irish, *glaschreamh*, or *creamh garraí*[10]), with three-angled stems and white flowers, is locally abundant on roadsides, hedgerows and roadsides. *Allium carinatum*, keeled garlic (in Irish, *gairleog chíleach*, or *farach-dubh* [*-duibh*], de Vere Coneys, 1849), is found only occasionally in the wild in Ireland as a garden escape.

Allium ursinum leaves.

Allium ursinum pesto.

Allium vineale

- Crow garlic (Webb, 1977); Wild onion (Scannell & Synnott, 1987)
- *Gairleog Mhuire* (Scannell & Synnott, 1987)

AMARYLLIDACEAE (Alliaceae, Liliaceae)—THE ONION FAMILY
Native (possible archaeophyte)

Allium vineale is a bulbous species that is found occassionally in the southern half of Ireland in fields and roadsides, as well as on sand dunes and dry banks, and in scrub and hedgerows.

Threlkeld (1726: unnumbered [8]) warned that frequent use of crow garlic is both 'troublesome and noxious, causing head-ach, thereby hurting the eyes, and all the organs of sense', perhaps suggesting that it was formerly used medicinally.

Peterson (1977) suggested that *Allium vineale* is edible and can be used as food, although I have not tried it. Rutty (1772: 36) said of crow garlic that 'the young chives shred are eaten with sallet or boiled in the pot as leeks or onions'.

The ordinary cultivated onion is *Allium cepa*, sometimes known as the 'garden onion'. It is known only in cultivation and possibly derived from wild species native to Central

9. Hoose is a bronchial infestation by nematodes (*Dictyocaulus, Metastrongylus, Protostrongylus*) seen in cattle, sheep, goats and pigs.
10. Uí Chonchubhair (1995) listed this name in *Flora of the Dingle Peninsula*.

Asia. Onions have been cultivated in Ireland since the earliest times and used for many culinary, medicinal and other purposes. A cut onion rubbed onto insect stings and bites was a popular remedy to remove itching and irritation (Vickery, 2010). Onion skins were used for dyeing textiles. In Irish, onion is *oinniún* (Uí Chanainn, 1962).

Rutty (1772: 107) described an intriguing use of onions—to make invisible ink. He said, 'If letters be written with the juice, they will continue invisible on the paper for many days; but if the paper be held to a hot fire, the letters become visible, appearing as if they had been burnt: by this means a secret correspondence may be held: and the same may be done with milk or juice of the tender branches of the fig-tree'.

Onions are an essential ingredient in many traditional recipes—Irish stew (mutton, potatoes and onions)[11]; Dublin coddle (ham, pork sausages, onions and potatoes); tripe (tripe and onions)[12]; champ (potatoes and onions); and colcannon (cabbage or kale, potatoes, onions or leeks).

Alnus glutinosa

- Alder (Scannell & Synnott, 1987)
- *Fearnóg* (Scannell & Synnott, 1987); *Fearn (-a)* (de Vere Coneys, 1849; Moloney, 1919; Dinneen, 1927); *Ruaim (-me) (-meacha)*[13] (de Vere Coneys, 1849); *Crann feárna* (Dinneen, 1927); *Crann fearnóige* (Dinneen, 1927)

BETULACEAE—THE BIRCH FAMILY
Native

Alnus glutinosa is a very common small tree throughout Ireland of wet woodlands, hedgerows, river and canal banks and lake-shores. The tallest known existing alder tree in Ireland is 19.5 m tall and found in Reenadinna Wood, Muckross, Co. Kerry, and one with a girth of 5.92 m is also known from Glaisin na marbh, Co. Kerry (Anonymous, 2005).

Alder was widely used in Ireland for many purposes. The bark of *Alnus glutinosa* is rich in tannins and formerly an important source for the tanning of leather. However, alder bark tends to make the leather more brittle and gives it a dark or reddish colour (Howes, 1974).

Alder wood was traditionally used for making shields, bowls, harps and containers, as well as charcoal (Mackay, 1836; Rimmer, 1969; MacCoitir, 2003), and was particularly used as a component of gunpowder (Mackay, 1836; Nelson & Walsh, 1993). In the early 1800s, charcoal made from alder (and willow) was used in the manufacture of gunpowder near Cork (Wakefield, 1812). The green wood of alder may require a long period of seasoning before it can be used for fuel (the green wood of most tree species requires seasoning and drying before it makes good fuel). Its soft timber was a positive asset for wood turners and other uses, such as the making of hat-blocks, which could be carved into the various shapes required by changing fashions (Edlin, 1973). Alder wood used for hat blocks was favoured because it retained its shape even when steamed (Arnold, 1968). For furniture, alder wood is an attractive, red-brown colour and has sometimes been called 'Irish mahogany' (John McLoughlin, pers. comm., 2008). Howes (1974) mentioned that this species has been used to make artificial limbs, although whether it was used in Ireland for such purposes is not known. Mitchell (1986) mentioned a shield in the National Museum of Ireland made

11. When I make Irish stew I generally add carrots and pearl barley, although these are not necessarily authentic 'original' ingredients.

12. Tripe is usually made from the stomachs of various farm animals, including cattle, sheep and pigs.

13. Another meaning for this word is 'fishing line'.

Alnus glutinosa tree in Lougher, Co. Kerry.

from alder wood that is 1 m in diameter, which would have been taken from a slice of an exceptionally large alder tree. He pointed out that in Ireland today, we could not find a single alder tree capable of supplying a blank for such a shield.

Alder timber is very durable when submerged in water but less so when exposed to the air (Nelson & Walsh, 1993). For that reason it has been used for many purposes where the wood would be subject to frequent wetting, such as for lock gates of canals, sluices and wooden piles sunk for various purposes in rivers and lakes (Hickie, 2002). One of the oldest objects made from wood known in Ireland is a box made of alder wood that was discovered in Killymoon, Co. Tyrone, in the early 19th century. The box contained a gold dress-fastener and is dated to approximately BC 700 (Teahan, 1994). A block cartwheel dated to about BC 400 and made of alder, with yew dowels, was found in a bog in Co. Roscommon in 1968–1969. This is the earliest evidence of wheeled transport in Ireland (Teahan, 1994). Leather clogs with soles made from alder wood were manufactured in Enniscorthy, Co. Wexford, and used by the gardeners at the National Botanic Gardens up to the 1950s. Wattles made of alder were used in building houses in ancient Ireland, of wattle and daub construction (Neeson, 1991). Mackay (1836) noted that alder timber was particularly useful for the piles of bridges because it will remain waterlogged for a long period without decaying.

Dutton (1808: 285–286) said that alder timber was 'generally despised' in Co. Clare, but if it is of sufficient age, it is 'little inferior to mahogany' and makes the best 'bolsters for cars and for bushing [pushing] the eye of the lower mill-stone round the spindle, as it never takes fire by friction. When used for handles for tools it does not blister the hands'.

Alder has been used to make a variety of dyes: the bark yields a tawny red dye that can be changed to black with the addition of copperas; a pinkish fawn dye is obtained from fresh wood; green dye from the catkins; and a yellow dye from young shoots (Edlin, 1973). Threlkeld (1726: unnumbered [8]) said that the bark of alder 'serves for dying instead of Galls'. Rutty (1772: 101) also mentioned such usage and said that the bark 'may supply the place of galls in making Ink'. A brown dye can also be obtained from this plant (Mahon, 1983). A description of how alder was used in dyeing is provided by Mahon (1983: 121–122), extracted from a manuscript held by the Irish Folklore Commission from Killorglin, Co. Kerry:

> They skimmed the iron mould from a stagnant pool and took the sediment from the bottom of the pool (a very dark substance)[14] and brought it home in timber cans and put it into timber barrels to store it up. Then they got the bark of the alder and broke it up and boiled it in some of the iron mould in a pot for two hours or so to soak. Then they put the articles to be dyed in a coil on the floor and poured the contents of the pot on the article and allowed it to rest for two hours or so to soak. Then they took it out and put it on the fence for two or three hours. They washed it in the river and in the cold water and when they had it drained the article was jet black.

Mairet (1916) suggested that 1 lb of wool should be dyed with 1 lb of alder bark. The wool was boiled with the bark for two hours, when it would be a dull, reddish brown colour, then, if half an ounce of copper was added, the wool would turn black.

K'Eogh (1735) noted alder bark was used medicinally to treat swellings of the throat and that it could be useful for healing or cauterizing sores and ulcers. Maloney (1972) recorded that the saplings of young alder trees are regarded as a very old and reliable cure for piles in Co. Cavan. Gallagher (2012) said that the bark contains anti-inflammatory properties. Rutty (1772: 102) stated that the leaves had a 'glutinous quality' and are sometimes worn next to the feet, 'in order to prevent their pain and sweating in travelling'. Rutty (1772: 102)

14. Sediments such as this containing iron were used as a mordant to fix dye.

also noted that 'they are also an effectual remedy against Fleas, by covering the floors of the rooms where they abound, with those leaves, while the dew hangs upon them, as they, especially when budding, contain a kind of tenacious humour, to which the fleas adhering, as little birds do to bird-lime, are surely detained and killed thereby.'

If one accepts that the expulsion of demons from the body is a medical matter, then alder was used medicinally for such purposes. Dianket, the great physician of ancient Irish legends, prescribed a decoction from the roots of the alder and the apple tree, to be boiled with the brains of a wild hog and drunk while fasting. The potion was to be administered until the demon was cast out of the stomach of the bewitched person (Wood-Martin, 1902).

Alders are occasionally grown as ornamental trees for parks and large gardens (Wyse Jackson, 1994). They have also been planted along stream and river-banks to help prevent erosion (Arnold, 1968). Several non-native species of alder are also occasionally grown in gardens, parks and demesnes. The most common of these is probably *Alnus incana*, the grey alder, which is quite a popular ornamental tree in Ireland.

**Alopecurus aequalis*

- Orange foxtail (Clapham et al., 1962)
- *Fiteog oráiste* (Parnell & Curtis, 2012)

POACEAE (Gramineae)—THE GRASS FAMILY

Native

Alopecurus aequalis is a rare grass in Ireland that is distributed only near the south coast in habitats such as wet meadows and the edges of ponds, lakes and ditches.

*No ethnobotanical uses of this plant have been noted in Ireland.

**Alopecurus geniculatus*

- Marsh foxtail (Scannell & Synnott, 1987)
- *Fiteog cham* (Scannell & Synnott, 1987); *Ulplosfhér glunlúbtha* (White, 1803)

POACEAE (Gramineae)—THE GRASS FAMILY

Native

Alopecurus geniculatus is common throughout Ireland in habitats such as wet meadows and the edges of ponds and ditches.

*No ethnobotanical uses of this plant have been noted in Ireland.

Alopecurus pratensis

- Meadow foxtail (Scannell & Synnott, 1987)
- *Fiteog léana* (Scannell & Synnott, 1987); *Ulplosfhear leana* (Wade, 1808a); *Ulplosfhér léna* (White, 1803)

POACEAE (Gramineae)—THE GRASS FAMILY

Native

Alopecurus pratensis is a common grass throughout Ireland of wet meadows, pastures and other grassy habitats.

It is included in some seed mixes for grasslands as a fodder grass.

Althaea officinalis

- Marsh-mallow (Scannell & Synnott, 1987)
- *Leamhach* (Scannell & Synnott, 1987); *Leamhach mhuighe* (Moloney, 1919; Dinneen, 1927); *Leamhadh* (de Vere Coneys, 1849); *Leamhach buí* (or *Leamaigh*) (Ó Dónaill, 1977); *Leamh* (Joyce, 1875)

MALVACEAE—THE MALLOW FAMILY
Probably Introduced

Althaea officinalis is a tall, perennial downy herb with large pink flowers that is closely related to the garden hollyhock. Although *A. officinalis* occurs in a variety of coastal habitats, such as wasteground and banks near the sea and along the upper edges of salt marshes, it is rather rare in Ireland. In most of its localities in Ireland it is obviously an escape from cultivation, but it occasionally has been found in localities far from human habitation.

Throughout Britain and Ireland, *Althaea officinalis* was formerly used medicinally in poultices and ointments to treat swellings, along with its two related genera in the Malvaceae, *Lavatera arborea* and *Malva sylvestris* (also probably introduced). Marsh-mallow was also taken internally for coughs, sore throats, asthma, colds and chest problems. Mackay (1836: 37) recorded that 'our indigenous species [of mallow] afford abundance of mucilage, and are employed for fomentations and cataplasing', and 'a decoction of it is of general use in England and Ireland. The common road-side mallow (*Malva sylvestris*) is often sold as Marsh Mallow, and has pretty much the same properties'. Wade (1804) noted its importance as a medicinal plant in Ireland, and according to Allen and Hatfield (2004), it was used in Ireland to treat rheumatism, still joints or backache and eye ailments. Allen and Hatfield (2004) also reported that a liquid obtained by boiling the leaves or roots was used to bathe a muscle sprain or bone fracture. Scully (1916) commented that its abundance around houses in Co. Kerry was probably due to its former cultivation as a medicinal plant.

According to Shaw (1998), *Althaea officinalis* was cultivated for its roots, which were traditionally used for sweet-tasting confectionary (candied root). Marsh-mallow sweets can be made from root extracts. The same author noted that its seeds are edible and nutty in flavour, and its leaves may be used as a vegetable, cooked like spinach, or chopped in salads, and were traditionally used to promote milk production in nursing mothers.

The Economic Botany Museum collections of the National Botanic Gardens include a beautiful polka jacket made from *Althaea officinalis* fibres. Marsh-mallow is still grown as an ornamental cottage-garden plant in some Irish gardens.

A jacket made from *Althaea officinalis.* From National Botanic Gardens, Glasnevin, Museum objects collection. © Department of Environment, Heritage and Local Government.

Ammophila arenaria (Arundo arenaria; Psamma arenaria)

- Marram (Scannell & Synnott, 1987); Bent (or Bint) (Colgan, 1911); Moorinyagh (Moolinyagh) (Hart, 1898); Miniheens (McCarthy & Hawkes, 2007)
- *Muiríneach* (Scannell & Synnott, 1987); *Cuile muiriunath* (Wade, 1808a); *Garbhán gaofar*[15] (Ó Dónaill, 1977); *Muiruidhneach* (Ó Criomhthain, n.d., ca. 1908); *Cuile muiriunadh* (White, 1803); *Muiriunach* (White, 1803); *Mehain*[16] (White, 1803)

POACEAE (Gramineae)—THE GRASS FAMILY
Native

Ammophila arenaria is an abundant grass species found all around the coasts of Ireland, particularly on sand dunes. This plant is among the first colonisers of sandy habitats and plays an important role in fixing and stabilising sand dunes with its spreading network of rhizomes, deep roots and tolerance for drought and low nutrients.

Marram was formerly harvested on sand dunes and used for thatching in some coastal areas, particularly for rope-thatched houses (Letts, 1997). Sometimes referred to as 'bent' or 'bent-grass', these vernacular names are generally applied to species of *Agrostis* today. When bent grass is mentioned as having been used in thatching, I believe in most cases authors are referring to the more robust *Ammophila arenaria*. I suspect that in coastal parts of Ireland, marram was very extensively used for thatching. Use of marram for thatching continued until relatively recent times, but such usage today is very rare. The marram was cut below the surface of the sand when being harvested, and in some parts of Ireland, such as Co. Donegal, a special toothed sickle was used for this purpose (Evans, 1957). Mackay (1806) commented that a thatch of marram with its tightly inrolled leaf blades and tough stems could last up to 20 years before it needed to be replaced. Smith ([1756] 1979: 176) recorded that in Kerry on the sand dunes of Inch Peninsula, local people would 'cut and carry away this sedge [marram is in fact a grass, not a sedge] as they frequently do for thatch, and other purposes'. O'Reilly (2004) noted that marram has been found as thatching material in Co. Donegal, and MacMahon (1994) wrote that it was used until relatively recently for thatching in Ballyheigue, Co. Kerry, to form a firm base for the scallops (flexible rods used to anchor the thatch in place) to hold. In his *Essay on the Indigenous Grasses of Ireland*, White (1803: 112) said that 'it makes an excellent thatch for houses, and will sometimes last for 20 years. Cattle will eat of it in winter, but in summer they leave it untouched'. White also said that 'the flowers and feed are possessed of an emetic quality, and may be used where ordinary emetics cannot be obtained. They vomit strongly, but are not dangerous in their operation'. Milliken and Bridgewater (2004) mention that in Scotland, marram was used in thatching in two distinct layers: the lower with the pointed leaf tips pointing upwards and the upper with the tips pointing downwards. There, the leaves of the wild yellow flag, *Iris pseudacorus*, were

Ammophila arenaria.

Ammophila at Strandhill, in Co. Sligo.

15. Ó Dónaill (1977) defined *garbhán gaofar* as 'marram (roots)'.
16. White (1803) gave the Irish names *cuile muiriunadh*, *muiriunach* and *mehain* for *Arundo arenaria*.

used underneath to pad out any depressions. I have not found records of this happening in Ireland. Rutty (1772) noted that in Co. Wexford, houses were thatched with it and that it was much more durable than straw. McCarthy and Hawkes (2007: 81) said that it is 'the best thatching material'.

In Scotland's Shetland Islands, *Ammophila arenaria* was used to make high-quality ropes and for weaving baskets (Vickery, 1995). Evans (1957) recorded its use in similar ways in Ireland. In the Outer Hebrides, marram was used to make a type of coiled grain basket (in Gaelic this is called a *ciosan*), as well as used to weave horse collars, cushions for pony pack saddles, hats and bags (Milliken & Bridgewater, 2004); the roots of marram were used to scrub floors (Milliken & Bridgewater, 2004). Evans (1957) wrote that a strong rope used for tethering sheep during shearing was made of marram or hay. Dutton (1808) mentioned that marram was used for making floor-mats in Dublin, as well as for thatching in Co. Clare, also noting that it would last on the roof for up to 20 years. It has been used for making brooms (Howes, 1974). Rutty (1772: 116) said that 'country people' cut and bleach it and then sell it in Dublin to make 'whisks and matting'. It was also used in parts of the west of Ireland as animal bedding, for example, in Co. Donegal (M. Sheehy Skeffington, pers. comm., 2007), and was formerly useful as winter fodder for cattle when other straw and hay were scarce (Mackay, 1806; Dutton, 1808).

Mackay (1836) noted that Lord Palmerston had used it extensively near Ballyshannon to protect the coastline from erosion. Wade (1808a: 34) mentioned that *Arundo arenaria* 'opposes and prevents the conflux of drifting sands, which threaten inundation—roots woody—collected for fuel—ruinous and melancholy consequences have ensued—the drifting sands meeting with no opposition—nearly a whole parish in Scotland has been destroyed … Acts passed to prevent its destruction—27th of Elizabeth—15th of George II—may be manufactured into mats and ropes'. Rutty (1772) also mentioned its use for such purposes as a sand-binder.

Anacamptis morio (*Orchis morio*)

- Green-winged orchid (Scannell & Synnott, 1987)
- *Magairlín féitheach* (Scannell & Synnott, 1987); *Magairlín an sciatháin ghlais* (Moloney, 1919); *Magairlín na stuaice* (Uí Chonchubhair, 1995)

ORCHIDACEAE—THE ORCHID FAMILY

Native

Anacamptis morio.

Anacamptis morio is a declining species in Ireland of coastal grassland habitats, sandhills, heaths, and rocky grassland and meadows. It is found mainly in the east and centre of the country, but also extends into west Co. Galway.

Green-winged orchid is reputed to be an aphrodisiac in Europe and particularly in the Orient. From there, its use spread to India where Ayurvedic physicians investigated its use and concluded that the tubers help to contribute to vigorous and healthy sexuality (Rätsch, 1997). Calling this orchid by the following names: female fool stones, *Satyrion faemina*, *Orchis femina* or *Morio femina*, K'Eogh (1735) noted that this species had the same properties as *Orchis mascula*.

In the Mediterranean region this species is a component of 'salep', a fine, white to yellowish-white powder that is obtained by drying the tuber and grinding it into a powder. Salep is said to be very nutritious and is made into a drink or can be added to cereals and used in making bread, etc. There are no records of *Anacamptis morio* being used traditionally in these ways in Ireland.

Anacamptis on dunes at Dog's Bay, Roundstone, Co. Galway.

Anacamptis pyramidalis.

**Anacamptis pyramidalis*

- Pyramidal orchid (Scannell & Synnott, 1987)
- *Magairlín na stuaice* (Scannell & Synnott, 1987)

ORCHIDACEAE—THE ORCHID FAMILY
Native

Anacamptis pyramidalis is an attractive orchid with densely packed spikes of pink flowers that can grow up to 50 to 70 cm tall. It is found in grasslands, particularly in the midlands, and also in sandhills all around the coasts of Ireland.

*No ethnobotanical uses of this plant have been noted in Ireland.

Anagallis arvensis

- Scarlet pimpernel (Scannell & Synnott, 1987); Shepherd's weather-glass (Moloney, 1919; Praeger, 1938)
- *Falcaire fiáin* (Scannell & Synnott, 1987); *Driofúr phine* (Ua Maoileoin, 2001); *Falcaire fuar* (Moloney, 1919; Dinneen, 1927); *Falcaire fiadhain* (de Vere Coneys, 1849; Moloney, 1919; Dinneen, 1927); *Luibh na muc* (Moloney, 1919; Dinneen, 1927; Uí Chanainn, 1962); *Ruinruise* (de Vere Coneys, 1849); *Glan-rosc*[17] (Dinneen, 1927); *Sailín cuach*[18] (Dinneen, 1927).

PRIMULACEAE—THE PRIMROSE FAMILY
Native

Anagallis arvensis is a common annual plant found throughout Ireland in a wide variety of dry calcareous or weedy habitats, such as sand dunes, roadsides, wasteground and cultivated fields.

It is recorded as having been used formerly in the treatment of chronic or muscular rheumatism (Allen & Hatfield, 2004). Threlkeld (1726: unnumbered [10]) noted its use to treat gripes in newborn children, as well as for 'Falling Sickness [epilepsy] and Madness.' It was used to treat toothaches and eye disorders (O'Rourke, 2006), and in the 18th century

17. According to Dinneen (1927), the name *glan-rosc* was used in West Cork.
18. Dinneen (1927) noted the word *sailín cuach* was generally applied to a violet or pansy.

Anagallis arvensis.

in Ireland was used to treat rabies (Williams, 1993). However, it does contain a poisonous glycoside similar to saponin (Heywood, 1978), so must be regarded as a poisonous plant. The Irish name of this plant, *falcaire*, is derived from the Irish word meaning 'to cleanse', and probably refers to its purgative properties. Writing about plants in Scotland, Clyne (1989) mentioned that it was used to remove obstructions of the bowels, liver and kidneys.

MacMahon (1994: 242) said that in Ballyheigue, north Co. Kerry, it was used 'to give second sight and hearing. Had power of moving against the current' [whatever that means!].

Gallagher (2012: 44) mentioned that scarlet pimpernel was 'better known as the poor man's barometer or the Shepherd's water glass, because its flowers always closed if rain was approaching'.

**Anagallis minima*

- Chaffweed (Scannell & Synnott, 1987); Bastard pimpernel (Harron, 1986)
- *Falcaire beag* (Scannell & Synnott, 1987); *Cáth-luibh* (Dinneen, 1927)

PRIMULACEAE—THE PRIMROSE FAMILY
Native

Anagallis minima is a rare species in Ireland that is confined to sandy coastal habitats and some lake-shores in north-east Ireland and in Co. Kerry.

*No ethnobotanical uses of this plant have been noted in Ireland.

**Anagallis tenella*

- Bog pimpernell (Scannell & Synnott, 1987)
- *Falcaire corraigh* (Scannell & Synnott, 1987); *Fruig na súl* (Ua Maoileoin, 2001)

PRIMULACEAE—THE PRIMROSE FAMILY
Native

Anagallis tenella is common only in the west and occurs less frequently elsewhere in Ireland. It grows in habitats such as bogs, wet acid meadows, lake-shores, fens and marshy places.

*No ethnobotanical uses of this plant have been noted in Ireland.

Anchusa arvensis

- Bugloss (Scannell & Synnott, 1987)
- *Boglas* (Scannell & Synnott, 1987)

BORAGINACEAE—THE BORAGE FAMILY
Probably Introduced (probable archaeophyte)

Anchusa arvensis is a straggling annual herb with sky-blue flowers that occurs in sandy waste places, arable fields and as a garden weed near the eastern and northern coasts. It is very rare elsewhere in Ireland.

Allen and Hatfield (2004) suggested that this species may have been used medicinally in Britain and Ireland, but because the common name 'bugloss' was applied variously to a number of different species, it is impossible to be certain.

**Andromeda polifolia*

- Bog rosemary (Scannell & Synnott, 1987); Marsh andromeda (Brunker, 1950)
- *Lus na móinte* (Scannell & Synnott, 1987); *Fraoch curraigh* (Moloney, 1919); *Fraoch corraigh* (Dinneen, 1927); *Andraiméid* (Ó Dónaill, 1977)

ERICACEAE—THE HEATHER FAMILY
Native

Andromeda polifolia is a dwarf shrub with pale pinky purple flowers that is found mainly in lowland bogs in the midlands and is rare elsewhere in Ireland.

Bog rosemary is said to be the county flower of Co. Offaly, in the midlands. However, I don't know of other so-called county flowers in Ireland, nor have I been able to find if this designation has any official basis. It is also the official county flower of Co. Tyrone. All counties have official flowers in the United Kingdom but not in the Republic of Ireland.

*No ethnobotanical uses of this plant have been noted in Ireland.

Anemone nemorosa

- Wood anemone (Scannell & Synnott, 1987); Nedcullion (Praeger, 1938)
- *Lus na gaoithe* (Scannell & Synnott, 1987); *Cogal dearg* (Moloney, 1919); *Nead coille* (Moloney, 1919; Ó Dónaill, 1977); *Nead chailleach*[19] (Cameron, 1883; Dinneen, 1927)

RANUNCULACEAE—THE BUTTERCUP AND CROWFOOT FAMILY
Native

Anemone nemorosa is a common perennial herb throughout Ireland of shady woodlands and hedgerows. It has attractive, white or pink-tinged flowers and is a conspicuous component of the spring woodland flora in most Irish broadleaved woodlands.

According to Moloney (1919), it was used in the treatment of chest infections and sores. Wood-Martin (1902) mentioned that it was used as a plaster for wounds. K'Eogh (1735: 5) suggested that its roots were useful for the 'expelling of phlegm', or boiled in wine and laid on the eyes, they are good to treat sight, and that a decoction of wood anemone mixed with a small amount of barley helps to promote milk production in nursing mothers.

Anemone nemorosa.

19. Cameron (1883) and Dinneen (1927) gave this word to mean 'old woman's nest'.

Allen and Hatfield (2004) noted its veterinary use as a remedy for sheep (sheep scab) in Co. Wicklow.

A range of attractive, Irish-raised cultivated forms (cultivars) of the wood anemone are grown as ornamental plants in Ireland with variously coloured blue, pink and white flowers. Ó Dónaill (1977) gave the Irish name *cloichín gorm* for 'blue anemone'.

This plant is regarded as poisonous.

Anethum graveolens

- Dill (Reynolds, 2002)
- *Dile* (Cameron, 1883); *Lus mín* (Ó Dónaill, 1977)

APIACEAE (Umbelliferae)—THE CARROT FAMILY

Introduced

Anethum graveolens is a short-lived perennial herb with strongly aromatic foliage that is frequently grown in Irish gardens as a culinary herb. It is native to Europe and the Mediterranean region but not to Ireland. Nevertheless, it probably has a long history of usage in Ireland for medicinal purposes. It occurs only very rarely in the wild as an escape or discard from cultivation and has been recorded in the wild only on one occasion, in Co. Waterford, where Paul Green found it growing in a wasteground in Ardmore in 2001 (Reynolds, 2002).

Angelica sylvestris

Angelica sylvestris.

- Wild angelica (Scannell & Synnott, 1987); Fifes, Flutes[20] (Colgan, 1911); Scout (or Scoot) (Hart, 1898)
- *Gallfheabhrán* (Scannell & Synnott, 1987); *Bliúcán* (Ó Criomhthain, n.d., ca. 1908; Uí Chonchubhair, 1995; Ua Maoileoin, 2001); *Geamhrán gléasta* (Ua Maoileoin, 2001); *Galluanán fiadhain* (Moloney, 1919; Dinneen, 1927); *Cuinneóg mhighe* (de Vere Coneys, 1849; Moloney, 1919); *Cuinneóg Mhuire* (Moloney, 1919); *Contran* (Cameron, 1883); *Aingealag* (Cameron, 1883); *Ainglice* (De Bhaldraithe, 1959); *Galluran (-ain)* (de Vere Coneys, 1849); *Gleóran*[21] (de Vere Coneys, 1849); *Gunna pléascáin* (Ó Dónaill, 1977); *Pléascán mór* (Ó Dónaill, 1977)

APIACEAE (Umbelliferae)—THE CARROT FAMILY

Native

Angelica sylvestris is a common perennial herb throughout Ireland in a variety of wetland habitats, such as ditches, stream and canal banks, and marshes and wet meadows. It is also found in woodlands, hedgerows and dune-slacks.

This plant was formerly widely used as a popular medicinal herb throughout Europe.

20. These English names were reported by Colgan (1911) from Clare Island, Co. Mayo, referring to the hollow stems of wild angelica.

21. Tóibín (1965) indicated the variant spelling *gleorán*.

This species can be used in salads, cooked as a vegetable, or chopped and added to stewed rhubarb. Its seeds are valuable culinary aromatic seeds.

The leaf stalks and stems of *Angelica sylvestris* can be candied (Phillips, 1983), but they are rather more tough and bitter than those of the related cultivated non-native angelica (*Angelica archangelica*). To crystallize angelica, choose the tenderest young shoots and wash and dry them. Put the dried shoots in a mixing bowl and pour boiling water over them (in which some salt has been dissolved—about 1 tablespoon of salt to 2 litres of water). Leave the stems for about a quarter of an hour to draw out moisture from the stems, then strain and rinse in cold water. When they are cool, peel the outer skin, leaving only the fleshy pulp behind, which you will then crystalize. For 1 lb/450 g of angelica shoots, 175 g of granulated sugar is needed. Dissolve the sugar in 300 mL/half a pint of water to make syrup. Spread the angelica in a shallow bowl, and pour the sugar syrup over it and leave for about 24 hours. Repeat this process four or five times more, adding more sugar to the syrup on subsequent days. Finally, drain any remaining liquid off, and dry the fruit in the oven. Place the angelica stems on a wire rack over a baking tray (to catch the drips) for about 45 minutes at 100°C/200°F until the outside of the angelica is softly crisp. Store the crystallised angelica dry in a sealed container.

In a manuscript list of Irish plant names compiled in the early 1900s (for the Norwegian scholar Carl Marstrander) by the Islandman Tomás Ó Criomhthain from Kerry's Great Blasket Island, Criomhthain mentioned that *Angelica sylvestris* (given under the name *bliúcán*) is 'a plant like the *molliona mear* [i.e., *Conium maculatum*, hemlock]—its roots are eaten and are sweet to the taste' (Ó Criomhthain, n.d., ca. 1908).

Angelica sylvestris was used medicinally in Ireland to treat lung and chest complaints (Allen & Hatfield, 2004). K'Eogh (1735: 4) said that the pulverized root taken internally 'causes sweat, expels all noxious humours and prevents any malignity that proceeds from the air'. He also suggested that it should be chewed in the morning before breakfast, and noted that a decoction of angelica 'helps to cure palpitations, oppressions of the heart and also helps to promote urine' (K'Eogh, 1735: 4). Angelica was also used to treat madness and falling sickness (epilepsy) (Wood-Martin, 1902).

In parts of the north-west of Ireland young boys used this species to make hollow tubes, called 'scouts', 'scoots', or 'squirts', which were used as 'pea-shooters' (Hart, 1898: 385).

**Anisantha sterilis (Bromus sterilis)*

- Barren brome (Scannell & Synnott, 1987)
- *Brómas aimrid* (Scannell & Synnott, 1987); *Brumfhér aimrid*[22] (White, 1803)

POACEAE (Gramineae)—THE GRASS FAMILY

Possibly Introduced (possible archaeophyte)

Anisantha sterilis is an annual, or short-lived, perennial grass that is found in most parts of Ireland except the north. It grows particularly in disturbed areas, such as in cities and town, on dry banks, roadsides and wastegrounds and around ruins.

*No ethnobotanical uses of this plant have been noted in Ireland.

Antennaria dioica

- Mountain everlasting (Scannell & Synnott, 1987); Mountain cudweed (Webb, 1977); Cat's foot (Webb & Scannell, 1983)

22. White (1803) applied this name to *Bromus sterilis*.

- *Catluibh* (Scannell & Synnott, 1987); *Cáth-luibh* (Dinneen, 1927); *Liath-lus* (Dinneen, 1927); *Cáithluibh*[23] (Ó Dónaill, 1977)

ASTERACEAE (Compositae)—THE DAISY FAMILY
Native

Antennaria dioica is abundant in some western and central regions, but rare in many other parts of Ireland. This low-growing, tufted and creeping perennial herb usually occurs on limestone soils or boulder clay in dry, rocky habitats, such as mountain pastures, heaths and sandhills.

It is a common plant in the Burren (*Boireann*, great rock) in north-west Co. Clare, an extensive area of limestone pavement and calcareous grassland. This area is among the most significant karst landscapes in Europe. The Burren is distinguished by its diverse and unusual plant assemblages found among the grikes or cracks of the rocky landscape.

This species was formerly chewed or mixed with cobwebs in Co. Clare to staunch bleeding from cuts (Allen & Hatfield, 2004).

Anthoxanthum odoratum

- Sweet-scented grass (Webb, 1977); Vernal-grass (Webb, 1977); Sweet vernal grass (Scannell & Synnott, 1987)
- *Féar cumhra* (Scannell & Synnott, 1987); *Earachfheur deadhbholadh* (Wade, 1808a); *Errachfhér dedhbholadh* (White, 1803)

POACEAE (Gramineae)—THE GRASS FAMILY
Native

Anthoxanthum odoratum is an abundant and widespread grass of pastures, meadows, scrub, heaths and bogs, and many other grassland habitats throughout Ireland.

It is used in some commercially available seed mixes as a fodder grass.

**Anthriscus caucalis*

- Bur chervil (Scannell & Synnott, 1987); Upright hedge parsley (Brunker, 1950)
- *Peirsil bhog* (Scannell & Synnott, 1987)

APIACEAE (Umbelliferae)—THE CARROT FAMILY
Native

Anthriscus caucalis was recorded in 1875 in the Aran Islands but has not been recorded there since. It is found elsewhere in Ireland, but only very rarely.

*No ethnobotanical uses of this plant have been noted in Ireland.

23. Ó Dónaill (1977) defined *cáithluibh* as 'chaffweed, cudweed'.

Anthriscus sylvestris

- Cow parsley (Scannell & Synnott, 1987); Hedge-parsley (Webb & Scannell, 1983); Queen Anne's lace (Webb & Scannell, 1983); Hi-how (Praeger, 1938); Keck (Harron, 1986)
- *Peirsil bhó* (Scannell & Synnott, 1987); *Cos uisce* (Moloney, 1919; Dinneen, 1927); *Cosuisge* (de Vere Coneys, 1849)

APIACEAE (Umbelliferae)—THE CARROT FAMILY
Native

Anthriscus sylvestris is an abundant and widespread plant of roadsides, hedgerows, waste places and some woodlands.

It has been used as a culinary herb for garnishing and has a mild flavour resembling chervil, to which it is closely related (Phillips, 1983). However, when cooked, it loses its flavour. Chopped leaves are a useful and tasty garnish for new potatoes, and in potato salads and tomato and cucumber dishes and with eggs (Hilton, 2007).

In Ireland it was used medicinally as a remedy for kidney trouble (Doolan, 1994; Allen & Hatfield, 2004), but its use in traditional medicines may have been limited, perhaps in part, because it closely resembles many related poisonous species.

Ashley (2001) described how sections of the hollow stems of this species were used as moulds to make candles. The pith from rush stems were inserted into sections of cow parley stems, which were then filled with tallow. When solid, these made useful (and very cheap) candles to light the poorer cottages.

Anthyllis vulneraria

- Kidney vetch (Scannell & Synnott, 1987); Lady's fingers (Brunker, 1950)
- *Méara Muire* (Scannell & Synnott, 1987); *Meóir Mhuire* (Moloney, 1919); *Cosán uain* (Moloney, 1919); *Luanfhiatghal coitchion* (Wade, 1808b); *Cúig Muire* (Dinneen, 1927)

FABACEAE (Leguminosae)—THE PEA AND CLOVER FAMILY
Native

Anthyllis vulneraria is a common perennial that is tufted and spreading, with yellow or pinky red flowers. It is found all around the coast of Ireland in dry grassland habitats as well as on sand dunes, coastal rocks and cliffs, and dry banks in parts of the midlands.

Threlkeld (1726: unnumbered [12]) noted that it was sold as an astringent in the markets in 18th-century Ireland under the name 'stanch', but Allen and Hatfield (2004) considered its use in folk medicines (to treat cuts and bruises) in Britain and Ireland questionable. However, Uí Chonchubhair (1995) suggested that it may have been used in the Middle Ages to heal wounds.

Mahon (1983) listed this species as the source of a yellow dye.

Antirrhinum majus

- Snapdragon (Scannell & Synnott, 1987)
- *Srubh lao* (Scannell & Synnott, 1987); *Scrubh na laoch* (Anonymous, 1978)

VERONICACEAE (Scrophulariaceae)—THE SPEEDWELL FAMILY

Introduced

Antirrhinum majus is a commonly grown ornamental plant in Irish gardens that is very frequently established in the wild as an escape from cultivation. It occurs in waste places and on walls. Although grown as an annual plant, it is a perennial plant in nature. It is native to and widespread in the Mediterranean region.

Aphanes species

- Parsley-piert (Scannell & Synnott, 1987)

ROSACEAE—THE ROSE FAMILY

Including:

Aphanes arvensis

- Parsley-piert (Scannell & Synnott, 1987)
- *Mionán Muire* (Scannell & Synnott, 1987)

Native

Aphanes australis (*Aphanes inexspectata*)

- Slender parsley-piert
- *Mionán Muire caol* (Scannell & Synnott, 1987)

Native

Both species of *Aphanes* are small, hairy annual plants of bare patches in fields, walls, roadsides, field margins and railway lines. Both species are common and found throughout Ireland. *Aphanes arvensis* generally occurs on limestone soils and *A. australis* on acid soils.

Aphanes arvensis was apparently used in Ireland as a traditional medicine, perhaps to treat kidney complaints (Allen & Hatfield, 2004). Mears and Hillman (2007) and Mabey (1972) noted that this species and *A. australis* (listed under a different name, *A. microcarpa*) have been used to make a pickle that was eaten raw in the British Isles.

Apium graveolens

- Wild celery (Scannell & Synnott, 1987)
- *Smaileog* (Scannell & Synnott, 1987); *Lus na smaileóg* (Moloney, 1919; Dinneen, 1927; Ó Dónaill, 1977); *Meirse* (Cameron, 1883; Ó Dónaill, 1977); *Peirsil mhór (-le moire)*[24] (de Vere Coneys, 1849)

APIACEAE (Umbelliferae)—THE CARROT FAMILY
Native

Apium graveolens is a biennial herb that smells strongly of celery and is a closely related form of cultivated celery. It is found occasionally around the coastline and occurs in salty ditches and salt marshes.

It was widely used, especially in Co. Dublin, as a medicinal plant (Allen & Hatfield, 2004) for a variety of health complaints, including rheumatism, boils, rashes and to cleanse the blood.

In Europe celery is one of the best-known aphrodisiacs. Apparently, the roots were eaten as celery salad to strengthen the body and help 'kindle the fires of love' (Rätsch, 1997: 47). Elsewhere, wild celery has been used as a pain killer and in the treatment of rheumatism, flatulence, blood pressure, arthritis and digestive problems (Reader's Digest Association, 2006).

It was recorded in excavated pits at Fishamble Street, Dublin, in Viking Age deposits dating back more than 1000 years and may have been used as food flavouring (Geraghty, 1996).

Mears and Hillman (2007) suggested that this species should not be used for food, as it can be toxic when growing in wet habitats.

**Apium inundatum*

- Lesser marshwort (Scannell & Synnott, 1987)
- *Smaileog bháite* (Scannell & Synnott, 1987)

APIACEAE (Umbelliferae)—THE CARROT FAMILY
Native

Apium inundatum is a poisonous species that is common throughout Ireland in lakes, marshes, slow streams, pools and bog-holes.

*No ethnobotanical uses of this plant have been noted in Ireland.

24. De Vere Coneys (1849) applied this name to a plant he called 'smallage; *Apium palustre*'.

**Apium nodiflorum*

- Fool's water-cress (Scannell & Synnott, 1987)
- *Gunna uisce* (Scannell & Synnott, 1987); *Fualactar*[25] (de Vere Coneys, 1849; Dinneen, 1927); *Biolar bréige* (Ó Dónaill, 1977)

APIACEAE (Umbelliferae)—THE CARROT FAMILY
Native

Apium nodiflorum is a poisonous species found frequently throughout Ireland in wet habitats, such as ditches, streams, marshes, mud and lake-shores.

*No ethnobotanical uses of this plant have been noted in Ireland.

Aquilegia vulgaris

- Columbine (Scannell & Synnott, 1987)
- *Colaimbín* (Scannell & Synnott, 1987); *Lus an choluim* (Moloney, 1919); *Crúba leisín* (Moloney, 1919; Dinneen, 1927); *Lus an cholmhain* (de Vere Coneys, 1849); *Lus an cholaim* (Dinneen, 1927); *Lus an choilm* (Ó Dónaill, 1977)

RANUNCULACEAE—THE BUTTERCUP AND CROWFOOT FAMILY
Possibly Introduced

Aquilegia vulgaris is a rare perennial plant of open scrub habitats, stony grasslands and roadsides that mainly occurs in the centre of Ireland. Its flowers are a purple colour, or more rarely, pink or white.

It is frequently grown as an ornamental garden plant in Ireland and is a common garden escape, although most escaped plants seen in the wild are large flowered and of hybrid origin.

Mears and Hillman (2007) suggested that the leaves of columbine can be eaten raw or steamed as a vegetable, but that the roots are toxic. I have not found records of its use in Ireland for food. Woodward (1985) lists it as a poisonous plant, so it is not a plant that I have tried eating.

**Arabidopsis thaliana*

- Thale cress (Scannell & Synnott, 1987)
- *Tailís* (Scannell & Synnott, 1987)

BRASSICACEAE (Cruciferae)—THE CABBAGE AND CRESS FAMILY
Native

Arabidopsis thaliana is a small, weedy annual species that is found occasionally throughout Ireland in habitats such as waste places, cultivated ground, tilled fields, gardens, dry banks and walls.

*No ethnobotanical uses of this plant have been noted in Ireland.

---

25. Dinneen (1927) defined *fualactar* as '*sium nodiflorum*, creeping water parsnip'.

*Arabis brownii

- Fringed rock-cress (Scannell & Synnott, 1987)
- *Gas caillí duimhche* (Scannell & Synnott, 1987)

BRASSICACEAE (Cruciferae)—THE CABBAGE AND CRESS FAMILY
Native

Arabis brownii is rare and grows only on sandhills in the west of Ireland. It is very similar to *A. hirsuta*, except that it is more or less without hairs (glabrous).

*No ethnobotanical uses of this plant have been noted in Ireland.

---

*Arabis hirsuta

- Hairy rock-cress (Scannell & Synnott, 1987)
- *Gas caillí giobach* (Scannell & Synnott, 1987); *Gas an chailligh (na caillighe)* (Moloney, 1919)

BRASSICACEAE (Cruciferae)—THE CABBAGE AND CRESS FAMILY
Native

Arabis hirsuta is found mainly in the midlands and in the west of Ireland in dry habitats, such as on walls and in banks and waste places.

*No ethnobotanical uses of this plant have been noted in Ireland.

---

*Arabis petraea (Cardaminopsis petraea)

- Northern rock-cress (Scannell & Synnott, 1987)
- *Gas caillí Artach* (Scannell & Synnott, 1987)

BRASSICACEAE (Cruciferae)—THE CABBAGE AND CRESS FAMILIY
Native

Arabis petraea is a rare perennial herb that is found in Ireland only in the south and north-west on mountain cliffs.

*No ethnobotanical uses of this plant have been noted in Ireland.

Arbutus unedo

- Strawberry-tree (Scannell & Synnott, 1987)
- *Caithne* (Scannell & Synnott, 1987); *Crann sugh-thalmhain* (Moloney, 1919)

ERICACEAE—THE HEATHER FAMILY
Native

Arbutus is a very long-lived native tree that is found exclusively in the south-west of Ireland, in Cos. Cork and Kerry, as well as very rarely around Lough Gill in Co. Sligo. It may have been more widespread in the past because the Irish name for *Arbutus*, which is *caithne*, occurs in several other places, such as in Co. Clare, where the strawberry-tree has never been recorded. It occurs on the margins of oak woods, particularly in the Killarney valley of Co. Kerry, and on rocky lake-shores and islands. It is one of the few species that occurs in Ireland but not in Britain.

The fruits of *Arbutus unedo* are edible and high in vitamin C, if not particularly sweet, except when very ripe. They seem to have been eaten regularly in the south-west of Ireland,

Arbutus unedo (above), *Arbutus* fruits (below).

at least by poor people. Carr (1806: 244) noted that 'the poor people eat it [*Arbutus* fruits] as a wild strawberry.... [It] requires to be eaten with caution on account of its producing a lethargic effect; and to qualify its juices, the country people generally drink a draught of water with it.' Their former use for food in the south-west is also indicated by Bush (1769: 124–125), who wrote in 1764, 'From the tempting beauty of its form and colour, 'tis not a little mortifying to find its taste so insipid. However, I know of no danger of eating more than one or two at a time, as is asserted by Mr Solomon, in his account of this tree. 'Tis certain, they are eat in great numbers by the people who live about the lake [in Killarney], without any sensible ill effect.' Mackay (1836: 182) described them as 'not ungrateful to the taste when fully ripe'. They are used in other European countries (but not much in Ireland) to make jelly, wines and in the production of liqueurs (Hora, 1981). Baskets of *Arbutus* fruits were sold in Killarney in the 19th century (Nelson & Walsh, 1993). However, in recent years there are very few records of *Arbutus* fruits being eaten in the south-west, although I have heard of the fruits being used to make wine (Cormac Foley, pers. comm., 2010). In Corsica, *Arbutus* fruits are cooked in water, then strained and squeezed, and the liquid boiled with sugar to prepare a popular preserve. I can find no record of *Arbutus* having been used for such purposes in Ireland. K'Eogh (1735: 6) suggested that the fruits of *Arbutus* are 'of a cold nature, hurts the stomach and causes headaches'.

I have made *Arbutus* jam from fruits in mid-November when the ripest fruits can be picked. The resulting jam tasted alright, although rather inferior and similar to a nutty strawberry jam. Jelly made from these fruits would probably have been a better option. A recipe for *Arbutus* jam from France suggests mixing the fruit with an equal quantity of sugar, leaving for several hours to draw out the juice, adding the juice of two lemons, and then boiling and simmering the mixture for two hours before putting it into warmed jars.

I've not seen *Arbutus* fruits eaten by birds, although I noted at one point that one of the most fruitful trees at the National Botanic Gardens was surrounded by piles of scat left by foxes who may have been feeding off the fallen fruits.

Killarney table.

Arbutus timber has been an important constituent of 'Killarney work', inlaid marquetry produced in the 19th and early 20th centuries in Co. Kerry. It was also used for making souvenirs for sale in Killarney as far back as Victorian times. *Arbutus* wood is reddish brown when polished. Larner (2005) described the development of this industry producing souvenirs for visitors in Killarney during the early part of the 19th century. During the famine years in the late 1840s, between 50 and 60 people were employed in this trade making wooden objects that were richly decorated with inlays depicting local scenes, such as Muckross Abbey, Ross Castle, Muckross House, and Glena Cottage, as well as shamrocks, harps, thistles, ferns, sprigs of *Arbutus*, ivy and oak leaves, acorns, round towers and other symbols of Ireland. The *Arbutus* industry declined in the 1880s but continued in Killarney until the early 20th century. Then, it moved to the Gap of Dungloe where the furniture workshop managed by the Kiernan family continued until 1952 when it was destroyed by fire.

In Cos. Cork and Kerry, *Arbutus* was felled for the production of charcoal in the 17th and 18th centuries, but today there is little *Arbutus* timber available for use (Smith, [1756] 1979; Nelson & Walsh, 1993; Larner, 2005). Charcoal samples of *Arbutus* wood dated back to ancient Ireland have apparently been found. *Arbutus* wood was also used for fuel and burns well, giving off excellent heat. In 1828 Prince Hermann von Pückler-Muskau, a German prince visiting Killarney, Co. Kerry, wrote: 'A fresh-caught salmon broiled on arbutus-sticks over the fire was an admirable specimen of Irish fare' (von Pückler-Muskau, 1833: 301). Skewers made from *Arbutus* wood were apparently popular with fishermen on Lough Leane in Killarney, on which fish were grilled over an open fire and gave an excellent flavour to the fish (Cormac Foley, pers. comm., 2010).

Nelson (in Nelson & Walsh, 1993) said that *Arbutus* wood is suitable for the making of drones for uilleann pipes. Box (*Buxus sempervirens*), black ebony (*Diospyros* spp.) and other non-native imported tropical hardwoods have been more commonly used for such purposes, though.

Arbutus unedo is a popular ornamental tree for Irish gardens. One of the oldest trees in the collections of the National Botanic Gardens at Glasnevin is a native *A. unedo*, which is said to have somewhat narrower leaves than the wild material and is named variety 'Glasnevin'. It was probably planted there in the early 1800s and has developed the large, solid root mass (lignotuber) so characteristic of this species, from which new shoots arise when older branches die off, thus ensuring that an individual *Arbutus* tree can constantly renew itself and live for centuries. Nelson (2009) mentioned that *A. unedo* was widely grown in Irish gardens in the 17th and 18th centuries.

Arctium lappa

- Great burdock (Scannell & Synnott, 1987)
- *Cnádán mór* (Scannell & Synnott, 1987); *Liodán liosta* (Ua Maoileoin, 2001); *Suirichean suirich* (Cameron, 1883); *Cocoil* (de Vere Coneys, 1849); *Crádán* (Colgan, 1911)

ASTERACEAE (Compositae)—THE DAISY FAMILY

Native

Arctium lappa is a rare biennial species throughout Ireland of waste places and roadsides.

Native to Europe, this species has been used medicinally to treat a range of gastrointestinal disorders (van Wyk & Wink, 2004). In Ireland it was used medicinally to treat sores, and according to Threlkeld (1726), the stalks make an excellent salad. O'Regan (1997) mentioned this species was used to treat ringworm, and was used internally to treat pimples (although I wonder if this may also apply to the next species, *Arctium minus*, which is much commoner). Dinneen (1927: 721) also refered to the use of this species to treat ringworm: '*meacan an táthabha*, root of hellebore, great common burdock, its roots are pounded, boiled and applied as a poultice for ringworm, the poultice (*ceiridhe*) being often prepared on the opposite side of a river, *an eagla go bhfaghadh an phiast boladh an mheacain is go n-aistreochadh sí* [lest the worm scent the root and shift its position].' Actually, ringworm is a fungal infection and has nothing to do with a worm.

The leaves and roots are also edible and can be used as a vegetable in the same way as *Arctium minus*. The name *A. lappa* was formerly used to include *A. minus*, so some of the Irish names and uses of *A. lappa* can equally be applied to *A. minus*.

Arctium minus

- Lesser burdock (Scannell & Synnott, 1987); Cuckel (Colgan, 1904); Cradan (Hart, 1898)
- *Cnádán* (Scannell & Synnott, 1987); *Liodán* (Ua Maoileoin, 2001); *Liost* (Ua Maoileoin, 2001); *Cradán* (*Cnadán*) (Moloney, 1919); *Meacan tobach* (Moloney, 1919); *Meacan dogha* (de Vere Coneys, 1849; Moloney, 1919); *Capóg tuaithil* (de Vere Coneys, 1849; Moloney, 1919); *Copóg thuathail* (Dinneen, 1927; De Bhaldraithe, 1959; Ó Dónaill, 1977; Uí Chonchubhair, 1995); *Meacan tuathail* (de Vere Coneys, 1849); *Meacan tuan* (de Vere Coneys, 1849); *Leadán* (Dinneen, 1927); *Tobha* (Dinneen, 1927); *Meacan tua* (Dinneen, 1927); *Tuabhal* (Dinneen, 1927); *Lideán liosta* (Ó Criomhthain, n.d., ca. 1908)

ASTERACEAE (Compositae)—THE DAISY FAMILY

Native

Arctium minus is a biennial plant of roadsides, wasteground, farm-yards and cultivated places that is widely distributed throughout Ireland.

This species has historically been used as food, with the roots, shoots and flower stalks all edible. Burdock roots, which should be peeled of their inedible rind before use, can be boiled as a vegetable and used to make beer. Young shoots can also be boiled as a vegetable (Phillips, 1983) and taste like celery. Flower stalks can be candied by boiling them and then simmering in sugar solutions. Rutty (1772) noted that the stems were peeled before the burrs appear and eaten raw or boiled with vinegar, oil or pepper and taste good. He also said that the roots were eaten. Lewis-Stempel (2012) noted that dandelion and burdock beer was reputedly invented by St Thomas Aquinas in the 13th century. As he walked from his place of prayer, he trusted God to provide for his needs and these were the first plants he encountered. Homemade dandelion and burdock beer is mildly alcoholic but is commercially available in Britain as a soft drink.

Arctium minus.

Arctium minus was formerly used medicinally in Ireland as a blood purifier; externally to treat glandular swellings (Moloney, 1919) and other ailments, such as burns, cuts and flatulence; as a poultice for boils, glandular swellings, convulsions, nervousness, jaundice, rheumatism, colds and respiratory problems and as a diuretic for the treatment of dropsy and kidney and urinary ailments (Allen & Hatfield, 2004). Uí Chonchubhair (1995) recorded that on the Dingle Peninsula, Co. Kerry, it was used in an infusion with other species (*Teucrium scorodonia* and *Scorzoneroides autumnalis* [*Leontodon autumnalis*]) to treat tuberculosis, as a laxative and for 'decay'. McGlinchey (2007) noted its use in Co. Donegal by people bothered with wind on the stomach. The roots were boiled and the infusion drunk. Lady Wilde (1890: 53) listed burdock roots as a most effective cure for scrofula ('King's evil'). Scrofula is tuberculosis of the neck, and was historically known as the King's Evil because of a belief that sufferers could be cured by the touch of the monarch, a belief which continued in England until the early 18th century.

The spiny, hooked bristles of the flower heads were often used in a popular game by children—particularly small boys—who would stick them onto the clothes or in the hair of their friends (usually girls!).

Arctostaphylos uva-ursi

- Bearberry (Scannell & Synnott, 1987); Blanchnog (Hart, 1898); Cowberry (Hart, 1898)
- *Lus na stalóg* (Scannell & Synnott, 1987); *Lus na m-breileóg* (de Vere Coneys, 1849); *Lus na geire boisnighe* (de Vere Coneys, 1849); *Stalóg* (Dinneen, 1927)

ERICACEAE—THE HEATHER FAMILY

Native

Arctostaphylos uva-ursi is a low-growing perennial shrub found mainly on mountain slopes and upland heaths, as well as near the coast. In Ireland, it can be found from Co. Clare, through the Connemara region and Co. Donegal, then north-east as far as Co. Antrim. Elsewhere, *A. uva-ursi* occurs in boreal montane regions across Europe and North America.

Its fruits are scarlet berries, which are both astringent and mucilaginous. There is no evidence that they were collected for food in Ireland, despite having been used as food in Scandinavia and North America (Mears & Hillman, 2007), but it seems likely that they have had some local usage. The fruits are a bit dry and best used when they have been boiled or stewed. I have come across recipes for bearberry wine but have not heard of it ever being made in Ireland.

In North America its dried leaves have been used as a tobacco substitute (Peterson, 1977), but I have not found records of such use in Ireland. The Chippewa people of North America burnt the leaves, smoked in a pipe, to attract game (Pennacchio et al., 2010).

Arctostaphylos uva-ursi (bearberry). Watercolour by Lydia Shackleton. Courtesy of the National Botanic Gardens, Glasnevin.

As a medicinal plant *Arctostaphylos uva-ursi* has been used to treat a variety of urinary complaints (Allen & Hatfield, 2004; MacCoitir, 2006). Wade (1804: 31) noted, 'It is much esteemed in nephritic complaints, and particularly serviceable in tanning leather, being highly astringent'. The leaves contain a phenolic glycoside known as arbutin, which is the basis for its medicinal use (van Wyk & Wink, 2004). Jude (1933: 46) said that the leathery leaves were used to make 'an aqueous infusion'.

Bearberry has also been used to obtain a blue dye for dyeing textiles (Mahon, 1983).

**Arenaria ciliata*

- Fringed sandwort (Scannell & Synnott, 1987)
- *Gaineamhlus gaelach* (Scannell & Synnott, 1987)

CARYOPHYLLACEAE—THE CARNATION OR PINK FAMILY

Native

Arenaria ciliata is a rare species in Ireland, found only on the mountain cliffs of Ben Bulben, Co. Sligo.

*No ethnobotanical uses of this plant have been noted in Ireland.

**Arenaria norvegica*

- Arctic sandwort (Scannell & Synnott, 1987)
- *Gaineamhlus Artach* (Scannell & Synnott, 1987)

CARYOPHYLLACEAE—THE CARNATION OR PINK FAMILY

Native

Arenaria norvegica is a rare species that was first found in Ireland in 1961 and is known only from limestone pavements in the Burren, Co. Clare, near Black Head. It was recently rediscovered in Ireland in 2008.

*No ethnobotanical uses of this plant have been noted in Ireland.

**Arenaria serpyllifolia*

- Thyme-leaved sandwort (Scannell & Synnott, 1987)
- *Gaineamhlus tíme* (Scannell & Synnott, 1987); *Luibh na duimhche* (Uí Chonchubhair, 1995)

CARYOPHYLLACEAE—THE CARNATION OR PINK FAMILY

Native

Arenaria serpyllifolia is a small annual plant of dry, rocky, stony and sandy places and walls that is distributed throughout Ireland, most commonly in the centre.

*No ethnobotanical uses of this plant have been noted in Ireland.

Armeria maritima

- Thrift (Scannell & Synnott, 1987); Sea pink (Moloney, 1919); Dead man's posy (Hart, 1898); Sea daisy (McCarthy & Hawkes, 2007); Lady's cushion (O'Rourke, 2006)
- *Rabhán; Nóiníní ruáin* (Ua Maoileoin, 2001); *Ruán* (Ua Maoileoin, 2001); *Nóinín an chladaigh* (O'Rourke, 2006); *Pabhsaer aille* (O'Rourke, 2006); *Tonn an chladaigh* (Moloney, 1919); *Cíp trágha* (Moloney, 1919). Probably also *Cíbt* (Moloney, 1919)

PLUMBAGINACEAE—THE THRIFT FAMILY

Native

Armeria maritima.

Armeria maritima is an attractive pink to red-flowered tufted plant in many coastal habitats around Ireland, such as in salt marshes and on cliffs, rocks and banks by the sea.

There are very few records of the use of sea pink in Ireland, although I have found a reference to its use for fuel and bedding on one of the islands off the south-west coast of Kerry. The following is taken from an account given by fishermen staying on Inis na Bró, one of the uninhabited islands of the Blasket Island group in Co. Kerry: 'We built a small four-sided stone wall and filled it up with soft sea-pink that we found growing on the cliffs. We set the sea-pink afire with a drop of paraffin and we soon had a fine fire blazing with the sea pink aflame because they were as dry as snuff'.... 'We gathered more of the sea-pink and scattered it over the floor of the cave and we spread the blankets down on them' (Ó Dubhshláine, 2009: 33). Dinneen (1927) also mentioned it was used for fuel.

It is occasionally grown as an ornamental plant in rock gardens (Wyse Jackson, 2006). A variety of flower colour forms occur from deep red to mauve to pink, and even very occasionally pure white.

Armoracia rusticana (*Cochlearia armoracia*)

- Horse-radish (Scannell & Synnott, 1987)
- *Meacan ragaim* (Scannell & Synnott, 1987); *Meacan each* (Moloney, 1919); *Meacan ruguim* (Moloney, 1919); *Uisce beatha na mbráthar* (Moloney, 1919; Dinneen, 1927); *Rácadal* (de Vere Coneys, 1849; Cameron, 1883; Dinneen, 1927; Uí Chanainn, 1962); *Raidis fhiadhain* (Dinneen, 1927; Ó Dónaill, 1977); *Reagam* (Dinneen, 1927; Uí Chanainn, 1962)

BRASSICACEAE (Cruciferae)—THE CABBAGE AND CRESS FAMILY

Introduced (possible archaeophyte)

This introduced plant, which is native to the Mediterranean region, is a robust and somewhat invasive perennial plant with large dock-like leaves arising from the tough taproot. Although found in some waste places and neglected gardens, it is rarely naturalised in Ireland. It does not produce mature fruits and so can spread only by vegetative means. It is cultivated in many larger and traditional vegetable gardens and sometimes survives as a remnant of cultivation.

The roots are used to make horse-radish sauce: roots are washed and cleaned and the brown outer layer removed. Peeling the horse-radish is not an easy task, however, in part because it is so knobbly but also because of its pungent roots. To make a delicious sauce to accompany roast beef, steaks and smoked fish, the white, inner part of the root is grated and mixed with cream, mayonnaise, cream cheese or yogurt, as well as vinegar, mustard and seasonings. The tender, young leaves can also be chopped and used (sparingly) in salads, and the roots can also be sliced and roasted as a vegetable. Beeton (1861: 215) suggested making horse-radish vinegar as 'an agreeable relish to cold beef.'

Horse-radish has also been used as a medicinal plant to treat bronchial conditions and infections of the urinary tract and used externally to treat rheumatism and inflammations (van Wyk & Wink, 2004). Moloney (1919: 15) also mentioned its use 'as a condiment externally applied as a counter-irritant.' Ui Sheasnáin (1982) mentioned that in Co. Cavan, brown sugar was put into the hollow in a piece of horse-radish and left overnight. Then, a teaspoon of the resulting liquid was taken twice daily to treat bronchitis. Maloney (1972) reported that it was used to treat a toothache in Co. Cavan: a piece of the root was put onto the gum near the painful tooth.

Arrhenatherum elatius.

Arrhenatherum elatius (*Arrhenatherum avenaceum*; *Avena elatior*)

• False oat (Scannell & Synnott, 1987); Pearl[26] (Hart, 1898)
• *Coirce bréige* (Scannell & Synnott, 1987); *Coirce fiadhain* (Moloney, 1919); *Coirce airde* (White, 1803; Wade, 1808a)
POACEAE (Gramineae)—THE GRASS FAMILY
Native

Arrhenatherum elatius is an abundant, tall perennial grass species throughout Ireland of rough grassland, hedges, banks, sand dunes and waste places.

Hart (1898) suggested that it was used medicinally to treat eye ailments.

According to Allen and Hatfield (2004), *Arrhenatherum elatius* was used in Co. Derry as a veterinary remedy for horses and dogs and to treat white 'scale'.

Wade (1808a) noted that *Avena elatior* is a good grass for hay production.

Artemisia absinthium

- Wormwood (Scannell & Synnott, 1987)
- *Mormónta* (Scannell & Synnott, 1987); *Bormónta* (Moloney, 1919); *Barramótar* (Dinneen, 1927)

ASTERACEAE (Compositae)—THE DAISY FAMILY
Introduced (possible archaeophyte)

Artemisia absinthium.

Artemisia absinthium is found occasionally in waste places, particularly in the east, south-east and south of Ireland. It was formerly grown and used as a medicinal plant, to treat stomach ailments, and perhaps was also used as a food or drink flavouring (Akeroyd, 1996).

This plant has been and continues to be used as a medicinal plant in various parts of the world. For example, in Uganda, fresh leaves, stems and roots are used to treat diarrhoea, abdominal pain and stomach aches, sore throats, fevers, worms, indigestion, convulsions, malaria and liver problems, and it is used as an appetite enhancer and insecticide (Adriaens, 2006). Wormwood contains active ingredients, including a compound called thujone, that enhance the effect of alcohol, but it is a convulsive poison at high doses. It can cause delirium, hallucinations and seizures (van Wyk & Wink, 2004). The traditional use of this species to flavour absinthe, vermouth and other alcoholic drinks has been banned in several countries because of the toxic effect of this plant extract, although occasional small doses may cause little harm. Moloney (1919: 30) listed it as having been used in traditional medicine in Ireland as 'one of the cures of falling sickness [epilepsy].'

26. According to Hart (1898: 274), *Arrhenatherum avenaceum* was known as 'Pearl', 'from the shining tuberous or knotted rootstock'.

Artemisia maritima (Seriphidium maritimum)

- Sea wormwood (Scannell & Synnott, 1987)
- *Liath na trá* (Scannell & Synnott, 1987); *Liath-lus trágha* (Dinneen, 1927)[27]

ASTERACEAE (Compositae)—THE DAISY FAMILY
Native

Artemisia maritima is a small, slightly shrubby, somewhat aromatic perennial that is found in parts of the east and west of Ireland in habitats close to sea, such as amongst seashore rocks and on salt marshes. It was formerly more widespread and today is very rare in Ireland. Its decline may well have been in part caused by its former collection and use.

Threlkeld (1726: unnumbered [2–3]) reported that this species was collected from the coast of Cos. Louth and Meath and taken to Dublin and used by 'Ale-house keepers making their Purl'. 'Purl' is a malted liquor, medicated or spiced; formerly, purl referred to ale or beer in which wormwood or other bitter herbs had been infused and was regarded as a tonic, which according to Threlkeld, 'great consumption of which is made in Winter mornings'. He described the virtues of this species as many, to treat weaknesses and disorders of the stomach, recover lost appetites, remove surfeits, strengthen the bowels, and it was also useful in cases of dropsy, jaundice, ague and against worms. The green leaves were made into an ointment with pig's lard for swellings. However, he noted drawbacks to the use of this plant medicinally and said that it can be blamed for causing 'Hurtfulness to the Eyes and blunting Cupid's Darts'. K'Eogh (1735) recommended that a decoction of this species drunk or applied to the stomach or navel as a poultice is useful to clear worms. Smith (1815: 361), quoting Ray, recommended 'its infusion in bad ale, in order to make it more wholesome'.

Artemisia maritima.

Artemisia vulgaris

- Mugwort (Scannell & Synnott, 1987); Muggart (Hart, 1898); Muggers (Hart, 1898)
- *Mongach meisce* (Scannell & Synnott, 1987); *Buachalán bán* (Ua Maoileoin, 2001); *Mugárd* (Moloney, 1919); *Liathlus* (Moloney, 1919; Dinneen, 1927); *Buachalán liath* (Moloney, 1919); *Mongach-measga* (Moloney, 1919); *Bofulan ban* (Cameron, 1883); *Buafannan ban* (Cameron, 1883); *Buafanán* (de Vere Coneys, 1849; Dinneen, 1927; Uí Chanainn, 1962; Uí Chonchubhair, 1995); *Buafanán bán* (de Vere Coneys, 1849); *Buafanán liath* (de Vere Coneys, 1849); *Mongach measca* (Dinneen, 1927); *Mugard* (Ó Dónaill, 1977); *Bothalán bán* (Colgan, 1911)[28]

ASTERACEAE (Compositae)—THE DAISY FAMILY
Native (possible archaeophyte)

Artemisia vulgaris is a robust, aromatic perennial herb found growing on roadside verges, waste places and disturbed ground throughout Ireland.

27. Dinneen (1927) defined *liath-lus trágha* as 'strand mugwort'.
28. According to Colgan (1911: 47) *Bothalán bán* was 'a common name for this species in Clare Island'.

Artemisia vulgaris.

Mugwort has been used for centuries as a medicinal plant throughout its native range in Europe and Asia, for example, to treat threadworm and roundworm, neuroses, depression, restlessness, insomnia, anxiety and irregular or painful menstruation (van Wyk & Wink, 2004). In more recent times it has been used to treat dyspepsia and lack of appetite. McGlinchey (2007) said that in Co. Donegal the leaves were smoked in a pipe like tobacco by people who had lost their appetites. In Ireland, the leaves were dried and smoked as a tobacco substitute by some people in parts of Fanet, Co. Donegal (Hart, 1898). Elsewhere in the world, the leaves were smoked for medicinal purposes also. For example, in China and Japan it was smoked to relieve asthma (Pennacchio et al., 2010). Threlkeld (1726: unnumbered [17]) said that it 'is a famous plant for disorders of the female sex; both given inwardly and used outwardly in Baths'. According to Allen and Hatfield (2004), it was formerly used in Ireland for the treatment of epilepsy.

Williams (1993) suggested that it may have been used in Ireland to treat drunkenness; hence, its name in Irish is *mongach meisce*. This same author noted it was used as an anti-epileptic, for cramps and female gynaecological problems, and to kill insects. It may also have been used as a cure for narcolepsy. Harrop (1977) indicated a strong infusion of the plant was dabbed on the skin and said to repel insects.

In parts of Europe it has been used as a culinary herb and aromatic tea and also to flavour beer (Phillips, 1983; Phillips said that it has a revolting flavour!). Shaw (1998) suggested that it can be added to stews and stuffing for meat dishes and promotes appetite. Hilton (2007) noted that it has been popular to aid digestion and for that reason has been used with fatty meats such as pork, goose, duck, eel and mutton.

Danaher (1972) noted that in East Cork it was kept in the house for luck or burnt in the fire for the same reason. Rutty (1772: 100) also mentioned that 'thrown upon coals, it banishes insects'. The leaves were burnt and the smoke inhaled as a remedy for unspecified ailments. O'Regan (1997) mentioned its use to fumigate a sick room. Elsewhere in the world, the leaves of this species have been burnt as an insect repellant against mosquitos, for example in Bolivia, central Asia and India (Pennacchio et al., 2010). Rutty (1772: 102) mentioned the use of mugwort leaves for tinder: 'The dry leaves being gently rubbed in the hands, until the external peel be separated, there appears internally a cottony substance, which serves some as tinder in striking fire'.

It was used as a veterinary remedy to treat gangrene in horses in Donegal (Allen & Hatfield, 2004).

In the Orkney Islands of Scotland, mugwort was used for basketry. These baskets were used by fishermen to hold fish and bait and were more suitable for use at sea than straw equivalents. These baskets were called 'huvies' (Milliken & Bridgewater, 2004). I have not found records of mugwort being used to make baskets in Ireland.

Moloney (1919) listed the following Irish names for another related species, *Artemisia abrotanum* (southernwood[29]): *lus an tsean-duine* (also given by Dinneen, 1927) and *surrabhán*, or *subharmant* (Dinneen, 1927). This species is sometimes grown as a shrub in gardens and has attractive grey-green foliage. Clyne (1989) indicated that the Irish name *lus an tsean-duine* comes from its use by old people to stay awake in church. Although native in parts of Europe, it is not found in Ireland, nor is it naturalised in the wild today. It may have been used formerly in Ireland as a medicinal plant, which is suggested by its having several vernacular names. Rutty (1772), in his *An Essay Towards a Natural History of the County of Dublin*, noted that it was spread between woollen clothes to defend

29. Hart (1898) gave the common name 'lad's love' for *Artemisia abrotanum*, from the Inishowen peninsula, Co. Donegal.

them from moths. It is strongly aromatic and bitter tasting and has had a long history as a medicinal plant in other parts of Europe as an antiseptic, a purgative, and particularly as an emmenagogue to bring on menstrual flow.

Arum maculatum

- Lords and ladies (Scannell & Synnott, 1987); Cuckoo pint (Hart, 1898; Colgan, 1904); Wake-robin (Praeger, 1938); Devil's bit (Hart, 1898; Colgan, 1904); Adder's tongue (Hart, 1898); Deadly nightshade[30] (Hart, 1898); Wild arum (Doogue & Krieger, 2010); Jack-in-the-box (Praeger, 1938); Jack-in-the-pulpit (Praeger, 1938)
- *Cluas chaoin* (Scannell & Synnott, 1987); *Bod gear* (Moloney, 1919); *Cluas chaoin* (Moloney, 1919); *Gaoicín chúthaigh* (Moloney, 1919; Dinneen, 1927); *Geathar* (Moloney, 1919; Dinneen, 1927); *Áram* (De Bhaldraithe, 1959; Ó Dónaill, 1977); *Gachar* (Cameron, 1883); *Cluas caicin* (de Vere Coneys, 1849; Dinneen, 1927)

ARACEAE—THE LORDS-AND-LADIES FAMILY
Native

Arum maculatum is a perennial herb that occurs widely throughout Ireland in woodlands, hedges and other shady habitats.

Baker (1999) mentioned that the roots of *Arum maculatum* produce a farinaceous powder like sago, which was distributed during the Irish potato famine in the 1840s for food. In Britain it was used on the Isle of Portland, Dorset, where it was known as Portland sago or Portland arrowroot. The starch it contains was also used to stiffen Elizabethan lace ruffs. I have also come across the roots of a related species, *A. italicum*, being used to make flour and baked as biscuits on the Portuguese island of Madeira.

Threlkeld (1726: unnumbered [19]) reported that the roots, mixed with powder of brimstone (sulphur), were used as a remedy for consumption (pulmonary tuberculosis). Threlkeld (1726: unnumbered [19]) also said that the roots boiled with honey can cure 'all pituitous diseases of the Breast', and that fresh roots bruised and distilled with cow's milk is a good remedy for cleaning the skin of wrinkles, scruff [?], freckles, spots or other blemishes. K'Eogh (1735: 6) wrote that the whole plant is 'very hot and biting, inflaming the mouth and throat for a long time', but nevertheless, he suggested that the roots can be pulverized and taken to induce vomiting and are good to clear obstructions from the lungs. K'Eogh (1735) also suggested that a poultice made from them, mixed with cow dung, helps to ease the pain of gout. Allen and Hatfield (2004) reported its use in Co. Wexford, where powdered root was placed under the tongue and swallowed with saliva to relieve palsy (paralysis). The roots were used to treat jaundice and, in Co. Offaly, given to boys as a cure for worms (Williams, 1993).

Some authors reported that all parts of this plant are poisonous, although in other parts of this and related species range, such as in the Middle East, it is eaten. I presume that the toxic aspect is removed by thorough cooking.

Although the red-orange berries of *Arum maculatum* are a common cause of poisonings, these are rarely fatal. The colourful berries and vegetation contain calcium oxalate crystals that intensely irritate the skin, the mouth, and any part of the body that comes in contact. There is a risk of severe allergic reactions from the minute crystals that derive from

30. Hart (1898) recorded the common name deadly nightshade from Favour Royal, Co. Tyrone. Of course, this name is more commonly applied to *Atropa belladonna*.

saponins, which are toxic. Individuals or animals may be attracted to the fruiting spike; however, the extreme tingling and irritation, plus the acrid taste of the berries, usually deter a person from eating enough berries to be seriously poisoned.

Asparagus officinalis subsp. *prostratus*

- Wild asparagus (Scannell & Synnott, 1987)
- *Lus súgach* (Scannell & Synnott, 1987); *Lus praicidh* (Moloney, 1919; Dinneen, 1927); *Creamh muc fiadhain* (Moloney, 1919); *Lus súch* (De Bhaldraithe, 1959); *Creamh muic fiadh* (Cameron, 1883); *Lus súghach* (Dinneen, 1927); *Asparagas* (Ó Dónaill, 1977); *Lus spreagtha* (Ó Dónaill, 1977); *Prachlus* (Uí Chanainn, 1962)[31]

ASPARAGACEAE—THE ASPARAGUS FAMILY
Native

The prostrate wild asparagus is a rare native plant in Ireland found in coastal habitats in the east and south. It is a very close relative (a subspecies) of the upright-growing garden asparagus. The Irish name for garden asparagus is *lus súgach garrai*.

There are no records of wild asparagus having been gathered and used in Ireland as a vegetable, although ordinary asparagus has been a popular vegetable grown in gardens in Ireland since the 16th century, most commonly used as a vegetable (lightly boiled, steamed or grilled and served with melted butter), in soup, stir-fried or included in egg dishes. Moloney (1919: 43) mentioned that asparagus was known as a vegetable in Ireland in the time of Jonathan Swift (the early 1700s) and noted 'it is said to have been introduced from Holland by his patron, Sir Wm. Temple'.

Asparagus officinalis (although not necessarily wild asparagus) was formerly used by herbalists in Ireland as a treatment for kidney ailments, gout and rheumatism (Threlkeld, 1726; Moloney, 1919) and reputed to be an aphrodisiac (Rätsch, 1997). Elsewhere in the world, asparagus has been used medicinally. For example in Tehran, Iran, the roots were burned to make smoke, which was inhaled for the relief of toothaches (Pennacchio et al., 2010).

Eating asparagus leads to a characteristic strong odour in the urine of some people due to the metabolic transformation of a compound in the asparagus into another (methyl-thiopropionate to methyl-mercaptan) (van Wyk & Wink, 2004).

Asperula cynanchica

- Squinancywort (Scannell & Synnott, 1987)
- *Lus na haincise* (Scannell & Synnott, 1987)

RUBIACEAE—THE BEDSTRAW FAMILY
Native

Asperula cynanchica is a small perennial plant of limestone grasslands, sand dunes and other lime-rich (calcareous) habitats. It is common from Cos. Kerry to Galway but is

31. Uí Chanainn (1962) gave the word *prachlus* for 'asparagus'.

absent from the rest of Ireland.

According to Uí Chonchubhair (1995) squinancywort was used medicinally to treat quinsy and other ailments. I have not been able to find any records of its use medicinally in Ireland. Quinsy is a serious throat infection, peritonsillar abscess, associated with acute tonsillitis.

The roots of *Asperula cynanchica* contain a red dye, but I have not found records of its use for dyeing purposes in Ireland.

Asperula cynanchica.

**Aster tripolium*

- Sea aster (Scannell & Synnott, 1987)
- *Luibh bhléine* (Scannell & Synnott, 1987); *Lus glotáin* (Moloney, 1919)

ASTERACEAE (Compositae)—THE DAISY FAMILY
Native

Aster tripolium is a common plant along the coasts of Ireland of salt marshes and other muddy coastal and estuarine habitats.

*No ethnobotanical uses of this plant have been noted in Ireland.

Aster tripolium.

Astragalus danicus (Astragalus hypoglottis)

- Purple milk-vetch (Scannell & Synnott, 1987)
- *Bleachtphiseán* (Scannell & Synnott, 1987); *Bliochtphiseáin* (Moloney, 1919)

FABACEAE (Leguminosae)—THE PEA AND CLOVER FAMILY
Native

Astragalus danicus is a very rare perennial in Ireland, known only from a limestone grassland habitat on cliff tops in the Aran Islands, Co. Galway.

Moloney (1919: 21) noted this plant (under the name *Astragalus hypoglottis*) was 'said to increase the milk supply in goats', hence milk-vetch is the common name. Moloney does not say whether it was used in Ireland for such purposes.

Astragalus danicus.

**Atriplex glabriuscula*

- Babington's orache (Scannell & Synnott, 1987)
- *Eilifleog chladaigh* (Scannell & Synnott, 1987)

AMARANTHACEAE (Chenopodiaceae)—THE GOOSEFOOT FAMILY
Native

Atriplex glabriuscula is found occasionally all around the Irish coast on sandy and shingle shores.

*No ethnobotanical uses of this plant have been noted in Ireland.

**Atriplex laciniata*

- Frosted orache (Scannell & Synnott, 1987)
- *Eilifleog phlúrach* (Scannell & Synnott, 1987)

AMARANTHACEAE (Chenopodiaceae)—THE GOOSEFOOT FAMILY
Native

Atriplex laciniata is found less frequently than *A. glabriuscula* and occurs along the coast on sandy shores.

*No ethnobotanical uses of this plant have been noted in Ireland.

**Atriplex littoralis*

- Grass-leaved orache (Scannell & Synnott, 1987)
- *Eilifleog thrá* (Scannell & Synnott, 1987)

AMARANTHACEAE—THE GOOSEFOOT FAMILY
Native

Atriplex littoralis occurs mainly on the eastern coast in Ireland in muddy and gravel shores.

Lewis-Stempel (2012: 36) said that *Atriplex littoralis* is inedible, having 'a foul smell, and fouler taste'.

*No ethnobotanical uses of this plant have been noted in Ireland.

Atriplex species

- Orache (Scannell & Synnott, 1987)
- *Eilifleog* (Scannell & Synnott, 1987); *Praiseach mhin*[32]; *Ceathramha caorach* (Colgan, 1911; Dinneen, 1927); *Meille* (Dinneen, 1927)[33]

AMARANTHACEAE (Chenopodiaceae)—THE GOOSEFOOT FAMILY

Including:

Atriplex patula

- Common orache (Scannell & Synnott, 1987); Lamb's quarters (Hart, 1898); Iron-root (Harron, 1986)
- *Eilifleog chaol* (Scannell & Synnott, 1987)

Possibly Introduced

Atriplex prostrata (Atriplex hastata)

- Spear-leaved orache (Scannell & Synnott, 1987); Lamb's quarters (Hart, 1898)
- *Eilifleog leathan* (Scannell & Synnott, 1987); *Praiseach mhín* (de Vere Coneys, 1849)

Native

Atriplex prostrata is a common plant near the coast and can be found along the seashore

32. Cameron (1883) said the name *praiseach mhin* was used for *Atriplex hastata* and *A. patula*. Uí Chonchubhair (1995) gave the Irish name *praiseach mhin* for *A. prostrata*. Ó Dónaill (1977) defined *praiseach mhin* as 'orache'.

33. Dinneen (1927) defined *meille* as 'atriplex, golden herb'.

and in waste places. *Atriplex patula* is also common and found in wastegrounds, cultivated fields and roadsides. *Atriplex patula* is a possible introduction, but *A. prostrata* is regarded as a native species.

According to Phillips (1983), the leaves of both species can be boiled and used as a vegetable or in soup. *Atriplex patula* was apparently sold in the 1700s in Ireland (Threlkeld, 1726), presumably as a vegetable. *Atriplex patula* and *Chenopodium album*, both of which have been called 'fat hen', were sold by hawkers and eaten as leafy vegetables in Dublin until the 18th century (Rutty, 1772; Geraghty, 1996). Geraghty (1996) recorded that seeds of fat hen were found in Viking remains at Fishamble Street in Dublin dating back more than 1000 years. According to Lewis-Stempel (2012), the roots of *A. patula* should not be used as food as they are mildly toxic.

Atriplex prostrata.

Uí Chonchubhair (1995) noted that oraches are mildly laxative, so are best not eaten in too much quantity at any one time.

In the north-east of Ireland, Praeger (1938) recorded the common name 'miles' for *Atriplex* species.

Atropa belladonna

- Deadly nightshade (Scannell & Synnott, 1987)
- *Lus na hoíche* (Scannell & Synnott, 1987); *Lus mór na coille* (Ó Dónaill, 1977; Anonymous, 1978); *Lus díoth mór* (Anonymous, 1978); *Miotóg bhuidhe* (Moloney, 1919); *Lus na h-oidhche* (de Vere Coneys, 1849); *Lus mór na coilleadh* (Dinneen, 1927); *Miotóg bhuí* (Ó Dónaill, 1977)[34]

SOLANACEAE—THE NIGHTSHADE FAMILY

Possibly/Probably Introduced (possible archaeophyte)

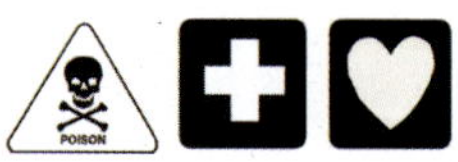

Atropa belladonna (deadly nightshade). Watercolor by Lydia Shackleton. Courtesy of the National Botanic Gardens, Glasnevin.

Atropa belladonna is a large perennial herb that can grow to as much as 2 m tall. It is rare in Ireland and when found, is usually as a relict of former cultivation. It has been recorded in waste places throughout but is often destroyed because it is quite widely recognised as a dangerously poisonous weed and is, in fact, one of the most toxic plants in the Western hemisphere. It is also strongly hallucinogenic, narcotic and has a long history of use for its supposed aphrodisiac and genuine medicinal properties. Moloney (1919) considered that it has had a long history of use in Ireland.

In modern medicine, alkaloid extracts of this species are used as a sedative to treat asthma and pain, especially abdominal spasms and pain in the area of the bile ducts (van Wyk & Wink, 2004). The drug atropine is derived from this species. The large, black glossy fruits might be regarded as tempting to eat, but as the tropane alkaloids they contain are very toxic, they should **not** be eaten and can lead to hallucinations, insomnia, confusion and death. Woodward (1985: 64) gives the symptoms of poisoning by this species as 'dilation of pupils, rapid heartbeat, cerebral excitement, trembling; followed by prostration, coma and death from asphyxia'.

K'Eogh (1735: 87) wrote of the fruit that 'it causes such as eat thereof to fall into a deep sleep, and when they awake they fall into an outrageous passion of anger, in which they

34. Ó Dónaill (1977) defined *miotóg bhuí* as 'nightshade'.

continue until they dye [die].'

Extracts of *Atropa belladonna* have been used in eye drops to dilate the pupils. It was thus formerly used during the Renaissance and even more recently by women to enhance their beauty and allure, hence its scientific name, *belladonna*.

Lady Wilde (1887: 250) mentioned an Irish superstition that 'the juice of deadly nightshade distilled, and given in a drink, will make the person who drinks believe whatever you will to tell him, and choose him to-believe'. It may work, but this treatment is one to avoid!

Pennacchio et al. (2010) described how in parts of the Ukraine, the leaves of this species were mixed with those of *Datura stramonium* (thorn-apple) and *Hyoscyamus niger* (henbane) and smoked to relieve asthma. All three are highly poisonous plants, fatally so, and this is not a treatment that is recommended.

**Avenula pubescens* (*Helictotrichon pubescens*; *Avena pubescens*)

- Downy oat-grass (Scannell & Synnott, 1987)
- *Coirce clumhach* (Scannell & Synnott, 1987); *Coirce cluimhadh* (White, 1803; Wade, 1808a)

POACEAE (Gramineae)—THE GRASS FAMILY
Native

Avenula pubescens is a common grass throughout Ireland that is found only on limestone soils in habitats such as pastures, roadsides, dry banks and rocky places.

*No ethnobotanical uses of this plant have been noted in Ireland.

**Baldellia ranunculoides*

- Lesser water-plantain (Scannell & Synnott, 1987)
- *Corrchopóg bheag* (Scannell & Synnott, 1987)

ALISMATACEAE—THE WATER-PLANTAIN FAMILY
Native

Baldellia ranunculoides occurs throughout Ireland and is found in the shallow margins of lakes and ponds.

*No ethnobotanical uses of this plant have been noted in Ireland.

Ballota nigra

- Black horehound (Scannell & Synnott, 1987)
- *Grafán dubh* (Scannell & Synnott, 1987); *Bréanlus* (Moloney, 1919; Dinneen, 1927)

LAMIACEAE (Labiatae)—THE DEAD-NETTLE FAMILY
Introduced (possible archaeophyte)

Ballota nigra is a rare and declining perennial weed in Ireland that is found mainly in the east. It occurs on wasteground and roadsides.

It is native to Europe and Asia and has been used in traditional medicine for centuries. According to van Wyk and Wink (2004), it has been used as a sedative and to treat nervous disorders, stomach pains, nausea, sleeplessness and coughs. Moloney (1919) said it was used in Ireland to make a poultice for skin ulcers, and Allen and Hatfield (2004) said it was used for colds, asthma, coughs and chest complaints. Dirrane (1997) mentioned that horehound was used in the Aran Islands in the late 1890s and early 1900s to treat blood disorders or in times of blood loss.

Remains of this plant have been found in Viking Age deposits in excavations from Fishamble Street, Dublin, where it was either a weed of wasteground or used as a medicinal plant, or both (Geraghty, 1996).

Barbarea vulgaris

- Winter-cress (Scannell & Synnott, 1987); Yellow rocket (Scully, 1916)
- *Treabhach* (Scannell & Synnott, 1987)

BRASSICACEAE (Cruciferae)—THE CABBAGE AND CRESS FAMILY
Native

Barbarea vulgaris is a common biennial plant throughout Ireland of cultivated ground, waste places and roadsides.

Threlkeld (1726) noted that it was used in Dublin as a cure for a sore leg.

Formerly, it was probably widely used as a green vegetable (Mears & Hillman, 2007). The young leaves may be used in salads, either finely chopped or whole. They have a pleasant but slightly bitter taste. They are best picked and used in the spring months of March and April. Older leaves can be boiled or steamed and used like spinach.

Bellis perennis

- Daisy (Scannell & Synnott, 1987); Nooneen (Hart, 1898)
- *Nóinín* (Scannell & Synnott, 1987); *Bó dhonn* (Dinneen, 1927); *Neóinín bán* (Ó Criomhthain, n.d., ca. 1908)

ASTERACEAE (Compositae)—THE DAISY FAMILY
Native

The daisy is abundant throughout Ireland in lawns, stream and river-banks, roadsides and short grasslands. It is a familiar element throughout temperate Europe and widely naturalised beyond its original range. Ornamental double-flowered forms are widely grown in gardens and sometimes survive in the wild from discarded plants.

Bellis perennis was formerly used medicinally in Ireland. According to Threlkeld (1726: unnumbered [23]), it 'helpeth Bones to knit again'. It was also used to make an ointment for burns, as an eye lotion, and for a variety of other ailments, including scrofula, consumption, pleurisy, coughs, colds, headaches, stomach and liver complaints, and skin troubles such as whitlows, ringworm, chilblains, erysipelas, and blasts or facial swellings (Uí Chonchubhair, 1995; Allen & Hatfield, 2004). Lady Wilde (1887: 247), mentioning the use of daisies to cure 'weak eyes', said 'A decoction of the flowers of daisies boiled down is an excellent wash, to be used constantly.'

Daisies were, and to a certain extent still are, popular plants for childhood games. To test the love of a sweetheart, the outer ray florets of the daisy inflorescence were plucked while alternately repeating, '(S)he loves me, (s)he loves me not', with the final floret indicating the answer. A similar game is played with other flowers, particularly other members of the Asteraceae, such as ox-eye daisies, (*Chrysanthemum leucanthemum* [*Leucanthemum vulgare*]). Making daisy chains was (and perhaps still is) a popular childhood pastime in Ireland. An alternative type of daisy chain that was made formerly was one in which the daisies were threaded onto a rush stem (MacCoitir, 2006) and worn as a necklace, bracelet or on the head.

Daisies are rich in vitamin C and have been used in salads or cooked and eaten, even though they are somewhat bitter. The unopened buds of *Bellis* can be pickled in vinegar-like capers.

Berberis vulgaris

- Barberry (Scannell & Synnott, 1987)
- *Barbróg* (Scannell & Synnott, 1987)

BERBERIDACEAE—THE BARBERRY FAMILY

Introduced

Berberis vulgaris is almost certainly an introduced plant in Ireland. This small shrub is seen occasionally in hedgerows and on rocky ground. Webb et al. (1996) noted that it was formerly more widespread but has been exterminated in many areas because it is a carrier of the wheat rust fungus (*Puccinia* species). It was prohibited in (the Republic of) Ireland in 1958 by the Noxious Weeds (Common barberry) Order.

Barberry bark was mixed with sulphur in stout as a treatment for jaundice in humans and animals (Moloney, 1919; Grigson, 1958)—a yellow bark mixed with yellow sulphur being used to treat a yellow disease. Logan (1972, 1981) reported that barberry was used to treat diarrhoea, liver diseases and jaundice. K'Eogh (1735: 10) said that 'the inner bark (which is of a deep yellow colour) is accounted a specific against yellow jaundice'. Roy Vickery included a record in the Plant-lore Archive (Vickery, 2010b, Plant-Lore.com) from T. C. Meelick in Clonlara, Co. Clare, who told of an old man who used barberry to cure jaundice: 'He would take a small sliver of the inner bark from barberry (*Berberis vulgaris*) and boil it in milk. This was given to the sufferer to drink and 'as soon as it hit the stomach, just as it was all drank', it caused fairly violent emesis, but 'the jaundice would be gone.'

During a visit to Vietnam in the early 1990s, I became violently ill with a stomach upset and fever that was treated very successfully by a local botanist who went to the Hanoi medicinal plant market and later fed me a treatment based on a *Berberis* species. I survived and so can probably vouch for its efficacy!

Mackay (1836: 11) noted that the berries of *Berberis vulgaris* are acid and astringent and when mixed with sugar can be made into 'an agreeable refreshing preserve'. Grigson (1958) suggested that the red fruits make a delicious jam or jelly, and that they can be candied and eaten as sweets, or used in a fever drink. The same author noted that punch was flavoured with barberry fruits, and that the bark was used to tan and dye leather. In her famous book of household management, published in 1859 to 1861, Mrs Beeton (1861: 631) provided a recipe for Barberry Tart and wrote that barberries are 'a fruit of such great acidity, that even birds refuse to eat it. In this respect, it nearly approaches the tamarind. When boiled with sugar, it makes a very agreeable preserve or jelly, according to the different modes of preparing it. Barberries are also used as a dry sweetmeat, and in sugarplums or confits; are pickled with vinegar, and are used for various culinary purposes. They are well calculated to allay heat and thirst in persons afflicted with fevers'.

Berberis vulgaris was used as a veterinary remedy to treat jaundice in cattle in Ireland (Allen & Hatfield, 2004).

---

*Berula erecta

- Lesser water-parsnip (Scannell & Synnott, 1987)
- *Ráib uisce* (Scannell & Synnott, 1987)

APIACEAE (Umbelliferae)—THE CARROT FAMILY
Native

Berula erecta is a perennial plant that is rather rare in Ireland, except quite locally in parts of the east and the midlands, in marshes, streams, ditches and canal banks.

*No ethnobotanical uses of this plant have been noted in Ireland.

Beta vulgaris subsp. *maritima*

- Sea beet (Scannell & Synnott, 1987); Sea kale (Colgan, 1904)
- *Laíon na trá* (Scannell & Synnott, 1987); *Biadhtas* (Moloney, 1919); *Bein-e fiadhain* (Moloney, 1919); *Biatas* (Cameron, 1883)

AMARANTHACEAE (Chenopodiaceae)—THE GOOSEFOOT FAMILY
Native

Beta vulgaris subsp. *maritima* is a common perennial plant of coastal locations throughout Ireland that grows on sandy and rocky seashores, coastal waste places, sand dunes and cliffs.

Beta vulgaris subsp. *maritima*.

Sea beet was traditionally collected from the seashore and cooked as spinach. It is still collected in Ireland as sea spinach and is a useful winter and spring vegetable (Mears & Hillman, 2007). I have seen it sold in organic and farmers' markets in Ireland, such as in Milltown, Co. Kerry, in April 2009. Mackay (1825a: 126; 1836: 229) recorded that it was often cultivated in gardens in Cork and other coastal places in Ireland as a vegetable 'as a good substitute for spinage (sic) in winter and spring'. Sea beet is an excellent and widely available wild vegetable, occurring in many Irish coastal localities. When cooked, it reduces considerably, so lots of leaves should be gathered. Before serving, remove the water in which they were boiled. The water is useful as a nutrient-rich vegetable stock. The younger leaves can be used in salads. Rutty (1772: 42) said that it is eaten 'as Cale, and the root is large, and of a saccharine sweetness'.

Beta vulgaris also includes a wide range of different varieties of beets used as a sugar source for food and fodder. Sugar beet is widely cultivated as a sugar source in Europe, Russia and the United States. Other varieties of *B. vulgaris* include beetroot, mangel-wurzels, Swiss chard, spinach beet and perpetual spinach.

Irish words for spinach are *spionáiste* (Ó Dónaill, 1977) and *spíonáiste* (Uí Chanainn, 1962).

Betula species

- Birch (Scannell & Synnott, 1987); Begh (Hart, 1898)
- *Beith* (Scannell & Synnott, 1987); *Beatha* (Cameron, 1883); *Beithe* (de Vere Coneys, 1849); *Cran-bheithe* (de Vere Coneys, 1849); *Beathóg* (Uí Chanainn, 1962)[35]

BETULACEAE—THE BIRCH FAMILY

Native

Including:

Betula pendula (*Betula verrucosa*)

- Silver birch (Scannell & Synnott, 1987)
- *Beith gheal* (Scannell & Synnott, 1987); *Beith charraigeach* (Moloney, 1919); *Beith chluasach* (Dinneen, 1927)

Betula pubescens

- Downy birch (Scannell & Synnott, 1987)
- *Beith chlúmhach* (Scannell & Synnott, 1987)

Birch bark.

The two native species of birch in Ireland are not always easy to distinguish. *Betula pubescens* is the more common of the two and is found in woodlands, rocky places, glens and heaths. *Betula pendula* is more frequent in the midlands and south and is found on lake-shores, rocky places and bogs. Birch can grow to be a substantial tree in Ireland, and indeed, the tallest native *B. pubescens* specimen known in Ireland is recorded from Derreen, Co. Kerry, and measured at 25 m tall (Anonymous, 2005). Birch trees with the largest known girth in Ireland are a specimen of *B. pendula* from Belcoo, Co. Fermanagh, that is 3.18 m in girth, and a specimen of *B. pubescens* from Tomies Wood, Killarney National Park, Co. Kerry, that is 3.37 m in girth (Anonymous, 2005).

Birch twigs were often used for making brooms, particularly for the garden. Birch brooms are traditionally used in Ireland and making them was a winter activity at the National Botanic Gardens in Glasnevin for the horticulture staff. Birch twigs have also been used to make ropes in Ireland (Evans, 1957). Although baskets made from birch are found throughout Scandinavia and also in parts of Scotland (although Milliken & Bridgewater [2004] say that it is not a traditional basketry material in Scotland), Hogan (2001) points out that they do not seem to have been made in Ireland, but there may also have been some knowledge of the use of birch bark in the past. Hogan also highlights, though, that an Irish-made milk strainer in the form of a funnel made from birch bark and filled with straw was mentioned in a 17th-century account.

Birch sap is collected in northern Europe and Russia to make wine, but traditional use of birch for this purpose in Ireland seems not to have been frequent, at least in recent times (although Mackay [1836] said that birch sap wine was made in Scotland, so it was undoubtedly also made in Ireland in times past). Threlkeld (1726: unnumbered [24]) noted that 'a juice drawn from this Tree…may be easily improved into Birch-wine, and is in great Vogue against the Stone'. He also mentioned that its juice was used in cheese-making. K'Eogh (1735: 14) said that 'the liquor that flows out of this tree (being bored with an augre

35. Uí Chanainn (1962) gave the word *beathóg* for 'birch tree'.

in the Spring) is accounted very good for the stone, gravel, strangury and pissing of blood' and 'a decoction of the leaves drunk is esteemed good for a dropsy, and scurvy'. Rutty (1772: 42) also mentioned that the trunks were bored in early spring for its juice, 'which is drank by many country people for ordinary drink, to quench their thirst'.

Birch sap can also be boiled to draw off much of the water and reduce it to a thick syrup, used in much the same way as maple syrup, on pancakes, waffles and ice-cream. I have bought commercially produced birch syrup in Finland and enjoyed it greatly. About 80 to 100 litres of sap are required to make 1 litre of birch syrup. Making birch sap is not easy, though, as the sugars it contains are prone to burn if boiled at too high a temperature, which results in a syrup with an unpleasant scorched taste. Queen Victoria is said to have drunk large quantities of birch sap while she was at Balmoral in Scotland in an attempt to halt the thinning of her hair—it was traditionally thought to help prevent baldness (Milliken & Bridgewater, 2004).

Phillips (1983) gives a recipe to make birch sap wine. The sap is collected by drilling a hole at about 18 in. from the base (upwards) into the birch tree and by inserting a tube, which feeds into a collecting bottle. After the required amount of sap is collected, the hole should be blocked to prevent bleeding. To make birch sap wine, the sap is boiled shortly after collecting and then sugar is added. A recipe for birch sap wine is as follows: 4.5 litres/8 pints of sap are used to 1 kg/2.25 lb of sugar; 200 g/0.5 lb of chopped raisins and the juice of two lemons are added to the boiling liquid and allowed to cool. Yeast is added and the mixture is left to ferment in a bucket for several days before it is put into a glass jar with an air lock and allowed to continue to ferment. When fermentation finishes, the wine is filtered and bottled and drunk after a couple of months.

Betula pendula syrup.

In Scotland, the traditional spout or tap (*bour*) inserted into the birch trunk when drawing off the sap was made from a hollow elder (*Sambucus nigra*) twig or a grooved peg (Milliken & Bridgewater, 2004).

Birch bark has been used to prepare a brown dye as well as for tanning leather and for fishing lines and nets (Lysaght, 1979; Mahon, 1983; MacCoitir, 2003). The inner bark of birch yields a red dye. The leaves of *Betula pubescens* have been used to provide a yellow dye (Mahon, 1983).

Rutty (1772: 105) noted other uses of birch in his *An Essay Towards a Natural History of the County of Dublin.*

> Our Fingallian[36] fishermen steep their lines, nets and sails, in a decoction of the barks of birch and oak, which not only tinges them brown, [whereas the fish would avoid them if they were white,] but preserves them a long time from rotting by the Mildew and the water, which they otherwise would not be able to resist but a short time; moreover, this perfectly tanns the Sail-cloth, and makes it firm like leather. The Birch bark also makes the fairest and best coloured leather for Booksellers, by tanning sheep skins dipt in an infusion of the same liquor.

---

36. Fingal is a region of north Co. Dublin where a strong fishing and farming tradition exists up to modern times. The heraldic crest for Fingal features a Viking longboat, representing the arrival and settlement of the Norse in Fingal. The region had its own language, Fingalian, which is now extinct. It was related to Old and Middle English and Old Norse and incorporated some elements of Old Gaelic and was spoken in Fingal until the mid-19th century.

Birch bark can be peeled easily in broad plates from a felled trunk and is waterproof, so it was no doubt used for a variety of purposes in early Ireland. Cameron (1883) mentioned the use of the smooth, inner birch bark in the Highlands of Scotland as a substitute for paper to write upon and to burn for light. According to Howes (1974), Native Americans used it for making canoes and as roofing materials.

Birch leaves have been used medicinally in various parts of the world, mainly in irrigation therapy against inflammations of the urinary tract, including cystitis (van Wyk & Wink, 2004), although I have not found any records of their use in such ways in Ireland. Birch bark was used medicinally in Ireland to treat eczema (Allen & Hatfield, 2004). In North America various species of birch have been used in traditional medicine by the Native Americans to treat stomach ailments (sap), burns and sores (boiled bark), catarrh (cones were heated over coals and the smoke inhaled) and to ease childbirth (tea) (King, 1984).

Drury (1984) mentioned that meal made from the bark of birch was used as a famine food in Ireland. The bark of other trees was also used in a similar way, such as Scots pine, beech, elm and lime, but not oak or willow, which contain too much tannin.

Birch timber is light, flexible and easy to work, and it is used commercially for plywood and pulp (Hickie, 2002). Mackay (1836) recorded that birch timber was much used (along with alder timber) to make charcoal, particularly in the manufacture of gunpowder. Birch (or holly) poles were used to make the two shafts of wheel-less sliding carts used to transport turf home from the bog (Bell & Watson, 2008).

Birch rods have traditionally been used as the weapon of choice for institutional corporal punishment in Britain and Ireland. Indeed, the use of the birch for minor breaches of the law by juveniles was only abolished in the Isle of Man in 1976. Those in favour of reintroducing such judicial penalties would often use the slogan 'bring back the birch'. To make a birch rod, a bundle of four or five rods (usually birch, but sometimes also willow) would be bound together to make a strong but flexible weapon. An example of a teacher's school cane (*cána múinteora*) from Castlebar, Co. Mayo, in the National Museum of Ireland (Museum of Country Life) in Castlebar, Co. Mayo (Ref. No: F2007.104), is made not from birch but from bamboo. When I was a child in primary school in Co. Limerick during the 1960s, the teacher's cane was also made from bamboo. Fortunately, I never had the opportunity to test whether it was an effective weapon or not. In his autobiography about growing up in the 1940s and 1950s in the West Kerry Gaeltacht, Maidhe Dainín Ó Sé (2007) wrote that the teacher's cane that he was familiar with was made of ash (*Fraxinus excelsior*). Corporal punishment in schools in Ireland was only banned in 1982.

Birch woods in the National Botanic Gardens, Glasnevin.

Birch twigs are still a traditional part of the sauna experience in Scandinavia and Russia. The skin is gently beaten with soaked birch twigs, still with their leaves on, which helps to open pores, improve blood circulation, and provide a mild massage. Such beating is seldom painful.

In parts of rural Scotland where heather was scarce, birch twigs were used for bedding. The twigs were cut (with leaves still attached) and stuffed tightly into a simple box frame (Milliken & Bridgewater, 2004). I have not found records of birch beds being made in Ireland. Rushes, bracken and heather were probably more usual plants used for bedding in this island.

Birch trees are frequently grown as ornamental trees in parks and gardens in Ireland. In addition to the two native species, a range of exotic species native to north temperate regions of the world are grown as ornamentals, including *Betula papyrifera* (paper-bark birch), from North America, and Asian species such as *B. utilis* var. *jacquemontii* (white-barked Himalayan birch). A stunningly white-barked variety of this Asian species was collected in the

Himalayas by Joseph Hooker, one-time director of the Royal Botanic Gardens, Kew, and grown at the Trinity College Botanic Gardens (originally located in Ballsbridge and more recently in the College's Garden in Dartry). This variety is aptly named 'Trinity College'.

**Bidens cernua*

- Nodding bur-marigold (Scannell & Synnott, 1987)
- *Scothóg Mhuire* (Scannell & Synnott, 1987)

ASTERACEAE (Compositae)—THE DAISY FAMILY
Native

Bidens cernua is a relatively common plant in Ireland that is found growing in habitats such as marshes, open mud, river-banks and bog-holes.

*No ethnobotanical uses of this plant have been noted in Ireland.

Bidens tripartita

- Trifid bur-marigold (Scannell & Synnott, 1987)
- *Scothóg leathan* (Scannell & Synnott, 1987); *Cnáib uisce* (Uí Chonchubhair, 1995)

ASTERACEAE (Compositae)—THE DAISY FAMILY
Native

Bidens tripartita is a rather rare plant in Ireland that is found mainly in the south in marshes, open mud, river-banks and bog-holes.

According to Uí Chonchubhair (1995), trifid bur-marigold has been used medicinally as a tonic and to cure liver, spleen and digestive ailments as well as genital and urinary problems in humans and animals.

Blackstonia perfoliata

- Yellow-wort (Scannell & Synnott, 1987)
- *Dréimire buí* (Scannell & Synnott, 1987); *Dreimire buidhe* (de Vere Coneys, 1849; Cameron, 1883)

GENTIANACEAE—THE GENTIAN FAMILY
Native

Blackstonia perfoliata is a small, but attractive, annual plant of lime-rich dry habitats, such as sand dunes and gravel banks throughout most of Ireland. It is rare in the south-west and absent from the north-east.

Mahon (1983) listed this plant as the source of a yellow dye.

**Blysmus rufus*

- Saltmarsh flat-sedge (Scannell & Synnott, 1987)
- *Searg rua* (Scannell & Synnott, 1987)

CYPERACEAE—THE SEDGE FAMILY
Native

Blysmus rufus is a perennial sedge of coastal salt-marshes that is rare in Ireland except in parts of the north.

*No ethnobotanical uses of this plant have been noted in Ireland.

Bolboschoenus maritimus (*Scirpus maritimus*)

- Sea club-rush (Scannell & Synnott, 1987)
- *Bogshifín mara* (Scannell & Synnott, 1987); *Bróbh* (de Vere Coneys, 1849; Cameron, 1883; Dinneen, 1927); *Broibhe* (de Vere Coneys, 1849)

CYPERACEAE—THE SEDGE FAMILY
Native

Bolboschoenus maritimus is found occasionally throughout Ireland in salt marshes, tidal river estuaries and at the edges of brackish pools and lakes. It is very rare inland.

Simpson and Inglis (2001) recorded that its rhizomes have been cooked and eaten as a vegetable and for thatching in Ethiopia and medicinal purposes in China. However, there are no records that I can find of its use in Ireland for such purposes.

Cameron (1883) noted that its roots are large and were used as a nutritious feed for cattle and in times of scarcity were ground down to make meal. This same author indicated its Irish name *bróbh* means a 'quern or hand-mill', which suggests that it was used in Ireland as a food source, although I have not found any specific references to its use. Clyne (1989) also mentioned such usage and how it would be ground by the quern-stone[37].

Huxley (1991) listed this plant as an ornamental used in horticulture, but I know of no records of its cultivation in Irish gardens.

Borago officinalis

- Borage (Scannell & Synnott, 1987)
- *Borráiste gorm* (Scannell & Synnott, 1987)

BORAGINACEAE—THE BORAGE FAMILY
Introduced

Borago officinalis is a perennial plant that is occasionally grown in Irish gardens as an ornamental and for culinary and medicinal purposes. It can also occasionally be found in the wild in waste places, rubbish dumps and near old ruins, particularly in the east.

37. A quern-stone is a stone tool used for the hand grinding of plant materials for food, particularly cereal grains and starch-rich roots, to make flour. It was a standard tool in Irish farms since Neolithic times.

Fresh leaves of borage have the taste of cucumber and can be used in salads and soups. Borage or cucumber have traditionally been used as a garnish for Pimms cocktails. The flowers are edible and can be used in salads, candied or made into fritters. It is not known to what extent it has been used medicinally in Ireland, but elsewhere it has been used as a diuretic, to treat fevers, and as a poultice to treat swellings.

Borage has also been used as an ingredient of cosmetics in Ireland.

**Brachypodium pinnatum* (*Bromus sylvaticus*)

- Tor-grass (Scannell & Synnott, 1987)
- *Brómas reatha* (Scannell & Synnott, 1987); *Brumfhér fiodhbhadh*[38] (White, 1803); *Brumfhér sgiathanach*[39] (White, 1803)

POACEAE (Gramineae)—THE GRASS FAMILY
Native

Brachypodium pinnatum is a tufted perennial grass of lime-rich grassland, such as railway embankments. Although it occurs throughout Ireland, it is somewhat rare, and mainly distributed in the south.

*No ethnobotanical uses of this plant have been noted in Ireland.

**Brachypodium sylvaticum*

- False brome (Scannell & Synnott, 1987)
- *Brómas bréige* (Scannell & Synnott, 1987); *Fescufhér cainnemhuil*[40] (White, 1803); *Brumfhér fiodhbhadh* (White, 1803)

POACEAE (Gramineae)—THE GRASS FAMILY
Native

Brachypodium sylvaticum is a common perennial grass throughout Ireland of woodlands and shady hedges and banks.

*No ethnobotanical uses of this plant have been noted in Ireland.

Brassica species

BRASSICACEAE (Cruciferae)—THE CABBAGE AND CRESS FAMILY

Including:

Brassica napus

- Rape (Scannell & Synnott, 1987); Swede (Hackney, 1992); Oil-seed rape (Forbes & Northridge, 2012); Shillogs (Hart, 1898); Yellow rape (Hart, 1898); Turnip (Forbes & Northridge, 2012)
- *Ráib* (Scannell & Synnott, 1987); *Praiseach bhuidhe* (de Vere Coneys, 1849)

38. White (1803) applied this name to *Bromus sylvaticus*.
39. White (1803) applied this name to *Bromus pinnatus*.
40. White (1803) applied this name to *Festuca calamaria*.

Brassica oleracea

- Wild cabbage (Scannell & Synnott, 1987); Cabbage; Kale
- *Cabáiste* (Scannell & Synnott, 1987); *Gabáiste* (Dinneen, 1927); *Praiseach bhaidhe*[41] (Cameron, 1883)

Introduced

Brassica rapa

- Wild turnip (Scannell & Synnott, 1987)
- *Tornapa fiáin* (Scannell & Synnott, 1987); *Neip* (Cameron, 1883)

Introduced (possible archaeophyte)

Cabbage masher.

Brassica oleracea 'Gortahork' cabbage.

Brassica rapa and *B. napus* are important economic plants of which various cultivated varieties are grown as oil-seed crops, turnips and other leaf vegetables. In the wild in Ireland they are found occasionally on roadsides and in the margins of fields and wastegrounds where they have escaped from cultivation.

Cabbage (in Irish, *gabáiste, cabáiste*) and kale are *Brassica oleracea* subsp. *oleracea*. *Brassica oleracea* is not native to Ireland but instead to western and southern Europe and the Mediterranean, reaching as far north as the coasts of England and Wales. Cabbage features prominently in early Irish diets, most notably in early Christian times and was particularly associated with the monasteries (Kelly, 1997). Cabbage is also one of the most traditional Irish vegetables, and a dish such as boiled bacon and cabbage is still regarded as a traditional and classic Irish dish. It was almost a staple vegetable in Ireland until relatively recently, along with carrots and onions, and I can well remember having to eat cabbage at least once or twice a week when I was a child growing up in Limerick and Kerry during the 1960s.

There are several traditional Irish varieties (cultivars) of cabbage that are still grown. The varieties vary according to shape and time of maturity. One of the most remarkable is a variety called 'Gortahork', a huge and vigorous drumhead cabbage named after the townland of Gortahork in north-west Co. Donegal and grown by a farming family there for many years. It was grown very successfully in recent years in the fruit and vegetable garden of the National Botanic Gardens in Glasnevin, Dublin, and made a spectacular display. It has now become quite an icon for those involved in conserving and promoting traditional vegetable varieties in Ireland. Another variety is Delaway cabbage, grown by the Hughes family in Co. Mayo for some generations. Both varieties are conserved and made available as seed for sale to the public by the Irish Seed Savers Association (see <http://irishseedsavers.ie>).

Cabbage features in traditional children's pastimes in Ireland and is documented in the archives of the National Museum of Ireland (Museum of Country Life) in Castlebar, Co. Mayo. The archive includes a letter dated 4 November 1951 from J. D. O'Leary of Ballyfainagh, Claremorris, Co. Mayo, to A. T. Lucas of the Museum describing a practice of children at Halloween of 'kicking heads of cabbage around the roads on that Holy Eve. You will see the roads around here littered with bruised battered cabbage taken in most cases from a neighbour's plot on Nov. 1st.'

Cabbage features in a traditional Irish remedy for a sore throat. A cabbage leaf was

41. Cameron (1883) gave the Irish name *praiseach bhaidhe* for *Brassica oleracea*, meaning the 'pot-herb of the wave'.

tied around the throat and cabbage juice taken with honey to relieve the symptoms of hoarseness and the loss of voice (Thiselton Dyer, 1889). Lady Wilde (1890: 35) listed 'fried cabbage-leaves' as a treatment for burns, 'used by people to deaden the pain'. Gallagher (2012: 17) said that a well-ripened white cabbage leaf can also help reduce blistering of the skin, and that cabbage 'clears the blood'. The same author recorded the use of the juice of boiled red cabbage in Co. Donegal to treat asthma and other chest problems.

Halloween cut turnip.

When I was growing up in Munster in the 1950s and 1960s, a large turnip was carved to make a Halloween lantern. The centre was hollowed out and a face was cut into one side with eyes, a nose, and a mouth with large teeth through which the light of a candle mounted inside could shine. In Irish this turnip was called a *móg* (or *móig*) (Dinneen, 1927). The collections of the National Museum of Ireland (Museum of Country Life) in Castlebar, Co. Mayo, include a 'ghost turnip' from Co. Donegal (Ref. No: F1943.350) that had been hollowed out as a lamp and carved with a ghostly face for Halloween. Pumpkins were little known in Ireland at that time but today have largely replaced the turnip for such Halloween festivities.

Turnips have also been used medicinally in Ireland. O'Regan (1997) wrote that in the 1930s in Ireland turnips were used to treat sore throats and colds. The centre of the turnip was cut out and 2 teaspoons of brown sugar were placed in it, with a few drops of lemon juice and glycerine also added. This was left overnight, and the next day the resulting syrup was used as a medicine. The same author recorded that a raw turnip was used to treat cracked chilblains and abscesses, and in Co. Wexford some farmers still use a turnip for drawing out an abscess on a bullock's hoof.

In Irish the word for turnip is *meacan ráibe* (Dinneen, 1927).

**Briza media*

- Quaking-grass (Scannell & Synnott, 1987); Shivery grass (Colgan, 1904); Shaky grass (Colgan, 1904); Shaky heads (Colgan, 1904); Fairy grass (Hart, 1898; Brunker, 1950); Doddering dillies (Harron, 1986)
- *Féar gortach* (Scannell & Synnott, 1987); *Conan* (Cameron, 1883); *Luascadhfhér medhon* (White, 1803)

POACEAE (Gramineae)—THE GRASS FAMILY
Native

Briza media is a slender and very attractive grass that is commonly found in the centre of Ireland in particularly dry, well-drained habitats, such as on sand dunes, gravel banks and wall tops and in grasslands. It can also be found in wetter grassland but is rarer in this habitat.

Clyne (1989) explained the Irish name *féar gortach* means the 'hungry grass', which is associated with the superstition that sudden weariness and extreme hunger can occur when walking on a particular fairy grass in the mountains. Other authors associate this superstition with *Agrostis stolonifera* (e.g., Allen & Hatfield, 2004).

*No ethnobotanical uses of this plant have been noted in Ireland.

*Bromopsis erecta (Bromus erectus)

- **Upright brome (Scannell & Synnott, 1987)**
- ***Brómas ceannard* (Scannell & Synnott, 1987); *Brumfhér sneigh*[42] (White, 1803); *Brumfhér giobach*[43] (White, 1803)**

POACEAE (Gramineae)—THE GRASS FAMILY
Native (possible archaeophyte)

Bromopsis erecta is a perennial grass of dry banks, sand dunes, railway embankments and pastures that is found mainly in the centre and south of Ireland. It is mostly rare except around Dublin.

*No ethnobotanical uses of this plant have been noted in Ireland.

---

*Bromopsis ramosa (Bromus ramosus)

- **Hairy brome (Scannell & Synnott, 1987)**
- ***Brómas giobach* (Scannell & Synnott, 1987)**

POACEAE (Gramineae)—THE GRASS FAMILY
Native

Bromopsis ramosa is a perennial grass throughout Ireland of woodlands, hedgerows, thickets and shady banks.

*No ethnobotanical uses of this plant have been noted in Ireland.

---

*Bromus commutatus

- **Meadow brome (Scannell & Synnott, 1987)**
- ***Brómas móinéir* (Scannell & Synnott, 1987)**

POACEAE (Gramineae)—THE GRASS FAMILY
Native (possible archaeophyte)

Bromus commutatus is a rare annual grass in Ireland of roadsides and meadows.

*No ethnobotanical uses of this plant have been noted in Ireland.

Bromus hordeaceus (Bromus mollis)

- **Soft brome (Scannell & Synnott, 1987); Sturdy (Hart, 1898)**
- ***Brómas bog* (Scannell & Synnott, 1987); *Brumfheur bog*[44] (Wade, 1808a); *Brumfhér bog*[45] (White, 1803)**

POACEAE (Gramineae)—THE GRASS FAMILY
Native

Bromus hordeaceus is a common annual grass throughout Ireland of meadows, pastures, roadsides and waste places.

Wade (1808a) noted that the seeds of this species are like corn, and some (presumably

42. White (1803) applied this name to *Bromus erectus*.
43. White (1803) applied this name to *Bromus asper*.
44. Wade (1808a) applied this name to *Bromus mollis*.
45. White (1803) applied this name to *Bromus mollis*.

of *Bromus*) are mixed with corn and made into bread, although they are apt to disagree with the stomach and head. They can easily be collected in large quantities.

Uí Chanainn (1962) gave the Irish word *broimfhéar* as a general name for 'brome-grass'. This same name was applied by Dinneen (1927) to scutch/couch-grass (*Elytrigia repens*).

**Bromus secalinus*

- Rye brome (Scannell & Synnott, 1987)
- *Brómas seagail* (Scannell & Synnott, 1987); *Brumfhér ségalamhuil* (White, 1803)

POACEAE (Gramineae)—THE GRASS FAMILY
Introduced

Bromus secalinus is a weed of cereal crops, dry marginal land and waste places. It was formerly common in Ireland but is now very rare and found only in scattered locations in the south and east.

*No ethnobotanical uses of this plant have been noted in Ireland.

Buddleja davidii

- Butterfly-bush (Scannell & Synnott, 1987)
- *Búidlia chorcra* (Anonymous, 1978)

SCROPHULARIACEAE (Buddlejaceae)—THE FIGWORT FAMILY
Introduced

Buddleja davidii, a native of China, is a common ornamental garden plant in Ireland with several cultivated varieties. It is widely established in the wild in waste places and especially urban settings, such as roadsides, walls, roof-tops and rubbish dumps.

Buddleja is a valuable nectar source in gardens for insects, particularly butterflies, hence butterfly-bush is its common name.

Butomus umbellatus

- Flowering rush (Scannell & Synnott, 1987)
- *Luachair dhearg* (Scannell & Synnott, 1987)

BUTOMACEAE—THE FLOWERING-RUSH FAMILY
Native

Butomus umbellatus is a rare plant throughout Ireland of river-banks and margins of canals and lakes.

Mabey (1972) noted that its roots are edible, but there are no records of it being used for food in Ireland. According to Heywood (1978), the rhizomes have been used as food in parts of Russia.

It is occasionally grown as an aquatic plant in the margins of decorative ponds and lakes.

Cakile maritima.

Cakile maritima

- Sea rocket (Scannell & Synnott, 1987); Purple sea rocket (Brunker, 1950)
- *Cearrbhacán mara* (Scannell & Synnott, 1987)

BRASSICACEAE (Cruciferae)—THE CABBAGE AND CRESS FAMILY
Native

Cakile maritima is a driftline annual plant of sandy and shingle shores all around the Irish coast.

Mears and Hillman (2007) suggested the leaves of this plant are edible when eaten raw but when cooked become very bitter.

**Calamagrostis epigejos*

- Wood small-reed (Scannell & Synnott, 1987)
- *Giolc* (Scannell & Synnott, 1987)

POACEAE (Gramineae)—THE GRASS FAMILY
Native

Calamagrostis epigejos is a very rare grass in Ireland, found only in rocky places in the north and west.

*No ethnobotanical uses of this plant have been noted in Ireland.

**Calamagrostis stricta*

- Narrow small-reed (Scannell & Synnott, 1987)
- *Giolc beag* (Scannell & Synnott, 1987)

POACEAE (Gramineae)—THE GRASS FAMILY
Native

Calamagrostis stricta is a very rare grass in Ireland, found only on the shores of Lough Neagh in Northern Ireland.

*No ethnobotanical uses of this plant have been noted in Ireland.

Calendula officinalis

- Garden marigold (Moloney, 1919); Pot marigold (Reynolds, 2002)
- *Blath Mhuire* (Moloney, 1919); *Liathán* (Moloney, 1919); *Ór Mhuire* (Moloney, 1919; Dinneen, 1927; Ó Dónaill, 1977)

ASTERACEAE (Compositae)—THE DAISY FAMILY
Introduced

Calendula officinalis is a commonly grown annual garden plant with very conspicuous, bright orange-yellow flowers. It occasionally occurs in the wild in Ireland as an escape or discard and grows in waste places and rubbish dumps.

Moloney (1919) noted that it was not only used in Ireland to give a deep colour to butter but also used in traditional medicine to make an external lotion for wounds, sprains and bruises, as well as internally as a uterine tonic.

Extracts of *Calendula officinalis* have also been used as an ingredient of cosmetics in Ireland in recent years.

Calendula officinalis.

Callitriche species

- Water-starwort (Scannell & Synnott, 1987)
- *Réiltín* (Scannell & Synnott, 1987)

PLANTAGINACEAE (Callitrichaceae)—THE WATER-STARWORT FAMILY
Native

There are seven freshwater aquatic species of *Callitriche* found as native plants in Ireland. Other than their occasional use as ornamental plants in coldwater aquaria, I have not recorded other uses for them in Ireland.

**Callitriche brutia*

- Pedunculate water-starwort (Scannell & Synnott, 1987)
- *Réiltín na ngasán* (Scannell & Synnott, 1987)

PLANTAGINACEAE (Callitrichaceae)—THE WATER-STARWORT FAMILY
Native

Callitriche brutia is an aquatic or semi-aquatic plant that is found throughout Ireland, most commonly in the north, in ditches, streams and lakes.

*No ethnobotanical uses of this plant have been noted in Ireland.

**Callitriche hamulata*

- Intermediate water-starwort (Scannell & Synnott, 1987)
- *Réiltín meánach* (Scannell & Synnott, 1987)

PLANTAGINACEAE (Callitrichaceae)—THE WATER-STARWORT FAMILY
Native

Callitriche hamulata is a rare aquatic or semi-aquatic plant throughout Ireland of ditches, streams and lakes.

*No ethnobotanical uses of this plant have been noted in Ireland.

**Callitriche hermaphroditica*

- Autumnal water-starwort (Scannell & Synnott, 1987)
- *Réiltín fómhair* (Scannell & Synnott, 1987)

PLANTAGINACEAE (Callitrichaceae)—THE WATER-STARWORT FAMILY
Native

Callitriche hermaphroditica is an aquatic or semi-aquatic plant throughout Ireland, most commonly in the north, of ditches, streams and lakes.

*No ethnobotanical uses of this plant have been noted in Ireland.

**Callitriche obtusangula*

- Blunt-fruited water-starwort (Scannell & Synnott, 1987)
- *Réiltín eangach* (Scannell & Synnott, 1987)

PLANTAGINACEAE (Callitrichaceae)—THE WATER-STARWORT FAMILY
Native

Callitriche obtusangula is an aquatic or semi-aquatic plant of muddy ditches, slow streams and ponds. It is occasionally found throughout Ireland.

*No ethnobotanical uses of this plant have been noted in Ireland.

**Callitriche platycarpa*

- Various-leaved water-starwort (Scannell & Synnott, 1987)
- *Réiltín leathan* (Scannell & Synnott, 1987)

PLANTAGINACEAE (Callitrichaceae)—THE WATER-STARWORT FAMILY
Native

Callitriche platycarpa is an aquatic or semi-aquatic plant of muddy ditches, slow streams and ponds. It is found throughout Ireland.

*No ethnobotanical uses of this plant have been noted in Ireland.

**Callitriche stagnalis*

- Common water-starwort (Scannell & Synnott, 1987)
- *Réiltín uisce* (Scannell & Synnott, 1987); *Luibh na luí* (Ua Maoileoin, 2001)

PLANTAGINACEAE (Callitrichaceae)—THE WATER-STARWORT FAMILY
Native

Callitriche stagnalis is an aquatic or semi-aquatic plant of muddy ditches, slow streams and ponds. It is found occasionally throughout Ireland.

*No ethnobotanical uses of this plant have been noted in Ireland.

*Callitriche truncata

- Short-leaved water-starwort (Scannell & Synnott, 1987)
- *Réiltín scoite* (Scannell & Synnott, 1987)

PLANTAGINACEAE (Callitrichaceae)—THE WATER-STARWORT FAMILY
Native

Callitriche truncata is only recorded from one site in Ireland in Co. Wexford.

*No ethnobotanical uses of this plant have been noted in Ireland.

Calluna vulgaris

- Heather (Scannell & Synnott, 1987); Ling (Scannell & Synnott, 1987); Heath broom (Colgan, 1904)
- *Fraoch mór* (Scannell & Synnott, 1987); *Craobh* (Ó Criomhthain, MSS, n.d., ca. 1908); *Fraoch* (Colgan, 1911); *Het* (McCarthy & Hawkes, 2007)

ERICACEAE—THE HEATHER FAMILY
Native

Calluna vulgaris is a low-growing evergreen shrub that is a characteristic component of mountain and bogland habitats and often the dominant plant found over large areas of the mountains. It also occurs in heaths, moors, open woodlands and on acid soils.

Calluna vulgaris.

Dried heather flowers can be used to make tea, and shoots have been used instead of hops (*Humulus lupulus*) to flavour beer (Phillips, 1983). Smith ([1756] 1979) recorded that a heath collected on the mountains was used in Co. Kerry to brew beer. It was probably *Calluna vulgaris*. Quoting from the *Dublin University Magazine* of 1854, O'Sullivan (1994: 136), described the use of heather for making beer: 'In Ireland, the invading Danes brewed from heath a strong and palatable ale, the recipe for which they so determinedly denied to the anxious Irish that often, when taken prisoner and offered life and benefits for the coveted secret, they preferred to die rather than gratify their enemies "desires".' Ó Dónaill (1977) gave the Irish name *leann fraoigh* for 'heather-ale'. The following is a recipe adapted from Macdonald Douglas (1982: 256) to make heather ale, who noted, 'At the end of the week the heather ale may be got rid of in the usual manner.'

> Collect the heather-bells when they are in full bloom. Cover them with water and boil for an hour. Strain the liquid into a clean wooden tub and for every dozen pints of water add half an ounce of hops, an ounce of ground ginger and a pound of treacle or honey. Boil the mixture again for twenty minutes and strain again. When almost cool add five tablespoons of barm (barm is defined as yeasty foam derived from fermenting beers and wines—I am sure that brewer's yeast could be used instead), cover and leave undisturbed for about 24 hours for the barm to work. Then strain carefully and pour gently into a tub, leaving all the barmy sediment behind. Put the liquid into bottles and store in a dark place for a week.

Another popular food use is as ling honey, or heather honey, which has a characteristic flavour and aroma. It is a popular product in Britain and Ireland and is derived from bees where a high proportion of their feed comes from this species.

In Co. Dublin *Calluna vulgaris* was used to make brooms for sweeping, which gave it the common name of heath broom (Colgan, 1904). Rutty (1772) also mentioned broom-making from this species and said that the brooms lasted a long time. Mackay (1836) also noted its use for making brooms in Ireland and as fuel where wood was scarce. The collections of the National Museum of Ireland (Museum of Country Life) in Castlebar, Co. Mayo, include a brush made from heather from Murher, Co. Kerry (Ref. No: F1964.88). Other objects made from heather in the museum include pot scrubbers or scourers (Ref. Nos: F1997.213 and F1973.664), one of them from Cornith Island, Co. Donegal. An example of a pot scrubber made from grass roots of some unspecified species is also included in the collections (Ref. No: F1955.9). Danaher (1964a) said that heather was used to make brooms and scrapers (although he does not mention what these scrapers were used for). A rough sieve made of cut shoots of heather (or alternatively from osier or hazel) was used to filter the sugar-rich wort drawn off from a vat of malted barley used in poteen making in Ireland (Ó Curraidhín, n.d., ca. 2009).

In Britain, heather branches have been used for thatching (such as for 'black houses' in Skye and the Hebrides in Scotland) as well as for packing pottery (Edlin, 1973). Dr Samuel Johnson (1775: 66) described huts thatched with heather in the Highlands and Western Islands of Scotland:

> The wall which is commonly about six feet high, declines from the perpendicular a little inward. Such rafters as can be procured are then raised as a roof, and covered with heath, which makes a strong and warm thatch, kept from flying off by ropes of twisted heath, of which the ends, reaching from the centre of the thatch to the top of the wall, are held firm by the weight of a large stone. No light is admitted but at the entrance, and through a hole in the thatch, which gives vent to the smoke. This hole is not directly over the fire, lest the rain should extinguish it; and the smoke therefore naturally fills the place before it escapes. Such is the general structure of the houses in which one of the nations of this opulent and powerful island has been hitherto content to live.

Heather ropes were made in Ireland, but probably less frequently so than in Scotland, where they were a standard in the construction of thatched roofs. They were used to tie down the thatch, and as they stand up to seawater well, they were used to haul home the harvest of seaweed (Milliken & Bridgewater, 2004). In Ireland ropes of straw were more frequently used to secure the thatch. Heather ropes are time-consuming to make but are strong and long-lasting.

Simple huts (often called 'cabins' in Ireland) were probably constructed very similarly to those described by Dr Johnson and thatched in the same way. In Ireland, it was also used for thatching, mainly for poorer cabins in the west and south, particularly when straw, rushes, reeds and other more traditional materials were not available (Sharkey, 2000). Rutty (1772) also mentioned its use in thatching cabins. I do not know of any houses thatched with heather that survive in Ireland today.

Larner (2005) provided a description of a poorer mud cabin from Co. Kerry in early times. A key piece of the furniture in these cabins consisted of a bag packed with *fraoch* (heather), probably mainly *Calluna*, which was used as a seat or sometimes as a bed. Heather beds were also constructed. These rude dwellings probably survived until the late 19th century in some parts of Ireland. Milliken and Bridgewater (2004: 109) describe these simple houses also in Scotland: 'The traditional Scottish heather bed was nothing more than an empty box frame into which heather bushes were stuffed, flowering ends upwards and slightly inclined towards the head, until they were dense enough to support the weight of the body. Few of us have ever slept on such a bed and few of us ever will, but anyone who has walked in the Scottish hills on a sunny day, and lain down in the heather for a rest, will already be aware of its comfortable springiness.' Rutty (1772: 111) also described these heather beds and said that 'for softness they may in some measure vie with feather beds, and greatly exceed them in wholesomeness, refreshing and strengthening the weary, so that they who lie on them rise lively and cheerful'.

Branches were also used to make a traditional woven heather lobster pot in Ireland. For this purpose the heather was pulled during its flowering time in late August or early September and allowed to season for a few days before use. The thicker branches were used for upright ribs and the thinner material to weave in and out between them (Hogan, 2001). An example of a heather lobster pot is included in the collections of the National Museum of Ireland (Museum of Country Life) in Castlebar, Co. Mayo. This pot was made by Joe Hogan, the renowned Irish basket-maker, who says that they are more effective in trapping lobster because they smell less than willow and look more natural in the setting of seaweed-covered rocks in which they are placed. Another example in the Museum was made by Richard Walsh in Achill Island, Co. Mayo (Ref. No: F1947.94). The heather cut for basketry needs to have six or seven years' growth before it can be used effectively (Hogan, pers. comm., 2010). Heather was used to make woven household or fireside seats (O'Súilleabháin, 1942).

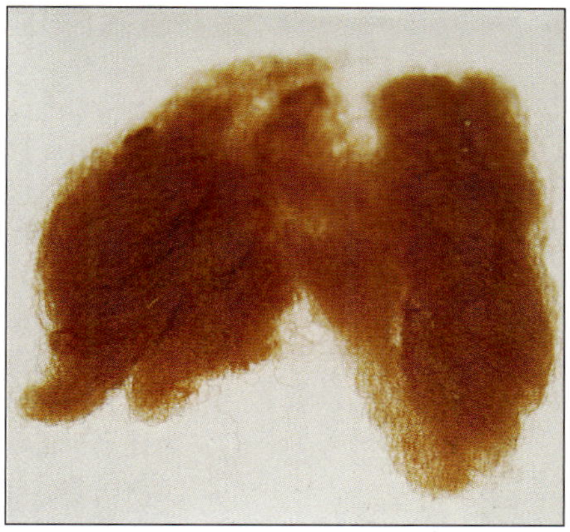

Wool dyed with *Calluna vulgaris*.

Heather yields dyes of various colours, from yellow to greenish-black and brown (Danaher, 1964a; Mahon, 1983; Williams, 1993). For the purposes of its use for dyeing, the tips of the shoots were gathered just before flowering and boiled in water for half an hour. The wool was first mordanted with alum or chrome, depending on the shade of yellow required, then put into a dye bath with the strained liquid and left overnight. Alternatively, the wool was boiled in the strained heather liquid until the desired colour was achieved (Mairet, 1916).

Smith (1815: 363) mentioned that it was used as a substitute for oak bark in tanning, 'where it answered tolerably well' for such a purpose. Rutty (1772) also mentioned experiments in Paris in 1762 where it was dried in an oven and powdered and used for tanning leather instead of oak bark.

According to Mitchell (1989), *Calluna* twigs were found in Valencia Island, Co. Kerry, in charcoal remains dating back more than 3000 years, suggesting that it was a fuel source in south-west Ireland, perhaps particularly when woodlands had been extensively cleared. It was also used as food for grazing sheep, and for that reason it was often burnt on the mountains to encourage tender young shoots that are more palatable.

A wide range of cultivated varieties (cultivars) of *Calluna vulgaris* have been raised in Ireland and are grown in Irish gardens (Nelson, 2000b). White-flowered forms are frequently found in the wild and are regarded as a good luck charm. It was traditionally given to brides after their marriage (Munnelly, 2005).

Caltha palustris

- Marsh-marigold (Scannell & Synnott, 1987); Kingcup (Webb & Scannell, 1983); Mayflower (Hart, 1898; Praeger, 1938)
- *Lus buí Bealtaine* (Scannell & Synnott, 1987); *Beánnán Bealtaine* (Moloney, 1919); *Plubairsín* (de Vere Coneys, 1849; Moloney, 1919); *Lus buidhe Bealtaine* (de Vere Coneys, 1849; Moloney, 1919; Dinneen, 1927); *Riasc-bláth Órdha* (Moloney, 1919; Dinneen, 1927);

Riasc-bhláth órtha (Uí Chanainn, 1962); *Plubairsin* (Cameron, 1883); *Lus Mairi* (Cameron, 1883); *Lus mhuire* (de Vere Coneys, 1849); *Pleibistín (buidhe)* (Dinneen, 1927; Uí Chanainn, 1962); *Plobaistín (-bairsín)* (Dinneen, 1927)

RANUNCULACEAE—THE BUTTERCUP AND CROWFOOT FAMILY

Native

Caltha palustris is very common throughout Ireland in marshy meadows and other wet places.

According to Allen and Hatfield, this plant was used in Irish folk medicine for the treatment of heart problems. It was also used in garlands as a talisman against the evil influence of fairies and witches (Lady Wilde, 1887; Moloney, 1919). Synnott (1979) noted a similar usage in Co. Louth in garlands to protect livestock. According to Uí Chonchubhair (1995), it is also a supposed cure for warts but is a strong irritant.

Uí Chonchubhair (1995) also recorded its use in Co. Kerry on May Day to protect houses, people, cattle and farm produce from evil and evil spirits. Singleton (1904: 457) described a similar custom from Co. Meath, enacted for protection:

> On May Eve the threshold must be strewn with 'May-flowers' (marsh-marigolds). On last May Eve, only a few days ago, I saw our cook coming in with a great bunch of May-flowers, which she told me she intended strewing on the thresholds of all the entrance doors of the house, as, being May Eve, the fairies would have great power, and the May-flowers are a potent charm to prevent their entering the house. 'Besides,' she said, 'whoever comes across the threshold, particularly that of the kitchen, must step on the flowers, and bring good luck and plenty of butter to the house.

Although poisonous if eaten raw, in North America this plant has been used as a food source, and the young leaves are collected before the plant has finished flowering (Peterson, 1977). The stalks are removed and the leaves boiled for about half an hour, in two or three changes of water. This same author suggested that the flowerbuds can also be pickled and eaten like capers. They are boiled for 10 minutes, in two or three changes of water, then pickled in hot vinegar. The juice in which they have been pickled should NOT be drunk, nor should any part of the plant be eaten raw.

Mahon (1983) listed this species as the source of a yellow dye.

Calystegia sepium

- Hedge bindweed (Scannell & Synnott, 1987); Robin-run-the-hedge (Colgan, 1904); Wild ivy (Colgan, 1904); Great bindweed[46] (Praeger, 1938; Brunker, 1950); Bellbine (Harron, 1986)
- *Ialus fáil* (Scannell & Synnott, 1987); *Currán carta* (Moloney, 1919); *Ladhlus* (Moloney, 1919); *Duil mhial* (Cameron, 1883); *Corrán casta, Craipleach*[47] (Ó Dónaill, 1977); *Féitheán talún* (Tóibín, 1965)

CONVOLVULACEAE—THE BINDWEED FAMILY

Native

46. Brunker (1950) gives this name for *Calystegia sepium*, not *C. silvatica*.
47. Ó Dónaill (1977) defined *craipleach* as 'bindweed'.

Calystegia sepium is a common perennial herb with large, white flowers that is a very persistent weed throughout Ireland of hedgerows and waste places. A variety with attractive striped pink and white flowers (subsp. *roseata*) is found near the western coasts.

This plant was used medicinally in Co. Fermanagh as a purgative and for kidney troubles (Allen & Hatfield, 2004).

A closely related introduced species, *Calystegia silvatica* (*C. sylvestris*), large bindweed (*ialus mór*; *iadh-lus*)[48], is also occasionally found in waste places.

Calystegia sepium subsp. *roseata*.

*Calystegia soldanella

- Sea bindweed (Scannell & Synnott, 1987)
- *Plúr an phrionsa* (Scannell & Synnott, 1987)

CONVOLVULACEAE—THE BINDWEED FAMILY

Native

Calystegia soldanella is found mainly in the south and east on sandy seashores.

*No ethnobotanical uses of this plant have been noted in Ireland.

*Campanula rotundifolia

- Harebell (Scannell & Synnott, 1987); Blue bell (Colgan, 1904); Blue-bonnet (Hart, 1898); Fairy thimble (Hart, 1898)
- *Méaracán gorm* (Scannell & Synnott, 1987); *Lus na gcoinnle gcarach* (Moloney, 1919); *Lus na gcoinnle gcorrach* (Dinneen, 1927)

CAMPANULACEAE—THE BELLFLOWER FAMILY

Native

Campanula rotundifolia is a small perennial herb of grasslands, limestone pavements, sand dunes, heaths, mountain cliffs and screes, lake-shores and rocky ground. It is more common in the north and west, although much rarer elsewhere in Ireland.

De Vere Coneys (1849) gave the Irish name *curach na cuaiche* for a plant called 'the flower blue bottle, small leafed bell flower', a *Campanula*. These names may refer to *C. rotundifolia*.

Campanula rotundifolia is the official flower of Co. Antrim.

*No ethnobotanical uses of this plant have been noted in Ireland.

Campanula rotundifolia.

*Campanula trachelium

- Nettle-leaved bellflower (Scannell & Synnott, 1987); Throatwort (Ó Dónaill, 1977)
- *Scornlus* (Scannell & Synnott, 1987)

CAMPANULACEAE—THE BELLFLOWER FAMILY

Native

48. Dinneen (1927) defined *craipleach* as 'great bindweed'.

Campanula trachelium is a rare plant in Ireland that is occasionally introduced as a garden escape throughout the country. It may be native in the south-east, where it is found on shady river banks, mainly on the River Nore.

The common name 'throatwort' comes from a belief that this plant could be used to provide a sore throat remedy. However, I have not found records of its use in traditional medicine.

*No ethnobotanical uses of this plant have been noted in Ireland.

Cannabis sativa

- Hemp (Reynolds, 2002); Indian hemp (Hackney, 1992); Cannabis (Parnell & Curtis, 2012); Marijuana[49]
- *Caineab* (Cameron, 1883); *Cnaib* (de Vere Coneys, 1849)

CANNABACEAE—THE HEMP FAMILY

Introduced

Cannabis sativa.

Cannabis sativa is rare in Ireland and is found mainly as a casual annual plant growing from discarded or spilled birdseed. It has been smoked throughout the world for centuries for its hallucinatory properties. As an illegal narcotic plant, its cultivation is not permitted in Ireland. However, individual plants, and even on occasion extensive crops, have been grown. Reynolds (2002) included about a dozen records of it growing in seven different botanical vice-counties in Ireland from the late 1800s up to the present day. Praeger (1909) said that it occurred as a casual plant in Limerick. Despite being illegal, it is still smoked widely in Ireland.

Rutty (1772) said that the seed fried is sometimes used in food and that hens fed with this seed lay eggs more plentifully, even in winter.

Although *Cannabis sativa* is regarded as an important medicinal plant in some cultures around the world, there is no history of its medicinal use in Ireland. Elsewhere, it has been used medicinally to treat pain, rheumatism and asthma, and it has been an important herb in Ayurvedic and traditional Chinese medicine (van Wyk & Wink, 2004). These same authors noted its use in modern medicine to treat nausea caused by chemotherapy, depression, lack of appetite in patients with AIDS, and to lower intra-ocular pressure in cases of glaucoma, and in Chinese medicine, to treat constipation in elderly people. Elsewhere, it has been used for a variety of medicinal purposes. For example, in Morocco, cannabis smoke was used to induce abortions in pregnant women wishing to terminate their pregnancies, and in the Democratic Republic of Congo, children with asthma were treated with cannabis smoke (Pennacchio et al., 2010).

Fibres from the stems of hemp have long been an important source of raw materials for many uses. It has been little grown in Ireland for such purposes, although in Britain it was widely cultivated as a crop to provide material for ropes used by Britain's navy (Lewington, 1990).

49. Dried plant material of this species used as a narcotic is often colloquially known by this name and many other slang terms in Ireland.

Capsella bursa-pastoris

- Shepherd's purse (Scannell & Synnott, 1987)
- *Lus an sparáin* (Scannell & Synnott, 1987); *Lus coscadh na fola* (Moloney, 1919); *Spáidín* (Moloney, 1919; Dinneen, 1927); *Luibh na nAosán* (Moloney, 1919; Dinneen, 1927); *Sraidin*[50] (de Vere Coneys, 1849; Cameron, 1883; Ó Dónaill, 1977); *Lus na fola* (de Vere Coneys, 1849; Ó Dónaill 1977); *Lus coiscthe na fola* (Dinneen, 1927); *Lus na n-aosán* (Ó Dónaill, 1977)

BRASSICACEAE (Cruciferae)—THE CABBAGE AND CRESS FAMILY
Probably Introduced (possible archaeophyte)

Capsella bursa-pastoris is an abundant annual weed of roadsides, cultivated ground, gardens, field margins and waste places.

This common cosmopolitan weed has been widely used as a traditional medicinal plant, generally taken as tea, to stop bleeding, treat heavy periods, diarrhoea and cystitis (van Wyk & Wink, 2004). In Ireland, it has been used as a treatment for haemorrhages and also as a diuretic and to promote menstrual flow (Moloney, 1919). According to Allen and Hatfield (2004), it was drunk as a tea or chewed to alleviate kidney troubles and has also been recorded in the treatment of rickets. Uí Chonchubhair (1995) also mentioned its use to control diarrhoea, to treat dysentery, spitting of blood, blood in urine and miscarriage.

Mears and Hillman (2007) suggested that the leaves can be eaten raw in salads and that the seeds are also edible. The leaves can also be cooked as a vegetable, like spinach. Apaches of North America winnowed the seeds of shepherd's purse and ground them into flour to make bread (Milliken & Bridgewater, 2004). Also in North America, the seed pods were dried and ground as a pepper-like seasoning (Peterson, 1977).

Other Irish names for this species include *luibh-na n-aosan* or *lus coscadh na fola* (O'Regan, 1997), which suggest its former use to staunch bleeding and as a remedy for rickets[51].

**Cardamine amara*

- Large bitter-cress (Scannell & Synnott, 1987)
- *Seilín cuaiche* (Scannell & Synnott, 1987); *Biolar searbh* (Moloney, 1919)

BRASSICACEAE (Cruciferae)—THE CABBAGE AND CRESS FAMILY
Native

Cardamine amara is a rare species in Ireland of meadows and river-banks that is found only in Ulster.

*No ethnobotanical uses of this plant have been noted in Ireland.

Cardamine flexuosa

- Wavy bitter-cress (Scannell & Synnott, 1987); Wood bittercress (Brunker, 1950)
- *Searbh-bhiolar casta* (Scannell & Synnott, 1987); *Biolar searbh casta* (Uí Chonchubhair, 1995)

BRASSICACEAE (Cruciferae)—THE CABBAGE AND CRESS FAMILY
Native

50. Or '*sráidín*'.
51. Ó Dónaill (1977) gave the Irish word *aos teannta* for 'rickets'.

Cardamine flexuosa is a common annual herb found throughout Ireland of damp, shady habitats, such as stream banks, woods and the lee sides of walls.

Mears and Hillman (2007) list bitter-cress as a delicious, mild mustard-flavoured herb for salads. There are no records traditional Irish use of this plant.

Cardamine hirsuta

- Hairy bittercress (Scannell & Synnott, 1987)
- *Searbh-bhiolar giobach* (Scannell & Synnott, 1987)

BRASSICACEAE (Cruciferae)—THE CABBAGE AND CRESS FAMILY
Native

Cardamine hirsuta is a common annual herb throughout Ireland that occurs in waste places and on cultivated ground, sand dunes and walls.

Phillips (1983) recommended this bitter-cress as an edible winter herb. I find no records of traditional Irish use of this plant.

**Cardamine impatiens*

- Narrow-leaved bitter-cress (Scannell & Synnott, 1987)
- *Searbh-bhiolar caol* (Scannell & Synnott, 1987)

BRASSICACEAE (Cruciferae)—THE CABBAGE AND CRESS FAMILY
Native

Cardamine impatiens is a very rare biennial herb that is known only as a native plant in Co. Westmeath in woodlands on an esker. As an introduced plant, it is more widespread.

*No ethnobotanical uses of this plant have been noted in Ireland.

Cardamine pratensis

- Cuckooflower (Scannell & Synnott, 1987); Lady's smock (Webb & Scannell, 1983); Bitter cress; Milkmaids (Hart, 1898)
- *Biolar gréagáin* (Scannell & Synnott, 1987); *Léine Mhuire* (Moloney, 1919; Dinneen, 1927; Ó Dónaill, 1977); *Seilín cuaiche* (Moloney, 1919; Dinneen, 1927); *Gleórán* (Moloney, 1919); *Biolar ghriagáin* (de Vere Coneys, 1849); *Lasair léana* (Dinneen, 1927); *Lus na mban* (Dinneen, 1927); *Léinteacha* (Ó Dónaill, 1977); *Léinidh* (Ó Dónaill, 1977)

BRASSICACEAE (Cruciferae)—THE CABBAGE AND CRESS FAMILY
Native

Cardamine pratensis is a common and conspicuous perennial herb with large pink, mauve or white flowers. It can be found throughout Ireland in wet meadows, marshes, lake-shores and stream banks.

It was used as an antispasmodic remedy for Saint Vitus dance (Sydenham's chorea—an acute neurologic disorder that can appear several months after a streptococcal infection) and epilepsy (Moloney, 1919). Uí Chonchubhair (1995) also recorded its medicinal use to treat scurvy, kidney stones, epilepsy, ulcers, and to whet the appetite and aid digestion.

Mears and Hillman (2007) suggested that over-wintering basal leaf rosettes that are just about edible can be eaten in salads, but later in spring, the leaves have an unpleasant oily flavour. On the other hand, Peterson (1977) said that the leaves are pungent but excellent to eat in salads. I am undecided!

Clyne (1989) said that in some parts of Ireland, the plant was given the name *clap-pede-pouch*, as the 'rattle-pouches', or a pouch used by a leper at cross-roads to receive alms, which was extended on a long pole with clappers or bells to draw attention. It is conceivable that this name could also have been applied to shepherd's purse (*Capsella bursa-pastoris*).

Carduus acanthoides (Carduus crispus subsp. *multiflorus)*

- Welted thistle (Scannell & Synnott, 1987)
- *Feochadán reangach* (Scannell & Synnott, 1987)

ASTERACEAE (Compositae)—THE DAISY FAMILY
Native (possible archaeophyte)

Carduus acanthoides is a rare plant in Ireland of dry banks and wasteground that is distributed mainly in the south and east.

Hart (1898: 375) recorded the common name 'Foughlan' for '*Carduus*' in Fanet, Co. Donegal.

*No ethnobotanical uses of this plant have been noted in Ireland.

Carduus tenuiflorus

- Slender thistle (Scannell & Synnott, 1987)
- *Feochadán caol* (Scannell & Synnott, 1987)

ASTERACEAE (Compositae)—THE DAISY FAMILY
Native

Carduus tenuiflorus is a rare plant in Ireland of dry banks and wasteground.

*No ethnobotanical uses of this plant have been noted in Ireland.

Carex species

- Sedge (Scannell & Synnott, 1987)
- *Cíb* (Scannell & Synnott, 1987); *Seisc*[52] (Ua Maoileoin, 2001); *Seasc* (Moloney, 1919; Ó Dónaill, 1977); *Ciab* (Moloney, 1919); *Seisg* (de Vere Coneys, 1849); *Seasclach*[53] (Uí Chanainn, 1962)

CYPERACEAE—THE SEDGE FAMILY

52. Dinneen (1927) defined the word *seisc* to mean 'a coarse grass or sedge that grows in a marsh, used for making grass ropes (súgáin), and thatching'. Ó Dónaill (1977) defined *seisc* as 'sedge'.

53. Uí Chanainn (1962) defined *seasclach* as 'sedge-grass'.

There are more than 50 *Carex* species in Ireland. All are perennial, and different species occur in a wide variety of habitats, both wet and dry, such as in marshes, wet fields, mountain-tops, bogs, heaths, woodlands and grasslands. Identification of the different species is not always easy, mainly because distinguishing characteristics are often based on small flower parts, and hybrids between species are also found with intermediate characteristics between the two parents.

Remarkably few *Carex* species have recorded traditional uses in Ireland, although many form the components of peat used for fuel and hay harvested as forage from damp meadows.

Carex acuta

- Slender tufted-sedge (Scannell & Synnott, 1987)
- *Cíb dhosach chaol* (Scannell & Synnott, 1987)

CYPERACEAE—THE SEDGE FAMILY
Native

Carex acuta is a relatively rare species in Ireland of marshes and wet grasslands, riverbanks and lake-shores. It occurs mainly in the north and is much rarer in the south.

Huxley (1992) listed this plant as an ornamental groundcover used in horticulture, but it is rarely, if ever, used for such purposes in Ireland.

Carex acutiformis

- Lesser pond-sedge (Scannell & Synnott, 1987)
- *Cíb locháin* (Scannell & Synnott, 1987)

CYPERACEAE—THE SEDGE FAMILY
Native

Carex acutiformis is found mainly in the west in Ireland in wet habitats, such as riverbanks, canal-banks, ponds, marshes, ditches and lake-shores.

Huxley (1992) listed this plant as an ornamental groundcover, but it is rarely, if ever, used for such purposes in Ireland.

**Carex appropinquata*

- Fibrous tussock-sedge (Scannell & Synnott, 1987)
- *Cíb schnáithíneach* (Scannell & Synnott, 1987)

CYPERACEAE—THE SEDGE FAMILY
Native

Carex appropinquata is found in bogs and marshes in Co. Westmeath; it is very rare elsewhere in Ireland.

*No ethnobotanical uses of this plant have been noted in Ireland.

*Carex aquatilis

- Water sedge (Scannell & Synnott, 1987)
- *Cíb uisce* (Scannell & Synnott, 1987)

CYPERACEAE—THE SEDGE FAMILY
Native

Carex aquatilis is a rare plant in Ireland of boggy river-banks, ditches and lake margins.

*No ethnobotanical uses of this plant have been noted in Ireland.

Carex arenaria

- Sand sedge (Scannell & Synnott, 1987); Black scutch (Colgan, 1904)
- *Cíb ghainimh* (Scannell & Synnott, 1987)

CYPERACEAE—THE SEDGE FAMILY
Native

Carex arenaria is a common plant throughout the Irish coastline of sand dunes and other sandy habitats by the sea.

In Scotland, Milliken and Bridgewater (2004) recorded that the tough roots of sand sedge were used to make woven restraints for animals being transported by boat from the islands to the mainland. I have not found records of its use for such purposes in Ireland, but with the close cultural linkages between Ireland and the Scottish Isles, such usage was probably also present in Ireland.

Simpson and Inglis (2001) listed *Carex arenaria* as having been used in Europe for unspecified medicinal purposes.

Huxley (1992) listed this plant as an ornamental groundcover used in horticultural, but I have found no records of its use in Ireland.

*Carex bigelowii

- Stiff sedge (Scannell & Synnott, 1987)
- *Cíb shléibhe* (Scannell & Synnott, 1987)

CYPERACEAE—THE SEDGE FAMILY
Native

Carex bigelowii is a rare plant in Ireland, found only on mountain tops and cliffs, mainly in the south and east.

*No ethnobotanical uses of this plant have been noted in Ireland.

*Carex binervis

- Green-ribbed sedge (Scannell & Synnott, 1987)
- *Cíb fhéithghlas* (Scannell & Synnott, 1987)

CYPERACEAE—THE SEDGE FAMILY
Native

Carex binervis is a common sedge throughout Ireland of wet peat, heaths and heathy grasslands, cliff-ledges and moorland.

*No ethnobotanical uses of this plant have been noted in Ireland.

**Carex buxbaumii*

- Club sedge (Scannell & Synnott, 1987)
- *Cíb ghorm* (Scannell & Synnott, 1987)

CYPERACEAE—THE SEDGE FAMILY
Native

Carex buxbaumii has been extinct since the late 1800s in the wild in Ireland, with its only known site on the shores of Lough Neagh, Co. Antrim. It is now maintained in cultivation at the National Botanic Gardens, Glasnevin, from that original source.

*No ethnobotanical uses of this plant have been noted in Ireland.

Carex caryophyllea

- Spring sedge (Scannell & Synnott, 1987)
- *Cíb earraigh* (Scannell & Synnott, 1987)

CYPERACEAE—THE SEDGE FAMILY
Native

Carex caryophyllea is a common plant throughout Ireland of dry unimproved grassland and heaths.

Huxley (1992) listed this plant as an ornamental used in horticulture, but it is rarely, if ever, used for such purposes in Ireland.

**Carex curta* (*Carex canescens*)

- White sedge (Scannell & Synnott, 1987)
- *Cíb bhán* (Scannell & Synnott, 1987)

CYPERACEAE—THE SEDGE FAMILY
Native

Carex curta is uncommon in Ireland and occurs mainly in the north and east in bogs and wet grassland.

*No ethnobotanical uses of this plant have been noted in Ireland.

**Carex demissa* (*Carex viridula* subsp. *oedocarpa*)

- Common yellow-sedge (Scannell & Synnott, 1987)
- *Cíb bhuí mhóna* (Scannell & Synnott, 1987)

CYPERACEAE—THE SEDGE FAMILY
Native

Carex demissa is common in most parts of Ireland in wet habitats on acidic soils, such as in bog-margins, moorland, marshes and wet acid grasslands, and on stony lake-shores and bare peat.

*No ethnobotanical uses of this plant have been noted in Ireland.

**Carex depauperata*

- Starved wood-sedge (Scannell & Synnott, 1987)
- *Cíb choille bhocht* (Scannell & Synnott, 1987)

CYPERACEAE—THE SEDGE FAMILY
Native

Carex depauperata is very rare in Ireland and known only from one woodland site in East Cork.

*No ethnobotanical uses of this plant have been noted in Ireland.

Carex diandra

- Lesser tussock-sedge (Scannell & Synnott, 1987)
- *Cíb thortógach bheag* (Scannell & Synnott, 1987)

CYPERACEAE—THE SEDGE FAMILY
Native

Carex diandra is found mainly in the centre of the country in wetland habitats, such as bogs, marshes, ditches and lake margins; it is very rare elsewhere in Ireland.

Huxley (1992) listed this plant as an ornamental used in horticulture, but it is rarely, if ever, used for such purposes in Ireland.

**Carex dioica*

- Dioecious sedge (Scannell & Synnott, 1987)
- *Cíb éagmaise* (Scannell & Synnott, 1987)

CYPERACEAE—THE SEDGE FAMILY
Native

Carex dioica is rare in most parts of Ireland and found mainly in the north in wet peaty habitats, such as bogs, marshes, fens and flushes in moorland.

*No ethnobotanical uses of this plant have been noted in Ireland.

**Carex distans*

- Distant sedge (Scannell & Synnott, 1987)
- *Cíb scrogallach* (Scannell & Synnott, 1987)

CYPERACEAE—THE SEDGE FAMILY
Native

Carex distans is a relatively common plant in Ireland of salt marshes, seepage areas and rocky and grassy places by the sea.

*No ethnobotanical uses of this plant have been noted in Ireland.

Carex disticha

- Brown sedge (Scannell & Synnott, 1987)
- *Cíb rua* (Scannell & Synnott, 1987)

CYPERACEAE—THE SEDGE FAMILY
Native

Carex disticha is a relatively common sedge throughout Ireland except in the south-west. It is found in wet habitats, such as marshes, fens, ditches, drains, wet grasslands and on lake-shores.

Simpson and Inglis (2001) listed this sedge as having been used to treat unspecified medicinal disorders in Europe. I have not found records of its use in Ireland.

**Carex divisa*

- Divided sedge (Scannell & Synnott, 1987)
- *Cíb ghabhlach* (Scannell & Synnott, 1987)

CYPERACEAE—THE SEDGE FAMILY
Native

Carex divisa is a sedge of estuarine marshes that is rare in Ireland and known only from the Barrow River estuary near New Ross, Co. Wexford.

*No ethnobotanical uses of this plant have been noted in Ireland.

**Carex divulsa*

- Grey sedge (Scannell & Synnott, 1987)
- *Cíb liath* (Scannell & Synnott, 1987)

CYPERACEAE—THE SEDGE FAMILY
Native

Carex divulsa is found occasionally in Ireland, mainly in the south and east, in woodlands, hedges and sandy roadsides.

*No ethnobotanical uses of this plant have been noted in Ireland.

**Carex echinata*

- Star sedge (Scannell & Synnott, 1987)
- *Cíb na réaltaí* (Scannell & Synnott, 1987)

CYPERACEAE—THE SEDGE FAMILY
Native

Carex echinata is common throughout Ireland in moorland, marshes and wet peaty habitats.

*No ethnobotanical uses of this plant have been noted in Ireland.

Carex elata

- Tufted-sedge (Scannell & Synnott, 1987)
- *Cíb dhosach* (Scannell & Synnott, 1987)

CYPERACEAE—THE SEDGE FAMILY
Native

Carex elata is found mainly in the centre of Ireland and is rather rare elsewhere. It grows in marshes, slow-moving streams and the margins of lakes.

Huxley (1992) listed this plant as an ornamental used in horticulture, but it is rarely, if ever, used for such purposes in Ireland.

*Carex elongata

- Elongated sedge (Scannell & Synnott, 1987)
- *Cíb ard* (Scannell & Synnott, 1987)

CYPERACEAE—THE SEDGE FAMILY
Native

Carex elongata is uncommon in Ireland and found in lake margins in the north.

*No ethnobotanical uses of this plant have been noted in Ireland.

*Carex extensa

- Long-bracted sedge (Scannell & Synnott, 1987)
- *Cíb fhada* (Scannell & Synnott, 1987)

CYPERACEAE—THE SEDGE FAMILY
Native

Carex extensa is an occasional sedge throughout the Irish coastline of salt marshes, rocky seashores and sandy habitats by the sea.

*No ethnobotanical uses of this plant have been noted in Ireland.

Carex flacca

- Glaucous sedge (Scannell & Synnott, 1987)
- *Cíb liathghorm* (Scannell & Synnott, 1987)

CYPERACEAE—THE SEDGE FAMILY
Native

Carex flacca is one of the commonest and most widespread sedges in Ireland of grasslands, banks, sand dunes, stony lake-shores and salt marshes.

Huxley (1992) listed this plant as an ornamental, but it is rarely used for such purposes in Ireland.

---

*Carex hirta

- Hairy sedge (Scannell & Synnott, 1987)
- *Cíb ghiobach* (Scannell & Synnott, 1987)

CYPERACEAE—THE SEDGE FAMILY
Native

Carex hirta is rare in Ireland, although it can be found throughout in dry grasslands and some wastegrounds.

*No ethnobotanical uses of this plant have been noted in Ireland.

---

*Carex hostiana

- Tawny sedge (Scannell & Synnott, 1987)
- *Cíb odhar* (Scannell & Synnott, 1987)

CYPERACEAE—THE SEDGE FAMILY
Native

Carex hostiana is found in rough grasslands, moors, marshes, upland flushes and other wet habitats, particularly in the west of Ireland.

*No ethnobotanical uses of this plant have been noted in Ireland.

---

*Carex laevigata

- Smooth-stalked sedge (Scannell & Synnott, 1987)
- *Cíb mhín* (Scannell & Synnott, 1987)

CYPERACEAE—THE SEDGE FAMILY
Native

Carex laevigata is an occasional sedge throughout Ireland of damp woodlands, streamsides and damp grasslands.

*No ethnobotanical uses of this plant have been noted in Ireland.

---

*Carex lasiocarpa

- Slender sedge (Scannell & Synnott, 1987)
- *Cíb chaol* (Scannell & Synnott, 1987)

CYPERACEAE—THE SEDGE FAMILY
Native

Carex lasiocarpa is an occasional sedge in most parts of Ireland except the south and east where it is rare. It grows in wetland habitats, such as lake margins, fens and bogs.

*No ethnobotanical uses of this plant have been noted in Ireland.

**Carex lepidocarpa* (*Carex viridula* subsp. *brachyrhyncha*)

- Long-stalked yellow-sedge (Scannell & Synnott, 1987)
- *Cíb bhuí chosach* (Scannell & Synnott, 1987)

CYPERACEAE—THE SEDGE FAMILY
Native

Carex lepidocarpa is found particularly in the west and the centre of Ireland in fens, calcareous (lime-rich) marshes and lake margins and is rare elsewhere on the island.

*No ethnobotanical uses of this plant have been noted in Ireland.

**Carex limosa*

- Bog-sedge (Scannell & Synnott, 1987)
- *Cíb chorraigh* (Scannell & Synnott, 1987)

CYPERACEAE—THE SEDGE FAMILY
Native

Carex limosa is a common sedge in the west of very wet bogs, pools, fens and lake margins; it is rare elsewhere in Ireland.

*No ethnobotanical uses of this plant have been noted in Ireland.

**Carex magellanica*

- Tall bog-sedge (Scannell & Synnott, 1987)
- *Cíb bhocht* (Scannell & Synnott, 1987)

CYPERACEAE—THE SEDGE FAMILY
Native

Carex magellanica is a rare sedge of mountain bogs in the north. It is unknown elsewhere in Ireland.

*No ethnobotanical uses of this plant have been noted in Ireland.

**Carex muricata*

- Prickly sedge (Scannell & Synnott, 1987)
- *Cíb dheilgneach* (Scannell & Synnott, 1987)

CYPERACEAE—THE SEDGE FAMILY
Native

Carex muricata is a rare sedge in most parts of Ireland of dry grassland.

*No ethnobotanical uses of this plant have been noted in Ireland.

Carex nigra

- Common sedge (Scannell & Synnott, 1987)
- *Cíb dhubh* (Scannell & Synnott, 1987); *Cíob dhubh* (Dinneen, 1927)

CYPERACEAE—THE SEDGE FAMILY
Native

Carex nigra is an extremely widespread and common sedge in most parts of Ireland of marshes, damp grasslands, river-banks and canal-banks.

It is listed by Huxley (1992) as an ornamental plant. Simpson and Inglis (2001) indicated that in parts of its range in Europe (Iceland and Russia), it is regarded as a useful forage plant for stock. I have not found records of its use in Ireland.

**Carex otrubae*

- False fox-sedge (Scannell & Synnott, 1987)
- *Cíb an mhadra rua* (Scannell & Synnott, 1987)

CYPERACEAE—THE SEDGE FAMILY
Native

Carex otrubae is an occasional plant in Ireland of ditches, drains and marshes especially close to the sea.

*No ethnobotanical uses of this plant have been noted in Ireland.

**Carex ovalis* (*Carex leporina*)

- Oval sedge (Scannell & Synnott, 1987)
- *Cíb mhaol* (Scannell & Synnott, 1987)

CYPERACEAE—THE SEDGE FAMILY
Native

Carex ovalis is found occasionally throughout most of Ireland in damp grasslands, upland areas, marshes and meadows and along roadsides.

*No ethnobotanical uses of this plant have been noted in Ireland.

**Carex pallescens*

- Pale sedge (Scannell & Synnott, 1987)
- *Cíb gheal* (Scannell & Synnott, 1987)

CYPERACEAE—THE SEDGE FAMILY
Native

Carex pallescens is found in marshes, damp heathy grassland and wet woods and scrub, particularly in the north and west; it is rare elsewhere in Ireland.

*No ethnobotanical uses of this plant have been noted in Ireland.

*Carex panicea

- Carnation sedge (Scannell & Synnott, 1987); Carnation-grass (Harron, 1986)
- *Cíb chruithneachta* (Scannell & Synnott, 1987)

CYPERACEAE—THE SEDGE FAMILY
Native

Carex panicea is a common sedge throughout Ireland of wet heaths, peaty grassland, bogs and moors.

Milliken and Bridgewater (2004) recorded that the roots of this sedge were woven to make a fish-trap in the Outer Hebrides. I have not found records of its use in this way in Ireland.

*No ethnobotanical uses of this plant have been noted in Ireland.

Carex paniculata

- Greater tussock-sedge (Scannell & Synnott, 1987)
- *Cíb thortógach* (Scannell & Synnott, 1987)

CYPERACEAE—THE SEDGE FAMILY
Native

Carex paniculata is an occasional plant throughout Ireland of marshes, bogs, canals, rivers and wet ditches and drains.

Simpson and Inglis (2001) listed this plant as having been used as a fibre source in the United Kingdom to make hassocks (kneeling stools used in churches) and brooms and in stables as a straw substitute. Huxley (1992) listed it as an ornamental sedge used in horticulture. I have found no records of its use in Ireland.

*Carex pauciflora

- Few-flowered sedge (Scannell & Synnott, 1987)
- *Cíb scáinte* (Scannell & Synnott, 1987)

CYPERACEAE—THE SEDGE FAMILY
Native

Carex pauciflora is known only on the Garron plateau in Co. Antrim in wet bogs and in one locality in Co. Down. It is otherwise unknown in Ireland.

*No ethnobotanical uses of this plant have been noted in Ireland.

Carex pendula

- Pendulous sedge (Scannell & Synnott, 1987)
- *Cíb chrom* (Scannell & Synnott, 1987)

CYPERACEAE—THE SEDGE FAMILY
Native

Carex pendula is a reasonably common plant throughout Ireland of moist shady habitats, such as damp hedgerows, woodlands, demesnes and gardens. It is often planted or survives from earlier cultivation as an ornamental plant or has escaped from cultivation.

It is widely grown ornamental plant for herbaceous borders or woodland gardens (Wyse Jackson, 2006). It is listed also by Huxley (1992) as an ornamental sedge.

Carex pilulifera

- Pill sedge (Scannell & Synnott, 1987)
- *Cíb na bpaidríní* (Scannell & Synnott, 1987)

CYPERACEAE—THE SEDGE FAMILY
Native

Carex pilulifera is an occasional plant throughout Ireland of heathy and stony grassland on acid soils.

It is listed by Huxley (1992) as an ornamental plant but is rarely used for such purposes in Ireland.

Carex pseudocyperus

- Cyperus sedge (Scannell & Synnott, 1987)
- *Cíb sheisceach* (Scannell & Synnott, 1987)

CYPERACEAE—THE SEDGE FAMILY
Native

Carex pseudocyperus is an occasional plant throughout Ireland of marshes, reed-swamps, ditches and fens.

Huxley (1992) listed this plant as an ornamental, but it is rarely, if ever, used for such purposes in Ireland.

**Carex pulicaris*

- Flea sedge (Scannell & Synnott, 1987)
- *Cíb na ndreancaidí* (Scannell & Synnott, 1987)

CYPERACEAE—THE SEDGE FAMILY
Native

Carex pulicaris is a common sedge throughout Ireland of marshes, bogs, heaths and mountain grasslands and cliffs.

*No ethnobotanical uses of this plant have been noted in Ireland.

*Carex punctata

- Dotted sedge (Scannell & Synnott, 1987)
- *Cíb bhallach* (Scannell & Synnott, 1987)

CYPERACEAE—THE SEDGE FAMILY
Native

Carex punctata is a coastal sedge that is found mainly in the south-west in Ireland, as well as in parts of West Galway and in one locality in Donegal, in grasslands and rocky habitats near the sea. It is otherwise unknown in Ireland.

*No ethnobotanical uses of this plant have been noted in Ireland.

*Carex remota

- Remote sedge (Scannell & Synnott, 1987)
- *Cíb scartha* (Scannell & Synnott, 1987)

CYPERACEAE—THE SEDGE FAMILY
Native

Carex remota is found quite commonly in Ireland in wet shady habitats, such as deciduous woodlands, scrub and hedgerows and in damp grassland.

*No ethnobotanical uses of this plant have been noted in Ireland.

Carex riparia

- Greater pond-sedge (Scannell & Synnott, 1987)
- *Cíb locháin mhór* (Scannell & Synnott, 1987)

CYPERACEAE—THE SEDGE FAMILY
Native

Carex riparia is found mainly in the east in habitats such as wet coastal marshes, ditches, drains, lake margins and river-banks; it is rare elsewhere in Ireland.

It is listed by Simpson and Inglis (2001) as used as a substitute for straw in stables. Huxley (1992) listed it as an ornamental plant in horticulture. There are no records of its use in Ireland.

*Carex rostrata

- Bottle sedge (Scannell & Synnott, 1987)
- *Cíb ghobach* (Scannell & Synnott, 1987)

CYPERACEAE—THE SEDGE FAMILY
Native

Carex rostrata is an occasional plant in most parts of Ireland of marshes, pools, bogs, ponds, lakeshores and other shallow-water habitats.

It is listed by Simpson and Inglis (2001) for its use in forage for cattle in parts of Europe (Iceland and Russia). There are no records of its use in Ireland.

**Carex serotina (Carex viridula subsp. viridula)*

- Small-fruited yellow-sedge (Scannell & Synnott, 1987)
- *Cíb bhuí bheag* (Scannell & Synnott, 1987)

CYPERACEAE—THE SEDGE FAMILY
Native

Carex serotina is found mainly in the west in habitats such as stony and muddy lake-shores and fens and wet grasslands; it is rare elsewhere in Ireland.

*No ethnobotanical uses of this plant have been noted in Ireland.

**Carex spicata*

- Spiked sedge (Scannell & Synnott, 1987)
- *Cíb spícíneach* (Scannell & Synnott, 1987)

CYPERACEAE—THE SEDGE FAMILY
Native

Carex spicata is rare in Ireland, but can be found throughout, in rough grasslands and on banks.

*No ethnobotanical uses of this plant have been noted in Ireland.

**Carex strigosa*

- Thin-spiked wood-sedge (Scannell & Synnott, 1987)
- *Cíb choille chaol* (Scannell & Synnott, 1987)

CYPERACEAE—THE SEDGE FAMILY
Native

Carex strigosa is rare in Ireland but can be found throughout in shady and damp habitats, such as woodlands and hedgerows.

*No ethnobotanical uses of this plant have been noted in Ireland.

Carex sylvatica

- Wood sedge (Scannell & Synnott, 1987)
- *Cíb choille* (Scannell & Synnott, 1987)

CYPERACEAE—THE SEDGE FAMILY
Native

Carex sylvatica is quite common throughout Ireland in shady places such as deciduous woodlands and hedgerows.

Huxley (1992) listed this plant as an ornamental, but it is rarely used for such purposes in Ireland.

*Carex vesicaria

- Bladder-sedge (Scannell & Synnott, 1987)
- *Cíb bhoilgneach* (Scannell & Synnott, 1987)

CYPERACEAE—THE SEDGE FAMILY
Native

Carex vesicaria is found occasionally throughout Ireland in marshes and shallow-water habitats and on lake-shores.

*No ethnobotanical uses of this plant have been noted in Ireland.

Carlina vulgaris

- Carline thistle (Scannell & Synnott, 1987)
- *Feochadán mín* (Scannell & Synnott, 1987)

ASTERACEAE (Compositae)—THE DAISY FAMILY
Native

Carlina vulgaris can be found in a variety of dry calcareous habitats, such as grassy banks, sand dunes and gravel banks, particularly in the centre of Ireland. It is abundant in the Burren, Co. Clare, but rarely seen in the north and south. The Burren is known for its diverse rare plant assemblages.

Allen and Hatfield (2004) recorded that in Limerick a preparation made from this species was rubbed onto the skin to treat an unspecified disease that was described as spreading over the body.

Carpinus betulus

- Hornbeam (Scannell & Synnott, 1987)
- *Crann sleamhain* (Scannell & Synnott, 1987); *Leamhan bog* (Dinneen, 1927)

BETULACEAE (Corylaceae)—THE BIRCH FAMILY
Introduced

Carpinus betulus is not a native tree in Ireland but is occasionally planted in woodlands and gardens and has perhaps gone wild and self-seeded in a few locations.

Hornbeam produces a very hard and heavy timber that is white in colour. It has been used widely for fine turnery, such as for the making of pulleys, wood-screws, tool handles, mallets and skittles (Howes, 1974). It is occasionally grown as a hedging plant in Ireland or as ornamental specimen trees.

Carum carvi

- Caraway (Scannell & Synnott, 1987)
- *Cearbhas* (Scannell & Synnott, 1987); *Carabhuadh* (Moloney, 1919); *Cearbhur* (Moloney, 1919); *Lus mhic chuimín* (Moloney, 1919; Dinneen, 1927); *Ainís* (Moloney, 1919)

APIACEAE (Umbelliferae)—THE CARROT FAMILY

Introduced

Although formerly more widespread as a casual plant, *Carum carvi* is an introduced, aromatic biennial herb that is now considered rare in the wild in Ireland. In the wild it is found particularly in the northern half of Ireland in waste places. It is regularly cultivated in gardens as a culinary herb.

Elsewhere in the world, caraway has been used as a medicinal plant. For example, in Hungary, it was burnt and the smoke inhaled to treat abdominal pains. It was used in the same way for babies who cried excessively after their birth (Pennacchio et al., 2010). In Ireland I have not found records of its use in this way.

**Carum verticillatum*

- Whorled caraway (Scannell & Synnott, 1987)
- *Cearbhas fáinneach* (Scannell & Synnott, 1987)

APIACEAE (Umbelliferae)—THE CARROT FAMILY

Native

Carum verticillatum is an uncommon plant in Ireland of damp grasslands that is found primarily in the north and west.

*No ethnobotanical uses of this plant have been noted in Ireland.

Castanea sativa

- Spanish chestnut (Scannell & Synnott, 1987); Sweet chestnut (Clapham et al., 1962)
- *Castán* (Scannell & Synnott, 1987); *Chraobh geann-chno* (Cameron, 1883)

FAGACEAE—THE BEECH, OAK AND CHESTNUT FAMILY

Introduced

The Spanish chestnut is native to the Mediterranean region of Europe and North Africa, and is distributed east as far as the Caucasus into Asia. It has been widely introduced and grown elsewhere for its edible nuts and useful timber and is a popular ornamental tree that gives excellent autumn colour. Its bark is also useful for tanning, and its timber has been used instead of oak. It grows to be a large deciduous tree in Ireland and is easily recognised by its spirally twisted trunk and long sharply toothed leaves. It is widely planted in Ireland but only very occasionally self-seeded and naturalised (Webb, 1977; Reynolds, 2002; Green, 2008). The largest girth of a single-trunked tree recorded in Ireland is a Spanish chestnut growing in Rossana, Ashford, Co. Wicklow, and in 2005, that measured 10.59 at 1.3 × 19 m tall. Another exceptional specimen, also in 2005, at Williamstown, Co. Meath, measured 7.7 × 27.5 m (Anonymous, 2005).

Castanea sativa has been used as a traditional medicinal plant elsewhere in Europe (although probably not in Ireland) to treat diarrhoea, coughs, whooping cough, bronchitis and asthma (van Wyk & Wink, 2004). A tea or decoction is made from dried chopped leaves.

Spanish chestnuts have been cultivated in the Mediterranean region for at least 3000 years. The Romans introduced them to Britain between AD 42 and 410 to provide a food source for their legionaries, but the cooler climate was not ideal for the development of a good chestnut crop (Ciesla, 2002). Chestnuts are highly nutritious and rich in carbohydrates, contain significant protein and also have a low fat content and no cholesterol. Roasted chestnuts are commonly roasted and sold on the streets in many parts of mainland Europe, often around Christmastime. This practice has also appeared in Ireland in recent decades too. Most of the chestnuts sold in Ireland seem to be imported (mainly from France), and I know of no commercial production of Spanish chestnuts in Ireland. In any case, the chestnuts rarely swell sufficiently in the Irish climate to make them worth using. Flour made from the seeds has been used in Mediterranean cooking, particularly in Italy. Pureed chestnuts (marron) are widely used in French (and other) cuisine and as stuffing for poultry. In France a traditional way to save chestnuts is to preserve them in syrup as *marrons glacés*.

De Vere Coneys (1849) gave the Irish word *cnobhfrancach* for 'chestnut' and Dinneen (1927) *cnó geanmnaidhe* or *geanm-chnó*.

*Catabrosa aquatica (Aira aquatica)

- Whorl-grass (Scannell & Synnott, 1987)
- *Casfhéar uisce* (Scannell & Synnott, 1987); *Gruagfheur uisce* (Wade, 1808a); *Gruagfhér uisge* (White, 1803)

POACEAE (Gramineae)—THE GRASS FAMILY
Native

Catabrosa aquatica is an occasional grass throughout Ireland of wetland habitats, such as marshes, ditches, drains, pools, ponds and open mud.

*No ethnobotanical uses of this plant have been noted in Ireland.

*Catapodium marinum (Desmazeria marina; Triticum loliaceum)

- Sea fern-grass (Scannell & Synnott, 1987)
- *Féar trá* (Scannell & Synnott, 1987); *Cruithneachd abhach*[54] (White, 1803)

POACEAE (Gramineae)—THE GRASS FAMILY
Native

Catapodium marinum is found throughout Ireland in waste places, on walls near the sea and along the seashore.

*No ethnobotanical uses of this plant have been noted in Ireland.

*Catapodium rigidum (Desmazeria rigida; Poa rigida)

- Fern-grass (Scannell & Synnott, 1987)
- *Féar raithní* (Scannell & Synnott, 1987); *Cuise cruidhata* (White, 1803)

POACEAE (Gramineae)—THE GRASS FAMILY
Native

54. White (1803) applied this name to *Triticum loliaceum*.

Catapodium rigidum can be found in waste places and on dry wall tops and rocky outcrops throughout Ireland, although it is rare in the north except along the coast.

*No ethnobotanical uses of this plant have been noted in Ireland.

Centaurea cyanus

- Cornflower (Scannell & Synnott, 1987); Corn bluebottle (Brunker, 1950); Bluebottle
- *Gormán* (Scannell & Synnott, 1987); *Lus gormáin* (Moloney, 1919; Dinneen, 1927; Ó Dónaill, 1977); *Coirce gorm*[55] (Moloney, 1919; Ó Dónaill, 1977); *Curac na cuig* (Cameron, 1883); *Gránphlúr* (Ó Dónaill, 1977)

ASTERACEAE (Compositae)—THE DAISY FAMILY
Native (or possible archaeophyte)

The cornflower was formerly a common annual weed of cereal and flax crops, but with improved seed cleaning and herbicide use, it has become very rare in Ireland and is almost extinct. Today, it is occasionally grown as a garden plant and included in wildflower seed mixtures. It is generally regarded as a native plant, but it was probably introduced to Ireland in ancient times with cereal crops by early farmers.

Centaurea cyanus has been widely used in traditional medicine in Europe and the Near East as a general tonic, to make a lotion to treat irritated eyes, for gall bladder, stomach, menstrual and kidney disorders, and to increase resistence to infections (van Wyk & Wink, 2004). Harrop (1977) mentioned that the dried flowers are useful to give colour to potpourri and also to make an infusion to bathe the eyes. I can find no records of its use as a source of traditional medicine in Ireland.

Centaurea nigra

- Common knapweed (Scannell & Synnott, 1987); Black heads (Colgan, 1904); Blackhead (Hart, 1898); Stickybacks[56] (Munnelly, 2005); Bachelor's buttons (Hart, 1898); Black knapweed (Brunker, 1950); Hardheads (Harron, 1986)
- *Mínscoth* (Scannell & Synnott, 1987); *Cnapán dubh* (Moloney, 1919); *Mullach dubh*[57] (Moloney, 1919; Dinneen, 1927; Ó Dónaill, 1977); *Nianscoth* (Moloney, 1919; Dinneen, 1927); *Niansgoth* (Cameron, 1883; de Vere Coneys, 1849); *Bodach dubh* (Dinneen, 1927); *Bata cogaidh* (McGlinchey, 2007); *Ceanna dubha* (Dinneen, 1927); *Cnaipe dubh* (Dinneen, 1927); *Tobac capaill* (Dinneen, 1927); *'Tobac an chapaill'* (Dinneen, 1927; Tóibín, 1965); *Brealla gorm* (Tóibín, 1965; Ó Dónaill, 1977)

ASTERACEAE (Compositae)—THE DAISY FAMILY
Native

55. Moloney (1919) and Ó Dónaill (1977) defined *coirce gorm* as 'blue-bottle'.
56. According to Munnelly (2005), the term 'stickybacks' was used for *Centaurea nigra* in Dublin.
57. Moloney (1919), Dinneen (1927) and Ó Dónaill (1977) defined *mullach dubh* as 'black knapweed'.

Centaurea nigra is an extremely common plant throughout Ireland of roadsides, banks, meadows and pastures.

According to Allen and Hatfield (2004), this plant was formerly used medicinally in Ireland for jaundice, liver ailments, scurvy, warts, 'pains in the bones', tuberculosis, asthma and as a cleansing tonic.

Other species of *Centaurea*, *C. solstitialis* (yellow star-thistle or St Barnaby's thistle), *C. calcitrapa* (common star-thistle) and *C. aspera* (rough star-thistle), are not native plants in Ireland or even found today as weeds. Nevertheless, Ó Dónaill (1977) gave the Irish name *plibisín* for 'star-thistle' and 'caltrop'. These species have been used in traditional medicine elsewhere in Europe for various purposes. In some parts of Europe, such as southern Italy and Crete, the leaves of *C. calcitrapa* have been eaten boiled as a vegetable. None of these species is known in the wild in Ireland today, although there are old records of *C. aspera* and *C. calcitrapa* occurring as rare casual weeds in several parts of Ireland, mainly around the major ports (Reynolds, 2002).

Munnelly (2005) mentioned that in West Clare, *Centaurea nigra* was used as a veterinary remedy to treat 'strangles' in horses. It was also used for a children's game, too, in Clare. Munnelly (2005: 166), quoting Francie Kennelly, a local farmer in Miltown Malbay, said, 'As young lads we used to be peltin' 'em at each other. An' especially peltin' 'em at girls' hair. 'Twould be very hard for a girl to take 'em out of her hair'.

Centaurea nigra.

Centaurea scabiosa (greater knapweed). Watercolour by Lydia Shackleton. Courtesy of the National Botanic Gardens, Glasnevin.

*Centaurea scabiosa

- Greater knapweed (Scannell & Synnott, 1987)
- *Mínscoth mhór* (Scannell & Synnott, 1987)

ASTERACEAE (Compositae)—THE DAISY FAMILY

Native

Centaurea scabiosa is an occasional plant in Ireland, particularly in the centre and south. It is much rarer elsewhere in Ireland. It occurs in habitats such as dry banks, rough grasslands, walls and rocky places.

*No ethnobotanical uses of this plant have been noted in Ireland.

Centaurium erythraea (*Erythraea centaurium*)

- Common centaury (Scannell & Synnott, 1987); Santry (Colgan, 1904); Dramwe na murrough (Hart, 1898); Feverwort (Reader's Digest Association, 2006)
- *Dréimire Mhuire* (Scannell & Synnott, 1987); *Cead-duilleach* (de Vere Coneys, 1849; Moloney, 1919; Dinneen, 1927); *Dréimire Mhuire* (Moloney, 1919); *Dréimire buidhe* (Moloney, 1919); *Ceadharlach* (Cameron, 1883); *Deagha*[58] (Ó Dónaill, 1977); *Dréimire bui* (or *dréimide* or *dréimre*) (Ó Dónaill, 1977); *Cróbán mhuire*[59] (Colgan, 1911)

58. Ó Dónaill (1977) gave the Irish word *deagha* for 'centaury'.
59. Colgan (1911) indicated *cróbán mhuire* was the common name for this plant in Clare Island, Co. Mayo.

GENTIANACEAE—THE GENTIAN FAMILY
Native

Centaurium erythraea is a frequent plant all around the coast of Ireland in dry or well-drained banks and pastures and on sand dunes. It is much rarer inland.

Sometimes called 'feverwort', this plant species has been used medicinally to help reduce fevers as well as loss of appetite, indigestion, flatulence, excessive tiredness, headlice and loss of hair (Reader's Digest Association, 2006). It has been used in traditional medicine throughout its native range in Europe and the Mediterranean for digestion, chronic dyspepsia, fevers and gastrointestinal, liver, bile, bladder and kidney complaints (van Wyk & Wink, 2004). Moloney (1919) indicated its use internally to treat rheumatism and as a tonic. Allen and Hatfield (2004) noted that it was also used in Ireland to treat indigestion, coughs and liver complaints. Williams (1993) recorded its use in Ireland for sore eyes, gravel, stomach worms, growing pains and muscular pains in children, and it is supposedly good for a voice when sweetened with honey.

**Centaurium littorale*

- Seaside centaury (Scannell & Synnott, 1987)
- *Dréimire trá* (Scannell & Synnott, 1987); *Dreimire buidhe* (Cameron, 1883)

GENTIANACEAE—THE GENTIAN FAMILY
Native

Centaurium littorale is a very rare plant in Ireland, known only from sandy coastal habitats in Co. Derry. It was recently found also near Dingle in Co. Kerry.

*No ethnobotanical uses of this plant have been noted in Ireland.

**Centaurium pulchellum*

- Lesser centaury (Scannell & Synnott, 1987)
- *Dréimire beag* (Scannell & Synnott, 1987)

GENTIANACEAE—THE GENTIAN FAMILY
Native

Centaurium pulchellum is a very rare plant in Ireland, known only on the east and south coasts. It grows on sand dunes and in other sandy habitats.

*No ethnobotanical uses of this plant have been noted in Ireland.

Centranthus ruber

- Red valerian (Scannell & Synnott, 1987); Pretty Betsy (Hart, 1898)
- *Slán iomaire* (Scannell & Synnott, 1987)

VALERIANACEAE—THE VALERIAN FAMILY
Introduced

Centranthus ruber is an introduced plant that is now invasive and spreading in Ireland. Its habitat is rocky outcrops, such as those in the Burren, Co. Clare, as well as old walls and ruins. It may have been introduced as an ornamental plant (white, pink and reddish flowered forms are seen), but its young leaves and shoots are also edible, although a bit bitter. It can be eaten raw in salads or used as a vegetable, lightly boiled and served with butter. The roots have also been used to flavour soups (Hilton, 2007).

*Cephalanthera longifolia

- Narrow-leaved helleborine (Scannell & Synnott, 1987)
- *Cuaichín caol* (Scannell & Synnott, 1987)

ORCHIDACEAE—THE ORCHID FAMILY
Native

Cephalanthera longifolia is a very rare orchid of woodlands and scrub that is found mainly in the southern half of Ireland.

*No ethnobotanical uses of this plant have been noted in Ireland.

*Cerastium arvense

- Field mouse-ear (Scannell & Synnott, 1987)
- *Cluas luchóige mhóinéir*[60] (Scannell & Synnott, 1987)

CARYOPHYLLACEAE—THE CARNATION OR PINK FAMILY
Native

Cerastium arvense is an occasional plant in Ireland, particularly of dry, rocky habitats and grasslands in the Burren, Co. Clare, and along the north Co. Dublin coastline. It is very rare elsewhere in Ireland.

*No ethnobotanical uses of this plant have been noted in Ireland.

*Cerastium diffusum

- Sea mouse-ear (Scannell & Synnott, 1987)
- *Cluas luchóige mhara* (Scannell & Synnott, 1987)

CARYOPHYLLACEAE—THE CARNATION OR PINK FAMILY
Native

Cerastium diffusum is an occasional plant in Ireland of coastal, sandy habitats; it is very rarely distributed inland.

*No ethnobotanical uses of this plant have been noted in Ireland.

*Cerastium fontanum

- Common mouse-ear (Scannell & Synnott, 1987); Common mouse-ear chickweed (Hackney, 1992)

60. Moloney (1919) gave the Irish names *cluas liath*, *cluas luch*, *lus midhe* and *lus na míol* for *Cerastium* species.

- *Cluas luchóige* (Scannell & Synnott, 1987); *Luibh na triucha*[61] (Dinneen, 1927); *Lus midhe* (Dinneen, 1927); *Lus na míol* (Dinneen, 1927)

CARYOPHYLLACEAE—THE CARNATION OR PINK FAMILY
Native

Cerastium fontanum is a common, and sometimes abundant, plant throughout Ireland of cultivated ground, waste places and grasslands.

According to Peterson (1977), *Cerastium fontanum* has been used for food in North America, as a vegetable.

*No ethnobotanical uses of this plant have been noted in Ireland.

---

*Cerastium glomeratum

- Sticky mouse-ear (Scannell & Synnott, 1987)
- *Cluas luchóige ghreamaitheach* (Scannell & Synnott, 1987)

CARYOPHYLLACEAE—THE CARNATION OR PINK FAMILY
Native

Cerastium glomeratum is a common plant throughout Ireland of wasteground, walls and cultivated ground.

*No ethnobotanical uses of this plant have been noted in Ireland.

---

*Cerastium semidecandrum

- Little mouse-ear (Scannell & Synnott, 1987)
- *Cluas luchóige bheag* (Scannell & Synnott, 1987)

CARYOPHYLLACEAE—THE CARNATION OR PINK FAMILY
Native

Cerastium semidecandrum is an occasional plant in Ireland of coastal sandy habitats; it is very rarely distributed inland but can be found in dry and sandy habitats.

*No ethnobotanical uses of this plant have been noted in Ireland.

---

*Ceratocapnos claviculata (Corydalis claviculata)

- Climbing corydalis (Scannell & Synnott, 1987)
- *Fliodh na tuí* (Scannell & Synnott, 1987)

PAPAVERACEAE (Fumariaceae)—THE POPPY FAMILY
Native

Ceratocapnos claviculata is a delicate herb that is found occasionally in Ireland, particularly in the south-east, in shady places.

*No ethnobotanical uses of this plant have been noted in Ireland.

---

*Ceratophyllum demersum

- Rigid hornwort (Scannell & Synnott, 1987)

61. Dinneen (1927) defined *luibh na triucha* as 'mouse ear, chickweed'.

- *Cornlach* (Scannell & Synnott, 1987)

CERATOPHYLLACEAE—THE HORNWORT FAMILY
Native

Ceratophyllum demersum is rather rare throughout Ireland and is an aquatic plant of ponds, lakes, canals and slow-moving streams.

*No ethnobotanical uses of this plant have been noted in Ireland.

**Ceratophyllum submersum*

- Soft hornwort (Reynolds, 2002)

CERATOPHYLLACEAE—THE HORNWORT FAMILY
Possibly Introduced

Ceratophyllum submersum is a very rare aquatic plant in Ireland that is known only from several ponds in Co. Down.

*No ethnobotanical uses of this plant have been noted in Ireland.

Chaerophyllum temulum

- Chervil (Webb, 1977); Rough chervil (Scannell & Synnott, 1987)
- *Camán gall* (Scannell & Synnott, 1987); *Lus na ríocach* (Moloney, 1919; Dinneen, 1927)

APIACEAE (Umbelliferae)—THE CARROT FAMILY
Introduced

Chaerophyllum temulum is a biennial herb that is found mainly in the southeast in hedgerows and on roadsides. It is considered rare, although it was formerly more common than it is today and was probably cultivated for its culinary or medicinal properties.

Chervil is a useful culinary herb that tastes mildly of liquorice and is widely used as a flavouring in salads, soups and omelettes. Chervil has been used for various medicinal purposes in Europe, but I have not found any records of its use in Ireland, although its Irish names suggests former usage.

Chamaemelum nobile (*Anthemis nobilis*)

- Chamomile (Scannell & Synnott, 1987)
- *Camán meall* (Scannell & Synnott, 1987); *Fíogadán* (Moloney, 1919; Dinneen, 1927; Ó Dónaill, 1977); *Camán (-áin) míonla* (*mínligh, míolla, meall*) (Dinneen, 1927); *Finéal madra*[62] (Ó Dónaill, 1977); *Coimeán meall* (Ó Criomhthain, MSS, n.d., ca. 1908)

ASTERACEAE (Compositae)—THE DAISY FAMILY
Native

62. Ó Dónaill (1977) defined *finéal madra* as 'dog's chamomile'.

Chamaemelum nobile is a creeping, aromatic perennial plant of roadsides, lake-shores, grasslands, heaths and waste places that is frequently found in the south-west but is rather rare elsewhere in Ireland.

Chamomile is a well-known herb used to aid digestion and to make a soothing tea. Despite being an Irish native plant, it was widely cultivated in gardens for its medicinal, aromatic and herbal qualities. Moloney (1919) noted that chamomile was formerly used to treat upset stomachs and wind in old people. The whole plant is intensely bitter, highly aromatic, and much used medicinally (Mackay, 1836). Chamomile is generally regarded as a soothing, slightly sedative tea, and Gallagher (2012: 18) mentioned that 'it is of great use for young girls who have an ailment with a tendancy for panic attacks'. Fresh or dried flowers can be used for this purpose. Boiling water is poured onto about a teaspoon full of the flowers and allowed to infuse for a few minutes. In Ireland, chamomile was formerly used to treat a number of ailments, including swellings, inflammations, boils, and jaundice (boiled in wine or porter and drunk over a period of 10 days), and as a relaxant and soporific (Allen & Hatfield, 2004). Chamomile is also said to be useful to treat sore throats (Harrop, 1977). Egan (1887) noted its use for cramps, flatulence, colic and indigestion.

Chamomile is also a mild bleach and has been used for lightening hair (Lewington, 1990; Shaw, 1998). It is also used in modern commercially available shampoos. The flower heads are steam-distilled to produce chamomile oil (Howes, 1974). Extracts of chamomile have also been used as ingredients in cosmetics in Ireland in recent years too.

Lawns composed of chamomile were popular as early as Elizabethan times as they give off a delicious fragrance when walked on. They are easy to establish and maintain and survive better than grass in times of drought. Sir Francis Drake is said to have played his famous game of bowls on a chamomile lawn as the Spanish Armada sailed towards Britain (Howes, 1974).

Moloney (1919) gave alternative Irish names for 'other varietieties' (sic) of chamomile: *camán míonla* and *finéal madra*. Related plants that are sometimes confused with *Chamaemelum nobile* include *Anthemis arvensis*[63] (corn chamomile) and *A. cotula* (stinking chamomile). Both of these species are rare, introduced casual plants of wasteground and cultivated fields and have no known ethnobotanical uses in Ireland.

Cheiranthus cheiri (*Erysimum cheiranthoides*)

- Wallflower (Scannell & Synnott, 1987)
- *Lus an bhalla* (Scannell & Synnott, 1987); *Lus leat an tsamhraidh* (de Vere Coneys, 1849; Moloney, 1919; Dinneen, 1927)

BRASSICACEAE (Cruciferae)—THE CABBAGE AND CRESS FAMILY
Introduced

Moloney (1919) recorded the use of *Cheiranthus cheiri* flowers in Ireland in traditional medicine, steeped in oil and rubbed on the skin as a liniment and as an infusion for 'nervous troubles'.

This popular garden plant, widely grown as an annual ornamental bedding plant for spring displays, is well established in the wild in Ireland on old walls, rocks, wasteground, and some coastal cliffs. Wild plants are generally yellow-flowered (Reynolds, 2002).

63. De Vere Coneys (1849) and Cameron (1883) gave the Irish name *coman mionla* for *Anthemis arvensis*.

Chelidonium majus

- Greater celandine (Scannell & Synnott, 1987); Sollendine (Hart, 1898)
- *Garra bhuí* (Scannell & Synnott, 1987); *Lacha cheannruadh* (Moloney, 1919); *Aonscoth* (Moloney, 1919; Dinneen, 1927); *Cliathach bhuidhe* (Dinneen, 1927)

PAPAVERACEAE—THE POPPY FAMILY
Introduced (possible archaeophyte)

Chelidonium majus is an occasional weed of hedges and walls, particularly in the east, but occasionally found elsewhere in Ireland.

Greater celandine is a well-known plant in traditional medicine throughout its natural range in Europe, Asia and North Africa, used to treat a variety of ailments including spasms of the gastrointestinal tract and bile duct, warts, eczema, ringworm and for eye complaints (van Wyk & Wink, 2004). It was used formerly in Ireland as a purgative and diuretic, an ingredient in cancer cures, and in the treatment of corns (Moloney, 1919). Its stems exude bright orange-yellow latex, which was formerly used as a cure for warts in Louth (Synnott, 1979). Allen and Hatfield (2004) also mentioned its use to treat warts on cattle. Hart (1898) and Gallagher (2012) mentioned its use in Co. Donegal to treat sore eyes, haemorrhoids and warts.

This plant is generally regarded as poisonous and causes gastroenteritis with diarrhoea if eaten, and can be fatal (Woodward, 1985).

Chenopodium album

- Fat hen (Scannell & Synnott, 1987); White goosefoot (Colgan, 1904); Lamb's quarters (Colgan, 1904); Dungweed (McDaeid, 1991); Muckweed (McDaeid, 1991); Dirtweed (McDaeid, 1991); Miles (Harron, 1986)
- *Blonagán bán* (Scannell & Synnott, 1987); *Lus coise gé (bán)* (Moloney, 1919; Dinneen, 1927); *Praiseach fhiadhain* (de Vere Coneys, 1849; Moloney, 1919; Dinneen, 1927); *Praiseach fhiáin* (Dinneen, 1927; Uí Chonchubhair, 1995); *Praiseach mhín* (Dinneen, 1927)

AMARANTHACEAE (Chenopodiaceae)—THE GOOSEFOOT FAMILY
Native (or possible archaeophyte)

Chenopodium album is a fast-growing and very common annual weed of waste places and cultivated fields.

The leaves of this species have been used for centuries in Ireland as a vegetable, like spinach. It was only after spinach was introduced that it faded from use (MacCoitir, 2006). *Chenopodium album* and *Atriplex patula*, both of which have been called fat hen, were sold by hawkers and eaten as leafy vegetables in Dublin until the 18th century (Rutty, 1772; Geraghty, 1996). Geraghty (1996) recorded that seeds of fat hen dating back more than 1000 years were found in Viking remains at Fishamble Street in Dublin. Mears and Hillman (2007) suggested that the seeds can be collected, processed and made into meal cakes. Seeds of fat hen were found in the stomach of Tollund Man, the 2000-year-old body found in a bog in Jutland, Denmark, in 1950. His last meal had been a gruel made from seeds of several species, including this one, barley, wild pansy and linseed (Mabey, 1998). In addition, a small amount of organic matter was found between the bones of a male grave occupant in excavations of medieval Dublin; his last meal included large numbers of seeds of *C. album* and *Persicaria/Polygonum* spp. (knotgrass) (Anonymous, 1973). The same seeds have been found in cesspits and demonstrate that these plants must have been a regular part of the diet of at least some of the city's occupants. Both fat hen and knotgrass would have been common and widespread weeds and may have formed part of the floras of cultivated ground that was being rested between crops, a standard practice in medieval and later agriculture (Anonymous, 1973).

The leaves of fat hen can also be used in salads, but only in moderation, as they contain saponins, which can be toxic if eaten in quantity. Leaves may also concentrate nitrates when growing in nitrogen rich places (Hilton, 2007). The seeds were gathered as a starch-rich source (Akeroyd, 1996) and can be ground and added to flour and used in baking. The seeds can also be sprouted and used in salads or stir-fried (Hilton, 2007).

According to Allen and Hatfield (2004), in Co. Dublin a decoction of the stems of *Chenopodium album* was drunk for rheumatism.

In North America, where it is a long-naturalised species, it was used by Native Americans to make pottage. It was also used by the Pawnee people for painting bows and arrows green (Gilmore, 1991). I have not found records of its use as a dye plant in Ireland.

De Vere Coneys (1849) gave the following Irish names for other *Chenopodium* species: *praiseach bhrathain*, meaning 'English mercury', 'wild spinage' for *Chenopodium*; *praiseach ghlas*[64], meaning 'fig-leaved goosefoot' for '*Chenopodium ficifolium*' and *praiseach na mara*, meaning 'annual sea side goosefoot' for '*Chenopodium maritimum*'. Dinneen (1927) gave the Irish names *lus coise ghé* or *cos ghé* for 'goosefoot' and Ó Dónaill (1977) gave the names *ceathrú chaorach* and *lus coise gé* for 'goosefoot'.

Chenopodium bonus-henricus

- Good King Henry (Scannell & Synnott, 1987); All-good (Harron, 1986); Mercury (Harron, 1986)
- *Praiseach bhráthar* (Scannell & Synnott, 1987)

AMARANTHACEAE (Chenopodiaceae)—THE GOOSEFOOT FAMILY
Introduced (possible archaeophyte)

Chenopodium bonus-henricus is an occasional plant throughout Ireland of waste places and roadsides. It was formerly used in salads or cooked as spinach and is reputed to taste similar to asparagus.

Chenopodium murale

- Nettle-leaved goosefoot (Scannell & Synnott, 1987); Sowbane (Colgan, 1904)
- *Praiseach an bhalla* (Scannell & Synnott, 1987); *Praiseach an fhalla* (Moloney, 1919); *Praiseach na m-balla* (de Vere Coneys, 1849)

AMARANTHACEAE (Chenopodiaceae)—THE GOOSEFOOT FAMILY
Introduced

64. Dinneen (1927) also gave *praiseach ghlas* for *Chenopodium*.

Chenopodium murale is a rather rare annual plant in Ireland of waste places near the coast, ports and dockyards. When found it is a casual plant that has usually been introduced.

Like other *Chenopodium* species its leaves are edible and can be cooked and used as a vegetable.

Chenopodium rubrum

- Red goosefoot (Scannell & Synnott, 1987)
- *Blonagán dearg* (Scannell & Synnott, 1987); *Lus coise gé (dearg)* (Moloney, 1919; Dinneen, 1927)

AMARANTHACEAE (Chenopodiaceae)—THE GOOSEFOOT FAMILY
Native

Chenopodium rubrum is a rather rare annual plant in Ireland of muddy waste places, generally in coastal localities.

Chenopodium species were said by Mackay (1836: 227) to be 'more or less employed as potherbs'. Its leaves can be boiled and used as a vegetable.

Chrysanthemum leucanthemum (*Leucanthemum vulgare*)

- Ox-eye daisy (Scannell & Synnott, 1987); Dog daisy (Colgan, 1904); White may (Colgan, 1904); Bishop's-posy (Hart, 1898); Espibawn (Hart, 1898; Praeger, 1938; Harron, 1986); Moon-daisy (Harron, 1986); Marguerite (Harron, 1986)
- *Noínín mór* (Scannell & Synnott, 1987); *Nóinín capaill* (Ua Maoileoin, 2001); *Súil daimh* (Moloney, 1919); *Nóinín mór* (Moloney, 1919); *Easbogán* (Moloney, 1919); *Easbuig speain* (Cameron, 1883); *Easpogán (-áin)* (Dinneen, 1927); *Noínín bán mór* (Dinneen, 1927)

ASTERACEAE (Compositae)—THE DAISY FAMILY
Native

Ox-eye daisy is a common plant in most rural parts of Ireland and can be found in meadows, roadsides, pastures and on walls.

Chrysanthemum leucanthemum was formerly used in Ireland as a treatment for coughs, chills and sore eyes (Allen & Hatfield, 2004), as well as jaundice, wounds, boils, ulcers and tuberculosis (Uí Chonchubhair, 1995). An infusion of the white ray-petals was formerly used as a favourite drink called 'Espibawn-tea', taken to ward off chills in Cos. Tyrone and Monaghan (Hart, 1898). It was also used in the form of a decoction to treat 'shortness of breath' (Threlkeld, 1726; unnumbered [23]). Williams (1993) also indicated it was used to treat abscesses, scrofula and asthma (Williams, 1993). Cameron (1883) said that the Irish name *easbogán* comes from it being an esteemed remedy for 'the King's-evil' (scrofula, a form of tuberculosis).

Mabey (1972) noted that John Evelyn, in his book *Acetaria: A Discourse of Sallets* (1699), reported the use of its roots as a salad vegetable in Spain. In North America, tender young leaves have been used in salads (Peterson, 1977). Rutty (1772: 42) said that the Irish sometimes eat it raw, and that it is of 'sweet taste and gently, warm'.

Ox-eye daisies have been used for the children's game 'She loves me, she loves me not' in the same way as daisies (see *Bellis perennis*) and also to make large-flowered 'daisy' chains.

The Economic Botany Collection at the Royal Botanic Gardens Kew includes specimens of fibres and lace made from *Chrysanthemum leucanthemum*, donated by Lady Doneraile (EBC 51474). Viscountess Doneraile, of Doneraile Court, Co. Cork, sent hand-made lace constructed from various plant fibres (including fibres of honeysuckle, nettle, marsh-mallow, bindweed, nasturtium, sweet pea and Solomon's seal) to Kew. In return, the Kew Director, Sir William Hooker, sent her various plants for her garden (Nelson, 2009). Hand-made 'botanical lace' was exhibited at the Dublin Exhibition of 1853 made in Bailieborough, Co. Cavan, where a Mrs Letitia Veevers had established a small lace-making enterprise as part of a famine relief scheme in the late 1840s and 1850s (her husband was Resident Magistrate at the time) (Nelson, 2009). Most of the lace items from Ireland sent by Lady Doneraile to Kew probably originated from Mrs Veevers.

Chrysanthemum segetum

- Corn marigold (Scannell & Synnott, 1987); Mogue tobin[65] (Booth, 1979); Geal gowan (Hart, 1898); Geal seed (Hart, 1898); Gilgowan (Praeger, 1938; Harron, 1986)
- *Buíán* (Scannell & Synnott, 1987); *Buachalán (Buafanán) buidhe* (Moloney, 1919; Dinneen, 1927); *Bile buidhe* (Moloney, 1919); *Íthbhláth órdha* (Moloney, 1919); *Liathán* (de Vere Coneys, 1849; Webb & Scannell, 1983); *Buafanán bridhe* (de Vere Coneys, 1849); *Ceannbhán buidhe* (Dinneen, 1927); *Comann míonla* (Dinneen, 1927); *Íth-bhláth* (Dinneen, 1927); *Buafanán buí* (Uí Chanainn, 1962)

ASTERACEAE (Compositae)—THE DAISY FAMILY
Introduced (possible archaeophyte)

Chrysanthemum segetum is an occasional weed of cornfields and disturbed ground in waste places, particularly in the east and north. It is rare elsewhere in Ireland. Corn marigold has been recorded in Ireland for many centuries, and seeds have been found dating to the 11th century in deposits excavated from Fishamble Street, Dublin (Kelly, 1997).

According to Moloney (1919), *Chrysanthemum segetum* was used medicinally and said to reduce sweating.

Chrysosplenium oppositifolium

- Golden saxifrage (Scannell & Synnott, 1987)
- *Glóiris* (Scannell & Synnott, 1987); *Luibh na luí* (Ua Maoileoin, 2001); *Clabrus* (Cameron, 1883); *Clabar* (Cameron, 1883); *Lus na laogh* (de Vere Coneys, 1849)

SAXIFRAGACEAE—THE SAXIFRAGE FAMILY
Native

---

65. In Carlow it is known by this name, after a farmer who was 'driven out of his farm when unable to grow anything else' (Booth, 1979: 110).

Chrysosplenium oppositifolium is a widespread plant throughout Ireland of damp woods, hedgerows, scrub and shady ditches.

Uí Chonchubhair (1995) recorded that a decoction of this species was used to treat kidney and urinary obstructions and that the crushed root, sweetened with honey or sugar, was used to ease flatulence or colic.

It can be a useful plant for salads (Mears & Hillman, 2007).

**Cicendia filiformis*

- Yellow centaury (Scannell & Synnott, 1987)
- *Deagha buí* (Scannell & Synnott, 1987)

GENTIANACEAE—THE GENTIAN FAMILY

Native

Cicendia filiformis is a tiny annual plant of wet sandy and peaty soils that is rare in Ireland and found only in parts of Cos. Kerry and Cork.

*No ethnobotanical uses of this plant have been noted in Ireland.

Cichorium intybus

- Chicory (Scannell & Synnott, 1987); Wild succory (Harron, 1986)
- *Siocaire* (Scannell & Synnott, 1987); *Lus an tsiúcaire* (Moloney, 1919; Dinneen, 1927); *Basruan mór* (Moloney, 1919); *Endivia siuchair* (Wade, 1808b); *Siocaire fiáin*[66] (Ó Dónaill, 1977)

ASTERACEAE (Compositae)—THE DAISY FAMILY

Introduced (possible archaeophyte)

Cichorium intybus is a blue-flowered perennial weed of roadsides and waste places that is rare in Ireland. It was formerly cultivated for use as a vegetable and probably occurred in Ireland in the 1800s as a remnant of that former cultivation (Moore & More, 1866).

Young leaves can be cooked and used as a vegetable if gathered early in the season before they become bitter. Peterson (1977) suggested using the white underground parts of the leaves in salads. The roasted roots of chicory are a well-known coffee substitute. The roots are washed well, then slow roasted until dark brown and dry, when they can be ground up and used like coffee.

De Vere Coneys (1849) gave the Irish name *eanach-garraidh*, meaning 'endive', for *Cichorium endivia*. Endive is a leaf vegetable that can be used raw in salads or eaten as a cooked vegetable.

In parts of Arabia the leaves were burnt as an insect repellant (Pennacchio et al., 2010), but in Ireland I have not found any records of this use.

---

66. Ó Dónaill (1977) defined *siocaire fiáin* as 'wild chicory'.

Cicuta virosa

- Water hemlock (Scannell & Synnott, 1987); Cowbane
- *Fealla bog* (Scannell & Synnott, 1987); *Tréanlus* (Moloney, 1919; Dinneen, 1927); *Dathabha bán* (Moloney, 1919; Dinneen, 1927; Ó Dónaill, 1977); *Feallair* (Cameron, 1883); *Forán (-áin)* (Dinneen, 1927)

APIACEAE (Umbelliferae)—THE CARROT FAMILY
Native

Cicuta virosa is a robust perennial herb of lake-shores, ditches, drains and riversides that is found in parts of the north, north-east and centre of Ireland.

This plant is poisonous. Moloney (1919: 25) suggested that this was the true poison that killed Socrates. He also noted its use as a medicinal plant in Ireland, 'as a curative poultice in glandular swellings which have suppurated'.

Cicuta virosa is the official county flower of Co. Armagh.

---

*Circaea ×intermedia

- Upland enchanter's-nightshade (Scannell & Synnott, 1987)
- *Lus na hóige* (Scannell & Synnott, 1987)

ONAGRACEAE—THE WILLOWHERB FAMILY
Native

Circaea ×*intermedia* is a hybrid between *C. lutetiana* and *C. alpina*, a plant that does not now occur in Ireland. It is a rare plant in Ireland, found only in parts of the north in rocky habitats.

*No ethnobotanical uses of this plant have been noted in Ireland.

---

*Circaea lutetiana

- Enchanter's-nightshade (Scannell & Synnott, 1987)
- *Fuinseagach* (Scannell & Synnott, 1987); *Fuinseach* (Moloney, 1919; Dinneen, 1927); *Fuinnseagal* (Moloney, 1919; Dinneen, 1927); *Lus na h'oidhnan* (Cameron, 1883)

ONAGRACEAE—THE WILLOWHERB FAMILY
Native

Circaea lutetiana is a common plant throughout Ireland of woodlands, hedges, shady gardens and roadsides. It is occasionally a pernicious weed of the herbaceous border.

*No ethnobotanical uses of this plant have been noted in Ireland.

Cirsium species

- Thistle (Scannell & Synnott, 1987); Thrissel (Hart, 1898)
- *Feochadán* (Scannell & Synnott, 1987); *Fothannán* (Colgan, 1911); *Deósadán* (Colgan, 1911)

ASTERACEAE (Compositae)—THE DAISY FAMILY
Native

Including:

Cirsium arvense

- Creeping thistle (Scannell & Synnott, 1987); Dawky thrissel (Colgan, 1904)
- *Feochadán reatha* (Scannell & Synnott, 1987)

Cirsium palustre (*Cnicus palustris*)

- Marsh thistle (Scannell & Synnott, 1987); Buck thrissel (Colgan, 1904); Black thistle (McCarthy & Hawkes, 2007)
- *Feochadán corraigh* (Scannell & Synnott, 1987)

Cirsium vulgare (*Cnicus lanceolatus*)

- Common thistle (Scannell & Synnott, 1987); Spear thistle (Colgan, 1904); Buck thrissel (Colgan, 1904); Dawky thrissel (Colgan, 1904); Thorn[67] (Colgan, 1911); White thistle (McCarthy & Hawkes, 2007); Spear plume thistle (Brunker, 1950)
- *Feochadán colgach* (Scannell & Synnott, 1987); *Feochadán capaill* (Ua Maoileoin, 2001); *Fothannán* (Moloney, 1919; Tóibín, 1965); *Greaseadán* (McCarthy & Hawkes, 2007)

Cirsium arvense.

Cirsium arvense is a very common, sometimes abundant, plant throughout Ireland of dry grasslands, roadsides and waste places. *Cirsium vulgare* is equally common and occurs in similar habitats but also in wetter habitats. *Cirsium palustre* is also extremely abundant and found in marshes, ditches, wet pastures and meadows and waste places.

According to Allen and Hatfield (2004), thistles have been used medicinally for various purposes in Ireland. These authors also note the use of *Cirsium* species in veterinary remedies: *Cirsium* species were used to staunch bleeding in cattle or horses in Limerick, and *C. vulgare*, in particular, was used to treat ailing cattle in Wexford.

The roots and flowering shoots of *Cirsium vulgare* are edible. Mears and Hillman (2007) suggested that they are excellent to eat roasted or steamed. The shoots can also be eaten raw. The shoots of *C. palustre* are also edible, but it does not have the thick taproot of *C. vulgare*. Vickery (1995: 357–358) noted '*Cirsium vulgare* was also known as Scotch thistle in Ulster, where children would pull off the florets and their young pappuses and peel off the spiny bractlets, so that you would find a nice little nutty core'. Milliken and Bridgewater (2004) mentioned that thistle flowers have a tiny edible heart, rather like an artichoke that is eaten by children in south-west Scotland. I have not come across records of children eating this flower in Ireland.

Cirsium vulgare.

Rutty (1772: 107) said that 'The Down [of *Cirsium vulgare*] which it yields in great plenty, may be used for stuffing pillows'.

Danaher (1964a) recorded that chickweed (*Stellaria media*), groundsel (*Senecio vulgaris*) and 'the big purple thistle' (probably *Cirsium vulgare*) were chopped up and added to chickens' food.

67. This English name was reported by Colgan (1911) from near the abbey, Clare Island, Co. Mayo.

Cirsium dissectum (Cnicus pratensis)

- Meadow thistle (Scannell & Synnott, 1987); Bog thistle (Brunker, 1950)
- *Feochadán móna* (Scannell & Synnott, 1987); *Cluthrán léana* (Moloney, 1919)

ASTERACEAE (Compositae)—THE DAISY FAMILY
Native

Cirsium dissectum is found in boggy and wet pastures, marshes, fens and bogs, particularly in the west, north-west and centre of Ireland and is rarely distributed elsewhere.

*No ethnobotanical uses of this plant have been noted in Ireland.

Cirsium heterophyllum (Cirsium helenioides)

- *Melancholy thistle* (Scannell & Synnott, 1987)
- *Cluasán fia* (Scannell & Synnott, 1987); *Cluas-an-fhéidh* (de Vere Coneys, 1849)

ASTERACEAE (Compositae)—THE DAISY FAMILY
Native

Cirsium heterophyllum is a very rare thistle found in only a few wet grassland sites in the north and north-west of Ireland.

*No ethnobotanical uses of this plant have been noted in Ireland.

Cladium mariscus

- Saw sedge (Scannell & Synnott, 1987); Prickly twigrush (Brunker, 1950); Twig-rush (Praeger, 1938)
- *Sábhsheisc* (Scannell & Synnott, 1987); *Colgrosc* (Moloney, 1919); *Colgróc* (Moloney, 1919)

CYPERACEAE—THE SEDGE FAMILY
Native

Cladium mariscus is a widespread plant throughout the west of Ireland of marshes and lake margins.

It was formerly, and still is, used for thatching houses and sheds in Connemara (M. Sheehy Skeffington, pers. comm., 2007) as well as in Europe and parts of North Africa (Heywood, 1978). In Britain it is still used to cover the ridges of thatched roofs because of its flexibility (Prendergast & Sanderson, 2004). Wade (1804) recorded that it was used for thatching in past times, too, instead of straw, as it can grow from 5 to 6 ft tall.

According to Mears and Hillman (2007), the seeds of this species are edible.

It is occasionally grown as an ornamental plant in horticulture (Huxley, 1992), but it is rarely cultivated in Irish gardens.

Clematis vitalba

- Traveller's joy (Scannell & Synnott, 1987); Old man's beard (Brunker, 1950; Clapham et al., 1962)
- *Gabhrán* (Scannell & Synnott, 1987)

RANUNCULACEAE—THE BUTTERCUP AND CROWFOOT FAMILY

Introduced

Clematis vitalba is occasionally established throughout Ireland as a garden escape in woodlands, scrub and hedgerows and also on sand dunes, roadsides and railway-banks. It is more common in the south than in the north, where it is rare.

Edlin (1973) indicated that it was used in Britain for weaving rustic baskets, and it may have been used similarly in Ireland.

Woodward (1985) lists this as a poisonous plant that should not be eaten.

Clinopodium ascendens (*Calamintha sylvatica*)

- Calamint (Scannell & Synnott, 1987)
- *Cailmint* (Scannell & Synnott, 1987)

LAMIACEAE (Labiatae)—THE DEAD-NETTLE FAMILY
Native (or possible archaeophyte)

Clinopodium ascendens is a rare plant in Ireland of limestone pavement, rocky ground, dry banks and gravelly soil that is found mainly in the south.

Mackay (1836: 219) said it could be 'employed in making herb tea.'

**Cochlearia anglica*

- English scurvygrass (Scannell & Synnott, 1987)
- *Carrán muirisce* (Scannell & Synnott, 1987)

BRASSICACEAE (Cruciferae)—THE CABBAGE AND CRESS FAMILY
Native

Cochlearia anglica is a perennial species that is rather rare in Ireland of muddy estuaries and salt marshes, particularly along the east and south coasts as well as in the River Shannon estuary.

*No ethnobotanical uses of this plant have been noted in Ireland.

Cochlearia danica

- Danish scurvygrass (Scannell & Synnott, 1987)
- *Carrán creige* (Scannell & Synnott, 1987)

BRASSICACEAE (Cruciferae)—THE CABBAGE AND CRESS FAMILY
Native

Cochlearia danica is a common annual plant of waste places near the sea, sand dunes and rocky places. It is more rarely distributed inland and can be found along some roadsides (where salt has been spread) and railway lines.

This is probably the species that Smith (1815: 362) described as *Cochlearia marina folio anguloso parvo* and said that 'the juice of this plant laid on the face for six hours, takes away freckles, but it must be afterwards washed off with a decoction of bran'.

The leaves are edible, although not tasty.

Cochlearia officinalis

- Common scurvygrass (Scannell & Synnott, 1987); Spoonwort (Moloney, 1919)
- *Biolar trá* (Scannell & Synnott, 1987); *Biolar trágha* (de Vere Coneys, 1849; Moloney, 1919; Dinneen, 1927); *Currán* (Moloney, 1919); *Amaraich (-che)* (de Vere Coneys, 1849; Dinneen, 1927); *Cairán* (de Vere Coneys, 1849); *Tráighbhiolar* (de Vere Coneys, 1849; Dinneen, 1927)

BRASSICACEAE (Cruciferae)—THE CABBAGE AND CRESS FAMILY
Native

Cochlearia officinalis is a common coastal plant all around Ireland on rocks, cliffs and moist banks and in salt marshes. It is also (*C. officinalis* subsp. *alpina*) found in mineral-rich flushes on rocks and stream-sides in high mountain locations, although its occurrence in these habitats is rare.

It was well known as a source of vitamin C for early travellers in northern regions and used for the treatment or prevention of scurvy (antiscorbutic). It was said to be useful in the treatment of 'spongy gums' (Moloney, 1919). Uí Chonchubhair (1995) recorded that the juice or an infusion of this plant was taken to purify the blood, and the plant was also macerated in vinegar to prepare a healing lotion for mouth ulcers and skin complaints. It was used in Inishtrahull Island off the Co. Donegal coast in an ointment to treat cuts (Sullivan & Stelfox, 1943). In Donegal it has also been used to make a mouthwash against bad breath or to treat gum problems. Boiling water is added to dried plants, then left to cool. After straining, the juice of an orange was added and the mouthwash was used two or three times a day until the problem disappeared (Gallagher, 2012).

The fleshy leaves are edible but not particularly tasty. I occasionally come across suggestions that the leaves can be added to salads and used in sandwiches. Despite studying *Cochlearia* in Ireland for more than five years for a research project, I never acquired a taste for this species.

**Coeloglossum viride*

- Frog orchid (Scannell & Synnott, 1987)
- *Magairlín an loscáin* (Scannell & Synnott, 1987); *Cuigeal an lorcáin* (Moloney, 1919)

ORCHIDACEAE—THE ORCHID FAMILY
Native

Coeloglossum viride is a widespread, but rare, orchid throughout Ireland of dry pastures.

*No ethnobotanical uses of this plant have been noted in Ireland.

Colchicum autumnale

- Meadow saffron (Harron, 1986; Scannell & Synnott, 1987); Autumn crocus (Clapham et al., 1962); Naked boys (Hart, 1898); Naked ladies (Clapham et al., 1962; Harron, 1986)

- *Cróch an fhómhair* (Scannell & Synnott, 1987); *Cróch léana* (Moloney, 1919)

COLCHICACEAE—THE MEADOW SAFFRON FAMILY

Native (or possible archaeophyte)

Colchicum autumnale is a very rare native plant found only in wet meadows in Co. Limerick and near the River Nore.

It is frequently grown as an ornamental plant in Irish gardens but (hopefully!) not derived from native stock.

Allen and Hatfield (2004) noted records of its use medicinally (called 'saffron'—although the genuine saffron is another plant, *Crocus sativus*, that does not grow as a native plant in Ireland) in Carlow and Tipperary to treat jaundice. It is a very poisonous plant. If eaten the symptoms of poisoning include abdominal pain and burning; diarrhoea, weakness, difficulty breathing and convulsions; coma and respiratory failure may occur (Woodward, 1985). Cattle may graze it and pass the toxins on in their milk, and the alkaloids it contains are resistant to drying and boiling too.

Mackay (1836: 283) said that it was used medicinally 'as a substitute for the famous Gout-medicine'. A powerful alkeloid, colchicine, is extracted from its seeds and corms and used medicinally for pain relief and to treat arthritis and gout (van Wyk & Wink, 2004). Colchicine has also been used in modern plant breeding to help induce genetic mutations such as chromosome doubling.

Comarum palustre (Potentilla palustris)

- Marsh cinquefoil (Scannell & Synnott, 1987); Bogberry (Hart, 1898)
- *Cnó léana* (Scannell & Synnott, 1987); *Cúig bileach uisce* (Moloney, 1919); *Seamróg chúig bhileach* (Moloney, 1919); *Cnámh léana* (Uí Chonchubhair, 1995)

ROSACEAE—THE ROSE FAMILY

Native

Comarum palustre is an occasional plant throughout Ireland of bogs, marshes and wet drains.

Rutty (1772: 122) said that 'it is used by the Irish to thicken their milk suddenly, by rubbing their vessels with it'. He also wrote: 'I confirmed this use by boiling both the roots and leaves separately in milk, and suspected that other astringents have the like effect, having observed it in the root of Tormentil [*Potentilla erecta*]'.

Conium maculatum

- Hemlock (Scannell & Synnott, 1987)
- *Moing mhear* (Scannell & Synnott, 1987); *Mil mhear* (Ua Maoileoin, 2001); *Muinmhear* (Moloney, 1919; Dinneen, 1927); *Cornán fáil* (de Vere Coneys, 1849; Moloney, 1919; Dinneen, 1927; Uí Chanainn, 1962); *Bainne-cíoch-éan*[68] (Moloney, 1919); *Milbhear (Tréanlus)* (Dinneen, 1927; Uí Chanainn, 1962; Uí Chonchubhair, 1995); *Milbhir* (or *Milmhear; Minmhear; Moingmhear*) (Dinneen, 1927); *Mongach mhear* (Dinneen, 1927); *Milliona mear*

68. This name is also given by the Islandman Tomás Ó Criomhthain, as '*Bainne-cíth-na-n-éun*', in a list of the flora and fauna of Kerry's Great Blasket Island, compiled for the Norwegian scholar, Carl Marstrander, in the early 1900s. Mairín Uí Chonchubhair (pers. comm., 2011) tells me that this name is applied to any plant with white milky latex, such as *Taraxacum officinale* (dandelion).

(Ó Criomhthain, n.d., ca. 1908); *Minmhear* (Tóibín, 1965); *Muing mhear* (Tóibín, 1965); *Mín mhara* (Colgan, 1911); *Muing mhara* (Colgan, 1911)
APIACEAE (Umbelliferae)—THE CARROT FAMILY
Possibly Introduced (possible archaeophyte)

A statue of Socrates at Glasnevin. Socrates was famously poisoned by hemlock.

It is not certain whether *Conium maculatum* is native to Ireland. If it is an introduction, then it was introduced in very ancient times and is now well established throughout Ireland, growing in waste places and on roadsides and farmland, particularly in coastal counties.

Hemlock is best known as the poison that the classical Greek philosopher Socrates is said to have drunk in AD 399 that caused his death. Ingestion of any part of the plant, in any quantity, is potentially fatal. Its pyridine alkaloids (coniine) are potent neurotoxins, with death from paralysis and respiratory failure. Nevertheless, it has been used medicinally (applied externally!) in Europe for thousands of years.

The leaves were used in combination with linseed meal externally as a pain-reliving poultice (Moloney, 1919). In the Belleek region of Donegal, Hart (1898) noted that it was regarded as good for sprains and bruises. It was used in a poultice to treat rheumatism, burns, swellings and sores in Ireland. It may have been used in the Aran Islands as a poultice for scrofula or abscesses, although the identity of the plant used is in question—the plant used may have been *Oenanthe crocata* (Allen & Hatfield, 2004). Small doses were used in Ireland as an anti-spasmodic, as well as against gout and arthritis (Williams, 1993). Lady Wilde (quoted in Wood-Martin, 1902: 188) said that it was used to treat epilepsy: 'Pour wine upon a pound of hemlock, fresh gathered, and let it be drank while the person is in the fit'. Lady Wilde (1887) said that 10 leaves of hemlock, dried and powdered and mixed in food, acts as a powerful love charm, making the person you like love you in return.

In Morocco, this species was included amongst a number of ingredients that were burnt, with the smoke used to help induce an abortion in pregnant women (Pennacchio et al., 2010). Other plants used in this recipe included *Cannabis sativa* (hemp) and *Datura stramonium* (thorn-apple). The smoke was directed into the vagina of the woman seeking an abortion. However, I have not found any evidence of such usage in Ireland.

Mackay (1836: 127) wrote that its name, given by Theophrastus, is derived from the Greek for a cone or top, 'whose whirling motion resembles the giddiness [it] produced on the human constitution' and that it was used to treat 'scrofulos and cancerous maladies and for the purpose of lowering the pulse'. He warned though that 'so powerful a plant should be carefully distinguished from its allies by its spotted stems and fetid smell…'.

Conium maculatum was used as a veterinary remedy for various ailments, such as liver fluke, sheep scab, cuts, swollen feet, sore breasts and farcy in various parts of Ireland, although some of these records may be of the use of other Umbellifers, such as cow parsley and hogweed (Allen & Hatfield, 2004). It was also used to treat dislocated joints in horses and as an ointment against sprains, but only if the skin was not broken (Williams, 1993). The same author noted that, made into an ointment, *C. maculatum* was used to treat swellings, such as on a cow's udder after milking.

Williams (1993) also noted that the hollow stems of this plant would sometimes be made into whistles and 'guns' (blow-pipes) by boys, but afterwards they would often fall ill.

Rutty (1772: 108) said that the ashes of this species (which he gave as '*Cicuta major*') burnt are very good for bleaching linen, and 'said by some to excel those of the thistle'.

Conopodium majus (*Bunium flexuosum*; *Conopodium flexuosum*)

- Pignut (Scannell & Synnott, 1987); Common earth nut (Moloney, 1919); Arnut (Hart, 1898); Earthnut (Harron, 1986)
- *Cúlarán* (Scannell & Synnott, 1987); *Prátaí luthragáin* (Ua Maoileoin, 2001); *Cnó talmhan* (Moloney, 1919); *Cnútharlán* (Moloney, 1919); *Cutharlán* (Moloney, 1919); *Coicearán muice* (Moloney, 1919); *Práitíní clutharacáin* (Moloney, 1919; Dinneen, 1927); *Caorthalmhuin (-ne)* (de Vere Coneys, 1849); *Coireanán-muice (-áin-muice)* (de Vere Coneys, 1849); *Cluathairicín*[69] (Dinneen, 1927); *Coireán muice* (Dinneen, 1927); *Práta clúracáin* (or *lurcáin*) (Dinneen, 1927); *Fáinne muice* (Ó Dónaill, 1977)
- *Harnothes* (in Yola)

APIACEAE (Umbelliferae)—THE CARROT FAMILY
Native

Conopodium majus is a relatively common perennial herb of woodlands, hedgerows and grasslands throughout Ireland.

Its roots were formerly gathered and roasted as a vegetable called 'fairy potatoes' (Moloney, 1919). Kelly (1997: 311–312) noted that in ancient Ireland 'the roots of a number of wild plants were eaten; of these the most prized seems to have been *curar*. This is probably to be identified with the common umbellifer *Conopodium majus* or pignut'. When Ireland was extensively covered with woodlands, pignut was probably much commoner than it is now and may have been a staple foodstuff for early people living in Ireland. Phillips (1983) suggested that pignut can be eaten raw in salads. Pignut tubers eaten raw taste like chestnuts and are crunchy. They may also be roasted. James Townsend Mackay, writing in 1836 in *Flora Hibernica*, the first comprehensive Irish Flora, said that the roots are 'much sought after by children and pigs' (Mackay, 1836: 122).

Conopodium majus was used in traditional medicine in Donegal to 'cleanse the blood' and as a tea substitute in Fermanagh (Allen & Hatfield, 2004). In Co. Cork, Paula O'Regan noted a recipe for the treatment of eczema from the mid-1900s where pignuts were mixed with sheep's fat or yellow wax, Stockholm tar (a high-grade pine tar) and dock root (O'Regan, 1997).

Convallaria majalis

- Lily-of-the-valley (Reynolds, 2002)
- *Fueraremore* (K'Eogh, 1735); *Lile na ngleanntán* (Ó Dónaill, 1977)

ASPARAGACEAE (Liliaceae)—THE ASPARAGUS FAMILY
Introduced

Convallaria majalis is a common and popular garden plant, with a delicious fragrance, that is occasionally established in the wild, mainly in woodlands or shady habitats, as an escape, a relic of cultivation or as a discard.

The lily-of-the-valley is a seriously poisonous plant. K'Eogh (1735: 71) described its use to treat apoplexy, palsy and gout and said that a dram of the root pulverized, taken in wine and vinegar is 'a powerful remedy against malignant or pestilential disorders'.

69. Dinneen (1927) defined *cluathairicín* as 'a pig-nut'.

*Convolvulus arvensis

- Field bindweed (Scannell & Synnott, 1987); Cornbine (Harron, 1986)
- *Ainleog* (Scannell & Synnott, 1987); *Eidhneán talon* (Dinneen, 1927); *Duillmheal* (Moloney, 1919); *Táthfhéithleann*[70] (Ó Dónaill, 1977); *Féitheán na mBánta* (Tóibín, 1965)

CONVOLVULACEAE—THE BINDWEED FAMILY
Native

Convolvulus arvensis is a widespread and persistent perennial weed of waste places throughout Ireland and is found on roadsides and railway lines and in cultivated fields and gardens, particularly on sandy, well-drained soils.

*No ethnobotanical uses of this plant have been noted in Ireland.

Cornus sanguinea

- Dogwood (Scannell & Synnott, 1987); Dogberry tree (K'Eogh, 1735); Gatter-tree (K'Eogh, 1735)
- *Conbhaiscne* (Scannell & Synnott, 1987); *Crann coirnéil* (Moloney, 1919; Dinneen, 1927); *Crann muchóra* (Moloney, 1919; Dinneen, 1927); *Coinbhile* (de Vere Coneys, 1849; Dinneen, 1927)

CORNACEAE—THE DOGWOOD FAMILY
Native

Cornus sanguinea is occasionally found in parts of the west, such as Clare, Galway and Limerick, and is rare in other parts of Ireland. It grows in hedgerows, scrub and rocky places.

Some consider dogwood an introduced tree in Ireland; however, there is no clear evidence for this, other than that it does not appear on the Old Irish tree-list that dates back to the early 8th century included in the Irish law text, *Bretha Comaithchesa.*

The fruits of the dogwood are berry-like but said to be bitter and inedible, although not poisonous. A useful oil has been obtained from the fruits.

It was formerly used in Britain for goads, skewers, ramrods, arrows and in making hoops (Edlin, 1973), and probably also in Ireland for the same purposes. The timber is tough and very durable. The 'Ice man' discovered on the border between Italy and Austria in 1991 was carrying arrows made from dogwood.

Cornus sanguinea has been used in the United Kingdom to prepare a pink dye (Prendergast & Sanderson, 2004) and a yellow dye in Ireland (Mahon, 1983).

Corylus avellana

- Hazel (Scannell & Synnott, 1987); Cob-nut (Harron, 1986)
- *Coll* (Scannell & Synnott, 1987); *Calltuinn* (Cameron, 1883); *Calldainn* (Cameron, 1883); *Callduinn* (Cameron, 1883); *Cailtin* (Cameron, 1883); *Colluinn* (Cameron, 1883); *Crann coilltín* (Dinneen, 1927)
- *Hawlse* (in Yola)

BETULACEAE (Corylaceae)—THE BIRCH FAMILY
Native

70. Ó Dónaill (1977) defined *táthfhéithleann* as 'convolvulus'.

Hazel was and remains one of the most important native plant resources in Ireland, used for food, timber, building, basketry, fuel, medicine and a multitude of other purposes. *Corylus avellana* is common throughout Ireland and temperate Europe in habitats such as woods, scrub, hedgerows and rocky places. The broadest known hazel tree in Ireland is a specimen from Cappawhite, Co. Tipperary, that is 3.12 m in girth (Anonymous, 2005).

The fruits of hazel (hazelnuts) (in Irish, *cnú*, *cnó*, *cnó gaedhealach*; Dinneen, 1927) were an extremely important food source in ancient Ireland and provided valuable carbohydrates and fats. Hazel was a common and recognisable tree in the wild, and hazelnuts could be stored for up to a year and used when other foods were scarce (Kelly, 1997). Kelly references early law texts in Old Irish that mention that a hungry person was entitled to gather a handful of hazelnuts from a privately owned wood. It was probably also a valuable famine food in Ireland and was certainly used as such in Scotland (Drury, 1984). Archaeological findings such as the excavations of a Stone Age (Neolithic) site at Ferriter's Cove on the Dingle Peninsula, Co. Kerry, confirm the use of hazelnuts as a food source in early Ireland. Burnt hazelnut shells were found in middens deposited between 5000 and 6000 years ago (Woodman et al., 1999). In addition, fragments of hazelnut shells were found in most layers excavated from Fishamble Street in early Dublin, possibly indicating that hazel provided not only an important food source for Viking Dublin, but also an important building material used as early as 1000 years ago (Anonymous, 1973; Geraghty, 1996). The hazelnut kernels would have been eaten whole as well as crushed and made into a nut meal that was eaten with milk as a form of muesli called '*maothal*'. They may also have been roasted and cooked in various ways—charred hazelnut shells found in archaeological excavations are evidence that they were cooked as food. Cooking makes them more palatable and easier to digest and also enabled the nuts to be stored and consumed over a period of months. Hazelnuts were not the only important nut species eaten in ancient Ireland; walnuts (*Juglans regia*), a non-native species, would also have been a food source.

Corylus avellana.

Corylus avellana (hazel). Watercolour by Lydia Shackleton. Courtesy of the National Botanic Gardens, Glasnevin.

Hazelnuts are collected on a small (noncommercial) scale from the wild in Ireland today but are not cultivated commercially for nut production. They are best collected in late September or October when their flesh becomes firmer and richer in flavour and when the leaves of the tree start to turn yellow. The kernel is oily and full of starch and can be eaten raw or roasted. Being rich in fat, the nuts can be liquidised and used as a substitute for milk or added to bread, cakes and biscuits (Hilton, 2007). They also can be used chopped or grated, such as in salads or added to homemade muesli.

According to Neeson (1991), hazel-mead was a popular drink in early Ireland. Rutty (1772: 108) mentioned that in Italy chips of hazel wood were used to clear 'turbid' wine.

Hazelnuts (unripe).

Also in some parts of England, according to Rutty, when barm[71] was needed, twisted twigs or rods of hazel were steeped in the dregs of ale at the end of fermentation, then hung up to dry and later added to a wort to be fermented. Rutty (1772: 109) said that in Ulster and 'several parts of Ireland, an Oak gad is used in the like manner for the same purpose'. 'Oak gad' may perhaps refer to oak gall.

Hazel was often managed as coppice and cut regularly to provide long, straight, flexible but durable rods for a variety of purposes. St Patrick supposedly used a hazel rod to drive the snakes out of Ireland (Baker, 1999). Hazel wood was used to make handles of implements, such as flails used for threshing. Dinneen (1927: 1148) gave the Irish word *súiste* for a 'flail for threshing', and said 'the best flails have a hazel handle (*colpán cuill*) and a holly striker (*buailteán cuilinn*)'. Hazel was also used for pikes, the main weapons used during the rebellion of the United Irishmen in 1798 (Lysaght, 1979). A pike was a spear-like weapon with a pointed iron blade and a sharp, hooked blade on the side that was mounted on a wooden shaft about 6 ft long and was used to cut the reins of horses.

Slender hazel branches, 'spars' or 'scallops' (or scollops) (*scolb* in Irish, *spraay* in Yola), were generally used to hold roof thatch in place, particularly in the midlands (Sharkey, 2000) (see also *Salix* species). Hazel scallops are made by twisting and bending short rods into a hoop shape. If the scallops are simply bent they often break, but if they are twisted at the same time they are more flexible and can be bent double. To make them more pliable, they were often soaked in water first. The use of hazel scallops for roofing goes back many centuries in Ireland, and remains of them also were found in excavations from Fishamble Street (Geraghty, 1996). In addition to hazel and willows, bog deal (bog pine, *Pinus sylvestris*) and brambles (*Rubus fruticosus*) were also used to make scallops (Evans, 1957).

Hazel rods (and other species too) were also used to make the frames of small dwellings in Ireland from ancient and medieval times up to the 16th and 17th centuries (Danaher, 1978). In *Ireland's Vernacular Architecture*, Danaher (1978) references small houses made from wooden frames covered with sods, straw or wattle, and he includes a photograph of the framework of a shelter for young calves near Sixmilebridge, Co. Clare, in 1968, that was awaiting a covering of sod and straw.

The medieval wattle and daub houses of Wood Quay in Dublin were constructed by using laths of hazel and ash (*Fraxinus excelsior*) (Nelson & Walsh, 1993), and hazel would have been a common species used for wattle throughout Ireland in building simple houses. In excavations of medieval Dublin, virtually all habitations were post and wattle constructions, with the walls consisting of lines of paired uprights in the round with a horizontal weft of stout rods—usually hazel, ash or elm—woven between them. The walls were originally plastered with mud to make them wind and rainproof (Anonymous, 1973). Wattlework was also laid on the floors too, some of which were found in an excellent state of preservation. Hazel wattle (or willow or briar rods) was used to make the smoke canopy on chimneys in some traditional Irish cottages, which was then plastered with a mixture of clay and cow dung or ashes and coated with limewash (Evans, 1957). In parts of the south of Ireland, hay or straw ropes were used instead of wattle. The ropes were soaked in mud (*putogs*) and twisted between a frame of horizontal rungs and then plastered with mud or limewash (Evans, 1957). Hazel twigs were also woven to make ropes (Evans, 1957), which were strong and long-lasting, but not very flexible.

71. Barm is a yeasty foam derived from fermenting beers and wines.

Hazel wood has been traditionally used to make furniture, fencing and wickerwork (MacCoitir, 2003), and the creel baskets used for carrying peat, fish and other materials were traditionally made of hazel (Edlin, 1973). The collections in the National Museum of Ireland (Museum of Country Life) in Castlebar, Co. Mayo, include a large number of traditional baskets in various shapes made from hazel. One such example is a D-shaped *parlóg*, a hinged bottom-opening creel used to transport seaweed or turf. It was made by the renowned contemporary Irish basket-maker, Joe Hogan, in 2001, and modelled after one collected in Tuam, Co. Galway. The Museum also contains a sheep-feeder made from hazel rods by John Slamon in Kilcormac, Co. Offaly (Ref. No: F1964.32). Sheep prefer to eat clean feed, and the woven sheep-feeder helps to keep their feed clean and dry. Wicker was used for many other purposes in Ireland, such as for doors, windows, cradles, beds and chairs. Wickerwork doors were used traditionally throughout Ireland from at least the 8th century until the 19th century (O'Dowd, n.d., ca. 2009); they were often lined with straw mats to act as draft excluders. The straw mats were as wide and high as the door itself and secured with sticks set between the jambs.

An Aran cradle. Reprinted with permission from Mason, Thomas H. 1936. *The Islands of Ireland.* Batsford, London.

Hazel rods were used to make hoops for butter-firkins (small, wooden vessels or casks) (Wakefield, 1812). Hazel rods were also used as cooking spits (Lysaght, 1979). Rutty (1772: 108) said of the hazel tree that 'It supplies us with Hoops, the greatest part of which are imported from England, where they have woods of hazel, and lopping the young shoots, nurse up the main stem, which therefore grows up smooth, the knots gradually disappearing; however the English hoops thus raised, have this disadvantage, that they are more spongy and tender, and do not bear the weather so well, nor last so long as the Irish, whose timber grows wild and more knotty'. Up until the 19th century the oak staves of barrels were bound into shape by hoops made from hazel or willow, after which they were replaced by metal hoops (McCracken, 1971). Gardey (1971) even described a pre-famine coffin made from hazel and sally (willow) rods found in a graveyard in Co. Down.

Hazel (and willow) rods were traditionally used in the making of coracles—small, oval boats used in particular for salmon fishing on the river Boyne in the east of Ireland. They were also used for the currachs that were traditionally used along much of the north-western, western and south-western coasts of Ireland. Similar boats were known to have been used on other Irish rivers, such as on the Blackwater at Rincrew near Youghal in Co. Cork, and on the rivers Shannon and Erne, as well as perhaps on the Bann and the Foyle in the north of Ireland (J. Hogan, in MacCárthaigh, 2008). To make a coracle, the hazel rods were stuck in the ground and bent over to form a dome. Other rods were then woven in and out of these to create an inverted basket that would be bound with hemp and covered with canvas (formerly cured and tanned cowhide was used) and waterproofed. The coracles were about 6 ft in length by 4 ft wide, which was small enough to enable them to be covered by a single cow or bullock hide. Hazel cross-pieces were also used in the making of rush rafts (see *Schoenoplectus lacustris*) in Co. Roscommon (MacPhilib, 2000). A small, ancient paddle for a boat made from hazel wood is included in the collections of the Kerry County Museum,

A coracle on the river Boyne, Co. Louth. Reprinted with permission from Mason, Thomas H. 1936. *The Islands of Ireland.* Batsford, London.

Illustration of a currach from Wakeman (1891).

Boat made from *Corylus avellana*.

having been found in a bog near Knocknagashel, Co. Kerry, in about 1988 (Murray, 2009). Giraldus Cambrensis in his *History and Topography of Ireland* described a wicker boat in Ireland in the 1100s that was 'narrow and oblong, made of wickerwork and covered on the outside with sewn hides of animals'; such a craft may have been commonly used for fishing, particularly in inland waters (O'Meara, 1982: 111). Hazel coppices were maintained in the Co. Donegal regions of Dunfanaghy, Downings and Fánaid until the mid-20th century, mainly for currach construction (H. C. Lönze, in MacCárthaigh, 2008). Hazel was apparently much used for fishing. Wicker made from hazel was used to make eel weirs, many of which were placed in major rivers through Ireland from the Middle Ages onward (Hogan, 2001). Hazel rods have also been used to make other artefacts for fishing, such as fish traps in rivers and lobster and crab pots (Arnold, 1968; Hogan, 2001). The Museum in Castlebar includes a traditional fishing pole made from a tall, slender hazel rod. Most fishing rods have traditionally been made from split cane, which is derived from the stems of tall perennial grasses with woody stalks, often from two grasses, *Arundo*, from Europe and Asia, and *Arundinaria*, from the Americas. *Arundo donax* is not found in the wild in Ireland. It is the preferred choice of musicians for reeds for woodwind instruments, such as clarinets, saxophones, bassoons and oboes, but the reed blanks come from Mediterranean fields, where warmer climes support the more melodic development of the lignified fibres. An example of a teacher's school cane (*cána múinteora*) from Castlebar, Co. Mayo, in the Museum (Ref. No: F2007.104) is made from bamboo. There are no native bamboos in Ireland, although diverse woody, cold-tolerant Bambuseae can form sizable clumps or groves that provide a rich resource for canes. Bamboos are frequently grown in Irish gardens. Most bamboos grown in Ireland and Britain are from cold-tolerant Asian species from the Himalayas across to Japan. Culms of *Chimonobambusa quadrangularis* from Japan can be as tall as 8 m, with populations established in the wild in Ireland in Cos. Kerry and Cork. A smaller bamboo, *Sasa palmata*, has stems only to 3 m and has become naturalised in Northern Ireland and Co. Cork. The arrow bamboo *Pseudosasa japonica* may be the best known bamboo in Ireland and Britain, readily establishing and even flowering in introduced groves. Other bamboos with scattered reports in Ireland include *Pleioblastus chino*, *P. simonii*, *Sasaella ramosa*, *Semiarundinaria fustuosa* and *Yushania anceps* (Stace, 2010).

The Castlebar Museum also includes an interesting artefact made from a hazel rod (Ref. No: F1956.103) from the Vale of Clara, Co. Wicklow, called a 'Vizor's staff'. This is an unpeeled hazel rod, 193 cm long, where the lower part is wrapped with coloured paper and held in place by a spiral string. The rod was used by a band of men and boys, locally called the 'Vizors', who went from house to house in disguise and with musical instruments and played for the people of the house on Halloween.

Other uses for hazel included for drawing and as dye. Vickery (1995) noted that homespun textiles in Donegal were dyed with hazel. Mackay (1836) noted that charcoal made from hazel was excellent for drawing. Rutty (1772: 109) also referred to this usage and described the preparation of hazel charcoal as follows:

> Painters, Engravers and Goldsmiths, prepare Coals for taking their draughts or delineating their respective designs on paper thus: they take wedges of the wood of the thickness of one's arm, and four or six inches long, dry them, then cleave them into boards of the thickness of a finger: these they put into a large pot of red sand, cover and secure from the air and flame, by stopping with clay. This they put into a potter's oven, until the pots be duly baked: the pot being taken out is not opened until cooled. The Coal thus prepared, is sharpened to a point for use, for drawing Plans on paper, which has this excellency, that if any alteration be wanted, the impression made by this is easily taken out, without leaving any mark or blackness.

Corylus avellana was used medicinally in Ireland to treat varicose veins, and burnt hazel stick ash was put on burns and bark was applied to boils and cuts (Allen & Hatfield, 2004). Corrigan (1984) mentioned that the legendary Irish druid and physician Diancnecht, who lived in around BC 470, had a recipe called 'Diancnecht's Porridge' that was used to treat colds, phlegm and sore throats, made from hazel buds, dandelions, chickweed and wood sorrel boiled in oatmeal. According to Pennacchio et al. (2010), in parts of Europe and Asia the leaves of hazel have been smoked, for unspecified purposes.

Hazel is also much associated with folklore. Forked hazel twigs (called *gabhlog* in Irish) were often used by dowsers or diviners for locating underground water (as a divining rod) in Ireland (Kinahan, 1881; Lysaght, 1979; Hickie, 2002) and in Britain (Vickery, 1995). A nut (perhaps a hazelnut) on a string, used as a pendulum, was also used for water divining (King, 2009).

Hazel is grown as a garden plant in some larger gardens in Ireland. Some cultivars have been selected for their nuts and others for their leaf colour and the form of their branches (such as the cultivar 'Contorta' with its dramatically twisted branches). Some of the cultivated forms are of hybrid origin, notably as crosses with the commercial filbert, *Corylus maxima*, a species native to south-east Europe and south-west Asia. Dinneen (1927: 426) gave the Irish name *faoisceog* for 'a nut-tree, a filbert'.

Crambe maritima

- Seakale (Scannell & Synnott, 1987); Sea colewort (Dinneen, 1927); Sea cabbage (O'Rourke, 2006)
- *Praiseach thrá* (Scannell & Synnott, 1987); *Praiseach na mara* (Moloney, 1919; Dinneen, 1927); *Praiseach trágha* (de Vere Coneys, 1849; Moloney, 1919); *Murán* (Moloney, 1919; Dinneen, 1927); *Raitheach trágha* (Dinneen, 1927)

BRASSICACEAE (Cruciferae)—THE CABBAGE AND CRESS FAMILY
Native

Crambe maritima.

Seakale (*Crambe maritima*) is a popular garden vegetable. This plant was probably once gathered in the wild for food in Ireland in its coastal habitats but is now rare and hopefully not gathered for domestic use anymore. *Crambe maritima* is a perennial plant that grows on shingle and sandy seashores. Presently, there are probably no more than 20 wild plants of the species known to survive in Ireland.

Seakale is a delicious vegetable when eaten in spring. I would pick the leaves in April and steam or boil them lightly for a few minutes and then serve them like asparagus with melted butter. Later in the season, the mature leaves are tough and leathery to eat. As a vegetable the leaves are often blanched in the garden (covered and forced in the spring, as one might blanch rhubarb) and eaten while they are white and soft. Seakale can also be pickled. The unopened flower heads can be eaten and taste somewhat like broccoli (Mears & Hillman, 2007). The roots are also edible and can be eaten raw or roasted. Rutty (1772: 44) mentioned that this species is found on Irish shores and (under the name *Brassica marina*) 'is in request as a Cale, and by some preferred in spring to the garden sort, as a potherb, which it resembles in taste, but is more succulent; and for this purpose the seed is sold in the shops in Dublin, in order to be propagated, and, after it has been sown in the ground, they throw stones over it, in order to make the artificial place of growth resemble the natural. If it be taken from the sea-shore, it requires to be boiled in two waters, to take away the saltness before it be eaten'.

Its Irish name, *praiseach thrá*, means the 'shore pot-herb', which suggests that it was gathered from the wild as food.

Seakale was a characteristic and traditional vegetable of larger, walled vegetable gardens in Ireland and was particularly associated with rural Church of Ireland rectory gardens.

Crataegus monogyna

- Hawthorn (Scannell & Synnott, 1987); Whitethorn[72] (Colgan, 1904); May bush (Colgan, 1904); Bread-and-cheese (Colgan, 1904); Drean bawn (Hart, 1898); Skeeoge (Hart, 1898); Skayug (Hart, 1898); Skaig (Hart, 1898)
- *Sceach gheal* (Scannell & Synnott, 1987); *Huath or Uath* (de Vere Coneys, 1849; Cameron, 1883; Dinneen, 1927; Uí Chanainn, 1962); *Coirr-sceach* (Dinneen, 1927); *Crann sceiche* (Dinneen, 1927; Ó Dónaill, 1977); *Scé* (Dinneen, 1927)

ROSACEAE—THE ROSE FAMILY

Native

Crataegus monogyna (hawthorn) is one of the most important small trees in Ireland and found abundantly throughout the country in hedges, fields, scrub, woodland edges and rocky places. Indeed, hawthorn is regarded as being the commonest tree in Ireland today. Some exceptional specimens may be found, however; the current girth champion in Ireland is a specimen measured at 2.65 m (12 m tall) at Springfield House, Bennekerry, Co. Carlow (Anonymous, 2005).

Hawthorn is economically important, particularly for its use for hedging, right up to the present day, as it provides an impenetrable barrier for stock, especially if regularly pruned. As well as growing plants, cut branches of hawthorn, blackthorn and holly are regularly used to fill gaps in hedges to keep cattle and sheep in their fields. As a widely used hedging plant in Ireland it is interesting to speculate whether it (and perhaps blackthorn and holly too) might once have been used in ancient times in Ireland as a defensive 'barrier-plant' around homesteads and forts. However, I have not found evidence that it ever had such a deliberate use in Ireland. A wide range of thorny and prickly plants have been used for defensive purposes in various parts of the world (Howes, 1946).

From the earliest of times, Ireland was a country of smallholdings and small field-plots (Evans, 1957). For example, before the famine, 80% of all Irish farms were less than 15 acres (6 ha) in size. Evans suggested that most hedgerows are of no great antiquity but were

72. Uí Chanainn (1962) also gave the word *uath* for whitethorn.

planted as a result of the agrarian revolution, in different parts of the country between 1750 and 1850, when thorn hedges were planted and stone walls were constructed on land where boundaries had been more temporary or ill-defined up until then. Some field boundaries are of course very ancient—stone walls have been found buried deep within peat bogs, such as the complex patterns of field boundaries, settlements and tombs dating from the Neolithic period at Céide Fields in Ballycastle, north Co. Mayo. The complex hedge management practices found throughout much of Britain, including hedge laying, have been but a little part of the Irish landscape or standard traditional farming practices. Hedgerows were much used to provide firewood, particularly in poorer parts of Ireland, and would probably have been hacked and harvested in a largely uncontrolled manner. In some areas too, *Crataegus* would have been little planted for hedging, with field boundaries mainly consisting of a ditch and earth bank or mound, topped with wild brambles and gorse. Lucas (1958) described the way a gorse (or hawthorn) hedge was made in Ireland, with the seeds inserted in a twisted *súgán* rope of hay or straw, which was then laid on the top of a field margin embankment and covered lightly with soil.

Crataegus monogyna.

Hawthorn trees growing by important religious or historic sites are often afforded important symbolic significance, for example, by holy wells and in ring forts, etc., where they are frequently decorated with ribbons, rags and other pieces of cloth, even small pieces of highly coloured plastics. Religious objects such as medals, crucifixes and rosary beads were sometimes also placed as offerings on these trees. These have been called 'Mass bushes,' and although they are generally hawthorns, other species can be found, sometimes ash and more rarely oak, willow, elder, holly, rowan, alder, elm, yew and [non-native] fir (Synnott, 1979). There are an estimated 3000 holy wells in Ireland (Lysaght, 1979). In 2006, I saw a large *Crataegus* growing on the Hill of Tara, Co. Meath, that had been decorated with ribbons, presumably for some personal symbolic purpose by a visitor. Cutting down or uprooting a lone hawthorn tree is still regarded as unlucky in Ireland as traditionally this was thought to have upset or annoyed the fairies (Kells, 1929; Synnott, 1979). Logan (1980) included stories of terrible consequences falling on those who cut down trees at holy wells, including bad luck, madness, strokes, facial palsy, pains in the limbs and stomach and even death.

Crataegus monogyna.

Some trees throughout Ireland have also been revered (and sometimes feared too) as 'fairy bushes' or 'fairy trees'. These are generally, but not always, *Crataegus* specimens. One of the best known is the so-called Fairy tree at Latoon that occurs on a roadside between Newmarket-on-Fergus and Ennis in Co. Clare. It was due to be removed as part of an expensive Newmarket-on-Fergus bypass scheme in the 1990s, but as a result of public pressure, and the fact that construction workers on the project were unwilling to cut it down, it was saved and incorporated into the scheme in 1999. A local storyteller and folklorist, Eddie Lenihan, said that the tree is a marker of a fairy path, a meeting point where the fairies of Kerry would gather as they prepared to do battle with the fairies of Connacht. Shortly after the tree was protected by a surrounding

Hawthorn on a hill near Tara.

The fairy tree (*Crataegus monogyna*) at Latoon, Co. Clare.

fence, it was vandalized in August 2002, when someone cut off all its branches with a chainsaw. The tree survived and has sprouted well since then (Markey, 2011).

The 19th-century Irish poet William Allingham wrote about thorn trees and the fairies in his poem 'The Fairies'.

By the craggy hill-side,
 Through the mosses bare,
 They have planted thorn-trees
For pleasure here and there.
 If any man so daring
 As dig them up in spite,
He shall find their sharpest thorns
 In his bed at night.

(Allingham, 'The Fairies', stanza 5)

Tradition suggests that trees growing around a holy well sprung from the walking stick or staff left by a visiting saint (Lysaght, 1979; Synnott, 1979), which may also have been an added reason why they were treated with reverence. Lysaght (1979: 35) described how willful damage of these trees was looked upon as an act of desecration and said that 'In some cases it was believed that the person who attempted to cut down the tree imagined that he saw his house on fire and ran to extinguish it. Finding he was mistaken, he returned to resume the felling only to experience a repetition of the illusion. He hurried home a second time, but again his house was intact, so he resumed his cutting. He saw his house in flames for a third time, but on this occasion he ignored the warning and finished cutting down the tree. When he returned home, he found his house had been burnt to the ground.'

The symbolic importance of hawthorn is also underlined by its central role in some traditional practices. For example, Wilde (1887) and Wood-Martin (1902) described an account of a marriage in Co. Kerry, in 1830, written about by Lady Wilde:

> A large Hawthorn tree that stood in the middle of a field, near a stream, was hung all over with bits of coloured stuff, while lighted rush candles were placed here and there amongst the branches, to symbolize, no doubt, the new life of brightness preparing for the bridal pair. Then came a procession of boys marching slowly with flutes and pipes made of hollow reeds, and one struck a tin can with a stick at intervals, with a strong rhythmical cadence.... A boy followed, bearing a lighted torch of bog-wood. After him came the betrothed pair hand-in-hand, a large square canopy of black stuff being held over their heads.... The procession then moved on to a bonfire, evidently the ancient altar, and having gone round it three times, the black shroud was lifted from the bridal pair, and they kissed each other before all the people, who shouted and waved their branches in approval (Wilde, 1887: 147).

This seems to be a fascinating remnant pagan ritual where the hawthorn was a central feature.

Hawthorn was central to the celebration of Bealtaine, the ancient Gaelic festival that takes place on 1 May and marks the beginning of the summer season. Bonfires were lit (and to an extent still are in some places) on Bealtaine night. Branches of hawthorn (May boughs) were cut and put as decorations around windows and doors and a 'May Bush' (usually hawthorn, but sometimes also mountain ash, *Sorbus aucuparia*, or gorse, *Ulex europaeus*) was erected in the farmyard and covered with ribbons, flowers and other decorations. Hawthorn is generally in bloom at that time and is often called the May Bush in Ireland.

The famous Beggar's Bush area in the south city region of Dublin is said to have been named after an ancient hawthorn tree under which beggars and perhaps highwaymen would shelter before descending on Dublin city. The Beggar's Bush was situated on the corner of Shelbourne and Lansdowne roads in Ballsbridge, immediately adjacent to what became the Trinity College Botanic Garden, established there in 1806. Up to the mid-1900s there was an old hawthorn tree growing in the Botanic Gardens that was said by some to have been the original Beggar's Bush.

Hawthorn bark was steeped in black tea and the resulting liquid used to treat toothaches in Co. Leitrim and also as a remedy for burns (Allen & Hatfield, 2004). Shaw (1998) noted its herbal use to treat heart ailments, including coronary artery disease, angina, hypertension and heart failure. Early European herbals recommend hawthorn for stomach colics and diarrhoea. The first mention of its effect on the heart is by Quercetanus, the personal physician of King Henry IV of France, who concocted an 'anti-age syrup' from the plant (Vogel/Bioforce AG, 2012). O'Regan (1997) reported a story of a medical doctor in the late 19th century in Co. Clare who treated hundreds of patients with a secret remedy that turned out after his death to be a tincture made from hawthorn berries. He had used it to treat angina, coronary thrombosis, palpitations, chest pains and other heart disorders.

In ancient Ireland hawthorn would have been used as a food source, even though its fruits (haws; in Irish, *scechóra* or *sceabhchóirí*)[73] are not particularly sweet or tasty. It crops heavily and is often widely available later in the autumn than many other native fruits. Geraghty (1996) noted that haw seeds have been found in excavations of Viking Dublin dating back more than 1000 years. The same author mentioned the old saying used in Ireland, 'When all fruit fails then welcome haw', suggesting that it was probably a useful 'last resort' fruit. In any case, haws are best collected late in the season when they become more palatable and somewhat sweeter.

Making haw jelly.

The berries of hawthorn have been used to make jellies and wine. Haw jelly is a delicious and delicately flavoured alternative to red currant jelly to serve with meat. To prepare jelly from the berries (haws), the fruits are trimmed of their stalks, washed, stewed and strained before being boiled with sugar and bottled. Haws can be mixed with other hedgerow fruits (rosehips, crabapples, blackberries and other fruits) to make 'hedgerow jam'. In Scotland, haws have been used to a limited extent for commercial purposes, for the preparation of jams and preserves, usually as part of fruit mixes (Milliken & Bridgewater, 2004), but I have not heard of any such commercial use in Ireland. Hilton (2007) suggested that dried fruits can be ground and added to flour used in baking. Rutty (1772: 85) mentioned such usage and said that, 'The ripe fruit is sold in our market [presumably in Dublin], and eaten by the poor, and reduced to meal, may be a substitute for bread in time of scarcity'.

Hawthorn leaves, 'bread-and-cheese'.

Hawthorn flowers are also edible and can be added to fruit salad or used to flavour puddings such as custard and junket. Hawthorn buds and young leaves can be included in salads (Phillips, 1983, recommends them in a salad with beetroot or potatoes). They have a nice, fresh flavour with very little bitterness if picked just as soon as they open from the buds in spring. Young hawthorn buds and leaves were once a favourite wild food for children, giving it an alternative common name, 'bread-and-cheese' (Colgan, 1904).

73. In Cork, a slang term for haws is 'skeory', no doubt derived from this Irish word (Beecher, 1984).

Mears and Hillman (2007) suggested that the flesh of haws can be processed and used as food by crushing the fruits and removing the skins and seeds to leave an edible gelatinous mush. The flowers can be used to make wine and a liqueur (Ó Céirín & Ó Céirín, 1978).

The leaves of hawthorn have been used to dye fabrics a dark blue colour (Danaher, 1964a). The branches of hawthorn were burnt and made excellent ashes for bleaching cotton (Rutty, 1772).

Haws were sometimes used as the ammunition for pea-shooters used by small boys in Ireland, as they were freely available in the hedgerows (rather than peas, which were harder to come by). O'Súilleabháin (1942) mentioned an old Irish traditional story of a man who shoots a fish using hawthorn seeds as bullets. The fish ends up sprouting a tree!

Hawthorn wood is hard and durable but seems to have been only little used in Ireland as a traditional source of timber in the past or today. It burns well even when freshly cut, so it makes a good fuelwood. Milliken and Bridgewater (2004) mentioned that its wood has been used to make durable cogs and teeth of mill-wheels and other machinery, and tool handles too, as it is tough and fine grained.

A cut branch of hawthorn was used in some parts of Ireland as a simple harrow, dragged along the ground to remove the weeds in a field, or to help sow seeds (Lucas, 1958).

Hawthorn was used as a veterinary remedy to treat diarrhoea in bullocks in Co. Longford (Allen & Hatfield, 2004).

The Irish word *sceath* translates as 'a bush' and is particularly referred to hawthorn or whitethorn (de Vere Coneys, 1849). *Sceach* is also given by Uí Chanainn (1962), meaning 'white-thorn, briar, bramble'. The Irish word *scairtín* is given by Dinneen (1927) and Uí Chanainn (1962), meaning 'a little white-thorn bush'.

**Crepis capillaris*

- Smooth hawk's-beard (Scannell & Synnott, 1987)
- *Lus cúráin mín* (Scannell & Synnott, 1987)

ASTERACEAE (Compositae)—THE DAISY FAMILY

Native

Crepis capillaris is a common plant throughout Ireland of grasslands, walls, heaths, banks and waste places.

*No ethnobotanical uses of this plant have been noted in Ireland.

**Crepis paludosa*

- Marsh hawk's-beard (Scannell & Synnott, 1987)
- *Lus cúráin corraigh* (Scannell & Synnott, 1987)

ASTERACEAE (Compositae)—THE DAISY FAMILY

Native

Crepis paludosa is a robust perennial plant of montane wet grasslands, stream banks and rocky habitats that is rare in most parts of Ireland except in parts of the north-west.

*No ethnobotanical uses of this plant have been noted in Ireland.

Crithmum maritimum

- Rock samphire (Scannell & Synnott, 1987); Samphire (Mackay, 1836; Smith, 1882)

- *Craobhraic* (Scannell & Synnott, 1987); *Greidhricí* (Ua Maoileoin, 2001); *Lus na gcnámh* (Moloney, 1919; Dinneen, 1927; Ó Dónaill, 1977); *Cnámhlus* (Moloney, 1919; Dinneen, 1927); *Grioloigin* (de Vere Coneys, 1849; Cameron, 1883); *Greidhri* (Uí Chonchubhair, 1995); *Géirgin (-ne)* (de Vere Coneys, 1849); *Lus na cnamh* (de Vere Coneys, 1849); *Cabáiste faille* (Dinneen, 1927); *Geirgín na trágha* (Dinneen, 1927); *Craobhracán* (Ó Dónaill, 1977)

APIACEAE (Umbelliferae)—THE CARROT FAMILY
Native

Crithmum maritimum.

Crithmum maritimum is a common plant of rocky cliffs, shores and seaside walls around the Irish coast, although it is more abundant in the south than the north.

Rock samphire continues to be used as a vegetable today in Ireland, but it is much more widely collected on the French coasts of Europe and pickled as a vegetable. It can be boiled as a vegetable or added to salads, although I don't much like the flavour. I have eaten pickled rock samphire in Greece, which was delicious. Uí Chonchubhair (1995) noted that it was used as a vegetable on the Blasket Islands in Co. Kerry, eaten raw or boiled. Mackay (1836: 117) reported that this species was made into a 'warm aromatic pickle and sold for that purpose in Dublin and elsewhere'. Smith (1815: 362) said that it was 'used as a pickle' in Co. Cork. J. O'Donovan (in O'Rourke, 2006) reported that it was regularly exported from the Aran Islands, Co. Galway, and sent to Dublin for use in pickles and preserves in the 19th century. I have also read that it was collected and used as a vegetable on Rathlin Island, off the north coast of Co. Antrim. Rutty (1772: 49) said that it was 'used as a potherb by the poor, and may answer the end of pepper, and is preserved in brine, and greened with vinegar, like other pickles, and is very grateful to the stomach'.

Samphire is mentioned in Shakespeare's *King Lear* (Act IV, Scene VI): 'Half way down, Hangs one that gathers samphire; dreadful trade', alluding to the often dangerous and difficult task of collecting samphire on rocky sea cliffs.

On the Greek island of Chios, I've seen *Crithmum maritimum* used for cosmetics, to make 'an anti-ageing, skin-rebirthing facial care cream'. I've not found it used as an ingredient in cosmetics in Ireland.

Crithmum maritimum leaves.

Crocosmia ×crocosmiiflora (*Tritonia ×crocosmiiflora*)

- Montbretia (Scannell & Synnott, 1987)
- *Feileastram dearg* (Scannell & Synnott, 1987); *Liostram dearg* (Ua Maoileoin, 2001)

IRIDACEAE—THE IRIS FAMILY
Introduced

Crocosmia ×crocosmiiflora is a widespread ornamental plant that has frequently escaped from cultivation to colonise roadsides and waste places throughout Ireland. It is a

Crocosmia ×crocosmiiflora.

particularly characteristic feature of many roadsides in the south-west of Ireland, mainly on acid soils, and is so common that local people and visitors often believe it to be an indigenous native plant. It is in fact a hybrid plant of garden origin derived from a cross between two South African species, *Crocosmia pottsii* × *C. aurea*.

In Larne, Co. Antrim, Vickery (1995) recorded that children have used its leaves as whistles, with different tones being produced depending on the width of the leaf.

Cuscuta epithymum

- Dodder (Scannell & Synnott, 1987)
- *Clamhán* (Scannell & Synnott, 1987); *Clúbhán (dearg)* (Moloney, 1919); *Clúmhán* (Uí Chonchubhair, 1995)

CONVOLVULACEAE—THE BINDWEED FAMILY

Possibly Introduced

Cuscuta epithymum is a parasitic plant that can be easily identified by its thin, reddish creeping stems. It is most often associated with small shrubs and herbs such as *Calluna vulgaris* (heather), *Thymus polytrichus* (wild thyme) and the gorses, *Ulex* species. It was possibly introduced into Ireland and can be found casually at scattered locations around the coastline on coastal dunes and sandy heaths.

Dodder was formerly used medicinally to treat skin, kidney and brain diseases, as well as to help relieve depression, heart palpitations and fainting, but its use for such purposes depended upon from which host plant it was derived, such as thyme, gorse, heather, etc. (Uí Chonchubhair, 1995).

A related dodder is *Cuscuta epilinum*, flax dodder, which was formerly a troublesome weed of cultivated flax fields in the early 19th century, but is now very rare in Ireland (Reynolds, 2002). Cameron (1883) gave its Irish name as *clamhainin lin*.

Cymbalaria muralis.

Cymbalaria muralis

- Ivy-leaved toadflax (Scannell & Synnott, 1987); Mother-of-thousands (Praeger, 1938)
- *Buaflíon balla* (Scannell & Synnott, 1987); *Lus lín an fhalla* (Uí Chonchubhair, 1995)

VERONICACEAE (Scrophulariaceae)—THE SPEEDWELL FAMILY

Introduced

Cymbalaria muralis was probably first grown as a garden ornamental that has spread widely to become an abundant weed of walls, gardens and waste places.

Phillips (1983) noted its use as an edible herb for salads. There are no records of traditional Irish use.

Cynoglossum officinale

- Hound's-tongue (Scannell & Synnott, 1987); Wild fire (Colgan, 1904)
- *Teanga chon* (Scannell & Synnott, 1987)

BORAGINACEAE—THE BORAGE FAMILY
Native

Cynoglossum officinale is a scarce plant in Ireland that is becoming much rarer. It is found in dry sandy coastal habitats, such as sand dunes and waste places, near the south and east coasts.

According to Allen and Hatfield (2004), *Cynoglossum officinale* was used medicinally in Ireland for arm rashes ('felons'), coughs, external and internal cancers and skin problems. Gallagher (2012: 34) described its use in Co. Donegal to treat an itchy scalp: 'Long ago young people who used to work at hay and corn would get lots of chaff on their hair by forking up hay and straw into the barns in autumn, causing it to itch. The grannies would use a remedy by going out to the fields and collect Hounds Tongue. Using its leaves and stem they put it into a jar in a pan of warm water and leave it to stew, do not strain, and leave to soften. Rub well into your scalp each night and this will help improve the complaint.' Wood-Martin (1902) mentioned that *Cynoglossum* powder had recently been advertised in the newspapers as a 'new cancer cure', even though it had been used for such purposes for centuries, dating back to the time of Dioscorides (ca. AD 40–90) and was included in a manuscript on medicinal plants held in the monastery at Mount Athos in Greece.

Cynosurus cristatus

- Dog's tail (Scannell & Synnott, 1987); Thraneen grass (Colgan, 1904); Crested dog's tail
- *Coinfhéar* (Scannell & Synnott, 1987); *Tráithnín* (Colgan, 1911; Moloney, 1919; Ó Dónaill, 1977); *Feur choinein* (Cameron, 1883); *Trathninfheur cirainach* (Wade, 1808a); *Trathnínfhér cirínach* (White, 1803); *Cuisiog* (White, 1803)

POACEAE (Gramineae)—THE GRASS FAMILY
Native

Cynosurus cristatus is a very common grass of grasslands and roadsides throughout Ireland. It is used in some commercial wildflower seed mixes.

Mackay (1825a: 114) commented that *Cynosurus cristatus* was 'found of late to be one of the best grasses for making ladies bonnets', and Howes (1974) recorded that mats and baskets have been made from it. The catalogue of exhibits from the Great Exhibition of 1851 held in London listed a series of objects made from traneen grass plait, such as hats, bonnets, flower-stands and baskets, exhibited by Louisa Maher (proprietor), of Ballinkeele, Enniscorthy, Co. Wexford.

The collections of the National Museum of Ireland (Museum of Country Life) in Castlebar, Co. Mayo, include a 'Dogtail grass besom', a broom made from this grass, from Canningstown, Co. Cavan (Ref. No: F1952.19). A 'besom' is an English word for a bundle of twigs or other material tied together and attached to a handle as a broom.

Cytisus scoparius (*Sarothamnus scoparius*)

- Broom (Scannell & Synnott, 1987); Green broom (Colgan, 1904)
- *Giolcach shléibhe* (Scannell & Synnott, 1987); *Giolcóg* (Moloney, 1919); *Brum* (Cameron, 1883); *Oir*[74] (de Vere Coneys, 1849; Uí Chanainn, 1962); *Beallaidhe* (Dinneen, 1927); *Giolcach* (Dinneen, 1927)

FABACEAE (Leguminosae)—THE PEA AND CLOVER FAMILY
Native

Cytisus scoparius is a shrub of dry and heathy habitats that is widespread in Ireland and relatively common except in parts of the centre.

Broom is frequently grown as an ornamental shrub in parks and gardens and has been used for basket-making. Broom twigs were used to make sweeping brooms, as their name suggests (Colgan, 1904). Burned broom twigs were also used as a fumigant (Moloney, 1919). Phillips (1983) noted that the bark of broom contains a useful fibre that has been used since ancient times, although I have not noted its use for fibre anywhere in Ireland.

In traditional medicine broom was used as a diuretic and to reduce swellings. Apparently, Henry VIII used a drink made from broom to treat his gout (MacCoitir, 2003). Elsewhere in Europe, broom has been used to treat heart and circulation disorders, acting to slow and regulate heartbeat, as well as, traditionally, for snakebite and as a diuretic (van Wyk & Wink, 2004). Egan (1887) and Lysaght (1979) recorded the use in Ireland of broom to treat dropsy and jaundice in which a decoction of the leaves and plant tops was drunk. Allen and Hatfield (2004) noted its widespread use for medicinal purposes in Ireland (it is a powerful diuretic) to treat rheumatism, neuralgia, as a liver remedy, for coughs and colds, asthma, abscesses, boils, erysipelas, rashes, sprains, heartburn and toothache. Mackay (1836: 75) noted that its use as a medicinal plant 'as a rustic remedy for dropsies' was still widespread at that time. McGlinchey (2007: 83) said that broom was used medicinally in Co. Donegal in the 1800s: 'If the tops of the branches are boiled they are a great cure for dropsy or swelling that would come from the kidneys'. Broom has been studied for its antioxidant properties and is now known to be rich in flavones, flavonals and isoflavones (Sundararajan et al., 2006). In ethnomedicine, this plant has been used for various purposes, such as for diuretics, as a hypnotic and sedative, and to treat diabetes and liver disease (Sundararajan et al., 2006). Smith (1815) said that in Co. Cork it was esteemed for the diuretic quality of its ash and was used also in preserving malt liquor when hops were scarce.

The unopened flowerbuds and young seedpods can be pickled, but they must first be soaked for about 24 hours in a strong solution of salt and vinegar, making sure that they do not float, and then rinsed well (Peterson, 1977). The flowerbuds should be collected at the beginning of May for such purposes (Milliken & Bridgewater, 2004). The seed pods can be roasted, ground and used as a coffee-like drink; however, eaten raw, the seed pods are poisonous.

Rutty (1772: 114) said that it is very useful for thatching, and 'most lasting for houses, and for binding and making dry hedges'. He also said that it provided excellent ash for bleaching.

Broom flowers have been used for salads and wine (N. B., <u>not</u> the flowers of *Spartium junceum*, the garden broom, which has poisonous flowers [Phillips, 1983]). In Scotland, broom was once added to beer made from barley, apparently as flavouring, although it is a

74. De Vere Coneys (1849), Uí Chanainn (1962) and Dinneen (1927) also gave the Irish word *oir* for 'furze' (*Ulex europaeus*) and the 'spindle-tree' (*Euonymus europaeus*).

poisonous plant if overused. Broom is reputed to have a narcotic effect by suppressing the nervous system, which leads to lack of coordination and periods of excitement, symptoms somewhat like alcohol poisoning (Milliken & Bridgewater, 2004). They say that in Scotland its dried flowers have been used as a marijuana substitute, presumably because of a narcotic effect, and that sheep eating broom that are unused to it become giddy, suggested because of the cystine content. Several quinolizidine alkaloids have been identified in broom, which are known to have neurologic effects on humans and animals, ranging from relaxation to euphoria to seizures at toxicological doses (Sundararajan et al., 2006).

Broom has been used as a veterinary remedy to treat various ailments (Lysaght, 1979), including gravel and broken wind in horses and fits in dogs in different parts of Ireland (Allen & Hatfield, 2004).

Mahon (1983) mentioned that broom yields a yellow dye formerly used for dyeing textiles in Ireland. A lime green–coloured dye is also obtained from broom.

Daboecia cantabrica

- St Dabeoc's heath (Scannell & Synnott, 1987)
- *Fraoch na haon choise* (Scannell & Synnott, 1987); *Fraoch galldha* (Moloney, 1919; Dinneen, 1927)

ERICACEAE—THE HEATHER FAMILY
Native

Daboecia cantabrica is found only in Connemara in Cos. Galway and Mayo and is unknown elsewhere in Ireland. It grows in rocky places and heathy habitats as well as in grassy banks and gorse thickets, and occasionally on rock-outcrops.

Allen and Hatfield (2004) noted that in 1700, Edward Llwyd (or Lhuyd) recorded that a sprig of this plant was carried by women in Mayo and Galway to protect them from mishap, or perhaps, 'incontinence' (Llwyd's manuscript is somewhat illegible on this). Nelson (2009: 53) wrote about the same reference ('sometimes ye women carry sprigs of it about them as a Preservative againsy Incontinency') and suggested that at that time of Llwyd's writing, the word incontinence had a slightly different meaning than today, and he clearly meant that it was believed to reinforce a young woman's self-restraint in sexual appetite.

St Dabeoc's heath is a common ornamental plant of gardens, best suited to acid soils. Instead of the usual purple flowers, white-flowered cultivated varieties are also grown. Albino white and reddish pink–flowered varieties are also found in the wild, some of which have been propagated and given cultivar names (such as cv. 'Praegerae', a pink form) (Webb & Scannell, 1983; Nelson, 2000b).

Dactylis glomerata

- Cock's-foot grass (Scannell & Synnott, 1987)
- *Garbhfhéar* (Scannell & Synnott, 1987); *Feur garbh* (Moloney, 1919; Dinneen, 1927); *Féar garbh* (Uí Chonchubhair, 1995); *Caolachosfheur coitchion* (Wade, 1808a); *Caileachas-fhéar coitcheann* (Dinneen, 1927); *Cailechosfhér coitchion* (White, 1803)

POACEAE (Gramineae)—THE GRASS FAMILY
Native

Dactylis glomerata is a common grass of pastures, meadows, roadsides and wasteground that is used in some commercial wildflower seed mixes.

Wade (1808a) noted that this grass is eaten greedily by horses and cattle.

Dactylorhiza species

- Spotted- and Marsh-orchids (Scannell & Synnott, 1987)

ORCHIDACEAE—THE ORCHID FAMILY

Native

Including:

Dactylorhiza fuchsii
- Common spotted-orchid (Scannell & Synnott, 1987)
- *Nuacht bhallach* (Scannell & Synnott, 1987)

Dactylorhiza incarnata
- Early marsh-orchid (Scannell & Synnott, 1987)
- *Magairlín mór* (Scannell & Synnott, 1987); *Magairlín mór an churraigh*[75] (Moloney, 1919; Dinneen, 1927)

Dactylorhiza kerryensis (*Dactylorhiza majalis* subsp. *occidentalis*; *Orchis latifolia*)
- Broad-leaved marsh-orchid (Scannell & Synnott, 1987); Western marsh-orchid; Brush sweet willie[76] (Hart, 1898)
- *Magairlín gaelach* (Scannell & Synnott, 1987); *Magairlín leathan* (Moloney, 1919; Dinneen, 1927)

Dactylorhiza maculata
- Heath spotted-orchid (Scannell & Synnott, 1987); Barney brack (Colgan, 1904)
- *Na circíní* (Scannell & Synnott, 1987); *Mangairlín meadhrach*[77] (Ua Maoileoin, 2001); *Úrach bhallach* (Moloney, 1919); *Cearc bhreach* (Moloney, 1919); *Cearc breac* (Uí Chonchubhair, 1995)

Dactylorhiza traunsteineri
- Narrow-leaved marsh-orchid (Scannell & Synnott, 1987)
- *Magairlín caol* (Scannell & Synnott, 1987)

Dactylorhiza species are widespread in Ireland in a variety of grassland and wetland habitats. *Dactylorhiza fuchsia* is common throughout Ireland in unimproved damp grasslands, marshes and river-banks. *Dactylorhiza maculata* occurs in more heathy and boggy habitats and wet grasslands. *Dactylorhiza incarnata*, *D. kerryensis* and *D. traunsteineri* are all marsh species occurring in marshes, fens, sand dunes (especially the margins of dune-slacks) and other wetland and wet meadow or grassland habitats.

Wade (1804: 64) suggested that 'it is highly probably that this, and every other species of our common *Orchis*, as well as the *O. mascula* may be used indiscriminately for making of salep, and affording an excellent nutriment in times of scarcity' (see also *Orchis mascula*

75. Dinneen (1927) defined *magairlín mór an churraigh* as 'common marsh orchid'.

76. Hart (1898: 369) reported this name from Ardara, Co. Donegal (for '*Orchis latifolia*'), and said that it was also applied to *O. mascula*.

77. Ua Maoileoin (2001) generally applied the Irish name *mangairlín meadhrach* to *Orchis mascula*, the early purple orchid.

Dactylorhiza species. A, *Dactylorhiza fuchsii* subsp. *okellyi*. B, *Dactylorhiza kerryensis*. C, *Dactylorhiza maculata*.

and *Anacamptis morio*).

Dactylorhiza fuchsii subsp. *okellyi* is a pretty pure white-flowered form of the common spotted orchid that was named by George Claridge Druce, the Oxford botanist, in honour of Patrick B. O'Kelly, an amateur botanist, plant collector and nurseryman from Ballyvaughan, Co. Clare. O'Kelly collected and sold wild plants from the Burren in the 1890s (Nelson, 2009). He sold a wide variety of wild plants, including this orchid, gentians (*Gentiana verna*) and many ferns to gardeners all over Ireland and Britain. His customers included the Royal (now the National) Botanic Gardens, Glasnevin, the Royal Botanic Gardens, Kew, and the Cambridge University Botanic Garden.

*Danthonia decumbens

- Heath grass (Scannell & Synnott, 1987)
- *Féar caoráin* (Scannell & Synnott, 1987); *Fescufhér sinte*[78] (White, 1803)

POACEAE (Gramineae)—THE GRASS FAMILY
Native

Danthonia decumbens is a common plant of heaths and pastures throughout Ireland.

*No ethnobotanical uses of this plant have been noted in Ireland.

Datura stramonium

- Thorn-apple (Scannell & Synnott, 1987)
- *Stoc an aingil* (Scannell & Synnott, 1987)

SOLANACEAE—THE NIGHTSHADE FAMILY
Introduced (possible archaeophyte)

Datura stramonium is a cosmopolitan annual weed that although widespread is rare in

78. White (1803) applied this name to *Festuca decumbens*.

Ireland. It is a seriously poisonous plant, like its near relative *Hyoscyamus niger* (henbane), and is well known for its hallucinogenic and euphorbic effects as well as for its use for medicinal purposes, although I have not found historic records of such usage in Ireland. However, Reynolds (2002: 200) recorded that it was grown as a crop near Lisnaskea, Co. Fermanagh, during World War I 'no doubt because of its medicinal properties'. Reynolds mentioned a remarkable record of this plant that was recorded by Austin O'Sullivan in the *Irish Naturalists' Journal* in 1979, when some 10,000 plants shaded out a fodder beet crop near Ballinspittle, West Cork. She noted that in Ireland it still turns up in late-maturing crops, but needs warm dry summers to thrive and was perhaps introduced with animal feed and slurry spread on tilled ground.

Datura stramonium contains anticholinergic tropane alkaloids such as hyoscyamine, hyoscine, atropine and scopolamine. These medically significant alkaloids impact smooth and cardiac muscle as well as the body's ability to sweat and salivate, acting as competitive antagonists to the neurotransmitter acetylcholine, which disrupts the parasympathetic nervous system. Thorn-apple can be considered one of the most toxic plants in the Western hemisphere.

This species was traditionally associated with witchcraft recipes due to its hallucinogenic properties that induce feelings of flight and surreal sensations (Pennacchio et al., 2010). The same authors suggested that the modern-day concept of witches riding brooms may have been derived from these feelings, although they noted that it may also be from broom handles being used to apply vaginal salves prepared from various *Datura* species. A particularly efficient means of self-administering the drug for women is through the moist tissues of the vagina (Davis, 1986). It was also used to dull the senses of people sacrificed during ceremonial executions (Rennacchio et al., 2010). Wade Davis wrote about this plant species in his book *The Serpent and the Rainbow* (1986) on Haiti, voodoo and zombies. He mentions that in Haiti it was called *concombre zombi* (the zombie's cucumber), and through his work, implicated this plant for the part it may have played in the creation of zombies on that island. He also pointed out that amongst the Yaquis of northern Mexico, sorcerers would anoint their genitals, legs and feet with a salve based on crushed *Datura* leaves and thus experience the sensation of flight. He, too, pointed out the purported link between medieval witches in Europe and the use of *Datura*, or its relatives *Atropa belladonna* (deadly nightshade), *Hyoscymus niger* (henbane) and *Mandragora officinarum*, (mandrake), suggesting that their broomstick journeys were not through space, but across the hallucinatory landscape of their own minds.

Rutty (1772: 175) said that this species is useful to make ointments for burns, but he tells a tale of 'several persons [in Naas, Co. Kildare] who had eaten of the leaves of this plant, imagining them to be Spinache, fell into a madness, attended with mirth and divers frolicksome tricks for a few days, and then recovered'.

Datura stramonium photographed in Cuba.

In various parts of the world the leaves of this species have been burned and the smoke inhaled as a treatment for asthma (Pennacchio et al., 2010). Reynolds (2002) mentioned that it was an occasional plant in wasteground near towns in early 19th-century Ireland, and prior to then, much used as an asthma cure but was considered extinct in Ireland by the 1890s. The leaves were rolled into cigarettes or smoked in a pipe. In fact its asthma-relieving properties come from the alkaloid, atropine, which impacts the pulmonary branches of the lungs, eliminating spasms during asthma attacks; therefore, it treats the symptoms of asthma rather than helping to cure it (Pennacchio et al., 2010).

This is a dangerous plant and should not be experimented with. Woodward (1985: 72) gave the symptoms of poisoning by *Datura stramonium* as 'unquenchable thirst, vomiting, enlarged pupils, delirium, jumbled speech, nervous twitches and convulsions; death follows in overdoses. Some symptoms may continue for several days.'

Its Irish name, *stoc an aingil*, means 'angel's trumpet'. Several *Datura* species grown as tropical and subtropical ornamental plants are given that name in English too.

Daucus carota

- Carrot; Wild carrot (Scannell & Synnott, 1987)
- *Mealbhacán* (Scannell & Synnott, 1987); *Bliúcán* (Anonymous, 1978); *Mil mhairt* (Ua Maoileoin, 2001); *Burgomán* (Moloney, 1919); *Mugomán* (de Vere Coneys, 1849; Moloney, 1919; Dinneen, 1927); *Miodhlucán* (Moloney, 1919; Uí Chonchubhair, 1995); *Curran bhuidhe* (Cameron, 1883); *Cairéad* (De Bhaldraithe, 1959); *Cearrac(h)án* (de Vere Coneys, 1849; Uí Chanainn, 1962); *Crallacóg*[79] (Dinneen, 1927); *Meacan dearg*[80] (Dinneen, 1927; Uí Chanainn, 1962; Ó Dónaill, 1977); *Meacan buidhe* (Dinneen, 1927); *Cuirdín* (or *Cuiridín*)[81] (Ó Dónaill, 1977); *Murlóg*[82] (Uí Chanainn, 1962)

APIACEAE (Umbelliferae)—THE CARROT FAMILY

Native

Daucus carota.

Wild carrot is a common plant of pastures, meadows, waste places, roadsides, sand dunes and banks, particularly near the sea. The cultivated carrot is *Daucus carota* var. *sativa*.

Carrots have been a well-known medicinal plant since antiquity as well as a highly praised aphrodisiac. In Ireland the wild species is distinguished as a medicinal plant from the cultivated garden vegetable varieties and said to be a more efficacious remedy for some ailments, such as earache, sores and wounds, and used as a diuretic (Allen & Hatfield, 2004). Rätsch (1997) reported that this root is said to have diuretic and abortifacient properties and, according to Dioscorides, it 'incites coitus'. The centres of the flower heads (the umbels) were used to treat epilepsy, but it was hard to collect large enough quantities in order to be effective (Williams, 1993). The same author recorded that a poultice was made from it to treat discharging ears. It was also boiled in wine and rubbed on the stomach to treat colic and for sore eyes. Carrot juice is an excellent health drink because it contains high levels of carotene that are converted into vitamin A in the liver (van Wyk & Wink, 2004). Wood-Martin (1902) listed its use as a blood purifier. In the Mizen region of West Cork, carrots were used to treat warts, and in Co. Donegal, it was used to treat gout (Gallagher, 2012). The core of the carrot was removed and the hole filled with salt, with the carrot left standing upright. The juice that seeps through the carrot into the salt was then collected and placed on warts (McCarthy & Hawkes, 2007).

The roots of wild carrots can be boiled and used as a vegetable (Peterson, 1977). According to Milliken and Bridgewater (2004), in the Outer Hebrides, Scotland, the

79. Dinneen (1927) defined *crallacóg* as 'a carrot'.
80. These authors defined *meacan dearg* as 'carrot'.
81. Ó Dónaill (1977) defined *cuirdín* (or *cuiridín*) as 'wild carrot'.
82. Uí Chanainn (1962) defined *murlóg* as 'carrot, turnip'.

roots of wild carrot were gathered and eaten raw as a traditional snack. These authors also mention a 19th-century account of young women gathering them 'for distribution among their acquaintances on Sundays, and at their dances' (Milliken & Bridgewater, 2004: 49).

On the Isle of Harris, Scotland, the seeds of wild carrot were used instead of hops for the brewing of beer during the late 17th century (Milliken & Bridgewater, 2004), but I have not found records of such usage in Ireland.

Daucus carota was used as a veterinary remedy for 'diseases in horses' in Carlow (Allen & Hatfield, 2004) and 'staggers' in cows (Williams, 1993).

McNeill (2010) mentioned that it has been included in wildflower seed mixes in Co. Tyrone, where it grows on the verges on a recently completed (2003) bypass of Newtownstewart.

Any tap-rooted plant in Irish is *meacan*, such as carrot or parsnip (de Vere Coneys, 1849; Dinneen, 1927).

Deschampsia cespitosa (*Aira cespitosa*)

- Tufted hair-grass (Scannell & Synnott, 1987)
- *Móinfhéar garbh* (Scannell & Synnott, 1987); *Gruagfheur sgrathogha* (Wade, 1803a); *Gruagfhér sgrathagach* (White, 1803)

POACEAE (Gramineae)—THE GRASS FAMILY
Native

Deschampsia cespitosa is a relatively common grass of woods, hedgerows, moist grasslands and river and stream banks throughout Ireland.

The seeds of *Deschampsia cespitosa* are edible, and Mears and Hillman (2007) suggested that early people in Britain and Ireland may have used this plant as a staple source of calories.

This grass was used in the north of England to stuff Church hassocks (kneelers) (Vickery, 1995), but I have not found records of such usage in Ireland.

---

*Deschampsia cespitosa subsp. *alpina*

- Alpine hair-grass (Scannell & Synnott, 1987)
- *Móinfhéar Alpach* (Scannell & Synnott, 1987)

POACEAE (Gramineae)—THE GRASS FAMILY
Native

Deschampsia cespitosa subsp. *alpina* is a plant of higher mountains that is rare in Ireland and mainly found in Kerry and West Mayo.

*No ethnobotanical uses of this plant have been noted in Ireland.

---

*Deschampsia flexuosa

- Wavy hair-grass (Scannell & Synnott, 1987)
- *Móinfhéar* (Scannell & Synnott, 1987); *Gruagfhér snímhain* (White, 1803)

POACEAE (Gramineae)—THE GRASS FAMILY

Native

Deschampsia flexuosa is a common grass in Ireland of heathlands and moors in the mountains and is rare in the lowlands.

*No ethnobotanical uses of this plant have been noted in Ireland.

Deschampsia setacea

- Bog hair-grass (Scannell & Synnott, 1987)
- *Móinfhéar seascainn* (Scannell & Synnott, 1987)

POACEAE (Gramineae)—THE GRASS FAMILY
Native

Deschampsia setacea is rare in Ireland and found only on lake margins and very wet bogs in southern and central Connemara.

*No ethnobotanical uses of this plant have been noted in Ireland.

Descurainia sophia (Sisymbrium sophia)

- Flixweed (Scannell & Synnott, 1987)
- *Finéal Muire* (Scannell & Synnott, 1987); *Fineul mhuire (-éil mhuire)* (de Vere Coneys, 1849)

BRASSICACEAE (Cruciferae)—THE CABBAGE AND CRESS FAMILY
Introduced

Descurainia sophia is a very rare weed of sandy fields and waste places. It was formerly more widespread in Ireland but is now found mainly in the east.

Moloney (1919) mentioned that this species was formerly used in Ireland to treat dysentery (called 'flix', or 'flux', hence its common name).

*Dianthus armeria

- Deptford pink (Akeroyd, 1996)
- *Lus gile Deptford* (Akeroyd, 1996)

CARYOPHYLLACEAE—THE CARNATION OR PINK FAMILY
Native

Dianthus armeria is a very rare plant in Ireland, known only from one grassy cliff-top on Horse Island, in West Cork, where it was discovered in 1992.

*No ethnobotanical uses of this plant have been noted in Ireland.

Digitalis purpurea

- Foxglove (Scannell & Synnott, 1987); Lady's fingers (Colgan, 1904); Foxlove (Colgan, 1904); Fairy thimble (Hart, 1898); Fox and leaves (Hart, 1898); Fox tree leaves (Hart, 1898); Lithmore (Hart, 1898); Shiggy[83] (or Shiggy thimble) (Hart, 1898)

83. This English name is from the Irish words *sighe* or *sidhe*, meaning 'fairy'.

- *Lus mór* (Scannell & Synnott, 1987); *Méaracáin*[84] *dhearga na bpúcaí* (Ua Maoileoin, 2001); *Lus mór baineann* (de Vere Coneys, 1849; Moloney, 1919; Dinneen, 1927); *Méaracán na mban sídhe* (Moloney, 1919; Dinneen, 1927); *Síodhán sléibhe* (Moloney, 1919); *Lus na mban sídhe* (Moloney, 1919; Dinneen, 1927); *Lus na mban sí* (Ó Dónaill, 1977); *Méaracán púca* (Webb et al., 1996); *An siothan* (Cameron, 1883); *Bolgan beic* (Cameron, 1883); *Lus na m-ban sighe* (de Vere Coneys, 1849); *Méaracán dearg* (Dinneen, 1927); *Méaracán daoine marbh* (Dinneen, 1927); *Méiríní púca* (Dinneen, 1927); *Digeatálas* (Dinneen, 1927); *Méaracán na mban sí* (Ó Dónaill, 1977); *Sian-sleibhe* (Joyce, 1875); *Maide méaracán* (Colgan, 1911); *Meirinighe phúca* (Colgan, 1911); *Mearacán sidhe* (Colgan, 1911)

VERONICACEAE (Scrophulariaceae)—THE SPEEDWELL FAMILY
Native

Digitalis purpurea.

Digitalis purpurea is a common plant throughout most of Ireland except in the centre. It occurs on acidic soils in woodlands, hedgerows and rough and rocky heathland.

Foxglove is well known in Ireland as a poisonous plant. It is the source of the heart drug digitalin. The action of the drug is to greatly increase the action of the heart and arterioles and all types of muscle tissue, which causes the heart and arteries to contract, raising the blood pressure. It is also used to slow the pulse and make it more regular. Large doses can cause death.

Historically, it was used in traditional medicine as an ingredient in ointments to reduce swellings (Moloney, 1919). McGlinchey (2007) mentioned its use in the 1800s in Co. Donegal to draw the poison out of a wound or bruise. The leaf was warmed over a coal fire and applied to the wound. Threlkeld (1726: Appendix 58–59) described how it was used medicinally in the Dublin region in the early 1700s: '...the People are often afflicted here with the King's-Evil, which is caused by their gross Food, eating too much butter, or ill-cured Fevers, which does corrupt the Glands'. To treat these ailments an ointment was made of the leaves of foxglove with 'May-Butter or Hogs-Lard'. Once made, some people stored it in a hole in the ground for 40 days or let it stand in the sun all summer. After this, it would be spread onto a linen cloth and applied to the tumour for a year, after which a cure would be affected. Threlkeld was encouraging about the use of such ointments and said, 'Tho the Ulcers become larger, be not afraid, since after the Ointment has wasted, and dryed up all the Humours, it will at last cure them, and draw a Skin over them'. Medieval Irish monks are known to have used foxglove as a diuretic to cure dropsy, and it was said to have been used throughout the British Isles to treat heart troubles (Allen & Hatfield, 2004), although Logan (1972) suggested that it was not used in traditional medicine in Ireland for heart ailments. Egan (1887: 12) mentioned its use to treat dropsy for which a tincture was made with gin and used in very small quantities on leaf sugar; it was known to be a dangerous medicine and 'though spoken of, seems little used'. Logan (1972, 1981) and K'Eogh (1735) mentioned that *Digitalis purpurea* was used to treat epilepsy in Ireland. It was also used (small quantities only) to treat yellow fever, epilepsy and skin ulcers (Williams, 1993). Maloney (1972) reported that in Co. Cavan the juice from 12 leaves of foxglove were used to treat a childhood ailment described as 'pining away of children'. This may be related to a use of this plant in Ireland to identify so-called changeling babies (healthy babies that had been stolen by the fairies and replaced by sickly fairy babies), which is described by Roy Vickery in his extremely useful and entertaining 2010 book, *Garlands, Conkers and Mother-Die*, on

84. Dinneen (1927) gave the Irish word *méaracán* for 'thimble'.

British and Irish plant lore. Various tests would be performed to try to determine whether a child was a changeling or not. In Co. Leitrim, Vickery reported that three drops of juice from a foxglove were placed on the tongue of the candidate child, and three in each ear. If the child died, it was obviously a fairy child, and if it survived, then it was clearly not a changeling. Vickery speculated that this may have been an acceptable form of infanticide for ailing children since the concentrated juice of foxglove contains digitalis, when about one-tenth of an ounce of dry leaf is fatal for an adult, and a child would be unlikely to survive a lesser dose (Vickery, 2010).

Another use was in snuff, and Colgan (1904: 301) reported that 'It's grand for sneezin', and in Dublin, old women would roast the leaves and grind them up for snuff.

Williams (1993) noted that foxgloves were used in veterinary medicine to treat staggers in sheep and swollen cow's udders.

Foxglove flowers have also been popular with children who would put the flowers on their fingers or pop unopened flowers, so they were sometimes called 'popgloves' (in England) or 'cracks' (in Galway) (Williams, 1993). O'Súilleabháin (1942) mentioned that it was a traditional pastime of children to capture and trap bees in the foxglove flowers.

Mahon (1983) listed *Digitalis purpurea* as providing a green dye for textiles.

Foxgloves are popular garden ornamental plants today and most frequently cultivated with a variety of pink, white and purple flower colours (Wyse Jackson, 2006).

Dipsacus fullonum

- Teasel (Scannell & Synnott, 1987)
- *Leadán úcaire* (Scannell & Synnott, 1987); *Leadán liosta* (Moloney, 1919; Dinneen, 1927); *Lus na leadán* (Moloney, 1919; Dinneen, 1927); *Lus an fhúcadóir* (Moloney, 1919); *Taga* (de Vere Coneys, 1849; Cameron, 1883); *Líodán an ucaire* (de Vere Coneys, 1849); *Lus an úcaire (-cadóra)* (Dinneen, 1927; Ó Dónaill, 1977); *Cnádán úcaire*[85] (Ó Dónaill, 1977); *Liosta* (Ó Dónaill, 1977)

DIPSACACEAE—THE SPEEDWELL FAMILY

Possibly Introduced (possible archaeophyte)

Dipsacus fullonum is found near the coast in waste places and on open ground and roadsides throughout Ireland; it is rare throughout except in the north and east.

The flower heads of teasel, which are covered with stiff prickly scales, were used to make simple carding boards that were used to tease out the strands of wool before it was spun (Sharkey, 2000) or to raise the nap on cloth (Howes, 1974). The heads of the teasel were held in a makeshift frame and the wool dragged across them to open out the fibres. Two separate subspecies are found: one, *Dipsacus fullonum* subsp. *sativus*, has hooked bracts, and the other, *D. fullonum* subsp. *fullonum*, has bracts that are not hooked.

Rutty (1772: 110) mentioned that a considerable quantity of this plant was imported from England with 'the hook of the English Teasel being better than ours'. Rutty also said that it was cultivated in Lusk and Balrothery in north Co. Dublin, where it was dried and supplied to 'clothiers' in Dublin.

Dipsacus fullonum.

Evans (1967) mentioned that dried stalks were used in thatching in locations

85. Ó Dónaill (1977) defined *cnádán úcaire* as 'fuller's teasel'.

where teasels were cultivated for the home woollen industry; they were placed over the roof beams of the house, and formed a base onto which the scraw and thatch would be laid.

Mahon (1983) listed this species as the source of a yellow dye.

**Draba incana*

- Hoary whitlowgrass (Scannell & Synnott, 1987)
- *Araflasach* (Scannell & Synnott, 1987)

BRASSICACEAE (Cruciferae)—THE CABBAGE AND CRESS FAMILY

Native

Draba incana is a rare plant in Ireland of sand-hills and rocks and found only in the north-west.

*No ethnobotanical uses of this plant have been noted in Ireland.

Drosera species

- Sundew (Scannell & Synnott, 1987); Oil plant (Hart, 1898)
- *Drúchtín móna* (Scannell & Synnott, 1987)

DROSERACEAE—THE SUNDEW FAMILY

Native

Including:

Drosera anglica

- Great sundew (Scannell & Synnott, 1987)
- *Cailís Mhuire mhór* (Scannell & Synnott, 1987); *Drúichtín móna* (Moloney, 1919; Dinneen, 1927); *Rós an tsoluis* (Moloney, 1919)

Drosera intermedia

- Oblong-leaved sundew (Scannell & Synnott, 1987)
- *Cailís Mhuire* (Scannell & Synnott, 1987)

Drosera rotundifolia

- Round-leaved sundew (Scannell & Synnott, 1987); Guller-na-gat (Hart, 1898)
- *Drúchtín móna* (Scannell & Synnott, 1987); *Dealtruaidh* (Moloney, 1919; Dinneen, 1927); *Lus na fearnaighe* (de Vere Coneys, 1849; Moloney, 1919; Dinneen, 1927); *Eil druich* (Cameron, 1883); *Drúichtín na móna* (Uí Chonchubhair, 1995); *Durachdín-móna* (de Vere Coneys, 1849); *Gealdruigh (-ghe)* (de Vere Coneys, 1849); *Cailís Mhuire* (Dinneen, 1927); *Rós an tsolais* (Dinneen, 1927)

These small perennial plants are found in boggy habitats. Worldwide, the sundews are found growing in acidic wetlands and *Sphagnum* bogs across temperate and montane Europe and North America. Sundews are well known for their ability to trap and digest insects to supplement the nutrients available to them in their peatland habitats. Their leaves are covered with sticky gland-tipped hairs that trap then dissolve their prey. *Drosera rotundifolia* is the commonest of the three species found in Ireland and occurs throughout much of Ireland where there are suitable bogland habitats. The other two species are locally frequent, mainly in bogs in the west, south-west and north-west.

All three native Irish species of sundew were formerly used as a treatment for asthma (Logan, 1981), whooping cough and jaundice (Allen & Hatfield, 2004). To treat asthma the leaves were chopped up finely and the juice squeezed out before being bottled and used when needed. The remedy was then taken as a few drops of the juice added to honey (Logan, 1972). According to Williams (1993), *Drosera* was also used in the treatment of lung ailments, and in the 1950s a herbalist near Dublin advertised that they would buy a pound weight of *Drosera* for £1 to make a cough remedy. This same author also noted it was used as a hair dye for women in the 18th century. Cameron (1883) said that it was much used by the 'Celtic tribes' for dyeing the hair. A yellow or purple/crimson clothes dye was also obtained from it (Williams, 1993). Mahon (1983) listed yellow and purple dyes as coming from *D. rotundifolia*.

Drosera intermedia (sundew). Watercolour by Lydia Shackleton. Courtesy of the National Botanic Gardens, Glasnevin.

Sundew was used as a love charm in the Isle of Man. A piece would be surreptitiously slipped into the clothing of the person who was to be attracted (Garrad, 1984). Wade (1804: 25) wrote that the *Drosera anglica* corrodes the skin, 'yet some ladies know how to mix the juice with milk, so as to form a safe cosmetic to remove freckles and sunburn'.

Cameron (1883) gave the Irish name *eil druich* as *eil*, meaning 'to rob', and *druich*, meaning 'dew'.

Dinneen (1927: 513) gave the Irish phrase *galar nag cat*, meaning 'a disease among lambs caused by eating sundew, where the tips of their ears fall off'. *Galar* means a 'disease, sickness or ailment' in Irish. Dinneen also gave the name *galar dubh* as the name of a poisonous herb dangerous to sheep but does not say what plant this refers to (this is also the Irish name given to cholera).

Dryas octopetala

- Mountain avens (Scannell & Synnott, 1987)
- *Leaithín* (Scannell & Synnott, 1987); *Machal sléibhe* (Moloney, 1919; Dinneen, 1927)

ROSACEAE—THE ROSE FAMILY
Native

Dryas octopetala is found in rocky grasslands on limestone, especially on the pavements of the Burren of Co. Clare. It also occurs on basic rock ledges in mountains in Cos. Sligo and Donegal, then to the north-east in Cos. Derry and Antrim. Mountain avens is an extremely decorative plant to grow on dry rockeries. It is grown and maintained easily and has been cultivated for many years in the Burren native plant section at the National Botanic Gardens in Glasnevin.

Dryas octopetala.

**Echium vulgare*

- Viper's-bugloss (Scannell & Synnott, 1987)
- *Lus nathrach* (Scannell & Synnott, 1987); *Lus na teangan* (Moloney, 1919; Dinneen, 1927); *Teanga na nathrach* (Moloney, 1919); *Lus na nathrach* (Moloney, 1919); *Lus nathrach* (Ó Dónaill, 1977)

BORAGINACEAE—THE BORAGE FAMILY
Native

Echium vulgare primarily occurs in the east in coastal habitats, such as grassy places and on cliffs, sand dunes and dry sandy and gravel soils; it is very rare elsewhere in Ireland.

*No ethnobotanical uses of this plant have been noted in Ireland.

**Elatine hexandra*

- Six-stamened waterwort (Scannell & Synnott, 1987)
- *Bosán na díge* (Scannell & Synnott, 1987)

ELATINACEAE—THE WATERWORT FAMILY
Native

Elatine hexandra is an annual of ponds, wet mud and shallow water in lakes and rivers that is distributed particularly in the north and south-west; it is very rare elsewhere in Ireland.

*No ethnobotanical uses of this plant have been noted in Ireland.

**Elatine hydropiper*

- Eight-stamened waterwort (Scannell & Synnott, 1987)
- *Bosán te* (Scannell & Synnott, 1987)

ELATINACEAE—THE WATERWORT FAMILY
Native

Elatine hydropiper is known in Ireland only from a few lake-shores in the north-east.

*No ethnobotanical uses of this plant have been noted in Ireland.

Eleocharis species

- Spike-rushes (Scannell & Synnott, 1987)
- *Spícíneach* (Scannell & Synnott, 1987)

CYPERACEAE—THE SEDGE FAMILY
Native

The spike-rushes are perennial sedges found in a range of marshy habitats in Ireland, both inland and coastal. I cannot find any recorded uses of any of the six species of *Eleocharis* in Ireland, although I suspect that the tallest and most robust species, *E. palustris*, may have been used for primitive thatching. It is a common plant of the shallow margins of lakes and ponds and also occurs in bogs and marshes.

Elsewhere in the world, species of *Eleocharis* have been used for food and as medicine. For example, in the Northern Territory of Australia, *E. dulcis*, the salt-water spike-rush, or water-chestnut, is used by Aboriginal people to treat wounds, and its corms are gathered for food. The young corms are eaten raw or older ones are roasted and ground down to make flour and baked as cakes (Isaacs, 1987). Apparently, these corms are also eaten by geese and when found in their crops are particularly tasty to eat (Wiynjorrotj et al., 2005). No Irish species of *Eleocharis* have corms or rhizomes thick enough to constitute a starch source worth gathering.

Eleocharis acicularis

- Needle spike-rush (Scannell & Synnott, 1987)
- *Spícíneach ghéar* (Scannell & Synnott, 1987)

CYPERACEAE—THE SEDGE FAMILY
Native

Eleocharis acicularis is a frequent plant of marshes, ponds, canals and lake-shores that is found particularly in the centre of Ireland.

Huxley (1992) listed this plant as an ornamental used in horticulture, but I have found no records of its use in Ireland.

**Eleocharis multicaulis*

- Many-stalked spike-rush (Scannell & Synnott, 1987)
- *Spícíneach thortógach* (Scannell & Synnott, 1987)

CYPERACEAE—THE SEDGE FAMILY
Native

Eleocharis multicaulis is a common plant of lake-shores and bogs, particularly in the west of Ireland.

*No ethnobotanical uses of this plant have been noted in Ireland.

Eleocharis palustris

- Common spike-rush (Scannell & Synnott, 1987)
- *Cíb dhéise* (Scannell & Synnott, 1987); *Luchair shínte*[86] (Moloney, 1919)

CYPERACEAE—THE SEDGE FAMILY
Native

Eleocharis palustris is a common plant of shallow water that is found in lake-shores, marshes, rivers and bogs.

Huxley (1992) listed this plant as an ornamental used in horticulture, but I have found no records of its use in Ireland.

86. Dinneen (1927) defined *luchair shínte* as 'creeping spike-rush'.

Eleocharis parvula

- Dwarf spike-rush (Scannell & Synnott, 1987)
- *Spícíneach bheag* (Scannell & Synnott, 1987)

CYPERACEAE—THE SEDGE FAMILY

Native

Eleocharis parvula is a very rare plant throughout Ireland of tidal mud that is known only in a few river estuaries.

Huxley (1992) listed this plant as an ornamental used in horticulture, but I have found no records of its use in Ireland.

**Eleocharis quinqueflora*

- Few-flowered spike-rush (Scannell & Synnott, 1987)
- *Spícíneach na gcúig bhláth* (Scannell & Synnott, 1987)

CYPERACEAE—THE SEDGE FAMILY

Native

Eleocharis quinqueflora is common throughout much of Ireland in wet habitats, such as lake margins, fens and marshes, except in the south where it is rare.

*No ethnobotanical uses of this plant have been noted in Ireland.

**Eleocharis uniglumis*

- Slender spike-rush (Scannell & Synnott, 1987)
- *Spícíneach chaol* (Scannell & Synnott, 1987)

CYPERACEAE—THE SEDGE FAMILY

Native

Eleocharis uniglumis is found mainly in coastal salt-marshes or streams near the sea.

*No ethnobotanical uses of this plant have been noted in Ireland.

**Eleogiton fluitans (Scirpus fluitans)*

- Floating club-rush (Scannell & Synnott, 1987)
- *Brobh ar snámh* (Scannell & Synnott, 1987)

CYPERACEAE—THE SEDGE FAMILY

Native

Eleogiton fluitans is occasionally found throughout Ireland and is an aquatic plant of bog-drains, bog-pools, streams and other peaty habitats.

*No ethnobotanical uses of this plant have been noted in Ireland.

Elodea canadensis

- Canadian waterweed (Scannell & Synnott, 1987); Canadian pondweed (Anonymous, 1978); Water thyme (Praeger, 1938; Anonymous, 1978); American weed (Hart, 1898); Swanweed (Hart, 1898); American water thyme (Brunker, 1950); Raave (Praeger, 1938)

- *Tím uisce* (Scannell & Synnott, 1987); *Gorlús* (Hart, 1898)

HYDROCHARITACEAE—THE FROGBIT FAMILY
Introduced

Elodea canadensis is a widespread and sometimes invasive aquatic plant of lakes, canals, rivers, ponds, turloughs and other watery habitats. It spreads perniciously by vegetative means. It was introduced first from Canada in the early to mid-1800s (Reynolds, 2002).

It is commonly grown in Ireland as an ornamental aquatic plant for artificial ponds and aquaria.

A close relative is *Elodea nuttallii* (Nuttall's waterweed; in Irish, *tím uisce chaol*), which also occurs as an invasive aquatic plant in similar habitats but is less common than *E. canadensis*. It has also been used as an ornamental aquatic plant for ponds and aquaria.

**Elymus caninus (Agropyron caninum)*

- Bearded couch (Scannell & Synnott, 1987)
- *Broimfhéar colgach* (Scannell & Synnott, 1987); *Fiorthan* (Moloney, 1919); *Aolfhér feusagach* (White, 1803)

POACEAE (Gramineae)—THE GRASS FAMILY
Native

Elymus caninus is a rare grass in Ireland and can be found in a range of woodland and rocky habitats.

*No ethnobotanical uses of this plant have been noted in Ireland.

**Elytrigia atherica (Elymus athericus; Elymus pycnanthus)*

- Sea couch (Scannell & Synnott, 1987)
- *Broimfhéar mara* (Scannell & Synnott, 1987)

POACEAE (Gramineae)—THE GRASS FAMILY
Native

Elytrigia atherica is a relatively rare grass in Ireland that is distributed mainly in the south on sandy seashores and in the upper margins of salt marshes.

*No ethnobotanical uses of this plant have been noted in Ireland.

**Elytrigia juncea (Elymus farctus; Elymus junceus; Elytrigia juncea* subsp. *boreoatlantica; Triticum junceum)*

- Sand couch (Scannell & Synnott, 1987)
- *Broimfhéar gainimh* (Scannell & Synnott, 1987); *Cruithneachd luachairamhuil*[87] (White, 1803)

POACEAE (Gramineae)—THE GRASS FAMILY
Native

Elytrigia juncea is a common grass of sandy seashores.

*No ethnobotanical uses of this plant have been noted in Ireland.

---

87. White (1803) applied this name to '*Triticum junceum*', 'sea wheat-grass'.

Elytrigia repens (*Agropyron repens*; *Elymus repens*; *Triticum repens*)

- Common couch (Scannell & Synnott, 1987); Scutch-grass (Webb & Scannell, 1983); Scutch (Colgan, 1904); Couch-grass (Colgan, 1904); Quitch (Hart, 1898; Praeger, 1938); Squitch (Hart, 1898; Praeger, 1938); Twitch (Harron, 1986)
- *Broimfhéar* (Scannell & Synnott, 1987); *Bruimfhéur (-éir)* (de Vere Coneys, 1849[88]); *Cruithneachd bruimsean*[89] (White, 1803; Wade, 1808a); *Cruithneacht bhruimsean* (Dinneen, 1927); *Féar gaoil* (Ó Dónaill, 1977); *Bruimsean* (White, 1803); *Bruimfheur* (White, 1803); *Fiothran* (White, 1803); *Urin* (White, 1803)

POACEAE (Gramineae)—THE GRASS FAMILY
Native

Elytrigia repens.

Elytrigia repens is a very common weedy grass of wasteground and cultivated fields. It is not a popular weed in gardens and can be troublesome to eradicate, particularly from herbaceous plantings.

It was formerly used medicinally in Ireland to treat kidney and genitourinary infections by making an infusion from the bruised roots (Allen & Hatfield, 2004). It was also used to treat bladder problems and to sooth irritated tissues (Moloney, 1919). Jude (1933) mentioned that during World War I in Ireland it was much in demand for medicinal purposes and high prices were paid for it (in 1933 it could be sold for between £40 and £50 a tonne).

O'Regan (1997) mentioned that it is well known in Ireland that pet dogs and cats will eat couch-grass when they have an upset stomach for its laxative effect, and that 'during the war years couch-grass was very much in demand and high prices were fetched for the dried rhizome. It was cultivated here in Ireland, collected in depots such as those that existed at University College, Cork, and fetched prices of £40 to £50 a ton' (O'Regan, 1997: 75). O'Regan also suggested that the creeping rhizomes are nutritious to eat, rather like parsley, but with a stronger taste. White (1803: 130) mentioned that during times of scarcity, the roots were dried and ground into meal and used as 'an agreeable substitute for bread'.

It was used as fodder for donkeys by the people of poorer districts in Ireland during the winter (Moloney, 1919).

Empetrum nigrum

- Crowberry (Scannell & Synnott, 1987); Berry heath (Colgan, 1904); Juniper[90] (Colgan, 1911); Curris-ran (Hart, 1898); She heather (Hart, 1898)
- *Lus na feannóige* (Scannell & Synnott, 1987); *Caor fionnóige* (Moloney, 1919); *Fraoch na bhfraochóg* (Dinneen, 1927); *Fraochóg* (Colgan, 1911); *Fraochán* (Colgan, 1911)

ERICACEAE (Empetraceae)—THE HEATHER FAMILY
Native

88. De Vere Coneys (1849) gave the Irish name *fiothrán (-áin)* for a plant called 'common wheat grass', *Triticum repens*, which is also this species.

89. Wade (1808a) applied this name to *Triticum repens*.

90. This English name was reported by Colgan (1911) from Clare Island, Co. Mayo.

Empetrum nigrum (crowberry). Watercolour by Lydia Shackleton. Courtesy of the National Botanic Gardens, Glasnevin.

Empetrum nigrum is a creeping, heather-like, small shrub that can grow to 120 cm tall that is found throughout the mountains of Ireland growing in acidic bogs and heaths. The black, berry-like (globose) fruits are edible but not tasty. It is widely used for food by local people in Scandinavia and North America and may have been widespread or common enough in parts of Ireland to be a staple food for the first hunter-gatherers. Vickery (1995: 96) mentioned that in Scotland 'the highlanders frequently eat the berries'.

Hart (1898: 377–378) recorded that 'On the first Sunday in August the country people from Derry, Buncrana, and all around gather and go up Slieve Snacht in Inishowen. Hundreds of them go. They gather the berries of *Empetrum nigrum* (not bilberries), and whiskey is not forgot.' They were probably gathered thus in other parts of Ireland, too, particularly as a winter food when other foods were scarce. The berries are retained on the plant throughout the winter and into the following spring. Peterson (1977) recommended that the flavour of these berries is improved by freezing or when cooked with a little lemon juice and served with cream and sugar.

If boiled in alum-water the berries yield a black dye suitable for yarn (Vickery, 1995; MacCoitir, 2006). Alum is an aluminium sulphate compound that is used as a chemical fixative for stains and dyes.

In the Northern Isles of Scotland, crowberry was used to make ropes (Milliken & Bridgewater, 2004), but I have not come across any records of its use in this way in Ireland.

*Epilobium alsinifolium

- Chickweed willowherb (Scannell & Synnott, 1987)
- *Saileachán sléibhe* (Scannell & Synnott, 1987)

ONAGRACEAE—THE WILLOWHERB FAMILY

Native

Epilobium alsinifolium is very rare plant in Ireland, known from only one mountain cliff in Glenade, Co. Leitrim.

*No ethnobotanical uses of this plant have been noted in Ireland.

Epilobium angustifolium (Chamaenerion angustifolium; Chamaerion angustifolium)

- Rose-bay willow-herb (Scannell & Synnott, 1987); Fireweed (Anonymous, 1978); Blooming willie (Hart, 1898); Blooming willow (Hart, 1898); Flowerin' sally (Hart, 1898)
- *Lus na tine* (Scannell & Synnott, 1987); *Rós saileóg na mbileóg chúmhang* (Moloney, 1919); *Saileachán francach* (Ó Dónaill, 1977)

ONAGRACEAE—THE WILLOWHERB FAMILY

Native

Epilobium angustifolium.

Epilobium angustifolium is a common wasteground and garden weed in most Irish towns and cities and found in the countryside in railway embankments, roadsides, bog-margins and waste places. It is also a very rare native plant in the mountains in Ireland.

The leaves and stems are edible and taste best when steamed or boiled (Milliken & Bridgewater, 2004). It is prepared in the same way that one would cook asparagus. The pith within the stems can also be eaten and has a mildly sweet taste of cucumber (Mears & Hillman, 2007). Young leaves can be boiled and eaten as spinach but become more bitter with age; mature leaves can be dried and used to make tea (Peterson, 1977).

Uí Chonchubhair (1995) gave the Irish name *saileachán sraoilleach* for the introduced creeping perennial New Zealand willowherb, *Epilobium brunnescens* (*E. nerter*[*i*]*oides*). It was originally introduced as a garden plant but is now a widespread weed.

Epilobium hirsutum.

*Epilobium hirsutum

- Great willowherb (Scannell & Synnott, 1987)
- *Lus na Tríoinóide* (Scannell & Synnott, 1987); *Saileachán mór* (Uí Chonchubhair, 1995); *Rás mór saileog* (Dinneen, 1927)

ONAGRACEAE—THE WILLOWHERB FAMILY

Native

Epilobium hirsutum is common throughout Ireland in wet habitats, such as ditches, stream sides and damp wastegrounds. It is also occasionally found in drier wasteground habitats.

*No ethnobotanical uses of this plant have been noted in Ireland.

*Epilobium montanum

- Broad-leaved willowherb (Scannell & Synnott, 1987)
- *Saileachán leathan* (Scannell & Synnott, 1987)

ONAGRACEAE—THE WILLOWHERB FAMILY

Native

Epilobium montanum is a common weed throughout Ireland of waste places, gardens, woodlands, hedges and roadsides.

Dinneen (1927) gave the Irish name *rás beag saileog* for 'smooth-leaved willow herb'. It is not clear which species this refers to, perhaps several, including *Epilobium montanum*.

*No ethnobotanical uses of this plant have been noted in Ireland.

*Epilobium obscurum

- Short-fruited willowherb (Scannell & Synnott, 1987)
- *Sealeachán caol* (Scannell & Synnott, 1987)

ONAGRACEAE—THE WILLOWHERB FAMILY

Native

Epilobium obscurum is a common weedy plant throughout Ireland of damp wasteground habitats as well as stream banks, wet rocks and marshes.

*No ethnobotanical uses of this plant have been noted in Ireland.

*Epilobium palustre

- Marsh willowherb (Scannell & Synnott, 1987)
- *Saileachán corraigh* (Scannell & Synnott, 1987); *Luibh na liaithe buí(the)* (Ua Maoileoin, 2001)

ONAGRACEAE—THE WILLOWHERB FAMILY
Native

Epilobium palustre is a common plant throughout Ireland of wet habitats, such as marshes, ditches, stream and river-banks and bogs.

*No ethnobotanical uses of this plant have been noted in Ireland.

*Epilobium parviflorum

- Hoary willowherb (Scannell & Synnott, 1987)
- *Saileachán liath* (Scannell & Synnott, 1987)

ONAGRACEAE—THE WILLOWHERB FAMILY
Native

Epilobium parviflorum is a common plant throughout Ireland of wet habitats, such as marshes and ditches.

*No ethnobotanical uses of this plant have been noted in Ireland.

*Epilobium roseum

- Pale willowherb (Scannell & Synnott, 1987)
- *Saileachán gasach* (Scannell & Synnott, 1987)

ONAGRACEAE—THE WILLOWHERB FAMILY
Probably Introduced

Epilobium roseum is a plant of damp waste places and wet, shady habitats that is rare in Ireland.

*No ethnobotanical uses of this plant have been noted in Ireland.

*Epipactis atrorubens

- Dark-red helleborine (Scannell & Synnott, 1987)
- *Cuaichín dearg* (Scannell & Synnott, 1987)

ORCHIDACEAE—THE ORCHID FAMILY
Native

Epipactis atrorubens is a beautiful orchid of the limestone habitats in the Burren (Cos. Clare and Galway); it is rare elsewhere in Ireland.

*No ethnobotanical uses of this plant have been noted in Ireland.

Epipactis atrorubens.

**Epipactis helleborine*

- Broad-leaved helleborine (Scannell & Synnott, 1987)
- *Ealabairín* (Scannell & Synnott, 1987); *Magarlín cruadh-lus an leathan* (Moloney, 1919)

ORCHIDACEAE—THE ORCHID FAMILY

Native

Epipactis helleborine is a woodland orchid frequently found in parts of the west and north and is rare elsewhere in Ireland.

*No ethnobotanical uses of this plant have been noted in Ireland.

**Epipactis palustris*

- Marsh helleborine (Scannell & Synnott, 1987)
- *Cuaichín corraigh* (Scannell & Synnott, 1987); *Magarlín cruadh-lus an churraigh* (Moloney, 1919)

ORCHIDACEAE—THE ORCHID FAMILY

Native

Epipactis palustris is found throughout Ireland in sand dune slacks, bogs, lake-shores, marshes and other wetland habitats.

*No ethnobotanical uses of this plant have been noted in Ireland.

**Epipactis phyllanthes*

- Green-flowered helleborine (Scannell & Synnott, 1987)
- *Cuaichín glas* (Scannell & Synnott, 1987)

ORCHIDACEAE—THE ORCHID FAMILY

Native

Epipactis phyllanthes is a very rare orchid of woodlands and dune-slacks in the south-east and in the north.

*No ethnobotanical uses of this plant have been noted in Ireland.

Erica species

- Heather; Heath (Scannell & Synnott, 1987)
- *Fraoch* (Scannell & Synnott, 1987); *Ur* (Dinneen, 1927; Uí Chanainn, 1962); *Oir* (Dinneen, 1927; Uí Chanainn, 1962)

ERICACEAE—THE HEATHER FAMILY

**Erica ciliaris*

- Dorset heath (Scannell & Synnott, 1987)
- *Fraoch frainseach* (Scannell & Synnott, 1987)

ERICACEAE—THE HEATHER FAMILY

Probably Introduced

Erica ciliaris is a very rare plant in Ireland, known from only one locality in Connemara near Roundstone, Co. Galway.

*No ethnobotanical uses of this plant have been noted in Ireland.

Erica cinerea

- Bell heather (Anonymous, 1978; Scannell & Synnott, 1987); Heath (Colgan, 1904); Purple heather (Hackney, 1992); Froogh[91] (Hart, 1898); Het[92] (McCarthy & Hawkes, 2007)
- *Fraoch cloigíneach* (Scannell & Synnott, 1987); *Fraoch fireann* (Moloney, 1919; Uí Chonchubhair, 1995); *Fraoch an fhrínse* (Dinneen, 1927); *Fraoch Lochlannach* (Ó Dónaill, 1977)

ERICACEAE—THE HEATHER FAMILY
Native

Erica cinerea is a common plant of dry heaths, peaty banks, moorlands and drier boglands.

This heather was used locally to treat a variety of ailments, including a weak heart, coughs, asthma, rheumatism, stomach troubles and heartburn, and was generally brewed as a tea (Allen & Hatfield, 2004).

'Heath' was used as a scouring agent when scrubbing wooden cottage furniture (Kinmouth, 1993). It can be assumed this common species was used, or *Calluna vulgaris* (ling). Bannister (1965: 541) gave an account of its use from G. Bonnier, 1919–1934, in *Flore complète illustrée en couleurs de France Suisse et Belgique, Vol VII*: 'Flowers of *Erica cinerea* have been used to prepare a yellow-brown dye which becomes dark brown after prolonged boiling and plant extracts have been used as perspirants and diuretic agents.' I have not found records of its use as a dye plant in Ireland.

Erica cinerea.

It is grown as an ornamental on acid soils in Irish gardens. A number of cultivated varieties are available, including several raised in Ireland (Nelson, 2000b).

De Vere Coneys (1849) gave the Irish word *ur* for 'heath', which may perhaps be applied to *Erica* species and *Calluna vulgaris*.

Erica erigena (*Erica mediterranea*; *Erica hibernica*)

- Irish heath (Scannell & Synnott, 1987); Mediterranean heath (Webb & Scannell, 1983)
- *Fraoch camógach* (Scannell & Synnott, 1987)

ERICACEAE—THE HEATHER FAMILY
Probably Introduced

Erica erigena is a rare heath in Ireland, found only in Connemara. Its distribution extends from west Co. Galway across Co. Mayo in wet boglands, stream banks and lake-shores.

A number of different cultivated varieties of this species are grown in Irish gardens (Nelson, 2000b).

91. Hart (1898: 375) recorded the name 'Froogh' for heather in Fanet, Co. Donegal.
92. This name was also used for *Erica tetralix* and *Calluna vulgaris*.

Erica mackayana (*Erica mackaiana*)

- Mackay's heath (Scannell & Synnott, 1987)
- *Fraoch Mhic Aoidh* (Scannell & Synnott, 1987)

ERICACEAE—THE HEATHER FAMILY

Native

Erica mackayana is very rare and found only in bogs and on lake margins associated with deep peat. Populations occur in west Co. Donegal and western parts of Cos. Mayo and Galway, with a small population also found in Co. Kerry.

A number of different cultivated varieties of this species are grown in Irish gardens (Nelson, 2000b).

Erica tetralix

- Cross-leaved heath (Scannell & Synnott, 1987)
- *Fraoch naoscaí* (Scannell & Synnott, 1987); *Fraoch an ruinnse* (Moloney, 1919); *Fraoch naoscaighe* (Moloney, 1919; Dinneen, 1927); *Het* (McCarthy & Hawkes, 2007)

ERICACEAE—THE HEATHER FAMILY

Native

Erica tetralix is a common species throughout Ireland and Britain of bogs, peaty marshes and wet heaths and acid grasslands.

Mahon (1983) recorded that this plant has been used in Ireland to obtain a yellow dye for dyeing textiles.

Cameron (1883) said that a bunch of plants of this species tied together make an excellent scouring brush, but that the other species of *Erica* are too coarse. I have not found records of its use in Ireland in this way although one of its Irish names, *fraoch an ruinnse*, meaning 'rinsing heath', suggests that it may have been used in this way.

A number of different named cultivated varieties of this species are grown in Irish gardens (Nelson, 2000b).

Erica vagans

- Cornish heath (Scannell & Synnott, 1987)
- *Fraoch gallda* (Scannell & Synnott, 1987)

ERICACEAE—THE HEATHER FAMILY

Possibly Native in Co. Fermanagh, but introduced elsewhere

Erica vagans is very rare in Ireland and only known as a possible native population from moorland near Belcoo in Co. Fermanagh. Other populations in Northern Ireland are likely introductions, naturalised from gardens in Cos. Antrim and Down and in scattered locations to the south.

A number of different cultivated varieties of this species are grown in Irish gardens (Nelson, 2000b).

**Erigeron acer*

- Blue fleabane (Scannell & Synnott, 1987)
- *Lus gorm na ndreancaidí* (Scannell & Synnott, 1987); *Lus gorm na ndeanathairtí* (Moloney, 1919; Dinneen, 1927)

ASTERACEAE (Compositae)—THE DAISY FAMILY
Native

Erigeron acer is a rare plant in Ireland of dry, sandy and gravelly banks and grasslands and found mainly in the south-east and centre of the country.

*No ethnobotanical uses of this plant have been noted in Ireland.

**Eriocaulon aquaticum* (*Eriocaulon septangulare*)

- Pipewort (Scannell & Synnott, 1987)
- *Píbín uisce* (Scannell & Synnott, 1987); *Píbín allaidh*[93] (Dinneen, 1927)

ERIOCAULACEAE—THE PIPEWORT FAMILY
Native

Eriocaulon aquaticum is a rare aquatic plant in Ireland and found mainly in shallow water in boggy pools and lake-shores in the west.

*No ethnobotanical uses of this plant have been noted in Ireland.

Eriophorum species

- Bog-cotton (Scannell & Synnott, 1987); Cotton-grass (Dinneen, 1927); Bog-down (Dinneen, 1927); Bent (Hart, 1898); Blackhead[94] (Hart, 1898); Cats' tails (Hart, 1898); Cottonsedge (Hart, 1898); Moss-coach (Hart, 1898); Flowan (Praeger, 1938)
- *Ceannbhán* (Scannell & Synnott, 1987); *Cona* (Cameron, 1883); *Ceannabhán* (Uí Chonchubhair, 1995); *Ceanabhán móna* (de Vere Coneys, 1849); *Sgathóg fhiadhain* (de Vere Coneys, 1849); *Scothóg* (or *Scathóg*) *fhiadhain* (Dinneen, 1927); *Scothóg fionnaidh* (Dinneen, 1927); *Buinneán bán* (Ó Dónaill, 1977); *Canach an tsléibhe* (Ó Dónaill, 1977); *Flocas* (Ó Dónaill, 1977); *Síoda móna* (White, 1803; Ó Dónaill, 1977); *Tóin fhionn* (Ó Dónaill, 1977); *Cennabhán monadh* (White, 1803)

CYPERACEAE—THE SEDGE FAMILY
Native

Including:

Eriophorum angustifolium

- Common cottongrass (Scannell & Synnott, 1987)

---

93. Dinneen (1927) defined *píbín allaidh* as 'jointed pipewort'.
94. Hart (1898: 366) said that 'Eriophora or Cottonsedges are called so about Killybegs [Co. Donegal], before the white plumes show'.

- *Ceannbhán* (Scannell & Synnott, 1987); *Ceannabhán sléibhe* (Ua Maoileoin, 2001); *Bláth portaigh* (Moloney, 1919)

Eriophorum gracile

- Slender cottongrass (Scannell & Synnott, 1987)
- *Ceannbhán caol* (Scannell & Synnott, 1987); *Ceannbhán cumhang-dhuilleach* (White, 1803; Dinneen, 1927)

Eriophorum latifolium

- Broad-leaved cottongrass (Scannell & Synnott, 1987)
- *Ceannbhán leathan* (Scannell & Synnott, 1987); *Cennabhán lethandhuillech* (White, 1803); *Keanuan ban* (White, 1803); *Scabog fiona* (White, 1803); *Ceanabhan mona* (White, 1803); *Ceanach na mona* (White, 1803); *Sioda mona* (White, 1803); *Siodha mona* (White, 1803); *Sgathoga fiona* (White, 1803)

Eriophorum vaginatum

- Hare's-tail cottongrass (Scannell & Synnott, 1987)
- *Ceannbhán gaelach* (Scannell & Synnott, 1987); *Tóin fhionn*[95] (Ó Dónaill, 1977)

Of the four *Eriophorum* species found in Ireland, two of them (*E. angustifolium* and *E. vaginatum*) are common and widespread throughout and grow in bogs, moorlands and other peaty habitats. The other two are much rarer: *E. gracile* is mainly found in Connemara and very rarely distributed elsewhere, and *E. latifolium*, although more widespread, is only very rarely found, in very wet swamps and bogs. Somewhat surprisingly, for such a conspicuous plant, there are few records of the use of bog-cotton in Ireland.

When mature, the fruiting heads of these species include a dense tuft of silky white tufts of cotton-like fibres. These fibres were said to have been collected in the poorest parts of Ireland and used for spinning simple threads and to stuff pillows (when goose-down [feathers] were not available) (White, 1803; Sharkey, 2000). They were also used for stuffing mattresses (O'Súilleabháin, 1942). Mattresses were traditionally stuffed with a range of other local materials, too, such as straw, chaff, hay, rushes, sawdust, feathers, hair and other materials enclosed in a cloth bag. Danaher (1964a) wrote that there is a tradition in Ireland that the fibres from *ceannabhán móna*, the bog cotton, were formerly spun and woven into cloth, and there is a long traditional story about a princess who made 12 shirts from bog-cotton for her brothers to free them from a spell. Simpson and Inglis (2001) indicated the fibres of *Eriophorum angustifolium* as having been used in the United Kingdom to make paper, candle wicks and to stuff pillows, and *E. vaginatum* to make yarn and fabric. It was also used to make candle wicks for poor people, 'but it becomes brittle when very dry' (White, 1803: 6).

Bog fibre derived from peat, and used to make cloth, was probably composed mainly of the stem and leaf fibres of *Eriophorum* species (see also *Sphagnum*). Milliken and Bridgewater (2004) pointed out that in the Great Exhibition of 1851 held in London, several items of clothing made from bog-cotton down were exhibited. These had been woven by women in Ross and Inverness-shire in Scotland and were said to have been much admired for their beauty and fine texture. Lady Doneraile, of Doneraile Court, Co. Cork (see entry for *Chrysanthemum leucanthemum*), sent insoles for shoes made from bog-cotton fibres to the Royal Botanic Gardens, Kew, in the 1850s for its Economic Botany Museum collections (Nelson, 2009).

Eriophorum angustifolium, *E. latifolium* and *E. vaginatum* were all listed by Huxley (1992) as ornamental plants used in horticulture, but I can find no records of their use in Irish gardens.

95. Ó Dónaill (1977) defined *tóin fhionn* as 'hare's-tail'.

Erodium cicutarium

- Common stork's-bill (Scannell & Synnott, 1987); Crow's toe (Colgan, 1904); Hemlock storksbill (Praeger, 1938)
- *Creagach* (Scannell & Synnott, 1987)

GERANIACEAE—THE GERANIUM FAMILY
Native

Erodium cicutarium is an occasional plant throughout Ireland of dry, sandy banks and waste places, and mainly distributed in coastal regions.

This species is edible, and young leaves may be used in salads or as a vegetable, lightly steamed (Mears & Hillman, 2007).

---

*Erodium maritimum

- Sea stork's-bill (Scannell & Synnott, 1987)
- *Creagach mara* (Scannell & Synnott, 1987)

GERANIACEAE—THE GERANIUM FAMILY
Native

Erodium maritimum is rare in Ireland, found near the sea on the south and east coasts in dry and sandy banks, waste places and short maritime grassland.

*No ethnobotanical uses of this plant have been noted in Ireland.

---

*Erophila verna

- Common whitlowgrass (Scannell & Synnott, 1987)
- *Bosán anagair* (Scannell & Synnott, 1987); *Biolar greagáin* (Moloney, 1919)

BRASSICACEAE (Cruciferae)—THE CABBAGE AND CRESS FAMILY
Native

Erophila verna is a small annual plant found throughout Ireland of dry banks, walls and waste places.

*No ethnobotanical uses of this plant have been noted in Ireland.

Eryngium maritimum

- Sea holly (Scannell & Synnott, 1987)
- *Cuileann trá* (Scannell & Synnott, 1987); *Cuileann duimhche* (Ó Dónaill, 1977; Ua Maoileoin, 2001); *Cuileann trágha* (Moloney, 1919)

APIACEAE (Umbelliferae)—THE CARROT FAMILY
Native

Eryngium maritimum is a frequent and conspicuous plant of sand dunes and sandy habitats all around the coast of Ireland.

Eryngium maritimum.

The roots can be eaten boiled, roasted or used to make jelly (Phillips, 1983) and were regarded as having useful medicinal properties (for flatulence). Mackay (1836: 129) recorded that the roots 'are well-tasted when candied'. Uí Chonchubhair (1995) also recorded their use candied as a sweetmeat, and that they were regarded as a remedy for impotence. Mabey (1972) mentioned that candied *Eryngium* was a vital ingredient of marrow-bone pie, an Elizabethan dish. Rutty (1772: 53) said that 'the tops before flowering, prepared like Asparagus, are of grateful taste, and eaten, not only candied as a restorative, but boiled, like skirrets [carrots] or parsnips; and Pies, of a pleasant taste and flavour, are made of it baked, and with apples'.

Allen and Hatfield (2004) recorded that Roderic O'Flaherty (in O'Síocháin, 1962) said it was used medicinally in the Aran Islands in 1684 to treat worms in children and as a general prophylactic, and was still in use in this way in the 1920s.

Euonymus europaeus (spindle). Watercolour by Lydia Shackleton. Courtesy of the National Botanic Gardens, Glasnevin.

Euonymus europaeus

- Spindle (Scannell & Synnott, 1987)
- *Feoras* (Scannell & Synnott, 1987); *Oir (Oire)*[96] (de Vere Coneys, 1849; Moloney, 1919); *Donn réidh*[97] (Dinneen, 1927); *Crann-feorais* (Ó Dónaill, 1977)

CELASTRACEAE—THE SPINDLE FAMILY

Native

Euonymus europaeus is a small tree of woodlands, thickets, hedgerows, lake-shores and rocky ground, usually on limestone soils, that is occasional and locally frequent throughout Ireland.

The spindle-tree was used in Irish folk medicine as a purgative, emetic and liver stimulant, and the bark of this tree was also used for a treatment of head lice in children (Moloney, 1919).

Spindle wood is hard and tough and used for making a variety of objects, such as shoemaker's pegs, toothpicks, knitting needles, skewers and, as its name suggests, spindles (Mackay, 1836; Edlin, 1973; MacCoitir, 2003). Spindle was used in Ireland for the manufacture of bows for violins, keys for virginals, and at least one harp made in Ireland is in part made from spindle wood (Rimmer, 1969; Nelson & Walsh, 1993). According to Mackay (1836), Linnaeus reported that it made the best charcoal for drawing. Rutty (1772: 112) said that it supplied skewers for butchers and made good 'Tooth-pickers' and was used by watchmakers for cleaning

96. De Vere Coneys (1849) and Dinneen (1927) also defined *oir* (*oire*) as 'broom' (*Cytisus*) and 'furze' (*Ulex*).
97. Dinneen (1927) defined *donn réidh* as 'peg wood, the spindle tree'.

watches, being tough and not easily broken, 'for which purpose it is cut when the tree is in flower, and then it will be tough, being otherwise too hard and brittle'.

Spindle-trees have been used for the preparation of a variety of dyes; red from the pink parts of the fruits, yellow from the orange seeds and green when the seeds are boiled in alum, which is an aluminium sulphate compound used to fix the colour of the dye (Edlin, 1973).

Woodward (1985) lists this plant as poisonous. No parts of the plant should be eaten. In particular, the brightly coloured fruits are attractive, but are rich in alkaloids such as theobromine and caffeine, and variably toxic by weight.

**Eupatorium cannabinum*

- Hemp agrimony (Scannell & Synnott, 1987); Crow-rocket (Hart, 1898)
- *Cnáib uisce* (Scannell & Synnott, 1987); *Scothóg Mhuire* (Dinneen, 1927; O'Rourke, 2006); *Marbhdhraighean* (Moloney, 1919)

ASTERACEAE (Compositae)—THE DAISY FAMILY

Native

Eupatorium cannabinum is a frequent plant throughout Ireland of river and stream banks, marshes and rocky places; it is more common in the south than the north.

*No ethnobotanical uses of this plant have been noted in Ireland.

Eupatorium cannabinum in Inisheer.

Euphorbia species

- Spurge (Scannell & Synnott, 1987)
- Moloney (1919) listed the following Irish names for spurges in Ireland but does not apply them to specific species (other than that they refer to *Euphorbia hyberna, E. helioscopia, E. peplus* and *E. exigua*): *bainne caoin; lus an leasaidh*[98]; *spuirse*[99]; *buidhe na ninghean*; *meacan buidhe an tsléibhe*[100]; *géarnabh; esbearna*; *dear-glaoich* (bog spurge); *bainne na néan*[101].

EUPHORBIACEAE—THE SPURGE FAMILY

Including:

Euphorbia helioscopia

- Sun spurge (Scannell & Synnott, 1987); Wart weed (Colgan, 1904); Fairy's milk (Colgan, 1904); Pig's milk (Colgan, 1904); Devil's churnstaff (Hart, 1898; Praeger, 1938); Seven sisters (Hart, 1898)
- *Lus na bhfaithní* (Scannell & Synnott, 1987); *Bainne cích na n-éan*[102] (Ua Maoileoin, 2001); *Fionneamh lus* (Cameron, 1883); *Buí na n-iníon* (Uí Chonchubhair, 1995)

Possibly Introduced (possible archaeophyte)

98. *Bainne caoin* and *lus an leasaidh* were also given by Ó Dónaill (1977) to mean 'spurge'.
99. *Spuirse* was also given by de Vere Coneys (1849), Dinneen (1927) and Ó Dónaill (1977).
100. *Meacan buidhe an tsléibhe* was given by Dinneen (1927) to mean 'mountain or knot-rooted spurge'.
101. *Buína n-iníon, na mbáb* was also given by Ó Dónaill (1977) to mean 'spurge'.
102. Mairín Uí Chonchubhair (pers. comm., 2011) tells me that this name is applied to any plant with white milky latex, such as *Taraxacum officinale* (dandelion).

Euphorbia hyberna

- Irish spurge (Scannell & Synnott, 1987); Barnicy Ann (McCarthy & Hawkes, 2007)
- *Bainne caoin* (Scannell & Synnott, 1987); *Buidhe na n-ingean* (de Vere Coneys, 1849; Dinneen, 1927); *Bainnicín* (Danaher, 1964a); *Meacan buidhe an t-sleibhe* (de Vere Coneys, 1849); *Lasair léana* (Dinneen, 1927)

Native

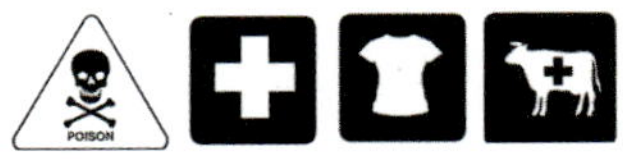

Euphorbia hyberna is a widespread plant of woodlands, damp banks, riversides and rocky habitats in the south and west of Ireland and is particularly common in Cos. Cork and Kerry. It is very rare elsewhere in Ireland.

It was used to knock out fish in Kerry and elsewhere (Mackay, 1836; Barrington, 1906). Scully (1916) reported that this practice continued in Co. Kerry during his time too, when it was still used to poach trout from the rivers. A basket was filled with bruised plants, which were then lowered into the river or pool. Fish are affected by the toxins in the water up to several miles down river. According to Danaher (1964a), another method was as follows: a canvas sack was filled with the plant and then pounded until the contents were beaten into a juicy mass and then dropped into the river when every fish—eels, minnows, sticklebacks, as well as trout and salmon—came floating dead to the surface for hundreds of yards downstream. Sometimes the poison affected miles of water. Danaher wrote that once in a while cases of rivers being poisoned with this plant were heard about, perhaps once every four or five years. He also noted that this was regarded as blackguardism and not as honest poaching. According to Clyne (1989), another species, *Euphorbia paralias*, the sea spurge, was also used in Ireland for such purposes.

The juice or latex of this and other *Euphorbia* species was sometimes applied to warts (Moloney, 1919). Allen and Hatfield (2004: 170) recorded that it is a powerful purgative—they noted that 'a medical correspondent of John Ray's, writing from Tipperary in 1697, reported that the root was sometimes boiled in Ireland to produce a purge, but was liable to give rise to severe convulsions, followed by early death'. It was also used to cure diarrhoea. Hart (1873: 339) noted that when it was given to a man in Gort, Co. Galway, he 'ran up and down the street like a madman, and swelled so big that his friends had to bind him round with hay-ropes lest he shall burst'.

Hart (1873) and Allen and Hatfield (2004) noted its use as a purgative in veterinary treatments for cattle and horses in Galway.

Barrington (1906) recorded its use as a black dye for wool.

Euphorbia helioscopia.

Euphorbia hyberna.

Also including:

Euphorbia paralias

- Sea spurge (Scannell & Synnott, 1987); Fairy's milk (Colgan, 1904)
- *Bainne léana* (Scannell & Synnott, 1987)

Native

Euphorbia peplus

- Petty spurge (Scannell & Synnott, 1987); Seven sisters (Hart, 1898)
- *Gearr nimhe* (Scannell & Synnott, 1987)

Possibly Introduced (possible archaeophyte)

Euphorbia helioscopia and *E. peplus* are both common annual weeds of cultivation occurring widely in tilled fields and waste places. *Euphorbia peplus* is also a common weed of gardens. *Euphorbia paralias* is a coastal plant found in sand dunes and sandy shores, especially in the east of Ireland.

Euphorbias are poisonous but formerly used medicinally in Ireland for their cathartic properties. Logan (1972, 1981) reported that *meacan buí tsléibhe* (the mountain spurge—perhaps *Euphorbia hyberna*) was used to treat diarrhoea. K'Eogh (1735: 117) said that it is a good, but dangerous, cathartic plant as 'it maketh the belly soluble, expelleth viscous phlegm and choleric humours' and 'dissolveth flatulences in the bowels, and cureth the hardness of the spleen'. The milky, white latex was applied to treat warts (Mackay, 1836; Colgan, 1904). *Euphorbia peplus* (petty spurge) was probably also widely used as a cure for warts.

Euphorbias were also used as a veterinary remedy to treat blackleg in Co. Cavan (Allen & Hatfield, 2004).

In the Isle of Man, Allen and Hatfield (2004) noted that Euphorbias were used as a sexual stimulant, giving rise to its occasional common name there, 'Saturday-night-pepper'. The juice was applied to the head of the penis, causing it to swell, to enhance the sexual pleasure and erotic experience. Vickery (1995: 362–363) mentioned that it was *Euphorbia helioscopia* that was used for such a purpose, called *lus y bwoid mooar* in the Manx language, which translated means 'the plant of the big knobs' or 'the plant of the big penis'. Apparently, if this caused excessive discomfort, the organ could be dipped in milk (soured milk was usually recommended).

Clyne (1989) said that *Euphorbia paralias*, the sea spurge, was used in Ireland to stupefy fish in rivers, as a means of poaching (see *E. hyberna*).

Housewives in Co. Kerry extracted a black dye from *Euphorbia* species in Co. Kerry (Williams, 1993).

Euphrasia officinalis agg.

- Eyebright (Scannell & Synnott, 1987); Fairy flax (Hart, 1898); Rock rue (Hart, 1898)
- *Glanrosc* (Scannell & Synnott, 1987); *Driofúr bhán* (Ua Maoileoin, 2001); *Líon radharc* (Moloney, 1919; Ó Dónaill, 1977); *Soillse na súl* (de Vere Coneys, 1849; Moloney, 1919; Dinneen, 1927); *Lus na leac* (de Vere Coneys, 1849; Moloney, 1919; Dinneen; 1927); *Radharcain (-aircín)* (de Vere Coneys, 1849; Moloney, 1919; Dinneen, 1927); *Caoimín* (de Vere

Euphrasia officinalis.

Coneys, 1849; Dinneen, 1927); *Linradharc (-radhairc)* (de Vere Coneys, 1849; Dinneen, 1927); *Lus an Phápa* (Dinneen, 1927); *Rinn roisc* (Ó Dónaill, 1977)

OROBANCHACEAE (Scrophulariaceae)—THE BROOMRAPE FAMILY

Native

Including:

Euphrasia anglica

Euphrasia arctica

Euphrasia confusa

Euphrasia frigida

Euphrasia micrantha

Euphrasia nemorosa

Euphrasia pseudokerneri

Euphrasia rostkoviana

Euphrasia salisburgensis

- Irish eyebright (Scannell & Synnott, 1987)
- *Glanrosc gaelach* (Scannell & Synnott, 1987); *Róisnín radhairc* (O'Rourke, 2006); *Lus na súl* (O'Rourke, 2006)

Euphrasia scottica

Euphrasia tetraquetra

There are 11 different species of eyebright recorded in Ireland: *Euphrasia anglica* is widespread and grows in rough glasslands, heaths and pastures; *E. arctica* is the commonest species in Ireland and occurs in disturbed grasslands such as field-margins; *E. confusa* is an occasional species of heaths and coastal sand-dune grasslands; *E. frigida* is a rare plant found on mountain cliffs, rock-ledges and exposed sites by the sea, mainly in the west; *E. micrantha* is widespread in Ireland, growing in heaths and moorland, and occasionally in bogs, most commonly in the north and west; *E. nemorosa* is a widespread species found in heaths, woods and pastures; *E. pseudokerneri* is found in limestone grassland in the west and nowhere else in Ireland; *E. rostkoviana* is a species of moist meadows that is widespread in Ireland but nowhere common; *E. salisburgensis* is found on coastal cliffs, amongst limestone rocks and sand dunes in the west of Ireland only (from Cos. Limerick to Donegal); *E. scottica* is occasionally found in Ireland, particularly in the west and north, growing in wet mountain moors, slopes and grasslands; and *E. tetraquetra* is most frequent in coastal localities such as sand-dune grasslands, cliff-tops and other grasslands near the sea, and rarely inland. Most occur in a variety of different grasslands where they are hemiparasitic annual plants.

The identification of Irish eyebrights to species is a difficult task even for an expert, and there has probably been little differentiation between the range of Irish species in folk medicine in Ireland.

Eyebrights have been widely used in Europe in traditional medicine, particularly to treat eye complaints, coughs, hay fever and sinusitis, as well as externally for skin

Euphrasia salisburgensis.

Eyebright (*Euphrasia officinalis*).

problems (van Wyk & Wink, 2004). Eyebrights provided an astringent used in the treatment of respiratory and eye problems and are particularly associated with the treatment of measles (Moloney, 1919).[103] Mackay (1836: 202) said that eyebright is much used 'in rustic practice as a remedy for diseases of the eye'. Uí Chonchubhair (1995) noted that eyebright (and chamomile) were applied to treat a sty and the species was also used to treat conjunctivitis and eye inflammations.

It has been used as a veterinary remedy to treat red-water fever in Co. Donegal (Allen & Hatfield, 2004). O'Regan (1997) mentioned that eyebright was used to treat weak kittens that were very sick with eyes and noses blocked with mucus. An infusion of eyebright was made and their eyes bathed with the lotion until the kittens were cured.

103. Another traditional remedy for measles in Ireland was the juice of sheep's droppings, steeped, which was given to children (*cróch na mbánta* or *cróch na gcaorach*; Dinneen, 1927).

Fagopyrum esculentum (Polygonum fagopyrum)

- Buckwheat (Scannell & Synnott, 1987)
- *Lus na gcearc* (Scannell & Synnott, 1987); *Gluneagh romhan* (Wade, 1808b); *Maol-chruithneacht* (Dinneen, 1927); *Ruadhán* (Dinneen, 1927)

POLYGONACEAE—THE KNOTWEED FAMILY
Introduced

Fagopyrum esculentum is a rare plant in Ireland and usually found as a rare casual plant or remnant of former cultivation. It was formerly grown as a crop, particularly for feeding game, such as pheasants (Reynolds, 2002). It is edible and used to make a porridge-like dish that is widely eaten in parts of northern Europe and Asia. (I had buckwheat for breakfast in Siberia.) Russia was formerly a major centre for its cultivation. Rutty (1772) said that in his time it had been cultivated near Tallow, Co. Waterford, by Dr Hoadley, late Archbishop of Dublin.

It is a fast-growing crop and so has been a popular choice in countries in northern latitudes where the summer and growing season is short. It has also been used as a green manure.

Fagus sylvatica

- Beech (Scannell & Synnott, 1987)
- *Feá* (Scannell & Synnott, 1987); *Fáibhile* (De Bhaldraithe, 1959; Ó Dónaill, 1977); *Feagha* (de Vere Coneys, 1849; Dinneen, 1927); *Fáigh* (-e) (Dinneen, 1927); *Fáidhbhile* (Uí Chanainn, 1962)[104]

FAGACEAE—THE BEECH, OAK AND CHESTNUT FAMILY
Introduced

Fagus sylvatica is widely grown in woods and hedges in Ireland and is often self sown and naturalised. It is not a native tree in Ireland but was probably introduced by the Normans (Nelson & Walsh, 1993). Many fine beech trees can be found throughout Ireland. The largest known is a specimen at Hilton Park, Scotstown, Co. Monaghan, that is 39 m tall and 5.13 m in girth (Anonymous, 2005). The current girth champion of this species in Ireland is a specimen from Stradbally Hall, Stradbally, Co. Laois, that has been measured at 7.9 m in girth (and 26 m in height) (Anonymous, 2005).

Young beech leaves are recorded as edible and can be used in salads. Beech nuts are edible too, but hard to peel and process; they can be toxic if eaten in large quantities, though. The beech nuts (called 'mast') are produced in quantity only every three or four years. In the tree's native range, beech mast was an important food source for pigs. A warning though, Woodward (1985: 64) lists this species as poisonous and gives the symptoms of poisoning as 'severe abdominal pain, nausea, diarrhoea; in extreme cases convulsions and death'.

In some parts of Europe beech nuts have been roasted, ground and used as a coffee substitute, and when crushed, they provide an oil similar to olive oil that is useful for

104. Uí Chanainn (1962) defined *fáidhbhile* as 'beech tree'.

salad dressings (Hilton, 2007). This oil was formerly made into butter and was apparently sold commercially in Britain (Ó Céirín & Ó Céirín, 1978). In his classic book *Food for Free*, Richard Mabey (1972) described how children were given the task of gathering beech mast during World Wars I and II in some areas of Germany as a source of domestic oil, which would be extracted in a communal village mill. He mentioned also that the oil can be used for salads and frying, and that beech nut butter is still made in some rural parts of the United States. It is not that difficult to extract the oil from beech nuts at home. The nuts can be boiled and the oil skimmed off the surface of the cooked mixture. I know of no such use of beech-derived products from domestic sources in Ireland.

Fagus sylvatica.

Ó Céirín and Ó Céirín (1978) provided a recipe for beech leaf liqueur that is probably English or French in origin. The young leaves are gathered in spring and stuffed into a glass jar or earthenware crock, which is then topped up with gin. After several weeks the liquid is strained and sugar dissolved in a little hot water or honey is added. I have not found evidence of this liqueur being made in Ireland.

Beech timber is strong, durable, fine-grained and widely used in cabinet making. In Britain beech has been traditionally used to make furniture such as chairs, as it is easy to bend and turn. Beech wood was also used to make butter-firkins in Co. Cavan (Ní Sheasnáin, 1984). Rutty (1772: 112) said that Irish beech timber is superior to English and foreign timber because of its hardness and 'variegation'.

According to Rutty (1772: 112), beech leaves make the best mattresses in the world 'to lay under the quilts instead of straw, because, beside their tenderness and loose lying together, they continue seven or eight years, before which straw becomes musty and hard'. The beech leaves have to be collected for such purposes before they become 'frost-bitten'.

Beech is also a popular hedge plant. It is slow growing and requires trimming only once a year. Beech hedges retain their dead leaves during the winter, providing privacy throughout the year. A fine beech hedge probably more than 100 years old surrounds the Director's Residence in the National Botanic Gardens, Glasnevin.

Fallopia convolvulus (Polygonum convolvulus)

- Black bindweed (Scannell & Synnott, 1987); Wild ivy (Colgan, 1904)
- *Glúineach dhubh* (Scannell & Synnott, 1987)

POLYGONACEAE—THE KNOTWEED FAMILY

Possibly Introduced (possible archaeophyte)

Fallopia convolvulus is a common weed of roadsides, wasteground, cultivated fields and hedgerows.

Mears and Hillman (2007) suggested that the seeds of this species are edible and can be made into dampers, a simple form of unleavened bread.

Fallopia japonica (Polygonum cuspidatum; Reynoutria japonica)

- Japanese knotweed (Scannell & Synnott, 1987); Hudsons[105]
- *Glúineach bhiorach* (Scannell & Synnott, 1987)

POLYGONACEAE—THE KNOTWEED FAMILY

Introduced

Fallopia japonica is native to eastern Asia (Japan, northern China, Taiwan and Korea) and is an extremely invasive and increasing weed of roadsides, waste places, woodlands, rubbish dumps and many habitats throughout Ireland. It is very difficult to control or eradicate. The roots can grow to 3 m in depth, and the plant will grow back from the tiniest fragment. It can even grow through tarmac and concrete! In Northern Ireland it is a prohibited species under the Wildlife (Northern Ireland) Order of 1985.

It is said to be a useful anti-oxidant plant. A colleague and friend in Germany recently told me that it is an edible species, and young shoots can be peeled and used, rather like rhubarb, to make chutney, jam, curries, soup and crumble, and as a vegetable (Frank Klingenstein, pers. comm., 2008). This may be one way of controlling it, to make extensive use of it as a wild food. (If you can't beat it, eat it!)

Jam from Japanese knotweed is made as follows: 3 kg of young shoots (from clean habitats) are combined with 2 kg sugar, 60 g of pectin and 150 mL of lemon juice. All the ingredients are left overnight, then boiled until setting point is reached. The shoots can also be pickled: 300 g of young shoots are mixed with 20 g of salt and left for four or five days; they are then washed and dried and placed in a jar with a few cloves, some mustard seeds, peppercorns and a bay leaf and filled with 80 mL of wine vinegar and 80 mL of white wine. The pickled shoots are ready to eat after a few weeks and go well with fish.

The young shoots picked when they are about 1 ft (30 cm) long can be steamed or boiled and served like asparagus. If the shoots are left too long, or the lower parts used, they can become tough and stringy. I generally cut away and discard the leaf nodes too. Peterson (1977) suggested that if they are a bit tart for some tastes, they can be sweetened with a little sugar.

Mícheál de Mórdha (pers. comm., 2011) told me that as a young boy growing up in West Kerry he would pick and eat raw the young shoots of this species.

A close relative and another species of giant knotweed native to Japan is *Fallopia sachalinensis*; it is also an invasive former garden plant. It, too, is prohibited in Northern Ireland under the Wildlife (Northern Ireland) Order of 1985.

**Festuca altissima*

- Wood fescue (Scannell & Synnott, 1987)
- *Feisciú coille* (Scannell & Synnott, 1987)

POACEAE (Gramineae)—THE GRASS FAMILY

Native

Festuca altissima is a rare grass in Ireland of woodlands and shady places.

*No ethnobotanical uses of this plant have been noted in Ireland.

105. *Fallopia japonica* is called 'Hudsons' in West Kerry in the Dunquin region. Apparently, it was introduced to the area as a garden plant by Mrs Hudson, the wife of a Kerry doctor who practised in the area in the early part of the 1920s (Dáithí de Mórdha & Mícheál de Mórdha, pers. comm., 2011).

Festuca arundinacea

- Tall fescue (Scannell & Synnott, 1987)
- *Feisciú ard* (Scannell & Synnott, 1987); *Fescufhér airde* (White, 1803)[106]

POACEAE (Gramineae)—THE GRASS FAMILY
Native

Festuca arundinacea is a common grass throughout Ireland of a wide variety of grassy habitats.

The seeds of this species are edible, and Mears and Hillman (2007) suggested that early people in Britain and Ireland may have used it as a staple source of calories.

**Festuca gigantea*

- Giant fescue (Scannell & Synnott, 1987)
- *Feisciú capaill* (Scannell & Synnott, 1987)

POACEAE (Gramineae)—THE GRASS FAMILY
Native

Festuca gigantea is a widespread grass throughout Ireland, although not commonly distributed everywhere, of woodlands and shady places.

*No ethnobotanical uses of this plant have been noted in Ireland.

Festuca species

- Fescue (Scannell & Synnott, 1987)
- *Feisciú* (Scannell & Synnott, 1987); *Féiste* (Cameron, 1883)

POACEAE (Gramineae)—THE GRASS FAMILY
Native

Including:

Festuca ovina

- Sheep's-fescue (Scannell & Synnott, 1987)
- *Feisciú caorach* (Scannell & Synnott, 1987); *Feur caorach* (Moloney, 1919; Dinneen, 1927); *Fescufheur caorach* (Wade, 1808a); *Féar caorach* (Ó Dónaill, 1977); *Fescufhér caorach* (White, 1803)

Festuca pratensis

- Meadow fescue (Scannell & Synnott, 1987)
- *Feisciú móinéir* (Scannell & Synnott, 1987); *Fescufheur leana* (Wade, 1808a); *Fescufhér léna* (White, 1803)

Festuca ovina is a common grass throughout Ireland of dry heaths and grasslands, seashores and mountains. *Festuca pratensis* is more frequent in damp grasslands and moist shady

106. White (1803) applied this name to *Festuca elatior*.

Festuca rubra.

Festuca vivipara.

habitats, such as roadside hedgebanks, and is most commonly found in the north-east and centre of Ireland.

Both species are used in some commercial wildflower seed mixes.

According to Clyne (1989), *Festuca ovina* is reputed to provide fine and sweet grazing for sheep, producing good mutton (hence its name, sheep's-fescue). There is no doubt that excellent lamb is produced on the mountains of Ireland, but whether this is thanks to their eating *F. ovina* or not is certainly debatable.

Wade (1808a) gave the Irish name *fescufheur durunta* for *Festuca duriuscula*, a synonym or variety of *F. ovina*.

Festuca rubra

- Red fescue (Scannell & Synnott, 1987); Creeping red fescue (Webb, 1977)
- *Feisciú rua* (Scannell & Synnott, 1987); *Fescufhér cruan* (White, 1803); *Fescufhér cruadh*[107] (White, 1803); *Fescufhér coilltemhuil*[108] (White, 1803)

POACEAE (Gramineae)—THE GRASS FAMILY

Native

Festuca rubra is an abundant grass in Ireland of dry grasslands, waste places, sand dunes, walls and lawns.

It is widely used in many commercial seed mixes as a lawn and fodder grass and in wildflower mixes.

**Festuca vivipara*

- Viviparous fescue (Scannell & Synnott, 1987)
- *Feisciú beobhreitheach* (Scannell & Synnott, 1987); *Féar caorach*[109] (Uí Chonchubhair, 1995); *Fescufhér beothuismighthach* (White, 1803)

POACEAE (Gramineae)—THE GRASS FAMILY

Native

Festuca vivipara is generally found in the mountains in Ireland except in some parts of the west where it is more widespread.

*No ethnobotanical uses of this plant have been noted in Ireland.

Ficaria verna (Ranunculus ficaria)

- Lesser celandine (Scannell & Synnott, 1987); Pilewort (Dinneen, 1927; Praeger, 1938; Ó Dónaill, 1977)
- *Grán arcáin* (Scannell & Synnott, 1987); *Searraigh* (de Vere Coneys, 1849; Moloney, 1919); *Gráin arcáin* (Moloney, 1919); *Na searraigh* (Uí Chonchubhair, 1995); *Lus na gcnapán* (Ó Dónaill, 1977)

107. White (1803) applied this name to *Festuca duriuscula*, which is a synonym of *F. rubra*.
108. White (1803) applied this name to *Festuca dumetorum*, which is a synonym of *F. rubra*.
109. This name is generally applied by other authors to *Festuca ovina*.

RANUNCULACEAE—THE BUTTERCUP AND CROWFOOT FAMILY
Native

Ficaria verna is a common woodland plant throughout Ireland. It also occurs in other damp, shady habitats, such as hedges and wild gardens.

It was used in the preparation of an ointment for the treatment of piles (Moloney, 1919; Maloney, 1972; Allen & Hatfield, 2004) and so has sometimes been called 'pile-wort' in Ireland (Dinneen, 1927; Ó Dónaill, 1977).

Milliken and Bridgewater (2004) speculated that in Scotland the tubers and bulbils of lesser celandine may have been used for food by some of the earliest settlers, as charred remains of these have been found in deposits in Oronsay dating to the Mesolithic period. Although most *Ranunculus* species contain ranunculin and are poisonous, cooking or drying may break down the toxic compounds. For native people in North America, most uses were medicinal, but some of the roots, plants and seeds of some species were eaten as well, after either boiling or pulverising dried material (Moerman, 1998). Livestock avoid the acrid taste of buttercups, which causes blisters and gastrointestinal problems, but dried hay containing them is safe. Contact dermatitis in both humans and animals can happen if too much fresh plant material is handled.

Ficus carica

- Fig (Reynolds, 2002)
- *Crann fíge* (*fíghis*) (Cameron, 1883); *Cran-fíge* (de Vere Coneys, 1849; Dinneen, 1927); *Crann figí* (Ó Dónaill, 1977)

MORACEAE—THE MULBERRY AND FIG FAMILY
Introduced

The fig is a common garden tree, grown traditionally in larger old Irish gardens against walls. Although relatively hardy, the figs don't often ripen sufficiently to make them sweet and edible during Ireland's short summers, and home-grown figs are not a regular treat.

Fig trees only rarely become established in the wild in Ireland but have been recorded on several occasions in the wild, such as near Mountjoy Jail on the banks of the Royal Canal in Dublin in 1980 (Wyse Jackson & Sheehy Skeffington, 1984); on sea cliffs at Tramore, Co. Waterford, in 2005 (Green, 2008); on wasteground in Cork city in 1990; and on a wall at Kingscourt, Co. Cavan, in 1980 (Reynolds, 2002). These plants probably all grew from seed. If climate change predictions are correct in Ireland, this species could very well become a much more frequent wild plant.

The seeds of figs have been excavated from ancient sediments in Dublin city dating back to the 13th century when they must have been a rare delicacy and a luxury import (Anonymous, 1973).

In parts of Europe, the leaves of *Ficus carica* have been smoked for recreational purposes (Pennacchio et al., 2010), but apparently not in Ireland.

**Filaginella uliginosa (Gnaphalium uliginosum)*

- Marsh cudweed (Scannell & Synnott, 1987)
- *Gnamhlus corraigh* (Scannell & Synnott, 1987); *Liathlus* (Uí Chonchubhair, 1995)

ASTERACEAE (Compositae)—THE DAISY FAMILY
Native

Filaginella uliginosa is found in moist places and shady ground throughout Ireland, except in parts of the midlands.

*No ethnobotanical uses of this plant have been noted in Ireland.

**Filago minima (Logfia minima)*

- Small cudweed (Scannell & Synnott, 1987)
- *Cáithluibh bheag* (Scannell & Synnott, 1987)

ASTERACEAE (Compositae)—THE DAISY FAMILY
Native

Filago minima is a rare plant in Ireland of gravel and sandy banks and other places that is distributed mainly in the south and east.

*No ethnobotanical uses of this plant have been noted in Ireland.

Filago vulgaris (Filago germanica)

- Common cudweed (Scannell & Synnott, 1987)
- *Cáithluibh* (Scannell & Synnott, 1987); *Cádhluibh* (Moloney, 1919); *Gnabhlus* (de Vere Coneys, 1849; Moloney, 1919); *Liath lus roid* (Cameron, 1883); *Gnamh-lus* (Dinneen, 1927)

ASTERACEAE (Compositae)—THE DAISY FAMILY
Native

Filago vulgaris is a very scarce plant in Ireland of dry habitats, such as sandy banks.

Rutty (1772: 115) said that the cottony leaves of this plant were 'antiently [anciently] used to stuff beds'. He gave the name of this plant as '*Gnaphalium* vulg. *majus*'.

Filipendula ulmaria (Spiraea ulmaria)

- Meadow-sweet (Scannell & Synnott, 1987); Bridewort (Tóibín, 1965); Sweet meadow (Colgan, 1904); Sweet William (Colgan, 1904); Arrigologher (Hart, 1898); Queen of the meadow (Hart, 1898)
- *Airgead luachra* (Scannell & Synnott, 1987); *Lus cneas* (Moloney, 1919; Dinneen, 1927); *Crios chonchulainn* (Moloney, 1919); *Crios Chu-chulainn* (Cameron, 1883); *Airgiod luchra* (de Vere Coneys, 1849); *Airgid luchra* (de Vere Coneys, 1849); *Cubhar léana* (Dinneen, 1927); *Airgidín na luachra* (Tóibín, 1965)

ROSACEAE—THE ROSE FAMILY
Native

Filipendula ulmaria is a common and conspicuous perennial herb with cream-coloured flowers of rough damp grasslands and fields, roadsides, marshes and other wet places throughout Ireland.

Filipendula ulmaria.

This plant has had a long history of usage in traditional medicine in Europe, particularly as an anti-inflammatory and analgesic, as well as a diuretic and antiseptic, and it was also used to treat arthritis and rheumatism (van Wyk & Wink, 2004). Like willows, meadow-sweet contains salicylate, from which aspirin was synthesized. As such, it has been used as a medicinal treatment for quite a wide range of ailments in Ireland, such as colds, diarrhoea, dropsy, kidney problems, scrofula and jaundice, and as a veterinary remedy in Co. Armagh to treat diarrhoea in calves (Allen & Hatfield, 2004). It was also made into a poultice to treat broken bones (Williams, 1993).

O'Regan (1997) gave a recipe for meadow-sweet hand cream: 4 oz meadow-sweet flowers; 2 oz horsetails; 1/2 pint of olive or almond oil and 1 oz beeswax. To make the hand cream, a jar is filled with the oil and meadow-sweet flowers and sealed. Then the jar is left in the sunshine for 10 days or beside a warm place. After that, the mixture is put in a saucepan and heated very slowly for half an hour to one hour, making sure that the oil does not burn. The flowers are then strained out using a muslin cloth. When the liquid is returned to the heat, the beeswax is grated and added to the mixture, and stirred well to prevent lumps from forming. A few drops of lavender oil may be added when the mixture is cooling.

Meadow-sweet is edible and all parts of the plant can be used as a vegetable or in soups or sauces (Hilton, 2007). Mabey (1972) suggested that the roots are rather bitter to eat; however, young leaves, flowerheads and roots have been used as a tea substitute. The dried leaves have a very sweet flavour and have been used as a sweetener in a variety of recipes. Dried leaves were used to add an aromatic bouquet to port, claret and Madeira (Mabey, 1972). Rutty (1772: 131) said that the flowers are 'of a grateful smell, cheer the heart without making the head heavy, and therefore in summer time are strewed as ornaments in dining-rooms and bed chambers'. Uí Chonchubhair (1995) also noted that its flowers were used to flavour drinks, a skin lotion was made by steeping meadow-sweet in rainwater or in dew, and the plants were strewn on floors to bring a fresh-smelling aroma to the house. Ó Céirín and Ó Céirín (1978) provided a recipe for meadow-sweet wine.

Filipendula ulmaria.

Meadow-sweet cordial is also used to make an excellent and refreshing honey-flavoured summer drink. It is made from the meadow-sweet flowers, sugar and lemon juice, in much the same way as elderflower cordial. I have seen it on sale in a farmers' market in Ballyvaughan, Co. Clare, in 2010, made by Peregringo's Kitchen in Fanore, Co. Clare. One recipe I have used to make meadow-sweet cordial is to put a good bunch of the flowers in about 2 litres of water and bring to a boil. Then, add 250 g of sugar and stir well until dissolved. Add the juice of a lemon and leave for a couple of hours until the flavours from the flowers have infused into the mixture. Then, strain the liquid through a fine sieve and bottle and seal in clean, sterilised bottles. Once opened, the cordial should be stored in the fridge and used within a week or two. The cordial is diluted in water according to taste and can also be used as an ingredient of cocktails.

The roots of meadow-sweet were used to prepare a black dye and a pale yellow dye (Mahon, 1983). Molony (1919) noted its use for cleaning milk churns, and MacMahon (1994) also mentioned that it was used for scouring vessels and for dyeing in Ballyheigue, Co. Kerry.

Filipendula vulgaris

- Dropwort (Scannell & Synnott, 1987)
- *Lus braonach* (Scannell & Synnott, 1987); *Greabhán* (Moloney, 1919; Dinneen, 1927)

ROSACEAE—THE ROSE FAMILY

Possibly Introduced (possible archaeophyte)

Filipendula vulgaris is a very rare plant in Ireland that is found in limestone grassland only in a few square kilometres in the eastern part of the Burren in Cos. Clare and Galway. It was first recorded in the Burren by K'Eogh (1735: 41) who said that 'it grows … wild in the barony of Burrin'.

As a medicinal plant it has been used in the treatment of epilepsy, kidney and bladder stones, genital discharges and intestinal worms.

It is grown as an ornamental plant in some Irish gardens (Wyse Jackson, 2006). Mabey (1972) noted that its roots have been eaten on continental Europe but are rather bitter. According to Cameron (1883: 18), Carl Linnaeus wrote, 'in a scarcity of corn the tubers have been eaten by men instead of food'.

Foeniculum vulgare.

Foeniculum vulgare

- Fennel (Scannell & Synnott, 1987)
- *Finéal* (Scannell & Synnott, 1987); *Macánleonta* (Dirrane, 1997); *Finéal mhuice* (Moloney, 1919; Dinneen, 1927); *Lus an tsaoi* (de Vere Coneys, 1849; Moloney, 1919; Dinneen, 1927); *Fineal chumhthra* (Cameron, 1883); *Finéal cumhra* (Dinneen, 1927; Ó Dónaill, 1977); *Lus na croiche* (Dinneen, 1927); *Meacan an leonta* (Ó Dónaill, 1977)

APIACEAE (Umbelliferae)—THE CARROT FAMILY

Introduced (possible archaeophyte)

Foeniculum vulgare is a tall, aromatic plant of waste places, roadside verges and dry banks that is particularly distributed in Ireland near the coast and in the southern half. It is almost certainly introduced but has occurred in Ireland for several centuries and has been used for a variety of culinary and medicinal purposes.

It has traditionally been used as a flavouring for fish and has a flavour of aniseed or liquorice. The young leaves and seeds can be used in salads and to make tea (Hilton, 2007). The roots are also edible and can be prepared and used as one would a carrot.

Emboden (1972) included it in his book of narcotic plants of the world, recording its use for eplileptiform convulsions and hallucinations. Dirrane (1997: 19) mentioned its use as a medicinal plant in the Aran Islands (Inishmore), where a sticky paste made by scraping off the plant is applied: 'One American visitor who fell in the graveyard, while visiting the grave of his ancestors, hurt his shoulder and elbow greatly. I applied the *macánleonta* sticky paste and when he got to Galway and was X-rayed, he was told the Aran remedy couldn't be bettered'.

Fragaria vesca

- Wild strawberry (Scannell & Synnott, 1987)
- *Sú talún fiáin* (Scannell & Synnott, 1987); *Súgh-thalmhan* (Moloney, 1919); *Tlachtshugh* (Moloney, 1919); *Súgh-láir* (Moloney, 1919); *Oighreóg*[110] (Moloney, 1919; Dinneen, 1927); *Sobha talmhuin*[111] (de Vere Coneys, 1849)

ROSACEAE—THE ROSE FAMILY
Native

Fragaria vesca.

Fragaria vesca is a common species throughout Ireland in shady habitats, such as woods and hedgerows.

Wild strawberries (*suba*) were certainly used as a food in ancient and medieval Ireland, and remains were found in excavations from Fishamble Street, Dublin, that date back to the Viking Age, more than 1000 years ago (Geraghty, 1996). The contents of drains and cesspits included seeds of strawberries, blackberries, apples, rowanberries, sloes and plums (Anonymous, 1973). In autumn, the dark red fruits of wild strawberries are sweet and delicious to eat. In some parts of Europe and Asia the fruits of wild strawberry are collected traditionally in autumn and form an important resource for local rural communities. I have bought baskets of delicious fresh wild strawberries on roadsides in Siberia collected by local people from the woods surrounding their villages. In Ireland the fruits were widely collected for food wherever they occurred in any abundance. Today, it is rare to find them in sufficient abundance to be worth collecting to take home. However, they can be made into jam (added pectin is needed to help the jam set).

The leaves have been used medicinally as a tonic. Allen and Hatfield (2004) noted that in Antrim there was a belief that 'excessive ardour' could be cooled with strawberry-leaf tea, although the leaves used may have been garden strawberries (*Fragaria ×ananassa* hybrids), not the wild species. An infusion of chopped leaves in boiling water makes a useful general medicinal herbal tea (van Wyk & Wink, 2004), and fresh leaves are rich in vitamin C (Peterson, 1977).

The rhizomes of wild strawberry have also been used as a coffee substitute (Hilton, 2007).

According to Allen and Hatfield (2004), *Fragaria vesca* was used as a veterinary remedy to treat red-water fever in Co. Cavan.

110. Dinneen (1927) defined *oighreóg* as 'strawberry'.
111. De Vere Coneys (1849) gave the Irish name *sobha talmhuin* for 'strawberries'.

Frangula alnus (Rhamnus frangula)

- Alder buckthorn (Scannell & Synnott, 1987); Black dogwood (Harron, 1986)
- *Draighean fearna* (Scannell & Synnott, 1987); *Ramhdhraighean mín* (Moloney, 1919)

RHAMNACEAE—THE BUCKTHORN FAMILY
Native

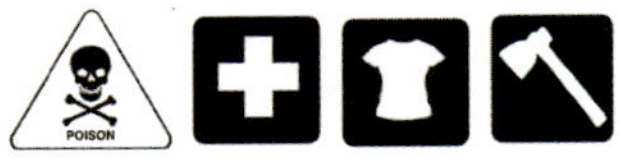

Frangula alnus is a small tree that is rare in Ireland and found in boggy or rocky habitats. It occurs on limestone pavements around turloughs and lakes in the Burren in north-west Co. Clare. Populations of *F. alnus* are also noted from Killarney, Co. Kerry, and in several midland counties in Ireland. The alder buckthorn can reach heights to 5 m. It deserves horticultural attention and is certainly deserving of cultivation in gardens.

Frangula alnus has long been used to treat constipation and is a powerful purgative, used in a similar way to *Rhamnus cathartica*, and it has also been used to treat liver and gall-bladder complaints such as jaundice, hepatitis and cirrhosis (Reader's Digest Association, 2006). It is said to cause acute abdominal pain or 'griping' (Moloney, 1919), especially if fresher bark is ingested. *Nature's Medicines* by the Reader's Digest Association (2006) warns that fresh bark is toxic, but van Wyk and Wink (2004) mentioned that dried bark is okay for use (the bark has to be stored for up to a year or heat dried before use). The bark is rich in antraquinones and it is these anthranoid laxatives that oxidize over time to reduce the toxicity of any herbal preparations (Duke et al., 2002: 310).

Frangula alnus.

Frangula alnus berries.

Frangula alnus (*Rhamnus frangula*; buckthorn). Watercolour by Lydia Shackleton. Courtesy of the National Botanic Gardens, Glasnevin.

MacCoitir (2003) noted that charcoal made from alder buckthorn was considered the best species for making charcoal for gunpowder (gunpowder being made from a combination of sulphur (brimstone), potassium nitrate (saltpetre) and charcoal, usually in the proportions 2:6:1). MacCoitir also said that its bark yields a yellow dye and its fruits a green dye.

Fraxinus excelsior

- Ash (Scannell & Synnott, 1987); Whinshag (Hart, 1898)
- *Fuinseog* (Scannell & Synnott, 1987); *Nuin* (Cameron, 1883; de Vere Coneys, 1849; Moloney, 1919; Dinneen, 1927; Uí Chanainn, 1962); *Uinseann* (*-sin*)[112] (de Vere Coneys, 1849; Cameron, 1883); *Uimhseann* (Cameron, 1883); *Cran-fuinseóige* (de Vere Coneys, 1849); *Crann fuinnse(oige)* (Dinneen, 1927); *Uinneas* (*-nse*)[113] (Dinneen, 1927); *Crann fuinseoige*[114] (Ó Dónaill, 1977); *Fuinse*[115] (Uí Chanainn, 1962)

OLEACEAE—THE ASH FAMILY

Native

Fraxinus excelsior is certainly one of Ireland's commonest native trees and is found abundantly in woodlands, farmland and hedgerows. Mature ash trees form some of the largest and most magnificent native trees in the island. Indeed, an ash specimen measured in 2005 in Marlfield Farm, Clonmel, Co. Tipperary, was the tallest native tree known in the country, measuring 2.7 m in girth and 40 m tall (Anonymous, 2005). The current ash girth champion is a specimen at Ballintray House, Youghal, Co. Waterford, that is 8.15 m in girth (and 14.5 m tall) (Anonymous, 2005).

Fraxinus excelsior.

Fraxinus excelsior fruits ('keys').

Ash is an important timber tree. Its timber is strong and rather hard, flexible, pale-coloured, straight-grained and widely used. It is light and very resistant to splitting, which has made it suitable for many diverse uses, such as bentwood, veneer, tool handles, ladders, turnery, solid furniture (Durand, 2008), wheel spokes, ploughs and spade handles (Lysaght, 1979). Mackay (1836: 179) noted that its wood is valuable for many purposes, 'especially for implements of husbandry'. Ash and oak wood were used in the making of wooden noggins in which milk and porridge were taken (Evans, 1957). Noggins were made from stakes of oak, with one projecting beyond the top of the vessel to act as a handle. The staves and a base piece were held together with a surrounding band of ash wood, joined by interlocking tongues. Ash wood was also used by local coopers to make

112. Cameron (1883) said that this also refers to the timber of the tree.
113. Dinneen (1927) defined *uinneas* (*-nse*) as 'an ash-tree (early form)'.
114. Ó Dónaill (1977) defined *crann fuinseoige* as 'ash-tree'.
115. Uí Chanainn (1962) defined *fuinse* as 'ash-tree'.

Fraxinus excelsior flowers.

Hurleys are most commonly made from *Fraxinus.*

An illustration of a faction fight. From Hall and Hall, 1841. *Ireland: Its Scenery, Character, etc.* Volume 1. How and Parsons, London.

butter-firkins in Co. Cavan. Ní Sheasnáin (1984) recorded that, at a local fair, a man buying a firkin of butter would ask what timber was used in the firkin so that he could subtract the weight of the empty firkin and only pay for the butter.

The best known and traditional use of ash wood in Ireland is for the making of hurleys. It has been widely used elsewhere for other sports equipment (oars, paddles, tennis rackets, billiard cues, croquet mallet handles, etc.) as well as, in the past, for spear shafts, lances and arrows (Edlin, 1973). The ideal tree to fell for making hurleys is one that is 25–30 years old; felled trees include the lower portion of the tree, or 'butt', so that the wider lower part of the hurley can be made from the wider part of the tree, following the natural grain of the tree and making for a stronger hurley (Durand, 2008). The same author noted that a cubic metre of good hurley ash should yield up to 200 planks. Two thirds of the planks would make top-quality senior hurleys, and the remaining third would be used for producing lighter, juvenile hurleys. Each hurley plank is sawn to follow the pattern of the grain and cut to be 2.5 cm in thickness and about 105 cm in length. In 2008 there were more than 100 hurley manufacturers in Ireland, with about 20 of them commercial producers, producing 10,000 to 20,000 hurleys each year. About half a million hurleys are needed each year to meet demand. The shape of the hurley can vary from county to county—one I bought in 2010 is the 'Kilkenny shape', made by Pat Campion of Borris-in-Ossory, Co. Laois.

Hurleys have also been used as weapons too, even up to the present day. Dillon (2006) described how they have been used in faction fights between rival families of Ireland's Traveller community, along with sticks, slash hooks and other weapons.

MacCoitir (2003) noted that ash timber was used for building (such as ridge-poles of houses), making fences, furniture (such as stools and chairs), spears and boat-building. McCracken (1971) recorded the widespread use of ash timber by wood turners to make small bowls and basins and that it would also have been used to make the staves of barrels (although much less frequently used for this purpose than oak). MacPhilib (2000) noted the use of ash timber to make frames for rush rafts in Ulster, used on Lough Erne (see *Schoenoplectus lacustris*), and Sharkey (2000) mentioned that ash, as well as willow and sycamore, were the timbers of choice for making shoulder-yokes used to carry milking pails. Ash was also used in the construction of currachs, the traditional boats of western Ireland, and used to make the ribs lining the skin or tarred canvas hulls (see *Quercus* species also).

A 'V'-shaped ash branch was also used in Co. Donegal to make a 'wooden horse', sometimes called a 'Gallen Maggie', to help in the transportation of manure from the manure pits to the fields on upland farms. One man filled a creel from the manure heap and carried it halfway to its destination. There, he placed it on the 'wooden horse' rest and picked up the emptied one returned by his partner from the field. The 'V'-shaped branch was bored with an inch and a half hole at the end of the 'V', and a similar hole put at the end of each arm. Wooden legs, about 3 ft long, were driven into the holes, and an upright one was driven into the top to prevent the creel from falling back when placed on the 'wooden horse' (Campbell, 1974).

Vickery (1995: 20) quoted from the archives of the Irish Folklore Commission's Schools' Scheme that ash sticks were used in Cos. Sligo and Longford as weapons: 'The Joyces are tinkers ... they are wary and row amongst themselves. They do have some fierce fights in which the women join in. When they have each others heads well cut with ash plants they settle down and are as friendly as ever'! In his autobiography about growing up in the 1940s and 1950s in the West Kerry Gaeltacht, Maidhe Dainín Ó Sé (2007) wrote that the teacher's cane that he was familiar with was made of ash (*Fraxinus excelsior*) (see also *Betula* species).

Ash and other timbers may have been used to make an interesting musical instrument called a 'bull-roarer'. It consisted of a short, flat slip of wood, sometimes notched along the sides, with a hole pierced at one end through which a string was threaded. The bull-roarer was twirled quickly in a circle and made an eerie humming sound. Bull-roarers have had important religious and ceremonial use in other cultures and for communication over distance (such as amongst Aboriginal Australians), but in Ireland they seem to have been used mainly, if not entirely, as playthings for children (McKean, 1933). Records of bull-roarers in use in Ireland come from Co. Cork, and other parts of Ireland, mainly in the north and north-east, in Cos. Antrim, Down, Donegal, Fermanagh and Tyrone (McKean, 1933). They were also in use by children in Co. Cavan in about 1904 (Benson, 1934). McKean mentioned that R. A. S. Macalister in his book, *Tara*, suggested that the old kings of Tara were instituted to the sound of bull-roarers, but I have not found any evidence documenting that they were widely used in ancient Ireland. Bull-roarers were sometimes made by children from their school rulers. The use of the bull-roarer in Ireland seems to have declined or disappeared in the early part of the 20th century. Several examples of bull-roarers are included in the collections of the National Museum of Ireland (Museum of Country Life) in Castlebar, Co. Mayo, with two of them from Cape Clear Island in Co. Cork (Ref. Nos: F1948.8 and F1934.17).

Ashwood is an important fuel. It burns well even when freshly cut and can be readily split into usable pieces (Durand, 2008). The fuel value of native Irish woods varies considerably. Those with close-grained, hard and heavy wood are generally superior to soft woods and burn longer and provide greater heat. Ash, oak, yew and hawthorn are all amongst the best Irish firewoods. Less useful are alder, poplar and willows, which produce less heat, burn poorly and are liable to spark. Woods of intermediate value include birch, pine, elm and sycamore (Howes, 1974). Seasoned wood is generally better for fuel than freshly cut sapwood.

Ash resin was made into pitch and used for caulking boats and as a protection for other objects from weathering (Neeson, 1991). An alkali used in the production of glass was extracted from ash trees in the 1600s (McCracken, 1971).

Ash bark has been used for tanning and also yields a yellow dye (Mahon, 1983). Rutty (1772) said that it was used for tanning calf skins. He also recorded that ash made from burning its branches were some of the best for bleaching linen.

Phillips (1983) noted a recipe for Ash Key Pickle, where young ash keys are picked very young and boiled to remove the bitterness, then added to a syrup made of white-wine vinegar, sugar and a little water, boiled, potted and cooled. The pickle takes on a green colour. Dried ash leaves have been used for livestock fodder (MacCoitir, 2003). Ash trees can be tapped to obtain a sugary syrup, used to make ash wine (Milliken & Bridgewater, 2004).

The ash is useful as an anti-inflammatory treatment for ailments such as rheumatism and has been used to treat kidney stones, gout, oedema, water retention, fevers and obesity (Reader's Digest Association, 2006). The bark was formerly recommended by herbalists for use in a footbath as a treatment for sore feet (Moloney, 1919). Allen and Hatfield (2004) noted it was formerly an important medicinal plant in Ireland, with various parts (sap, bark, leaves, buds) used to treat ailments such as fevers, warts, heartburn, ringworm, burns and earache (a bud would be heated over a fire and the liquid that emerges put hot into the ear). They recorded that the leaves were used (fresh or boiled) for rheumatism and gout. K'Eogh (1735) suggested its use as a treatment for liver complaints and dropsy; the inner bark was used for fevers, and the ash resulting from burning the wood was used to cure scabs and ringworm. In Co. Antrim, the smoke from smouldering ash twigs was inhaled as a treatment for worms (Vickery, 1995; Pennacchio et al., 2010).

The ash is a popular tree in gardens in Ireland despite being one of the last native trees to produce leaves in spring and one of the first to lose them in autumn. Various cultivated varieties are also grown, such the weeping ash (*Fraxinus excelsior* var. *pendula*; in Irish, *fuinseog shilte*) with pendulous branches.

Fuchsia magellanica

- Fuchsia (Scannell & Synnott, 1987)
- *Fiúise* (Scannell & Synnott, 1987); *Deora dé*[116] (Ó Dónaill, 1977; O'Rourke, 2006); *Mileanna dearg* (O'Rourke, 2006); *Na fiúsaíos* (Uí Chonchubhair, 1995); *Siogairlíní*[117] (Dinneen, 1927); *Singirlín*[118] (Dinneen, 1927; Uí Chanainn, 1962)[119]

ONAGRACEAE—THE WILLOWHERB FAMILY

Introduced

Fuchsia magellanica.

Fuchsia magellanica is native to temperate South America, from Peru, south through Chile to Tierra del Fuego. It is widely grown as an ornamental and hedge plant, particularly in the south-west of Ireland, where exuberant flowering fuchsia hedges are a characteristic feature of the landscape in several parts of Cos. Cork and Kerry. Its fruits are blackish and succulent and said to be edible, although they may have diuretic properties. However, the commonest variety grown in Ireland, *F. magellanica* var. *riccartonii*, which is distinguished by its spherical buds, rarely produces fruits. It is most commonly grown from cuttings, although there are certainly occasions in Ireland where plants have become established from self-sown seed (e.g., see McNeill, 2010).

The Irish Folklore Commission holds records of fuchsia flowers being used to produce a dark red dye in Ireland.

According to Joe Hogan (pers. comm., 2010), *Fuchsia* is not used for basketry in Ireland, although its timber has been used for various craft purposes. I have also used fuchsia wood for fuel and as tinder. It burns quickly and easily when dry and is easy to cut. McCarthy and Hawkes (2007) said that in the Mizen region of West Cork it was (or is) 'used for field boundaries, fuel and the hay rick berth'. The hay was stored in the haggard on a platform made from round beach pebbles (called 'pounders') and fuchsia bushes.

116. Ó Dónaill (1977) defined *deora dé* as 'fuchsia'.

117. Included as a manuscript note '*fide* Mrs Michael Long, Dingle', Co. Kerry, in a copy of Dinneen (1927) in the Library of the National Botanic Gardens, Glasnevin.

118. Dinneen (1927) defined *singirlín* as 'fuchsia globosa'.

119. Uí Chanainn (1962) defined *singirlín* as 'fuchsia'.

Fumaria officinalis

- Common fumitory (Scannell & Synnott, 1987)
- *Camán searraigh díge* (Scannell & Synnott, 1987); *Dearg talmhan* (Moloney, 1919); *Gidiríam* (Moloney, 1919; Dinneen, 1927); *Deatach thalmhuin* (de Vere Coneys, 1849; Cameron, 1883)[120]; *Giodairiam* (Dinneen, 1927; Ó Dónaill, 1977)

PAPAVERACEAE (Fumariaceae)—THE POPPY FAMILY
Introduced (possible archaeophyte)

Fumaria officinalis is an occasional weed of tilled fields, cultivated ground, field margins, roadsides and wasteground, particularly near the east coast of Ireland and occasionally elsewhere. It is almost certainly an introduced non-native plant.

It was used in Ireland as a decoction in the treatment of skin problems, especially scalp problems in infants (Moloney, 1919). Elsewhere, it has been used in traditional medicine as a diuretic and laxative, for intestinal problems, and externally to treat psoriasis and chronic eczema (van Wyk & Wink, 2004). In Co. Cavan inhaling the smoke from burning a *Fumaria* species was regarded as an excellent treatment for stomach troubles (Pennacchio et al., 2010).

**Galeopsis angustifolia*

- Red hemp-nettle (Scannell & Synnott, 1987)
- *Ga corcra* (Scannell & Synnott, 1987)

LAMIACEAE (Labiatae)—THE DEAD-NETTLE FAMILY
Possibly Introduced (possible archaeophyte)

Galeopsis angustifolia is a rare plant in Ireland in the midlands and east of limestone gravel on old esker ridges.

*No ethnobotanical uses of this plant have been noted in Ireland.

**Galeopsis tetrahit*

- Common hemp-nettle (Scannell & Synnott, 1987)
- *Ga buí* (Scannell & Synnott, 1987); *Gath bhuidhe* (Moloney, 1919)

LAMIACEAE (Labiatae)—THE DEAD-NETTLE FAMILY
Possibly Introduced (possible archaeophyte)

Galeopsis tetrahit is an occasional weed in Ireland of cultivated land that occurs mainly on peaty soils.

*No ethnobotanical uses of this plant have been noted in Ireland.

Galinsoga species

- Gallant-soldier (Reynolds, 2002)

ASTERACEAE (Compositae)—THE DAISY FAMILY
Introduced

120. Uí Chonchubhair (1995) gave *deatach talún*, not for *Fumaria officinalis*, but instead for *F. capreolata*.

Two *Galinsoga* species (*G. parviflora* and *G. quadriradiata*) are found in Ireland as rare annual weeds, mainly as garden weeds. They are native to South America and during visits to Colombia I have eaten *Galinsoga* leaves cooked in a traditional potato soup, to which they added a delicious and distinctive flavour. They can also be used as a green vegetable, boiled or steamed, and are worth gathering as food when seen in Ireland.

Galium aparine (cleavers). Watercolour by Lydia Shackleton. Courtesy of the National Botanic Gardens, Glasnevin.

Galium aparine

- Cleavers (Scannell & Synnott, 1987); Robin-run-the-hedge (Scannell & Synnott, 1987); Goose grass (Colgan, 1904; Moloney, 1919); Sticky-back (Doogue & Krieger, 2010); Catchweed; Jack at the hedge (Hart, 1898); Sticky billy (Praeger, 1938); Devil's garter (Praeger, 1938)
- *Garbhlus* (Scannell & Synnott, 1987); *Garbh(l)ach* (Ua Maoileoin, 2001); *Uirlis chríonna* (Ua Maoileoin, 2001); *Lus garbh* (de Vere Coneys, 1849; Moloney, 1919; Dinneen, 1927; Ó Dónaill, 1977); *Airmeirg* (Cameron, 1883; Ó Dónaill, 1977); *Briosclán (-áin)(-cán)*[121] (Dinneen, 1927; Uí Chanainn, 1962); *Caol-fhail* (Dinneen, 1927)

RUBIACEAE—THE BEDSTRAW FAMILY
Native

Galium aparine is an extremely familiar and common annual weed in Ireland of waste places, gardens, cultivated ground, thickets and hedges. It also occurs on shingle beaches, although less frequently.

According to Moloney (1919), the leaves and stems were used externally in the treatment of ulcers and cancerous growths, and its juice was used internally for the same purposes. The plant would be pounded well and squeezed to extract the juice. In Donegal it was pounded and squeezed and given to children to 'strike out hives from the stomach' (Hart, 1898: 379). In Derry it was used as a cancer cure (Hart, 1898). Threlkeld (1726) noted that it was much praised as a treatment for kidney stones (gravel) and simple gonorrhoea. Allen and Hatfield (2004) recorded its former use medicinally in Ireland to treat tumours, burns, whooping cough, swellings, inflammations of the bowels in children, stomach-ache and 'softening of the joints', and in Kerry, it was mixed with *Geranium* species in a preparation drunk for kidney trouble. Williams (1993) recorded that it was used as an ingredient of a purgative and also used to treat skin complaints such as impetigo.

Phillips (1983) and Mears and Hillman (2007) noted that the soft, young leaves and tender tips can be used in soups or as a vegetable. Mackay (1836) recorded that its seeds were recommended as a substitute for coffee (which is interesting because it is a member of the Rubiaceae, as are *Coffea arabica* and *C. robusta*, the genuine coffee species).

121. The same name is also given for silverweed (*Potentilla anserina*).

The ripe seeds are slow-roasted then ground down and used like coffee. Uí Chonchubhair (1995) said that it was fed to geese (hence one of its common names, goosegrass) and added to soups and stews, with a reputation for helping to make the obese lean. John Gerard's *Herball*, published first in 1597, said that 'Women do usually make pottage of Clevers with a little mutton and Otemeale, to cause lanknesse, and keep them from fatnesse'. For use as a vegetable, the young shoots are boiled for 10 to 15 minutes and served with butter (although if one is hoping to address 'fatnesse' it is better to leave out the butter!). Peterson (1977) suggested that chilled, cooked shoots can be used also in salads. Rutty (1772: 38), in his *An Essay Towards a Natural History of the County of Dublin*, said that 'the tops are boiled in Lenten broth, chopt and mixt with bran and oatmeal, fattens geese effectually, and it is eaten by cows, goats, sheep and horses, and the seeds roasted, have supplied the place of Coffee'.

This is a favourite plant for children to play with, sticking on the clothes or in the hair of their friends, and is probably where it gets its common names, 'sticky willie' or 'sticky-backs'.

According to Sharkey (2000), it was also boiled and used to make an effective hair tonic.

Galium aparine.

Galium boreale

- Northern bedstraw (Scannell & Synnott, 1987)
- *Rú crua* (Scannell & Synnott, 1987)

RUBIACEAE—THE BEDSTRAW FAMILY

Native

Galium boreale is a perennial plant of rocky lake-shores and stream-sides with patchy distribution in Ireland; it is very common in parts of the west, rarer in the north and absent from much of the south and east.

Wade (1804) recorded that the roots of this species yield a red dye.

Galium odoratum

- Woodruff (Scannell & Synnott, 1987); Hayplant (Brunker, 1950)
- *Lus moileas* (Scannell & Synnott, 1987); *Lus molach* (Moloney, 1919; Dinneen, 1927; Ó Dónaill, 1977); *Moleas* (Moloney, 1919)

RUBIACEAE—THE BEDSTRAW FAMILY

Native

Galium odoratum.

Galium odoratum is a widespread perennial plant in Ireland of woodlands and other shady habitats.

Moloney (1919) listed it (under the name *Asperula odorata*) as having been used in Ireland to flavour wine.

According to Threlkeld (1726), it was used medicinally to treat jaundice and for tumours (externally).

This strongly scented plant has a sweet smell, and Rutty (1772: 103) said that the flowers were put among clothes, 'both for the grateful smell and to banish insects'.

**Galium palustre*

- Common marsh-bedstraw (Scannell & Synnott, 1987)
- *Rú corraigh* (Scannell & Synnott, 1987)

RUBIACEAE—THE BEDSTRAW FAMILY
Native

Galium palustre is a widespread perennial plant throughout Ireland of muddy streams, marshes, fens and ditches.

*No ethnobotanical uses of this plant have been noted in Ireland.

**Galium saxatile*

- Heath bedstraw (Scannell & Synnott, 1987)
- *Luibh na bhfear gonta* (Scannell & Synnott, 1987); *Lus na bhfear mbeag* (Moloney, 1919; Dinneen, 1927; Ó Dónaill, 1977); *Madar fraoigh* (Moloney, 1919; Dinneen; 1927; Ó Dónaill, 1977); *Luibh na bhfear ngonta* (Dinneen, 1927; Ó Dónaill, 1977)

RUBIACEAE—THE BEDSTRAW FAMILY
Native

Galium saxatile is a widespread perennial of rocky places and heaths that is common throughout Ireland except in some parts of the centre.

*No ethnobotanical uses of this plant have been noted in Ireland.

**Galium sterneri*

- Limestone bedstraw (Scannell & Synnott, 1987)
- *Rú beag* (Scannell & Synnott, 1987)

RUBIACEAE—THE BEDSTRAW FAMILY
Native

Galium sterneri is a common perennial plant in Ireland that occurs on the limestone soils of the west, such as on cliffs and rocky slopes and in short grassland, and on the basalts of the north-east.

*No ethnobotanical uses of this plant have been noted in Ireland.

**Galium uliginosum*

- Fen bedstraw (Scannell & Synnott, 1987); Rough bog bedstraw (Brunker, 1950)
- *Rú eanaigh* (Scannell & Synnott, 1987)

RUBIACEAE—THE BEDSTRAW FAMILY
Native

Galium uliginosum is a perennial plant of marsh habitats that is found mainly in the west in Ireland.

*No ethnobotanical uses of this plant have been noted in Ireland.

Galium verum

- Lady's bedstraw (Scannell & Synnott, 1987)
- *Boladh cnis* (Scannell & Synnott, 1987); *Rú Mhuire* (Uí Chonchubhair, 1995; O'Rourke, 2006); *Baladh chnis*[122] (Moloney, 1919); *Madadh (-aidh) fraoigh*[123] (Dinneen, 1927); *Rú* (Dinneen, 1927)

RUBIACEAE—THE BEDSTRAW FAMILY
Native

Galium verum is a common plant of dry banks, hedges and fixed sand dunes throughout most parts of Ireland except in the most acid of soils.

This plant was used as rennet to curdle milk in cheese-making (Smith, [1756] 1979), and the flowering tips can be distilled as a summer drink (Phillips, 1983). Mackay (1836) recorded that in the Scottish Highlands it was combined as a rennet with nettle leaves and a little salt to curdle milk.

Galium verum in sand dune grassland.

It was a source of a grey dye (Kelly, 1997) and a red or crimson dye from the stolons (Clapham et al., 1962; Mahon, 1983). Cameron (1883: 35) described how the dye is extracted: 'The bark is stripped off the roots, in which the virtue principally lies. They then boil the roots thus stripped in water, to extract what little virtue remains in them; after taking them out, they last of all put the bark into the liquor, and boil that and the yarn they intend to dye together, adding alum to fix the colour'.

Galium verum was formerly used in traditional medicine in Ireland, and the crushed flowers were applied to burns to help alleviate inflammation, heal wounds and stop haemorrhages, including nosebleeds (K'Eogh, 1735). Elsewhere, it has been used as a diuretic (van Wyk & Wink, 2004).

It was also used as a veterinary remedy to treat fits in dogs in Co. Westmeath (Allen & Hatfield, 2004).

Gentiana verna

- Spring gentian (Scannell & Synnott, 1987)
- *Ceadharlach Bealtaine* (Scannell & Synnott, 1987); *Bláth (Pabhsaer) Mhuire* (O'Rourke, 2006); *Lus an chrúbáin*[124] (Ó Dónaill, 1977)

GENTIANACEAE—THE GENTIAN FAMILY

122. Cameron (1883) suggested that this name is more likely to have been used for *Galium odoratum*.
123. Dinneen (1927) defined *madadh (-aidh) fraoigh* as 'yellow lady's bedstraw', although later in the same work the name *madar (-air) fraoigh* is given for 'heath bedstraw' (*Galium saxatile*).
124. Ó Dónaill (1977) defined *lus an chrúbáin* as 'gentian'.

Gentiana verna growing in the Burren, Co. Clare.

Native

Gentiana verna is a widespread perennial herb in the west of Ireland, from north Co. Clare to south Mayo Co. It is particularly abundant in parts of the karst topography of the well-known Burren in Co. Clare, where it grows in rocky limestone grassland and stabilised sand dunes. It is probably the most famous of the Burren plants, and a field of bright blue-flowered gentians in spring is a wonderful sight to see.

This plant is occasionally cultivated in Irish gardens, mainly by keen alpine plant enthusiasts.

---

*Gentianella amarella

- Autumn gentian (Scannell & Synnott, 1987); Autumn felwort (Brunker, 1950)
- *Muilcheann* (Scannell & Synnott, 1987)

GENTIANACEAE—THE GENTIAN FAMILY

Native

Gentianella amarella is mainly distributed in the centre of Ireland, such as around Galway Bay, in limestone grasslands and in sandy and gravelly habitats and on dry banks and sand dunes. It is rare elsewhere in Ireland.

*No ethnobotanical uses of this plant have been noted in Ireland.

Gentianella campestris

- Field gentian (Scannell & Synnott, 1987)
- *Lus an chrúbáin* (Scannell & Synnott, 1987)

GENTIANACEAE—THE GENTIAN FAMILY

Native

Gentianella campestris occurs in heathy pastures along the coast and in the mountains, particularly in the west and north; it is rare elsewhere in Ireland.

Clyne (1989) said that it was used to treat a disease of cattle called *crúbain*. This disease causes cattle to draw their hind legs forward towards their fore legs, as if pulled by a rope, and they become lean and weak.

---

*Geranium columbinum

- Long-stalked crane's-bill (Scannell & Synnott, 1987)
- *Crobh coilm* (Scannell & Synnott, 1987); *Crobh colaim* (Dinneen, 1927); *Crobh an colúr* (Dinneen, 1927)

GERANIACEAE—THE GERANIUM FAMILY

Native

Geranium columbinum is an occasional plant of disturbed limestone soils, rocky ground and roadsides throughout most of Ireland, except the centre and north-west where it is rare.

*No ethnobotanical uses of this plant have been noted in Ireland.

---

*Geranium dissectum

- Cut-leaved crane's-bill (Scannell & Synnott, 1987)
- *Crobh giobach* (Scannell & Synnott, 1987); *Luibh an neorbhais* (Ua Maoileoin, 2001); *Crobh choluim* (Moloney, 1919); *Creagrach* (Moloney, 1919); *Crobh gioblach* (Dinneen, 1927)

GERANIACEAE—THE GERANIUM FAMILY
Possibly Introduced (possible archaeophyte)

Geranium dissectum is a common plant in Ireland of cultivated ground, roadsides, pastures, field margins, stony ground and waste places.

*No ethnobotanical uses of this plant have been noted in Ireland.

---

*Geranium lucidum

- Shining crane's-bill (Scannell & Synnott, 1987)
- *Crobh geal* (Scannell & Synnott, 1987)

GERANIACEAE—THE GERANIUM FAMILY
Native

Geranium lucidum is an occasional plant in Ireland of limestone walls, pavement and waste places that is distributed particularly in parts of the west and centre; it is rare elsewhere in Ireland.

*No ethnobotanical uses of this plant have been noted in Ireland.

---

*Geranium molle

- Dove's-foot crane's-bill (Scannell & Synnott, 1987)
- *Crobh bog* (Scannell & Synnott, 1987)

GERANIACEAE—THE GERANIUM FAMILY
Native

Geranium molle is a small annual plant of sand dunes, roadsides, dry grassland and banks, waste places and limestone pavement that is common in most areas of Ireland except the north-west.

*No ethnobotanical uses of this plant have been noted in Ireland.

---

*Geranium pratense

- Meadow crane's-bill (Scannell & Synnott, 1987)
- *Crobh gorm* (Scannell & Synnott, 1987)

GERANIACEAE—THE GERANIUM FAMILY
Possibly Introduced

Geranium pratense is found throughout Ireland in coastal grassland and sand dunes on the north coast; it is occasionally found elsewhere as a garden escape in waste places and roadsides.

*No ethnobotanical uses of this plant have been noted in Ireland.

*Geranium purpureum

- Little-robin (Scannell & Synnott, 1987)
- *Eireaball rí* (Scannell & Synnott, 1987)

GERANIACEAE—THE GERANIUM FAMILY
Native

Geranium purpureum is a very rare plant in Ireland of walls and rocks near the south coast and is possibly now extinct.

*No ethnobotanical uses of this plant have been noted in Ireland.

*Geranium pusillum

- Small-flowered crane's-bill (Scannell & Synnott, 1987)
- *Crobh beag* (Scannell & Synnott, 1987)

GERANIACEAE—THE GERANIUM FAMILY
Possibly Introduced

Geranium pusillum is a rare weed in Ireland of cultivated ground, tilled fields, roadsides and wasteground that occurs mainly in the east and south.

*No ethnobotanical uses of this plant have been noted in Ireland.

Geranium robertianum

Geranium robertianum.

- Herb-Robert (Scannell & Synnott, 1987); Wire fire (Colgan, 1904); Bloodweed (Hart, 1898); Dog's toe (Hart, 1898); Elfshot cow (Hart, 1898); Stinking roger (Hart, 1898)
- *Ruithéal rí* (Scannell & Synnott, 1987); *Luibh an chreatha* (Ua Maoileoin, 2001; Uí Chonchubhair, 1995); *Crobhdearg* (Moloney, 1919; Dinneen, 1927); *Ruithéal ríog* (Moloney, 1919); *Rian ríogh* (Moloney, 1919); *Earball rí* (Webb et al., 1996; O'Rourke, 2006); *Righeal cuil* (Cameron, 1883)

GERANIACEAE—THE GERANIUM FAMILY
Native

Geranium robertianum is an extremely common plant throughout Ireland of roadsides, hedgerows, walls, woodlands, scrub, waste places and other habitats.

Herb-Robert has a long history of use in traditional medicine in Europe and Asia and was typically used to treat diarrhoea and infections of the urinary tract (van Wyk & Wink, 2004). It was used for a number of medicinal purposes throughout Ireland, such as to treat 'bad eyes' and to help make women fertile, as a cure for blood complaints, and to treat bleeding piles, liver complaints and kidney disease (Williams, 1993). It has also been used in Co. Dublin (where it was sometimes called wild-fire). Herb-Robert was used to treat skin ailments, and according to Colgan (1904: 309), 'Its grand for the felon, a sore that breaks out on a man. If you've got a round ring on your arrum, a sort of an itch, rub the Wild Fire on it and that'll backen … (keep back or retard) it'. Allen and Hatfield (2004) noted its medicinal use in Ireland to treat back and stomach pains, sore throats and coughs, diabetes, gallstones, urinary complaints, and to stop bleeding and for rheumatism.

Geranium robertianum was widely used in Ireland as the standard treatment for 'red-water' (blood in the urine), a disease of cattle, and apparently in the treatment of diabetes (Moloney, 1919; Allen & Hatfield, 2004). This species was also called 'Bloodweed' in some parts of the country, as it was said to be good for treating cattle that are passing blood (Hart, 1898). According to Allen and Hatfield (2004), it has been used in other veterinary treatments in Ireland, such as for worms in horses and cattle, and constipation, diarrhoea and tuberculosis in dogs.

**Geranium rotundifolium*

- Round-leaved crane's-bill (Scannell & Synnott, 1987)
- *Crobh cruinn* (Scannell & Synnott, 1987)

GERANIACEAE—THE GERANIUM FAMILY
Native

Geranium rotundifolium is a rare plant in Ireland of stone walls and roadsides that is found mainly in the south.

*No ethnobotanical uses of this plant have been noted in Ireland.

Geranium sanguineum

Geranium sanguineum.

- Bloody crane's-bill (Scannell & Synnott, 1987)
- *Crobh dearg* (Scannell & Synnott, 1987); *Créachtach-dearg* (*-aighe-déirge*) (de Vere Coneys, 1849); *Creachtach dearg, Crobh preacháin, Earball riogh*[125] (Moloney, 1919)[126]

GERANIACEAE—THE GERANIUM FAMILY
Native

Geranium sanguineum is an attractive and conspicuous perennial of limestone pavements, rocky grassland and dry banks that is common in the karst topography of the Burren in Co. Clare and Co. Galway; it is very rare elsewhere in Ireland.

Bloody crane's-bill is occasionally grown as an ornamental garden plant in a rockery or herbaceous border (Wyse Jackson, 2006).

According to O'Rourke (2006), it has been used in folk medicine to treat urinary disorders in humans and animals.

125. Moloney (1919) gave the English name 'stinking crane's-bill' for this Irish name.
126. Moloney (1919) gave these names for both *Geranium sanguineum* and *G. molle*.

**Geranium sylvaticum*

- Wood crane's-bill (Scannell & Synnott, 1987)
- *Crobh coille* (Scannell & Synnott, 1987)

GERANIACEAE—THE GERANIUM FAMILY
Native

Geranium sylvaticum is a perennial plant of woods and rough grasslands along the coasts of Northern Ireland. It is unknown elsewhere in Ireland.

*No ethnobotanical uses of this plant have been noted in Ireland.

Geum rivale

- Water avens (Scannell & Synnott, 1987)
- *Machall uisce* (Scannell & Synnott, 1987); *Machall-uisge* (*-aill-uisge*) (de Vere Coneys, 1849); *Macall uisce* (Ó Dónaill, 1977)

ROSACEAE—THE ROSE FAMILY
Native

Geum rivale can be found in the northern half of Ireland in wet habitats in woods and stream banks and occasionally on mountain cliffs.

Although no ethnobotanical uses of this plant have been noted in Ireland, in North America the root stocks have been boiled to make a drink rather like hot chocolate, to which milk and sugar may be added (Peterson, 1977).

Geum urbanum

- Wood avens (Scannell & Synnott, 1987); Herb bennet (Harron, 1986)
- *Machall coille* (Scannell & Synnott, 1987); *Machal coilleadh* (Moloney, 1919); *Machall-fiadhain* (*-aill-fiadhain*) (de Vere Coneys, 1849; Dinneen, 1927); *Traithnín*[127] (de Vere Coneys, 1849); *Beinidín* (Ó Dónaill, 1977); *Macall coille* (Ó Dónaill, 1977)

ROSACEAE—THE ROSE FAMILY
Native

Geum urbanum is a common perennial herb throughout Ireland of hedges, woodlands and other shady places.

This plant occurs widely in Europe, Asia and North America where its roots and rhizomes have been used in traditional medicine for diarrhoea, dyspepsia and appetite loss, as a gargle for throat and mouth infections, and to treat chilblains, haemorrhoids and other skin problems (van Wyk & Wink, 2004). These same authors mentioned that it has been used as an additive to liqueurs, brandies, toothpastes and mouthwashes, but I have not found records of such usage in Ireland. It was also formerly used in traditional medicine in Ireland in the treatment of chills (Moloney, 1919) and for kidney trouble (Allen & Hatfield,

127. De Vere Coneys (1849) listed *Traithnín* as the Irish name for 'the herb bennet'.

2004). Mackay (1836) reported that this species has aromatic roots. It was widely grown in Britain in the 16th century as a pot herb and added to broths and soups (Mabey, 1972).

**Glaucium flavum*

- Yellow horned-poppy (Scannell & Synnott, 1987)
- *Caillichín na trá* (Scannell & Synnott, 1987); *Bárróg ruadh* (Moloney, 1919); *Cailcín na trágha* (Dinneen, 1927)

PAPAVERACEAE—THE POPPY FAMILY

Native

Glaucium flavum is a plant of sandy and shingle beaches that was formerly more widespread in Ireland. It is now found only occasionally around the coasts, except in the northwest between Cos. Antrim and Mayo where it is absent.

The seeds of this species have been used in the manufacture of soap (Heywood, 1978), but I have not found records of its use in Ireland.

Rutty (1772: 172) gave an account of it being poisonous:

> 'a certain person made a pye of the roots of this plant, supposing them to be the roots of the Eryngo [*Eryngium maritimum*], of which he had before eaten pies which were very pleasant, and eating it while it was hot, became delirious, and having voided a stool in a white chamber pot, fancied it to be gold, breaking the pot in pieces, and desiring what he imagined to be gold, might be preserved as such. Also his man and maid servant eating of the same pye, fancied most of what they saw to be gold; and a child in the cradle, having tasted of the same pye, was much dozed, and turned its mouth to and fro: thus they continued for some days, and then recovered'.

Glaux maritima

- Sea-milkwort (Scannell & Synnott, 1987)
- *Lus an tsaillte* (Scannell & Synnott, 1987); *Lus bainne mara* (Uí Chonchubhair, 1995)

PRIMULACEAE—THE PRIMROSE FAMILY

Native

Glaux maritima is a common plant all around Ireland of several coastal habitats, including salt marshes and muddy, sandy and gravel shores.

It has been eaten by native people in North America with the roots boiled and used as a vegetable and young shoots eaten raw or pickled (see Pfaf.org).

Used medicinally, the plant is reputed to promote sleep.

I can find no records of its use for food or medicine in Ireland, but as a common coastal plant, it would undoubtedly have been gathered and used.

Glechoma hederacea (Nepeta glechoma)

- Ground ivy (Scannell & Synnott, 1987)
- *Athair lusa* (Scannell & Synnott, 1987); *Eidhneán talún* (Ua Maoileoin, 2001); *Eidhneán*

talmhán (de Vere Coneys, 1849; Moloney, 1919; Dinneen, 1927); *Aignean thalmhuin* (Cameron, 1883); *Athair lus* (de Vere Coneys, 1849; Dinneen, 1927); *Athair talmhan* (Dinneen, 1927)

LAMIACEAE (Labiatae)—THE DEAD-NETTLE FAMILY

Native

Glechoma hederacea is a common perennial herb throughout Ireland of woodlands, hedgerows and shady grasslands.

Ground ivy has been used medicinally in an infusion for bronchitis and also as a tonic and diuretic (Moloney, 1919). Allen and Hatfield (2004) noted its use in Ireland to treat skin complaints and ulcers, flush out the kidneys, stimulate menstruation in cases of chlorosis (an iron deficiency, especially in adolescent girls), and heal sores and blisters. Maloney (1972) noted its use in Co. Cavan to treat asthma. Lady Wilde (1890: 57) also mentioned this usage and said it was made into a potion. Lady Wilde also described its use to cure epilepsy, whereby it was mixed with 'Mandragore' (*Mandragora* [mandrake], which is not an Irish plant), and the two plants were boiled together to make a plaster that was then applied to the head: 'If the patient sleeps he will do well, and if not, he will not' (in Wood-Martin, 1902).

It was used as a veterinary remedy and Allen and Hatfield (2004) noted its use to treat kidney trouble in cattle in Kilkenny (Allen & Hatfield, 2004). Drury (1985) described its use in Britain as a herbal remedy for livestock and to treat eye troubles in horses; powdered ground ivy was mixed with ale and honey and then strained before use. Trailing shoots of ground ivy were also given to horses to rid them of worms and other parasites.

Although bitter, ground ivy is edible. Mears and Hillman (2007) suggested that it is a useful addition to nettle soup, giving it more 'bite'. In North America, it has been used to make tea (Peterson, 1977). In Scotland it was used instead of hops to make ale bitter (Clyne, 1989). Rutty (1772) also mentioned this use and said that it was put into the hogshead to clear the ale, before the introduction of hops.

**Glyceria declinata*

- Small sweet-grass (Scannell & Synnott, 1987)
- *Milseán beag* (Scannell & Synnott, 1987)

POACEAE (Gramineae)—THE GRASS FAMILY

Native

Glyceria declinata is mainly found in the south and west in muddy habitats.

*No ethnobotanical uses of this plant have been noted in Ireland.

**Glyceria fluitans*

- Floating sweet-grass (Scannell & Synnott, 1987); Flote grass (Brunker, 1950)
- *Milseán uisce* (Scannell & Synnott, 1987); *Glyceria snamhuighill*[128] (White, 1803)

POACEAE (Gramineae)—THE GRASS FAMILY

Native

128. White (1803) applied this Irish name to *Festuca fluitans*.

Glyceria fluitans is quite common throughout Ireland in ditches, ponds, pools, drains and slow-moving streams.

*No ethnobotanical uses of this plant have been noted in Ireland.

Glyceria maxima (Glyceria aquatica; Poa aquatica)

- Reed sweet-grass (Scannell & Synnott, 1987); Reed manna (Brunker, 1950) • *Milseán mór* (Scannell & Synnott, 1987); *Féur milis* (Moloney, 1919; Dinneen, 1927); *Cuise cuilcamhuil*[129] (Wade, 1808a; White, 1803)

POACEAE (Gramineae)—THE GRASS FAMILY
Native

Glyceria maxima is a common plant in the centre and south-east in river-banks, canals and ditches.

Mackay (1836: 303) said *Glyceria maxima* is a useful grass to plant on river-banks as grazing for horses, cows and goats, 'who are exceedingly fond of it'. Wade (1808a) indicated this grass is sweet, nutritious and liked by cattle. White (1803: 45) said that it is a useful grass to sow on the banks of rivers, 'as it would prevent them from being worn away by the rapidity of the water'.

**Glyceria plicata (Glyceria notata)*

- Plicate sweet-grass (Scannell & Synnott, 1987)
- *Milseán géagach* (Scannell & Synnott, 1987)

POACEAE (Gramineae)—THE GRASS FAMILY
Native

Glyceria plicata is a common grass throughout Ireland of ditches and slow streams.

*No ethnobotanical uses of this plant have been noted in Ireland.

**Gnaphalium sylvaticum (Omalotheca sylvatica)*

- Heath cudweed (Scannell & Synnott, 1987)
- *Gnamhlus móna* (Scannell & Synnott, 1987); *Liathlus beág*[130] (Moloney, 1919); *Lochall caol* (Moloney, 1919)

ASTERACEAE (Compositae)—THE DAISY FAMILY
Native

Gnaphalium sylvaticum is a rare and decreasing plant in Ireland of damp sandy places and upland pastures.

De Vere Coneys (1849) gave the Irish word *cadhluibh* (*-bhe*) for 'the herb cudweed, gnaphalium', which may refer to this species, or to a species of *Filago*.

*No ethnobotanical uses of this plant have been noted in Ireland.

129. Wade (1808a) gave the Irish name *cuise cuilcamhuil* for *Poa aquatica*.
130. Dinneen (1927) defined *liathlus beág* as 'upright cudweed'.

Groenlandia densa

- Opposite-leaved pondweed (Scannell & Synnott, 1987)
- *Líobhógach dlúth* (Scannell & Synnott, 1987)

POTAMOGETONACEAE—THE PONDWEED FAMILY
Native

Groenlandia densa is a rare aquatic pondweed of streams and canals that occurs mainly in the centre and south in Ireland. It is a protected species in Ireland under the Wildlife Act of 1976.

*No ethnobotanical uses of this plant have been noted in Ireland.

Gunnera tinctoria

- Giant-rhubarb (Scannell & Synnott, 1987)
- *Gunnaire* (Scannell & Synnott, 1987)

GUNNERACEAE—THE GUNNERA FAMILY
Introduced

Gunnera tinctoria is a South American plant species, with huge, spreading leaves that can grow up to 2 m across that is now well-established in Ireland and spreading as an invasive alien plant, particularly in parts of the west, such as in Achill Island, Co. Mayo, where it is a serious pest. It was originally introduced into Ireland as an attractive herbaceous plant for large gardens, particularly in stream banks, and has become well adapted to Irish conditions and is spreading widely.

In its native Chile it is used for several ethnobotanical purposes, including as a dye plant (hence its specific name 'tinctoria'). It is also regarded as edible there. Apart from its use in ornamental horticulture, I know of no other ethnobotanical use to which the plant has been put in Ireland.

Despite its common name, it is unrelated to garden rhubarb (in Irish, *biabhóg* or *bióbhóg*, Uí Chanainn, 1962), which belongs to a different genus, *Rheum*.

Gymnadenia conopsea.

Gymnadenia conopsea (Habenaria conopsea)

- Fragrant orchid (Scannell & Synnott, 1987)
- *Lus taghla* (Scannell & Synnott, 1987); *Lus taighte* (Moloney, 1919; Dinneen, 1927)

ORCHIDACEAE—THE ORCHID FAMILY
Native

Gymnadenia conopsea is an orchid of grasslands, sand dunes, fens and damp peaty places that is found mainly in the centre of Ireland.

*No ethnobotanical uses of this plant have been noted in Ireland.

Halimione portulacoides (Atriplex portulacoides)

- Sea purslane (Scannell & Synnott, 1987)
- *Lus an ghaill* (Scannell & Synnott, 1987)

AMARANTHACEAE (Chenopodiaceae)—THE GOOSEFOOT FAMILY

Native

Halimione portulacoides is a low-growing shrub of salt marshes and muddy seashores that is mainly found on the east and south coasts.

Mears and Hillman (2007) and Phillips (1983) said this saltmarsh plant is useful as a vegetable or pickled. Mabey (1972) suggested that the fleshy leaves are a useful addition for salads.

It was formerly used for medicinal purposes in Ireland, and according to Threlkeld (1726: unnumbered [21]), it 'has obtained a great reputation among the good Women both in Decoction and Conserve for Uterine Disorders, Hysteric Fits, and Menstrual Evacuations'.

**Hammarbya paludosa (Malaxis paludosa)*

- Bog orchid (Scannell & Synnott, 1987)
- *Magairlín na móna* (Scannell & Synnott, 1987); *Magairlín an churraigh* (Moloney, 1919); *Magairlín an chorraigh* (Ó Dónaill, 1977)

ORCHIDACEAE—THE ORCHID FAMILY

Native

Hammarbya paludosa is a small orchid that grows amongst *Sphagnum* moss in the wettest bogs; it is rare in Ireland.

*No ethnobotanical uses of this plant have been noted in Ireland.

Hedera helix

- Ivy (Scannell & Synnott, 1987); Eenagh (Hart, 1898)
- *Eidhneán* (Scannell & Synnott, 1987); *Feidhneán* (Moloney, 1919); *Eidheann* (Moloney, 1919); *Faithlah* (Cameron, 1883); *Aighneann* (Cameron, 1883); *Gort* (de Vere Coneys, 1849; Cameron, 1883; Uí Chanainn, 1962); *Ialluin* (Cameron, 1883); *Iadh-shlat* (Cameron, 1883; Dinneen, 1927); *Gath* (Dinneen, 1927); *Íodha*[131] (Dinneen, 1927; Uí Chanainn, 1962); *Aidhneán* (Ó Criomhthain, MSS, n.d., ca. 1908)

ARALIACEAE—THE IVY FAMILY

Native

Hedera helix is a very common woody climber in woodlands, scrub and hedgerows and can be found on rocky cliffs and walls throughout Ireland.

It was formerly recommended for use in a foot bath for 'soft' or perspiring feet (Moloney, 1919) and to treat corns, bunions or verrucas (K'Eogh, 1735; Ui Sheasnáin, 1982; MacMahon, 1994; Allen & Hatfield, 2004). Ivy leaves were soaked in vinegar to soften them and made into a poultice that was bound onto the foot or boiled, and the resulting liquid rubbed on (Logan, 1981). A simpler method noted by Allen and Hatfield (2004) was to wear the leaves inside the sock and the corn was reputed to drop off without any pain some days later. Ivy leaves steeped in vinegar and left for 48 hours were also used in traditional medicine

131. Dinneen (1927) defined this word as a 'yew tree'.

Hedera helix (ivy). Watercolour by Lydia Shackleton. Courtesy of the National Botanic Gardens, Glasnevin.

Hedera helix leaves.

Hedera helix berries.

in Co. Cavan to treat a varicose ulcer (Ui Sheasnáin, 1982): the solution was painted on the ulcer. This solution was also used to treat burns and scalds. Boiled ivy leaves were used to treat sore eyes (Munnelly, 2005). According to Allen and Hatfield (2004), the plant was used in Ireland to treat coughs, colds and bronchitis, burns, eczema in children, jaundice and kidney problems, thorns in fingers, boils and abscesses, inflammations and cuts, and its berries were used as a cure for aches and pains. Bridget Dirrane (1997: 2–3) described ivy leaves being used to treat a bad abscess when growing up as a child in Inishmore, Aran Islands: water and ivy leaves were collected from a holy well in Corough and mixed with some earth and applied as a poultice to her abscess. She wrote, 'Believe me that did the trick. Early next morning I woke up soaking wet and wakened my sister Julia, who slept beside me, to show her all the pus and badness that had seeped out. It ran for some time afterwards but eventually dried up. In the end, a snail was put on my neck to dry it up completely. It did the work of a leech and the wound healed up in time'. Ivy leaves were also chewed and the juice swallowed to 'clear' the chest (Logan, 1972). Williams (1993) noted that it was used to treat headaches and earaches in Ireland, and that a tea made from the plant was useful for lung complaints.

Allen and Hatfield (2004) also noted its use in Ireland as a veterinary remedy to treat various ailments: eye troubles in cows and sheep, warts in cattle, liver fluke, digestive troubles, mastitis and expelling afterbirth. Ui Sheasnáin (1982) also notes its use for expelling afterbirth in cows in Co. Cavan; chopped ivy leaves were put in a mash and given to the cow to eat.

In mythology Bacchus wears a garland of ivy leaves around his head because in some countries it was believed that ivy can overcome drunkenness. Similarly, describing a cure that involved the wearing of ivy, Hart (1898) said that in Donegal a ring of leaves sewn together made from ivy was put around an arm or leg to help cure erysipelas.

Kelly (1997) recorded that ivy and holly were used as a source of winter fodder for livestock in early Ireland.

Ivy has been used as part of Irish Halloween traditions and as a symbol of eternal life. Each member of the family would place an ivy leaf in a cup of water overnight. If it was still perfect and without spots the following day, the person could be assured good health for the following year. Ivy is also traditionally associated with Christmas for decorative purposes and was used widely in Ireland. Eibhlís Ní Shúilleabháin wrote in 1931 that on the Blasket Islands ivy was used to decorate the walls indoors during the Christmas season (holly does not grow on the Blaskets) (Ní Shúilleabháin, 1988). In recent years it has largely been replaced in the modern Irish home with shop-bought decorations and holly.

In former times during the Puck Fair in Killorglin, Co. Kerry, sprigs of ivy were given out from a float on which The

Puck (a large male goat) was carried through the streets in a procession. The float was decorated with coloured paper and wreathed with ivy. Murray (1953: 352) described the fair: 'Six young boys, dressed in green and wearing green cloaks lined with yellow, stood three on each side of the shrine [which held the goat]. Every now and then young men came to the lorry and were given sprigs of ivy by one of the boys'.

The woody stems of ivy have been used to make walking sticks and for smaller household objects, and ivy roots were once used by leather workers to sharpen their knives (Edlin, 1973). Rutty (1772: 117) mentioned such use and said that wood taken from the root was used by shoemakers 'after they have sharpened their knives by a whetstone, they then take off the asperities left, and smoothed the edge by drawing the knife to and fro upon the wood; which by reason of its softness is peculiarly fit for this purpose'.

The hard, black fruits of ivy have been used as ammunition for children's pea-shooters. Gerry Sharkey (pers. comm., 2009) tells of them being used effectively in Co. Dublin to attack the boys of the local FCA (part-time Civil Defence Forces) by other boys, the self-styled 'Revolutionary Socialists'!

Various cultivated varieties of *Hedera helix* are grown as ornamental plants in Irish gardens.

Ivy is generally regarded as a poisonous plant. Woodward (1985: 14) gives the symptoms of poisoning by ivy as 'sap may cause dermatitis; ingestion of berries leads to strong purgation, in large quantities to vomiting, diarrhoea, coma'.

**Helianthemum nummularium*

- Common rock-rose (Scannell & Synnott, 1987)
- *Grianrós* (Scannell & Synnott, 1987)

CISTACEAE—THE ROCK-ROSE FAMILY
Native

Despite its common name, *Helianthemum nummularium* is a very rare species in Ireland that is known only from East Donegal, near Ballintra.

*No ethnobotanical uses of this plant have been noted in Ireland.

Helianthemum oelandicum subsp. *piloselloides* (*Helianthemum canum*)

- Hoary rock-rose (Scannell & Synnott, 1987)
- *Grianrós liath* (Scannell & Synnott, 1987); *Plúr na gréine* (Moloney, 1919; Dinneen, 1927); *Rós na carraige* (Moloney, 1919)

CISTACEAE—THE ROCK-ROSE FAMILY
Native

Helianthemum oelandicum subsp. *piloselloides.*

Helianthemum oelandicum subsp. *piloselloides* is found on limestone pavement and grasslands in the Burren region of Co. Clare.

Moloney (1919) indicated this plant (as *Helianthemum vineale*) was 'said to have been used in the treatment of shingles or wild fire (Herpes zoster)'.

Moloney (1919) gave the Irish name *plúr na gréine* for 'rock rose', but it is uncertain to me whether this name refers to a *Rosa* or *Cistus* species or to *Helianthemum*.

The likelihood that *Helianthemum oelandicum* was ever used for medicinal purposes must therefore be in some doubt.

Heracleum mantegazzianum

- Giant hogweed (Scannell & Synnott, 1987)
- *Feabhrán capaill* (Scannell & Synnott, 1987); *Ollfheabhrán* (Anonymous, 1978)

APIACEAE (Umbelliferae)—THE CARROT FAMILY
Introduced

Heracleum mantegazzianum is a very tall and robust perennial herb that can be found in an increasing number of waste places, roadsides and river-banks around Ireland. It is a dangerous invasive, not only because it can grow to dominate the vegetation in which it occurs, but also because its sap can cause a serious photoallergic reaction (phytophotodermatitis), causing skin blistering whenever touched. It is so seriously regarded at the National Botanic Gardens that any plants that arise from seed washed down along the banks of the River Tolka in the Gardens in Glasnevin are immediately eradicated, and it is not included at all in the collections.

O'Regan (1997) mentioned that a poultice of mashed comfrey leaves has been used to treat the severe burns, rashes, blistering and swellings caused by giant hogweed.

Giant hogweed was originally introduced from the Caucasus Mountains in south-west Asia as an ornamental plant. Despite its potential to cause blistering to hands and face, its hollow stems have been used in recent years by children in Co. Dublin, at Cardiff's Bridge, to make pea-shooters (Gerry Sharkey, pers. comm., 2009).

Heracleum sphondylium

- Hogweed (Scannell & Synnott, 1987); Cow-parsnip (Colgan, 1904); Hefferaun (or Heffrane) (Colgan, 1904); Shooraun (Colgan, 1904); Singer (Uí Chonchubhair, 1995); Orn (or Oran) (Hart, 1898); Pipers (McCarthy & Hawkes, 2007); Keck (Harron, 1986)
- *Feabhrán* (Scannell & Synnott, 1987); *Amhrán* (Ua Maoileoin, 2001); *Odharán* (de Vere Coneys, 1849); *Faidhf na mada í* (Dinneen, 1927); *Gleorán* (Dinneen, 1927); *Odhrán* (Dinneen, 1927; Tóibín, 1965); *Pleascán* (Dinneen, 1927); *Faidhf* (Ó Dónaill, 1977); *Pléascán mór*[132] (Ó Dónaill, 1977); *Feabhránach* (Tóibín, 1965); *Odhránach* (Tóibín, 1965); *Pléasgán* (Colgan, 1911); *Fuarán* (Colgan, 1911); *Siurán* (Colgan, 1911); *Uaran* (Colgan, 1911)

APIACEAE (Umbelliferae)—THE CARROT FAMILY
Native

Heracleum sphondylium is a very common robust perennial herb of neglected pastures, roadsides, hedgerows and waste places.

Young shoots can be cooked as a vegetable in spring, fried in butter or boiled. I

---

132. *Pléascán mór* was also given by Ó Dónaill (1977) for 'wild angelica'.

sometimes pick and eat the raw young shoots in spring as a snack while on fieldwork, peeling off the tougher outer skin first. Mabey (1972) described the shoots as 'marvellously fleshy'.

According to Allen and Hatfield (2004), *Heracleum sphondylium* was used medicinally in Ireland to stop bleeding (Allen & Hatfield, 2004). Uí Chonchubhair (1995) recorded that this plant is called 'The Singer' in the Tralee area of Co. Kerry because of the noise that it makes as the wind whistles through its dried stems in winter.

Hollow stems of hogweed have been used by children to make pea-shooters. Cameron (1883) gave a Scots Gaelic name for the plant, *gunnachan sputachain*, meaning 'squirt-guns', used because of its hollow stems. Colgan (1911: 4.13) said that its Irish name, *pléasgán*, is 'a contraction of *gunna pléasgán*, pop-gun, for which the hollow stems are used by boys'.

It was used as a veterinary remedy to treat rheumatism in cattle in Ireland (Allen & Hatfield, 2004). Roy Vickery also included this use of *Heracleum sphondylium* in the Plant-lore Archive (Vickery, 2010), and mentioned that one of his correspondents used to boil it up in water and give it to cattle to keep rheumatism at bay. Cameron (1883) said that the whole plant is wholesome and nourishing for cattle.

Heracleum sphondylium.

Hieracium species

- Hawkweed (Scannell & Synnott, 1987)
- *Lus na seabhac* (Scannell & Synnott, 1987); *Caisearbhán na muc* (Dinneen, 1927)

ASTERACEAE (Compositae)—THE DAISY FAMILY

As many as 50 different wild hawkweed species can be found in Ireland in habitats such as rocky places, walls, cliffs, dry banks, woodlands, heaths and river-banks. Hawkweeds are extremely difficult to identify to species level, even for an expert. They produce seed without pollination and so any plant that has developed individual characteristics can pass these on to all following generations, in effect, creating a new species. Dandelions (*Taraxacum*) reproduce in a similar way and are represented by thousands of 'microspecies'.

I have not found ethnobotanical uses for hawkweed in Ireland except for *Hieracium scullyi*, which is a robust and showy yellow-flowered species known only from Co. Kerry. It has been grown at the National Botanic Gardens in Dublin and is a promising and rewarding ornamental plant. *Hieracium scullyi* was named in honour of Reginald Scully, author of the *Flora of County Kerry*, published in 1916.

Cameron (1883) gave the Irish name *srubhan na muc* for one species, *Hieracium murorum*. Dinneen (1927: 1115) gave the same name for 'great hawkweed'.

Hieracium scullyi.

**Hieracium umbellatum*

- A hawkweed

ASTERACEAE (Compositae)—THE DAISY FAMILY

Native

Hieracium umbellatum is a rare species in Ireland of shady river-banks and other similar habitats throughout.

*No ethnobotanical uses of this plant have been noted in Ireland.

**Hierochloe odorata*

- Holy-grass (Scannell & Synnott, 1987)
- *Féar Muire* (Scannell & Synnott, 1987)

POACEAE (Gramineae)—THE GRASS FAMILY

Native

Hierochloe odorata is a very rare grass in Ireland that is known only from one marsh near Lough Neagh in Northern Ireland.

*No ethnobotanical uses of this plant have been noted in Ireland.

Hippophae rhamnoides (sea buckthorn). Watercolour by Lydia Shackleton. Courtesy of the National Botanic Gardens, Glasnevin.

Hippophae rhamnoides

- Sea buckthorn (Scannell & Synnott, 1987); Biddy bush (Colgan, 1904)
- *Draighean mara* (Scannell & Synnott, 1987)

ELAEAGNACEAE—THE SEA-BUCKTHORN FAMILY

Introduced

Hippophae rhamnoides is a dense, spiny shrub that has been planted in several parts of Ireland to help stabilise sand dunes. In some areas it has become dangerously invasive, such as in Magherabeg, north of Castlegregory, Dingle Peninsula, Co. Kerry.

It produces orange-yellow berries that are widely used in other countries for food, cordials and conserves and made into fruit leathers (see *Prunus spinosa*). Male and female flowers occur on separate plants and so good crops of fruit will only occur when both are present. The flowers are wind pollinated. In some colonies of this plant in Ireland the fruits are produced only sparingly, which makes their collection tedious and hardly worth the effort. Mears and Hillman (2007) suggested that the fruits are strongly purgative for some people. The seeds can be roasted and ground down to make flour.

Hippophae fruits are internationally important as a source of vitamins for the pharmaceutical industry (Ciesla, 2002) and contain vitamins such as provitamin A, vitamins B, B_2 (riboflavin), E (okoferol), K (phyllokinon), P and PP. Their use has been the subject of considerable research in Russia and Eastern European countries, and it is recorded that they formed part of the diet of Russia's cosmonauts and were made into a cream to protect against cosmic rays (Ciesla, 2002). In Ireland, as yet, they are little used.

It has been grown in some Irish gardens as an attractive grey-leaved ornamental plant.

Hippuris vulgaris

- Mare's-tail (Scannell & Synnott, 1987)
- *Colgrach* (Scannell & Synnott, 1987); *Cáití collagán* (Moloney, 1919); *Brobh i dtóin* (Ó Dónaill, 1977)

PLANTAGINACEAE (Hippuridaceae)—THE MARE'S-TAIL FAMILY
Native

Hippuris vulgaris is an aquatic plant of ponds, slow-moving streams, lake margins, drains, ditches and canals that is found mainly in the centre of Ireland and is less common elsewhere.

According to Wade (1804), the whole plant was considered a weak astringent.

Mahon (1983) mentioned a yellow dye was obtained from this plant.

Holcus lanatus

- Yorkshire fog (Scannell & Synnott, 1987)
- *Féar an chinn bháin* (Scannell & Synnott, 1987); *Minfheur fadchalgach* (Wade, 1808a); *Minfhér birchalgach* (White, 1803)

POACEAE (Gramineae)—THE GRASS FAMILY
Native

Holcus lanatus.

Holcus lanatus is an abundant perennial grass throughout Ireland of pastures, meadows, roadsides and waste places.

In Scotland, Milliken and Bridgewater (2004) recorded that it has been used to make tethers for animals and reins in Shetland and Orkney. I have not found records of its use for such purposes in Ireland.

It is used in some commercial wildflower seed mixes.

Peterson (1977) listed this species as poisonous.

**Holcus mollis*

- Creeping soft-grass (Scannell & Synnott, 1987)
- *Mínfhéar reatha* (Scannell & Synnott, 1987); *Minfheur birchalgach* (Wade, 1808a); *Minfhér fadchalgach* (White, 1803)

POACEAE (Gramineae)—THE GRASS FAMILY
Native

Holcus mollis is a perennial grass throughout much of Ireland of woodlands, hedgerows, heathy places and wasteground.

*No ethnobotanical uses of this plant have been noted in Ireland.

Honckenya peploides

- Sea sandwort (Scannell & Synnott, 1987)
- *Gaineamhlus mara* (Scannell & Synnott, 1987); *Lus na gainimhe* (Uí Chonchubhair, 1995)

CARYOPHYLLACEAE—THE CARNATION OR PINK FAMILY
Native

Honckenya peploides (sea sandwort). Watercolour by Lydia Shackleton. Courtesy of the National Botanic Gardens, Glasnevin.

Honckenya peploides is a coastal plant all around Ireland of sandy and shingle seashores.

Mears and Hillman (2007) noted that this succulent plant can be eaten as a crunchy green vegetable or in salads.

---

*Hordelymus europaeus

- Wood barley (Scannell & Synnott, 1987)
- *Eorna choille* (Scannell & Synnott, 1987)

POACEAE (Gramineae)—THE GRASS FAMILY
Possibly Introduced

Hordelymus europaeus is a very rare grass known only from one shady woodland habitat in the Glens of Antrim in Northern Ireland.

*No ethnobotanical uses of this plant have been noted in Ireland.

Hordeum murinum

- Wall barley (Scannell & Synnott, 1987)
- *Cuiseogach fhionn* (Scannell & Synnott, 1987); *Eorna balla* (White, 1803; Wade, 1808a)

POACEAE (Gramineae)—THE GRASS FAMILY
Introduced

Hordeum murinum is a common annual weed, particularly in the Dublin region, that grows in waste places and roadsides. The flowerheads have been used by children for play darts.

White (1803: 124) mentioned that 'before the flower spike appears [this grass species] is relished much by cattle; but afterwards is much disagreeable and injurious to them, from the awns or beard of the ears which stick in their mouths and cause a soreness that renders them incapable of eating for some time' and 'while it is in foliage it is generally kept down by cattle, but when shooting for flower is afterwards left untouched'.

---

*Hordeum secalinum

- Meadow barley (Scannell & Synnott, 1987)
- *Eorna mhóinéir* (Scannell & Synnott, 1987); *Eorna leana*[133] (Wade, 1808a); *Eorna léna*[134] (White, 1803); *Eorna mara*[135] (White, 1803)

POACEAE (Gramineae)—THE GRASS FAMILY
Native (or possible archaeophyte)

Hordeum secalinum is a rare grass found in grassland near the sea, mainly in the southern half of Ireland.

*No ethnobotanical uses of this plant have been noted in Ireland.

133. Wade (1808a) applied this name to *Hordeum pratense*, a synonym of *H. secalinum*.
134. White (1803) applied this name to *Hordeum pratense*.
135. White (1803) applied this name to *Hordeum maritimum*, 'sea barley'.

*Hottonia palustris

- Water violet (Scannell & Synnott, 1987)
- *Cleiteán uisce* (Scannell & Synnott, 1987)

PRIMULACEAE—THE PRIMROSE FAMILY
Possibly Introduced

Hottonia palustris is a very rare aquatic plant of ponds, marshes and ditches that is likely to have been introduced in most of its localities and may not be native in Ireland.

*No ethnobotanical uses of this plant have been noted in Ireland.

Humulus lupulus

- Hop (Scannell & Synnott, 1987)
- *Hopa* (Scannell & Synnott, 1987); *Sail-bhraich* (Moloney, 1919); *Lionnluibh* (Moloney, 1919); *Seanbhlus* (Moloney, 1919; Dinneen, 1927); *Lus an leanna* (Moloney, 1919; Dinneen, 1927; Ó Dónaill, 1977); *Brach* (*-aiche*)[136] (Uí Chanainn, 1962)

CANNABACEAE—THE HEMP FAMILY
Introduced

Humulus lupulus is a widespread, but rare, introduced perennial climbing plant. It was widely cultivated in Ireland for its use in beer-making and is now established in the wild throughout, although never abundantly. Hops are an essential ingredient in the production of Guinness stout (and other Irish beers too of course). Those hops used in brewing Guinness stout in Ireland are imported.

Hops are sometimes cultivated as a garden plant and can escape into hedgerows, roadsides and waste places. Today, it is only a rarely grown crop in Ireland. Emboden (1972) included it amongst his list of narcotic plants of the world, listing it as a sedative and soporific plant. In traditional medicine it has been used in Europe, Asia and North America as a bitter tonic and diuretic, and in more recent times for mood and sleep disorders, anxiety and restlessness (van Wyk & Wink, 2004). Hop extracts are used in a number of sleep-promoting teas and sedatives.

A famous stand of this plant survives growing up the back steps of the Cashel Palace in Cashel, Co. Tipperary. These plants are said to be descended from those grown by a former Archbishop of Cashel, Archbishop Arthur Price. He was godfather to Arthur Guinness, who at the age of 27 in 1752 was left 100 pounds in the Archbishop's will, which he invested in a brewery in Leixlip. He later set up his famous brewery at St James's Gate in Dublin. His father Richard Guinness (or Guinis) was land steward to the Archbishop of Cashel. As a child I spent my first years living in the Cashel Palace, where my father was Dean of Cashel. In our day the house was the Deanery. I remember being told by my father that these hops were the original Guinness hops.

Young hop shoots are edible and can be cut in spring and cooked like asparagus (boiled or steamed) and are best served with Hollandaise sauce.

Male plants are prohibited in the Republic of Ireland under the Noxious Weeds (Male Wild Hop Plant) Order, 1965.

Hops were also used to provide a brown dye for dyeing textiles (Mahon, 1983). Rutty (1772) wrote that in Sweden cloth was made from its stalks. For this purpose it is harvested

136. Uí Chanainn (1962) defined *brach* (*-aiche*) as 'the hop plant; hops'.

in autumn, soaked in water all winter and then dried in a stove in March. The fibres can then be extracted like those of flax and yield strong white filaments that can be spun and woven.

Hyacinthoides non-scripta (Endymion non-scriptus; Scilla non-scripta; Scilla nutans)

- Bluebell (Scannell & Synnott, 1987); Bummuck (Hart, 1898); Burmuck (Hart, 1898); Crow-picker (Hart, 1898); Wild hyacinth (Praeger, 1938)
- *Coinnle corra* (Scannell & Synnott, 1987); *Bugha* (Moloney, 1919); *Butha muc* (Moloney, 1919); *Fuath muice* (de Vere Coneys, 1849; Moloney, 1919); *Connle corra* (Moloney, 1919); *Bú muc* (Ó Dónaill, 1977); *Cloigín gorm* (Ó Dónaill, 1977); *Coinnle coradh* (Ó Criomhthain, MSS, n.d., ca. 1908)

ASPARAGACEAE (Hyacinthaceae)—THE ASPARAGUS FAMILY

Native

Hyacinthoides non-scripta.

Hyacinthoides non-scripta is native to western Europe and common throughout most of Ireland. This attractive plant occurs in woodlands and other shady habitats, such as grassy banks, hedges and thickets. It is also a characteristic plant of some coastal heathy habitats and grows amongst bracken (*Pteridium aquilinum*).

Its close relative the Spanish bluebell (*Hyacinthoides hispanica*) is the bluebell most frequently seen growing in gardens where it is a persistent and weedy species that has also become naturalised quite frequently in the wild. This species hybridises easily with the native bluebell.

Hyacinthoides non-scripta is little used medicinally in Ireland, although Allen and Hatfield (2004) noted records of it being boiled in Cavan for throat ailments and applied to whitlows in Monaghan. Williams (1993) recorded the use of its roots in Co. Donegal to treat coughs.

According to Williams (1993), a glue in its roots was used as an adhesive. MacCoitir (2006) mentioned the use of this glue for purposes such as bookbinding and setting the tail feathers on arrows and as a useful substitute for starch. To make the glue, sometimes the bulb would be chewed, then spat out, but as this plant is poisonous, care must have been taken not to swallow! Rutty (1772: 117) also referred to the bluebell as being a source of 'glutinous juice whereof paste is made'.

The bluebell is regarded as a poisonous plant if eaten, and Woodward (1985) listed the symptoms of poisoning as abdominal pain, diarrhoea and bloody urine.

In Irish mythology Gráinne used it (mixed with tormentil, *Potentilla erecta*) at a great betrothal feast at Tara to put her husband-to-be, Fionn MacCumhaill, and all his companions asleep before she escaped with her lover-to-be Diarmaid (MacCoitir, 2006). During the early 20th century the Norwegian scholar Carl Marstrander asked the Islandman Tomás Ó Criomhthain to compile a list of the flora and fauna of Kerry's Great Blasket Island for him. This list is preserved in the University of Oslo in Norway. Amongst the list is a name given as *néunartaċ*, which is annotated as 'a kind of bulb with a few bare leaves and a bluish flower, it was formerly used for tanning leather'. This may refer to this species, although the list also includes an entry for '*Coinnle coradh*', which is certainly bluebell and is described as 'a bulb growing in bogs; the stem is long and its blossom is a lovely blue'.

The other possibility is that *néunartaċ* refers to a blue-flowered species of *Allium*, but the Irish naturalist R. M. Barrington (1881) does not list any such species as occurring on the Blasket Islands. The other possibility is that this is an error for a yellow-flowered plant, *Potentilla erecta* (which has the Irish name *néanthartach*), which was formerly quite widely used for tanning leather.

Hydrilla verticillata

- Hydrilla (Scannell & Synnott, 1987)
- *Ilphéistein fáinneach* (Scannell & Synnott, 1987)

HYDROCHARITACEAE—THE FROGBIT FAMILY
Native

Hydrilla verticillata is a very rare aquatic plant in Ireland that is known only from one small lake close to the seashore at Renvyle, West Galway.

Hydrilla is occasionally grown as an ornamental plant in home aquaria.

Hydrocharis morsus-ranae

- Frogbit (Scannell & Synnott, 1987)
- *Greim an loscáin* (Scannell & Synnott, 1987); *Lus loscáin* (Moloney, 1919; Dinneen; 1927); *Bior-rós* (Moloney, 1919)

HYDROCHARITACEAE—THE FROGBIT FAMILY
Native

Hydrocharis morsus-ranae is a rare aquatic plant in Ireland that is found mainly in the east in ditches, marshes, bog-pools, swamps, canals and slow-moving streams.

Writing about folk medicine in Co. Donegal, Gallagher (2012) said that 'frog bit' 'was always used as a great medicine long ago, but if you have any breakout or troublesome pimples, gather a handful of frog bit, wash it thoroughly. Place into a jar with half a pound of clarified lard. Stand the jar in a pan of water and simmer for at least two hours, stirring frequently. An ointment is made that can be used for all types of cuts and bruises'.

Hydrocotyle vulgaris

- Marsh pennywort (Scannell & Synnott, 1987); White-rot (Harron, 1986)
- *Lus na pingine* (Scannell & Synnott, 1987); *Lus pingne (Curraigh)* (Moloney, 1919); *Lus na pinginne curraigh* (Dinneen, 1927)

APIACEAE (Umbelliferae)—THE CARROT FAMILY
Native

Hydrocotyle vulgaris is found frequently throughout Ireland in habitats such as marshes, fens, wet meadows, bog-margins, ditches and lake-shores.

*No ethnobotanical uses of this plant have been noted in Ireland.

Hyoscyamus niger

- Henbane (Scannell & Synnott, 1987)
- *Gafann* (Scannell & Synnott, 1987); *Bainne cích na n-éan* (Ua Maoileoin, 2001); *Caoch na gcearc* (de Vere Coneys, 1849; Moloney, 1919); *Deódha gafann* (Moloney, 1919); *Crann gafainn* (Moloney, 1919, Dinneen, 1927); *Beann mhear* (Dinneen, 1927); *Deodha* (Dinneen, 1927)

SOLANACEAE—THE NIGHTSHADE FAMILY
Native or Possibly Introduced (possible archaeophyte)

Hyoscyamus niger is a very rare plant in Ireland today of stony and sandy seashores and is a very rare and fleeting weed of cultivated ground and gardens. Occasional plants are seen in some localities where it has been absent for years, no doubt springing from seed lying dormant in the sand for years and revealed when disturbed by a winter storm.

Webb (1996) regarded henbane as a native species in Ireland, although other authors have suggested that it may be an introduction. If it is introduced, then it was a very early introduction indeed. In all fairness, though, it is probably indigenous in many of its coastal localities.

Henbane has been found in deposits excavated from Viking Dublin dating back more than 1000 years ago (Geraghty, 1996). The same author reported that it was used by the Vikings as a medicinal/magical plant and could have been used in Dublin for medicinal or other purposes.

Despite being toxic, even when taken in low doses, *Hyoscyamus niger* has had a long history of use throughout its native range in Europe and Asia. Smoke from roasting henbane seeds was inhaled as a treatment for toothache in some European countries. It is toxic when taken fresh, dried or cooked and must be regarded as a lethally dangerous plant. It has been used as a sedative and to promote sleep and as a pain killer and anaesthetic. In Co. Cavan, Maloney (1972) reported that it was used to treat whooping cough. The leaves were boiled and the juice then drunk. According to Rätsch (1997), its seeds were roasted in medieval European bathhouses to create an erotic mood amongst the guests. It is rich in physiologically active compounds and has a long history of medicinal use for its hypnotic and sedative properties. Henbane has been smoked for its hallucinogenic properties in various parts of the world (Pennacchio et al., 2010) and contains a number of tropane alkaloids such as hyoscyamine, scopolamine and atropine (Emboden, 1972), which are used today in modern medicine. Apparently, henbane produces visual hallucinations and a sensation of flight when ingested, which may explain its use by witches in potions to aid in flying on their broomsticks. In Germany it was used to flavour beer up until 1516 when its use was banned. In 2008 the British 'celebrity' chef Anthony Worrell Thompson suggested that its leaves would be a useful addition to summer salads. He was quickly forced to retract his advice with a public apology that was featured on national news bulletins in Britain and Ireland. He said that he had meant to recommend 'fat hen' (*Chenopodium album*). Kelly (1997) said that henbane is notorious as a poisoner of hens and other poultry: in French the plant is called *mort aux poules*, or 'death to hens', and there are references to the ill effects of this plant on people in early Irish texts. Perhaps as early as the 10th century there is reference to henbane which implies that drinking it causes madness.

In Ireland henbane seems to have been only a little used medicinally, although there are some records of its use in Co. Cavan to treat various ailments including whooping cough in children (Allen & Hatfield, 2004). According to Williams (1993) it was also used by doctors in the Middle Ages in Ireland to treat gout, arthritis and for pain relief, and it has also been used to reduce the swelling of breasts after childbirth and for coughs and toothache and in cooling ointments. It is exceedingly poisonous and lethally toxic in low doses, so the plant must have been used as a medicinal only with the utmost care. A little poem quoted by Wood-Martin (1902: Vol. 2, 177) demonstrates that its dangers were well known:

> Fair to the sight, but by the smell
> Unprized, the henbane's straw-ting'd hell
> With danger pregnant.

Despite its rarity in Ireland, henbane is still recognised by people locally as a poisonous plant, and I have come across its deliberate extermination for that reason on at least one occasion when it was growing on a stony seashore of Illauntannig, in the Magharee Islands, Co. Kerry, in the 1990s. This may be one reason why it is such a rare plant in Ireland. Jude (1933) mentioned that formerly 'large tracts of Henbane existed in the neighbourhood of Fenit and Castlegregory [Co. Kerry] over a century ago, but these have since been plucked up by herbalists, and only a few scattered plants remain'.

Cameron (1883) translated its Irish name *gafann* (or *gabhann*) as 'the dangerous one'.

Hypericum androsaemum

- Tutsan (Scannell & Synnott, 1987)
- *Meas torc allta* (Scannell & Synnott, 1987); *Úlla an mhadaruaidh* (Ua Maoileoin, 2001); *Súile an mhadaruaidh* (Ua Maoileoin, 2001); *Meastuirc allta* (de Vere Coneys, 1849); *Meas allaidh* (Dinneen, 1927)

HYPERICACEAE (Guttiferae)—THE ST JOHN'S WORT FAMILY
Native

Hypericum androsaemum is a small shrub occurring throughout Ireland of woodlands, hedgerows and other shady habitats.

Hypericum androsaemum.

According to Smith ([1756] 1979) and Allen and Hatfield (2004), this plant was formerly used medicinally in Ireland in an ointment with lard for dressing cuts, wounds and carbuncles. It has been used by herbalists for a variety of medical purposes, including both externally to clean wounds and internally as a diuretic (Moloney, 1919). Its common name, tutsan, comes from the French word *tout sain* or 'cure-all'.

Uí Chonchubhair (1995) noted that in Ireland it is particularly associated with Mary and Colm Cille, as well as St John the Baptist, and has both important magical and religious significance.

Hart (1898) gave the name 'butternut' from Derry for the fruit of *Hypericum androsaemum* and other larger garden *Hypericum* species grown in shrubberies.

It is occasionally used as an ornamental plant in gardens and is particularly suitable for shady situations.

Hypericum species

- St John's-wort (Scannell & Synnott, 1987)
- *Luibh eoin bhaiste* (Ó Dónaill, 1977; Webb & Scannell, 1983); *Lus choluim chille* (de Vere Coneys, 1849); *Lus Colaim Chille* (Dinneen, 1927); *Beathnua* (Ó Dónaill, 1977)
- Other Irish names for *Hypericum* species include *Luibh eoin baiste, Lus choluimcille, Allus Mhuire, Beathnuadh boineann* and *Beathnuadh fireann* (Moloney, 1919)

HYPERICACEAE (Guttiferae)—THE ST JOHN'S WORT FAMILY

Including:

Hypericum canadense
- Irish St John's-wort (Scannell & Synnott, 1987)
- *Beathnua gaelach* (Scannell & Synnott, 1987)

Possibly Introduced

Hypericum elodes
- Marsh St John's-wort (Scannell & Synnott, 1987); Mioullough (Hart, 1898)
- *Luibh na chiorraithe* (Scannell & Synnott, 1987); *Luibh an choirithe* (Uí Chonchubhair, 1995; Ua Maoileoin, 2001); *Luibh eoin an uisce* (Dinneen, 1927)

Native

Hypericum hirsutum
- Trailing St John's-wort (Scannell & Synnott, 1987)
- *Lus an fhógra* (Scannell & Synnott, 1987)

Possibly Introduced (possible archaeophyte)

Hypericum humifusum
- Hairy St John's-wort (Scannell & Synnott, 1987)
- *Beathnua sraoilleach* (Scannell & Synnott, 1987)

Native

Hypericum maculatum (Hypericum dubium)
- Imperforate St John's-wort (Scannell & Synnott, 1987)
- *Beathnua gan smál* (Scannell & Synnott, 1987)

Native

Hypericum perforatum
- Perforate St John's-wort (Scannell & Synnott, 1987)
- *Lus na Maighdine Muire* (Scannell & Synnott, 1987); *Eala bhuidhe* (Cameron, 1883)

Native

Hypericum pulchrum
- Slender St John's-wort (Scannell & Synnott, 1987); Rock rue (Hart, 1898)
- *Beathnua baineann* (Scannell & Synnott, 1987)

Native

Hypericum tetrapterum
- Square-stalked St John's-wort (Scannell & Synnott, 1987)
- *Beathnua fireann* (Scannell & Synnott, 1987)

Native

Hypericum canadense is a very rare plant in Ireland of wet peaty habitats and stream banks in the west and south-west. *Hypericum elodes* grows in bogs, marshes and streams and is particularly distributed in the west and south-east and is much rarer elsewhere. *Hypericum hirsutum* is also a very rare plant in Ireland that only occurs in any quantity in the Liffey Valley in Co. Dublin, in woodlands and shady places. *Hypericum humifusum* is a relatively frequent species throughout of peaty, sandy and heathy places. *Hypericum maculatum* is particularly distributed in the southern half of Ireland and occurs in damp places on woodland edges and shady roadsides and in hedgerows. *Hypericum perforatum* grows in more open situations, such as in woodlands, and on dry banks and hedgerows and in grasslands throughout Ireland, although it is nowhere common except in the centre of the country. *Hypericum pulchrum* is much more common and grows in heaths, dry banks, and grassy places on non-calcareous soils. *Hypericum tetrapterum* is one of the commonest St John's-worts in Ireland and is found in wet habitats, such as stream banks, ditches, drains, damp meadows and marshes.

Hypericum species have a long history of use in traditional medicine throughout their range. In particular, *Hypericum* species were used to treat mild cases of depression, anxiety, nervous conditions and mood problems, as well as wounds and burns and stomach and intestinal inflammations and worms (van Wyk & Wink, 2004). *Hypericum* species were also important medicinal plants in Ireland and used to treat a variety of ailments. For example, Egan (1887) recorded the use of St John's-wort to treat gravel, taken as a decoction of leaves and flowers, but does not mention which species were used. Williams (1993) recorded that they were used to treat diarrhoea, sciatica, blood clots and worms and gangrene in old people, as well as 'hard udder' in cows. According to Moloney (1919), a Co. Waterford herbalist recommended St John's-wort for the treatment of depression and insanity. *Hypericum pulchrum* was used as a red-water cure for cattle (Hart, 1898). Allen and Hatfield (2004) noted that *H. elodes* was used in Ireland to treat diarrhoea in cows. O'Regan (1997) mentioned the use of *H. perforatum* as an infusion taken three times daily to treat bed-wetting in old people and children.

Mackay (1836: 38) recorded that *Hypericum* species in Ireland yield 'a resinous and slightly purgative febrifugal juice'. Of *H. perforatum* he said, 'this plant was formerly of great repute for its supposed tonic or stimulating qualities; it is also esteemed as a vulnerary, and became the 'balm of the warrior's wound" (Mackay, 1836: 39). Hart (1898) mentioned that in Ramelton, Co. Donegal, the flowers of *H. perforatum* were steeped in olive oil and used to treat bruises.

Hypericum pulchrum and *H. elodes* were reputedly of value against spells, demons and evil spirits and gave protection from the 'evil eye' (Uí Chonchubhair, 1995).

Uses for some of the species above are not known or documented, but most (except the rarest species) were probably used for medicinal purposes.

Extracts from the flowers of St John's-wort have been used as an ingredient of cosmetics made in Ireland.

Mahon (1983) listed *Hypericum* species as the source of a yellow dye.

Hypochaeris species

ASTERACEAE (Compositae)—THE DAISY FAMILY
Native

Including:

Hypochaeris glabra

- Smooth cat's-ear (Scannell & Synnott, 1987)
- *Cluas chait mhín* (Scannell & Synnott, 1987)

Hypochaeris radicata

- Cat's-ear (Scannell & Synnott, 1987); Long-rooted catsear (Brunker, 1950)
- *Cluas chait* (Scannell & Synnott, 1987); *Castarbhan* (Danaher, 1964a)

Hypochaeris radicata is an extremely common plant in Ireland of grasslands, sand dunes, heaths and roadsides. Its close relative *H. glabra* is much rarer and occurs only along the north coast on sand dunes.

The leaves of cat's ear are edible, if a bit bitter, and widely used in Mediterranean countries for food (Mears & Hillman, 2007) and probably were also used in the past in Ireland. Mears and Hillman report that the leaves can be used in salads, steamed as a vegetable or used in omelettes, pancakes and cheese dishes. *Hypochaeris glabra* is probably also edible but it is very rare in Ireland and is found only on sand dunes on the north coast.

Hypochaeris radicata in acid grassland.

Danaher (1964a) recorded that you often saw 'children and girls at the pleasant task of picking castarbhan (*Hypochaeris radicata*) on ditches and dry banks', along with chickweed (*Stellaria media*), groundsel (*Senecio vulgaris*) and 'the big purple thistle' (probably *Cirsium vulgare*), which were all chopped up and added to chicken feed. He wrote that 'in the spring before it has flowered this little plant looks like a small green rosette, but it was very difficult to pull up as it has a long root, and the gatherers cut it with a small knife, usually a broken table knife kept for the purpose, and this, with a canvas sack, was the gatherer's equipment. Often you saw two children, one on each side of a ditch working away with two knives and one bag, but it was a nice job on a warm day, and could be varied by picking primroses and cowslips and making daisy chains'.

**Hypopitys monotropa (Monotropa hypopitys)*

- Yellow bird's-nest (Scannell & Synnott, 1987)
- *Buíán sailí* (Scannell & Synnott, 1987)

ERICACEAE—THE HEATHER FAMILY
Native

Hypopitys monotropa is a rare plant in Ireland, found only in old woodlands.

*No ethnobotanical uses of this plant have been noted in Ireland.

Ilex aquifolium

- Holly (Scannell & Synnott, 1987); Cullen (Hart, 1898)
- *Cuileann* (Scannell & Synnott, 1987); *Cran-cuilin* (de Vere Coneys, 1849); *Cuilion* (de Vere Coneys, 1849; Ó Dónaill, 1977); *Crann cuilinn* (Dinneen, 1927)

AQUIFOLIACEAE—THE HOLLY FAMILY
Native

Holly is one of the few evergreen trees native to Ireland, and it is found commonly in woodlands (particularly as an understory tree in oak woodlands), glens and hedgerows. Holly is traditionally associated with Christmas for decorative purposes. Such use probably dates back to pre-Christian times when the evergreen holly was regarded as a symbol of continued life and hope in midwinter.

It is a useful timber tree providing hard, dense, light-coloured wood for furniture making and household implements. Holly can grow to form a large tree and some exceptional specimens are known in Ireland. The tallest known holly in Ireland is recorded from Red House, Ardee, Co. Louth, which measures 21.3 m in height and 2.31 m in girth (Anonymous, 2005). MacCoitir (2003) noted that its timber was used to make spears and chariot poles, and that the fresh shoots of holly were used as a fodder for livestock. Kelly (1997) noted this was particularly done in winter, and in Co. Kerry, the upper less-spiny leaves of the holly bush were fed to cattle until relatively recently. Young shoots of holly have been used for basketry, although not commonly (Joe Hogan, pers. comm., 2010). Supplies of home-grown holly timber are very limited in Ireland.

Nelson and Walsh (1993) recorded that holly wood, when stained black, has been used as a substitute for ebony to make piano keys, mathematical instruments, knife handles and bobbins. It has also been used for making inlays in marquetry furniture. Mackay (1836) noted that its timber was used to make veneers and for knife handles and that birdlime can be made from the mucilaginous bark. In Britain, holly wood was also used to make butter moulds and whip stocks (Edlin, 1973). McGlinchey (2007) mentioned that holly (or apple wood) was used to make the shuttle of a spinning loom.

Holly (or birch) poles were used to make the two shafts of wheel-less sliding carts used to transport turf home from bogs (Bell & Watson, 2008) and for other purposes, such as removing boulders from fields and transporting apples from orchards to their stores in Co. Antrim (Evans, 1957). Evans (1957) mentioned that holly was used for such purposes because of its strength and its reputed special sliding qualities. In Ulster such 'slide-carts' were distinguished from those in most other parts of Ireland, as well as from Welsh and Scots examples, by having detachable shoes or separate wooden runners that were pegged to the ends of the shafts and could be removed and replaced when they had worn out (Evans, 1957). Holly was also used to provide the keel for some traditional boats in Ireland, before iron keel-banks became commonplace (according to Evans, 1957, and Donal MacPolin, who wrote about a Mourne 'Norway Yawl' or 'Drontheim', a traditional boat used on the north and north-east coasts of Ireland, in MacCárthaigh, 2008). When this holly wood keel was worn down it could easily be removed and replaced by a new one nailed to the bottom of the boat.

In Co. Offaly, in central Ireland, part of a flail used in the threshing of corn was made from holly wood (Delaney [1980], see also *Salix cinerea*).

Arnold (1968) noted that holly wood burns to give a hotter fire than any other native

Ilex aquifolium (holly). Watercolour by Lydia Shackleton. Courtesy of the National Botanic Gardens, Glasnevin.

Ilex aquifolium flowers.

Ilex aquifolium berries.

wood and so was popular on the Christmas fire.

According to Allen and Hatfield (2004), it was used medicinally in Ireland to treat burns (in Meath), and for a stiff neck in Waterford (the neck was beaten with a sprig from the tree!). Traditional uses in Ireland include treatment for burns and scalds, colic and gall bladder complaints, and to relieve a stiff neck (Allen & Hatfield, 2004). Medicinal uses are limited by its toxicity, but holly has also been used as a diuretic, febrifuge and laxative (Wren, 1988). The berries of holly are not edible and could be fatal to children, but their bitter taste deters most. The toxic compounds include saponins as well as the alkaloid theobromine and ilexanthin, a yellow pigment. Saponins are amphipathic glycosides that are quite bitter in taste and are foaming surfactants of water. Saponins have ethnobotanical uses as fish poisons, since the saponins are water soluble and particularly toxic to fish.

Threlkeld (1726) also recorded that holly bark was used to make birdlime, and its leaves used medicinally to treat colic and gripes. Rutty (1772: 100) described the making of birdlime from the holly: 'The tree is barked in June or July, and the bark is boiled seven, eight, or even twelve hours, viz. until it be very tender, and the ash-coloured bark be separated from the green bark. Having boiled it, and carefully separated the water from it, place it in a heap on the ground, with alternate layers of the bark thus prepared, and fern: let it ferment or rather putrefy for two or three weeks until it turn to a mucilage, then bruise in a mortar until it can be formed like dough: wash the mass, holding and kneading it in the fingers in a running stream; by which means the impurities will be separated, and pure Viscus will be left behind: put in to an earthen vessel four or five days, that it may be further cleaned by despumation, and then put it up in a new vessel for use'. Birdlime was made by boiling the bark of the holly tree as described above and putting this onto twigs or branches of trees or posts, onto which the birds would stick (Lysaght, 1979). The trapping of birds with birdlime is now illegal in Ireland.

The foliage of holly has been used in some European countries to sweep chimneys (Hora, 1981) and probably in Ireland too.

Holly features prominently in a former folklore practice on St Stephen's Day, 26 December. This was the day when it was traditional to display a poor, unfortunate wren, hunted down by village boys. The dead bodies of any wrens caught would be tied to a large holly bush, which itself would sometimes be decorated with rags or, more recently, with strips of coloured paper, which would be carried in procession through the streets on St Stephen's Day to the tunes of 'The wran, the wran, the King of all birds, on St Stephen's Day, was caught in the furze' (Danaher, 1972). 'Furze' is an Irish term for gorse.

Evans (1957) reported the use of holly for a children's game called 'building the house' in which 12 pairs of holly twigs are arranged in a circle, pushed into the ground and tied together at the apex. A live turf representing a fire is placed in the centre and each pair of twigs is named after the girls and boys present. The pair of twigs that catch on fire first represent the couple that will be first to marry. Kinahan (1881) described the use of a holly leaf for another children's game: 'A man's or woman's love may be tested on a holly leaf by going around the thorns, saying "Loves me, loves me not". A head of any kind of corn may be similarly used, or a handful of pebbles'.

Wood-Martin (1902) mentioned that local people around Lough Neagh in Northern Ireland[137] believed that holly wood became fossilised in the lake, and that the process took seven years only. Such fossil wood was used to make whetstones (a hard, fine-grained flat stone used for sharpening tools, weapons or knives), and sold with the cry:

137. Lough Neagh is a large freshwater lake in Northern Ireland; it is the largest lake in Ireland or Britain with an area of 392 square kilometres (151 sq mi). It is located 30 km (20 miles) west of Belfast and borders five of the six counties of Northern Ireland (Cos. Derry, Antrim, Down, Armagh and Tyrone).

'Lough Neagh bones, Lough Neagh bones
You put them in sticks, you take them out stones'.

Decorating the house at Christmas time with sprigs of holly is a continuing tradition in Ireland, although unsustainable cutting of female holly trees for their berries in parts of Ireland is resulting in damage to the populations in some areas, leaving only male trees in hedgerows. Berried holly in Irish is called *cuileann na gcaortha*. Withered Christmas holly is *cuileann Muire* (Ó Dónaill, 1977). Bringing evergreen plants into the house to celebrate winter festivals is a custom and practice that goes back many centuries in Ireland and Britain. Evergreen plants such as holly, ivy and mistletoe were regarded as representing undying life as they remained green at a time when most other plants had withered away, and in addition, they bear their fruits in winter (Drury, 1987).

Mrs Beeton (1861) gave useful directions on how to make frosted holly leaves for decoration and as Christmas garnishes. Pick some holly leaves and wipe them clean of all dust or moisture. Dry them near a fire, but not so much that they shrivel. Dip them into oiled butter then sprinkle them with coarse sugar and dry again in front of the fire. They should be stored in a dry place and although attractive, are not intended to be eaten.

Holly is commonly grown as an ornamental plant in parks and gardens, with female trees (for its decorative winter berries) or cultivated varieties usually preferred. A range of quite different, but very ornamental, cultivated varieties have been developed that have variegated leaf margins, different leaf shapes and variously coloured yellow and orange berries.

**Inula crithmoides*

- Golden samphire (Scannell & Synnott, 1987)
- *Ailleann Pheadair* (Scannell & Synnott, 1987); *Geirgín* (Moloney, 1919); *Grioloigín* (Moloney, 1919; Dinneen, 1927; Ó Dónaill, 1977)

ASTERACEAE (Compositae)—THE DAISY FAMILY
Native

Inula crithmoides is a rare plant in Ireland of rocky cliffs and shores that is distributed mainly along the east and south coasts.

*No ethnobotanical uses of this plant have been noted in Ireland.

Inula helenium

- Elecampane (Scannell & Synnott, 1987)
- *Meacan aillinn* (Scannell & Synnott, 1987); *Aillean* (Moloney, 1919; Dinneen, 1927); *Meacan uillean* (or *uilleann*) (de Vere Coneys, 1849; Dinneen, 1927)

ASTERACEAE (Compositae)—THE DAISY FAMILY
Introduced (possible archaeophyte)

Inula helenium is a large herbaceous plant that is rather rare in Ireland. It was formerly grown for medicinal use to treat coughs, chest complaints and skin complaints and now mainly occurs as a remnant of former cultivation along roadsides and around ruins and old buildings. It is also grown as an occasional garden plant in herbaceous borders.

Drury (1985) described the importance of this plant in Britain as a herbal remedy for

Inula helenium.

Inula helenium.

livestock (horses, sheep and cattle), where it was sometimes called 'horse-heal'. The bitter leaves were fed to horses to improve their appetites, and it was also used to treat various skin conditions and farcy (with its associated skin lesions). In sheep it was used to treat sheep scab.

Peterson (1977) suggested that its roots can be candied to make a bittersweet, aromatic and addictive delicacy.

**Inula salicina*

- Irish fleabane (Scannell & Synnott, 1987)
- *Lus gréine gaelach* (Scannell & Synnott, 1987); *Lus dréimreach na ndeanathairtí*[138] (Moloney, 1919; Dinneen, 1927)

ASTERACEAE (Compositae)—THE DAISY FAMILY
Native

Inula salicina is a very rare species in Ireland, known only from the shores of Lough Derg in Co. Tipperary. It was formerly in south-east Galway and is now almost extinct. It is not known elsewhere in Ireland or Britain.

*No ethnobotanical uses of this plant have been noted in Ireland.

Iris foetidissima

- Stinking iris (Scannell & Synnott, 1987)
- *Glóiriam* (Scannell & Synnott, 1987)

IRIDACEAE—THE IRIS FAMILY
Introduced (possible archaeophyte)

Iris foetidissima is a rare plant in Ireland of banks, scrub, woodlands, seashores and sea-cliffs that has escaped from cultivation and is occasionally grown in gardens. It is probably not native to Ireland.

Walter Wade, in his *Plantae Rariores in Hibernia Inventae* (1804: 5), said that 'swine

138. Dinneen (1927) defined *lus dréimreach na ndeanathairtí* as 'willow-leaved inula'.

bitten by a mad dog have escaped the disease, by taking the fresh roots mixed with their food, whilst others bitten by the same dog died raving mad'. He also noted that it was used to provide a black dye.

Iris pseudacorus

- Yellow flag (Scannell & Synnott, 1987); Flaggers (Joyce, 1875; Hart, 1898; Colgan, 1904; 1914); Wild iris (Uí Chanainn, 1962); Marsh flag (Uí Chanainn, 1962); Fellistrums (Doogue & Krieger, 2010); Yellow Iris (Doogue & Krieger, 2010); Saggons (Hart, 1898; Harron, 1986); Saggon (Praeger, 1938); 'The May Flower'[139] (McCarthy & Hawkes, 2007)
- *Feileastram* (Scannell & Synnott, 1987); *Liostram* (Ua Maoileoin, 2001; Ó Dónaill, 1977); *Soileastar* (Moloney, 1919; Uí Chanainn, 1962; Ó Dónaill, 1977); *Curcais* (Moloney, 1919); *Seilistrom* (Moloney, 1919); *Seilisdear* (Cameron, 1883); *Seilistring* (Dinneen, 1927); *Feileastar* (Ó Dónaill, 1977); *Feileastrom* (Ó Dónaill, 1977); *Eileastar* (or *Eileastrom* or *Eileastron*) (Ó Dónaill, 1977); *Glóiriam*[140] (Ó Dónaill, 1977); *Eileastram* (Tóibín, 1965); *Soileastrach* (Ó Dónaill, 1977); *Uilliostrum* (Criomhthain, MSS, n.d., ca. 1908); *Sellistring* (Colgan, 1911); *Illistrum* (Colgan, 1911); *Selistrum* (Colgan, 1911); *Dileastrum* (Colgan, 1911); *Solastrum* (Colgan, 1911); *Sellistroch* (Colgan, 1911)
- *Levers* (in Yola—also used for any sword-bladed plant)

IRIDACEAE—THE IRIS FAMILY
Native

Iris pseudacorus is a common and conspicuous plant throughout Ireland, especially when in flower, of wet fields, ditches, drains and marshes.

The roots were formerly used to produce a black dye (Mackay, 1836; Moloney, 1919; Mahon, 1983; Sharkey, 2000). Williams (1993) noted that blue-green and green dyes and ink were also obtained from this species. Mahon (1983) also mentioned that the roots of this species provide a blue dye.

A marshy field of *Iris pseudacorus*.

Wool dyed with *Iris pseudacorus*.

Allen and Hatfield (2004) noted that it was used in Ireland medicinally to treat toothache and was included in a mixture of yellow flowers in one area as a treatment for jaundice. The rhizomes are strongly purgative. Wood-Martin (1902) said that pounded flag-root was used traditionally for dressing cuts and wounds. Mackay (1836) recorded that Sir William J. Hooker (once Director of the Royal Botanic Gardens, Kew) recommended the roasted seeds as a coffee substitute and said that a piece held between the teeth is said to cure toothache. It was mixed with pig's lard as a cure for burns. Warmed leaves were put around the neck in Co. Kildare to treat mumps or throat pains (Williams, 1993).

A thatch of *Iris* leaves was placed on the top of a haycock, presumable to keep off the rain (Williams, 1993). The same author noted that its dried leaves were used in Galway for

139. The yellow iris is used for 'May Day decoration' (McCarthy & Hawkes, 2007: 83).
140. Ó Dónaill (1977) defined *glóiriam* as 'iris'.

cattle bedding and in Cork for pigs, and sometimes as fodder for cows and sheep. During the Middle Ages in Dublin, the leaves were used as bedding for people. Amhlaoibh Ó Súilleabháin, an early 19th-century historian and diarist, called the festival of Corpus Christi[141] *Díardaoin Álainn na mBínsí Breátha* (Beautiful Thursday of the Fine Benches), on account of a custom around Callan, Co. Kilkenny, to spread rushes (*Juncus* species) and leaves of *I. pseudacorus* on benches outside houses as a mark of respect (O'Dowd, 2011).

Iris leaves were probably rarely, if ever, used for thatching houses. In part, this was because they seldom grow in the abundance necessary to thatch a house, and in part, because they rot quickly and become very brittle and disintegrate after a short time.

Toy boats were made by children from its leaves, and gillies (fishing guides) would also wrap fish in its leaves (Williams, 1993).

Yellow flag is regarded as a poisonous plant and should not be eaten. It is also dangerous to animals. Nevertheless, Rutty (1772) said that he has seen a horse eating the leaves green, and that the 'Fingallians' of Portrane, Co. Dublin, save, dry and stack it like hay for their cows and horses.

A close but non-native relative that is occasionally found as a garden escape is *Iris foetidissima*, the stinking iris[142]. It is also poisonous. Praeger (1938: 211) said that 'it is an old medicinal plant, inclined to be persistent when it escapes'. Its capsules contain bright orange-red fruits which should not be eaten.

Isatis tinctoria

- **Woad (Reynolds, 2002)**
- ***Glaisean* (Dinneen, 1927); *Glas lus* (Cameron, 1883); *Glaisín* (Dinneen, 1927; Ó Dónaill, 1977); *Goirmín*[143] (Ó Dónaill, 1977); *Vód* (Ó Dónaill, 1977)**

BRASSICACEAE (Cruciferae)—THE CABBAGE AND CRESS FAMILY
Introduced (Archaeophyte)

Woad (sometimes also called 'dyer's weed') is a very rare casual plant in Ireland that was formerly cultivated in Ireland in the 19th century and earlier (Reynolds, 2002). Colgan (1904) mentioned that it was cultivated in Dublin in Rutty's time (ca. 1772) but had not become established (as a wild plant). It has long been prized in Europe and the Mediterranean regions for the striking and long-lasting blue dye provided by the leaves. The image of fierce warriors with blue-stained bodies going into battle is one that we associate with early impressions of battling tribes in Britain. However, it was also much used in Ireland and is greatly linked with the Celts (Kelly, 1997). Kelly reported that there are numerous references to woad in Old Irish texts and it was clear that it was a widely grown crop in ancient Ireland.

In one 8th-century tale, the queen of Tara is reported to have owned a private woad-garden (*glaisengort*). Its Irish name, *glaisean*, is derived from the word *glas*, meaning 'grey, green and blue', but the colour derived from woad was the Irish word *gorm*, which can be more narrowly translated as 'blue' (Kelly, 1997). After the late Middle Ages its use as a dye was largely replaced by indigo (*Indigofera tinctoria* and other related

141. The Festival of Corpus Christi, observed mainly by Roman Catholics, celebrates the Eucharist and the transubstantiation of bread and wine into the actual body and blood of Christ during Mass. Corpus Christi falls between late May and mid-June, on the first Thursday after Trinity Sunday (60 days after Easter).

142. Praeger (1938) recorded the common name 'gladdon' for this species in the north-east of Ireland.

143. Ó Dónaill (1977) defined *goirmín* as '(dyer's) woad'.

species), a tropical plant group that provides brilliant blue dyes.

To make the dye, the leaves were harvested and crushed, rolled into balls and dried. Before use these were mixed with water and fermented (Howes, 1974). Woad dye was produced commercially in England up to the 1930s, but I know of no such commercial production in Ireland in recent times.

Isolepis cernua (Scirpus cernuus)

- Slender club-rush (Scannell & Synnott, 1987)
- *Crualuachair chaol* (Scannell & Synnott, 1987)

CYPERACEAE—THE SEDGE FAMILY

Native

Isolepis cernua is a frequent plant all around the Irish coastline of marshy habitats and lake-shores near the sea.

Huxley (1992) listed this plant as an ornamental used in horticulture, but I can find no records of its use in Irish gardens.

Isolepis setacea (Scirpus setaceus)

- Bristle club-rush (Scannell & Synnott, 1987)
- *Crualuachair bhiorach* (Scannell & Synnott, 1987)

CYPERACEAE—THE SEDGE FAMILY

Native

Isolepis setacea is a frequent plant in Ireland of damp bare ground in marshes and drains and in waste places, such as on roadsides on sandy and gravelly soils.

Huxley (1992) listed this plant as an ornamental used in horticulture, but I can find no records of its use in Irish gardens.

**Jasione montana*

- Sheep's-bit (Scannell & Synnott, 1987); Sheep's scabious (Brunker, 1950; Anonymous, 1978)
- *Duán na gcaorach* (Scannell & Synnott, 1987); *Luibh an ghalair cham* (Ua Maoileoin, 2001); *Cab an deasáin* (Moloney, 1919); *Cab an ghasáin* (Moloney, 1919); *Dubhán na gcaorach* (Dinneen, 1927)

CAMPANULACEAE—THE BELLFLOWER FAMILY

Native

Jasione montana is a common herb in Ireland of sandy and grassy habitats, such as heaths, dry banks and cliffs, particularly those close to the sea. It also occurs more rarely inland, particularly in the mountains.

*No ethnobotanical uses of this plant have been noted in Ireland.

Juglans regia

- Walnut (Clapham et al., 1962)
- *Cran-gall-chno* (de Vere Coneys, 1849; Dinneen, 1927; Ó Dónaill, 1977); *Gallchno*[144] (Ó Dónaill, 1977); *Gallchnó* (Uí Chanainn, 1962); *Cnó francach*[145] (Uí Chanainn, 1962)

JUGLANDACEAE—THE WALNUT FAMILY

Introduced

Juglans regia.

Walnuts in sugar syrup (in Greece).

The common walnut (*Juglans regia*), a non-native species, is widely planted as an ornamental tree in Ireland in parks, gardens and demesnes, but it is rarely, if ever, established in the wild in this country. Walnuts have been imported into Ireland as a food source for centuries. Evidence of this has been found in deposits excavated in Viking Dublin dated to about 1000 years ago (Geraghty, 1996). The fruits ripen during long, hot summers and have been collected for food. It was widely cultivated in Europe in early times, although it is native in much of temperate Asia, from western China as far west as the Balkans (Mitchell, 1974). Another species, black walnut (*J. nigra*), a North American species, is also grown in some gardens and plant collections in Ireland but is much more rarely seen. In North America the nuts of this species were used as food by Native Americans (eaten raw, served with honey, or made into soup) and a black dye was made from the roots (Gilmore, 1991).

Walnut timber is highly prized hardwood used for fine cabinet-making and to make veneers. It is a hard and heavy timber. Walnut was particularly popular before the introduction of mahogany (mahogany is a general name used for a variety of similar tropical hardwoods, of which the main species in commerce was formerly *Swietenia mahagoni* from the West Indies) (Howes, 1974). Howes also mentioned that walnut wood was much used for gun and rifle stocks as it does not warp and will stand up to rough usage.

Walnuts are commonly sold and eaten around Halloween time in Ireland, all from imported sources. Walnuts are good to eat straight from the shell or can be used in recipes grated or chopped, such as in walnut cakes and bread, muesli and in tarts. Walnut oil is also sometimes available and a useful substitute for olive oil in cooking. The oil has also been used to make soap.

Pickled walnuts are also good to eat and can be prepared from home-grown Irish walnuts. These are prepared from unripe, green walnuts. The walnuts are pricked with a large knitting or darning needle, in part to check that the nut has not formed inside, then put into cold water and brought to the boil to scald them, when the skins are rubbed off with a cloth. Then they are put into a pan of strong brine. The walnuts are left to soak in the brine for about a week, stirred occasionally. After that the brine is replaced and they are soaked again for a further three days before being spread out to dry in the sun, when they will turn black. After that, they are bottled in vinegar flavoured with peppercorns, cloves, allspice and seasoned with a little salt. Green unripe walnuts can also be made into ketchup (walnuts boiled in vinegar, with chopped shallots or onions, a little horse-radish, anchovies

144. Ó Dónaill (1977) defined *gallchno* as 'walnut'. A walnut is also *cnó franncach*, according to Dinneen (1927) and Ó Dónaill (1977).

145. Uí Chanainn (1962) defined *cnó francach* as 'walnut'.

and spices such as mace, nutmeg, ground cloves, ground ginger, salt and pepper). I have also eaten the most delicious sweet dessert: preserved walnuts in Turkey and Greece, prepared in sticky sugar syrup.

A nut butter can be made from walnuts too. The walnuts are smashed and boiled in water until the meat of the nuts and the oils they contain rise to the surface, while the broken pieces of shell settle and sink to the bottom of the pot. The oil and walnut meat can then be combined to make nut butter, or used separately as oil and flour.

Preparing walnuts that you have collected from the tree can be a time-consuming task. The walnut fruits are gathered when ripe in autumn, then their fleshy outsides are removed (use gloves as the juice will stain your hands). Afterwards, the hard nuts are washed to remove more of the fleshy outside and then they are left to dry. The nuts need to be stored where air can circulate freely amongst them and left to cure for a few weeks before use. When they are cured, the nuts are cracked (I use a hammer, particularly for the nuts of *Juglans nigra*, the black walnut, which are very hard), and the meat extracted.

Rutty (1772: 117) provided interesting information on a former use of *Juglans regia*: 'The decoction of the outward peels of the nuts, or the husks and leaves macerated in warm water, and that water poured on the carpet of walks and Bowling greens, kills worms without endangering the grass; and it is said that the leaves rubbed on horses, or rather a Sponge dipt in a decoction of the leaves, prevents them from being molested by flies or other insects'. Rutty also said that oil from the nuts was sometimes used in lamps and is more durable than olive oil, and is preferred by painters to linseed oil because it does not change in colour over time, and 'renews the lustre of old paint'.

Mackay (1825a) recorded that this tree was planted on the banks of the Grand Canal to bind soils.

Juncus species

- Rush (Scannell & Synnott, 1987); *Fiaha*[146] (Hart, 1898)
- *Luchair* (Moloney, 1919); *Úr-luachair* (Moloney, 1919); *Brobh* (Moloney, 1919); *Bogbhuinne*[147] (de Vere Coneys, 1849; Moloney, 1919; *Bruadhna* (pl) (Dinneen, 1927); *Fiag* (Dinneen, 1927; Uí Chanainn, 1962); *Leaglaidh*[148] (Dinneen, 1927); *Feag* (-*gh*) (Ó Dónaill, 1977); *Ruathar rábach*[149] (Ó Dónaill, 1977); *Geitirí*[150] (Ó Criomhthain, MSS, n.d., ca. 1908)
- In Yola, the word for rushes is *rusheen.*
- In Shelta, the word for rush is *snail* (or *snēl*) (*snailee* is a bundle of rushes).

JUNCACEAE—THE RUSH FAMILY

Native

Including:

Juncus acutiflorus

- Sharp-flowered rush (Scannell & Synnott, 1987)
- *Fiastalach* (Scannell & Synnott, 1987); *Crocán* (Colgan, 1911)

146. Rushes.
147. De Vere Coneys (1849) gave the Irish word *bogbhuinne* for 'bulrush'.
148. Dinneen (1927) defined *leaglaidh* as 'a rush, rushes'.
149. Ó Dónaill (1977) defined *ruathar rábach* as 'wild rush'.
150. Ó Criomhthain (MSS, n.d., ca. 1908) gave this name for 'rushes'.

Juncus acutus

- Sharp rush (Scannell & Synnott, 1987); Great sea rush (Brunker, 1950)
- *Luachair bhiorach* (Scannell & Synnott, 1987); *Meithan* (Cameron, 1883)

Juncus articulatus

- Jointed rush (Scannell & Synnott, 1987)
- *Lachán na ndamh* (Scannell & Synnott, 1987); *Crocán*[151] (Dinneen, 1927)

Juncus bufonius

- Toad rush (Scannell & Synnott, 1987)
- *Buafluachair* (Scannell & Synnott, 1987)

Juncus bulbosus

- Bulbous rush (Scannell & Synnott, 1987)
- *Luachair bhleibeach* (Scannell & Synnott, 1987)

Juncus compressus

- Round-fruited rush (Scannell & Synnott, 1987)
- *Luachair chruinn* (Scannell & Synnott, 1987)

Possibly Introduced

Juncus conglomeratus

- Compact rush (Scannell & Synnott, 1987)
- *Luachair dhlúth* (Scannell & Synnott, 1987)

Juncus effusus

- Common rush (Scannell & Synnott, 1987); Soft rush (Anonymous, 1978)
- *Geataire* (Scannell & Synnott, 1987); *Feath* (Cameron, 1883); *Brógbraidhe*[152] (de Vere Coneys, 1849); *Fidheog* (*-eag*) (Colgan, 1911; Dinneen, 1927)

Juncus foliosus

- Leafy rush (Scannell & Synnott, 1987)
- *Buafluachair dhuilleach* (Scannell & Synnott, 1987)

Juncus gerardii

- Saltmarsh rush (Scannell & Synnott, 1987)
- *Luachair mhuirisce* (Scannell & Synnott, 1987); *Luachsir sáile* (Uí Chonchubhair, 1995)

Juncus inflexus

- Hard rush (Scannell & Synnott, 1987)
- *Luachair chrua* (Scannell & Synnott, 1987)

Juncus maritimus

- Sea rush (Scannell & Synnott, 1987); Lesser sea rush (Brunker, 1950)
- *Meathán mara* (Scannell & Synnott, 1987); *Meithan* (Cameron, 1883)

Juncus ranarius

- *Buafluachair bheag* (Scannell & Synnott, 1987)

Juncus squarrosus

- Heath rush (Scannell & Synnott, 1987)
- *Luachair chaoráin* (Scannell & Synnott, 1987)

---

151. A name used on Clare Island.

152. De Vere Coneys (1849) applied this name to *Juncus conglomeratus*, the 'common rush'; however, *J. effusus* is known as the common rush, not *J. conglomeratus*.

Juncus subnodulosus

- Blunt-flowered rush (Scannell & Synnott, 1987)
- *Luachair gheal* (Scannell & Synnott, 1987)

Juncus effusus.

There are about 15 native species of rush in Ireland (as well as two non-native species, *Juncus planifolius* from Australia and South America, and *J. tenuis* from North America). Most rush species in Ireland are perennial and occur in inland wetland habitats or on the coast. *Juncus maritimus* is found commonly on Irish coastlines in salt marshes and sandy seashores. *Juncus gerardii* is another coastal species found in wet places near the sea, such as brackish marshes and other salt marshes. *Juncus acutus* is a striking and robust tufted plant that is found on seashores in the south and east of Ireland. *Juncus acutiflorus* occurs in wet fields, moist woodlands, bog-margins and acid marshes throughout Ireland. *Juncus articulatus* is an extremely common species found in wet fields, marshes, ditches, roadsides and lake-shores throughout Ireland. *Juncus bufonius* is a small annual rush found commonly in a wide variety of muddy and wet habitats. It is closely related to *J. foliosus*, which occurs in similar habitats in the west and south, and *J. ranarius*, which is found in more coastal habitats, such as salt marshes and brackish marshes. *Juncus bulbosus* occurs in semi-aquatic habitats, such as pools, ditches, lake margins and bogs, usually on peaty soil. *Juncus compressus* occurs in damp grassland and is very rare and known only from sites in Cos. Louth and Roscommon. *Juncus conglomeratus* is common in wet upland habitats, such as on wet moors and damp mountain pastures. *Juncus squarrosus* is another rush found in upland habitats in Ireland. It occurs in moors and wet bogs. *Juncus effusus* is one of the commonest rushes found in a wide variety of wet habitats, such as wet meadows, roadsides, ditches, stream banks and marshes. It is similar to *J. inflexus*, which is particularly abundant in the centre of Ireland and a common plant in wet fields and ditches. *Juncus subnodulosus* is quite rare and occurs mainly in the west and centre of Ireland in fens and calcareous marshes but also occasionally in acid bogs.

Rushes were used for a variety of purposes, such as thatch for houses and corn stacks, matting, hats and caps, for making and stuffing saddles and mattresses, making ropes, rattles, fishing nets, baskets, bracelets, belts, whips and rush candles.

Rush matting was an important component of many traditional cottages in Ireland. Sharkey (2000) recorded that whole families would sleep on a single woven rush mat. According to an account from Mayo in 1799, the family slept together lying on rushes, naked with their feet towards the fire, with a covering of blankets or day-clothes (Evans, 1957: 86). The same author mentioned a mid-18th century account from Co. Donegal: 'All the family lay together in one bed, and if any visitors came in the evening, they too slept with them; for they set no bounds to their hospitality . . . heath [probably *Calluna vulgaris*] or bent bushes were spread across the floor, to a length sufficient for the number present, and in breadth about 6 ft; over this litter, the mistress of the house laid part of a long plaid or blanket.' In *An Béal Boċt* (*The Poor Mouth*), Brian O'Nolan's (Myles na gCopaleen) 1941 novel in Irish, a parody on Gaeltacht autobiographies, he described a poverty stricken family in a remote part of Ireland called Corca dorca (or Corkadorkey) where it always rains and they sleep on rushes on the floor in front of the fire.

A baby's rattle made from rushes.

A Frenchman's account from 1644 published in the *Irish Penny Journal* of 1840 mentioned that 'The castles or houses of the nobility consist of four walls extremely high, thatched with straw... They have little furniture, and cover their rooms with rushes, of which they make their beds in summer, and straw in winter. They put the rushes a foot deep on the floors, and on their windows, and many of them ornament the ceilings with branches' (Anonymous, 1840). Preserved remains of rushes spread on the floors of houses in medieval Dublin have been uncovered in excavations undertaken by the National Museum of Ireland in the 1960s and 1970s, suggesting that this practice was widespread throughout Ireland, in country and town houses (Anonymous, 1973). 'If I knew you were coming I'd shake green rushes for you' was an Irish greeting for someone who had not visited your home for a long time that was in common usage in Ireland up to the 1960s (O'Dowd, n.d., ca. 2009, 2011). The same author noted that it is recorded in medieval documents in Ireland that spreading rushes on the floors of churches was usual too. Rushes seem to have been widely and generally used to provide a rough floor covering for many purposes. For example, Lysaght (2003) described a wake in the midlands of Ireland in the late 17th century. The corpse was set up in the barn, laid out on an old door mounted on barrels and with green sods set around as seats and the floor strewn with rushes. Soft boggy ground on the bog was sometimes spread with rushes to provide a firm pathway over which a wheelbarrow laden with turf could be wheeled (O'Dowd, 2011).

Sharkey (2000) mentioned that rush mats were used as sieves too, for example to strain a thirst-quenching drink called 'sowans'. This drink was made by soaking the husks of oats and some whole oats too in an earthenware pot of water for up to a week.

Rushes were also woven to provide seats for chairs in simple country houses, although straw was more commonly used for such purposes. Some seats were made entirely of coiled rush or straw ropes, both stools and high-backed armchairs (Evans, 1957). Rutty (1772: 119) said that the Dublin Society encouraged the making of rush mats by offering 'premiums'.

Rushes were very important for thatching in some parts of Ireland. O'Hare (1993) noted that in one part of Co. Kerry, Knocknagoshel, they were the main material used, although they were perhaps not as waterproof as other thatching materials such as reeds and straw. The same author mentioned that potato stalks were even used sometimes to thatch poorer cabins (and also Dutton [1808] in Co. Clare where he noted that they form a wretched covering and last only one winter). Tomás O'Crohan (*The Islandman*) described how rushes were used for thatching on the Blasket Islands, Co. Kerry, but when the rushes started to decay, hens would disturb the thatch, hunting for worms or laying eggs in them. Rushes are common on the Blasket Islands and would have been the most easily available material for thatching there. Robin Flower wrote in 1944 that 'In the memory of the old men they [the houses] were thatched with rushes which rotted easily in the rains and let through the constant drip—"*an braon anuas*"—of which so much is said in proverbs and traditional poems' [later the roofs were timber covered with tarred felt] (Flower, 1944: 41). Rutty (1772: 119) too referred to rushes being used for thatching. He said that 'In some of the moist bottoms among the mountains, these are separated from the grass, and exposed to bleach in stacks in order for the thatching the Corn and humble cottages of the mountaineers, and make a more durable Thatch than straw: some of the poor make use of small Rushes for stuffing beds'.

A cottage, perhaps thatched with rushes, on the Blasket Islands. Courtesy Ionad an Bhlascaoid Mhóir/The Blasket Centre.

Haycocks were also thatched with rushes to protect them from wind and rain (Evans, 1957; Campbell, 1974). Native rushes are little used today in Ireland for thatching. Throughout the country there are now probably close to 1500 traditional thatched cottages, of which no more than a handful would be thatched with native rushes. Imported reeds and straw are now the most commonly used materials.

Rushes were also an important component in the building of mud houses in Ireland. The majority of the Irish population up to the 18th century probably lived in houses with mud walls. To build with mud, the mud was kneaded into a dough-like consistency with rushes, straw, grass and even sometimes hay, moss or gorse added to help bind the mud (Evans, 1957; O'Hare, 1993). Then the mud was dried for a day or two before being shovelled in layers to form the walls. Layers of rushes were placed between each of the layers of mud, which apparently helped not only to bind the mud but also to absorb water from the wall.

Rushes were probably one of the most widely used materials for making ropes in ancient and medieval Ireland. Rush ropes are flexible and strong. The rushes were crushed with a mallet or beetle after harvesting then dried, before being twisted and woven to make ropes. In Scotland, Milliken and Bridgewater (2004) recorded that *Juncus acutiflorus* (sharp-flowered rush) was used to make tethers for cattle.[153]

St Brigid's Crosses.

In Scandinavian countries St Brigid's Crosses were traditionally made from rushes (or straw). The crosses were hung in houses and outhouses to protect them and those using the houses (people and livestock) from harm and misfortune. In Ireland, St Brigid's crosses have traditionally been made on 1 February, usually made of *Juncus* spp. (Hogan, 2001). In the Aran Islands these were placed inside the house on the underside of the thatch to protect the family within for the coming year (O'Sullivan, 1977). Some houses could be aged by the number of St Brigid's crosses they contained, with a new one added each year. The rushes had to be pulled (not cut—otherwise their protective charms would not work) on the eve of St Brigid's Day (Sharkey, 2000). Various forms of St Brigid's crosses were made in different parts of Ireland, some with three legs. In some parts of Ireland a potato was incorporated into the St Brigid's cross. The potato, having been blessed by St Brigid, was often used as the first seed potato planted that year. An example of a St Brigid's cross with an attached potato is included in the collections of the National Museum of Ireland (Museum of Country Life) in Castlebar, Co. Mayo.

Small crosses often made from rushes were formerly worn on St Patrick's Day (MacCoitir, 2006). Summer sun-hats woven from rushes were popularly worn by young girls (Sharkey, 2000). A variety of other ceremonial and traditional objects were made from rushes (and straw), such as harvest knots, rush ladders at Halloween, masks, hats, imitation flowers and birds' feet and toy dolls. Evans (1957) described butterfly cages made of rushes from Co. Down.

Rushes were used for plaiting and weaving objects, such as a variety of containers and baskets, in particular when straw was not available. O'Dowd (n.d., ca. 2009) described the preparation of rushes for use thus: the rushes were cut and crushed with a mallet or 'beetle', then steeped in a bog hole for some days before being retrieved, spread out and dried. When dry they became a light white or brown colour and could be used for weaving. The rushes were then pleated into long strips using between five and 21 stalks to make narrow or wide lengths. These lengths were then sewn together (sometimes using stems of blackberry or bramble, *Rubus fruticosus*) to make whatever sized receptacle was needed.

The collections of the National Museum of Ireland (Museum of Country Life) in Castlebar, Co. Mayo, includes a strawboy's hat made from rushes (made in Turlough, Co. Mayo), as well as a small basket woven from rushes and used to catch pinkeens or minnows in Delvin, Co. Westmeath.

---

153. The Vikings used ropes made from the inner bark of the linden tree, *Tilia petiolaris*, but *Tilia* is not native to Ireland. The Irish name for the lime tree is *crann teile*, a corruption of the generic name *Tilia* (de Vere Coneys, 1849; Cameron, 1883; Scannell & Synnott, 1987).

Rush light holders. The top example is a contemporary one made in the Muckross Estate, Killarney, in the 2000s. It is embedded into a piece of *Arbutus* wood (from the author's collection).

Rushes provided an important everyday household use for many centuries in Ireland as they were used to make rush lights (in Irish, *fewoge*; *fáideog*; *ádhann*; *ádhann ithearnach*; *trillseán*; *páideóg*; *íotharna*, Moloney, 1919; Dinneen, 1927). Dinneen (1927: 1150) also gave the Irish word *sútróg* for 'a dip-rush candle'. Rush lights have been described as a poor man's candle (Ashley, 2001). The pith found in the stems was used to make the wicks for candles (Mackay, 1836; Uí Chonchubhair, 1995). The wicks (or in Irish, *geataire* or *fiag(a)*—a peeled rush (Moloney, 1919), derived from dried rushes, were dipped in tallow (*geir*) (or lard) repeatedly until a layer of fat had built up on them to form the rush lights (Sharkey, 2000). Drury (1925) said that mutton fat was used in preference as hog's lard was too soft and beef suet was not as readily obtainable. Fish oil was also used instead of tallow in some communities in western Ireland (Ashley, 2001). Oil from other sources would also have been used to make rush lights too. For example, Richard Pococke, writing in 1758 about the seabird, the gannet, said that in Kerry, 'they take the young ones and sell 'em dear for their fat, which the people use to make rush lights' (McVeagh, 1995). On the Great Blasket Island of the west coast of the Dingle Peninsula, Co. Kerry, oil from seals was used in the making of rush lights (Ní Dhálaigh-Uí Chonchubhair, 1999). Kelly (1997) said that in ancient Ireland candles were made as much as 2.5 ft long and in thickness they tapered from the width of a grain to the width of an inch.

Drury (1925) said that rush lights were used in the houses of poorer people, as well as a taper (a source of feeble light or a long narrow candle or wax-coated wick used to light candles or gas lamps) in the homes of the more well-to-do. Rush light tapers were used in churches to light altar candles and chandeliers too. The same author mentioned that they were used in Ireland since the earliest times and included a mention of them in use in about AD 200 when Cormac mac Airt, King of Ireland, met a young maiden cutting rushes with a sharp hook. She selected the long ones for making rush lights and the short ones were to scatter on the floor. When questioned, she said that it was the work of the women to cut and peel the rushes, as women were the light of the house. Impressed by her charm, answers and beauty, Cormac mac Airt made her his queen. The rushes were collected in late summer or autumn when mature but still green. When they become brown they are more difficult to peel. The rushes were gathered mainly by elderly women or old men, who cut them with a reaping hook (*corrán*), tying them into bundles like sheaves of corn. They were then stored in a cool place until needed and then peeled, mainly by old women and children. The tip of the rush was cut off and the skin peeled off from top to base.

A single rush light would burn and give light for about half an hour, depending on its length. *Juncus effusus* (soft rush or common rush) was regarded as the best species for making rush lights (Drury, 1925; Sharkey, 2000). One or two thin strips of the outer skin were left on each side of the rush to ensure that it remained rigid and strong enough to stand upright. 'Two strip lights' were used when watching over a sick person at night because they gave less light.

Gilbert White described the preparation of rushes to make rush lights in *The Natural History of Selborne* (1789): 'These rushes are in the best condition in the heights of summer. As soon as they are cut they must be flung into water and kept there, otherwise they will dry and shrink and the peel will not run. At first a person would find it no easy matter to divest a rush of its peel or rind so as to leave one, regular, narrow even rib from top to bottom that may support the pith; but this, like other feats, soon becomes familiar even to children . . .. When the junci are thus prepared they must lie out on the grass to be bleached, and take the dew for some nights, and afterwards be dried in the sun. Some address is required in dipping these rushes in the scalding fat or grease, but this knack also is to be attained by practice.' In Ireland the water treatment of rushes used for rush lights was not employed (Drury, 1925).

The rush lights were dipped or soaked in the hot melted grease or oil in a canoe-shaped iron vessel, called a 'grisset', 'cam' or 'sleig', pointed at both ends and about 10 in. long (Drury, 1925). The grisset usually had three short feet and a handle projecting from the side. It was made heavy with metal so that it could be removed from the fire and would retain heat long enough for lots of rushes to be soaked. The rushes were drawn slowly through the grisset one by one. Drury noted that the absorption potential of the rush pith is extraordinary: a peeled rush 1 ft in length weighed more than 22 times its own weight when soaked in mutton fat. When completed, the rush lights were stored in long, hollowed-out wooden troughs until they were needed, at least in some parts of Ireland. A drying trough for rush lights made from wood is included in the collections of the National Museum of Ireland, from Achill Island, Co. Mayo, held at the Museum of Country Life in Castlebar, Co. Mayo (F1957.102).

The rush lights would have been used to provide a portable light source for cottages that could be used away from the fire, which would have been the primary source of illumination in the poorest households. Wax candles were used more often in richer households (Ashley, 2001). Other materials were also used as wicks for simple rush lights, including flax, yarn and, later on, cotton. The use of rush lights in preference to wax candles was encouraged also when a tax on wax candles was imposed by the government of these islands in 1709, which remained in force until 1831 (Ashley, 2001). The tax imposed the sum of fourpence on a pound weight of wax candles and a halfpence on tallow candles per pound. The tax was then doubled in 1711. Small homemade rush lights were free of this tax.

I've experimented making rush lights, collecting and peeling the thick stems of *Juncus effusus* in late summer. After passing the rush through melted beef fat a couple of times the rush light was ready. I then placed it in the grip of a special rush light holder and lit it. My rush light holder was made recently by a blacksmith working in the old forge at the Muckross Traditional Farms museum at the Killarney National Park in Co. Kerry. However, old rush lights holders of similar design are occasionally seen in museums and antique shops around Ireland. Once dried, my rush light burned well and steadily for about 40 minutes before it needed to be replaced by the next one. The light it provided was very poor though, much less than an ordinary wax candle and was barely enough to do more than faintly illuminate the room. It burnt best if held at about 45° away from the vertical. 'Burning the candle at both ends' is possible too when the rush light is held horizontally. Rutty (1772: 118) said that sometimes these rushes were twisted in great numbers together, sometimes 'to the bulk of a man's arm, and even of a man's middle', which must have made for impressive lights!

Baskets made from rushes were widespread in Ireland. Blaeberry baskets were made from rushes in Co. Down (Evans, 1957). These were used to carry bilberries (*Vaccinium myrtillus*) on Garland Sunday (or Lammas Sunday), at the end of July. Baskets were widely used in the Hebridean Islands for carrying fish or fishing bait (Evans, 1957). Examples are known of bog butter finds in Irish bogs contained within rush baskets (Synnott, 2008). According to Rutty (1772), rushes were used instead of strings for tying up bunches of flowers.

Logan (1981) reported that disused sweat-houses are frequent in some parts of Ireland, particularly in the north-west such as in Co. Leitrim, but they are also known in Cos. Cork, Kerry and Wicklow (Harte, 2008). The sweat-houses were low buildings built of corbelled stone, 4 to 7 ft in diameter inside, with a narrow, low entrance and a stone slab, cobbled or earthen floor (Evans, 1957). In these a turf fire would be lit and kept burning for a few days to warm the house and the stones with which it was built. Green rushes would be brought in and spread on the floor to protect the user's feet from burning or while sitting. Water was often sprinkled on the flagstone or cobble floors to make the steam bath. Inside between two and half a dozen people would sit naked sweating profusely for half an hour or so before emerging and cooling off in a nearby stream or with a bucket of water. The sweathouse was used to treat rheumatic pains, and Evans (1957) reported that on Rathlin Island in Co. Antrim, young girls would take a sweat bath to clear their complexions before the annual Lammas Fair at Ballycastle on the mainland nearby.

Gardey (1971) noted that in Co. Westmeath rushes have been used to line early coffins made from hazel and willow rods. Rushes were also used to line pits in which potatoes were stored (McCarthy & Hawkes, 2007).

Juncus inflexus was used in parts of Ulster to treat jaundice, and in other parts of Ireland it was burnt and the ashes used to treat ringworm, or mixed with lard as an ointment to treat 'wildfire'—probably shingles. It was also used to rub onto chickenpox pustules to prevent scarring (Allen & Hatfield, 2004). An ointment made from the ashes of burnt rushes mixed with gunpowder was used to treat pocks on the face during chicken pox to prevent scarring (Logan, 1972).

Mahon (1983) listed *Juncus* species as having been used to obtain a green dye.

An interesting form of *Juncus effusus*, forma *spiralis*, the 'Corkscrew rush' with spirally twisted leaves, was introduced to cultivation by David Bishop, who was Curator of Belfast Botanic Gardens at one time, for cultivation as a garden ornamental plant (Nelson, 2000b).

Kickxia elatine (Linaria elatine)

- Sharped-leaved fluellen (Scannell & Synnott, 1987)
- *Buaflíon breatnach* (Scannell & Synnott, 1987); *Seamar criadh* (*cré*) (Moloney, 1919); *Lus cré* (Moloney, 1919; Dinneen, 1927)

VERONICACEAE (Scrophulariaceae)—THE SPEEDWELL FAMILY
Introduced

Kickxia elatine is a very rare annual weed in Ireland, particularly in the south and east, of tilled fields, cultivated ground and waste places.

Allen and Hatfield (2004) discussed its credentials as a species used in folk medicine in Britain and Ireland. They noted that in Gerard's *Herball*, *Veronica officinalis* is mentioned as a long, esteemed plant in Wales for its healing virtues, under the name 'fluellen'. But Gerard also mentioned two British species of *Kickxia* under this name too. It is uncertain therefore whether this species has been used medicinally in Ireland, although Chevallier (1996) mentioned its use as a medicinal herb used externally to staunch herbs and bleeding. I have not found records of its use in Ireland, although Gallagher (2012: 15) mentioned the use of one or two herb species called 'fluellen' for indigestion, stomach disorders and headaches in Co. Donegal.

Knautia arvensis (Scabiosa arvensis)

- Field scabious (Scannell & Synnott, 1987)
- *Cab an ghasáin* (Scannell & Synnott, 1987); *Bodach gorm* (Moloney, 1919); *Cloigne gorma* (Moloney, 1919); *Caba deasain* (de Vere Coneys, 1849; Cameron, 1883); *Bréanla gorm* (Dinneen, 1927; Ó Dónaill, 1977)

DIPSACACEAE—THE TEASEL FAMILY
Native

was either drunk by itself or mixed with ale and 'it was held in high esteem before the days of whisky' (Clyne, 1989: iii).

Bitter-vetch was used in the mid-19th century in Donegal to make 'an excellent stomach drink' (Hart, 1898; Allen & Hatfield, 2004). It is also said to have been used in parts of Ireland, perhaps in Connemara even up to recent times, as an appetite suppressant and against hunger and thirst during Lent and other times of religious fasting. Colgan (1911) said that the tubers were dug up and eaten by children in Clare Island, Co. Mayo, as in many other parts of Ireland.

**Lathyrus palustris*

- Marsh pea (Scannell & Synnott, 1987)
- *Peasairín corraigh* (Scannell & Synnott, 1987)

FABACEAE (Leguminosae)—THE PEA AND CLOVER FAMILY
Native

Lathyrus palustris is a rare plant in Ireland and is mainly found in marshy fields in the midlands.

*No ethnobotanical uses of this plant have been noted in Ireland.

Lathyrus pratensis

Lathyrus pratensis.

- Meadow vetchling (Scannell & Synnott, 1987); Meadow pea; Mouse-pea (Hart, 1898)
- *Peasairín buí* (Scannell & Synnott, 1987); *Peasain bhuidhe* (Moloney, 1919); *Pis bhuidhe* (Moloney, 1919; Dinneen, 1927); *Pis bhuí* (Uí Chonchubhair, 1995; Ó Dónaill, 1977); *Pis bhuidhe* (de Vere Coneys, 1849); *Fiatghail leana* (Wade, 1808b); *Peasair(e) bhuidhe* (Dinneen, 1927); *Piseánach (-aigh) na gcapall* (Dinneen, 1927)

FABACEAE (Leguminosae)—THE PEA AND CLOVER FAMILY
Native

Lathyrus pratensis is a common plant throughout Ireland of hedges, banks and grasslands.

Mears and Hillman (2007) recorded that this plant has edible seeds. There are no Irish records of its use for food, however.

Laurus nobilis

- Bay laurel (Smith, 1882); Bay (Reynolds, 2002); Sweet bay (Smith, 1882)
- *Labhras* (de Vere Coneys, 1849; Dinneen, 1927; Uí Chanainn, 1962; Anonymous, 1978); *Labhróg* (Dinneen, 1927); *Labhraidh* (Dinneen, 1927)

LAURACEAE—THE LAUREL FAMILY
Introduced

This evergreen tree is very commonly grown in Ireland in tubs, trained and clipped into various shapes and cultivated as a garden shrub. It occasionally grows to be a large tree and sometimes self-seeds in the wild. Native to the Mediterranean region, the bay laurel has been a symbol of honour for centuries in Greece and Rome.

Bay leaves are a commonly used as a flavouring in Ireland as an ingredient of bouquet garni and added to stews, marinades, soup and mulled wine.

Lavatera arborea.

Lavatera arborea

- Tree mallow (Scannell & Synnott, 1987); Holly hawk (Colgan, 1904)
- *Hocas ard* (Scannell & Synnott, 1987); *Crann liúrach* (Uí Chonchubhair, 1995)

MALVACEAE—THE MALLOW FAMILY

Native

Lavatera arborea is a large biennial herb with rather woody stems that is widespread around the Irish coasts, growing on cliffs and other rocky habitats and in waste places.

Like *Althaea officinalis*, the tree mallow was formerly used medicinally throughout the British Isles to treat swellings or taken internally for coughs, sore throats, asthma, colds and chest problems (see *Althaea officinalis*) (Allen & Hatfield, 2004).

It is grown as an ornamental in some Irish gardens and has probably been grown in cottage gardens for centuries.

Lemna species

- Duckweed (Scannell & Synnott, 1987)
- *Ros lachan* (Scannell & Synnott, 1987)

LEMNACEAE—THE DUCKWEED FAMILY

Native

Including:

Lemna gibba

- Fat duckweed (Scannell & Synnott, 1987)
- *Ros lachan dronnach* (Scannell & Synnott, 1987)

Lemna minor

- Common duckweed (Scannell & Synnott, 1987); Duck's meat (Hart, 1898; Harron, 1986)
- *Ros lachan* (Scannell & Synnott, 1987); *Grán tonnóg* (Moloney, 1919); *Grán tonóg*[156] (Ó Dónaill, 1977) *Lus gan athair gan mháthair* (Moloney, 1919; Dinneen, 1927; Ó Dónaill, 1977); *Grán lachan* (de Vere Coneys, 1849; Ó Dónaill, 1977); *Grán tonnóige* (Dinneen, 1927)

156. Ó Dónaill (1977) defined *grán tonóg* as 'duckweed'.

Lemna trisulca

- Ivy-leaved duckweed (Scannell & Synnott, 1987); Ivy duckweed (Harron, 1986)
- *Ros lachan eidhneach* (Scannell & Synnott, 1987); *Grán tonnóg* (Moloney, 1919); *Lus gan athair gan mháthair* (de Vere Coneys, 1849; Cameron, 1883; Moloney, 1919); *Abhran donog* (de Vere Coneys; 1849; Cameron, 1883)

All three native Irish duckweed species are tiny aquatic plants that are generally found in habitats such as ponds, pools, canals, lakes and still or slow-moving water. *Lemna minor* is common throughout Ireland, *L. trisulca* is found occasionally in most parts of Ireland except the west where it is rare and *L. gibba* is rare and found mainly in the north and east of Ireland.

K'Eogh (1735: 41) mentioned the use of duckweed for medicinal purposes: 'It is of a refrigerating nature, mixt with wheaten flowers, or Barley Meal very much pravails against Phlegmons [an inflammation of connective tissues], the Erysipelas, Gout, and all kinds of Inflammations'. Lady Wilde (1890: 28) said that 'Duckweed boiled down, and the liquid drank three times a day, is an excellent potion for the sick'. According to Clyne (1989), it was also used to treat headaches in Scotland.

Allen and Hatfield (2004) indicated it was used in Co. Wicklow as a veterinary remedy to treat diarrhoea in animals.

Leontodon hispidus

- Rough hawkbit (Scannell & Synnott, 1987)
- *Crág phortáin gharbh* (Scannell & Synnott, 1987)

ASTERACEAE (Compositae)—THE DAISY FAMILY
Native

Rough hawkbit is found occasionally in most parts of Ireland in habitats such as dry pastures and banks. It is absent in the north.

Mears and Hillman (2007) said that the leaves of this species are edible, if a bit bitter.

Leontodon hispidus.

**Leontodon taraxacoides (Leontodon saxatilis)*

- Lesser hawkbit (Scannell & Synnott, 1987)
- *Crág phortáin bheag* (Scannell & Synnott, 1987)

ASTERACEAE (Compositae)—THE DAISY FAMILY
Native

Lesser hawkbit is found in the south and centre of Ireland in habitats such as dry banks and pastures; it is rarely distributed in the north.

Thomas de Vere Coneys (1849) gave the Irish name *caisearbhán* (*-áin*) for 'dandelion, leontodon taraxacum', but this probably refers to *Taraxacum officinale* and not to this species.

*No ethnobotanical uses of this plant have been noted in Ireland.

Leonurus cardiaca

- Motherwort (Scannell & Synnott, 1987)
- *Lus na clainne* (Scannell & Synnott, 1987)

LAMIACEAE (Labiatae)—THE DEAD-NETTLE FAMILY
Introduced

Leonurus cardiaca is now very rare if not extinct in Ireland.

K'Eogh (1735: 83) described a range of ailments that this introduced plant was used for in traditional medicine in Ireland and said that it was a 'powerful remedy to facilitate the birth'; hence its common name motherwort. In K'Eogh's time he described it as growing in untilled places and lanes and on old walls.

**Lepidium coronopus (Coronopus squamatus; Senebiera coronopus)*

- Swine's cress (Scannell & Synnott, 1987)
- *Cladhthach* (Scannell & Synnott, 1987); *Slánlus na muc* (Moloney, 1919; Dinneen, 1927; Ó Dónaill, 1977); *Fochla* (Moloney, 1919); *Cladhthach buidhe* (Moloney, 1919)

BRASSICACEAE (Cruciferae)—THE CABBAGE AND CRESS FAMILY
Possible Archaeophyte

Lepidium coronopus is an occasional weed of waste places, especially trampled ground, gateways and roadsides.

A related species, *Lepidium didymium* (*Coronopus didymus*)[157], lesser swine's cress, is an introduced annual that is a widespread weed of waste places and roadsides, particularly in the south and especially in coastal areas. Uí Chonchubhair (1995) gave the Irish name *cladthach mhín* for *L. didymium*.

*No ethnobotanical uses of this plant have been noted in Ireland.

**Lepidium heterophyllum*

- Smith's pepperwort (Scannell & Synnott, 1987)
- *Piobar an duine bhoicht* (Scannell & Synnott, 1987)

BRASSICACEAE (Cruciferae)—THE CABBAGE AND CRESS FAMILY
Native

Lepidium heterophyllum is found occasionally in the south and east but is rare elsewhere in Ireland. It grows on walls and in dry heaths and acid sandy ground.

*No ethnobotanical uses of this plant have been noted in Ireland.

Lepidium latifolium

- Dittander (Scannell & Synnott, 1987); Pepperwort (Scully, 1916)
- *Piobracas* (Scannell & Synnott, 1987); *Lus an phiobair* (Moloney, 1919; Dinneen, 1927); *Piobarcas*[158] (Dinneen, 1927)

157. Brunker (1950) gave the common name 'lesser watercress' for this species.
158. Dinneen (1927) defined *piobarcas* as 'pepperwort, garden cress'.

BRASSICACEAE (Cruciferae)—THE CABBAGE AND CRESS FAMILY
Probably Introduced (possible archaeophyte)

Lepidium latifolium is a rare plant in Ireland that is found mainly in waste places along the coast. In the 1980s I remember encountering this striking plant in some abundance along a section of the banks of the Grand Canal in Dublin where it was well established as a long surviving alien plant (Wyse Jackson & Sheehy Skeffington, 1984). It was formerly used for medicinal and culinary purposes, similar to horse-radish. O'Mahony (2009) mentioned that it was formerly used in Co. Cork for gynaecological purposes, to hasten childbirth.

Leucojum aestivum

- Summer snowflake (Scannell & Synnott, 1987); Loddon lily (Harron, 1986)
- *Plúirín samhraidh* (Scannell & Synnott, 1987)

ALLIACEAE (Liliaceae)—THE ONION FAMILY
Possibly Introduced

Leucojum aestivum is a rare plant of marshes and wet meadows. It is a beautiful plant cultivated in quite a wide range of gardens in Ireland, although probably very few if any of these plants are derived from wild Irish sources. It is a rare plant in Ireland and considered by Webb (1977) as probably native in the south, in its typical marshy meadow habitat, although it has escaped from cultivation in many of the localities in which it occurs.

This plant is reputed to be poisonous.

**Leymus arenarius (Elymus arenarius)*

- Lyme-grass (Scannell & Synnott, 1987)
- *Táithean* (Scannell & Synnott, 1987); *Aolfheur mara*[159] (Wade, 1808a)

POACEAE (Gramineae)—THE GRASS FAMILY
Native

Leymus arenarius is a rare coastal grass in Ireland of sandy shores in the north and east.

Wade (1808a) said this grass was a more effective resister of drifting sands than *Ammophila arenaria*, but I have not found records of it being planted in Ireland for such erosion-control reasons.

*No ethnobotanical uses of this plant have been noted in Ireland.

Ligusticum scoticum

- Lovage (Scannell & Synnott, 1987); Scottish lovage
- *Sunais* (Scannell & Synnott, 1987); *Lus an liagaire* (Moloney, 1919; Dinneen, 1927); *Lus an lugaire* (Moloney, 1919; Dinneen, 1927); *Luáiste*[160] (Ó Dónaill, 1977)

159. Wade (1808a) applied this name to *Elymus arenarius*.
160. Ó Dónaill (1977) defined *luáiste* as 'lovage'.

Apiaceae (Umbelliferae)—THE CARROT FAMILY
Native

Ligusticum scoticum is a rare plant in Ireland and found only on cliffs and rocky shores on the north coast.

Allen and Hatfield (2004) recorded its use as a medicinal plant for human use, to treat flatulence, as an aphrodisiac, as a sedative and 'against consumptions'. Eaten in the morning it was believed to keep a person safe from infection for the rest of the day. It was also used to treat various ailments in cattle (used as a purge) and sheep (to treat a cough). As it is known to have been used in the Western Isles of Scotland it was undoubtedly used in Northern Ireland too, which is the only part of Ireland to which it is native.

It is used today as a culinary herb but not derived from wild Irish sources. The leaves and seeds have a strong flavour and should be used sparingly but can be used to flavour soups, stews, sauces and added to salads. Cameron (1883) mentioned that it was sometimes eaten raw as a salad, or boiled as greens in the Western Isles of Scotland.

Common lovage is a different plant species, *Levisticum officinale* (in Irish, *lus an liagaire*), that is much used in southern European cookery.

De Vere Coneys (1849) gave the Irish name *sunais* (*-se*) also for '*Ligusticum leviaticum*, the herb lovage'.

Ligustrum vulgare

- Privet (Scannell & Synnott, 1987)
- *Pribhéad* (Scannell & Synnott, 1987); *Primhéad* (Moloney, 1919; Dinneen, 1927); *Tor luathfás* (Moloney, 1919; Dinneen, 1927); *Crannla* (Dinneen, 1927)

OLEACEAE—THE ASH FAMILY
Native

As a native plant, *Ligustrum vulgare* is very rare in Ireland and only found on a few cliffs and rocky places. Scannell and Synnott (1987) suggested that it is native only in Waterford, north Tipperary, south-east Galway and Dublin, but even in some of those counties its native status must be considered suspect. It is widely naturalised throughout Ireland, having spread from gardens or having 'gone wild' from specimens that were originally planted.

Privet is a common ornamental and hedge plant, which if left untrimmed, will grow to become a small tree. The Irish name, *pribhéad*, comes from the Irish word *priobháideach*, meaning 'private' (*privatus*, in Latin), suggesting its use for hedging.

In Britain its young twigs have been used for basketry (Edlin, 1973), but its use for such purposes has not been recorded in Ireland.

Privet leaves have been used to prepare yellow, green and brown dyes and its berries a blue-green dye (with alum and salt) (Mairet, 1916; Edlin, 1973; Mahon, 1983).

A decoction of the berries was used in Ireland to treat earache and an infusion was drunk for sore throats (Allen & Hatfield, 2004).

Privet is generally regarded as a poisonous plant and its black berries should not be eaten.

*Limonium binervosum agg.

- Rock sea-lavender (Scannell & Synnott, 1987)
- *Lus liath aille* (Scannell & Synnott, 1987)

PLUMBAGINACEAE—THE THRIFT FAMILY
Native

Plants recorded as *Limonium binervosum* in the past in Ireland are probably all *L. recurvum*.

*No ethnobotanical uses of this plant have been noted in Ireland.

Limonium humile (Statice rariflora)

- Lax-flowered sea-lavender (Scannell & Synnott, 1987)
- *Lus liath na mara* (Scannell & Synnott, 1987)

PLUMBAGINACEAE—THE THRIFT FAMILY
Native

Limonium humile.

Limonium humile is an occasional plant of salt marshes and muddy seashores around much of the Irish coast except the north where it is rare.

The roots and seeds were used in the treatment of diarrhoea, dysentery and excessive menstrual discharges, and the root held and chewed in the mouth helps to cure a toothache (K'Eogh, 1735).

*Limonium procerum

- *Lus liath ard* (Scannell & Synnott, 1987)

PLUMBAGINACEAE—THE PLUMBAGO OR THRIFT FAMILY
Native

Limonium procerum is distributed on the east coast of Ireland only and occurs on coastal rocks and in other seashore habitats.

*No ethnobotanical uses of this plant have been noted in Ireland.

*Limonium recurvum

- *Lus liath* (Scannell & Synnott, 1987)

PLUMBAGINACEAE—THE THRIFT FAMILY
Native

Limonium recurvum is found on coastal limestone rocks and salt marshes in several parts of Ireland. It is represented by a number of distinct subspecies outlined below.

*No ethnobotanical uses of this plant have been noted in Ireland.

Limonium recurvum* subsp. *humile (Limonium paradoxum)

- Paradoxical sea-lavender (Scannell & Synnott, 1987)
- *Lus liath Albanach* (Scannell & Synnott, 1987)

PLUMBAGINACEAE—THE THRIFT FAMILY
Native

Limonium recurvum subsp. *humile* occurs in Co. Donegal only.

*No ethnobotanical uses of this plant have been noted in Ireland.

Limonium recurvum* subsp. *portlandicum

PLUMBAGINACEAE—THE THRIFT FAMILY
Native

Limonium recurvum subsp. *portlandicum* occurs in Co. Kerry only.

*No ethnobotanical uses of this plant have been noted in Ireland.

Limonium recurvum* subsp. *pseudotranswallianum

- Western sea-lavender (Scannell & Synnott, 1987)
- *Lus liath na Boirne* (Scannell & Synnott, 1987)

PLUMBAGINACEAE—THE THRIFT FAMILY
Native

Limonium recurvum subsp. *pseudotranswallianum* occurs in Co. Clare only.

*No ethnobotanical uses of this plant have been noted in Ireland.

****Limosella aquatica***

- Mudwort (Scannell & Synnott, 1987)
- *Lus lathaí* (Scannell & Synnott, 1987)

SCROPHULARIACEAE—THE FIGWORT FAMILY
Native

Limosella aquatica is a plant of wet mud at lake edges and shallow ponds in the west of Ireland.

*No ethnobotanical uses of this plant have been noted in Ireland.

****Linaria vulgaris***

- Common toadflax (Scannell & Synnott, 1987)
- *Buaflíon corcra* (Scannell & Synnott, 1987); *Lorcán-líon buidhe* (Moloney, 1919)

VERONICACEAE (Scrophulariaceae)—THE SPEEDWELL FAMILY
Possibly Introduced

Linaria vulgaris is found occasionally in Ireland, mainly in the south and east, in waste places and on roadsides and railway lines.

According to Moloney (1919), this plant was not used medicinally in Ireland.

*No ethnobotanical uses of this plant have been noted in Ireland.

*Linum bienne

- Pale flax (Scannell & Synnott, 1987)
- *Líon beag* (Scannell & Synnott, 1987)

LINACEAE—THE FLAX FAMILY
Possibly Native (or possible archaeophyte)

Linum bienne is an occasional but declining plant in the south and east of Ireland of pastures and dry banks. It was formerly also distributed in the south-west but is no longer there.

*No ethnobotanical uses of this plant have been noted in Ireland.

Linum catharticum

- Fairy flax (Scannell & Synnott, 1987); Cathartic flax (Anonymous, 1978); Purging flax (Praeger, 1938; Anonymous, 1978)
- *Lus na mban sí* (Scannell & Synnott, 1987); *Lus na mban sidhe* (Moloney, 1919); *Míosach* (de Vere Coneys, 1849; Moloney, 1919; Dinneen, 1927; Ó Dónaill, 1977); *Caolach* (de Vere Coneys, 1849; Moloney, 1919; Dinneen, 1927); *Mionnuach* (Moloney, 1919); *Líon beag* (Moloney, 1919); *Ceolagh* (Cameron, 1883); *Líon na m-ban sighe* (de Vere Coneys, 1849; Dinneen; 1927); *Líon purgóideach*[161] (Dinneen, 1927); *Míonbhaigh* (Dinneen, 1927); *Míonsach (-aighe)* (Dinneen, 1927); *Líon na mban sí* (Ó Dónaill, 1977)

LINACEAE—THE FLAX FAMILY
Native

Linum catharticum is a very common delicate annual plant throughout Ireland of short acid grasslands, bogs, fens and heaths.

It was formerly used in Ireland to promote vomiting or as a purgative. When boiled in beer it has been used as a treatment for jaundice (Moloney, 1919). It was also formerly used to treat urinary complaints (in Cavan) (Allen & Hatfield, 2004). Munnelly (2005) mentioned the use of fairy flax in West Clare for the treatment of constipation in human and veterinary medicine. He quotes a local man, Francie Kennelly, from Miltown Malbay, who said one would use 'as much that'd go in a woman's ring, they used say, for a person. And as much that'd go in there (makes circle with thumb and forefinger), like that, for an animal' (Munnelly 2005: 168).

Linum usitatissimum

- Flax (Reynolds, 2002)
- *Lín* (Dinneen, 1927); *Líon* (Dinneen, 1927; Ua Maoileoin, 2001)

LINACEAE—THE FLAX FAMILY
Archeophyte and Introduced

161. Dinneen (1927) defined *líon purgóideach* as 'purging flax'.

Linum usitatissimum (flax). Watercolour by Lydia Shackleton. Courtesy of the National Botanic Gardens, Glasnevin.

Flax fibres.

Linum usitatissimum is a tall annual herb with a single stem to as much as 85 cm topped with blue flowers. Flax was formerly widely grown in Ireland, particularly in the north-east of the country, first as a fibre crop and increasingly for the oil extracted from its seeds (linseed oil). It is now very rarely seen. In the wild, *L. usitatissimum* occurs only as a casual plant and is a remnant of cultivation. The stem fibres (called 'lint', particularly in the north of Ireland) were used to make linen and for thatching.

Flax has been grown in Ireland since early times and it is well represented in archaeobotanical finds from the early Christian period (Kelly, 1997). Kelly mentioned that it was used to make a linen tunic (*léine*) that was worn under the cloaks (*brat*) by the early Irish. It is not a native plant in Ireland but was probably first grown as a crop during the Neolithic period, derived from a wild flax native to an area from the eastern Mediterranean to India, where *Linum* was probably first domesticated.

A traditional flax spinning wheel from the United States (author's collection).

When the flax crop was ready for harvesting, the tall stalks were pulled by hand, not cut. To extract the fibres for use in the manufacture of textiles after harvesting, the plants would be decomposed (called 'retting') by placing them in a pond, pool or drain that helps the fibres separate from the other decayed parts of the plant. McGlinchey (2007) wrote that, after soaking, the flax fibres were separated from the stem ('scutched') with a stick on a block; later there were scotch-mills. The women would take their bundles of lint for sale in the local towns. If the lint was to be spun into thread at home instead of being sold, the flax fibres were 'heckled' after being scutched by being pulled across a board with nails sticking up in it. Flax lint was then 'cloved' on a cloving stick. The cloving stick had a blade along its side and the lint was drawn through between the blade and the stick, when it would be ready for spinning. The same spinning wheel would be used for spinning flax as for wool. McGlinchey described the way people came together for a communal night of scutching or cloving the flax where there would be 20 or 30 people present. The work was carried out in a barn or an out building, or in the kitchen of the house. Afterwards they would have a dance. He said that the old people all had a supply of linen sheets that would last a lifetime. Dinneen (1927) gave the word *gairmint* for cleaning or scutching flax, and *tlúgh gairminte* is a clove or tongs used in cleaning flax from woody fibre.

In parts of the world the ripe dried seeds of flax are used medicinally to treat constipation and the symptoms of irritable colon (van Wyk & Wink, 2004). The seeds are eaten crushed or whole as a bulk-forming laxative. Moloney (1919: 18) listed flax as having been used by bone-setters because it 'makes excellent splinting material when supported with leather'. O'Regan (1997) recorded that linseed was used in Ireland to feed calves and also medicinally to draw out bites, poisons and thorns. The seeds were cooked and applied hot over the affected area as a poultice. It was also used to treat pleurisy.

A cottage with flax thatching. Ulster Museum.

McGlinchey (2007: 35) wrote that in the part of north Donegal in which he lived in the mid-1800s, flax was an annual and widely grown cash crop. 'Nearly everyone in the parish has a barrel of lint [flax] sown'. It was used as part of a regular crop rotation, with cabbage, potatoes and cereals.

In Northern Ireland, in particular, *Linum usitatissimum* has been used for thatching when sale of the crop for its fibre has not been achieved. McGlinchey (2007) mentioned that flax thatch was very long-lasting. Evans (1967) wrote that 'old houses in the north-east sometimes have a thick layer of flax waste or "shows" under the thatch'.

The waste product from the flax scutching mills was sometimes used in Northern Ireland to fuel the bonfires of 'The Twelfth' of July during Orangemen's celebrations. Hedges were often cut back before 11 July to provide fuel also (Gailey & Adams, 1977).

Rutty (1772) mentioned that flax was used instead of rushes to make the wicks of candles.

Lithospermum arvense (Buglossoides arvensis)

- Field gromwell (Green, 2008); Rough gromwell (Brunker, 1950)
- *Gormail* (Scannell & Synnott, 1987)

BORAGINACEAE—THE BORAGE FAMILY

Probably Introduced (possible archaeophyte)

Lithospermum arvense is now very rare but was formerly widespread throughout Ireland, mainly as a weed of cultivated ground.

Wade (1804) said that the bark of the roots of this species tinge wax and oil a beautiful red.

**Lithospermum officinale*

- Common gromwell (Scannell & Synnott, 1987)
- *Gormail* (Scannell & Synnott, 1987)

BORAGINACEAE—THE BORAGE FAMILY

Native

Lithospermum officinale is a rare plant in Ireland that is found throughout in hedges, sandy banks and roadsides.

*No ethnobotanical uses of this plant have been noted in Ireland.

**Littorella uniflora*

- Shoreweed (Scannell & Synnott, 1987)
- *Lus an chladaigh* (Scannell & Synnott, 1987)

PLANTAGINACEAE—THE PLANTAIN FAMILY
Native

Littorella uniflora can be found on lake-shores in the north and west and is rare elsewhere in Ireland.

*No ethnobotanical uses of this plant have been noted in Ireland.

**Lobelia dortmanna*

- Water lobelia (Scannell & Synnott, 1987)
- *Plúr an locháin* (Scannell & Synnott, 1987)

CAMPANULACEAE—THE BELLFLOWER FAMILY
Native

Lobelia dortmanna is rare in Ireland, except in the west, where it grows in shallow water at the edges of lakes.

*No ethnobotanical uses of this plant have been noted in Ireland.

Lolium perenne

- Perennial rye-grass (Scannell & Synnott, 1987); Ray-grass (Harron, 1986)
- *Seagalach buan* (Scannell & Synnott, 1987); *Coirce préacháin* (Moloney, 1919; Ó Dónaill, 1977; Uí Chonchubhair, 1995); *Roille marthanach* (White, 1803; Wade, 1808a); *Seagalach ilbhliantúil* (Ó Dónaill, 1977); *Roille caol*[162] (White, 1803)

POACEAE (Gramineae)—THE GRASS FAMILY
Native

Perennial rye-grass is one of the most abundant and important meadow grasses in Ireland. It occurs in meadows, on roadsides, lawns and waste places and is widely sown as an agricultural grass. It occurs both as a naturally occurring native plant and is included in some commercially available seed mixes as a lawn and fodder grass. Perennial rye-grass is often the grass species of choice for playing fields as it is robust, fast-growing and very resistant to trampling.

Of course, as a common meadow grass, it is also an important constituent of hay. Hay was harvested (generally called 'saving the hay') to provide fodder for stock and to supplement the food available for animals during the winter months when growth is slow in the meadows. Farmers cut the meadow usually in mid-June or later when a period of dry weather is expected, during which time the hay was dried in the fields and collected into bales (or formerly into haycocks). Farmers would try to cut the hay before the seed heads had become ripe, when the leaves were still green and lush and while they were at their most nutritious point. Choosing a time when the hay could be harvested was always one of the great gambles of Irish farming. If a field was cut and then the following days were very wet, the hay would rot or develop mould before it could be dried. In those conditions the mould can be toxic for animals. In Ireland, the closing of schools in early June (compared with late July in England) was partly due to the need for children to be available on the

162. White (1803) applied this name to *Lolium tenue*.

farm to help with the hay preparation. This was a time of feverish activity for farmers: to cut the hay, turn it in the fields to help drying, rake and gather it all in and store it safely before wet weather returned. Today hay is much less frequently made and alternative winter animal feed is available (silage and other commercially available foodstuffs). Modern mechanised balers are now available to automate aspects of hay harvesting. Kelly (1997) pointed out that the practice of making and storing hay as winter fodder for livestock may have been a Norman farming innovation as it seems not to date from much before the Norman invasion in the 1100s. The mild winters and continued growth of grass throughout each month of the year (perhaps combined with the unpredictable summer weather) may have discouraged the practice of hay-making until the Normans' arrival.

The constituents of hay are those species that occur in the field but generally include a variety of different grass species, such as *Lolium perenne* and the non-native *L. multiflorum* (Italian rye-grass; in Irish, *seagalach Iodálach*); various species of *Festuca* (fescue; in Irish, *feisciú*), such as *F. pratensis* (meadow fescue) and *F. rubra* (red fescue); *Poa* species (meadow-grass); *Dactylis glomerata* (cock's-foot grass); and *Phleum pratense* (cat's-tail or Timothy); as well as a range of other wildflowers, such as legumes (especially clovers [*Trifolium* species]).

Apart from its use as animal feed, hay had many other uses. It was used to provide bedding for animals and fowl. Mattresses and pillows were stuffed with it. It has been used as mulch around fruit and vegetable crops. Ropes were made from hay. Hay or straw placed around strawberries (note the name!) keeps the developing fruits from becoming spoilt in contact with the soil. Hay has also been used in cooking (hay-box cooking). It works on the same principle as a slow cooker, insulating and keeping the pot and its contents hot or cold as required.

A variety of grass species seed mixes is widely available today at hardware shops and agricultural suppliers throughout the country for reseeding tilled fields, often including a large proportion of *Lolium perenne*. Pat O'Shea from Killorglin, Co. Kerry (in King 2009), said that in former times when new grasslands were being sown, the seed would be gathered from underneath the previous year's haystack in the haggards of local farmers.

An example of a pot scrubber made from grass roots of some unspecified species is included in the collections of the National Museum of Ireland (Museum of Country Life), Castlebar, Co. Mayo (Ref. No: F1955.9).

Lolium temulentum

- Darnel (Scannell & Synnott, 1987)
- *Roille* (Scannell & Synnott, 1987); *Breallán* (de Vere Coneys, 1849; Moloney, 1919; Dinneen, 1927); *Baoidhealán* (Moloney, 1919); *Dithean* (Moloney, 1919); *Raidhleadh* (Cameron, 1883); *Roille breoillean* (White, 1803; Wade, 1808a); *Ruinhelais* (White, 1803); *Bod a rinker* (White, 1803); *Roille branar*[163] (White, 1803)

POACEAE (Gramineae)—THE GRASS FAMILY

Possibly Introduced (possible archaeophyte)

Lolium temulentum is a possible introduction that is now very rare and only known from the Aran Islands in the 1980s and 1990s and East Mayo (recorded there in the 1960s, Reynolds, 2002). It was formerly a widespread weed of cereal crops throughout Ireland.

163. White (1803) applied this name to *Lolium arvense*.

The seeds or grains of *Lolium temulentum* were used medicinally by the ancient Greeks and Romans. Gerard's *Herball* (1597) recorded 'the new bread wherein Darnel is eaten hot causeth drunkenness'. Mackay (1836: 317) noted that it was a common weed of corn-fields in several parts of Ireland, and 'the seeds mixed with wheat and made into bread have proved highly injurious to those who have eaten it'. Wade (1804: 7) wrote that it was sometimes common in cornfields at that time too and swine that ate it have been known to be 'seized with temporary blindness and drunkenness, when it was mixed with their food'. White (1803: 119) mentioned that brewers and distillers did not mind buying barley that contained small quantities of the seed of this species mixed in as a weedy contaminant as it had an 'inebriating quality'. The symptoms of darnel poisoning are giddiness, staggering and loss of sight.

Rutty (1772: 69) said, though, that 'in times of scarcity of corn, life has been sustained by this feed; for tho' it be intoxicating in drink, this quality is in some measure lost in bread made of it'.

Lonicera periclymenum (honeysuckle; woodbine). Watercolour by Lydia Shackleton. Courtesy of the National Botanic Gardens, Glasnevin.

Lonicera periclymenum

- Honeysuckle (Scannell & Synnott, 1987); Woodbine (Colgan, 1904; Anonymous, 1978); Irish vine (Hart, 1898); Irish ivy (O'Rourke, 2006)
- *Féithleann* (Scannell & Synnott, 1987); *Milleanna gabhar* (Ua Maoileoin, 2001); *Táithfhéithleann* (Ua Maoileoin, 2001); *Bainne gabhan* (O'Rourke, 2006); *Táith-fheileann* (Moloney, 1919); *Féithleóg* (de Vere Coneys, 1849; Moloney, 1919); *Lus na meala* (de Vere Coneys, 1849; Moloney, 1919; Dinneen, 1927); *Iadh-shlat* (Cameron, 1883; Dinneen, 1927); *Cas fa chrann* (de Vere Coneys, 1849; Cameron, 1883); *Duilliur-féithlean* (de Vere Coneys, 1849; Colgan, 1911); *Duilleabhar féithle(ann)* (Dinneen, 1927); *Duilleabhar féarána* (Dinneen, 1927); *Feath (Féith)*[164] (Dinneen, 1927; Uí Chanainn, 1962); *Féithle*[165] (Dinneen, 1927; Ó Dónaill, 1977); *Seamar* (Dinneen, 1927); *Tromán na habhna* (Dinneen, 1927); *Uilleann (-linne)* (Dinneen, 1927); *Táthfhéithleann* (Ó Dónaill, 1977); *Luibh féarána*[166] (Uí Chanainn, 1962); *Feithleóg faoi crann* (Colgan, 1911)

CAPRIFOLIACEAE—THE HONEYSUCKLE FAMILY
Native

Lonicera periclymenum is a very common woody climber throughout Ireland of hedgerows, woods, walls, scrub and rocky ground. It is native to Britain and throughout Europe, as far north as Norway.

Lonicera periclymenum has been used to treat inflammations of the mouth, or herpes, and perhaps for the fungal infection of the mouth known as thrush (Allen & Hatfield, 2004). Juice from the plant was used as a purgative or rubbed on

164. Uí Chanainn (1962) defined this word as 'fibre, twig, honeysuckle, osier'.
165. This word is normally used for clover species, but Dinneen (1927) also defined it as 'honeysuckle'.
166. Uí Chanainn (1962) defined *luibh féarána* as 'woodbine'.

Lonicera periclymenum.

Lonicera periclymenum.

the skin, chest and under the arms against jaundice. Leaves could be chewed for bad breath and toothaches, and honeysuckle was also used against tuberculosis (Williams, 1993). Lady Wilde (1890) quoted a cure for epilepsy, from an old manuscript dated ca. 1450 in the Library of the Royal Irish Academy, where salt and white snails were put into a vessel for three nights, to which 7 lb of woodbine leaves were added and mixed all together into a paste. The paste was applied as a poultice that cures in nine days (see also Wood-Martin, 1902). Lady Wilde (1890) also gave a recipe for 'woodbine' (*Lonicera periclymenum*) and 'maiden-hair' (*Adiantum capillus-veneris*) in milk with oatmeal to treat dysentery. The woody parts of the stems were used to make aromatic mouthparts for smoking pipes (Williams, 1993).

As a child in Limerick, sucking the sweet nectar from the flowers was a popular summer pastime for us. Lucas (1960) recorded that the fruits of woodbine are mentioned 'in a poem of the 15th/16th century date as edible dainties'.

Uí Chonchubhair (1995) noted that this plant was believed to be able to protect the wearer against the 'evil eye'.

According to Allen and Hatfield (2004), *Lonicera periclymenum* was used as a veterinary remedy to treat farcy in Cos. Tipperary and Limerick.

Lotus corniculatus

- Common bird's-foot-trefoil (Scannell & Synnott, 1987); Crowfoot (Hart, 1898); Crawfoot (Hart, 1898); Croobeen cut (Hart, 1898); Cuckoo-flower (Hart, 1898); Eggs-and-bacon (Harron, 1986)
- *Crobh éin* (Scannell & Synnott, 1987); *Sicín* (O'Rourke, 2006); *Pabhsaer na cuaiche* (O'Rourke, 2006); *Bárr an mhisléin* (Moloney, 1919); *Ungeun coitchion* (Wade, 1808b)

FABACEAE (Leguminosae)—THE PEA AND CLOVER FAMILY

Native

Lotus corniculatus is a very common plant of calcareous grasslands, sand dunes and roadsides throughout Ireland.

Quoting from *A Dictionary of English Plant-names* by James Britten (1846–1924) and Robert Holland (1829–93), Vickery (1995: 33) said that 'in the south of Ireland children gather this plant to take with them to school in the belief that the possession of it will save them from punishment; and it is said they will go miles out of their way to obtain it'.

Lotus corniculatus.

Lotus pedunculatus.

Lotus pedunculatus (Lotus uliginosus)

- Greater bird's-foot-trefoil (Scannell & Synnott, 1987)
- *Crobh éin corraigh* (Scannell & Synnott, 1987); *Barr an mhisléin* (Uí Chonchubhair, 1995)

FABACEAE (Leguminosae)—THE PEA AND CLOVER FAMILY
Native

Lotus pedunculatus is a common plant of ditches and wet meadows throughout most of Ireland except the centre and in some regions of the west.

*No ethnobotanical uses of this plant have been noted in Ireland.

Lotus subbiflorus

- Hairy bird's-foot-trefoil (Scannell & Synnott, 1987)
- *Crobh éin mosach* (Scannell & Synnott, 1987)

FABACEAE (Leguminosae)—THE PEA AND CLOVER FAMILY
Native

Lotus subbiflorus is a very rare plant in Ireland of coastal grasslands in West Cork and Co. Wexford only.

*No ethnobotanical uses of this plant have been noted in Ireland.

Luronium natans

- Floating water-plantain (Scannell & Synnott, 1987)

ALISMATACEAE—THE WATER-PLANTAIN FAMILY
Possibly Native

Luronium natans is a very rare and recently discovered aquatic plant of lake margins in Co. Galway.

*No ethnobotanical uses of this plant have been noted in Ireland.

Luzula campestris

- Field wood-rush (Scannell & Synnott, 1987)
- *Giúnach léana* (Scannell & Synnott, 1987)

JUNCACEAE—THE RUSH FAMILY
Native

Luzula campestris is a common plant throughout Ireland of dry banks and grasslands.

*No ethnobotanical uses of this plant have been noted in Ireland.

Luzula multiflora

- Heath wood-rush (Scannell & Synnott, 1987)
- *Giúnach caoráin* (Scannell & Synnott, 1987)

JUNCACEAE—THE RUSH FAMILY

Native

Luzula multiflora is a common plant throughout Ireland of the uplands, bogs, heaths, marshes and fens.

*No ethnobotanical uses of this plant have been noted in Ireland.

*Luzula pilosa

- Hairy wood-rush (Scannell & Synnott, 1987); Spring woodrush (Brunker, 1950)
- *Giúnach coille* (Scannell & Synnott, 1987)

JUNCACEAE—THE RUSH FAMILY
Native

Luzula pilosa can be found in woodlands and shady habitats mainly in the north and east and is much rarer elsewhere in Ireland.

*No ethnobotanical uses of this plant have been noted in Ireland.

*Luzula sylvatica

- Great wood-rush (Scannell & Synnott, 1987); Bendle (Hart, 1898); Mad ye rall (Hart, 1898)
- *Giúnach mór* (Scannell & Synnott, 1987); *Giuineach (-nigh)*[167] (Dinneen, 1927)

JUNCACEAE—THE RUSH FAMILY
Native

Luzula sylvatica is a common plant throughout Ireland of woodlands, hedgerows, river-banks, scrub, mountain ledges, slopes and glens.

*No ethnobotanical uses of this plant have been noted in Ireland.

Luzula sylvatica.

Lychnis flos-cuculi

- Ragged robin (Scannell & Synnott, 1987)
- *Lus síoda* (Scannell & Synnott, 1987); *Plúr na cuaiche* (Moloney, 1919; Dinneen, 1927; Uí Chonchubhair, 1995)

CARYOPHYLLACEAE—THE CARNATION OR PINK FAMILY
Native

Lychnis flos-cuculi is a common plant throughout Ireland of wet meadows and marshes.

It is said to be unlucky to pick the flowers of this species and take them indoors as bad weather (thunder) may follow (Uí Chonchubhair, 1995).

Lycopus europaeus

- Gipsywort (Scannell & Synnott, 1987); Water-horehound (K'Eogh, 1735; Wade, 1794)
- *Feorán corraigh* (Scannell & Synnott, 1987)

167. Dinneen (1927) applied this name to *Luzula maxima*.

LAMIACEAE (Labiatae)—THE DEAD-NETTLE FAMILY
Native

Lycopus europaeus is an occasional perennial herb throughout Ireland of ditches, marshes, lake margins and river-banks; it is most commonly found in the north and south-west.

It is well known as a traditional medicinal plant but has also been used to treat an over-active thyroid gland and associated symptoms, such as heart palpitations and nervousness (van Wyk & Wink, 2004). Moloney (1919) indicated it was used in traditional medicine in Ireland as an ingredient in cough remedies.

According to Mackay (1836), it yields a black dye, which gypsies used to darken the skin.

Lysimachia species

PRIMULACEAE—THE PRIMROSE FAMILY
Native

Including:

Lysimachia nemorum

- Yellow pimpernel (Scannell & Synnott, 1987); Wood pimpernel (Brunker, 1950)
- *Lus Cholm Cille* (Scannell & Synnott, 1987); *Seamair Mhuire* (de Vere Coneys, 1849; Ua Maoileoin, 2001); *Lus Coluimcille* (Moloney, 1919; Dinneen, 1927); *Lusaire na hurchóide* (Moloney, 1919); *Lus na Maighdeana Muire* (Moloney, 1919; Dinneen, 1927); *Lusradh na h-urchóide* (Dinneen, 1927); *Lusnadh na hurchóide* (Dinneen, 1927)

Lysimachia nummularia

- Creeping-Jenny (Scannell & Synnott, 1987); Moneywort (Anonymous, 1978); String of sovereigns (Hart, 1898)
- *Lus an dá phingin* (Scannell & Synnott, 1987)

Lysimachia nemorum is quite a common plant throughout Ireland of woods, scrub, hedges, moist banks and grassland and ditches. *Lysimachia nummularia* is less frequent and found in damp and shady grassland and muddy habitats on lake-shores and in woods and river-banks. In many of its habitats in Ireland it is certainly an escape from cultivation, but it is probably native in some places.

Lysimachia nemorum was formerly used in Ireland as a treatment for gallstones (Allen & Hatfield, 2004). Smith ([1756] 1979: 380) recorded that '*Nummularia* (moneywort), or herb two pence' was 'common in the marshy ground near Lixnaw', Co. Kerry, and said that 'the country people esteem it in decays, coughs, dysenteries, and spitting of blood'.

Lysimachia vulgaris

- Yellow loosestrife (Scannell & Synnott, 1987)
- *Breallán léana* (Scannell & Synnott, 1987); *Lus na síothchána (buidhe)* (Moloney, 1919; Dinneen, 1927); *Lus na síochána buí* (Ó Dónaill, 1977; Uí Chonchubhair, 1995); *Seamar Mhuire*

(Dinneen, 1927)
PRIMULACEAE—THE PRIMROSE FAMILY
Native

Lysimachia vulgaris is found in wet habitats, such as fens, marshes, river-banks, lake-shores and woodlands. It is mainly distributed in the centre of Ireland and is much rarer elsewhere.

It is occasionally grown as an ornamental garden plant in Ireland (Wyse Jackson, 2006).

Lythrum portula (Peplis portula)

- Water purslane (Scannell & Synnott, 1987)
- *Puirpín uisce* (Scannell & Synnott, 1987)

LYTHRACEAE—THE LOOSESTRIFE FAMILY
Native

Lythrum portula is an especially common annual plant in the south-west of ditches and the muddy margins of pools and puddles.

Mears and Hillman (2007) noted it is edible with a pleasant, mild flavour and useful in salads or it can be eaten on its own.

Lythrum salicaria

- Purple loosestrife (Scannell & Synnott, 1987); Brian braw (Hart, 1898); Foxtail (Hart, 1898); Garyagh (Hart, 1898); Redshanks (Hart, 1898)
- *Créachtach* (Scannell & Synnott, 1987); *Earbail caitín* (O'Rourke, 2006); *Bealán léana* (Moloney, 1919); *Brian breágh* (Moloney, 1919); *Brian breá* (Ó Dónaill, 1977); *Lus na síothchána* (de Vere Coneys, 1849; Moloney, 1919; Dinneen, 1927); *Conaire* (Moloney, 1919); *Earball cuitín* (Moloney, 1919); *Eireball caitín* (Dinneen, 1927; Uí Chonchubhair, 1995); *Cró lus* (Dinneen, 1927)

LYTHRACEAE—THE LOOSESTRIFE FAMILY
Native

Lythrum salicaria.

Lythrum salicaria is a common plant throughout Ireland of wet meadows, marshes, swamps, ditches and lake-shores.

It was formerly used in Ireland as an 'astringent' to cure diarrhoea (Allen & Hatfield, 2004). Threlkeld (1726) recorded that he used it to treat a patient suffering from dysentery. Moloney (1919) mentioned that he had not met any herbalists in Ireland that used it internally or externally at that time, although its Irish name, *creachtach*, meaning 'wound plant', suggests former medicinal usage, and another Irish name, *bealán*, means a 'knob' or 'gland', perhaps suggesting that it was used to treat glandular diseases.

MacMahon (1994: 241) said it was 'supposed to banish discord in a house' and used for dyeing. Dickie (1864) recorded that it was used in 'Glen Columbkil' (Glencolumbkille), Co. Donegal, to yield a brown dye. Hart (1898) also mentioned that it was used to supply a red dye in Donegal. Mahon (1983) listed it as providing a purple dye.

Malus sylvestris

- Crab-apple (Scannell & Synnott, 1987)
- *Crann fia-úll* (Scannell & Synnott, 1987); *Ubhall fiadhain* (Moloney, 1919; Dinneen, 1927); *Abhlacán* (Moloney, 1919; Dinneen, 1927); *Gortóg* (Moloney, 1919); *Ceirt*[168] (Dinneen, 1927); *Corr-abhall* (Dinneen, 1927); *Cumhnóg* (*-óige*); *Fiadh-ubhall* (Dinneen, 1927); *Goirteog* (Dinneen, 1927; Uí Chanainn, 1962); *Fia-úll* (Uí Chanainn, 1962)

ROSACEAE—THE ROSE FAMILY
Native

Malus sylvestris flowers.

Crab-apple is an occasional small- to medium-sized tree found in old woods and hedgerows throughout Ireland. The tallest crab-apple known in Ireland is an exceptional specimen recorded from Kissawamy, Carbury, Co. Kildare, that is 14.5 m tall and 2.25 m in girth (Anonymous, 2005). The girth champion is a tree at Coolusty, Ataleague, Co. Roscommon, that is 3.35 m in girth (and 12.5 m tall) (Anonymous, 2005).

The crab-apple tree (or wild apple, *fiaduball*) was an extremely important tree in ancient Ireland and much featured in folklore. It is included amongst a list of the most valuable trees (the 'seven nobles of the wood') (Kelly, 1997) probably largely because of its fruits. It grows to become a medium-sized tree and is often mistaken for self-sown apple trees (*Malus domestica*; in Irish, *abhaill*, Ó Dónaill, 1977) derived from the pips from discarded apple cores. Hart (1898: 370) gave a local English name for the apple tree from Fanet, Co. Donegal, as 'Cran ooly'.

Geraghty (1996) described finds of apple pips and endocarp (core) in sediments excavated from Fishamble Street, dating back more than 1000 years ago. Other species found in the same excavations include blackberries (*Rubus fruticosus*), rowans (*Sorbus aucuparia*), sloes (*Prunus spinosa*), haws (*Crataegus monogyna*) and bilberries (*Vaccinium myrtillus*).

Crab-apple trees provide a hard and durable, close-grained reddish-brown timber, which has been used to a limited extent for furniture making and woodcarvings. In Britain, crab-apple was used by foresters to make their mallet heads and was used also for golf club heads (Edlin, 1973). In the Aran Islands, apple wood was used to make the 'thwart knees', or internal corner braces within the hull of the traditional currach boats. McGlinchey (2007) mentioned that apple wood (or holly wood) was used to make the shuttle of a spinning loom. In Co. Offaly part of a flail used in the threshing of corn was sometimes made of crab-apple wood (see also sally; *Salix cinerea*) (Delaney, 1980).

Its bark provides a dye that has been used to dye wool yellow (Mahon, 1983; MacCoitir, 2003). The bark is boiled to extract the dye. Rutty (1772) wrote that the bark of an apple tree tans leather well too.

It was formerly used medicinally, mixed with buttermilk to treat sore throats (Moloney, 1919) and internal cancer (Allen & Hatfield, 2004). In Co. Antrim, rotten apples (of *Malus*

168. Dinneen (1927) defined *ceirt and cumhnóg* (*-óige*) as 'an apple-tree'.

domestica) were used to treat chilblains on the toes to ease the itching (Vickery, 2010). Chilblains were more commonly treated though by soaking them in urine.

It is still used to make delicious country preserves (jams and jellies) and is a useful source of pectin for setting preserves made from other fruits. Crab-apple jelly can be simply made with the following recipe: 2 lb/900 g of crab-apples (unbruised) are roughly cut and simmered with 600 mL/1 pint of water until soft, squeezing occasionally to ensure that all the juice is extracted. Don't remove the cores, peels or pips as these provide both pectin (to help set the jelly) and flavour. The mixture is then strained through a jelly bag or muslin and 450 g/1 lb of sugar is added to each 600 mL/1 pint of juice. The apple pulp can be boiled again to extract more juice, but less water is needed the second time. The sugar is dissolved slowly before the liquid is subsequently boiled rapidly until setting point is reached. After that the jelly is poured into sterilised jam jars, sealed and allowed to cool. Crab-apple jelly provides a useful base for other wild collected flavours and plants, such as *Origanum vulgare* (marjoram), *Mentha aquatica* (water mint) and *Thymus polytrichus* (wild thyme). Crab-apples can also be mixed with other wild fruits to make jelly, including berries of *Sorbus aucuparia* (rowanberries), *Rosa* species (rosehips), *Rubus fruticosus* (blackberries), *Crataegus monogyna* (hawthorn) and *Prunus spinosa* (sloes).

Crab-apples can also be made into fruit cheeses, ideally mixed with blackberries or other wild fruits. Cheeses are like a stiff jelly that can be cut into slices. The fruit is cooked until very stiff, then sieved or liquidised to make a puree. The peel, core and pips should be removed before it is liquidised and a non-metal sieve should be used if possible as the metal may taint or colour the mixture. To make crab-apple and blackberry cheese use 2 lb/900 g of crab-apples and 1.5 lb/675 g of blackberries and simmer together with a few tablespoons of water until they become a thick pulp. The pulp is then sieved to remove pips, seeds and the skin before being measured and heated again with sugar. To each 600 mL/pint of puree add about 400 g/14 oz of sugar (depending on taste) and a few knobs of butter. Cook these together over a low heat until the sugar has dissolved, then boil steadily until setting point is reached (be careful to stir the mixture to prevent it from sticking to the pan and burning). Fruit cheeses are delicious with roast duck, goose or game, or served with cheese or as part of a pudding or dessert. Crab-apples also can be stewed and baked.

Crab-apples have been used to make wine and cider. The making of cider has been much less a part of traditional activity in Ireland than in Britain, although Ó Céirín and Ó Céirín (1978) mentioned that there was a strong tradition of cider-making in Cos. Limerick and Clare until recent times. Clonmel, Co. Tipperary, is also the home of Bulmer's Cider (sold as Magners Cider overseas), which has been made there since 1935.

Crab-apple vinegar of 'verjuice' can also be made. Verjuice, a slightly fermented, very bitter cider, can be made by collecting the fruits late and storing them until they begin to sweat. After this the bad fruit is removed and the remaining ones crushed with a wooden mallet. The juice is then collected, bottled and stored for some weeks before use (Ó Céirín & Ó Céirín, 1978).

Crab-apples are occasionally grown in Irish gardens, both for their fruit and as ornamental plants. Nelson (2000b) documented a wide range of cultivated varieties of *Malus sylvestris* and *M. domestica* raised in Ireland.

Malva neglecta (Malva rotundifolia)

- Dwarf mallow (Scannell & Synnott, 1987)
- *Hocas francach* (Scannell & Synnott, 1987); *Ucas franncach* (Moloney, 1919)

MALVACEAE—THE MALLOW FAMILY

Introduced (possible archaeophyte)

Malva neglecta is an introduced species on roadsides and wasteground, particularly in the east of Ireland. Like its near relative, the common mallow (*Malva sylvestris*), it has been used as a medicinal plant. Its leaves have been used to treat coughs. Moloney (1919: 18) mentioned that *M. neglecta* and *M. sylvestris* were used in Ireland 'internally as demulcent remedies for coughs and hoarseness and that an ointment made from the roots was still in demand as an emollient for rubbing into painful and stiff joints'.

Peterson (1977) described its use as an edible plant in North America, where it has been used as a thickener, in salads and cooked as a green vegetable. I have not found records of its use for food in Ireland.

The Irish word *hocas* or *ucas* means 'breast', suggesting that the mucilage in this plant was used to make a skin preparation for the breasts.

Malva sylvestris

- Common mallow (Scannell & Synnott, 1987); Hawk (Colgan, 1904); Marsh mallow (Colgan, 1904)
- *Lus na meall Muire* (Scannell & Synnott, 1987); *Milmheacan* (de Vere Coneys, 1849; Moloney, 1919; Dinneen, 1927); *Lus na moil mór* (Moloney, 1919); *Lus na míol mór* (de Vere Coneys, 1849; Dinneen, 1927); *Ucas* (*Ocos*) (Moloney, 1919); *Ucas fiadhain* (de Vere Coneys, 1849; Moloney, 1919); *Hocas* (Dinneen, 1927; Webb et al., 1996); *Grosis* (*-se*)[169] (de Vere Coneys, 1849); *Mealbhacán* (*-áin*)[170] (Dinneen, 1927)

MALVACEAE—THE MALLOW FAMILY

Possibly Introduced

Malva sylvestris is widespread and common throughout Ireland on roadsides and wasteground.

It has been used as a medicinal plant throughout Britain and Ireland for centuries, as a poultice for cuts, to treat blood-poisoning, urinary problems, coughs, and rheumatism and as a veterinary remedy for a range of ailments in cattle and horses in Ireland (Allen & Hatfield, 2004). Mallow contains a high percentage of mucilage, or slime (10% dry weight of flowers and 8% of leaves), which provides the medicinally useful ingredient (van Wyk & Wink, 2004).

Young mallow leaves and flowers are edible and can be used in salads. The leaves can be used too as a vegetable to help thicken soups and stews or deep-fried as a snack food (Hilton, 2007). The dried leaves can also be used to make tea, and the small seeds, known as cheeses, have an interesting nutty flavour.

Marrubium vulgare

- White horehound (Scannell & Synnott, 1987)
- *Orafunt* (Scannell & Synnott, 1987); *Grafán bán* (de Vere Coneys, 1849; Moloney, 1919;

169. De Vere Coneys (1849) gave this name for 'mallows, *Malva*'.
170. Dinneen (1927) defined *mealbhacán* (*-áin*) as 'a mallow'.

Dinneen, 1927; Ó Dónaill, 1977)
LAMIACEAE (Labiatae)—THE DEAD-NETTLE FAMILY
Introduced

Marrubium vulgare is a very rare weed in Ireland and has probably not been recorded for about the last 50 years (Reynolds, 2002).

White horehound is well known throughout its native range in southern Europe and Asia as an important plant in traditional medicine for digestive complaints, such as indigestion, lack of appetite, dyspepsia and flatulence, and for dry coughs and bronchitis (van Wyk & Wink, 2004). It was used medicinally in Ireland to treat earache and headache and to 'clean out the valves of the heart; it was also drunk as a tea for rheumatism (Moloney, 1919; Allen & Hatfield, 2004). Wade (1804) and Moloney (1919) also noted it was used to treat colds, coughs and asthma.

Matricaria discoidea (Matricaria matricarioides; Chamomilla suaveolens)

- Pineapple-weed (Scannell & Synnott, 1987); Rayless mayweed (Anonymous, 1978); American mayweed (Brunker, 1950)
- *Lus na hiothlann* (Scannell & Synnott, 1987); *Lus anainn* (Ó Dónaill, 1977)

ASTERACEAE (Compositae)—THE DAISY FAMILY
Introduced

Matricaria discoidea.

Matricaria discoidea is an annual weed that was introduced near Dublin in 1894 and is now widespread throughout Ireland (Scannell & Synnott, 1987). It is reputed to have been introduced from North America as a seed contaminant of corn used as poultry feed in Ireland (Reynolds, 2002). It has a delicate aroma that resembles the fragrance of pineapples and is now an extremely common weed of waste places, farmyards, roadsides and cultivated ground.

Although I have not found records of its use in Ireland, Peterson (1977) mentioned its flowers have been used fresh or dried in North America to make a pale-golden, mildly pineapple-scented tea.

**Matricaria maritima (Tripleurospermum maritimum)*

- Sea mayweed (Scannell & Synnott, 1987)
- *Lus Bealtaine* (Scannell & Synnott, 1987)

ASTERACEAE (Compositae)—THE DAISY FAMILY
Native

Matricaria maritima.

Matricaria maritima is a common plant all around the coasts of Ireland of shingle and rocky shores and waste places.

*No ethnobotanical uses of this plant have been noted in Ireland.

Matricaria perforata (Matricaria inodora; Tripleurospermum inodorum)

- Scentless mayweed (Scannell & Synnott, 1987); Dog daisy (Hart, 1898)
- *Meá drua* (Scannell & Synnott, 1987); *Meadh-duach* (Moloney, 1919); *Meadh druadh* (Moloney, 1919); *Lus deartan*[171] (Moloney, 1919; Dinneen, 1927)

ASTERACEAE (Compositae)—THE DAISY FAMILY
Probably Introduced

Matricaria perforata is an annual plant that is found occasionally throughout Ireland as a weed in waste places.

According to Moloney (1919), it was formerly used to treat uterine problems and to prevent insect bites.

Meconopsis cambrica

- Welsh poppy (Scannell & Synnott, 1987)
- *Poipín Breatnach* (Scannell & Synnott, 1987)

PAPAVERACEAE—THE POPPY FAMILY
Native

Meconopsis cambrica is a perennial yellow-flowered poppy that is occasionally grown as a garden plant in Ireland. It more often survives as a garden weed or in waste places having escaped from cultivation. It is a very rare native plant of rocky damp shady habitats in the mountains.

Medicago lupulina

- Black medick (Scannell & Synnott, 1987); Black hay (Harron, 1986)
- *Dúmheidic* (Scannell & Synnott, 1987); *Medicaga dubhsiolach* (Wade, 1808b)

FABACEAE (Leguminosae)—THE PEA AND CLOVER FAMILY
Native

Medicago lupulina is a very common annual herb of dry grasslands, sand dunes, roadsides and waste places throughout Ireland, except the north-west where it is surprisingly rather rare.

In North America, the ripe seeds have been gathered as food and ground down to make flour (Peterson, 1977). I have not found any records of such usage in Ireland.

This plant has occasionally been used in Ireland as shamrock and worn to celebrate St Patrick's Day (Nelson, 1991).

---

171. Dinneen (1927) defined *lus deartan* as 'scentless feverfew'.

Medicago sativa subsp. *sativa*

- Lucerne (Scannell & Synnott, 1987); Alfalfa (Anonymous, 1978)
- *Meidic na mbó* (Scannell & Synnott, 1987); *Medicaga treabhamtach* (Wade, 1808b)

FABACEAE (Leguminosae)—THE PEA AND CLOVER FAMILY
Introduced

Medicago sativa subsp. *sativa* is occasionally grown in Ireland as a fodder crop and is naturalised in some places throughout, such as in waste places and on roadsides, sand dunes and banks.

It has been used in traditional medicine as a general tonic and dietary supplement and has been considered useful in helping to lower cholesterol levels (van Wyk & Wink, 2004). I have not found records of its usage this way in Ireland as a medicinal plant.

The dried and powdered flower heads and young leaves are nutritious and can be made into tea or added to breakfast cereals or stews (Peterson, 1977). Soft young leaves can also be added to salads.

**Melampyrum pratense*

- Common cow-wheat (Scannell & Synnott, 1987); Yellow cow-wheat (Brunker, 1950)
- *Lus an tsagairt* (Scannell & Synnott, 1987)

OROBANCHACEAE (Scrophulariaceae)—THE BROOMRAPE FAMILY
Native

Melampyrum pratense is an annual herb throughout Ireland of woodlands, stony lake-shores, bogs and mountain slopes.

*No ethnobotanical uses of this plant have been noted in Ireland.

**Melampyrum sylvaticum*

- Small cow-wheat (Scannell & Synnott, 1987)
- *Lus sagairt beag* (Scannell & Synnott, 1987)

OROBANCHACEAE (Scrophulariaceae)—THE BROOMRAPE FAMILY
Native

Melampyrum sylvaticum is a rare species in Ireland and found only in woods in Cos. Derry and Antrim in Northern Ireland.

*No ethnobotanical uses of this plant have been noted in Ireland.

**Melica uniflora*

- Wood melic (Scannell & Synnott, 1987)
- *Meilic* (Scannell & Synnott, 1987); *Meloigfhér aonbhlátha* (White, 1803)

POACEAE (Gramineae)—THE GRASS FAMILY
Native

Melica uniflora is a grass species that can be found occasionally in woodlands, hedges and other shady places.

*No ethnobotanical uses of this plant have been noted in Ireland.

Melilotus species

- Melilot (Scannell & Synnott, 1987)
- *Crúibín* (Scannell & Synnott, 1987)

FABACEAE (Leguminosae)—THE PEA AND CLOVER FAMILY

Introduced

Including:

Melilotus altissimus

- Tall melilot (Scannell & Synnott, 1987)
- *Crúibín cait mór* (Scannell & Synnott, 1987)

Melilotus officinalis

- Ribbed melilot (Scannell & Synnott, 1987); Common melilot (Colgan, 1904); Strongbow (Colgan, 1904)
- *Crúibín cait* (Scannell & Synnott, 1987)

Melilotus altissimus and *M. officinalis* are tall, weedy legumes of dry waste places, such as railway banks and coastal sandy habitats, that were introduced in Ireland. *Melilotus altissimus* is mostly widespread in the Dublin region and *M. officinalis* in eastern Co. Antrim; both are rarely distributed elsewhere in Ireland.

Melilot species have been used in traditional medicine in parts of their range to treat a variety of ailments, such as cramps, haemorrhoids and problems with the circulatory system, as well as gastrointestinal disorders (van Wyk & Wink, 2004). *Melilotus officinalis* has also been used to treat varicose veins and eczema (Ody, 1997b), but I have not found evidence of its use to treat such problems in Ireland. Moloney (1919) listed *M. officinalis* as a homeopathic remedy for epilepsy and nervous headache.

Melilot species could probably be used in salads or cooked as a green vegetable, although I have not tried them, nor have I found any records of this use in Ireland. Peterson (1977) suggested that the pea-like seeds of *Melilotus officinalis* are a good source of protein and can be used to flavour soups and stews, and the crushed dried leaves can be used as a vanilla-like flavouring for pastries.

Mentha species

- Mint (Scannell & Synnott, 1987)
- *Mismín* (Scannell & Synnott, 1987); *Miontas*[172] (Dinneen, 1927)

LAMIACEAE (Labiatae)—THE DEAD-NETTLE FAMILY

Mints are extremely important plants worldwide and provide aromatic essential oils for a wide range of products and uses, such as toothpaste, sweets and chewing-gum. Four different mints are used commercially and industrially: *Mentha* ×*piperita*, peppermint[173]; *M. arvensis* and two species of spearmint[174], *M. spicata* and *M. cardiaca*. Of these, only *M. arvensis*, corn mint[175], is native to Ireland, and it has never been used commercially there. As far as I can tell it has been little used for folk or traditional purposes. *Mentha*

172. Dinneen (1927) defined *miontas* as 'mint, spearmint'.

173. Moloney (1919) and Dinneen (1927) gave the Irish word *lus an phiobair* for 'peppermint'.

174. De Vere Coneys (1849) gave the Irish word *méantus gairdín* for 'spearmint', Dinneen (1927) *glais-léana*, and Uí Chonchubhair (1995) *misímín garraí*.

175. Dinneen (1927) and Ó Dónaill (1977) gave the Irish word *miontas arbhair* for 'cornmint'.

×*piperita* is a hybrid between *M. aquatica* and *M. spicata* and is found as an occasional introduction on river-banks and ditches in Ireland. It is also commonly grown as a garden herb.

Mentha aquatica

- Water mint (Scannell & Synnott, 1987)
- *Mismín mionsach* (Scannell & Synnott, 1987); *Pisimín* (Moloney, 1919); *Cairteal* (Moloney, 1919); *Misimín dearg* (de Vere Coneys, 1849; Colgan, 1911; Moloney, 1919); *Cartloin* (Cameron, 1883); *Cartláid*[176] (Dinneen, 1927); *Miontas uisce* (Dinneen, 1927)

LAMIACEAE (Labiatae)—THE DEAD-NETTLE FAMILY

Native

Mentha aquatica is a common perennial herb that is found in a variety of wet habitats, such as swamps, marshes, lake-shores, stream-sides, ditches and drains.

Water mint is occasionally collected and used as an aromatic herb for culinary use (e.g., for mint sauce as an accompaniment for roast lamb). Another mint species, *Mentha spicata*, spearmint, is a non-native plant that was mentioned as an ingredient of 'lamb-sauce' by James Townsend Mackay in *Flora Hibernica* (1836: 212). It has probably been used like this for 200 years or more. For mint sauce I prefer species with hairless (glabrous) leaves, and *M. aquatica* is a bit hairy for my taste. It is also used to make mint tea.

Mint jelly is made by adding finely chopped mint leaves to apple jelly or crab-apple jelly during cooking, together with several tablespoons of malt or wine vinegar (see *Malus sylvestris*). It is delicious with roasted lamb, cheese and pâtés. Mint jelly is usually not green coloured when cooked (unless a little green colouring is added, such as in commercially available varieties). Mint marmalade (chopped leaves added to a lemon marmalade base during cooking) is also a delicious and interesting alternative to chutney to accompany cold meat and cheeses.

Mentha aquatica.

Mentha aquatica was formerly used in Ireland for a variety of medicinal purposes, such as indigestion and stomach pains, headaches, deafness, measles, nettle stings and jaundice, and was occasionally used to treat colds, coughs and whooping cough (Allen & Hatfield, 2004). Williams (1993) noted its use as a cure for deafness. The plant was boiled, then strained through linen, and a thimble-full of the resulting liquid was poured into the ears. It was also thought to be useful to place mint leaves in your socks to protect against infection when visiting infectious patients. The same author noted its use to treat nettle stings in Cavan and as a tea for menstrual pain and throat and stomach ailments.

Allen and Hatfield (2004) recorded the use of *Mentha* species, 'peppermint', to treat fits in dogs and pigs in Co. Leitrim.

Williams (1993) noted it was used to keep away fleas and protect clothes from moths. Maurice Maxwell (pers. comm., 2008) relayed the following story to me: 'In the early 1980s while I worked as a gardener for the Daughters of Charity at Knockmaroon Hill, Chapelizod, Co. Dublin, one of the elderly nuns would insist that a garland of fresh mint

176. Dinneen (1927) gave the Irish word *cartláid* for spearmint.

was placed all around the windows of the kitchen and dining area to keep off flies. The mint had to be replaced as soon as it became dry.' *Mentha spicata* was probably the species used.

Williams (1993) also noted it was used in the storage of smoked trout, and in Kerry it was sometimes smoked instead of tobacco.

Mentha arvensis

- Corn mint (Scannell & Synnott, 1987); Field mint (Anonymous, 1978)
- *Mismín arbhair* (Scannell & Synnott, 1987)

LAMIACEAE (Labiatae)—THE DEAD-NETTLE FAMILY
Native

Mentha arvensis is found only occasionally throughout Ireland in habitats such as waste places, roadsides and grasslands.

An essential oil is distilled from the leaves and stems of corn mint, mainly to produce pure menthol. The oil has been used to treat a variety of ailments, such as flatulence, gastrointestinal problems and catarrh, as well as for oral hygiene (van Wyk & Wink, 2004). I have not found records of its use in traditional medicine in Ireland.

Hart (1898) mentioned that this plant may have been dried and used in Co. Donegal as a tobacco substitute.

Mentha pulegium

- Penny-royal (Scannell & Synnott, 1987)
- *Borógach* (Scannell & Synnott, 1987); *Arán glas* (Moloney, 1919)

LAMIACEAE (Labiatae)—THE DEAD-NETTLE FAMILY
Possibly Introduced (possible archaeophyte)

Mentha pulegium occurs in wet habitats on sandy soil. It was formerly much more common in Ireland and abundant in Co. Kerry but is now declining and has become rare.

Mackay (1836) recorded that this species was much used (presumably in Ireland) for medicinal purposes. Scully (1916) recorded that in Co. Kerry it was used medicinally by peasants for infusions. Wade (1804) reported that it was mixed with honey or sugar to treat children's whooping cough. Penny-royal soaked in vinegar and held in the mouth was used to relieve toothache (Baker, 1999). Rutty (1772: 74) said that it was sometimes used in broths ('but gives way to thyme and savory') and boiled with black puddings and 'lamb's puddings'.

According to Vickery (1995, 2010) this plant was widely used as an abortifacient in Britain, but I have not found records of such usage in Ireland.

Menyanthes trifoliata

- Bogbean (Scannell & Synnott, 1987); Bog hop; Marsh trefoil; Boughram (Hart, 1898); Baughram (Hart, 1898); Buckbean (Praeger, 1938; Brunker, 1950)
- *Báchrán* (Scannell & Synnott, 1987); *Pónaire chapail(l)* (de Vere Coneys, 1849; Moloney,

1919; Dinneen, 1927); *Pónaire chunnaigh* (Moloney, 1919); *Bearnán lacháin* (Moloney, 1919); *Pacharán* (de Vere Coneys, 1849; Moloney, 1919); *Pór an capaill* (Moloney, 1919; Dinneen, 1927); *Pónaire churraigh* (De Bhaldraithe, 1959); *Pónaire chorraigh*[177] (Dinneen, 1927)

MENYANTHACEAE—THE BOGBEAN FAMILY

Native

Menyanthes trifoliata is an aquatic and bogland perennial that is found occasionally throughout Ireland in ponds, lake margins, wet bogs and fens.

It was an important and widely used medicinal plant in Europe. In traditional medicine, fresh or dried bogbean leaves have been used to stimulate the appetite and for digestion (Reader's Digest Association, 2006). According to Moloney (1919), it was used as a tonic and purgative and recommended for rheumatism. Hart (1898) recorded that in the Ardara region of Co. Donegal it was picked in summer and dried and used in March for blood purification. It was also used in Ireland for boils and skin troubles, digestion, liver complaints, stomach upsets, constipation, rheumatism, coughs and colds, heart disease and dropsy (Uí Chonchubhair, 1995; Allen & Hatfield, 2004). Lady Wilde (1890) listed it as an effective cure for scrofula ('King's evil'). Bogbean contains a bitter glucoside compound called menyanthin, which has been used widely in medicine as a tonic and to bring down fevers (Heywood, 1978).

Mackay (1836) recorded that it was used to cure a disease of cattle called 'darn', and in Scotland it was sometimes used as a substitute for hops (Milliken & Bridgewater, 2004).

Allen and Hatfield (2004) recorded that it has been used as a veterinary remedy to help remove afterbirth in cattle in Co. Leitrim and to treat a trace-element deficiency called 'pine' in cattle and sheep in Co. Mayo.

Peterson (1977) listed it as having been used for food in North America where the rootstock was harvested to make flour and as a vegetable. The long rootstock was boiled in several changes of water for about half an hour, or it was dried, crushed and well leached to remove some of the bitterness to make unpalatable but nutritious flour. I have not found records of its use for food in Ireland.

Hart (1898) noted that it yields a useful dark brown dye. I have come across references in several works to a plant used for medicinal purposes in Ireland called 'bog bine' or 'bogbine', but I have not been able to identify with any certainty what species is being referenced. It may, in fact, be a variant of 'bogbean'. Of bogbine, McGlinchey (2007: 84) wrote, 'In my young days every house pulled bogbine in the spring-time. It was for the bogbine roots, and they were boiled and mixed with treacle and sulphur. This was a well-tried remedy for purifying the blood. The blood gets out of order in the spring-time. Bogbine is used to the present day [he wrote this in the mid-1900s]'. On the Glens of Antrim Historical Society website, bogbine is also mentioned in an oral history by Annie McKillop: 'They used to give bogbine, for instance, to teenage girls, to build up their iron content, you know, so that they wouldn't go anaemic when they were teenagers, when they started their young life. You see, bogbine was a very hard thing to take, for it was very, very bitter. You would get it in the mountain. There is places it grows in the bog, in the moss, and they would have to take it, and boil it and then bottle it. Aye, just like taking an iron tonic, you know. I never could have it. I had to do without it. I couldn't take it!' (Antrimhistory.net, n.d.). Another reference to bogbine is found on the website of Clonmany in Co. Donegal (Charles McGlinchey's county) (Clonmany.com, n.d.): 'Bogbine is another great thing. It's like the root of a plant, white and solid with broad green leaves something like a palm. You pulled it up and saw its stretchy white stem. So you boiled it and strained it and bottled

177. Dinneen (1927) defined *pónaire chorraigh* as 'marsh bean'.

it and you could keep it in the fridge and every morning before you broke your fast, you would take it by the spoonful and it killed all the harmful bacteria in your blood. People don't know how to identify it any more. The only man could give it to you now is a man up in Dunaff. It's a pity it doesn't come back into circulation' (Margaret Farren).

**Mercurialis annua*

- Annual dog's mercury (Scannell & Synnott, 1987); Annual mercury (Brunker, 1950)
- *Lus glinne beag* (Scannell & Synnott, 1987); *Praisneach larla* (Moloney, 1919)

EUPHORBIACEAE—THE SPURGE FAMILY
Introduced

Mercurialis annua is a common annual weed, particularly in urban areas, of cultivated ground, waste places and muddy roadsides.

It is poisonous if eaten raw, but the leaves are said to be edible if cooked (like spinach) (see Pfaf.org).

*No ethnobotanical uses of this plant have been noted in Ireland.

**Mercurialis perennis*

- Perennial dog's mercury (Scannell & Synnott, 1987)
- *Lus glinne* (Scannell & Synnott, 1987); *Praiseach bhráthar* (Moloney, 1919; Ó Dónaill, 1977)

EUPHORBIACEAE—THE SPURGE FAMILY
Possibly Introduced

Mercurialis perennis is rare in Ireland and occurs in woodlands, hedgerows and other shady habitats. In most of its localities it is certainly introduced but could possibly be native in a very few places, such as in Co. Clare.

This plant is poisonous.

*No ethnobotanical uses of this plant have been noted in Ireland.

Mertensia maritima

- Oysterplant (Scannell & Synnott, 1987); Sea gromwell (Brunker, 1950)
- *Lus na sceallaí* (Scannell & Synnott, 1987)

BORAGINACEAE—THE BORAGE FAMILY
Native

Mertensia maritima occurs along the coast from Cos. Donegal to Louth. It was formerly found more often along the Irish coastline but is now declining and confined to a small number of populations in the north and north-west on shingle shores.

Mabey (1972) recorded that the fleshy leaves of oysterplant have been eaten raw or cooked as a vegetable and are reputed to taste of oysters.

**Mibora minima*

- Early sand-grass (O'Mahony, 2006)
- *Taenfhér robhég*[178] (?) (White, 1803)

POACEAE (Gramineae)—THE GRASS FAMILY
Native

Mibora minima is a very rare grass in Ireland that was found only in 2005 by Tony O'Mahony in Cannawee dune system in West Cork.

*No ethnobotanical uses of this plant have been noted in Ireland.

**Milium effusum*

- Wood millet (Scannell & Synnott, 1987); Millet-grass (Moloney, 1919)
- *Muiléad coille* (Scannell & Synnott, 1987); *Meanbhphreasair* (Moloney, 1919); *Miledfhér coitchion* (White, 1803)

POACEAE (Gramineae)—THE GRASS FAMILY
Native

Milium effusum is a relatively rare grass throughout Ireland of mountain woodlands and glens.

*No ethnobotanical uses of this plant have been noted in Ireland.

**Minuartia recurva*

- Recurved sandwort (Scannell & Synnott, 1987)
- *Gaineamhlus cuar* (Scannell & Synnott, 1987)

CARYOPHYLLACEAE—THE CARNATION OR PINK FAMILY
Native

Minuartia recurva is a very rare plant in Ireland of rocky areas in the mountains in the south.

*No ethnobotanical uses of this plant have been noted in Ireland.

**Minuartia verna*

- Spring sandwort (Scannell & Synnott, 1987)
- *Gaineamhlus earraigh* (Scannell & Synnott, 1987)

CARYOPHYLLACEAE—THE CARNATION OR PINK FAMILY
Native

Minuartia verna grows amongst rocks and screes and on some sand dunes in only a few places in Ireland in Co. Clare, the Aran Islands, Antrim and Derry.

*No ethnobotanical uses of this plant have been noted in Ireland.

178. White (1803) applied this name to *Agrostis minima*. The Plant List (see Theplantlist.org) gives *A. minima* as a synonym of *Mibora minima*; however, the application of this Irish name to this species should be regarded as very tentative.

*Misopates orontium

- Lesser snapdragon (Scannell & Synnott, 1987); Weasel's-snout
- *Srubh lao beag* (Scannell & Synnott, 1987)

VERONICACEAE (Scrophulariaceae)—THE SPEEDWELL FAMILY
Possibly Introduced (possible archaeophyte)

Misopates orontium is a very rare annual weed of cultivated fields that is found mainly in the south of Ireland.

*No ethnobotanical uses of this plant have been noted in Ireland.

Moehringia trinervia

- Three-nerved sandwort (Scannell & Synnott, 1987)
- *Gaineamhlus féitheach* (Scannell & Synnott, 1987)

CARYOPHYLLACEAE—THE CARNATION OR PINK FAMILY
Native

Moehringia trinervia is a woodland herb that is found occasionally throughout Ireland.

Mears and Hillman (2007) noted that this plant is edible and is best eaten steamed or boiled. I have not found records of its use for food in Ireland.

Molinia caerulea (Melica caerulea)

- Purple moor-grass (Scannell & Synnott, 1987); Keib (or Kabe) (Hart, 1898)
- *Fionnán* (Scannell & Synnott, 1987); *Préamh(acha) earraigh* (Ua Maoileoin, 2001); *Mealoigfér corcuir* (Cameron, 1883); *Béitean* (Uí Chonchubhair, 1995); *Punglas (-ais)*[179] (de Vere Coneys, 1849; Dinneen, 1927); *Meloigfheur corcuir* (Wade, 1808a[180]); *Féar boirne*[181] (Ó Dónaill, 1977); *Meloigfhér corcuir* (White, 1803)

POACEAE (Gramineae)—THE GRASS FAMILY
Native

Molinia caerulea is a common, sometimes abundant, grass throughout Ireland of moorland, bogs, limestone pavements and heaths. *Molinia* meadows growing on the lower slopes of western mountains are a conspicuous feature of the Irish landscape.

Maura Scannell (pers. comm., 2008) told me that there are literature references to this grass being used to make creel handles in Co. Kerry and that it was known locally as 'fedget grass'.

Richard M. Barrington, writing in the *Irish Naturalist* in 1906, reported that in Sneem, Co. Kerry, he saw tenant farmers drying *Molinia caerulea*, locally called *finnaun*, to make ropes for thatching. Apparently, no other grass would make ropes as strong and lasting as

179. De Vere Coneys (1849) and Dinneen (1927) applied the name *punglas (-ais)* to *Melica caerulea*.
180. Wade (1808a) applied this name to *Melica caerulea*.
181. *Féar boirne* is translated as 'blue moor-grass'.

this species (Barrington, 1906). Nathaniel Colgan (author of the *Flora of Co. Dublin*) is recorded in the same article as having seen the grass used in the same way in Co. Mayo, gathered during the autumn. Wade (1808a) said that this grass was made into twine and much valued by Continental fishermen as it is not much affected by water. In the Scottish island of Skye, *M. caerulea* was twisted by fishermen to make anchor ropes and cables (Milliken & Bridgewater, 2004). According to Clyne (1989), these ropes last well in water without rotting. McCarthy and Hawkes (2007) said that it was used traditionally in the Mizen region of West Cork as a fire lighter.

Molinia caerulea.

Dinneen (1927: 457) included the Irish name *fionnán* as a 'kind of long coarse white grass which grows on marshy land, used for making grass ropes (*súgáin*) and as bedding for cattle'. Dinneen (1927: 458) also listed another Irish name, *fionntarnach*, as 'a coarse brown sedge-like grass that grows on mountains and swamps, sometimes used for thatch'. It is not clear if this name also refers to *Molinia caerulea*.

The Islandman Tomás Ó Criomhthain gave the name *fionán* for 'bog-grass—long white grass used for making *súgaín* or hay-ropes' in a list of the flora and fauna of Kerry's Great Blasket Island that he compiled during the early part of the 20th century for the Norwegian scholar Carl Marstrander. This list is preserved by the University of Oslo in Norway.

According to Uí Chonchubhair (1995), purple moor-grass is resistant to heath fires and tends to replace heather with frequent burning.

Montia fontana

- Blinks (Scannell & Synnott, 1987); Water chickweed (Webb, 1977; Anonymous, 1978; Webb & Scannell, 1983)
- *Fliodh uisce* (Scannell & Synnott, 1987)

MONTIACEAE (Portulacaceae)—THE BLINKS FAMILY
Native

Montia fontana is found throughout Ireland in wet habitats, such as stream banks, wet meadows, marshes and mountain flushes.

Remains of this plant dating back to the Viking Age have been found in grassland sods excavated from houses in Fishamble Street, Dublin (Geraghty, 1996) and may indicate some former usage in ancient Ireland.

*Myosotis species

- Forget-me-not (Scannell & Synnott, 1987)
- *Driofúr ghorm* (Ua Maoileoin, 2001); *Lus míonla* (Dinneen, 1927; Ó Dónaill, 1977; Scannell & Synnott, 1987) *Ceotharnach* (Uí Chanainn, 1962)

BORAGINACEAE—THE BORAGE FAMILY

*No ethnobotanical uses of this plant have been noted in Ireland.

*Myosotis arvensis

- Field forget-me-not (Scannell & Synnott, 1987); Common forget-me-not (Anonymous,

1978); Field scorpion grass (Brunker, 1950; Anonymous, 1978)
- *Lus míonla goirt* (Scannell & Synnott, 1987)

BORAGINACEAE—THE BORAGE FAMILY
Native

Myosotis arvensis is a small annual plant throughout Ireland of dry habitats, such as roadsides, sparse grasslands, cultivated fields and sand dunes.

*No ethnobotanical uses of this plant have been noted in Ireland.

Myosotis discolor

- Changing forget-me-not (Scannell & Synnott, 1987)
- *Lus míonla buí* (Scannell & Synnott, 1987)

BORAGINACEAE—THE BORAGE FAMILY
Native

Myosotis discolor is a widespread and common plant in the north of cultivated ground, roadsides and dry, well-drained habitats, such as sandy and stony banks. It can also be found in marshy grasslands, although it is rare in this habitat.

*No ethnobotanical uses of this plant have been noted in Ireland.

Myosotis laxa

- Tufted forget-me-not (Scannell & Synnott, 1987)
- *Ceotharnach beag* (Scannell & Synnott, 1987)

BORAGINACEAE—THE BORAGE FAMILY
Native

Myosotis laxa is a common plant throughout Ireland of wet habitats, such as ditches, stream-sides and marshes.

*No ethnobotanical uses of this plant have been noted in Ireland.

Myosotis ramosissima

- Early forget-me-not (Scannell & Synnott, 1987)
- *Lus míonla luath* (Scannell & Synnott, 1987)

BORAGINACEAE—THE BORAGE FAMILY
Native

Myosotis ramosissima occurs mainly along the east coast in dry, sandy and stony habitats; it is very rare elsewhere in Ireland.

*No ethnobotanical uses of this plant have been noted in Ireland.

Myosotis scorpioides (Myosotis palustris)

- Water forget-me-not (Scannell & Synnott, 1987)
- *Ceotharnach uisce* (Scannell & Synnott, 1987); *Súil an chuir* (Moloney, 1919); *Lus mínle* (Moloney, 1919); *Lus míola* (Moloney, 1919); *Cotharach* (Cameron, 1883)

BORAGINACEAE—THE BORAGE FAMILY
Native

Myosotis scorpioides can be found throughout Ireland and is a frequent plant of marshy places, such as lake-shores, river-banks, muddy soils and wet fields.

Rutty (1772) said this plant is poisonous to sheep.

*No ethnobotanical uses of this plant have been noted in Ireland.

**Myosotis secunda*

- Creeping forget-me-not (Scannell & Synnott, 1987)
- *Ceotharnach reatha* (Scannell & Synnott, 1987)

BORAGINACEAE—THE BORAGE FAMILY
Native

Myosotis secunda is found throughout Ireland and is most common in the north in wet acid habitats such as ditches, marshes and stream sides.

*No ethnobotanical uses of this plant have been noted in Ireland.

Myrica gale

- Bog myrtle (Scannell & Synnott, 1987); Sweet-willow (Dinneen, 1927); Gale (Hart, 1898); Golden willow (Hart, 1898); Raidsogagh (Hart, 1898); Rideokut (Hart, 1898)
- *Raideog* (Scannell & Synnott, 1987); *Raidhleóg* (Moloney, 1919); *Railleóg* (Webb & Scannell, 1983); *Rileóg* (Colgan, 1911; Danaher, 1964a); *Ruid-eagach* (McGlinchey, 2007); *Roid*[182] (Dinneen, 1927); *Roideog* (*-eioge*) (Dinneen, 1927) (or *Raidleog*; *Rudóg*; *Roidleog*; *Roilleog*; *Reileog*; *Roithleog* (Dinneen, 1927); *Raideogach* (Ó Dónaill, 1977); *Releog*, *Rideóg* (Colgan, 1911)

MYRICACEAE—THE MYRTLE FAMILY
Native

Myrica gale (bog myrtle). Watercolour by Lydia Shackleton. Courtesy of the National Botanic Gardens, Glasnevin.

Myrica gale is a common shrub of bogs, fens and lake-shores, particularly in the western half of Ireland, and is rarely found in the east.

This aromatic shrub traditionally had many uses. In ancient Ireland bog myrtle was considered semi-sacred. However, historically its main use was as an insecticide to repel fleas and other insects, and there are records of its use to treat kidney complaints, measles and sore throats (Allen & Hatfield, 2004). McGlinchey (2007) also noted its use to treat kidney ailments in Co. Donegal. Hilton (2007) suggested that this plant contains compounds that may cause

182. Dinneen (1927) defined *roid* as 'a dye-plant cut into bundles called *trillsíní* or *scriplíní*, or prepared as a powder; applied to bog myrtle'.

miscarriages, so it should not be used during pregnancies.

Bog myrtle was also used in parts of Ireland as a veterinary remedy to treat various ailments of animals, such as red-water fever and collar sores in horses, liver fluke in calves, fits in dogs and gravel in horses (Allen & Hatfield, 2004). It was used to cure worms in cattle (and boys!) and as a cure for red murrain and colic, and the ashes of burnt leaves were put in the eyes of sheep, with a goose feather, to treat blindness (Williams, 1993).

The leaves of *Myrica gale* were used to flavour beer (Phillips, 1983). Leafy branches are added to the hot liquid in the early stages of the beer-making process (Mabey, 1998). There are no particular records of its use in beer making in Ireland, however. In Belgium this plant was one of the constituents of gruit, a mixture of herbs for bittering and flavouring beer, and was widely used until the late 16th century when it was superseded by hops. In addition to *M. gale*, gruit usually included yarrow (*Achillea millefolium*) and marsh rosemary (*Ledum palustre*). (*Ledum palustre* does not occur in Ireland.) In recent years a Belgian brewery revived its use to make an organic 'Gageleer' beer using this species collected from one of Belgium's important conserved wetland sites near Antwerp (Anonymous, 2008). Milliken and Bridgewater (2004) reported that bog myrtle beer is said to produce heavy intoxication and particularly bad hangovers.

Branches of bog myrtle were cut and placed under beds to keep away vermin (Hart, 1898; Vickery, 2010) and its dried leaves placed in linen presses to keep away moths (Danaher, 1964a; Uí Chonchubhair, 1995). Rutty (1772: 121) said that the leaves and branches were brought into bedrooms ('chambers') in summer for ornament and their 'grateful smell' and were put into chests to drive away moths. In Scotland, fishermen would wear a sprig in a button-hole to keep away midges. A commercially available midge-repellent was prepared from bog myrtle in Scotland in the 1990s (Mabey, 1998), made for a few years by a company called Scotia Pharmaceuticals that operated on the Isle of Skye (Milliken & Bridgewater, 2004).

Bog myrtle was also collected and used as kindling and brought a lovely scent into the house (Williams, 1993). Mackay (1836) recorded that in Scotland beds were made of *Myrica gale* twigs. Uí Chonchubhair (1995) suggested its culinary use to flavour soups and stews. The leaves are aromatic and fragrant and are best used as one would use bay leaves for flavouring, removing the leaves before serving. Dried fruits have also been used as a spice (Hilton, 2007) and added to soup (Williams, 1993). The leaves were dried and smoked in West Galway instead of tobacco (Williams, 1993).

The bark of bog myrtle was also regarded as excellent for tanning leather and the tips of branches provided a yellow dye (Dinneen, 1927; Mahon, 1983; Uí Chonchubhair, 1995). In the 18th century the Dublin Society (later the Royal Dublin Society) offered a prize for the best cured leather using *Myrica gale* (Williams, 1993). Rutty (1772: 121) said that the prize (a 'premium') was for tanning 'veal' skins.

Bog myrtle was used locally as a palm substitute on Palm Sunday (Dinneen, 1927).

In Scotland, Edlin (1973) recorded that candle wax was extracted from the shoots of *Myrica gale*. According to Sharkey (2000) *M. gale* was also used for the same purposes in Ireland. The shoots were washed in warm salted water and the wax skimmed off from the surface like scum. The candles derived from bog myrtle give off a delicate perfume. Milliken and Bridgewater (2004) said that wax can be extracted from bog myrtle catkins by boiling them in water, but because such tiny quantities are yielded this way, they were probably little used.

*Myriophyllum alterniflorum

- Alternate water-milfoil (Scannell & Synnott, 1987)
- *Líonán crom* (Scannell & Synnott, 1987); *Snaithe bhatheadh* (Cameron, 1883)

- Dinneen (1927) gives the Irish name *Fliodh uisce* and Ó Dónaill (1977) gives *Snáth báite* for 'Water milfoil'.

HALORAGACEAE—THE WATER-MILFOIL FAMILY
Native

Myriophyllum alterniflorum is a widespread aquatic plant of lakes, streams, ponds and rivers throughout Ireland except in the south-east where it is rare.

*No ethnobotanical uses of this plant have been noted in Ireland.

---

*Myriophyllum spicatum

- Spiked water-milfoil (Scannell & Synnott, 1987)
- *Líonánach* (Scannell & Synnott, 1987)

HALORAGACEAE—THE WATER-MILFOIL FAMILY
Native

Myriophyllum spicatum is an aquatic plant of lakes, ponds, pools, springs and rivers throughout Ireland.

*No ethnobotanical uses of this plant have been noted in Ireland.

---

*Myriophyllum verticillatum

- Whorled water-milfoil (Scannell & Synnott, 1987)
- *Líonán fáinneach* (Scannell & Synnott, 1987); *Fliodh uisce* (Moloney, 1919); *Snáithe báidhte* (Moloney, 1919); *Linnearach*[183] (Moloney, 1919; Dinneen, 1927)

HALORAGACEAE—THE WATER-MILFOIL FAMILY
Native

Myriophyllum verticillatum is an aquatic plant of ponds, lakes, canals and ditches in limestone areas, particularly in the centre of Ireland.

*No ethnobotanical uses of this plant have been noted in Ireland.

Myrrhis odorata

- Sweet cicely (Scannell & Synnott, 1987)
- *Lus áinleoige* (Scannell & Synnott, 1987)

APIACEAE (Umbelliferae)—THE CARROT FAMILY
Introduced

Myrrhis odorata is a perennial species that was formerly widely used as a culinary herb and medicinal plant, particularly in Northern Ireland, where it is a frequent garden escape or relict of cultivation (Reynolds, 2002). It is mainly found on roadsides and banks, and although it is frequently found in the north, it is very rare elsewhere in Ireland (Webb et al., 1996). Praeger (1915) considered its introduction in the north via Scottish settlers who held its medicinal properties in high repute.

183. Dinneen (1927) gave this name for 'water milfoil'.

It is highly aromatic and smells strongly of aniseed. The leaves can be used in salads or boiled in soups and stews (Phillips, 1983). It has a thick root that can be used as a vegetable like parsnip. Phillips (1983: 41) noted sweet cicely was mentioned in an old text and said to be 'very good for old people that are dull and without courage; it rejoiceth and comforteth the heart and increaseth their lust and strength'.

**Najas flexilis*

- Slender najad (Scannell & Synnott, 1987)
- *Síofróg uisce* (Scannell & Synnott, 1987)

HYDROCHARITACEAE (Najadaceae)—THE FROGBIT FAMILY

Native

Najas flexilis is a rare aquatic plant in Ireland that is found in lakes in the west.

*No ethnobotanical uses of this plant have been noted in Ireland.

**Nardus stricta*

- Mat-grass (Scannell & Synnott, 1987)
- *Beiteán* (Scannell & Synnott, 1987); *Luachair ghabhair* (Moloney, 1919); *Fitigh direch* (White, 1803)

POACEAE (Gramineae)—THE GRASS FAMILY

Native

Nardus stricta is a grass species found mainly in the west and north in habitats such as mountain moors and heaths.

*No ethnobotanical uses of this plant have been noted in Ireland.

Narthecium ossifragum.

Narthecium ossifragum

- Bog asphodel (Scannell & Synnott, 1987); Bent[184] (Hart, 1898); Cruppany-grass (Hart, 1898); Glasharlana (Hart, 1898); Goose-grass (Hart, 1898)
- *Sciollam na móna* (Scannell & Synnott, 1987); *Blioch* (Cameron, 1883); *Bliochan* (de Vere Coneys, 1849; Cameron, 1883)

NARTHECIACEAE—THE BOG ASPHODEL FAMILY

Native

Narthecium ossifragum is a common bog species throughout Ireland.

In the Shetland Islands, Scotland, it has been used as a substitute for saffron for medicine and dyeing (Grigson, 1958). In summer it has bright yellow flowers, and in autumn the dried flower spikes and stems are a striking orange colour. Mahon (1983) noted its use in Ireland to provide a yellow dye for dyeing fabrics.

184. According to Hart (1898: 366) this name was applied to bog asphodel in Co. Tyrone and was 'a term usually applied to the old stalks of various grasses'.

Nasturtium (Rorippa) species

- Water-cress (Scannell & Synnott, 1987)
- *Biolar* (Scannell & Synnott, 1987); *Dur-lus* (Dinneen, 1927); *Dobharlus* (Ó Dónaill, 1977)

BRASSICACEAE (Cruciferae)—THE CABBAGE AND CRESS FAMILY
Native

Including:

Nasturtium microphyllum (Rorippa microphylla)

- Narrow-fruited water-cress (Scannell & Synnott, 1987)
- *Biolar mion* (Scannell & Synnott, 1987)

Nasturtium officinale (Rorippa nasturtium-aquaticum)

- Water-cress (Scannell & Synnott, 1987); Water-grass (Colgan, 1904); Biller (Hart, 1898; Praeger, 1938)
- *Biolar* (Scannell & Synnott, 1987); *Biorar* (de Vere Coneys, 1849; Moloney, 1919)

Nasturtium microphyllum and *N. officinale* are both common throughout Ireland in muddy slow-moving streams and drains.

Water-cress was probably an important wild food plant in ancient Ireland. It grows quickly and thrives throughout the year. Kelly (1997) noted that it was a feature of ascetic diets and probably formed an important part of the diet of those living in early Irish monasteries. Joyce (1875) pointed out that in the *Book of Rights*[185], access to the cresses of the river Brosna in Co. Westmeath is the prerogative of the King of Ireland. Drury (1984) highlighted the importance of water-cress as a famine food in Ireland. Other important famine foods were charlock (*Sinapis arvensis*), nettles, sorrel, blackberries and probably hazelnuts.

Nasturtium species were formerly used in Ireland for the treatment of scurvy and scrofula (a form of tuberculosis affecting the lymph nodes). Scrofula is also called 'King's Evil' in some old medicinal texts (e.g., Egan, 1887). Allen and Hatfield (2004) and Uí Chonchubhair (1995) recorded that in Ireland *Nasturtium* species were used as a purifying tonic and remedy for colds, coughs and chest complaints, as well as used locally for stomach, heart and kidney troubles, rashes, headaches, scurvy, bleeding, rheumatism, cuts, scrofula, swellings, sore eyes, worms, blackheads and melancholia. Uí Chonchubhair (1995) recorded that it was believed to be effective for 'women's ailments', and in Dunquin, Co. Kerry, it was used as a poultice to ease labour pains. Ui Sheasnáin (1982) said large amounts of water-cress eaten raw were used in Co. Cavan for the treatment of anaemia. Boiled in whiskey and sugar, water-cress was also used as a cure for bronchitis (Wood-Martin, 1902). McGlinchey (2007: 86) wrote about the value of water-cress to treat heart ailments, and that it was 'good for man and beast':

> I heard of a stirk [a 1- or 2-year-old bullock or heifer] that was poorly one time and couldn't go to the hill with the other cattle. It grazed about the river and ate a lot of the wild *biolar* and soon mended up and went with the rest. Some short time after it was got drowned in a bog-hole. When they skinned and opened it up, a new heart was found growing beside the old one, about the size of a small scoiltín of a potato. People ate *biolar*, too, and it gave many a body a new life.

185. The *Book of Rights* (*Lebor na gCeart*) outlines the rents and taxes paid by the King of Cashel. A copy of the *Book of Rights* is included in the *Book of Ballymote* (*Leabhar Bhaile an Mhóta*), written in 1390 or 1391.

Glannagalt in Co. Kerry.

There is a beautiful valley on the Dingle Peninsula, Co. Kerry, called 'Glannagalt', the Glen of the Madmen, where afflicted people from all over Ireland would come to be cured by drinking the waters from a well and eating the cresses that grow nearby (Wood-Martin, 1902). The well survives today and is a celebrated folklore site, although no species of *Nasturtium* are now to be found growing near it.

Water-cress is an excellent culinary herb, being useful in salads, boiled or steamed as a vegetable, or used to make water-cress soup. The leaves and stems have a delicious peppery taste. To make watercress soup the leaves and tender stems are added to stock, along with seasoning and other vegetables, such as potatoes or root vegetables. To make the soup really rich, mix in a liberal amount of double cream before serving. In the early 1900s, Moloney (1919) reported that watercress was used as a decoration for blancmange made from Carrageen moss (the red marine alga, *Chondrus crispus*). It was mixed with alexanders (*Smyrnium olusatrum*) and nettles (*Urtica dioica*) to make a 'Lenten pottage' (soup) (Threlkeld, 1726: unnumbered [111]).

Mears and Hillman (2007) wisely noted that care should be taken when using this plant, firstly so that it is not confused with *Apium nodiflorum* (fool's water-cress), which is poisonous, and secondly because it often grows well in ditches and drains that are contaminated by sewage, or where there are water snails present that may carry liver fluke. Caution should therefore be exercised when collecting watercress from the wild and any leaves used should also always be washed really well.

Neotinea maculata.

**Neotinea maculata (Neotina intacta; Habenaria intacta)*

- Dense-flowered orchid (Scannell & Synnott, 1987)
- *Magairlín glas* (Scannell & Synnott, 1987); *An magairlín tiugh* (Moloney, 1919)

ORCHIDACEAE—THE ORCHID FAMILY

Native

Neotinea maculata is an orchid of rocky limestone grasslands of only a few western and midland counties. There is a single 1966 collection from the Isle of Man and a handful scattered to the north as far as Co. Donegal. *Neotinea maculata* is best known from the Burren flora in Co. Clare, where it is frequently seen blooming among the limestone pavements, grikes and calcareous grasslands of this striking karst landscape.

*No ethnobotanical uses of this plant have been noted in Ireland.

**Neottia cordata (Listera cordata)*

- Lesser twayblade (Scannell & Synnott, 1987)
- *Dédhuilleog bheag* (Scannell & Synnott, 1987); *Dá-dhuilleóg* (*An chroidhe*) (Moloney, 1919)

ORCHIDACEAE—THE ORCHID FAMILY

Native

Neottia cordata is a rather rare orchid in Ireland that is usually found growing amongst moss under heather.

*No ethnobotanical uses of this plant have been noted in Ireland.

*Neottia nidus-avis

- Bird's-nest orchid (Scannell & Synnott, 1987)
- *Magairlín neide éin* (Scannell & Synnott, 1987); *Magairlín nid éin* (Dinneen, 1927)

ORCHIDACEAE—THE ORCHID FAMILY
Native

Neottia nidus-avis is a rare saprophytic orchid that is lacking green chlorophyll. It is found only rarely in woodlands throughout Ireland.

*No ethnobotanical uses of this plant have been noted in Ireland.

*Neottia ovata (Listera ovata)

- Common twayblade (Scannell & Synnott, 1987); Wild tulip (Hart; 1898)
- *Dédhuilleog* (Scannell & Synnott, 1987); *Dá-dhuilleóg* (*An uibh*) (Moloney, 1919; Dinneen, 1927)

ORCHIDACEAE—THE ORCHID FAMILY
Native

Neottia ovata is found occasionally throughout Ireland in damp meadows and woodlands.

*No ethnobotanical uses of this plant have been noted in Ireland.

Nuphar lutea

- Yellow water-lily[186] (Scannell & Synnott, 1987); Brandy-bottle (Harron, 1986)
- *Cabhán abhann* (Scannell & Synnott, 1987); *Bileóg bháidhte bhuidhe* (Moloney, 1919); *Liath laghar* (Cameron, 1883); *Liach loghar* (*-air*) (de Vere Coneys, 1849); *Liach lobhar* (Dinneen, 1927)

NYMPHAEACEAE—THE WATERLILY FAMILY
Native

Nuphar lutea is a frequent plant of lakes, canals, ponds, bog-pools and slow streams in most parts of Ireland. It is most common in moderately acidic to basic water.

Mears and Hillman (2007) reported archaeological evidence that yellow water-lily seeds were cooked and eaten in Europe, as they were by indigenous people in North America. The seeds are quite nutritious and most palatable when gathered ripe and allowed to rot in water for a couple of weeks before having their seed coats removed, winnowed, parched and ground to make flour. Peterson (1977) mentioned that the seeds can be fried and used to make a popcorn-like snack. The rootstock can be eaten as

Nuphar lutea (yellow water-lily). Watercolour by Lydia Shackleton. Courtesy of the National Botanic Gardens, Glasnevin.

186. Ó Dónaill (1977) gave the Irish words *bior-rós*, *bual-lile* and *póicíní locha* for water-lilies, and Uí Chanainn (1962) gave the word *buail-lile*.

a vegetable, although they may need to be boiled in several changes of water to make their flavour less strong. Although I have not tried them, Peterson (1977) suggested that they have been a useful wild food source in North America.

It is occasionally used as an ornamental in ponds and small lakes. Nelson (2000b) mentioned that a red-flowered form of *Nuphar lutea* was collected by Robert Lloyd Praeger in 1940 for cultivation in the National Botanic Gardens at Glasnevin, but it is not in cultivation there now.

Nymphaea alba

- White water-lily (Scannell & Synnott, 1987)
- *Bacán bán* (Scannell & Synnott, 1987); *Dhuilleóg (Bileóg) bhádhte bán* (Moloney, 1919); *Cuirinín bán* (Moloney, 1919); *Dhuilleóg bháite* (Webb et al., 1996); *Buillite* (Cameron, 1883); *Duilleog bháite bhán* (Uí Chonchubhair, 1995); *Pónaire chorraigh* (Dinneen, 1927)

NYMPHAEACEAE—THE WATERLILY FAMILY
Native

Nymphaea alba is found frequently in lakes, ponds and slow-moving rivers throughout much of Ireland except the east where it is rare. It prefers alkaline water but can tolerate acidic conditions too.

Mackay (1806, 1836) noted that in Connemara the roots of *Nymphaea alba* were used for dyeing wool black.

Charred seeds have been found at Mount Sandel archaeological site in Northern Ireland, suggesting that like *Nuphar lutea*, the seeds of white water-lily may have been eaten (Mears & Hillman, 2007). The same authors recorded that the unopened flower buds may be roasted and eaten but are a bit bitter. Apparently, the fleshy rhizome can be eaten (Uí Chonchubhair, 1995).

This plant is occasionally used as an ornamental in ponds and small lakes.

Nymphaea species are used as food elsewhere in the world, such as in northern Australia, where the seeds, bulbs and stems are eaten as a high energy and carbohydrate source (Isaacs, 1987). Peterson (1977) listed *Nymphaea* species in North America as providing edible young leaves, flower buds, seeds and tubers.

**Odontites vernus*

- Red bartsia (Scannell & Synnott, 1987); Red rattle (Harron, 1986)
- *Hocas tae* (Scannell & Synnott, 1987); *Hocas té* (Dinneen, 1927)

OROBANCHACEAE (Scrophulariaceae)—THE BROOMRAPE FAMILY
Native

Odontites vernus is a common annual plant of roadsides, grassland and stony waste places throughout Ireland.

*No ethnobotanical uses of this plant have been noted in Ireland.

**Oenanthe aquatica (Oenanthe phellandrium)*

- Fine-leaved water-dropwort (Scannell & Synnott, 1987); Horsebane (Moloney, 1919)
- *Dathabha uisce* (Scannell & Synnott, 1987); *Nimh capaill* (Moloney, 1919); *Greabhán mín* (Moloney, 1919; Ó Dónaill, 1977); *Táthabha bán* (Dinneen, 1927)

APIACEAE (Umbelliferae)—THE CARROT FAMILY
Native

Oenanthe aquatica is found occasionally in most parts of Ireland except the centre. It occurs in wet ditches and drains, canals, pools and lakes, and on river-banks and pond-edges. It is poisonous.

*No ethnobotanical uses of this plant have been noted in Ireland.

Oenanthe crocata

- Hemlock water-dropwort (Scannell & Synnott, 1987)
- *Dathabha bán* (Scannell & Synnott, 1987); *Tréanlus braonach an churraigh* (Moloney, 1919; Dinneen, 1927); *Táthabha bán* (Dinneen, 1927); *Tréanlus*[187] (Ó Dónaill, 1977)

APIACEAE (Umbelliferae)—THE CARROT FAMILY
Native

Oenanthe crocata is a widespread plant in Ireland except the west and centre where it is rare. It occurs in wet habitats, such as river and canal banks, wet ditches and drains, lake margins and alongside ponds and pools.

It is an extremely poisonous plant. Threlkeld (1726: unnumbered [113]) provided an early warning of how dangerous it is:

> The very aspect of it seemed to me always grim and dismal . . .. A tragic story of its pernicious nature, you have set down by Mr Vaughan. Eight lads about thirty years ago, went a fishing near Clonmell in Ireland, and there meeting a great parcel of *Oenanthe* [*crocata*] they mistook the roots of it for *Sium aquaticum* [an early Latin name for watercress] roots, and did eat a great deal of them; about four or five hours after going home, the eldest of them, who was almost of Man's stature, without the least previous disorder, or complaint, on a sudden fell down backwards, and lay kicking and sprawling on the ground, his countenance soon turned very ghastly, and he foamed at the mouth. Soon after four more were seized the same way, and they all died before morning, not one of them having spoken a word from the moment in which the venenate particles surprised the Genus Nervosum. Of the other three, one ran stark mad, but came to right reason next morning, another his hair and nails fallen off; the third who is my brother-in-law, and from whom I had my account, only escaped without any harm, he either eat less, or his athletic constitution overcame it; he run home about two miles, drunk warm milk, which caused a diaphoresis.

187. Ó Dónaill (1977) defined *tréanlus* as 'water-hemlock'.

Despite its poisonous nature, according to Allen and Hatfield (2004), it was used in Ireland against tumours and to treat external swellings.

Oenanthe fistulosa

- Tubular water-dropwort (Scannell & Synnott, 1987); Wild rue (Hart, 1898)
- *Dathabha giolcaí* (Scannell & Synnott, 1987); *Tréanlus braonach* (Moloney, 1919; Ó Dónaill, 1977)

APIACEAE (Umbelliferae)—THE CARROT FAMILY
Native

Oenanthe fistulosa is rare in Ireland and found mainly in the east and centre in marshes, pools and canals.

According to Allen and Hatfield (2004) it was used medicinally in Wicklow as a treatment for rheumatism.

It is poisonous.

**Oenanthe fluviatilis*

- River water-dropwort (Scannell & Synnott, 1987)
- *Dathabha abhann* (Scannell & Synnott, 1987)

APIACEAE (Umbelliferae)—THE CARROT FAMILY
Native

Oenanthe fluviatilis is a rare plant in Ireland and is mainly found in rivers and canals in the centre of the island.

It is poisonous.

*No ethnobotanical uses of this plant have been noted in Ireland.

**Oenanthe lachenalii*

- Parsley water-dropwort (Scannell & Synnott, 1987)
- *Dathabha peirsile* (Scannell & Synnott, 1987)

APIACEAE (Umbelliferae)—THE CARROT FAMILY
Native

Oenanthe lachenalii is an occasional plant of salt marshes all around the Irish coast.

It is poisonous.

*No ethnobotanical uses of this plant have been noted in Ireland.

**Oenanthe pimpinelloides*

- Corky-fruited water-dropwort (Scannell & Synnott, 1987)
- *Dathabha ainíse* (Scannell & Synnott, 1987)

APIACEAE (Umbelliferae)—THE CARROT FAMILY
Possibly Introduced (possible archaeophyte)

Oenanthe pimpinelloides is a very rare plant in Ireland and found in damp grasslands in a few localities in the west and south-west only.

It is poisonous.

*No ethnobotanical uses of this plant have been noted in Ireland.

Oenothera species

- Evening-primrose (Scannell & Synnott, 1987)
- *Coinneal oíche* (Scannell & Synnott, 1987)

ONAGRACEAE—THE WILLOWHERB FAMILY
Introduced

Including:

Oenothera biennis

- Common evening-primrose (Scannell & Synnott, 1987)
- *Coinneal oíche* (Scannell & Synnott, 1987)

Oenothera erythrosepala

- Large-flowered evening-primrose (Scannell & Synnott, 1987)
- *Coinneal oíche mhór* (Scannell & Synnott, 1987)

Evening-primroses are widespread plants of waste places, sand dunes and roadsides throughout Ireland. Some are also grown in gardens and escape from there into the wild.

Elsewhere in the world they are cultivated as a source of evening-primrose oil that is used as a nutrition supplement and to treat skin disorders, such as eczema, as well as irritable bowel syndrome, rheumatism, and circulation and menopausal problems (van Wyk & Wink, 2004). Evening primrose oil has been used in the making of cosmetics in Ireland.

Young leaves can be cooked and used as a (slightly bitter) green vegetable (Peterson, 1977) or as a peppery addition to salads. The first-year roots can also be cooked and eaten. These are gathered from young non-flowering rosettes, peeled and boiled for about half an hour in several changes of water, and served with melted butter. Peterson (1977) described them as 'mildly pungent, excellent'.

Ononis repens

- Common restharrow (Scannell & Synnott, 1987); Bowstring (Anonymous, 1978)
- *Fréamhacha tairne* (Scannell & Synnott, 1987); *Sreang bogha* (Cameron, 1883; Dinneen, 1927; Webb & Scannell, 1983); *Pnéamh tharraingthe* (Dinneen, 1927); *Préumh fhairge* (Ó Criomhthain, MSS, n.d., ca. 1908)

FABACEAE (Leguminosae)—THE PEA AND CLOVER FAMILY
Native

Ononis repens is a low-growing perennial herb of sand dunes and dry grasslands that is found mainly in the south and east.

According to Moloney (1919) this plant was used medicinally, and juice extracted from the leaves was used to treat chapped and rough hands. Decoctions from the roots were apparently used to treat kidney complaints (Threlkeld, 1726).

Even though its root is tough to eat, Mabey (1972) recorded that children in northern England would dig it up and chew it, from whence it acquired alternative English names of wild liquorice and Spanish root. I have not found records of its use in this way in Ireland.

The Irish name, *sreang bogha*, meaning 'bowstring', suggests that the strong fibres from this plant may have been used for such purposes, but I have not found any direct references to such usage. Its English name, restharrow, also alludes to the strength of the fibres it contains, suggesting that the harrow has to be rested and perhaps untangled when plants of this species are encountered during harrowing.

**Ophrys apifera*

- Bee orchid (Scannell & Synnott, 1987)
- *Magairlín na mbeach* (Scannell & Synnott, 1987); *Magairlín na beiche* (Uí Chonchubhair, 1995)

ORCHIDACEAE—THE ORCHID FAMILY
Native

Ophrys apifera is a rare but widespread orchid in Ireland of limestone grasslands, dry banks and sand dunes.

*No ethnobotanical uses of this plant have been noted in Ireland.

**Ophrys insectifera*

- Fly orchid (Scannell & Synnott, 1987)
- *Magairlín na stuaice* (Scannell & Synnott, 1987); *Magairlín na cuileóige* (Moloney, 1919)

ORCHIDACEAE—THE ORCHID FAMILY
Native

Ophrys insectifera is a rare orchid in Ireland of limestone grasslands and fens that is found mainly in the centre and west.

*No ethnobotanical uses of this plant have been noted in Ireland.

Orchis mascula

- Early purple orchid (Scannell & Synnott, 1987); Blue rocket (Hart, 1898); Brush sweet willie (Hart, 1898)
- *Magairlín meidhreach* (Scannell & Synnott, 1987); *Magairlin meireach* (Cameron, 1883); *Magairlín meadhrach* (Dinneen, 1927)

ORCHIDACEAE—THE ORCHID FAMILY
Native

Orchis mascula.

Orchis mascula is found occasionally throughout Ireland but seems to be declining. It is present in unimproved moist grasslands, shady banks and open woodland and scrub.

It is reputed to be an aphrodisiac and is mentioned as such in Brian Merriman's famous Irish poem 'Cuirt an Mheán Oíche' [The Midnight Court]. K'Eogh (1735: 49) described its use (under the name fool stones, dog stones or male *Satyrion*) as an aphrodisiac, thus, 'A decoction of the roots drunk with goat's milk excites sexual desire, aids conception and strengthens the genitals . . . It is said that if men eat the largest roots, they will beget sons, and if women eat the smallest roots they shall bring forth daughters'.

Williams (1993) recorded that it was used as a love potion and its roots were used to treat tuberculosis, diarrhoea and sterility. Kinahan (1881) said that in Co. Wicklow it was called *mogra-myra* and believed to be effective as a love potion. Ó Dubhshláine (1996: 93–94, translated from the Irish by M. De Mórdha) also noted its use in a love potion in Dunquin (Dún Chaoin), Co. Kerry:

> Some people in the parish had the power to concoct a drink, and whosoever drank it, he or she, would fall in love with the person who prepared the drink and there would be a strong urge to have sex. It was associated very strongly with Dún Chaoin. The drink was prepared with the tuber of the herb, *an magairlín meidhreach* ('merry diminutive testicle') (*Orchis mascula*) being boiled in a black saucepan. It is said here to this day regarding any person who falls in love with a man or woman from Dún Chaoin: Yes, indeed, did he not get a drink from the black saucepan.

In Scotland, it was eaten to soothe or provide a protective lining for the alimentary canal (Allen & Hatfield, 2004). Mackay (1836: 275) indicated 'salep is made from *Orchis mascula* and other terrestrial European species', but whether salep was ever made in Ireland is uncertain. Salep is a starchy meal ground from the roots of dried orchids and used as a food and to make a milky beverage (see also *Anacamptis morio*).

The Irish word for orchid, *magairlín*, means 'little testicle'. Ó Dónaill (1977) gave *lus na magairlí* as a general name for 'orchid'.

Origanum vulgare

- Marjoram (Scannell & Synnott, 1987)
- *Máirtín fiáin* (Scannell & Synnott, 1987); *Oragán* (de Vere Coneys, 1849; Moloney, 1919; Dinneen, 1927); *Seathbhóg* (*-óige*) (*-óga*) (de Vere Coneys, 1849)

LAMIACEAE (Labiatae)—THE DEAD-NETTLE FAMILY

Native

Origanum vulgare is an occasional plant throughout Ireland except in the north and south-west. It grows on limestone (calcareous) soils on dry banks, grassland, roadsides and stony, sandy places, such as sand pits.

Marjoram is a commonly used perennial aromatic herb for culinary purposes and is usually available only in dried form. Nevertheless, it is a widespread native wild Irish plant that can be collected and used for a variety of purposes. It has had considerable use in traditional medicine to treat bronchitis, coughs, colds, colic, dyspepsia, catarrh and nasal congestion (van Wyk & Wink, 2004). In Ireland, Moloney (1919) reported that oil extracted from this species was used as a treatment for rheumatism, rubbed onto joints. Similarly, Egan (1887) noted its use for cramps and indigestion.

It has been used widely as flavouring for stews, sauces, soups and meat dishes and its strong flavour combines well with other flavourings, such as garlic, onions or chillies (Hilton, 2007). Phillips (1983) provided a recipe for marjoram jelly, in which it is used to flavour apple jelly (like mint jelly). Marjoram jelly can be made by adding finely chopped leaves to apple jelly or crab-apple jelly during cooking, together with several tablespoons of malt or wine vinegar. It is delicious with roasted meat and cheeses.

Origanum tea can also be made from dried leaves and flower stems.

**Ornithopus perpusillus*

- Bird's-foot (Scannell & Synnott, 1987)
- *Crúba éin* (Scannell & Synnott, 1987)

FABACEAE (Leguminosae)—THE PEA AND CLOVER FAMILY
Native

Ornithopus perpusillus is a very rare annual plant in Ireland of dry and sandy places, such as wasteground near the south and east coasts only.

*No ethnobotanical uses of this plant have been noted in Ireland.

**Orobanche alba*

- Thyme broomrape (Scannell & Synnott, 1987)
- *Múchóg dhearg* (Scannell & Synnott, 1987)
- *Síor*; *Síorlach*; *Muchóg*[188] Moloney (1919)

OROBANCHACEAE (Scrophulariaceae)—THE BROOMRAPE FAMILY
Native

Orobanche alba is a rare parasitic plant found mainly in the north and west of Ireland growing on the roots of thyme.

*No ethnobotanical uses of this plant have been noted in Ireland.

**Orobanche hederae*

- Ivy broomrape (Scannell & Synnott, 1987)
- *Múchóg mhór* (Scannell & Synnott, 1987); *Síor* (*-orr*) (or *Luibh-shíor*)[189] (Dinneen, 1927)

OROBANCHACEAE (Scrophulariaceae)—THE BROOMRAPE FAMILY
Native

Orobanche hederae is a parasitic plant found growing on the roots of ivy and is widespread in Ireland except in the north where it is rare.

*No ethnobotanical uses of this plant have been noted in Ireland.

188. Moloney (1919) listed these names for *Orobanche alba* and *O. hederae*.
189. Dinneen (1927) also gave this name for 'red broom rape'.

**Orobanche minor*

- Common broomrape (Scannell & Synnott, 1987); Lesser broomrape (Anonymous, 1978); Clover broomrape (Brunker, 1950)
- *Múchóg bheag* (Scannell & Synnott, 1987); *Siorra-lach* (Cameron, 1883)

OROBANCHACEAE (Scrophulariaceae)—THE BROOMRAPE FAMILY
Probably Introduced

Orobanche minor is a parasitic plant found growing on the roots of a number of different plants, including clovers, mainly in the south and east of Ireland.

Orobanche species, and probably this species too, have been used in North America by native people as food, where the underground portions of the stems have been cooked like asparagus (Peterson, 1977).

*No ethnobotanical uses of this plant have been noted in Ireland.

**Orobanche rapum-genistae*

- Greater broomrape (Scannell & Synnott, 1987)
- *Múchóg chapaill* (Scannell & Synnott, 1987); *Srathman* (Moloney, 1919)

OROBANCHACEAE (Scrophulariaceae)—THE BROOMRAPE FAMILY
Native

Orobanche rapum-genistae is a very rare parasitic plant in Ireland, found mainly near the east and south coasts growing on the roots of broom and gorse.

*No ethnobotanical uses of this plant have been noted in Ireland.

**Orthilia secunda*

- Serrated wintergreen (Scannell & Synnott, 1987)
- *Glasluibh fhiaclach* (Scannell & Synnott, 1987)

ERICACEAE—THE HEATHER FAMILY
Native

Orthilia secunda is a very rare plant in Ireland found only in Cos. Antrim and Fermanagh growing on rocky ledges.

*No ethnobotanical uses of this plant have been noted in Ireland.

**Otanthus maritimus*

- Cottonweed (Scannell & Synnott, 1987)
- *Cluasach mhara* (Scannell & Synnott, 1987)

ASTERACEAE (Compositae)—THE DAISY FAMILY
Native

Otanthus maritimus is a very rare and declining plant in Ireland and is found only on a sand bank on the south coast of Co. Wexford.

*No ethnobotanical uses of this plant have been noted in Ireland.

Oxalis acetosella

- Wood sorrel (Scannell & Synnott, 1987); Cuckoo-sorrel (Hart, 1898)
- *Seamsóg* (Scannell & Synnott, 1987); *Samhadh coilleadh* (Moloney, 1919); *Seamróg coilleadh* (Moloney, 1919); *Sealbhóg fiodha* (de Vere Coneys, 1849; Moloney, 1919); *Bileóg na néan* (Moloney, 1919); *Biadh* (or *Bia*) *éanáin* (*na-éanán*) (Dinneen, 1927; Ó Dónaill, 1977); *Samhadh coille* (Dinneen, 1927; Ó Dónaill, 1977); *Seamar coille* (Dinneen, 1927); *Seamar ghéar* (Dinneen, 1927); *Seamróg* (Dinneen, 1927); *Siomsán* (Dinneen, 1927); *Seamair choille* (Ó Dónaill, 1977)

OXALIDACEAE—THE WOOD-SORREL FAMILY
Native

Oxalis acetosella is a common plant of woodland, scrub, heathy ground amongst bracken, stream-sides, shady banks, hedgerows, mountain cliffs and flushes and other shady places throughout Ireland.

It is recorded as an edible plant in ancient Ireland (Kelly, 1997). Mears and Hillman (2007) suggest that the leaves be stripped from their inedible stems and are then ideal in salads. They have a sour lemony flavour and may also be used as a stuffing for fish. These same authors warn that the leaves of *Oxalis acetosella* should be used in moderation as they contain oxalic acid. Peterson (1977) mentioned that excessive consumption of a related North American species, *O. montana*, is toxic if used over a long period of time as it may inhibit the absorption of calcium by the body. Grigson (1958) mentioned, too, that the leaves were used in a sauce, in a similar way as *Rumex acetosa*. Williams (1993) suggested that the leaves are good to chew on a hot day.

Williams (1993) reported that it has been used medicinally as an ancient cure in Ireland for sore throats, colds, blood weaknesses, drawing the pus out of boils and heart complaints. Egan (1887) recorded its use in Ireland to treat cancer; an ointment prepared from the leaves was applied or the leaves were eaten. Allen and Hatfield (2004) noted its use medicinally in Ireland to treat diarrhoea, cancer and palsy and as a heart and blood tonic, as well as a veterinary remedy to treat ticks and lice in sheep. Also, for veterinary medicine, it was mixed with buttermilk to cure mange in sheep and to treat lice and ticks (Williams, 1993).

It is occasionally used as shamrock (Nelson, 1991) and was probably the template for the shamrock logo of the airline Aer Lingus.

**Oxyria digyna*

- Mountain sorrel (Scannell & Synnott, 1987)
- *Samhadh sléibhe* (Scannell & Synnott, 1987); *Sealbhaig nan fiadh* (Cameron, 1883)

POLYGONACEAE—THE KNOTWEED FAMILY
Native

Oxyria digyna is a rare mountain plant in Ireland of cliffs, ledges, screes and other wet habitats in the west and the Galtee Mountains.

*No ethnobotanical uses of this plant have been noted in Ireland.

Papaver species

- Poppy (Scannell & Synnott, 1987); Blind-eye (Hart, 1898)

- *Poipin*[190] (Webb et al., 1996; Dinneen, 1927); *Cromlus* (de Vere Coneys, 1849; Uí Chanainn, 1962); *Paipín* (de Vere Coneys, 1849); *Barróg ruadh* (Dinneen, 1927)

PAPAVERACEAE—THE POPPY FAMILY
Probably Introduced (probable archaeophyte)

Including:

Papaver dubium

- Long-headed poppy (Scannell & Synnott, 1987)
- *Cailleach fhada* (Scannell & Synnott, 1987)

Papaver rhoeas

- Common poppy (Scannell & Synnott, 1987); Corn poppy (Anonymous, 1978); Scabby hands (Colgan, 1904)
- *Cailleach dhearg* (Scannell & Synnott, 1987); *Bláth na mbodhach* (Moloney, 1919); *Cailicín* (Moloney, 1919); *Caithleach dearg* (Dinneen; 1927; Webb et al., 1996); *Blath na mbodaigh* (de Vere Coneys, 1849); *Loirgneach dhearc* (de Vere Coneys, 1849); *Cáithligh dheirg* (de Vere Coneys, 1849)
- *Copprouse* (in Yola) (Dolan & Ó Muirithe, 1979)

These poppies are common weeds of roadsides, field margins and wasteground throughout Ireland. Both *Papaver dubium* and *P. rhoeas* are probably introduced and may have been introduced as a weed of crops with the first or early farmers.

The petals were used as a colouring agent and food flavouring and to make a syrup formerly used as an anodyne expectorant. A drink made from the flowers of *Papaver rhoeas* or *P. dubium* was used as a sedative, anti-spasmodic or diaphoretic and given in cases of anxiety, colic, tonsillitis or bronchitis (Uí Chonchubhair, 1995). The corn poppy is said to be edible and the leaves can be eaten raw in salads, cooked in the same way as spinach, but it is best used early in the spring in March and April when the leaves are soft and before flower buds are formed (Hilton, 2007). The seeds are the most widely used part of the plant for culinary purposes, particularly used in baking. Rutty (1772) said that the tender young leaves could be boiled for food, mixed with cheese and butter.

Vickery (2010) noted that poppies were used to make dolls and included a record from Northern Ireland. The petals were turned back and two pieces of stalk were used to make arms and a second leg was added to the existing piece of stalk as the first leg. A piece of wool was wrapped around to give the waist and a face was drawn on the exposed seed capsule.

These two wild poppy species are widely used in commercial 'wildflower seed mixes', which have been used to revegetate roadsides and other waste places.

Papaver somniferum

- Opium poppy (Scannell & Synnott, 1987)
- *Codlaidín* (Scannell & Synnott, 1987); *Cromlus* (Moloney, 1919; Dinneen, 1927); *Paipín* (Moloney, 1919); *Lus an chodail* (Moloney, 1919; Dinneen, 1927; Ó Dónaill, 1977)

PAPAVERACEAE—THE POPPY FAMILY
Introduced (possible archaeophyte)

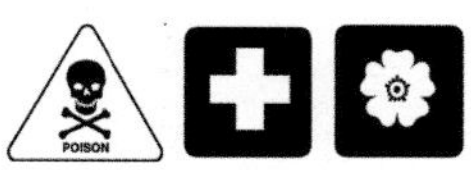

190. Dinneen (1927) spelled as '*poipín*.'

Papaver somniferum is an occasional annual weed of cultivated ground and waste places, particularly on dry or sandy soils. It has been a long established weed in Ireland. It is native to Asia Minor. Despite its extensive use worldwide as a medicinal and narcotic plant, it has probably been little grown in Ireland for such purposes. Its active compounds include a range of alkaloids, such as morphine, codeine and many others (Emboden, 1972), which are amongst the greatest painkillers in the world. It is an extremely valuable drug to treat intense pain, such as suffered by some cancer patients. Unfortunately it is highly addictive. Morphine was used as early as BC 6000 by Swiss Lake Dwellers and by the Sumerians (BC 4500) as small balls of opium that were eaten or mixed with wine (Lewington, 1990). The Irish name *lus an chodail* means the 'herb of sleep'.

Its seeds have been widely used in Ireland for culinary purposes, particularly in baking, but probably rarely from Irish-grown stock. The seeds of the opium poppy do not contain opium (Heywood, 1978). It is also a widely grown ornamental garden plant in Ireland of which a range of different forms with double flowers or various colour varieties can be seen.

Opium poppy is regarded as a poisonous plant and should not be eaten. Woodward (1985: 34) gave the symptoms of poisoning as 'depression of central nervous system; headache, giddiness, vomiting, drowsiness, weakness; irregular breathing and coma in severe cases'.

**Parapholis incurva (Rottboellia incurvata)*

- **Curved hard-grass (Scannell & Synnott, 1987)**
- ***Cruafhéar cuar* (Scannell & Synnott, 1987); *Durfhér fairge*[191] (White, 1803)**

POACEAE (Gramineae)—THE GRASS FAMILY
Native

Parapholis incurva is a very rare annual grass in Ireland that is found in only a few coastal sites in Cos. Dublin and Cork.

*No ethnobotanical uses of this plant have been noted in Ireland.

**Parapholis strigosa*

- **Hard-grass (Scannell & Synnott, 1987)**
- ***Cruafhéar* (Scannell & Synnott, 1987)**

POACEAE (Gramineae)—THE GRASS FAMILY
Native

Parapholis strigosa is an occasional plant of salt marshes and seashores all around the Irish coast.

*No ethnobotanical uses of this plant have been noted in Ireland.

**Parentucellia viscosa*

- **Yellow bartsia (Scannell & Synnott, 1987)**
- ***Hocas tae buí* (Scannell & Synnott, 1987)**

OROBANCHACEAE (Scrophulariaceae)—THE BROOMRAPE FAMILY
Native

191. White (1803) applied this name to '*Rotbollia incurvata*'. It is more likely that it refers to the more common related species *Parapholis strigosa* than to *P. incurva*, which is rare and a relatively recent discovery in Ireland.

Parentucellia viscosa is a hemi-parasitic annual plant of grasslands, particularly in the south-west.

*No ethnobotanical uses of this plant have been noted in Ireland.

Parietaria judaica (Parietaria diffusa; Parietaria officinalis)

- Pellitory-of-the-wall (Harron, 1986; Scannell & Synnott, 1987); Peniterry (Thiselton Dyer, 1889); Pellitory (Webb, 1977); Wall pellitory (Mackay, 1836)
- *Feabhraíd* (Scannell & Synnott, 1987); *Lus an bhalla* (de Vere Coneys, 1849; Moloney, 1919; Dinneen, 1927); *Miontas (Meantus) chaisil* (de Vere Coneys, 1849; Moloney, 1919; Dinneen, 1927; Ó Dónaill, 1977); *Peileatra*[192] (Ó Dónaill, 1977)

URTICACEAE—THE NETTLE FAMILY

Native

Parietaria judaica.

Parietaria judaica is a common plant of walls, particularly old mortared walls, throughout much of Ireland. It is particularly common in the south and also on boulder- and shingle-beaches mainly in the west.

This plant was probably important in traditional medicine in ancient times in Ireland. This species has several uses as a medicinal plant. According to Moloney (1919), it was used as a laxative and diuretic. It was also noted by Uí Chonchubhair (1995) to have been crushed and applied as a poultice to treat fevers and boils or used in an ointment to clean and help heal piles. This same author recorded that various preparations of this plant were taken internally to treat chest and urinary ailments, menstrual disorders and skin problems. A powder made from the dried plant was used to relieve wheezing, dry coughs and shortness of breath, and the whole plant was heated on cinders and applied to the back, side or stomach to relieve pains.

Rutty wrote in 1772 that the fresh leaves are excellent for cleaning bottles and 'even Urinals fouled by what adheres to them' (Rutty, 1772: 122).

Thiselton Dyer (1889: 238) mentioned that this plant was used by children in Ireland as a talisman against a beating at school: 'A weed called, locally at least, peniterry, to which the suddenly terrified (schoolboy) idler might run in his need, grasping it hard and threateningly, and repeat the following "words of power":'

> Peniterry, peniterry, that grows by the wall,
> Save me from a whipping, or I'll pull you, roots and all.

Wood-Martin (1902) also mentioned this folklore use and repeated Dyer's quote.

**Parnassia palustris*

- Grass-of-parnassus (Scannell & Synnott, 1987)
- *Fionnscoth* (Scannell & Synnott, 1987); *Fionnán bán* (Moloney, 1919; Dinneen, 1927)

PARNASSIACEAE—THE GRASS-OF-PARNASSUS FAMILY

192. Ó Dónaill (1977) defined *peileatra* to mean 'pellitory', although the name *peileatra* may also apply to other species sometimes knows as pellitory, such as *Achillea ptarmica* or *Anacyclus pyrethrum*, a non-native member of the Compositae family used in the past to treat toothaches.

Native

Parnassia palustris is a beautiful white-flowered perennial plant of marshy habitats, such as fens, lake-shores, mountain cliffs and damp grassland, that is common in the west and midlands. It is rare elsewhere in Ireland.

*No ethnobotanical uses of this plant have been noted in Ireland.

Pastinaca sativa (Peucedanum sativum)

- Wild parsnip (Scannell & Synnott, 1987)
- *Cuirdín bán* (Scannell & Synnott, 1987); *Bliúcán* (Ua Maoileoin, 2001); *Geamhrán* (Ua Maoileoin, 2001); *Meacan riogh fiadhain* (de Vere Coneys, 1849; Moloney, 1919; Dinneen, 1927); *Meacan buidhe*[193] (de Vere Coneys, 1849)

APIACEAE (Umbelliferae)—THE CARROT FAMILY
Introduced

Pastinaca sativa is a wild ancestor of the cultivated parsnip and is found in a few places in Ireland today in habitats such as roadsides, waste places, quarries, and sandy soils. Its roots are edible and can be steamed, boiled, sautéed or roasted as a vegetable.

Contact with the leaves of this plant can cause an allergic reaction on the skin, especially if the skin is sweaty or wet, which can enhance a photosensitive reaction (phytophotodermatitis). This is a similar, although probably less severe, reaction than that caused by contact with the giant hogweed, *Heracleum mantegazzianum*.

According to Cline (1989), on the Scottish island of Harris, the seeds of this species were used instead of hops in brewing beer.

**Pedicularis* species

OROBANCHACEAE (Scrophulariaceae)—THE BROOMRAPE FAMILY
Native

Moloney (1919) listed a number of Irish names for the two species of *Pedicularis* (see below) but does not distinguish between them: *míl-ghabhair*, *lus an giolla*, *lus riabhach*, *bodán cloigín* and *maothlán móna*. He also applied these names to *Rhinanthus christi-galli* (i.e., *R. minor*). Dinneen (1927: 743) gave the name *milseán móna* for 'bog honeysuckle'.

**Pedicularis palustris*

- Marsh lousewort (Scannell & Synnott, 1987); Red rattle (Anonymous, 1978); Honey cup (Hart, 1898); Wild honeysuckle (Hart, 1898); Marsh redrattle (Brunker, 1950)
- *Milseán móna* (Scannell & Synnott, 1987); *Lus riabhach* (Uí Chonchubhair, 1995)

OROBANCHACEAE (Scrophulariaceae)—THE BROOMRAPE FAMILY
Native

Pedicularis palustris is an occasional plant in the west and north in marshes, wet meadows and fens.

*No ethnobotanical uses of this plant have been noted in Ireland.

193. The ordinary parsnip, the vegetable, is called *meacan bán* or *meacan ríogh* (Dinneen, 1927).

**Pedicularis sylvatica*

- Lousewort (Scannell & Synnott, 1987)
- *Lus an ghiolla* (Scannell & Synnott, 1987); *Artha na ndreancaidí* (Ua Maoileoin, 2001); *Riabhach* (de Vere Coneys, 1849); *Lus riabhach* (Dinneen, 1927); *Rianaitín*[194] (Dinneen, 1927)

OROBANCHACEAE (Scrophulariaceae)—THE BROOMRAPE FAMILY
Native

Pedicularis sylvatica is an occasional plant throughout Ireland of mountain and moorland heaths, dry bogs and boggy grasslands.

*No ethnobotanical uses of this plant have been noted in Ireland.

**Persicaria amphibia (Polygonum amphibium)*

- Amphibious bistort (Scannell & Synnott, 1987)
- *Glúineach uisce* (Scannell & Synnott, 1987); *Gluineach dhearg* (Cameron, 1883)

POLYGONACEAE—THE KNOTWEED FAMILY
Native

Persicaria amphibia is a common perennial plant throughout Ireland of wet habitats, such as river-banks, ponds, canals, streams, ditches and slow-moving rivers.

*No ethnobotanical uses of this plant have been noted in Ireland.

Persicaria bistorta (Polygonum bistorta)

- Common bistort (Scannell & Synnott, 1987); Snakeweed; Easter-ledges (Harron, 1986); Snake-root (Harron, 1986)
- *Stóinse* (Scannell & Synnott, 1987); *Copóg nimhe* (Moloney, 1919); *Stóinire* (Moloney, 1919); *Bilur* (Cameron, 1883); *Lus na h-imirche* (Dinneen, 1927)

POLYGONACEAE—THE KNOTWEED FAMILY
Introduced

Persicaria bistorta is a rare perennial herb throughout Ireland of damp grasslands.

Moore and More (1866) said that it was formerly much cultivated in Ireland but they do not say for what purposes. Allen and Hatfield (2004) mentioned its popularity as an ingredient of soup and spring or Easter puddings and said that it has a root that is rich in tannins and is a powerful astringent. I have not found records of its use in traditional medicine in Ireland.

Persicaria hydropiper (Polygonum hydropiper)

- Water pepper (Scannell & Synnott, 1987)
- *Biorphiobar* (Scannell & Synnott, 1987); *Cochall an uisce* (Ua Maoileoin, 2001); *Glúineach mhór* (Moloney, 1919); *Glúineach theith* (Moloney, 1919); *Glúineach the* (Webb et al., 1996; Uí Chonchubhair, 1995)

194. Dinneen (1927) defined *rianaitín* as 'red rattle'.

POLYGONACEAE—THE KNOTWEED FAMILY
Native

Persicaria hydropiper is a very common annual plant of muddy habitats and wet places, such as ditches, streams, pools and ponds and the edges of lakes. It is widespread and common throughout most of Ireland except parts of the centre where it is rare.

K'Eogh (1735) suggested this plant was used medicinally to treat swellings, kidney and bladder disorders and as a decoction in water for aches, pains, gout and arthritis. K'Eogh also noted that the dried powdered leaves of this species could be used with meat instead of pepper. Mears and Hillman (2007) suggested that the peppery taste actually comes from compounds that are toxic, so perhaps it should be used cautiously. The seeds can be ground to make flour.

Uí Chonchubhair (1995) recorded that the green plant was put into beds to repel fleas. Rutty (1772: 122) said that it was of great use to keep flies off cattle. He wrote that 'if the juice of this be rubbed to any of their wounds and sores, no Flies will attack them in the hottest summer; and a handful of it put under a bed is said to banish Fleas effectually'.

Mahon (1983) listed this species as the source of a yellow dye.

Ordinary pepper is a spice derived from an unrelated tropical climbing plant, *Piper nigrum*, which is native to India. It has been imported into Ireland since the earliest of times as a condiment and seasoning spice. Kelly (1997) reported that as early as the 5th century it was being exported from India to the Mediterranean region, but we do not know at what date it was first used in Ireland for flavouring food. Pepper comes in two forms: 'black pepper', which is derived from ground immature berries, and 'white pepper', where the seeds are ground after the outer skin of the berries has been removed.

Persicaria lapathifolia (Polygonum lapathifolium)

- Pale persicaria (Scannell & Synnott, 1987); Bloodweed[195] (Hart, 1898)
- *Glúineach bhán* (Scannell & Synnott, 1987)

POLYGONACEAE—THE KNOTWEED FAMILY
Possibly Introduced (possible archaeophyte)

Persicaria lapathifolia is an occasional annual plant of cultivated fields and wasteground throughout Ireland. It is also found on river-banks on gravel.

According to Allen and Hatfield (2004), it was used medicinally in Ireland to stop bleeding.

The seeds can be ground and used as flour, and the whole plant can be chopped up and used in salads or boiled or steamed as a vegetable (Mears & Hillman, 2007).

Persicaria maculosa (Polygonum persicaria)

- Redshank (Scannell & Synnott, 1987); Spotted persicaria (Webb & Scannell, 1983)
- *Glúineach dhearg* (Scannell & Synnott, 1987); *Gluineach mhór* (Cameron, 1883)

---

195. Hart (1898: 367) said that in Belleek, Co. Fermanagh, 'the dark spot on the leaf is supposed to be blood, and the plant was at the foot of the Cross'.

POLYGONACEAE—THE KNOTWEED FAMILY
Native

Persicaria maculosa is a very common and widespread annual plant throughout Ireland of cultivated ground, roadsides and waste places.

Like *Persicaria lapathifolia*, *P. maculosa* has been used medicinally. Juice from the leaves was used to cool ulcers and inflammations and to dissolve congealed blood, cleanse putrid wounds, relieve ear-aches and as a vermifuge (Uí Chonchubhair, 1995). The same author recorded that a poultice of crushed leaves was applied to whitlows, and a decoction of the plant was given to people or animals with kidney stones or urinary infections.

The young leaves can be used as a vegetable, boiled like spinach, or served raw in salads (Peterson, 1977).

According to Mahon (1983), it was used to obtain a yellow dye.

**Persicaria minor (Polygonum minus)*

- Small water-pepper (Scannell & Synnott, 1987)
- *Biorphiobar beag* (Scannell & Synnott, 1987)

POLYGONACEAE—THE KNOTWEED FAMILY
Native

Persicaria minor is a common annual plant in the north of Ireland of lake-shores and land subject to flooding.

*No ethnobotanical uses of this plant have been noted in Ireland.

Persicaria vivipara (Polygonum viviparum)

- Alpine bistort (Scannell & Synnott, 1987)
- *Glúineach shléibhe* (Scannell & Synnott, 1987)

POLYGONACEAE—THE KNOTWEED FAMILY
Native

Persicaria vivipara is a very rare plant in Ireland that is found only on the highest mountain ridges and summits in the north-east and in Co. Kerry.

In Scandinavia and North America the roots of this plant have been eaten raw and cooked as a vegetable (Peterson, 1977; Mears & Hillman, 2007). It has also been used for food in Scotland, where Milliken and Bridgewater (2004) recorded that in the 18th century poor people on the Duke of Gordon's estate in Morayshire prepared meal from its dried roots and seeds. In Ireland, it was probably always too rare during historic times to ever have been a widely used food source.

Petasites fragrans

- Winter heliotrope (Scannell & Synnott, 1987); Docken (Hart, 1898)
- *Plúr na gréine* (Scannell & Synnott, 1987); *Meacan an phubóil* (Moloney, 1919; Dinneen, 1927; Ó Dónaill, 1977); *Gréinseiceach* (Moloney, 1919; Dinneen, 1927)

ASTERACEAE (Compositae)—THE DAISY FAMILY
Introduced

Petasites fragrans is an introduced perennial herbaceous plant that is widely naturalised on shady banks and roadsides throughout Ireland. It was originally introduced and grown as a garden plant, in particular because it flowers in winter. However, it is invasive and aggressive and no longer rated as an ornamental in Ireland.

Petasites hybridus

- Butterbur (Scannell & Synnott, 1987); Gallon (Praeger, 1938; Harron, 1986)
- *Gallán mór* (Scannell & Synnott, 1987); *Borrán* (Moloney, 1919; Ó Dónaill; 1977); *Meacan an phobóil* (Ó Dónaill, 1977)

ASTERACEAE (Compositae)—THE DAISY FAMILY
Native

Petasites hybridus is an occasional perennial plant throughout Ireland of wet river-banks.

Butterbur has been used as a plant of traditional medicine in Europe and north-western Asia; its rhizomes were used to treat spasms of pain in the head and the gastrointestinal and urinary tracts, and the leaves were used as a traditional remedy for nervous cramps, headaches, wounds, colic and skin disorders (van Wyk & Wink, 2004). It was used medically in Ireland for several purposes, similar to the uses of *Tussilago farfara* (Uí Chonchubhair, 1995).

Petroselinum crispum (Carum petroselinum; Petroselinum sativum)

- Parsley (Scannell & Synnott, 1987)
- *Peirsil gharraí* (Scannell & Synnott, 1987); *Bliúcán* (Ua Maoileoin, 2001); *Fionnas garraidhe* (Moloney, 1919; Dinneen, 1927); *Pearsáil* (de Vere Coneys, 1849; Dinneen, 1927); *Fionnas garraí* (Uí Chanainn, 1962)

APIACEAE (Umbelliferae)—THE CARROT FAMILY
Introduced (possible archaeophyte)

Parsley is a biennial herbaceous plant in Ireland of old walls, castles and ruins. It is usually found growing as a wild plant but only very rarely, and probably survives as a relict of former cultivation. It also occurs as a discarded plant on rubbish dumps and disturbed ground, banks and some rocky places. Of course it is extremely commonly cultivated and used as a culinary herb and has probably been grown and used in Ireland for more than 300 years.

The great Irish botanist Robert Lloyd Praeger (1909) mentioned that parsley grew 'amongst the picturesque ruins' at Moyne Abbey between Killala and Ballina in West Mayo where its occurrence 'still tell[s] of the old monkish kitchen-garden'. Such survivals on castles and monastic ruins elsewhere are probably also relics of its use both as a culinary herb and medicinal plant in former times. Donal Synnott (1979: 38) mentioned that there is also a particular group of medicinal or culinary herbs that are associated with old Norman castles and abbeys in Ireland 'with striking regularity' and he suggested that 'although there are many difficulties in the way of interpreting this phenomenon, it seems not unreasonable to suggest that many of these species were originally introduced by the Normans or at least their culinary or medicinal use was begun in Ireland by the Normans'. Although parsley is not included in the list of plants he gives for this group, it may well be a Norman introduction too because of its occurrence in such sites. Amongst this group of plants he included are *Silybum marianum* (milk thistle), *Conium maculatum* (hemlock), *Malva sylvestris* (mallow), *Hyoscyamus niger* (henbane), *Parietaria judaica* (wall pellitory), *Urtica urens* (annual nettle) and the thistles *Carduus acanthoides* and *C. tenuiflorus*. In 1977, 18 out of 20 Norman castles he visited from Tipperary to North Louth had plants of hemlock and mallow. He also noted that older habitations (pre-Norman sites) do not have such plants associated with them. He suggested that traditions associated with these plants are probably Norman and date from about the 12th century at the earliest.

I am not aware of parsley being collected from the wild in Ireland for use as a culinary herb or medicinally, although it is a staple plant of most kitchen and herb gardens. Its leaves, roots and aromatic seeds have all been used for medicinal purposes. Parsley tea can be made and acts as a diuretic (Harrop, 1977). Egan (1887: 13) recorded its use in Irish traditional medicine to treat gravel and other 'slight disorders of kidneys', taken as a strong decoction. O'Regan (1997) mentioned a case of a decoction of the leaves and roots being used in Co. Mayo to treat back pain. Vickery (2010: 120) provided information on an unusual gynaecological use of parsley recorded in London in 1982 'to bring on your period put a sprig of parsley inside your vagina for 12 hours—your period should start 24 hours later'.

Moutray Read (1916: 257–258) described some Irish folklore associated with garden parsley, saying 'Parsley should be sown on Good Friday, because it goes down into Hell seven times before it comes up'. This probably refers to the difficulties that are sometimes encountered in germinating parsley seed. (The secret to good germination of parsley seed is never to let it dry out after sowing.)

Phalaris arundinacea (Arundo colorata)

- Reed canary-grass (Scannell & Synnott, 1987)
- *Cuiscreach* (Scannell & Synnott, 1987); *Biorrach lachan*[196] (Moloney, 1919; Dinneen, 1927); *Siosgalach* (Moloney, 1919); *Cuile ioldathach*[197] (White, 1803)

POACEAE (Gramineae)—THE GRASS FAMILY

Native

Phalaris arundinacea is a common and robust grass of river-banks, stream-sides, ditches and lake-shores.

It was planted for its use in stabilising river-banks (Mackay, 1836).

Rutty (1772) said that the stalks were used to make corn pipes.

**Phleum arenarium*

- Sand cat's-tail (Scannell & Synnott, 1987)
- *Tiomóid duimhche* (Scannell & Synnott, 1987); *Sleghfhér traghainmhech*[198] (White, 1803)

196. Dinneen (1927) defined *biorrach lachan* as 'common reed-grass'.
197. White (1803) applied this name to *Arundo colorata*.
198. White (1803) listed *sleghfhér traghainmhech* as '*Phalaris*', 'sea canary-grass'.

POACEAE (Gramineae)—THE GRASS FAMILY
Native

Phleum arenarium is an occasional plant in all parts of the Irish coast in sand-dune grassland.

*No ethnobotanical uses of this plant have been noted in Ireland.

Phleum pratense

- Timothy (Scannell & Synnott, 1987); Cat's-tail (Anonymous, 1978)
- *Tiomóid* (Scannell & Synnott, 1987); *Bodán* (Moloney, 1919); *Feur capaill* (Moloney, 1919); *Féar capaill* (Dinneen, 1927; Ó Dónaill, 1977; Uí Chonchubhair, 1995); *Loscatfheur leana* (Wade, 1808a); *Loschaitfhér léna* (White, 1803)

POACEAE (Gramineae)—THE GRASS FAMILY
Native

Phleum pratense is an abundant meadow grass and generally important as a fodder grass and constituent of hay. It is widespread in Ireland and found in meadows, pastures and roadsides. It is used in some commercial wildflower seed mixes.

Allen and Hatfield (2004) recorded its use medicinally in Co. Waterford where a leaf was wound around a finger to keep cloth from sticking to a wound.

Phormium tenax.

Phormium tenax

- New Zealand flax (Reynolds, 2002)
- *Líon na Nua-Shéalainne* (Uí Chonchubhair, 1995)

XANTHORRHOEACEAE (Phormiaceae)—THE ASPHODEL FAMILY
Introduced

Phormium tenax is a popular ornamental garden plant that has long been established in the wild in several parts of Ireland. For example, in West Kerry on the Dingle Peninsula it was the basis of a local flax fibre industry established by the 5th Baron Ventry alongside the Burnham Estate. The industry operated for only a few short years in the 1900s before production of ropes and twine ceased in about 1920 (Siobhán Ní Chorcora, pers. comm., 2011). Local stories survive in Kerry that the mill was run by a manager by the name of Jim Crow, who was brought in from the tiny island of St Helena by Lord Ventry (St Helena is famous as the place to which Napoleon Bonaparte was exiled in 1815 and where he died in 1821). New Zealand flax still grows in abundance on that South Atlantic island. Baron Ventry served in the British Army in South Africa before returning to Ventry, and as St Helena was on the Britain-to-South Africa route, he may have got the idea to start the industry during a stop at that island (Siobhán Ní Chorcora, pers. comm., 2011). The ruined remains of the flax mill survive today, and scattered New Zealand flax plants grow along field boundaries and embankments nearby, no doubt ancestors of those introduced for the flax industry. In parts of the west and south-west of Ireland it has been planted on field margins to provide a shelterbelt for fields and gardens.

In its native New Zealand it has been widely and traditionally used by the indigenous Maoris and European settlers to make a variety of objects, such as woven baskets, mats, ropes, nets and fabrics. Its fibres are strong and very durable. In Ireland it is little if ever used today as a fibre source, and other than the Dingle example, I have not found other references to former usage in Ireland.

In 2010 reports of birds with bright orange heads, otherwise looking like common starlings and house sparrows, were received by the organisation Birdwatch Ireland. These turned out to be starlings and sparrows that had been feeding on the rich, sweet nectar supplies in the flowers of the New Zealand flax; an interesting example of native species that have found a useful new food source from an exotic plant. In New Zealand this plant is particularly pollinated by a native bird called a 'Tui'.

A range of cultivated varieties are grown as ornamental garden plants.

Phragmites australis (Phragmites communis; Arundo phragmites)

- Reed (Scannell & Synnott, 1987); Sregs (Colgan, 1904); Cushkina (Hart, 1898); Queel rods (or Quill rods) (Hart, 1898); Wheel rod (Hart, 1898)
- *Giolcach* (Scannell & Synnott, 1987); *Biorrach* (Moloney, 1919; Ó Dónaill, 1977); *Cruisgiornach* (or *Cruisigh*) (White, 1803; de Vere Coneys, 1849; Cameron, 1883); *Giolcuigh* (de Vere Coneys, 1849); *Biorrach laighean* (Dinneen, 1927); *Cuile cruisgiornadh*[199] (Wade, 1808a); *Clabhascar*[200] (Dinneen, 1927); *Cuiscireach* (Dinneen, 1927); *Lachán*[201] (Dinneen, 1927); *Coiscreach*[202] (Ó Dónaill, 1977); *Cuiscreach* (or *Cuiscear(n)ach, Cuiscireach*) (Ó Dónaill, 1977); *Gáinne* (Ó Dónaill, 1977); *Cuile cruisgiornagh*[203] (White, 1803); *Biorach lachan* (White, 1803); *Birah laghon* (White, 1803); *Coisgearach* (Colgan, 1911)

POACEAE (Gramineae)—THE GRASS FAMILY
Native

Phragmites australis is a tall grass species that is common throughout Ireland in marshes, wet ditches and lake-shores.

Perhaps surprisingly, all or most parts of reeds are edible but are a much overlooked wild food. Young roots can be dried and ground to obtain a flour or make a porridge and can be boiled and used as a vegetable. Young shoots and leaves can be dried and ground down to make a flour and made into dumplings. (For a description of the way the flour is harvested, see *Typha* species.) The stems contain a sugar-rich sap, which can be collected and eaten or cooked until it becomes like a brown toffee. The seeds are also edible and can be eaten raw or boiled (Hilton, 2007) or ground and made into gruel. It is not known whether reeds were widely used for food in early times in Ireland.

Reeds were formerly widely harvested and used for thatch and weaving. Danaher (1964) noted that they were especially prized for first-class thatching. Colgan (1911) mentioned that he had seen this species cultivated and even manured in a fenced field on Achillbeg,

199. Wade (1808a) applied this name to *Arundo phragmites*.

200. Dinneen (1927) defined *clabhascar* to mean 'reed' and also 'long seedless straw'.

201. Dinneen (1927) defined *lachán* as 'common reed'.

202. Ó Dónaill (1977) gave the Irish words *coiscreach, cuiscreach* (or *cuiscear(n)ach, cuiscireach*) for reeds; *gáinne* for 'reed'.

203. White (1803) applied this name to '*Arundo phragmitis*'.

Co. Mayo, which were being grown for thatching. Water reeds are generally harvested between January and March; reeds cut later in the year are more brittle (O'Reilly, 2004). O'Reilly (2004) recorded that a majority of the thatching of vernacular buildings was carried out by using oats, barley, rye and wheat straw.

In a survey of almost a 1000 thatched buildings in 11 counties, water reed had been used in 39% of buildings. Modern Irish roof thatching uses cereal (oat and rye) straw (as well as a variety of imported materials, including water-reeds, from Turkey, Hungary, France and England), whereas older roofs were often primarily wheat (or barley) straw and *Phragmites australis* (more rarely *Juncus* spp. also).

A 1000 sq ft property thatched with water-reeds was estimated to cost in the region of €15,000 in 2007 (Niall Toner, *The Sunday Times*, 8 April 2007). A large new thatched roof made from Turkish reeds that I saw being completed in Anascaul, Co. Kerry, in 2010 was reported to me as having cost as much as €60,000. It is no wonder that people are reluctant today to use thatch for the roof of a new house or to replace older slated-roof houses again with thatch. My own cottage on the Dingle Peninsula in Co. Kerry is recorded in an earlier Irish Census as having once been thatched, but today it has a slate roof. Even though this roof needs to be renewed, it is unlikely that our finances would allow us to thatch it once again. Ironically, while today having a thatched roof on one's house in Ireland suggests that its owners are probably relatively wealthy, in the past it would have been seen as a sign of poverty.

O'Sullivan (1977) described the way in which traditional houses were built in the Aran Islands in the 1900s. When the walls and roof beams were in place a specialist thatcher would be called in to complete the roof, using cheap local materials such as reeds (but more usually rye straw, which was also grown extensively). A thatch roof needed to be repaired every year and replaced completely every eight years. Before the roof was thatched a layer of sod was spread on it (scraws). Scraws were cut off a good thick bog, roughly 0.76–0.91 m (ca. 2.5–3 ft) wide and ca. 80 mm (ca. 3 in) thick (O'Hare, 1993). They were then rolled up and left to dry to reduce the weight. Then when they were needed they would be lifted by inserting a stout stick or spade handle through the middle of the roll. They would then be transported by cart to where they were needed before being unrolled onto the roof of the house. Grassland sods were also used; these were generally grasslands that were well grazed so that a dense and thick mat of roots would have developed (Geraghty, 1996). Onto that a layer of straw was spread about 10 cm thick along the eaves at the base of the roof. Onto this another layer of straw would be spread that overlapped the first layer. Further layers would be spread until finally the roof ridge was reached, where a final layer of straw would be spread over the ridge. The completed thatch was secured using grass ropes (called *súgán*) every 30 cm along the ridge of the roof, which were secured onto pegs in the walls below the eaves or weighed down with stones. Other ropes were placed cross-ways on the roof and secured onto the gable ends of the house. The ropes were made from twisted straw. In other parts of Ireland where the thatched roofs did not have to withstand the same regular gales, ropes were less used to anchor the thatch but instead the thatch would be fixed with hazel or willow spars (or scallops) rooted into the scraws below.

Spars showing through thatch and scraw construction.

Maura Scannell (2007) described a chromolithograph of a painting in her possession, which she recently (2008) donated to the National Botanic Gardens, Glasnevin, entitled 'Rush Gatherers, Lough Corrib, Connemara, Ireland' by J. J. Hill. The original painting was painted in about 1859 and depicts three people (two women and a man) harvesting *Phragmites australis* for thatch.

Along the banks of the river Shannon in Cos. Clare and Limerick, the vast reed beds just west of Limerick city were traditionally harvested for thatching reeds. A traditional flat-bottomed wooden boat, called a 'gandelow' (the name is probably derived from the word 'gondola') was used for this purpose. It carried the bundles of harvested reed (called 'hundreds') back to the shore, where they were stacked by the roadsides and from where they were collected by the thatchers (McInerney, 2005; J. McInerney, in MacCárthaigh, 2008). I remember well these elegant boats when as a child growing up in Limerick city in the 1960s they were moored all along the shore of the river Shannon. They were constructed from pine or spruce planks, timber which was frequently imported into Ireland through the port of Limerick. They were generally painted a grey colour. The gandelows were also used for fishing (they were regarded as ideal for salmon fishing) and many other purposes. Their flat bottoms meant that they were particularly convenient for shallow water and for dragging over the mudflats. Similar boats were used in the Cashen and Feale river estuaries. Small flat-bottomed wooden boats called 'cots' were, and to some extent still are, used on the river Suir, Co. Waterford, to cut and carry reeds for thatching or willows for basketry (D. Tully, in MacCárthaigh, 2008).

A gandelow (traditional river boat) at Bunratty Castle, Co. Clare, used for harvesting *Phragmites* and as a fishing boat.

Phragmites reeds for thatching.

MacGowan (2003: 3–4) described the traditional cutting of reeds for thatching in part of Co. Donegal in the late 1800s and early 1900s:

> All the houses in the area were thatched. Reeds were used to thatch our house and the other houses in the area. They were plenty down by the sand-banks at Magheraroarty and everyone's land had a strip by the sea both there and at Derryconor. The bogs and their limits are the same now as they were hundreds of years ago. This is the way the limits are fixed: as the reeds are cut, those along the edges are left uncut so that the owner can come along and tie what remains into knots. This is done year after year and in this way everyone knows his own. When the men are cutting reeds as I was young they had a kind of toothed sickle—a sickle with an edge something like a saw. The reeds had to be cut under the sand and they would go about the cutting on soft days when there would be nothing much doing on the land. They would often be wet to the skin at this work. Then they would sell a load of reeds for a couple of shillings and well might it be said that this was money earned the hard way.

Reeds were probably never used for basketry, as being hollow-stemmed they are too brittle (Joe Hogan, pers. comm., 2010).

Reed beds have also become an increasingly popular means for environmentally friendly sewage and waste water treatment in Ireland. (See also *Juncus* and *Scirpus* spp.)

Wade (1808a) said that starlings have an astonishing and violent passion for its seeds.

Clyne (1989) said that this species was formerly used to make reeds for bagpipes in Scotland.

*Picris echioides

- Bristly oxtongue (Scannell & Synnott, 1987)

- *Teanga bhó gharbh* (Scannell & Synnott, 1987); *Boglus* (Moloney, 1919; Dinneen, 1927); *Teanga bó* (Moloney, 1919)

ASTERACEAE (Compositae)—THE DAISY FAMILY
Possibly Introduced

Picris echioides is a very rare plant in Ireland that occurs mainly in the east in waste places and tilled fields.

*No ethnobotanical uses of this plant have been noted in Ireland.

Pilosella officinarum (Hieracium pilosella)

- Mouse-ear hawkweed (Scannell & Synnott, 1987)
- *Searbh na muc* (Scannell & Synnott, 1987); *Lus na seabhac* (Moloney, 1919); *Searbhán na muc* (Moloney, 1919); *Cluasluch* (de Vere Coneys, 1849; Dinneen, 1927); *Cluas liath* (Ó Dónaill, 1977)

ASTERACEAE (Compositae)—THE DAISY FAMILY
Native

Pilosella officinarum is a relatively common plant throughout Ireland of dry habitats, such as esker ridges, sand dunes, dry banks, wall-tops and grasslands.

Allen and Hatfield (2004) noted its former use for the treatment of coughs and whooping cough in Ireland. These authors also noted it was used for urinary troubles, whitlows, sores and as a salve for burns.

Pimpinella major

- Greater burnet saxifrage (Scannell & Synnott, 1987)
- *Coll an dromáin* (Scannell & Synnott, 1987); *Clabhras mór* (Uí Chonchubhair, 1995)

APIACEAE (Umbelliferae)—THE CARROT FAMILY
Native

Pimpinella major is common in parts of the south and west only in Ireland and is absent from most of the country. It occurs on roadsides and in woodland margins, hedges and rocky pastures.

In parts of its natural range in Europe and western Asia, the roots and rhizomes of this species have been used in traditional medicine to treat respiratory infections and to prepare a cough medicine (van Wyk & Wink, 2004). Whether it was used for such purposes in Ireland is not known.

According to Wade (1804: 22), 'The root is very acrid, burning the mouth like pepper, and affords a blue oil'.

Pimpinella saxifraga

- Burnet saxifrage (Scannell & Synnott, 1987)

- *Ainís fhiáin* (Scannell & Synnott, 1987); *Clabhrus* (Moloney, 1919)

APIACEAE (Umbelliferae)—THE CARROT FAMILY
Native

Pimpinella saxifraga is found in parts of the centre and south-east only in Ireland and is rare elsewhere. It occurs on dry calcareous habitats, such as dry banks and pastures.

Smith (1815: 360) described its use as a medicinal plant with 'anti-scorbutic' properties and noted that it was 'found in great plenty about Cork'. Rutty (1772) said that the root was dried and powdered and used instead of pepper in salads.

Its scientific name *saxifraga*, meaning 'breaker of rocks or stones', may relate to medicinal use in the treatment of kidney or bladder stones.

Pinguicula species

- Butterwort (Scannell & Synnott, 1987)
- *Liath uisce* (Webb et al., 1996); *Bodán meascáin*[204] (Ó Dónaill, 1977)

LENTIBULARIACEAE—THE BLADDERWORT FAMILY
Native

Including:

Pinguicula grandiflora

- Large-flowered butterwort (Scannell & Synnott, 1987)
- *Leith uisce* (Scannell & Synnott, 1987)

Pinguicula lusitanica

- Pale butterwort (Scannell & Synnott, 1987)
- *Leith uisce beag* (Scannell & Synnott, 1987)

Pinguicula vulgaris

- Common butterwort (Scannell & Synnott, 1987); Steep-grass (Praeger, 1938)
- *Bodán meascáin* (Scannell & Synnott, 1987); *Léith uisce* (Ua Maoileoin, 2001); *Liaith* (Colgan, 1911; Moloney, 1919); *Meascán* (Moloney, 1919); *Bróga na cumhaigh* (Moloney, 1919)

The three species of butterwort in Ireland all occur in boggy habitats. *Pinguicula grandiflora* is very common in parts of the south-west, such as in Co. Kerry and West Cork, where it more or less replaces *P. vulgaris*. It grows in bogs and fens and on wet and dripping rocks and in heaths in the mountains. *Pinguicula vulgaris* grows in similar habitats, mainly in the west, north and centre of Ireland. *Pinguicula lusitanica* is the smallest and most inconspicuous of the three and grows by mountain streams and in bogs, wet heaths and flushes, mainly in the west and is rare elsewhere.

It is often stated that the leaves of butterworts were used to curdle milk (e.g., Moloney, 1919), although my limited experiments in this regard of adding leaves to milk have been unsuccessful in causing any perceptible change. Nevertheless, according to Williams (1993), it was used as a rennet in Ulster. Cameron (1883) said it was believed that the

---

204. Ó Dónaill (1977) defined *bodán meascáin* as 'butterwort'.

consistency of milk could be changed and thickened to cream by straining it through the leaves of *Pinguicula*.

There are no known records of its use in folk medicine in Ireland (Allen & Hatfield, 2004).

Threlkeld (1726: unnumbered [124]) said that it was used to treat swellings and 'clefts' in cow's udders. Moloney (1919) said that the name *léith uisce* denotes the disease of liver fluke in sheep. Drury (1985) mentioned the use of butterworts in Britain to make an ointment to cure wounds and bites in cattle.

Plantago species

- Plantain (Scannell & Synnott, 1987)

PLANTAGINACEAE—THE PLANTAIN FAMILY

Including:

Plantago coronopus

- Buck's-horn plantain (Scannell & Synnott, 1987)
- *Adharca fia* (Scannell & Synnott, 1987)

Native

Plantago lanceolata

- Ribwort plantain (Scannell & Synnott, 1987); Narrow-leaved plantain (Anonymous, 1978); Rubbart (Colgan, 1904); Ribgrass (Hart, 1898; Colgan, 1904); Cocks (Hart, 1898; Praeger, 1938); Cruk slanlish (Hart, 1898); Ripple-grass (Hart, 1898)
- *Slánlus* (Scannell & Synnott, 1987); *Slándus* (Ua Maoileoin, 2001); *Deideog* (Cameron, 1883); *Cruachphadruic slanlus* (Wade, 1808b); *Lus an teanchais*[205] (Dinneen, 1927); *Slándás* (Ó Criomhthain, MSS, n.d., ca. 1908)

Native

Plantago major

- Greater plantain (Scannell & Synnott, 1987); Way bread (Colgan, 1911; Praeger, 1938); Plantain leaf (Colgan, 1904); Rat's-tail (Praeger, 1938); Rat-tail plantain (Harron, 1986)
- *Cuach Phádraig* (Scannell & Synnott, 1987); *Copóg Phádraig* (Moloney, 1919); *Cró Phádraig* (Moloney, 1919); *Slánlus mór* (Uí Chonchubhair, 1995); *Cál Phádraig*[206] (Ó Dónaill, 1977)

Native

Plantago maritima

- Sea plantain (Scannell & Synnott, 1987)
- *Slánlus mara* (Scannell & Synnott, 1987)

Native

Plantago media

- Hoary plantain (Scannell & Synnott, 1987); Lamb's tongue (Praeger, 1938)
- *Slánlus liath* (Scannell & Synnott, 1987)

Introduced

---

205. Dinneen (1927) defined *lus an teanchais* as 'rib-wort'.
206. Ó Dónaill (1977) defined *cál phádraig* as 'plantain', and did not specifically refer to *Plantago major*.

Plantago lanceolata (ribwort plantain). Watercolour by Lydia Shackleton. Courtesy of the National Botanic Gardens, Glasnevin.

Plantains grow in a very wide variety of habitats throughout Ireland. *Plantago coronopus*, or buck's-horn plantain, is common all around the Irish coastline in habitats such as salt marshes, cliffs, banks, rocks and wasteground near the sea. *Plantago maritima*, or sea plantain, is also coastally distributed and one of the dominant plants of many Irish salt marshes. It grows on seashores and more rarely inland in Ireland in the mountains of the west and north in rocky habitats and on lake-shores. *Plantago lanceolata*, or ribwort plantain, is an extremely common plant of weedy habitats, such as waste places, roadsides and cultivated ground as well as pastures and meadows. *Plantago major*, or greater plantain, is similarly abundant and occurs on roadsides, waste places, pastures and dune-slacks and as a weed of lawns and gardens. The hoary plantain, or *P. media*, is introduced and is occasionally found growing as a weed of lawns and waste places, and is neutral to basic grasslands.

Plantains were important medicinal plants in Ireland in historic times, indeed throughout Europe, and have been used for a variety of treatments and ailments. Elsewhere in other parts of the world, *Plantago major* is a cosmopolitan weed and in the Caribbean was used by the native Caribs to treat eye inflammations. Its leaves were used in the Caribbean islands as a cooling tea, with mint, thyme and salt to counteract shock; the leaves have been used in Martinique to promote menstruation (Honeychurch, 1986). In Ireland, plantain leaves were applied to wounds and sores (Colgan, 1904, 1911; Moloney, 1919; Logan, 1981; Vickery, 2010). Colgan (1911) said that in Clare Island, Co. Mayo, the leaves were pounded up to make a poultice for cuts and sores, and, according to one informant, mixed with the root-leaves of *P. major* for this purpose. Hart (1898) noted that the poor people of Derry applied the leaves of *P. major* to bruises and to wounds to stop bleeding. Plantain leaves are rich in tannins, which help in the healing of grazes and wounds (Williams, 1993). K'Eogh (1735: 96) wrote that plantain is 'exceedingly good against haemorrhages and all kinds of fluxes, also useful to conglutinate and heal green wounds'. Egan (1887) recorded the use of plantain in Ireland to treat cuts. Bruised leaves, while still moist with the sap, were applied to the wound to diminish pain, stop profuse bleeding and prevent festering. Dirrane (1997: 17–18) wrote that ribwort plantain was extremely effective for treating cuts and was used in Inishmore in the Aran Islands. The rough side of the leaf was put next to the cut or ulcer and kept there until the cut was healed: 'Of course you needed a bandage to keep it in place. I remember splitting my index finger while opening a tin of sardines in Aran and the *slánlus* did the job for me in seven days. Cobwebs were often used in the old days too to stop bleeding. There was no scarcity of cobwebs at the time but that particular usage is largely a thing of the past'. In Ireland, Allen and Hatfield (2004) reported that a decoction of plantain juice was drunk for a cough, applied to burns, and used to draw pus out of boils, wounds or swellings. It was also used externally to treat lumps and swellings, pimples, chapped hands or legs, corns, warts, sore eyes, headaches and gout, and internally for jaundice and other liver complaints. The sea plantain, *P. maritima*, was used in Co. Cork also to staunch bleeding and to treat wounds (Allen & Hatfield, 2004). The ribwort plantain, *P. lanceolata*, was used as a healing herb for consumption (tuberculosis), to heal whitlows, as a purgative for worms, in ointment for burns and was also good for 'weak blood' (Williams, 1993).

Plantago species: A, *Plantago lanceolata;* B, *Plantago major;* C, *Plantago maritima.*

This same author recorded that its leaves were eaten for gout. *Plantago coronopus* and *P. lanceolata* were both used medicinally to 'cure the bite of a mad dog' (Smith 1815: 363). Three or four spoonfuls of the juice were taken three or four times every morning as soon as possible after the bite was suffered. This treatment was repeated 'for two or three next changes of the moon'. In the treatment of whitlows in the Aran Islands, the leaves were washed and soaked in hot water, then wrapped around the affected finger (O'Regan, 1997).

The name *slánlus* is derived from '*slán*', meaning 'whole', or '*sláinte*', meaning 'health' and '*lus*', a herb (Maírín Uí Chonchubhair, pers. comm., 2011). She also tells me that seven-ribbed leaves, rather than the more usual three to five, were regarded as being more potent for medicinal purposes.

In veterinary medicine, *Plantago lanceolata* was used to treat red murrain and 'red milk' in cows and to cure cow's udders (Williams, 1993). Ribwort plantain was mixed with butter and given to a dog to prevent it from going mad; it was also used to cure bites in cattle (Smith, 1815).

Lewis-Stempel (2012: 36–37) said that the young leaves of ribwort plantain are edible but need to be steamed to 'within an inch of their life'.

Plantago lanceolata was, and remains to some extent, a popular plant for childhood country games. Children would try to knock the flowerheads off each other's plants. Colgan (1904) mentioned that this game (and the plant itself) was named 'cocks' or 'fighting cocks' in Co. Dublin and elsewhere in Ireland, but it is also called 'soldiers'. Vickery (2010) gave other names for this game in England and Scotland, such as 'blackmen', 'cocks and hens', 'hard heads', 'knights' and 'carl doddies'[207]. O'Regan (1997) gave the Irish name *lus na Saighduir* for this plant, either because of its use in children's games or perhaps because it was used for healing wounds. The flowerheads are called *ceann-táthair* in Irish (Dinneen, 1927).

In Irish a name for *Plantago major* is *cuach Phaidraig* (the drinking vessel of Patrick) (Synnott, 1979), although it is not known if this name suggests a particular religious or symbolic usage in Ireland.

The leaves of all native plantains in Ireland are edible and may be cooked as a pot herb. The seed husks are mucilaginous and can be ground down to make flour to make dampers (Mears & Hillman, 2007). The chopped leaves can be added to salads (Peterson, 1977).

**Platanthera bifolia*

- Lesser butterfly-orchid (Scannell & Synnott, 1987)
- *Magairlín beag an fhéileacáin* (Scannell & Synnott, 1987)

ORCHIDACEAE—THE ORCHID FAMILY

Native

Platanthera bifolia is an occasional orchid of damp peaty grassy hillsides, open woods, bog-margins and cut-away bogs. It is widespread throughout most of Ireland except the east and south.

*No ethnobotanical uses of this plant have been noted in Ireland.

207. Vickery (2010) and Hart (1898) gave the name 'carl doddie or 'carl doddy' as a name for devil's-bit scabious (*Succisa pratensis*) in Co. Donegal and Co. Derry.

**Platanthera chlorantha*

- Greater butterfly-orchid (Scannell & Synnott, 1987)
- *Magairlín mór an fhéileacáin* (Scannell & Synnott, 1987)

ORCHIDACEAE—THE ORCHID FAMILY
Native

Platanthera chlorantha is an orchid that grows in similar habitats as *P. bifolia*, such as rough grassland and heaths. It is widespread throughout most of Ireland except the south-west where it is rare.

*No ethnobotanical uses of this plant have been noted in Ireland.

Platanus ×hispanica (Platanus ×hybrida; Platanus occidentalis × Platanus orientalis)

- London plane (Scannell & Synnott, 1987)
- *Plána Londan* (Anonymous, 1978)

PLATANACEAE—THE PLANE FAMILY
Introduced

Platanus ×hispanica, the London plane, is a widely planted street tree in Dublin city. Its flaking bark makes it particularly useful for polluted city environments and it grows vigorously even in what might be regarded as unfavourable city conditions with compacted soils. It is little troubled by pests and diseases and rarely blows down (Mitchell, 1974).

Although little used for its wood, London plane timber is very similar in colour and texture to that of beech.

**Poa alpina*

- Alpine meadow-grass (Scannell & Synnott, 1987)
- *Cuisce Alpach* (Scannell & Synnott, 1987); *Cuise ailp* (White, 1803; Dinneen, 1927)

POACEAE (Gramineae)—THE GRASS FAMILY
Native

Poa alpina is a rare grass in Ireland and found only on the highest parts of Mount Brandon in Co. Kerry and the Ben Bulben range in Co. Sligo.

*No ethnobotanical uses of this plant have been noted in Ireland.

**Poa annua*

- Annual meadow-grass (Scannell & Synnott, 1987)
- *Cuise bliantúil* (Scannell & Synnott, 1987); *Glas léana* (Moloney, 1919); *Móinfhéur* (Moloney, 1919); *Ceanna dubha* (Moloney, 1919); *Brobh i dtóin* (Moloney, 1919); *Cuise bliadhainta* (Wade, 1808a); *Cuise bliadhantamhail* (Dinneen, 1927); *Cuise blidhainta* (White, 1803)

POACEAE (Gramineae)—THE GRASS FAMILY
Native

Poa annua is a very common annual grass of waste places, grasslands, roadsides, walls and a weed of gardens and cultivated land.

*No ethnobotanical uses of this plant have been noted in Ireland.

**Poa nemoralis*

- Wood meadow-grass (Scannell & Synnott, 1987)
- *Cuise coille* (Scannell & Synnott, 1987); *Cuise coiltemhuil* (White, 1803)

POACEAE (Gramineae)—THE GRASS FAMILY

Possibly Introduced

Poa nemoralis is an occasional plant of shady habitats, such as woodlands, mainly in the east of Ireland.

*No ethnobotanical uses of this plant have been noted in Ireland.

Poa pratensis

- Smooth-stalked meadow-grass (White, 1803); Smooth meadow-grass (Scannell & Synnott, 1987)
- *Cuise mín* (Scannell & Synnott, 1987); *Cuise mionghasach* (Wade, 1808a; Dinneen, 1927); *Cuise mínghasach* (White, 1803); *Cuise cumhangdhuillech*[208] (White, 1803)

POACEAE (Gramineae)—THE GRASS FAMILY

Native

Poa pratensis is a common grass throughout Ireland of meadows, pastures, sand-hills and wasteground.

It is used in some commercial wildflower seed mixes as a useful fodder grass.

**Poa subcaerulea (Poa humilis)*

- Spreading meadow-grass (Scannell & Synnott, 1987)
- *Cuise reatha* (Scannell & Synnott, 1987)

POACEAE (Gramineae)—THE GRASS FAMILY

Native

Poa subcaerulea is a common small grass found in a variety of habitats particularly close to the sea.

*No ethnobotanical uses of this plant have been noted in Ireland.

**Poa trivialis*

- Rough meadow-grass (Scannell & Synnott, 1987)
- *Cuise garbh* (Scannell & Synnott, 1987); *Cuise garbghasach* (Wade, 1808a); *Cuise garbhghasach* (White, 1803)

POACEAE (Gramineae)—THE GRASS FAMILY

Native

Poa trivialis is an abundant grass of all types of grasslands throughout Ireland.

*No ethnobotanical uses of this plant have been noted in Ireland.

---

208. White (1803) applied this name to *Poa angustifolia*, 'narrow-leaved meadow-grass'.

Polygala species

- Milkwort (Scannell & Synnott, 1987)
- *Glúineach* (Webb et al., 1996); *Lusan baine* (Cameron, 1883); *Na deirbhshiuiríní* (Dinneen, 1927); *Lus an bhainne* (Dinneen, 1927; Ó Dónaill, 1977); *Lus na seacht mbuadh* (Dinneen, 1927)
- Moloney (1919) gives the following Irish names for *Polygala* but does not differentiate between those names applied to the two Irish species: *Lus an bhainne*; *Glúineach*; *Na deirbhshiúiríní.*

POLYGALACEAE—THE MILKWORT FAMILY
Native

Including:

Polygala serpyllifolia
- Heath milkwort (Scannell & Synnott, 1987)
- *Na deirfiúirīní* (Scannell & Synnott, 1987)

Polygala vulgaris
- Common milkwort (Scannell & Synnott, 1987); Fairy soap (Hart, 1898)
- *Lus na bhainne* (Scannell & Synnott, 1987); *Driféaracha* (Ua Maoileoin, 2001)

Polygala vulgaris (common milkwort). The blue colour in this image may have faded somewhat as *P. vulgaris* generally has a strong blue, violet or purple flower colour. Watercolour by Lydia Shackleton. Courtesy of the National Botanic Gardens, Glasnevin.

There are two species of *Polygala* in Ireland: *Polygala serpyllifolia* and *P. vulgaris*. Both are perennial herbs and widespread throughout. *Polygala serpyllifolia* is common in heaths and grassy places, such as pastures in limestone-free (non-calcareous) areas. *Polygala vulgaris* occurs in heaths, sand dunes and rocky places.

Dioscorides said that 'it is believed to make milk more abundant', presumably in nursing mothers (rather than in grazing cows) (Grigson, 1958), and accordingly milkwort was used by nursing women who wanted to increase the flow of milk (Moloney, 1919; Uí Chonchubhair, 1995). Allen and Hatfield (2004) considered its credentials as a folk remedy in Ireland to be questionable, but MacMahon (1994) said that in Ballyheigue, Co. Kerry, milkwort was used to cure warts. Wade (1804) reported its use as an infusion to treat coughs, although not necessarily in Ireland.

In Donegal it was called 'fairy soap' because the fairies were believed to make lather from its leaves and roots (Hart, 1898; Grigson, 1958).

Polygonum species

- Knotgrass (Scannell & Synnott, 1987)
- *Glúineach bheag*

POLYGONACEAE—THE KNOTWEED FAMILY

Including:

Polygonum arenastrum

- Equal-leaved knotgrass (Scannell & Synnott, 1987)
- *Glúineach ghainimh* (Scannell & Synnott, 1987)

Native or Possibly Introduced (archaeophyte)

Polygonum aviculare

- Knotgrass (Scannell & Synnott, 1987); Stanch (Colgan, 1904)
- *Glúineach bheag* (Scannell & Synnott, 1987)

Native

Polygonum maritimum

- Sea knotgrass (Scannell & Synnott, 1987)
- *Glúineach mhara* (Scannell & Synnott, 1987)

Native

Polygonum oxyspermum subsp. *raii*

- Ray's knotgrass (Scannell & Synnott, 1987)
- *Glúineach ghlé* (Scannell & Synnott, 1987)

Native

The knotweed *Polygonum aviculare* is a common annual weed throughout Ireland of waste places and roadsides and a significant agricultural weed on cultivated ground, indeed throughout much of the world. The equal-leaved knotweed, *P. arenastrum*, occurs in similar habitats but is a more prostrate plant and can tolerate trampling so is a common weed of paths. Ray's knotweed, *P. oxyspermum* subsp. *raii*, is another procumbent knotweed but found much more rarely in Ireland and is restricted to sandy seashores. The seeds of these species can be gathered and ground as a flour (Mears & Hillman, 2007). *Polygonum maritimum* could probably also be harvested for its seeds to make a flour, but it is such a rare plant in Ireland that it is unlikely that it has been used for any purpose in the past. Green (2008) suggested that the sea knotweed may be extinct in Co. Waterford, the only county from which it is known, although it is possible that it has been overlooked elsewhere. This prostrate knotgrass grows on sandy seashores just above the high-tide mark and can be found along Atlantic shorelines into the Mediterranean.

In parts of Europe *Polygonum aviculare* has been used in traditional medicine to treat coughs, bronchial catarrh and mouth and respiratory inflammations (van Wyk & Wink, 2004). I have not found any records of its use for medicinal purposes in Ireland.

According to Allen and Hatfield (2004), *Polygonum aviculare* has been used as a veterinary treatment for swine fever in Kilkenny.

Populus nigra

- Black poplar (Scannell & Synnott, 1987)
- *Poibleog dhubh* (Scannell & Synnott, 1987); *Crann pobail*[209] (Dinneen, 1927; Ó Dónaill, 1977)

SALICACEAE—THE WILLOW FAMILY

Possibly Native

209. Dinneen (1927) defined *crann pobail* as 'poplar' and Ó Dónaill (1977) as 'poplar-tree'.

Although *Populus nigra* is generally an introduced tree in Ireland, there seems to be a strong case for considering it native in parts of the midlands, where it occurs in hedges, river-banks and on roadsides. Naturally regenerating populations have been found, for example, on the shores of Lough Ree, Co. Westmeath, and Lough Allen, Co. Leitrim (Reynolds, 2002). The largest known black poplar in Ireland is a specimen from Gloster, Roscrea, Co. Offaly, that is 30.5 m in height (and 4.46 m in girth) (Anonymous, 2005). The current Irish girth champion of this species is a tree growing at Organic Galloway Farm, Lecarrow, Drumkeeran, Co. Leitrim, that is 6.07 m in girth (and 22.3 m tall) (Anonymous, 2005).

Black poplar provides a useful timber but because of its rarity in the wild in Ireland it has probably only been used locally.

It is occasionally grown as an ornamental tree in parks and gardens.

Populus tremula

- Aspen (Scannell & Synnott, 1987); Trembling poplar; Quiggen espy (Hart, 1898); Quakin' esp (Praeger, 1938)
- *Crann creathach* (Scannell & Synnott, 1987); *Eabhadh* (Moloney, 1919); *Crann critheach* (Moloney, 1919; Dinneen, 1927; Uí Chanainn, 1962); *Cran-chrith* (de Vere Coneys, 1849); *Crain-cretha* (de Vere Coneys, 1849); *Eadha* (de Vere Coneys, 1849); *Crann crithir* (Dinneen, 1927); *Crann criotháin* (Dinneen, 1927); *Crithean* (Dinneen, 1927; Uí Chanainn, 1962); *Eadhadh* (Dinneen, 1927); *Eabhóg* (Dinneen, 1927; Uí Chanainn, 1962); *Fiodhad* (Dinneen, 1927; Uí Chanainn, 1962); *Poibleog chreathach* (Ó Dónaill, 1977)

SALICACEAE—THE WILLOW FAMILY

Native

Populus tremula.

Populus tremula is not a common tree everywhere in Ireland, although it is quite frequently found in some regions, particularly in rocky places and glens in the west and north, as well as in some woods, scrub and hedges. It rarely grows to be a tree of major size and is often accompanied by numerous suckers that grow around it.

This tree species provides useful light-coloured and light-weight timber that has several uses (although not necessarily in Ireland), including for clogs, floorboards, paddles and oars (Hickie, 2002). The soft timber is not used much in Ireland. Aspen timber has traditionally been used to make matches; it burns slowly and is apparently able to withstand rough usage without splintering, so it was formerly used for the bottoms of lorries, carts and barrows, and for brake blocks (Anonymous, 1941). Poplar wood, in general, is one of the few native timbers that is not recommended as firewood because of its poor burning characteristics (Warren, 2006). However, Rutty (1772: 123) wrote that 'the bark of young trees is sometimes used for Torches in the night'.

The large Iron Age pot, called the Altartate Cauldron, that is included in the collections of the National Museum of Ireland was found in a bog near Clones, Co. Monaghan, in around 1933. It is made from poplar wood (if it is of Irish origin, it is probably *Populus tremula*, aspen), with handles of yew and has been dated to approximately the 2nd century BC.

Aspen foliage was formerly used for fodder in Ireland.

K'Eogh (1735: 8) suggested the use of juice from aspen leaves to treat earaches and 'healeth ulcers of the same, being dropt into them. A decoction of the bark, provokes urine, relieves strangury [restricted urine flow, sometimes caused by kidney stones] and them that have the sciatica'.

Populus tremula is occasionally grown as an ornamental tree in parks and gardens. Non-native poplars grown in Ireland, mainly as ornamental trees, include:

Populus alba

- White poplar (Scannell & Synnott, 1987)
- *Poibleog gheal* (Scannell & Synnott, 1987); *Poibleag* (Cameron, 1883)

Populus ×canescens (Populus alba × Populus tremula)

- Grey poplar (Scannell & Synnott, 1987)
- *Poibleog liath* (Scannell & Synnott, 1987)

Populus deltoides

- *Poibleog ghallda* (Scannell & Synnott, 1987)

Populus ×canadensis (Populus deltoides × Populus nigra)

- Italian poplar (Scannell & Synnott, 1987); Black Italian poplar (Webb et al., 1996)

All of these non-native poplars occurring in Ireland were originally planted and may have become established by suckering into wild places.

Potamogeton species

- Pondweed (Scannell & Synnott, 1987)
- *Líobhógach* (Scannell & Synnott, 1987); *Duileasc na habhann* (Moloney, 1919); *Duilleóg féithe* (Moloney, 1919)

POTAMOGETONACEAE—THE PONDWEED FAMILY

Native

Despite the considerable diversity and relative abundance of some species of native pondweeds in Ireland I can find few references to the use of these aquatic plants. Elsewhere in the world *Potamogeton* species have been used for medicinal purposes: *Potamogeton natans* has been used in China to reduce fevers and against tumours and *P. pectinatus* to treat liver problems (Duke & Ayensu, 1985). Some *Potamogeton* species are also reputed to be edible but I have never tried them.

Preston (1995) provided a comprehensive review of the taxonomy and distribution of the group in Britain and Ireland.

**Potamogeton alpinus*

- Red pondweed (Scannell & Synnott, 1987)
- *Líobhógach dhearg* (Scannell & Synnott, 1987)

POTAMOGETONACEAE—THE PONDWEED FAMILY

Native

Potamogeton alpinus is mainly distributed in the northern half of Ireland and is an aquatic plant of slow-moving rivers and lakes that usually occurs in relatively shallow water.

*No ethnobotanical uses of this plant have been noted in Ireland.

**Potamogeton berchtoldii*

- Small pondweed (Scannell & Synnott, 1987)
- *Líobhógach bheag* (Scannell & Synnott, 1987)

POTAMOGETONACEAE—THE PONDWEED FAMILY
Native

Potamogeton berchtoldii is an aquatic plant of slow-moving rivers and lakes throughout Ireland.

*No ethnobotanical uses of this plant have been noted in Ireland.

**Potamogeton coloratus*

- Fen pondweed (Scannell & Synnott, 1987)
- *Liach eanaigh* (Scannell & Synnott, 1987)

POTAMOGETONACEAE—THE PONDWEED FAMILY
Native

Potamogeton coloratus is an aquatic plant of pools and ditches in limestone areas and is particularly and mainly distributed in the midlands.

*No ethnobotanical uses of this plant have been noted in Ireland.

**Potamogeton crispus*

- Curled pondweed (Scannell & Synnott, 1987)
- *Líobhógach chatach* (Scannell & Synnott, 1987)

POTAMOGETONACEAE—THE PONDWEED FAMILY
Native

Potamogeton crispus is an aquatic plant throughout Ireland of slow-moving rivers, lakes and other still water.

*No ethnobotanical uses of this plant have been noted in Ireland.

**Potamogeton filiformis*

- Slender-leaved pondweed (Scannell & Synnott, 1987)
- *Líobhógach chaol* (Scannell & Synnott, 1987)

POTAMOGETONACEAE—THE PONDWEED FAMILY
Native

Potamogeton filiformis is a rare aquatic plant in Ireland of lakes that is mainly distributed in the north, north-west and west.

*No ethnobotanical uses of this plant have been noted in Ireland.

Potamogeton friesii

- Flat-stalked pondweed (Scannell & Synnott, 1987)
- *Líobhógach leata* (Scannell & Synnott, 1987)

POTAMOGETONACEAE—THE PONDWEED FAMILY
Native

Potamogeton friesii is a rare aquatic plant in Ireland of canals and lakes that is mainly distributed in the centre, west and north-west of Ireland.

*No ethnobotanical uses of this plant have been noted in Ireland.

Potamogeton gramineus

- Various-leaved pondweed (Scannell & Synnott, 1987)
- *Líobhógach fhéarúil* (Scannell & Synnott, 1987)

POTAMOGETONACEAE—THE PONDWEED FAMILY
Native

Potamogeton gramineus is a relatively frequent aquatic plant of rivers and lakes throughout much of the northern half of Ireland; it occurs elsewhere, although rarely.

*No ethnobotanical uses of this plant have been noted in Ireland.

Potamogeton lucens

- Shining pondweed (Scannell & Synnott, 1987)
- *Líobhógach lonrach* (Scannell & Synnott, 1987)

POTAMOGETONACEAE—THE PONDWEED FAMILY
Native

Potamogeton lucens is an aquatic plant of pools, lakes and streams that is mainly distributed in the north and centre of Ireland.

*No ethnobotanical uses of this plant have been noted in Ireland.

Potamogeton natans

- Broad-leaved pondweed (Scannell & Synnott, 1987); Blackweed (Hart, 1898); Doolaman (Hart, 1898); Lady wrack (Hart, 1898); Slanlis (Hart, 1898)
- *Liach Bhríde* (Scannell & Synnott, 1987); *Liachroda* (Cameron, 1883); *Duileasg na h-abhan* (de Vere Coneys, 1849; Dinneen, 1927); *Liachbrighide* (de Vere Coneys, 1849; Dinneen, 1927)

POTAMOGETONACEAE—THE PONDWEED FAMILY
Native

Potamogeton natans is probably the commonest pondweed in Ireland and is found in many aquatic habitats, such as ponds, pools, canals, lakes, ditches and slow-moving streams and rivers.

In Irish a name for *Potamogeton natans* is *liach Bhríde* (Brigid's spoon) (Synnott, 1979), although it is not known if the name suggests a particular religious or symbolic usage in Ireland.

Hart (1898) noted it was used in Donegal to treat burns.

In other countries its starchy rootstock and turions (short, scaly branches off the rhizomes) have been used as food (Heywood, 1978), but there are no records of its use this way in Ireland.

**Potamogeton obtusifolius*

- Blunt-leaved pondweed (Scannell & Synnott, 1987)
- *Líobhógach mhaol* (Scannell & Synnott, 1987)

POTAMOGETONACEAE—THE PONDWEED FAMILY
Native

Potamogeton obtusifolius is an occasional aquatic plant of ditches, bog-pools, drains and ponds that occurs throughout Ireland and is more common in the north.

*No ethnobotanical uses of this plant have been noted in Ireland.

**Potamogeton pectinatus*

- Fennel pondweed (Scannell & Synnott, 1987)
- *Líobhógach fhinéiliúl* (Scannell & Synnott, 1987); *Duilleog féithe* (Uí Chonchubhair, 1995)

POTAMOGETONACEAE—THE PONDWEED FAMILY
Native

Potamogeton pectinatus is an aquatic plant of rivers, canals, streams, lakes and ponds, including near the coast and sometimes brackish water.

*No ethnobotanical uses of this plant have been noted in Ireland.

**Potamogeton perfoliatus*

- Perfoliate pondweed (Scannell & Synnott, 1987)
- *Dréimire uisce* (Scannell & Synnott, 1987)

POTAMOGETONACEAE—THE PONDWEED FAMILY
Native

Potamogeton perfoliatus is found quite commonly throughout Ireland and is an aquatic plant of streams, rivers, lakes and canals.

*No ethnobotanical uses of this plant have been noted in Ireland.

**Potamogeton polygonifolius*

- Bog pondweed (Scannell & Synnott, 1987)
- *Liach mhóna* (Scannell & Synnott, 1987); *Duileasc abhann móna* (Uí Chonchubhair, 1995)

POTAMOGETONACEAE—THE PONDWEED FAMILY
Native

Potamogeton polygonifolius is a common aquatic plant throughout Ireland of acid waters and other peaty habitats, such as bog-pools, drains, lakes and wet peat.

*No ethnobotanical uses of this plant have been noted in Ireland.

Potamogeton praelongus

- Long-stalked pondweed (Scannell & Synnott, 1987)
- *Líobhógach fhada* (Scannell & Synnott, 1987)

POTAMOGETONACEAE—THE PONDWEED FAMILY
Native

Potamogeton praelongus is a rare aquatic plant in Ireland of deep water in lakes and rivers that is mainly distributed in the northern half.

*No ethnobotanical uses of this plant have been noted in Ireland.

Potamogeton pusillus

- Lesser pondweed (Scannell & Synnott, 1987)
- *Líobhógach mhion* (Scannell & Synnott, 1987)

POTAMOGETONACEAE—THE PONDWEED FAMILY
Native

Potamogeton pusillus is a rare aquatic plant throughout Ireland of slow-moving rivers and lakes; it is more common in the north than in the south.

*No ethnobotanical uses of this plant have been noted in Ireland.

Potentilla anglica

- Trailing tormentil (Scannell & Synnott, 1987)
- *Néalfartach shraoilleach* (Scannell & Synnott, 1987)

ROSACEAE—THE ROSE FAMILY
Native

Potentilla anglica is a common plant of roadsides and banks throughout Ireland.

*No ethnobotanical uses of this plant have been noted in Ireland.

Potentilla anserina

- Silverweed (Scannell & Synnott, 1987); Tansy (Colgan, 1904); Brisken (Hart, 1898); Silver feather (Hart, 1898); Mashcorn (Praeger, 1938)
- *Briosclán* (Scannell & Synnott, 1987); *Bliogarán* (Ua Maoileoin, 2001); *Briosglán* (de Vere Coneys, 1849); *Fearbán*[210] (Dinneen, 1927; Uí Chanainn, 1962); *Blioscán* (Colgan, 1911); *Briosc* (Colgan, 1911); *Brioscán* (Colgan, 1911); *Brioslóg* (Colgan, 1911)

ROSACEAE—THE ROSE FAMILY
Native

Potentilla anserina is an extremely common plant throughout Ireland of roadsides, grasslands, seashores and waste places.

210. Dinneen (1927) defined *fearbán* as 'silverweed' in Clare and Dublin.

In the past it was an important wild food source, particularly during periods of famine and shortage. Akeroyd (1996) suggested that it may even have been a cultivated crop in Ireland. Mabey (1972) recorded that in parts of Britain the roots were cultivated as a crop from late prehistoric times, and in upland areas they were used up until the introduction of the potato. The plant has both fibrous roots and thicker storage roots, the latter of which are edible and which are an excellent source of starch and good to eat. They can be steamed, roasted or boiled. Rutty (1772: 39) noted they are of a sweet taste 'and said to be as grateful as the parsnip'. They can also be added to soups or stews and when dried can be ground to make a flour and baked. They are best collected in autumn but can be gathered throughout the winter into the spring. According to Hart (1898), the roots of this species were eaten by poor people in Donegal. In Scotland this plant was a particularly important carbohydrate source in coastal areas and used at the times of the year before the new potato crop could be harvested and when all of the previous year's grain had been used up (Milliken & Bridgewater, 2004). Cameron (1883: 19) wrote that its succulent roots were 'not unfrequently used by the poorer people in some parts of the Highlands [of Scotland] for bread'. Mears and Hillman (2007) noted that silverweed was a staple food source for a wide range of hunter–gatherer peoples in north-western North America and in Central Asia. Dinneen (1927) gave the Irish word *braoscán* for silverweed roots. Colgan (1911) gave the name *blucán* for the roots, from Clare Island, Co. Mayo, where he said they were roasted and eaten by boys.

Potentilla anserina (silverweed). Watercolour by Lydia Shackleton. Courtesy of the National Botanic Gardens, Glasnevin.

Potentilla anserina.

Potentilla anserina has a long history of use in traditional medicine in Europe, Asia and North America. According to van Wyk and Wink (2004), it has been used in an astringent to stop bleeding sores and for haemorrhoids, eczema, diarrhoea, menstrual problems, cramps, spasms and colic and for mouth and throat inflammations. Allen and Hatfield (2004) noted it was used in Ireland to treat diarrhoea, bleeding piles and heart troubles. Threlkeld (1726: unnumbered [17]) said that it was 'good against spitting of Blood and Diarrhaeas, the distilled Water is good against Freckles'. An infusion of the leaves was used to help relieve bleeding, ulcers and sore throats (Uí Chonchubhair, 1995). The same author noted that fresh leaves were used to line the boots of travellers to soothe their feet. It was used to treat epilepsy and sore eyes, and a tea made from it was useful against toothache and sunburn (Williams, 1993).

Potentilla erecta (Potentilla tormentilla)

- Tormentil (Scannell & Synnott, 1987); Nyamman (Hart, 1898); Nyammany (Hary, 1898); Sheep's napperty (Praeger, 1938); Tormenting root (Praeger, 1938)

- *Néalfartach* (Scannell & Synnott, 1987); *Néanthartach* (Ó Dónaill, 1977; Ua Maoileoin, 2001); *Benedín* (Moloney, 1919); *Leanartach* (Moloney, 1919; O'Regan, 1997); *Neamhnaidh* (Cameron, 1883; Moloney, 1919); *Lus na* (or *an*) *codlata* (Moloney, 1919; Dinneen, 1927); *Bárr braonán na gcon* (Moloney, 1919); *Mianfhartach* (Uí Chonchubhair, 1995); *Neamhain* (*Neamhna*) (de Vere Coneys, 1849; Dinneen, 1927); *Nealfhartach* (Dinneen, 1927); *Niamhnaid* (Ó Dónaill, 1977); *Néunartach* (Ó Criomhthain, MSS, n.d., ca. 1908)

ROSACEAE—THE ROSE FAMILY
Native

Potentilla erecta is a common plant of grasslands, heaths and banks throughout Ireland and is particularly abundant in the mountains and on higher ground.

This plant has been used in traditional medicine in parts of Europe to treat diarrhoea, dysentery and gastroenteritis; as a gargle to treat mouth and throat inflammations and externally to treat wounds and sores (van Wyk & Wink, 2004). According to Moloney (1919), it was used in Ireland with *Lysimachia nemorum*, 'yellow pimpernel', as a hypnotic for insomnia. It was also used traditionally in Ireland to treat liver problems and for burns and scalds (Egan, 1887; Allen & Hatfield, 2004). Uí Chonchubhair (1995) noted that the plant was used for its astringent and anti-inflammatory properties in treatments for humans and animals. Decoctions of the plant were used to treat foot rot in animals and as a gargle for a sore throat and sore gums and to help fasten loose teeth, as well as to soothe piles, a cracked anus or nipples. It was mixed with *Hypericum* species to help promote sleep and was also used to treat gluttony or morbid hunger (Williams, 1993). The same author recorded that it was used to make a drink given to patients; staunch blood flow in wounds; treat measles, diarrhoea and toothache; lessen 'gaga' (chapped skin) on hands and when its roots were boiled in milk it was a useful general medicine for children. In Donegal the whole plant was given as a decoction to treat diarrhoea (Hart, 1898).

According to Allen and Hatfield (2004), *Potentilla erecta* was used as a veterinary remedy to treat various ailments in animals in Ireland, such as diarrhoea in cattle and horses and red-water fever and white scour (Allen & Hatfield, 2004).

In the early 1700s, when the clearance of Ireland's forests was resulting in a shortage of trees for many purposes, the Irish Parliament awarded £200 to William Maple who discovered that leather could be tanned using tormentil roots[211]. In 1750 the Dublin Society (later the Royal Dublin Society) offered a premium of £10 to the person who tanned the most hides with tormentil roots and £8 to whomever gathered and sold the most tormentil roots to tanners (McCracken, 1971).

Remains of *Potentilla erecta* have been found in excavations at Fishamble Street, Dublin, that date back more than 1000 years, which may suggest that the species was used for some purpose because it would not have been a common weed at that time on the calcareous land surrounding Dublin (Geraghty, 1996).

Mahon (1983) listed this species as the source of a yellow dye and a red dye from the roots.

Potentilla erecta.

211. Maple's pamphlet was entitled *A Method of Tanning without Bark* (1729).

Potentilla fruticosa

- Shrubby cinquefoil (Scannell & Synnott, 1987)
- *Tor cúigmhéarach* (Scannell & Synnott, 1987)

ROSACEAE—THE ROSE FAMILY
Native

Potentilla fruticosa is a rare native shrub found growing on lake-shores and rocky habitats that are regularly flooded, such as turloughs. Shrubby cinquefoil occurs as part of the flora in the Burren, Co. Clare, and is also known from Connemara, Cos. Galway and Mayo. Elsewhere in the world, this hardy shrub thrives in cold temperate to Arctic regions throughout the Northern hemisphere.

I have seen dried leaves and stems of *Potentilla fruticosa* for sale in local markets in parts of Siberia where they are used to make tea. The species has also been used for such purposes in North America (Peterson, 1977).

Shrubby cinquefoil is commonly grown as an ornamental shrub in rock gardens and borders (Wyse Jackson, 2006), and a range of different cultivated varieties (cultivars) have been raised and grown in Ireland with variously coloured flowers, from white to rose-pink (Nelson, 2000b).

Potentilla reptans

- Creeping cinquefoil (Scannell & Synnott, 1987)
- *Cúig mhéar Mhuire* (Scannell & Synnott, 1987); *Cúig-mhéarach* (Dinneen, 1927; Webb et al., 1996)

ROSACEAE—THE ROSE FAMILY
Native

Potentilla reptans is common on roadsides and wasteground in the south and central parts of Ireland. It also occurs in the north but is rare.

This plant was formerly used to treat 'ague' (typically malaria) in Ireland (Maloney, 1972; Allen & Hatfield, 2004).

**Potentilla sterilis*

- Barren strawberry (Scannell & Synnott, 1987)
- *Sú talún bréige* (Scannell & Synnott, 1987); *Súgh-thalmhan seasc* (Moloney, 1919)

ROSACEAE—THE ROSE FAMILY
Native

Potentilla sterilis is a common plant of hedges, woodlands, and damp shady walls and banks.

*No ethnobotanical uses of this plant have been noted in Ireland.

Primula veris

- Cowslip[212] (Scannell & Synnott, 1987); Paigle (Harron, 1986)
- *Bainne bó bleachtáin* (Scannell & Synnott, 1987); *Bainne-bó-báine* (Moloney, 1919; Dinneen, 1927); *Fleasc i mbeal bó* (Moloney, 1919; Dinneen, 1927); *Seichearlan* (Cameron, 1883); *Seicheirghin* (Cameron, 1883); *Seicheirghlan* (Cameron, 1883); *Bainne bó bleacht* (Ó Dónaill, 1977; Uí Chonchubhair, 1995); *Baine bó bleacht* (de Vere Coneys, 1849); *Bó muc* (Dinneen, 1927); *Múiseán* (Dinneen, 1927); *Lusna múiseán* (Dinneen, 1927); *Seimhirín* (Dinneen, 1927)

PRIMULACEAE—THE PRIMROSE FAMILY

Native

Primula veris.

Primula veris is an occasional plant in the central plain of Ireland of old unimproved grasslands and especially roadsides. It is much rarer in the north-east and south-west than in the rest of the country. *Primula veris* is grown as an ornamental plant in Irish gardens and wildflower meadows (Wyse Jackson, 2006).

Cowslip is said to be narcotic and was used to treat insomnia (Moloney, 1919). It was formerly used for other ailments in Ireland too, such as palsy and dropsy. It was also used as a blood cleanser as well as in ointment for wrinkles and spots and even to cure deafness (Allen & Hatfield, 2004). Wood-Martin (1902: 187) quoted Lady Wilde (1890: 61) on the use of cowslip as a deafness cure: 'Take the cowslip roots, blossom and leaves, clean them well, then bruise and press them in a linen cloth, add honey to the juice thus pressed out, put it in a bottle, and pour a few drops into the nostrils and ears of the patient, he lying on his back. Then, after some time, turn him on his face, till the water pours out, carrying away whatever obstructives lay on the brain. This must be repeated for three days'. Elsewhere in Europe it has been used to treat colds and coughs, headaches and as a heart tonic (van Wyk & Wink, 2004). Boiled cowslip made into an ointment was also useful for sunburn and to remove freckles, as well as for the spots and wrinkles mentioned previously (Clyne, 1989).

Cowslip can also be made into wine. The following is a recipe from a 17th-century herbal:

> Having boil'd your Water and Sugar together, pour it boiling hot upon your Cowslips beaten, stir them well together, and let them stand in a Vessel close cover'd till it be almost cold; then put into it the Ye[a]st beaten with the Juice of Lemons; let it stand for two days, then press it out with as much speed as you can, and put it up into a Cask, and leave a little hole open, for the working; when it hath quite done working stop it up close for a Month or Six Weeks, then Bottle it. Cowslip Wine is very Cordial, and a glass of it being drank at night Bedward, causes sleep and rest. . . .' (Hartman, 1696).

Ó Céirín and Ó Céirín (1978) provided recipes for cowslip tea, wine and syrup (which

212. Moloney (1919) said that *Primula veris* and *P. vulgaris* have been 'indiscriminately styled' the following Irish names: *bainne-bó-bleacht*, *bainne-bó-buidhe*, *peacán buidhe* and *múiseán*. Dinneen (1927) gave these Irish names for primrose or cowslip: *péacán*, *béacán*, *peidhceán*, *peidreacán*, *padhacán*, and *buidheachán*.

they say tastes like honey and is delicious spread on bread or hot scones). The syrup is made by pouring a litre of boiling water over 700 g of cowslip flowers and leaving the mixture to stand for a day. After that, 400 g of sugar is added and it is boiled until a syrup consistency has been obtained. Young leaves are edible and can be used in salads and stuffing or to make tea (Hilton, 2007).

There are records of cowslip balls being made by children in Ireland and used as toys or decorations. Vickery (2010) mentioned that in Co. Louth these were called 'dodge-balls', where children would see how long they could keep them up in the air with their hands.

Primula vulgaris (Primula acaulis)

- Primrose (Scannell & Synnott, 1987); White summer (Hart, 1898; Colgan, 1904); May flower (Colgan, 1911); Fearkan (Hart, 1898); White may (Hart, 1898)
- *Sabhaircín* (Scannell & Synnott, 1987); *Ba bleachtas* (Ua Maoileoin, 2001); *Samhaircín* (Moloney, 1919); *Padhcán buidhe* (Moloney, 1919); *Buidheacán* (Moloney, 1919); *Soghradhach* (Cameron, 1883); *Baine bó bleacht* (de Vere Coneys, 1849; Dinneen, 1927); *Buidheachán bó bleacht* (Dinneen, 1927); *Seimhirín* (Dinneen, 1927); *Sobhrach* (*-bhairche*) (Dinneen, 1927); *Buíoc(h)án* (Uí Chanainn, 1962; Ó Dónaill, 1977); *Neóinín buidhe*[213] (Ó Criomhthain, MSS, n.d., ca. 1908); *Péacán buí* (Tóibín, 1965); *Sobhaircín* (Uí Chanainn, 1962)

PRIMULACEAE—THE PRIMROSE FAMILY
Native

Primula vulgaris × Primula veris.

Primula vulgaris is a common plant of hedges, shady banks, woodlands, coastal slopes and shaded montane cliffs in most parts of Ireland. This plant reaches its highest altitude (850 m) on Mount Brandon, in south Co. Kerry.

Roy Vickery (2010) included the English name 'goslings' for primrose flowers from Clonlara, Co. Clare, as they were in bloom when goslings hatched (see Plant-lore.com). Wade (1804) mentioned that the leaves of the 'common primrose' were used to feed silkworms, although presumably not in Ireland.

Primroses were formerly used in Ireland for a variety of purposes, such as an ointment for burns, cuts and skin ailments. It has been used to treat jaundice as well as stomach and toothache pains (to treat toothache the leaves were rubbed against the offending tooth) and also for insomnia (Allen & Hatfield, 2004). It has also been used to treat various nervous conditions (O'Rourke, 2006). Primrose tea traditionally has medicinal uses as well. In Co. Kildare the leaves were made into an ointment and rubbed on piles and in Co. Kerry they were rubbed on the face to cure razor cuts (Williams, 1993). The same author recorded that in Co. Kildare the roots were boiled to cure tuberculosis, and in Co. Westmeath the flowers were made into a paste with pig fat to treat burns.

In veterinary medicine, primrose has been used to treat coughs in horses. Crushed primrose roots were strained in breast milk or in the whey of goat's milk and then put into the horse's nose (Williams, 1993; MacCoitir, 2006).

Phillips (1983) suggested that primrose leaves can be eaten in salads or boiled as a green

213. *Neóinín buidhe* was given by Ó Criomhthain (MSS, n.d., ca. 1908) as 'yellow daisy (primrose)'.

pot herb. The flowers may also be fermented to make wine or ground with rice, almonds, honey and saffron to make a 'primrose pottage'. O'Rourke (2006) reported that it was used in the Aran Islands to prevent the fairies spoiling butter and milk, and on 1 May it was scattered on the threshold or tied to the tails of livestock to ward off evil spirits. O'Dowd (2011) refers to a similar custom from Kilmore, Co. Down, where up to the 1960s rushes were spread in front of the byre on May Eve, mixed with 'mayflowers'[214], *Primula vulgaris*, to bring good luck and a blessing on the cows and their milk for the coming year.

Primula vulgaris is still grown as an ornamental plant, particularly in woodland gardens (Wyse Jackson, 2006). Occasionally, plants with pink-coloured flowers are found in the wild, but a soft yellow is the usual primrose colour. Moutray Read (1916: 263) mentioned that it had become a custom in the early part of the 20th century for flower girls to sell primrose flowers on 19 April in Dublin and to wear them on that day, which he described as 'an instance of Calendar Custom in the making'. The anniversary of the death of the British politician Benjamin Disraeli is 19 April. The primrose was his favourite flower, and he was commemorated on what was called Primrose Day. The custom does not continue up to the present day. It is not clear to me when this practice died out.

Prunella vulgaris.

Prunella vulgaris

- Selfheal (Scannell & Synnott, 1987); Kinnaman beg (Colgan, 1904); Heart's ease (or Hearts' ease) (Hart, 1898; Colgan, 1904); Keanadha-hassog (Hart, 1898); Shashagh na creeagh (Hart, 1898); Touch-an'-heal (Praeger, 1938)
- *Duán ceannchosach* (Scannell & Synnott, 1987); *Coinnleoir* (Ua Maoileoin, 2001); *Lus an chroidhe* (Moloney, 1919; Dinneen, 1927); *Ceannabhán beag* (de Vere Coneys, 1849; Moloney; 1919; Dinneen, 1927); *Dubháinín an tseanchuis* (Moloney, 1919); *Dubhán ceannchasach* (Colgan, 1911; Moloney, 1919); *Té na ngarrdhanta* (Moloney, 1919); *Duainín an tseanchais* (Uí Chonchubhair, 1995); *Dubhan-ceanchosach* (*-ain-ceanchosaigh*) (de Vere Coneys, 1849; Dinneen, 1927) *Té na ngarrantaí* (or *na gcailleach*) (Dinneen, 1927); *Ténagarantai* (Colgan, 1911)

LAMIACEAE (Labiatae)—THE DEAD-NETTLE FAMILY

Native

Selfheal is a very common plant found throughout Ireland in pastures, meadows, lawns, gardens, roadsides, woods, hedges and waste places.

Allen and Hatfield (2004) recorded that selfheal was an important and widely used medicinal herb in Ireland for the treatment of respiratory problems, heart complaints, and sometimes to staunch bleeding, treat worms in children, cure eczema, back pain, piles and for other purposes. Various authors have written that a tea made from this species was recommended for 'weak' hearts (e.g., Hart, 1898—used to treat heart palpitations; Colgan, 1911; Moloney, 1919; Uí Chonchubhair, 1995). MacMahon (1994) recorded that a drink was made from it in Ballyheigue, Co. Kerry. Williams (1993) noted that it was used in Ireland to treat any ailment that happens quickly, like a stroke, and was also useful against

214. O'Dowd (2011) said that 'mayflowers' refer to *Primula vulgaris*. However, this name is more usually applied to *Caltha palustris*, the marsh-marigold, and I have some doubts as to whether it is correct to attribute this usage to the primrose.

fevers in children—the leaves were rubbed on their palms or the soles of their feet. Hart (1898) suggested that the name 'hearts' ease' was used in Co. Donegal for *Prunella vulgaris* (not for *Viola*), where an infusion of that plant was valued for the heart. Dinneen (1927) mentioned that it was used as a sedative.

Prunus avium

- Wild cherry (Scannell & Synnott, 1987); Gean (Harron, 1986)
- *Crann silíní fiáin* (Scannell & Synnott, 1987); *Sirís* (de Vere Coneys, 1849; Moloney, 1919; Dinneen, 1927); *Seilín fhiadhain* (Moloney, 1919); *Cran-siris* (de Vere Coneys, 1849); *Crann sirise*[215] (Dinneen, 1927); *Silín* (Dinneen, 1927)

ROSACEAE—THE ROSE FAMILY
Native

Prunus species are economically important worldwide and grown for their ornament, timber and fruits, such as almonds, apricots, cherries, damsons, nectarines, plums and peaches. Cultivated cherries are derived from *P. avium*, which gave rise to sweet or dessert cherries, and another species, *P. cerasus* (in Irish, *seilín*, Moloney, 1919), from which culinary or cooking cherries, such as morello cherries, have been selected (Howes, 1974).

The wild cherry is an occasional tree occurring in old hedges and woodlands throughout Ireland. Wild cherries generally form relatively small to medium-sized trees in Ireland, although the largest known specimen of *Prunus avium* is 28.5 m tall (and 2.37 in girth), from Gurteen le Poer, Kilsheelan, Co. Waterford (Anonymous, 2005). The current Irish girth champion is a specimen from Ardmulchan, Co. Meath, that is 4.15 in girth (and 15 m tall) (Anonymous, 2005).

It was an important wild fruit tree in early Ireland that is now more or less neglected in favour of cultivated forms with large, fleshy fruits. Wild cherry stones have been found, mostly from the Late Bronze Age, in various archaeological excavations (Kelly, 1997). Phillips (1983) provided recipes for wild cherry brandy, wild cherry soup and cherry wine. Homemade wild cherry brandy is made by half filling a bottle with cherries, adding some sugar and topping up the bottle with brandy. After a few months stored in a dark cupboard the cherry brandy is ready to drink.

Cherry jam can also be made with wild cherry fruits. The ripe fruits can be sieved to remove the stones or stewed gently until soft and then sieved. Don't overcook the mixture though. Although it is a bit tedious to do, the stones can be removed without sieving if a more chunky cherry jam is preferred. After this, the mixture is combined with some lemon juice (to help set the jam) and sugar. Once the sugar has dissolved, the mixture should be boiled rapidly until setting point has been reached. Then the jam is spooned into sterilised jam jars (see *Rubus fruticosus*, blackberry) and sealed.

Wild cherries that taste very bitter should be avoided as they may have a high content of the poison hydrogen cyanide, the chemical that gives almonds (*Prunus dulcis*) their particular flavour and smell.

Cherry wood is used in the making of furniture and is a dark reddish brown and very durable (MacCoitir, 2003). The wood is fine-grained and ideal for making turned objects or musical instruments, smoking pipes and for delicate cabinet-making purposes (Edlin, 1973). Rutty (1772: 123) said that 'the Wood is in request for handles of knives, and its

---

215. Dinneen (1927) defined *crann sirise* as 'cherry tree' and *silín* as 'a cherry, a cherry tree'.

Prunus avium.

colour is improved to an elegant redness by boiling in lye, or with quick-lime and urine'.

The bark was used to obtain a brown dye (Mahon, 1983).

A gum extracted from the bark is said to be useful for the complexion (John McLoughlin, pers. comm., 2008). Rutty (1772: 108) mentioned this property too and said that '*Cerasus sativa*. . .The Cherry-tree. . .yield Gum in great abundance'.

Prunus avium is grown as an ornamental tree for parks and gardens in Ireland generally as a cultivar with variously coloured, large or double flowers. According to Allen and Hatfield, it was little used in traditional medicine in Ireland.

Prunus padus

- Bird cherry (Scannell & Synnott, 1987)
- *Donnroisc* (Scannell & Synnott, 1987)

ROSACEAE—THE ROSE FAMILY

Native

Prunus padus is a rare small tree or shrub of damp woods and rocky places that occurs particularly in the north-west in Ireland. It is easily distinguished from other native cherries by having flowers in long drooping heads (racemes) and by its small black fruits. Its timber was probably used for various purposes, including furniture, veneers, wood turning and decorative panels (Hickie, 2002).

It is occasionally grown as an ornamental tree in parks and gardens. The commonest cultivated variety is cv. 'Watereri', with larger and denser drooping flower racemes.

Prunus spinosa

- Blackthorn (Scannell & Synnott, 1987); Sloe (Anonymous, 1978); Drean dhu (Hart, 1898); Slac (Hart, 1898)
- *Draighean* (Scannell & Synnott, 1987); *Áirne* (de Vere Coneys, 1849; Moloney, 1919; Ó Dónaill, 1977); *Droíghion* (de Vere Coneys, 1849); *Droíghneán* (de Vere Coneys, 1849); *Áirneog* (Dinneen, 1927; Ó Dónaill, 1977); *Cosán* (Dinneen, 1927); *Máthair an áirne* (Dinneen, 1927); *Straith*[216] (Dinneen, 1927); *Draighean dubh* or *donn* (Ó Dónaill, 1977); *Tom draighin*[217] (Ó Dónaill, 1977)

ROSACEAE—THE ROSE FAMILY

Native

Prunus spinosa is a small tree or shrub of woods, hedgerows, scrub, ditches and rocky

216. Dinneen (1927) defined *straith* as 'a sloe-bush'.
217. Ó Dónaill (1977) defined *tom draighin* as 'blackthorn-bush'.

places that can grow up to 4 m tall and is common throughout Ireland and Great Britain. *Prunus spinosa* also occurs throughout temperate Europe and has widely naturalised beyond its native range. Plants sucker readily and can sometimes form dense, almost impenetrable thickets with their much-branched spiny stems. In Ireland, it is one of the most important woody plants used for hedging (together with *Crataegus monogyna*, hawthorn). An ancient type of Irish field boundary, called a *nochtaile*, a type of fence, is described by Kelly (1997) and was known to have been constructed as early as the 7th century. The fence was constructed with a series of 4 ft stakes driven into the ground with a mallet, set about 8 in. apart. Three bands of rods were then woven between the stakes, one at the bottom, one at the middle and a third at the top. At the very top of the fence a band of blackthorn was woven into the fence, like an ancient barbed wire to make the fence stock-proof.

Prunus spinosa.

The fruits of *Prunus spinosa.*

Blackthorn is well known for its hard wood and was much used in the past for fuel and is still used today. It is still preferred for walking sticks and saw former uses as shillelaghs (*sail éille*) and cudgels (Lysaght, 1979). In Irish, *draighean duibh* refers to a blackthorn stick (Dinneen, 1927). In 19th-century Ireland, blackthorn sticks were favoured as weapons in faction fights (Vickery, 1995), but today shillelaghs are only made and sold as souvenirs. Traditionally the blackthorn stick was coated in butter and placed in the chimney to season, which is how it gained its usual black colour. Sometimes the thick end of the staff was hollowed out and filled with molten lead to make them heavier and a more powerful striking weapon (known as a 'loaded stick'). In some cases, a leather strap would be attached so that the shillelagh could be attached around the wrist and swung more effectively. Shillelaghs were also made from oak.

Blackthorn sticks were also used in the practice of 'stick-fencing', the practice of which survived in Ireland up until the 1840s, at least in southern Tipperary. Lyons (1943) described how medium-sized blackthorn sticks were used for duel-type fencing, which he suggested may have been a last survivor of an ancient Irish military exercise. There had even been a stick-fencing school in Cahir, Co. Tipperary, during the first quarter of the 19th century. In fencing, the blackthorn sticks were held at a quarter of their length from the handle end, the spare portion of the stick being a shield to the elbow in action.

Blackthorn walking sticks remain popular. The sticks are normally cut in winter, leaving a knuckle at the top to fit comfortably into the hand (Shaw-Smith, 1984). While clearing blackthorn scrub in Kerry some years ago I noticed that some of its stems arise vertically from thick horizontal roots, which can be cut to make a convenient handle, requiring very little moulding or bending to fit snugly in the hand. The sticks can also be steamed and made supple and then tied to a straight piece of timber and left to harden straight. It takes about a month to season such a walking stick before personal use. This species was also used in Britain for whip stocks (Edlin, 1973) and as the durable teeth of hay-rakes (Howes, 1974). A cut branch of blackthorn was used in some parts of Ireland as a simple harrow and dragged along the ground to remove the weeds in a field, or to help sow seeds (Lucas, 1958).

Ó Céirín and Ó Céirín (1978) and Mahon (1983) recorded that it was used to dye linen, yielding a light blue colour.

Prunus spinosa has been used in traditional medicine as a diuretic and astringent (van Wyk & Wink, 2004). Lysaght (1979) mentioned that its leaves were used in Ireland to treat indigestion and summer fever. It has also been used as a veterinary remedy to treat diarrhoea in animals in Wicklow (Allen & Hatfield, 2004).

In autumn, *Prunus spinosa* produces a crop of bluish black, bitter and acidic fruits (*grán muine*, Ó Dónaill, 1977); each drupe is covered with a blue bloom, looking like small plums. Moloney (1919) reported that sloes are considered unwholesome by Irish herbalists, as they are powerfully astringent.

In ancient Ireland Kelly (1997) suggested that sloes may have been used as an emergency food in times of hunger. Although the fruits are strongly astringent when eaten fresh, Milliken and Bridgewater (2004) mentioned that if dried and rehydrated they become much more palatable, and taste like stewed plums. The stony hearts of sloe fruits have been found as remains in excavations in Viking Dublin dating back at least 1000 years ago (Geraghty, 1996). The same author suggested that they may have been fermented to make fruit vinegar, or verjuice, which is generally made from crab-apples (see *Malus sylvestris*) or unripe grapes. In Viking Dublin the excavated contents of drains and cesspits included seeds of strawberries, blackberries, apples, rowan berries, sloes and plums (Anonymous, 1973). Sloes are occasionally collected today and used to flavour alcoholic spirits (sloe gin and port) or to make jellies and jam. Sloe jelly can be delicious but is best used in combination with a settling agent, such as crab-apples, which contain plenty of pectin. Sloe gin is a particularly good and warming after-dinner liqueur for winter consumption. To make this, sugar and sloes (with their skins pricked with a pin) are added to gin and left for some months to mature. Phillips (1983) provided a recipe for sloe wine and mentioned that its leaves were used to make tea in Ireland. Geraghty (1996) mentioned that in the 1980s a commercial firm making sloe gin in Co. Cork bought its sloes gathered by local schoolchildren. In Ireland, traditionally, sloe gin was considered good for the kidneys and used to make a decoction to treat worms in children (Allen & Hatfield, 2004).

In the Basque Navarra region of Spain sloes are used to make a liqueur called patxaran (Hilton, 2007). The liqueur is made by soaking the sloes in an aniseed-flavoured spirit (flavoured with a Mediterranean plant, *Pimpinella anisum*), with several coffee beans and a vanilla pod. The liqueur produced is a light reddish-brown colour, which is served cold as a digestif.

Ó Céirín and Ó Céirín (1978) also provided two recipes for sloe wine, one of them from West Cork that resembles a sweet port. One gallon (4.5 litres) of boiling water is poured over 1 gallon (4.5 litres) of sloes, which are then mashed, covered and left for a week and stirred occasionally before being strained. After resting and straining, 4–5 lb (2–2.5 kg) of sugar is dissolved into some of this must, which has been heated, and it is then added to the rest of the liquid. Once this mixture has cooled, 1 oz (25 g) of yeast is added and then it is placed in a fermentation jar with a fermentation lock that is fitted and left in a warm place for a month or so to ferment. More sugar can be added if it is too bitter. The wine should be left to clear for a month more before being bottled. The wine needs to mature for a year or more before it is ready to drink.

Vickery (1995) noted records of the Irish Folklore Commission that document the use of dried blackthorn leaves, known as Irish tea, as a tobacco substitute in Co. Carlow. Lysaght (1979) also mentioned such usage in Ireland.

Mears and Hillman (2007) described how sloes and other native stone-fruits can be made into 'fruit leathers'. The fruit is collected and crushed into a paste, then spread thinly onto sheets of bark or other material and allowed to dry. This process also cuts down the astringency and bitterness of the fruit and allows it to be preserved for storage. In this way such fruits would have been available in ancient times for use in winter when other food may have been scarce. If the leather is heated it helps to release the natural sugars in the

Prunus domestica.

The fruits of *Prunus domestica* subsp. *insititia*.

fruit and makes them sweeter to eat.

Other native species with fruits that can be preserved in this way include *Prunus avium* and *P. padus* as well as the non-native *P. domestica* (wild plum, Scannell & Synnott, 1987; bullace, Webb, 1977, Webb & Scannell, 1983; wild damson, Webb & Scannell, 1983) (in Irish, *boláiste*, Dinneen, 1927). Wild plums are very variable and found occasionally throughout Ireland, usually naturalised close to houses, often in hedges or the margins of woods. The wild plum is thought to have arisen originally as a hybrid between *P. spinosa* and *P. cerasifera*, the cherry plum (Sell, 1991). The fruits are excellent for making jam, wine, cordials, chutney, syrup and can be added to gin or vodka to make a liqueur (like sloe gin). The flowers of the wild plum are edible and have been used as a tea substitute. Flower buds, which taste mildly of almond, have also been added to salads. The seeds should not be eaten as they contain cyanogenic glycosides, which in the human intestine produce hydrogen cyanide and can be lethal (Hilton, 2007). The damson is closely related, known either as a variety of *P. domestica*, as var. *damascena*[218] or is recognised as a subspecies of the wild plum, as *P. domestica* subsp. *insititia* (Stace, 2010: 197).

Byrne (1907) described a tradition involving the use of sloes in Co. Roscommon on All Hallows Eve (October 31st). A cake was made in nearly every house and a ring, a coin, a sloe and a chip of wood were put into it, which caused great excitement. The coin meant riches and the one who got the ring would be married first, whoever got the chip of wood (which stands for a coffin) would be the first to die, and the sloe denotes the person who would live the longest, because the fairies are supposed to blight the sloes and haws and other berries on November night, so this would be the last eatable sloe of the year. Today the tradition lives on in Ireland with the barm brack, a fruit loaf that is traditionally made at Halloween and containing a ring. When I was growing up in the 1960s in Limerick and Kerry our favourite barm bracks were made containing, in addition to a ring (the finder would be wed within a year), a stick (to beat the wife—suggesting an unhappy marriage), a pea (the finder would not marry within a year) and a small rag (indicating poverty or bad luck). In the 1940s, Brian O'Nolan (Myles na gCopaleen) wrote an amusing piece of mock folklore about an old Gaelic custom for the *Irish Times* about barm bracks and their contents (translated from the original Irish into English by John Wyse Jackson and reproduced with his kind permission):

> Halloween is a great night of merry-making for the Irish. The Way of Knowledge. Especially in Corcadorky. With the coming of twilight the woman of the house

218. Dinneen (1927) and Uí Chanainn (1962) gave the word *daimsín* for a damson plum.

> places her husband's trousers into a pot of water and puts the pot down onto a large turf fire. She leaves it alone and begins making the barm brack for the supper. She adds to it a large variety of different objects: for example, a baby's button, pens, pencils, old spent matches, clay pipes, whole crow's eggs in the shell, small knobs of coal, brass nails, hazelnuts, postage stamps, income-tax forms, a needle, shreds of tobacco, ladies' lipsticks, sodlets of turf, horse's teeth, robin's feathers, Indian meal, potato sprouts, and many other things that are not those things. The woman of the house places the barm brack in another pot and puts that onto the fire as well. At midday the youngest son of the family lies flat on the floor at the lower end of the house and starts to screech at the top of his voice, uttering these words as follows: "Go out, O Hunger, go out, O Hunger, and stay from this house!" Outside, the eldest son is on top of the dung-heap. In through the window he throws dung onto the floor, and responds to the presence within: "Come out, O Hunger, come out, O Hunger, and get into this heap!" At two o'clock the man of the house gets up and asks his wife for tidings of his trousers. She recites a lengthy Gaelic lay, takes the trousers out of the pot with a pair of tongs, and offers them to her husband. He examines them to make sure they have been well boiled. If they have, he puts them on (after they have cooled down), and says: "Begone, O Hunger, begone, O Hunger, and come not close to my trousers." At midday, the younger generation packs into the kitchen with the older generation at its heels, and there is dancing, merriment and high roistering until morning. I myself have never been in Corcadorky during Halloween. Thanks be to God.

In earlier times the brack would also include a small coin (originally a silver sixpence), which suggested that the finder would be rich and have good fortune.

**Pseudorchis albida*

- Small-white orchid (Scannell & Synnott, 1987)
- *Magairlín bán* (Scannell & Synnott, 1987); *Magairlín bán an tsléibhe* (Moloney, 1919)

ORCHIDACEAE—THE ORCHID FAMILY
Native

Pseudorchis albida is a rare orchid in Ireland that is found only in upland heaths and grasslands.

*No ethnobotanical uses of this plant have been noted in Ireland.

**Puccinellia distans*

- Reflexed saltmarsh-grass (Scannell & Synnott, 1987)
- *Féar muirisce crom* (Scannell & Synnott, 1987); *Cuise sgaoltebhlathach*[219] (White, 1803)

POACEAE (Gramineae)—THE GRASS FAMILY
Native

Puccinellia distans is a coastal grass of salt marshes and bare ground near the sea. It is rare on the coasts of the north and west but locally common elsewhere.

*No ethnobotanical uses of this plant have been noted in Ireland.

---

219. White (1803) applied this name to '*Poa distans*'.

*Puccinellia fasciculata

- Borrer's saltmarsh-grass (Scannell & Synnott, 1987)
- *Féar muirisce triopallach* (Scannell & Synnott, 1987)

POACEAE (Gramineae)—THE GRASS FAMILY
Native

Puccinellia fasciculata is a very rare grass in Ireland of brackish and saline muddy habitats in the south and east.

*No ethnobotanical uses of this plant have been noted in Ireland.

Puccinellia maritima (Poa maritima)

- Common saltmarsh-grass (Scannell & Synnott, 1987); Murvagh (Hart, 1898); One-pointed grass (Hart, 1898)
- *Féar muirisce* (Scannell & Synnott, 1987); *Cuise muruisc* (White, 1803)

POACEAE (Gramineae)—THE GRASS FAMILY
Native

Puccinellia maritima is a common coastal grass of salt marshes and muddy and rocky seashores all around the Irish coast.

It was used as a veterinary remedy as a mild purge for cattle in Co. Donegal (Allen & Hatfield, 2004).

Pulicaria dysenterica

- Fleabane (Scannell & Synnott, 1987)
- *Lus buí na ndreancaidí* (Scannell & Synnott, 1987); *Lus buidhe na ndeanathairtí* (Moloney, 1919); *Lus na gcoscart* (Moloney, 1919; Dinneen, 1927); *Dreancaid díbeartach* (Moloney, 1919); *Lus na ndreancaidí* (Uí Chonchubhair, 1995); *Luibh* (or *Lus*) *na ndeargnataí* (Dinneen, 1927)

ASTERACEAE (Compositae)—THE DAISY FAMILY
Native

Pulicaria dysenterica is an occasional plant throughout Ireland that is commonest in the south. It occurs in wet meadows, ditches and marshes and on roadsides.

The name *Pulicaria* is derived from the Latin word *pulex*—a flea (the human flea is *Pulex irritans*). Mackay (1836: 144) stated that 'on the authority of General Keith, that it cured the Russian army of dysentery—hence the specific name'. If it is an effective deterrent for fleas, I imagine it was formerly collected and included in bedding, although I have not found direct evidence of this use in Ireland.

*Pyrola media

- Intermediate wintergreen (Scannell & Synnott, 1987)
- *Glasluibh bheag* (Scannell & Synnott, 1987)

ERICACEAE—THE HEATHER FAMILY
Native

Pyrola media is mainly distributed in the north in woods, moors and heaths.

*No ethnobotanical uses of this plant have been noted in Ireland.

Pyrola minor

- Common wintergreen (Scannell & Synnott, 1987)
- *Glasluibh mheánach* (Scannell & Synnott, 1987)

ERICACEAE—THE HEATHER FAMILY
Native

Pyrola minor is mainly distributed in the north in woods, moors and heaths.

Mabey (1972) noted that in America the young leaves of this plant are used as a spicy aromatic salad vegetable and in tea. The fruits are also edible and taste similarly spicy. There are no records of it being used for food in Ireland, but the fruits in particular would no doubt have been gathered as valuable winter food by early Irish inhabitants whenever they occurred in any abundance.

**Pyrola rotundifolia* subsp. *maritima*

- *Glasluibh chruinn*

ERICACEAE—THE HEATHER FAMILY
Native

Pyrola rotundifolia subsp. *maritima* is very rare in Ireland and distributed in Co. Wexford only in damp patches on sand dunes.

*No ethnobotanical uses of this plant have been noted in Ireland.

**Pyrola rotundifolia* subsp. *rotundifolia*

- Round-leaved wintergreen
- *Glasluibh chruinn*

ERICACEAE—THE HEATHER FAMILY
Native

Pyrola rotundifolia subsp. *rotundifolia* is rather rare in Ireland and distributed mainly in the centre in fens and wet bogs.

*No ethnobotanical uses of this plant have been noted in Ireland.

Quercus species

- Oak (Scannell & Synnott, 1987)
- *Dair* (Scannell & Synnott, 1987); *Rí na coille* (Moloney, 1919); *Cran-daraigh* (de Vere Coneys,

1849); *Crann darach (daraighe)(daróige)* (Dinneen, 1927); *Railach*[220] (Dinneen, 1927); *Tuilm*[221] (Dinneen, 1927)

FAGACEAE—THE BEECH, OAK AND CHESTNUT FAMILY

Native

Including:

Quercus petraea (Quercus sessiliflora)

- Sessile oak (Scannell & Synnott, 1987); Irish oak[222]
- *Dair ghaelach* (Scannell & Synnott, 1987)

Quercus robur

- Pedunculate oak (Scannell & Synnott, 1987); English oak
- *Dair ghallda* (Scannell & Synnott, 1987)

Quercus species.

There are two native oak species in Ireland, *Quercus petraea* and *Q. robur*. *Quercus petraea* was a dominant tree of western and mountainous areas and is now frequently found in most regions of the country. *Quercus robur* was also a dominant tree of woodlands in Ireland but occurs more frequently in midland and lowland areas today. Indeed, hybrids between the two oaks are extremely frequent and often intermediate in morphology between the two (*Q. ×rosacea*).

Oak woodland was the most important and widespread vegetation in Ireland before its clearance began through human settlement. Significant forest clearance apparently began in Ireland with the arrival of Neolithic man about 5700 years ago, and by AD 300 at the latest, most of the land was permanently deforested (Kelly & Fuller, 1988). The woodlands provided rich natural resources for Ireland's first human inhabitants—timber for dwellings, habitats for animals and birds, fuel for fires and a diversity of plant-derived products for food, fibres, weapons, medicines and many other uses. The destruction of Ireland's native forests continued through the Middle Ages up to the 17th century when most of the extensive remaining tracts of native woodland had been cleared. E. Estyn Evans (1957: 13) noted the resulting loss of oak forests in *Irish Folk Ways*: 'With the passing of the oak forests a wealth of woodcraft and tree lore perished and a wide range of skills declined.' Without woodlands to provide the natural resources they had formerly enjoyed, rural people turned instead to exploiting timber preserved in the bogs (mainly bog oak, pine and yew). After centuries of felling for timber, tanning, smelting ores and clearing oakwoods for farmland, extensive woodlands of these two native oak species are now few and far between in Ireland. Very few sites remain in Ireland that can be considered 'ancient woodland' (sites continuously wooded since AD 1600 or earlier) (Kelly & Fuller, 1988). Nevertheless, oak remains one of the most highly prized and most widely used trees for timber. Its wood is hard, heavy and long-lasting and polishes well to make attractive and robust furniture. Because it is so hard, particular techniques

Quercus petraea.

220. Dinneen (1927) defined *railach* as 'an oak tree'.

221. Dinneen (1927) defined *tuilm* as 'vagina, or oak or elm'.

222. The name 'Irish oak' is sometimes applied to this species (Nelson & Walsh, 1993), especially in the context of it being designated as the national tree. Strictly speaking there is no oak that can be defined as being exclusively Irish.

A traditional spinning wheel made of oak.

A reconstructed crannog at Craggaunowen, Co. Clare.

have been used for constructing with oak, especially when it is 'green' or unseasoned. Oak wood is generally too hard to be suitable for the use of nails, which will often corrode due to the high content of tannins in the wood. As such, oak objects were traditionally put together with joints and pegs rather than nails.

Some of the earliest surviving traditional Irish furniture was made from oak, although according to Evans (1957: 85), not much native oak furniture was being produced by the end of the 17th century: 'Good craftsmanship in wood is so rarely seen in the Irish peasant house that one easily notices a piece of quality furniture. The chances are that it has come out of 'the big house' or, near the coast, from a wrecked ship. Very little oak furniture survived from the days before the famine, and what there is likely to be bog oak, that is timber dug out of the bogs. The truth seems to be that by the time improved styles of furniture and settled conditions of life came about, in the second half of the 17th century, the native oak woods had been almost everywhere destroyed, and the common furniture came to be made of bog wood or imported soft woods. The gentry favoured mahogany'.

Other household items, such as wooden noggins in which milk and porridge were taken, were made from oak timber (Evans, 1957). Noggins were made from stakes of oak, with one projecting beyond the top of the vessel to act as a handle. The staves and a base piece were held together with a surrounding band of ash wood, joined by interlocking tongues. They were made by skilled craftsmen called noggin-weavers and used instead of earthenware pottery, for which in Ireland there is little tradition (Evans, 1957). Early milk churns were made from oak staves, similar to narrow barrels, held together with hoops of hazel, willow or rowan (*Sorbus aucuparia*; rowan hoops were considered lucky) (Evans, 1957). Oak wood was also used to make butter-firkins in Co. Cavan (Ní Sheasnáin, 1984). In addition, thin sheets of split oak timber were used to make baskets, although there is currently no one in Ireland making such objects with oak (Joe Hogan, pers. comm., 2010).

Oak was also used for early habitations for some of Ireland's first inhabitants. Wakeman (1891) described early lake-dwellings, called *crannógs* (in English, crannogs), which have been found in lakes and bogs throughout Ireland. These dwellings were essentially small defensive islands built slightly offshore, constructed of a series of oak posts driven into the lake bed on which platforms of oak logs were built, surrounded by oak plank walls and covered by a thatched roof. Sometimes a series of crannogs were constructed, linked by raised walkways between them. Today the remains of crannogs can still be seen in some places, such as Lough Gur in Co. Limerick. As Wakeman (1891: 241) noted, 'A slight elevation on the surface of a bog, some bleached sprays of birch, ash or sallow, or the appearance of a few gray or white stones—reed surrounded, and rising a few inches above the surface of a loch—will often to the practised eye indicate the position of a "crannog"'. Although originally built over water, many crannogs have become submerged and buried in the bogs that grew up in these basins.

The remains of great oak logs sunk into the bed of the River Shannon were found recently at Clonmacnoise. These once formed a long bridge spanning the river at Clonmacnoise that was built from oak trees dated by dendrochronology to about AD 800 (Moore,

1996). Trackways and causeways of round logs (often oak) laid side by side on a bog surface have also been found in several locations, built as, in effect, floating tracks, roads or walkways through wet bogs. In Irish these causeways are called *tóchar* (Dinneen, 1927), a word that is quite common in placenames in Ireland. Kelly (1997) reported such a trackway discovered during turf cutting at Corlea, Co. Longford, that is 1 km long and composed of oak logs laid on top of brushwood; it has been dated to 150 BC. Before widespread draining of Ireland's bogs, some of the wettest bogs, particularly in the Midlands, would have been exceedingly dangerous to cross and were significant barriers to the movement of people, vehicles and stock. The oldest major roads are generally recognised as having followed the great esker ridges of the Midlands, which are raised gravel ridges, deposited as glacial stream-beds under the ice. The great highway, or Esker Riada (*Slí Mór*), is reputed to have followed the line of eskers across the central plain of Ireland from Dublin to Galway (Kelly, 1997).

A 'Galway hooker' in Inisheer, Aran Islands.

Other common uses for oak timber included ship-building and making barrels (McCracken, 1971). The curved staves of barrels were made from oak, the number of staves was constant depending on the type of barrel and its size: a pipe (105 Imperial gallons) had 26 staves, a hogshead (52 Imperial gallons) had 25 staves and a barrel (32 Imperial gallons) had 23 staves. The staves were bound into shape by hoops made from hazel or willow until the 19th century, after which they were replaced by metal hoops. Oak was an ideal timber for ship-building, not only because it is very strong and long-lasting but also because the growth form of its branches allows timber to be cut in curved shapes following the grain and giving even stronger timber. It is widely prized for boat-building and used in the construction of a great variety of traditional boats all around the Irish coasts (MacCárthaigh, 2008). Oak would often be used to make the boat frames and then the lighter planking fashioned from 'deal' (pine), spruce or larch. Larger boats such as the 'Galway hookers' were generally constructed of oak timbers, although even as early as the 1820s much of this timber was being imported, presumably because Irish oak timber had become scarce. Specifications for hookers qualifying for a £10 bounty[223] issued by the Fisheries Bounties Commission, 1828–1832, said that white oak and black birch, both timbers from North American trees, should be used in their construction (P. Barry & D. Scott, in MacCárthaigh, 2008).

Oak ship construction in the Lelystad shipyards, The Netherlands.

Oak would also have been frequently used to make dug-out canoes (logboats), which were commonly used in Ireland from the earliest times, probably from at least BC 2000 onward (Teahan, 1994) until the 17th century (Lucas, 1957). In 1978 a large dug-out oak boat of unknown age was discovered in the mud in the Cashen River estuary in Co. Kerry.

223. These bounties were government incentives to promote boat-building, thus helping fisheries development in Ireland.

An illustration of a dugout canoe from a *crannóg* in Ballinderry, Co. Westmeath. Reprinted with permission from Mason, Thomas H. 1936. *The Islands of Ireland*, Batsford, London.

A Lurgan logboat. Illustration from Wakeman, William F. 1898. *Archaeologica Hibernica. A Hand-Book of Irish Antiquities*. Hodges, Figgis and Co., Dublin.

Construction of a currach. Courtesy Ionad an Bhlascaoid Mhóir/The Blasket Centre.

It was 5.97 m long and up to 86 cm wide, hewn from an oak log that had been split longitudinally and which may have been used as a ferryboat (Kelly, 1981). A remarkable logboat is included in the collections of the National Museum of Ireland in Kildare Street. It was found in 1902 in a raised bog during drainage operations in the townland of Addergoole, Lurgan, Co. Galway. Over 15 m in length and more than a metre wide, this is the largest intact logboat ever discovered in Ireland and has been dated to 3940 ±25 BP, a period that was transitional between the Neolithic and Bronze Ages (Mitchell, 1986). A 15 m length is a truly remarkable size when one considers that the trunk of many mature oak trees in Ireland makes up approximately half a tree's stature. Thus, the Lurgan logboat would have been cut from a tree that was probably more than 30 m in height. The current champion oak trees are, for *Quercus robur*, 37 m in height and 5.88 m in girth from Tynan Abbey, Tynan, Co. Armagh, in Northern Ireland and, for *Q. petraea*, 37 m in height and 6.54 m in girth from Coolattin Golf Club, Shillelagh, Co. Wicklow (Anonymous, 2005). The current girth champion tree of *Q. petraea* is a specimen from Northern Ireland on the Abercorn Estates, Barenscourt, Newtownstewart, Co. Tyrone, that is 8.35 m in girth and 24.2 m tall, and for *Q. robur*, a specimen in Charleville Forest, Tullamore, Co. Offaly, that is 8.82 m in girth and 20 m tall (Anonymous, 2005). An exceptional specimen of *Q. robur* measured in 2005 in Mote Demesne, Co. Roscommon, was 6 m in height, with a girth of 7.18 m (Anonymous, 2005).

Some of Ireland's most ancient boats were simple craft constructed with woven lengths of willow (*Salix* spp.) or hazel (*Corylus avellana*) as ribs and laths (stringers), covered with animal skins. Early records from Britain and Ireland indicate that boats covered with stitched skins were used for various purposes. For example, Geraldus Cambrensis described such boats in his description of Ireland, *The History and Topography of Ireland*, written ca. 1185 (O'Meara, 1982). These boats are called currachs throughout much of Ireland (alternatively spelled curachs or curraghs, and known as coracles in Wales from the Welsh *corwgl*), and called *naomhógs* in the Dingle Peninsula, Co. Kerry. While many of these boats were small and used for in-shore fishing and other purposes, and on inland waters, larger sea-going vessels were also constructed. One such boat is described in St Brendan's 6th-century account of his voyage in search of a promised land, *Navigatio Sancti Brendani*. This voyage took him at least as far as Iceland and the Faroe Islands, and perhaps even to North America, in a large currach capable of carrying more than a dozen companions. St Brendan's boat was made with wickerwork sides and ribs, covered with cowhide and tanned with oak bark.

There seems to have been an unbroken tradition in using such boats in Ireland from earliest times. Nevertheless, when wood was scarce in coastal areas there may also have been a move toward the construction of these small inexpensive boats for in-shore fishing. While they were traditionally covered with animal skins (cow, horse and perhaps deer hides) more recently these have been replaced by tarred canvas and even fibreglass reinforced polyester resin, with timber only being used for the frames. Oak would have been commonly used for the frames of such boats. The shapes and styles of currachs varied by region along the coasts of Ireland but all were characterised by having a rounded hull without a keel. These were very effective seaworthy craft for local people to use to transport themselves, supplies and stock along the coast and between islands and the mainland, as well as for fishing. They are light (at least lighter than dug-out canoes or clinker-built timber boats), manoeuvrable and stable in heavy seas. They are also fast and, depending on their size, they could accommodate between one and four oarsmen. However, great skill is required to manage a currach safely and effectively. Split oakwood was used to make narrow beams for use in the currach construction. The oakwood was steamed to make it flexible so that it could then be bent into hoops or ribs to form the hull. Other timbers would also have been used in currach construction, particularly for the ribs, such as ash (*Fraxinus excelsior*), hazel (*Corylus avellana*), willow (*Salix* spp.) and even driftwood. 'Deal' (Scots or bog pine, *Pinus sylvestris*) was generally used for the gunwales (Clare Champion, 2002).

A currach and typical oars at Brandon, Co. Kerry.

Carved bog wood box.

Oak was probably the main timber used in Ireland to make charcoal, and woodlands in some parts of Ireland were either clear-felled or heavily harvested to feed the supply of charcoal needed for iron smelting. Charcoal is one of the most ancient wood-derived products. It was widely used for metallurgical and chemical industries as a fuel as it burns to give an intense heat. It has also been used for its absorptive properties and as such can be eaten as a treatment for poisoning. Larner (2005) mentioned that 25-year-old oaks make the best charcoal and described the damage to Killarney woodlands in Co. Kerry from the extensive ironworks in that region. Other tree species including alder, arbutus, ash, birch, hazel, holly and yew were also used for charcoal making. Charcoal is made by heating timber in the absence of air to burn off impurities and leave a heap of carbon. The wood is cut in winter or early spring and split into lengths up to 1 m long and left to dry. It is then placed in a tight stack about 2 m high and 5 m wide before being covered with sods and sealed with damp clay and ashes. Burning the wood stack took several days during which time it was carefully monitored and damped down whenever necessary to ensure that no flames were produced while at the same time making certain that it continued to burn slowly. Larner (2005) recorded that a tonne of wood would produce about a hundredweight of charcoal and that 1.3 tonnes of charcoal was needed to reduce 2.5 tonnes of iron ore into a tonne of cast iron.

Bog oak timber, preserved sometimes for thousands of years in peat, has been a valuable and valued timber resource for centuries in Ireland. Sharkey (2000) noted that bog oak was used to make rafters for thatched cottages in Ireland, as it is durable and strong. Dinneen (1927) gave the Irish word *círín* for the 'trunk of an oak tree embedded in a bog'. In addition, carved objects made from bog oak (and other bog timbers) were made as early tourist souvenirs, which today sell for high prices in antique shops. Crane (1907) said that Lady Kenmare had established a flourishing wood carving school in Killarney, Co. Kerry. The National Botanic Gardens' museum collections include bog oak jewellery, hand-carved tobacco pipes and even a decorative skull and crossbones carved from bog oak. The trade in objects made from bog oak for the tourist market thrived particularly from the 1820s up to about 1920, but reached its peak in the 1850s and 1860s (Teahan, 1994). These objects included bracelets, brooches, earrings and buttons, as well as ornaments, such as miniature carved Irish harps; also functional objects, such as work boxes, card cases, candle sticks and book stands. Almost all of them included shamrocks as part of their carved decorations. Bog oak was also used to make furniture, cranes for the fire (Evans, 1957) and sometimes for inlays. In Co. Armagh, bog oak was commonly used to make fire cranes (Lysaght, 1979). Bog oak often has an attractive black colour similar to that seen in tropical ebony woods.

To locate bog timber (such as bog oak, pine and yew), one probed the bog with rods. Timber was easiest to find on a dewy morning; dew would not be formed on the spot above where the bog timber would be found (O'Hare, 1993). Bog timber was most widely used for making furniture during the 19th century; bog oak was said to become very dark when exposed to the air and be both very hard and possess great strength (Teahan, 1994).

Oak was also used for food, and acorns were dried and ground into flour for baking, with the flour first washed in hot water to remove the tannins and bitter flavour. Eaten raw, the acorn kernels are very bitter and barely edible, unless the tannins they contain are leached from them. Mabey (1972) described how acorns were buried, sometimes with ash or charcoal, and watered from time to time until they became sweet after the bitter tannins leach. Acorns can also be used as a coffee substitute if roasted, and for such purposes Mabey noted that the kernels were chopped and roasted until they were light brown in colour, then ground up and roasted again. Chopped and roasted they can also be used as a substitute for almonds (Mabey, 1972). The tannins can be removed from crushed acorn kernels by placing them in a porous bag and leaving them for a week or two in running water (such as a small stream). A quick method to remove bitterness is to grind the acorns and then repeatedly run hot water through the grinds in a coffee percolator or filter machine. Oil can be obtained from acorns. In North America, many tribes of Native Americans used acorns extensively for food, including oil extracted from them. To obtain the oil the acorns were pounded into flour and boiled in water containing maple-wood ashes, after which the oil was skimmed off the surface. The remaining flour was used in breads and cereals (Kavasch, 2005). According to Phillips (1983), oak leaves can be used to make a white wine.

Old Irish literature contains numerous references to pigs fattened on acorns (*dercain*, Kelly, 1997; *darcain*, Dinneen, 1927; *meas na daraighe*, Moloney, 1919; *dearcán*, Moloney, 1919) in woodlands. They were probably also used as food for other domestic animals too. However, if eaten in excess acorns (and oak leaves) can poison horses and cattle, but not pigs (Howes, 1974).

Oak bark in Ireland was an extremely important source of tannins for the tanning of leather, and the tannins also occur in the leaves and acorns. Good quality oak bark contains 12% to 14% tannin, by air-dried weight (Howes, 1974). Tannins in the bark combine with the collagen proteins in the animal skins to produce leather. Larner (2005) noted that in 1736 the cost of bark harvested in Killarney, Co. Kerry, was £2–10 shillings a tonne, and the bark from 320 twenty five-year-old oak trees went to make up a tonne of bark. It is no wonder that so many of Ireland's native oak woods disappeared quickly under the onslaught of harvesting for timber, charcoal-production and tanning. Oak bark was particularly good for producing heavy leathers, such as for the soles of shoes. The main drawback of vegetable tanning with tannins was the slowness of the process, measured in weeks in comparison to mineral or chrome tanning of less than a day. Oak bark was also used to treat sails of sailing boats (C. Ó Siocháin, in MacCárthaigh, 2008).

After oak was used for tanning, the waste bark was then used for fuel. According to Rutty (1772: 124),'What remains in the pit after the bark has been used for leather, being trod fine, and made up into a solid mass like turf, and dried, makes a sweet winter firing, and is well known in Dublin and Paris as such'.

Because of the tannins it contains, oak bark has been used medicinally to treat inflamed or chapped skin, chilblains, sprains, diarrhoea, toothache, neuralgia, ringworm and pin-worms, as well as in foot baths to reduce foot odour and as a gargle or mouthwash (Allen & Hatfield, 2004; Reader's Digest Association, 2006). McGlinchey (2007) mentioned that oak bark was a great cure for sore feet. The bark was steeped in water for a day or two and the water was good for hardening the skin of the feet in the summer.

Oak also was used as a veterinary remedy to treat harness sores in horses and diarrhoea in cattle in different parts of Ireland (Allen & Hatfield, 2004).

Oak bark can be use to produce a black dye (Mairet, 1916; Mahon, 1983; MacCoitir, 2003), and when combined with various metallic salts it will yield a variety of shades from brown, purple, yellow and black (Edlin, 1973). Sharkey (2000) mentioned that wool can be dyed a glossy black through the addition of oak chips to the dyeing pot. A dye extracted from oak galls has been used as a hair dye in Europe since antiquity and Pliny recorded its use for this purpose in about AD 50 (Lewington, 1990). Oak galls (or 'oak apples') contain large quantities of tannins and are formed by oak trees around the grubs of cynipid wasps, which attack their buds. Oak galls have been used to make inks used in ancient Irish manuscripts. An early Christian Psalter found recently in Faddan More bog in Co. Tipperary had been written with inks made of oak galls and iron on vellum. The oak gall-based ink had acted as a tanning agent to preserve the written text left for hundreds of years in the bog (Leavy, 2007).

These stately trees are much used as ornamental species planted in parks and larger gardens throughout Ireland. Oaks and other trees are important too for providing shelter, both to afford protection from wind and rain and as shelter-belts, planted to protect a house, garden or crops from wind, which in Ireland is often the prevailing wind from the south-west.

Apparently, this is a traditional autumn children's game in Ireland, although it is not one I remember from my own childhood in Limerick or Kerry. My son Thomas told me that he and his friends have amused themselves trying to catch leaves falling from an oak tree, as it was said to be good luck to catch one.

Cameron (1883) gave *craobh thuilm*[224] or *darach sior-uaine* as Irish names for the non-native *Quercus ilex*, the evergreen oak or holm oak. He also gave *crann airc* as an Irish name for *Q. ilex*, the cork oak. Ó Dónaill (1977) gave the Irish words *crann thoilm*, *craobh thoilm* or *dair thoilm* for 'holm-oak'.

**Radiola linoides*

- Allseed (Scannell & Synnott, 1987)
- *Gathán lín* (Scannell & Synnott, 1987)

LINACEAE—THE FLAX FAMILY

Native

Radiola linoides is a tiny annual plant that is rare in Ireland except in parts of the west, growing in open habitats on damp, sandy or peaty ground.

*No ethnobotanical uses of this plant have been noted in Ireland.

Ranunculus species

- Buttercups[225] and Crowfoots[226]

RANUNCULACEAE—THE BUTTERCUP AND CROWFOOT FAMILY

Native

224. Dinneen (1927) also gave this name for the holm oak.

225. In Irish, 'buttercup' is *dúil-in-im* (Ua Maoileoin, 2001) or *fearbán* (de Bhaldraithe, 1959; Uí Chanainn, 1962).

226. McGlinchey (2007) gave the name *fearaban* for 'crowfoot'. Dinneen (1927) gave the Irish words *cearbán* and *cos cromáin* for 'crowfoot'.

Ranunculus bulbosus (buttercup). Watercolour by Lydia Shackleton. Courtesy of the National Botanic Gardens, Glasnevin.

Various species of buttercup and crowfoot have been used in Irish folk medicine for a variety of ailments (Allen & Hatfield, 2004). It is not always clear which *Ranunculus* species may have been used for particular treatments. It was used in the treatment of 'stitches' and St Anthony's fire (a skin disease caused by ergot infection) (Moloney, 1919). Logan (1981) reported that 'wild buttercups' were used to make two cures for piles—the roots were ground up, boiled with lard and made into an ointment, or the leaves were boiled in water and the patient drank the water. Wood-Martin (1902) mentioned that 'crowfoot' was pounded up with butter and used to treat erysipelas.

Buttercups have been used in veterinary remedies to treat blackleg and red-water fever in various parts of Ireland (Allen & Hatfield, 2004). Another remedy was noted in Ireland by Baker (1999) in which buttercups were rubbed onto cow's udders on May Day, apparently to increase the milk yield and richness of their butter.

Buttercup flowers have long been used in Ireland to amuse children, where the flowers are put under or close to a child's chin to see if there is a reflected yellow colour on the skin. If there is then the child is said to 'love butter'.

Ó Dónaill (1977) gave the Irish word *gairgín* for 'crowfoot'. Munnelly (2005: 165) gave a common name from West Clare for *Ranunculus* species as 'the lion's paw', and quoting a local farmer from Miltown Malbay, Francie Kennelly, said that it was used to cause blistering: 'You'd break it up, like, with a stone or with a hammer or anything. An' if you had lumbago in your back or a bad hip or anything you'd put it to it an' 'twould blister it and 'twould bring the water out of the joints, they used to say, like'.

Including:

Ranunculus acris

- Meadow buttercup (Scannell & Synnott, 1987); Farrabun (Hart, 1898); Meadow crowfoot (Brunker, 1950)
- *Fearbán feír* (Scannell & Synnott, 1987); *Cearbán feóir* (Moloney, 1919); *Fearbán feoir* (Uí Chonchubhair, 1995); *Cam an ime* (Tóibín, 1965)

Ranunculus bulbosus

- Bulbous buttercup (Scannell & Synnott, 1987); Bulbous crowfoot (Moloney, 1919)
- *Tuile thalún* (Scannell & Synnott, 1987); *Fuil talmhan* (Moloney, 1919); *Tuile talmhan*[227] (Moloney, 1919; Dinneen, 1927); *Tuile-talmhuin* (de Vere Coneys, 1849)

Ranunculus acris is a common plant throughout Ireland of meadows and other grasslands, such as roadsides and hedgerows. *Ranunculus bulbosus* is similarly distributed but

227. Dinneen (1927) defined *tuile talmhan* as 'a buttercup, bulbous crowfoot'.

occurs in dry calcareous grasslands, for example, on sand dunes and esker ridges, in most parts of Ireland.

These species were used in Irish folk medicine for the treatment of various ailments, including warts, heartburn, kidney trouble, mumps and swollen glands and jaundice (Allen & Hatfield, 2004), as well as piles (Ui Sheasnáin, 1982). In Co. Cavan this author recorded that the roots of buttercups were boiled and mixed well with lard to make into an ointment and then applied to external piles. Ui Sheasnáin (1982) also mentioned that buttercups (as well as the leaves of *Iris pseudacorus* and *Sinapis arvensis*) were boiled and the water drunk to treat jaundice, although only one of these species at a time and never mixed. Uí Chonchubhair (1995) mentioned that in common with other *Ranunculus* these two species contain a poisonous juice. Before being used medicinally the plant material would be boiled to neutralise the acrid juice. These two species were probably not distinguished from each other in folk medicine. Uí Chonchubhair (1995) noted that *R. acris* was used as a rubefacient, placed on the skin to raise blisters and draw out disease. A poultice was also made from the crushed leaves and applied to rashes, scabs and warts, and the crushed roots were held under the nose to induce sneezing and help to clear the head. The same author also noted that it was believed that the scent of the meadow buttercup could cause madness. According to Moloney (1919), *R. acris* and *R. bulbosus*, as well as *R. repens*, were used in the treatment of 'stitches' and St Anthony's fire.

**Ranunculus aquatilis*

- Common water-crowfoot (Scannell & Synnott, 1987)
- *Néal uisce coiteann* (Scannell & Synnott, 1987); *Neul uisge* (Cameron, 1883); *Fleanuisge*[228] (de Vere Coneys, 1849); *Niúl uisce* (Dinneen, 1927); *Fliodh uisce*[229] (Ó Dónaill, 1977); *Líon abhann* (Ó Dónaill, 1977)

RANUNCULACEAE—THE BUTTERCUP AND CROWFOOT FAMILY
Native

Ranunculus aquatilis is an aquatic buttercup or crowfoot of streams, ditches and lake margins that is distributed throughout most of Ireland.

*No ethnobotanical uses of this plant have been noted in Ireland.

**Ranunculus auricomus*

- Goldilocks buttercup (Scannell & Synnott, 1987); Goldilocks (Anonymous, 1978; Webb & Scannell, 1983)
- *Gruaig Mhuire* (Scannell & Synnott, 1987); *Garbhlus na Móna* (Moloney, 1919; Dinneen, 1927); *Foloscain* (Cameron, 1883)

RANUNCULACEAE—THE BUTTERCUP AND CROWFOOT FAMILY
Native

---

228. De Vere Coneys (1849) applied this name to '*Ranunculus aquaticus*'. Dinneen (1927) listed a variant of this Irish word, *fleann uisce*.

229. Ó Dónaill (1977) defined *fliodh uisce* and *líon abhann* as 'water-crowfoot'.

Ranunculus auricomus is distributed mainly in the east in shady grasslands, woods and hedges.

*No ethnobotanical uses of this plant have been noted in Ireland.

**Ranunculus baudotii*

- Brackish water-crowfoot (Scannell & Synnott, 1987)
- *Néal uisce sáile* (Scannell & Synnott, 1987)

RANUNCULACEAE—THE BUTTERCUP AND CROWFOOT FAMILY
Native

Ranunculus baudotii is an occasional plant of ponds, pools and ditches near the coast, and sometimes brackish water, and is distributed particularly in the west and south.

*No ethnobotanical uses of this plant have been noted in Ireland.

Ranunculus bulbosus

- Bulbous buttercup (Scannell & Synnott, 1987)
- *Tuile thalún* (Scannell & Synnott, 1987); *Fuil talmhan* (Moloney, 1919); *Tuile talmhan*[230] (Moloney, 1919; Dinneen, 1927); *Tuile-talmhuin* (de Vere Coneys, 1849)

See discussion under *Ranunculus acris*.

**Ranunculus circinatus*

- Fan-leaved water-crowfoot (Scannell & Synnott, 1987)
- *Néal uisce righin* (Scannell & Synnott, 1987); *Flidh uisce* (Moloney, 1919); *Fleann uisce* (Moloney, 1919); *Líon na hAbhann* (Moloney, 1919)

RANUNCULACEAE—THE BUTTERCUP AND CROWFOOT FAMILY
Native

230. Dinneen (1927) defined *tuile talmhan* as 'a buttercup, bulbous crowfoot'.

Ranunculus circinatus is an aquatic plant of canals, rivers and lakes that is mainly in the centre of Ireland and is rare elsewhere.

*No ethnobotanical uses of this plant have been noted in Ireland.

Ranunculus flammula

- Lesser spearwort (Scannell & Synnott, 1987)
- *Glasair léane bheag* (Scannell & Synnott, 1987); *Lasair léana* (Dinneen, 1927; Uí Chonchubhair, 1995; Moloney, 1919); *Luibh na Franncac*[231] (Borland, 1920); *Glasiar léun* (Ó Criomhthain, MSS, n.d., ca. 1908)

RANUNCULACEAE—THE BUTTERCUP AND CROWFOOT FAMILY
Native

Ranunculus flammula can be found throughout Ireland in wetland habitats, such as marshes, ditches and lake-shores.

Uí Chonchubhair (1995) described the use of a poultice made from the crushed roots of this species with salt for use as a rubefacient to produce skin irritations and blisters, which it was believed would draw out disease from the body. It was used extensively in this way for the treatment of bubonic plague in the 16th century and to help heal scabs and running sores. In Fermanagh and Tyrone the flowers and leaves were used to treat tuberculosis and, mixed with wine, were used as a cure for diarrhoea and colic (Williams, 1993). Mrs John Borland (1920: 236) wrote about folklore in Co. Kerry and said that 'a handful of Spearwort, called locally *Luibh na Franncac* (rat herb), placed in a stack of oats will keep away rats. The leaves chewed are a cure for toothache'.

**Ranunculus fluitans*

- River water-crowfoot (Scannell & Synnott, 1987); Eelweed (Hart, 1898)
- *Néal uisce abhann* (Scannell & Synnott, 1987)

RANUNCULACEAE—THE BUTTERCUP AND CROWFOOT FAMILY
Native

Ranunculus fluitans is a very rare plant in Ireland, known only from one river in Co. Antrim.

*No ethnobotanical uses of this plant have been noted in Ireland.

**Ranunculus hederaceus*

- Ivy-leaved crowfoot (Scannell & Synnott, 1987)
- *Néal uisce eidhneach* (Scannell & Synnott, 1987)

RANUNCULACEAE—THE BUTTERCUP AND CROWFOOT FAMILY

231. This name has also been applied to *Stellaria media*, chickweed.

Native

Ranunculus hederaceus is a reasonably common plant of mud, marshes, ditches, streams and pools and can be found in most parts of Ireland except the centre.

*No ethnobotanical uses of this plant have been noted in Ireland.

Ranunculus lingua

- Greater spearwort (Scannell & Synnott, 1987)
- *Glasair léana mhór* (Scannell & Synnott, 1987)

RANUNCULACEAE—THE BUTTERCUP AND CROWFOOT FAMILY
Native

Ranunculus lingua is a large wetland buttercup of marshes, fens and canals that is mainly distributed in the centre of Ireland.

According to Moloney (1919), it was used in traditional medicine in the treatment of swine fever and was occasionally the cause of poisoning in cattle.

**Ranunculus omiophyllus*

- Round-leaved crowfoot (Scannell & Synnott, 1987)
- *Néal uisce cruinn* (Scannell & Synnott, 1987)

RANUNCULACEAE—THE BUTTERCUP AND CROWFOOT FAMILY
Native

Ranunculus omiophyllus is found occasionally only in the south and south-east in Ireland in wetland habitats, such as marshes, ditches, shallow ponds and pools and on wet mud.

*No ethnobotanical uses of this plant have been noted in Ireland.

**Ranunculus peltatus*

- Pond water-crowfoot (Scannell & Synnott, 1987)
- *Néal uisce scéithe* (Scannell & Synnott, 1987)

RANUNCULACEAE—THE BUTTERCUP AND CROWFOOT FAMILY
Native

Ranunculus peltatus is a widespread aquatic plant of rivers throughout Ireland.

*No ethnobotanical uses of this plant have been noted in Ireland.

**Ranunculus penicillatus*

- Stream water-crowfoot (Scannell & Synnott, 1987)
- *Néal uisce bréige* (Scannell & Synnott, 1987)

RANUNCULACEAE—THE BUTTERCUP AND CROWFOOT FAMILY
Native

Ranunculus penicillatus is a widespread aquatic plant of rivers and streams in Ireland and is reasonably frequent.

*No ethnobotanical uses of this plant have been noted in Ireland.

Ranunculus repens

- Creeping buttercup (Scannell & Synnott, 1987); Crawtee (Hart, 1898); Setfast (or Sitfast) (Hart, 1898)
- *Fearbán* (*reatha*) (Scannell & Synnott, 1987); *Bairghin* (Cameron, 1883); *Crobh préacháin*[232] (Dinneen, 1927); *Fearbóg* (Ó Dónaill, 1977); *Gurradh préachán* (Ó Criomhthain, MSS, n.d., ca. 1908)

RANUNCULACEAE—THE BUTTERCUP AND CROWFOOT FAMILY
Native

Ranunculus repens is a very common buttercup throughout Ireland of meadows, pastures, roadsides, damp waste places, ditches and drains.

According to Allen and Hatfield (2004) *Ranunculus repens* was used in Irish folk medicine for the treatment of various ailments. It was mentioned by Moloney (1919) in the treatment of 'stitches' and St Anthony's fire (a skin disease caused by ergot infection).

**Ranunculus sceleratus*

- Celery-leaved buttercup (Scannell & Synnott, 1987)
- *Toircheas fiáin* (Scannell & Synnott, 1987); *Torachas biadhain* (de Vere Coneys, 1849; Dinneen, 1927; Webb et al., 1996)

RANUNCULACEAE—THE BUTTERCUP AND CROWFOOT FAMILY
Native

Ranunculus sceleratus is mainly distributed around the coast in marshy habitats.

232. Dinneen (1927) defined *crobh préacháin* as 'crowfoot' (*Ranunculus repens*).

Cameron (1883) said that the name *Torachas biadhain* probably means 'food of which one would be afraid'.

*No ethnobotanical uses of this plant have been noted in Ireland.

**Ranunculus trichophyllus*

- Thread-leaved water-crowfoot (Scannell & Synnott, 1987); Water fennel (Brunker, 1950)
- *Néal uisce ribeach* (Scannell & Synnott, 1987); *Líon na habhann* (Uí Chonchubhair, 1995)

RANUNCULACEAE—THE BUTTERCUP AND CROWFOOT FAMILY
Native

Ranunculus trichophyllus is mainly found in the centre and south-east in wet habitats, such as ditches, drains, shallow ponds and pools, lake margins and open mud; it is much rarer elsewhere in Ireland.

*No ethnobotanical uses of this plant have been noted in Ireland.

Ranunculus tripartitus.

**Ranunculus tripartitus*

- Three-lobed crowfoot (Scannell & Synnott, 1987)
- *Néal uisce na dtrí chluas* (Scannell & Synnott, 1987)

RANUNCULACEAE—THE BUTTERCUP AND CROWFOOT FAMILY
Native

Ranunculus tripartitus is a very rare aquatic species in Ireland and found only in a small pond near Baltimore in West Cork.

*No ethnobotanical uses of this plant have been noted in Ireland.

Raphanus raphanistrum

- Wild radish (Scannell & Synnott, 1987); White charlock (Harron, 1986); Runch (Harron, 1986)
- *Meacan raidigh* (Scannell & Synnott, 1987); *Currán dearg* (Moloney, 1919); *Fiadh roidis* (Cameron, 1883); *Corrán dearg*[233] (Dinneen, 1927)

BRASSICACEAE (Cruciferae)—THE CABBAGE AND CRESS FAMILY
Native (or possible archaeophyte)

233. Dinneen (1927) also defined *meacan ráidigh* to mean 'radish'.

Raphanus raphanistrum occurs in Ireland as an occasional annual weed of cultivated fields and waste places, mainly on sandy or peaty soils.

Mears and Hillman (2007) suggested using immature seed capsules in salads, which they say are fleshy, juicy and delicious.

Remains of seed pods of wild radish have been found in pits containing faecal material excavated from Fishamble Street, Dublin, dating back to the Viking Age, more than 1000 years ago when they may have been eaten. Alternatively, they may have been eaten involuntarily when harvested as contaminants of cereal crops (Geraghty, 1996).

Raphanus raphanistrum subsp. *maritimus*

- Sea radish (Scannell & Synnott, 1987)
- *Meacan mara* (Scannell & Synnott, 1987)

BRASSICACEAE (Cruciferae)—THE CABBAGE AND CRESS FAMILY
Native

Raphanus raphanistrum subsp. *maritimus* is an occasional plant around the Irish coast on sandy beaches and sandy waste places near the sea.

Mears and Hillman (2007) reported that the leaves and flowers of this coastal plant can be eaten in salads. The leaves are chopped finely or young flowers and buds are used.

**Reseda alba*

- White mignonette (Scannell & Synnott, 1987)
- *Buí geal* (Scannell & Synnott, 1987)

RESEDACEAE—THE MIGNONETTE FAMILY
Introduced

Reseda alba is a casual introduction in Ireland that persists at coastal sites around Dublin and in Co. Waterford. *Reseda alba* is native along Mediterranean coasts and sparingly naturalised elsewhere in Europe.

*No ethnobotanical uses of this plant have been noted in Ireland.

Reseda luteola

- Weld (Scannell & Synnott, 1987); Yellow-weed (Anonymous, 1978); Dyer's rocket (Brunker, 1950; Anonymous, 1978); Dyer's weed (Praeger, 1938)
- *Buí mór* (Scannell & Synnott, 1987); *Ruadhchán* (Moloney, 1919); *Ruadhchán buidhe* (Moloney, 1919); *Buidhe mór* (Dinneen, 1927; Webb et al., 1996); *Buidhemór* (de Vere Coneys, 1849)

RESEDACEAE—THE MIGNONETTE FAMILY
Introduced (possible archaeophyte)

Reseda luteola is an occasional weed throughout most of Ireland of waste places, roadsides and dry, sandy and gravelly habitats, such as sand-pits and quarries, and particularly on limestone substrates. It is almost certainly not a native plant but may have been introduced as a weed of cultivation in past times.

This species is well known for providing a particularly bright and brilliant yellow colour when dyeing textiles. Mairet (1916) regarded it as the best and fastest of the natural yellow dyes, and in Ireland *Reseda luteola* was indeed the most commonly used plant yielding a yellow dye and called *buí mór* (Mahon, 1983). The same author noted that not only wild plants were used, but it was also sown as a dye crop under barley and pulled out during the second year of growth. The plant was gathered in June and July and dried in the shade before being tied in bundles. When it was needed for dyeing it was broken into pieces or chopped finely. The roots were discarded. After that, a decoction of the plant was made by boiling it in water for up to an hour (Mairet, 1916). Used with different mordants, it yields various colours. With alum or tartar it gives a bright yellow colour; with chrome it yields a gold colour; with tin, an orange-yellow colour and with copper and iron, an olive-yellow colour.

Mackay (1836) noted that the name *Reseda* comes from the Latin word *resedo*, meaning 'to calm', from its supposed medicinal qualities. He also recorded that it was used to dye woollens yellow. Scully (1916) recorded that in Co. Kerry it mainly occurred near villages and houses; he speculated that it was a relict of cultivation for dyeing purposes. Between 3 and 6 pounds weight of plant material was needed to dye a pound of clothes (Williams, 1993).

Another species, *Reseda lutea* (wild mignonette[234]), also occurs quite frequently throughout much of Ireland as an introduced weed.

Rhamnus cathartica

- Buckthorn (Scannell & Synnott, 1987); Purging buckthorn (Brunker, 1950)
- *Paide bréan* (Scannell & Synnott, 1987); *Ramhdhraighean* (Moloney, 1919; Dinneen, 1927; Uí Chanainn, 1962); *Maide bréan* (Moloney, 1919); *Bréan-ubhall*[235] (Dinneen, 1927)

RHAMNACEAE—THE BUCKTHORN FAMILY

Native

Rhamnus cathartica is a small to medium-sized tree of woods, scrub, river-banks and rocky lake-shores that occurs mainly in the centre and west of Ireland.

The fruits are powerful purgatives and a laxative, and the tree is generally regarded as poisonous. A tea was made from dried berries and drunk morning and evening to promote the production of soft stools (van Wyk & Wink, 2004). Its fruits were gathered and sold in Dublin markets in the early 18th century for medicinal purposes (MacCoitir, 2003).

According to Edlin (1973), buckthorn berries have been used to produce a green dye. Mackay (1836) noted that unripe fruits yield a yellow dye and the bark produces a green dye. Mahon (1983) listed that the berries and bark provide a yellow dye. Wade (1804) said that the saffron-coloured juice of the unripe berries were used for staining maps or paper and mentioned that they were found in 'druggists shops' in his time. Wade also noted that the juice of ripe berries mixed with alum was used to create a green paint for

234. Moloney (1919) gave the Irish words *buí beag* and *buidhe mór* to mean 'wild mignonette' and Ó Dónaill (1977) *buí cumhra*.

235. Dinneen (1927) defined *bréan-ubhall* as 'purging blackthorn'.

artists (alum is an aluminium sulphate compound used as a chemical fixative).

Allen and Hatfield (2004) recorded that it was used as a veterinary remedy to treat 'dry murrain' in cattle in Co. Mayo.

**Rhinanthus minor*

- Yellow-rattle (Scannell & Synnott, 1987); Rattlers (Colgan, 1904); Rattle boxes (Colgan, 1904); Bull's peas (Hart, 1898)
- *Gliográn* (Scannell & Synnott, 1987); *Boilgeáin* (Ua Maoileoin, 2001); *Maothlán móna* (Moloney, 1919; Uí Chonchubhair, 1995); *Bodach gliogair* (Dinneen, 1927); *Gliogarán* (Dinneen, 1927); *Míl-ghabhair* (Moloney, 1919); *Lus an giolla* (Moloney, 1919); *Lus riabhach* (Moloney, 1919); *Bodán cloigín* (Moloney, 1919)

OROBANCHACEAE (Scrophulariaceae)—THE BROOMRAPE FAMILY

Native

Rhinanthus minor is a very common annual plant of grasslands throughout Ireland.

*No ethnobotanical uses of this plant have been noted in Ireland.

Rhodiola rosea (Sedum rosea)

- Roseroot (Scannell & Synnott, 1987)
- *Lus na laoch* (Scannell & Synnott, 1987); *Rós lus*[236] (Uí Chanainn, 1962)

CRASSULACEAE—THE STONECROP FAMILY

Native

Rhodiola rosea is a relatively rare mountain plant in Ireland of cliffs and mountain ledges that is distributed mainly in the west and north. It also occurs on some sea-cliffs too.

It is a valued medicinal plant in northern and eastern Europe and Russia, used to combat depression. It was widely used by the Vikings and Siberians to increase their physical strength and endurance, as well as to provide them with resistance to cold and disease, and to enhance fertility and longevity (van Wyk & Wink, 2004). In Nordic countries it is being produced commercially to meet a growing demand (Wagner, 2006). There is apparently little if any history of its use in Ireland as a medicinal plant, which is perhaps surprising when Ireland was so heavily colonised by the Vikings.

Mackay (1825b) said that the roots have a fragrance of roses when bruised, thus giving its name 'roseroot'.

The young leaves and stems can be eaten in salads and cooked as a green vegetable or made into pickles (Peterson, 1977).

Rhododendron ponticum

- Rhododendron (Scannell & Synnott, 1987)
- *Róslabhras* (Scannell & Synnott, 1987); *Ródaideandrón fiáin* (Ua Maoileoin, 2001)

ERICACEAE—THE HEATHER FAMILY

Introduced

---

236. Uí Chanainn (1962) defined *rós lus* as 'rose wort'.

Rhododendron ponticum is a very common large shrub throughout Ireland on acid soils, sometimes growing abundantly, in woodlands and bogs and on mountainsides. It was planted originally as an ornamental plant and to provide cover in woodlands for game birds, particularly pheasants (Cross, 1975), but has since become a seriously invasive plant whose control represents one of the greatest nature conservation challenges in Ireland. Recent work on *R. ponticum* naturalised in Britain and Ireland suggests that it may not be the pure *R. ponticum* that occurs, but instead is a more vigorous hybrid (Milne & Abbott, 2000).

Honey made from the nectar of *Rhododendron ponticum* flowers contains a toxin called grayanotoxin that is toxic to some people if they eat too much of it, but there are no recorded cases of death in Ireland from honey poisoning, or so-called honey-madness, caused by eating this honey (Milliken & Bridgewater, 2004).

Rhododendron is generally regarded as a poisonous plant and should not be eaten. Woodward (1985: 42) gave the symptoms of poisoning as 'vomiting, vertigo, delirium'.

**Rhynchospora alba*

- White beak-sedge (Scannell & Synnott, 1987)
- *Gobsheisc* (Scannell & Synnott, 1987)

CYPERACEAE—THE SEDGE FAMILY
Native

Rhynchospora alba is a sedge of lowland wet bogs in the west and the midlands mainly; it is very rare elsewhere in Ireland.

*No ethnobotanical uses of this plant have been noted in Ireland.

**Rhynchospora fusca*

- Brown beak-sedge (Scannell & Synnott, 1987)
- *Gobsheisc rua* (Scannell & Synnott, 1987)

CYPERACEAE—THE SEDGE FAMILY
Native

Rhynchospora fusca is a sedge of wet bogs in the west and the midlands mainly; it is very rare elsewhere in Ireland.

*No ethnobotanical uses of this plant have been noted in Ireland.

Ribes nigrum

- Black currant (Scannell & Synnott, 1987)
- *Cuirín dubh* (Scannell & Synnott, 1987); *Raosar dubh*; *Srionán dubh*, *Dearcóg dubh* (Moloney, 1919).

GROSSULARIACEAE—THE GOOSEBERRY FAMILY
Introduced

Ribes nigrum is widely grown as a fruit bush in Irish gardens. It can also be found growing throughout Ireland in damp woods, hedgerows and river-banks, presumably from discarded plants as well as bird-sown seed, although this occurrence is rare. Black currants are used to make jams and sweet sauces and used in puddings, and they are a useful source of vitamin C. Its dried leaves can be used as tea.

This species has been used in parts of Europe in traditional medicine as a diuretic, and the leaves and sometimes the buds too were taken against arthritis, coughs, diarrhoea, rheumatism and urinary problems (van Wyk & Wink, 2004).

Cameron (1883) gave the Irish name *spiontog* for 'currant, gooseberry'. De Vere Coneys (1849), Ó Dónaill (1977) and Uí Chanainn (1962) gave the Irish name *spionán* for 'gooseberry'.

Ribes rubrum

- Red currant; White currant (Scannell & Synnott, 1987)
- *Cuirín dearg* (Scannell & Synnott, 1987); *Raosar dearg* (Moloney, 1919); *Srionán dearg* (Moloney, 1919); *Dearcóg dearg* (Moloney, 1919)

GROSSULARIACEAE—THE GOOSEBERRY FAMILY
Introduced

Ribes rubrum is widely grown as a fruit bush in Irish gardens. Like *R. nigrum*, it can also be found growing in hedgerows, roadsides and waste places, presumably from discarded plants as well as from bird-sown seed, although this occurrence is rare. Red and white currants are used to make jelly and jams and used in puddings.

Moloney (1919) listed the following Irish names for white currant: *raosar bán*, *srionán bán* and *dearcóg bán*.

Ribes uva-crispa

- Gooseberry (Scannell & Synnott, 1987)
- *Spíonán* (Scannell & Synnott, 1987); *Spiúnán* (Ua Maoileoin, 2001); *Spriúnán* (Moloney, 1919); *Grosaid* (Moloney, 1919); *Spionóg* (Moloney, 1919); *Ifin* (de Vere Coneys, 1849); *Crann spíonáin* (Dinneen, 1927); *Groiseog* (Dinneen, 1927; Uí Chanainn, 1962); *Iphin*[237] (Dinneen, 1927); *Iphín* (Uí Chanainn, 1962); *Sceach spíonán* (Dinneen, 1927); *Tor spíonán*[238] (Ó Dónaill, 1977)
- *Kraanberry* (in Yola; this name was also used for cranberry)

GROSSULARIACEAE—THE GOOSEBERRY FAMILY
Introduced

Ribes uva-crispa.

237. Dinneen (1927) defined *iphin* as 'the gooseberry tree'.
238. Ó Dónaill (1977) defined *tor spíonán* as 'gooseberry bush'.

Ribes uva-crispa is widely grown as a fruit bush in Irish gardens. It also can be found growing throughout Ireland, mainly in hedgerows, presumably from discarded plants as well as from bird-sown seed, although this occurrence is rare. Gooseberries are used to make jams and puddings.

McNeill (2010: 200) said that in Co. Tyrone the berries of gooseberry are known as 'goosegabs', especially by children. I have come across them being known as 'goosegogs' or 'goosegobs' elsewhere and very commonly in Ireland.

Roy Vickery in the Plant-lore Archive (Vickery, 2010) included a treatment for a stye in the eye by old folk in the 1920s. The stye was pricked every morning with a thorn from a gooseberry bush. In Cos. Tipperary and Monaghan this practice was well known up until at least the 1960s, when the stye would merely be touched with the thorn but not pricked (Con Cummins & Geraldine Lambe, pers. comm., 2011).

**Rorippa amphibia*

- Great yellow-cress (Scannell & Synnott, 1987)
- *Biolar buí mór* (Scannell & Synnott, 1987); *Biolar curraigh* (Moloney, 1919)

BRASSICACEAE (Cruciferae)—THE CABBAGE AND CRESS FAMILY

Native

Rorippa amphibia is found mainly in the centre and east of Ireland in wet habitats, such as ditches, drains, marshes, streams, lake-shores, turloughs and river-banks; it is rare elsewhere in Ireland.

*No ethnobotanical uses of this plant have been noted in Ireland.

**Rorippa islandica*

- Northern yellow-cress (Scannell & Synnott, 1987)
- *Biolar buí na Boirne* (Scannell & Synnott, 1987)

BRASSICACEAE (Cruciferae)—THE CABBAGE AND CRESS FAMILY

Native

Rorippa islandica is a very rare plant in Ireland that is mainly distributed in the west in wet habitats, such as lakes, ponds, turloughs, marshes and wet mud.

*No ethnobotanical uses of this plant have been noted in Ireland.

**Rorippa palustris*

- Marsh yellow-cress (Scannell & Synnott, 1987)
- *Biolar buí corraigh* (Scannell & Synnott, 1987)

BRASSICACEAE (Cruciferae)—THE CABBAGE AND CRESS FAMILY

Native

Rorippa palustris is a rare plant throughout Ireland of wet habitats, such as lakes, ponds, turloughs, marshes and wet mud.

*No ethnobotanical uses of this plant have been noted in Ireland.

**Rorippa sylvestris*

- Creeping yellow-cress (Scannell & Synnott, 1987)
- *Biolar buí reatha* (Scannell & Synnott, 1987); *Biolar coilleadh* (Moloney, 1919)

BRASSICACEAE (Cruciferae)—THE CABBAGE AND CRESS FAMILY
Native

Rorippa sylvestris is widespread throughout Ireland except in the south-west and mid-west where it is rare. It occurs in marshy habitats, such as river-banks, lake-shores and turloughs.

*No ethnobotanical uses of this plant have been noted in Ireland.

Rosa species

- Rose; Buckie-briar (Praeger, 1938)
- *Rós* (Scannell & Synnott, 1987); *Conrós* (Moloney, 1919); *Cocán* (Moloney, 1919); *Plúirín* (Moloney, 1919); *Fírdhris* (Wild rose tree) (Moloney, 1919; Dinneen, 1927)

ROSACEAE—THE ROSE FAMILY
Native

Including:

Rosa agrestis

- Small-leaved sweet-briar (Scannell & Synnott, 1987)
- *Dris chumhra gheal* (Scannell & Synnott, 1987)

Rosa arvensis

- Field-rose (Scannell & Synnott, 1987); Trailing rose (Brunker, 1950)
- *Rós léana* (Scannell & Synnott, 1987)

Rosa caesia (Rosa coriifolia)

- Hairy dog-rose (Scannell & Synnott, 1987)
- *Feirdhris bhog* (Scannell & Synnott, 1987)

Rosa canina

- Dog-rose (Scannell & Synnott, 1987); Dog briar (Colgan, 1904); Buckey brier (Colgan, 1904); Buckie (Hart, 1898); Buckie-berry (Hart, 1898); Corragy (Hart, 1898)
- *Feirdhris* (Scannell & Synnott, 1987); *Rós cú* (Cameron, 1883); *Sgeach mhadra* (de Vere Coneys, 1849; Cameron, 1883); *Con-rós* (or *Conrós*) (Dinneen, 1927; Ó Dónaill, 1977); *Oireachas* (Dinneen, 1927); *Sceach mhadraidh* (Dinneen, 1927); *Coindris*[239] (Ó Dónaill, 1977)

Rosa micrantha

- Small-flowered sweet-briar (Scannell & Synnott, 1987)
- *Dris chumhra bheag* (Scannell & Synnott, 1987)

Rosa mollis

- Soft downy-rose (Scannell & Synnott, 1987); Soft rose (Brunker, 1950)
- *Rós bog* (Scannell & Synnott, 1987)

Rosa obtusifolia

239. Ó Dónaill (1977) also gave an alternate spelling: *coin-dris*.

- Round-leaved dog-rose (Scannell & Synnott, 1987)
- *Feirdhris mhaolduilleach* (Scannell & Synnott, 1987)

Rosa pimpinellifolia (R. spinosissima)

- Burnet rose (Scannell & Synnott, 1987); Scotch rose (Webb, 1977; Webb & Scannell, 1983)
- *Briúlán* (Scannell & Synnott, 1987); *Rós fiáin* (O'Rourke, 2006); *Stancán* (Dinneen, 1927)

Rosa rubiginosa

- Sweet briar (Scannell & Synnott, 1987)
- *Dris chumhra* (Scannell & Synnott, 1987); *Sgeach-chumhra* (de Vere Coneys, 1849; Cameron, 1883); *Forrdris* (de Vere Coneys, 1849); *Foirdhris* (Dinneen, 1927); *Fordhris*[240] (Uí Chanainn, 1962)

Rosa sherardii

- Sherard's downy-rose (Scannell & Synnott, 1987)
- *Rós Shioraird* (Scannell & Synnott, 1987)

Rosa stylosa

- Short-styled field-rose (Scannell & Synnott, 1987)
- *Rós stíleach* (Scannell & Synnott, 1987)

Rosa tomentosa

- Harsh downy-rose (Scannell & Synnott, 1987); Corragy (Hart, 1898)
- *Rós clúmhach* (Scannell & Synnott, 1987)

Ireland is home to a wide variety of wild rose species. At least 10 native roses are known, as well as numerous hybrids between them (Doogue et al., 1998). Their identification is rarely easy and sometimes only possible for experts. Most occur in hedgerows, scrubs, thickets and waste places, although one or two have more specialised habitats, such as *Rosa pimpinellifolia*, which grows in sand dunes, rocky heaths and on limestone pavement. *Rosa canina*, the dog-rose, is the commonest species in Ireland and widespread throughout the country. Other widespread species are *R. sherardii* and *R. rubiginosa*, but neither is common. *Rosa arvensis* is found most commonly in the west and south-east and is of scattered distribution elsewhere. Several species are rare or confined largely to specific parts of Ireland, such as *R. stylosa*, which is rare and mainly found in the southern half of Ireland; *R. obtusifolia*, which is known only from Cos. Laois and Kildare; *R. mollis*, which is rather rare and mainly found in the north; and *R. micrantha*, which is rare and occurs in hedges in Cos. Cork, Kerry, Kildare, Kilkenny and Wexford. *Rosa caesia* is widespread but very rare. *Rosa tomentosa* is a species of uncertain status in Ireland, and records of it may instead be of hybrids such as *R. canina* × *R. tomentosa* and *R. canina* × *R. sherardii* (Webb et al., 1996). Cultivated roses are generally complex hybrids and highly bred selections, mainly from Europe and Asia, and are almost unrecognisable from the wild roses that were their ancestors.

Rosa pimpinellifolia.

Apart from the Irish names for the various rose species listed above, Moloney (1919) gave the following names: corn rose (*cailicín*; *cailleach dhearg*, *bláth na mbodach*, *luirgneach dhearg*); damask rose (*cocán liath*, *rós leathbhán*); and wild

240. Uí Chanainn (1962) defined *fordhris* as 'sweet briar, bramble'.

rose (*rós fiadhain, rós mhadra ruaidh*). Moloney also suggested that an Irish name for *Rosa rubiginosa*, sweet briar, is probably the 'Rose of Jerusalem', *rós na bainríoghna Muire*. Uí Chonchubhair (1995) gave the Irish name *rós Bhaile an Chalaidh* for *R. rugosa*, an introduced species.

Rosa canina.

All the above-mentioned roses produce rosehips (fruits) (*mucóra*), and probably all have similar traditional uses in Ireland. Rosehips were a significant wild fruit in ancient Ireland. They have been used to make syrup, jams, pie fillings, jellies and wine. They are extremely rich in vitamin C, and according to Mears and Hillman (2007), contain greater concentrations of vitamin C than any other wild fruit in these islands (with levels of up to 2000 mg per 100 g of fruit). Roses have also been used in traditional medicine and their hips were used to treat a range of ailments, such as intestinal ailments and as a diuretic (van Wyk & Wink, 2004). Rutty (1772) said that the ripe fruits of *Rosa pimpinellifolia* were eaten by children. Rosehip syrup was often given to children in Ireland during World War II and afterwards into the 1950s and 1960s as an easily available vitamin C supplement. In the 1970s an overseas aid charity, Concern, organised the collecting of rosehips in Ireland for vitamin C manufacture (Geraghty, 1996).

Rosehip syrup is made by boiling rosehips in water and straining the resulting mixture through a jelly bag (being careful to remove all of the hairs that surround the seeds). The strained liquid is then boiled again with sugar and bottled. Smith ([1756] 1979) recorded that the hips were used in Co. Kerry to make a conserve (conserves are similar to jams but include larger pieces of fruit when cooked). Stewed rosehips are also good to eat. Rose leaves and rosehips have been used to make tea. The hips are best collected after frost when they have the most flavour. In cooking the hips, only stainless-steel utensils should be used to keep their colour.

A delicious and delicately flavoured rose petal jam can be made easily: 1 lb/450 g of sugar, 300 mL/1/2 pint of water and 3 tablespoons of lemon juice are simmered together until a syrup-like consistency is achieved. After that, about 600 mL/1 pint of (unbruised) rose petals are added to the mixture, stirred gently and cooked until the petals are tender and coated with syrup. After that the mixture is spooned into sterilised jam jars and sealed. The darker coloured the petals, the better the colour of the jam.

Rosehips can also be made into tea. Crushed or minced hips are simmered in boiling water for about half an hour and strained carefully to remove the hairs that surround the seeds. The tea can be bottled and stored, sealed well until needed. Before use the tea is reheated, it is usually sweetened with sugar or honey (Ó Céirín & Ó Céirín, 1978). Tea can also be made from rose petals and leaves (Peterson, 1977).

When I was a child it was common practice for children to use the seeds and hairs as a substitute for itching power to place down the backs of their friends. Colgan (1904: 303) described its use in this way as 'an instrument of refined torture'.

The collections of the National Museum of Ireland (Museum of Country Life) in Castlebar, Co. Mayo, include a necklace made of rosehips from South Ulster from the 1940s (Ref. No: F1944.59).

O'Súilleabháin (1942) mentioned that the wood of the dog-rose was used in Ireland for 'clams', the castration of animals, although he does not specifically describe how this was done (see also *Sambucus nigra*).

Briar rose shoots have been used for basket-making and provide very strong ribs useful in the construction of various types of baskets (Joe Hogan, pers. comm., 2010). They have the advantage also that they don't taper towards the ends like willows.

Sharkey (2000) noted that rose petals were mixed with *Tussilago farfara* (colt's foot) for use as a tobacco substitute for poor people. Briar pipes for smoking tobacco are not made from rose wood as one might imagine but instead carved from the roots of an unrelated Mediterranean species, *Erica arborea*, a tree heath (or *bruyere*) that is occasionally grown in gardens in Ireland but is not native. Tree heath forms large woody nodules on the roots and base of the stems, which are the parts used (Howes, 1974). In Dublin, Kapp and Peterson have made briar pipes for more than a century from this imported species (Shaw-Smith, 1984). The firm was established in Grafton Street in 1865 and today makes about 100,000 pipes each year.

In the east of Ireland, rose hips have been known by an unusual common name, 'Johnnie Magories' (or 'Johnny Magoreys'). It has been said that they may also (or alternatively) refer to the fruits of hawthorn (*Crataegus monogyna*)—the fruits of the hawthorn are called *sceachóir* in Irish (or skeehories), which may be the derivation of this name, like a rhyming slang. Colgan (1904: 298) said that the fruits of *Rosa canina* are always called 'Johnny McGorey' in Dublin. Hogan and O'Neill (1947: 274) in their north county Dublin glossary of traditional words said that 'johnnie magories' referred to hips of the wild rose and gave a rhyme associated with them: 'I'll tell you a story of Johnnie Magory; He went to the wood and shot a Tory'[241]. Growing up in Limerick in the 1960s we heard a variant of this rhyme 'I'll tell you a story of Johnnie Magory; And now my story's done'. We didn't know any more of the rhyme, nor did we associate it with rose hips. I seem to remember that the rhyme was used when teasing a child who wanted to be told a story. James Joyce (1922: 166) included Johnny Magories in *Ulysses*: 'If you didn't know risky putting anything into your mouth. Poisonous berries. Johnny Magories. Roundness you think good. Gaudy colour warns you off. One fellow told another and so on. Try it on the dog first. Led on by the smell or the look. Tempting fruit'.

Roses are amongst the most commonly grown garden ornamental plants in Ireland. They thrive in the mild, moist Irish climate. As such, they have become a major part of Irish modern-day lore and are widely used as cut flowers and to create both private and public rose gardens (such as the beautiful rose gardens at the National Botanic Gardens, Glasnevin and in St Anne's Park, Clontarf, Co. Dublin). Cut red roses are today the universal symbol for love, and each year hundreds of thousands of red rose blooms are bought and given to loved ones on St Valentine's Day in Ireland. Extracts of roses (rose oil and rose water, in particular) are used also in the making of cosmetics and perfumes in Ireland.

Rubia peregrina.

Rubia peregrina

- Wild madder (Scannell & Synnott, 1987)
- *Garbhlus na Boirne* (Scannell & Synnott, 1987); *Madar* (Moloney, 1919; Dinneen, 1927; Ó Dónaill, 1977)

RUBIACEAE—THE BEDSTRAW FAMILY

Native

Rubia peregrina is a plant of the Mediterranean and Atlantic coasts of Europe that reaches the northern limit of its distribution in Ireland. Here, it occurs in the west as far north as Co. Mayo and is common in the karst topography

241. A 'tory' is a robber or bandit, probably from the old Irish word *tóraidhe*, meaning 'robber', derived from the word *tóir*, meaning 'pursuit'.

of the Burren in Co. Clare as well as in parts of the south-west but is rarely encountered in the Irish midlands. Wild madder grows on walls and limestone pavement and in scrub and hedges.

Wade (1804), Mahon (1983) and Sharkey (2000) recorded that this species yields a red dye. Wade suggested that it could be used for dyeing purposes in preference to the more costly non-native madder (*Rubia tinctorum*). However, the colour that wild madder produces is more so a rose-pink and not the brilliant red of its relative. Mahon (1983) recorded that the dye was obtained from the roots.

Rubia tinctorum (madder, *roid* or *madar* [*-air*] in Irish) has never been reported as a wild plant in Ireland but was probably fairly widely cultivated as a source of dye (Kelly, 1997). Kelly reported that it is known from one early Christian site in Ireland, at Boho, Co. Fermanagh, and madder persists in scattered locations across Britain (Stace, 2010). It is the polyphenols obtained from the roots (alizarin, purpuroxanthin, quinizarin) that yield the typical red colorant that was used for various dyeing purposes as well as in artists' colours such as 'rose madder' and 'madder brown' (Howes, 1974).

Rubus caesius

- Dewberry (Scannell & Synnott, 1987)
- *Eithreog*[242] (Moloney, 1919; Scannell & Synnott, 1987); *Gorm dhearc* (Moloney, 1919)

ROSACEAE—THE ROSE FAMILY

Native

Rubus caesius occurs frequently in parts of the south and west in stony and sandy habitats, especially on limestone. It is much less common in other parts of Ireland and is not as common as its close relative, the blackberry, *R. fruticosus*. The dewberry has edible fruits that can be delicious to eat, but fruits rarely occur in abundance and are not collected in the same way as blackberries. Dewberry jam is excellent if sufficient fruits can be gathered. The leaves can be used fresh or dried as a tea substitute.

Rubus chamaemorus

- Cloudberry (Scannell & Synnott, 1987)
- *Eithreog shléibhe* (Scannell & Synnott, 1987); *Lus na nEithreóg* (Moloney, 1919; Dinneen, 1927); *Eireag* (Cameron, 1883)

ROSACEAE—THE ROSE FAMILY

Native

Rubus chamaemorus jam from Scandinavia.

Rubus chamaemorus is a low-growing herbaceous plant of bogs, acid woodlands and tundra in the north of Asia, Scandinavia and North America and Arctic regions. It is a very rare plant in Ireland that is found only in one small

242. Moloney (1919) listed this Irish name for *Rubus chamaemorus*.

patch in the Sperrin Mountains of Co. Tyrone, on the north slope of the summit of Mullaghdoo. It was first discovered in Ireland in 1826 by Professor Murphy and Admiral Jones. After that it was not found again until 1892. Apart from the 1892 sighting, it has never been recorded in flower or fruit (McNeill, 2010) and seems to be barely hanging on as a native Irish plant. It evidently occurs at the southern limit of its climatic tolerance and survives as a remnant of the last Ice Age. If predicted changes in Ireland's climate occur over the coming century, this species is one that may disappear quickly and easily. It is interesting that it has a well-established Irish name, *eithreog*, and is included in several dictionaries (e.g., Uí Chanainn, 1962), since it is so rare in Ireland.

Although it has edible fruits that are tasty (especially with cream and sugar) and is much used in Scandinavia for jam and drinks, it was probably never gathered in Ireland for such purposes due to its rarity. I particularly enjoy cloudberry jam on toast at breakfast, instead of marmalade, sourced from my local Ikea shop and imported from Scandinavia. Cloudberry seeds have been found in an Iron-Age crannog (lake-dwelling) on Loch Tay in Scotland and may have been used for food there in ancient times (Milliken & Bridgewater, 2004).

Rubus fruticosus fruits.

Rubus fruticosus agg.

- Bramble (Blackberry) (Scannell & Synnott, 1987); Blackberry bush (Anonymous, 1978); Brier (Colgan, 1904,1911); Drisag (Hart, 1898)
- *Dris* (Scannell & Synnott, 1987); *Sceach* (Ó Dónaill, 1977; Uí Chonchubhair, 1995; Ua Maoileoin, 2001); *Smear dubh* (the fruit) (Moloney, 1919; Dinneen, 1927); *Driseóg* (Colgan, 1911; Moloney, 1919; Dinneen, 1927); *Dreasóg* (de Vere Coneys, 1849; de Bhaldraithe, 1959); *Smeur* (de Vere Coneys, 1849); *Sméire* (de Vere Coneys, 1849); *Foirdhreis*[243] (Dinneen, 1927); *Sméir-dhris* (Dinneen, 1927); *Sméardhris*[244] (Ó Dónaill, 1977); *Sméirdhris*[245] (Uí Chanainn, 1962); *Sgeach* (Ó Criomhthain, MSS, n.d., ca. 1908); *Fordhris*[246] (Uí Chanainn, 1962)
- *Brimel*; *Brimelès* (in Yola)

ROSACEAE—THE ROSE FAMILY
Native

Rubus fruticosus is a ubiquitous plant of hedgerows, banks, woodlands, hedgerows, scrub and many other habitats throughout Ireland. *Rubus fruticosus* is a so-called aggregate species made up of more than 80 different micro-species that have arisen as a result of brambles reproducing largely by asexual means. For that reason, whenever new mutations occur, a new species may be born. Hybrids also occur occasionally, and the identification of most micro-species in Ireland is the preserve of a very small number of experts. For traditional users of blackberries, however, their complex reproductive systems are very largely irrelevant, other than the fact that some micro-species have larger fruits or produce ripe fruits earlier or later than others. Several varieties of this species are cultivated in fruit and vegetable gardens, particularly thornless forms or those with large fruits. One species,

243. Dinneen (1927) also gave this word as a name for 'sweet briar'.
244. Ó Dónaill (1977) defined *sméardhris* as 'blackberry-bush'.
245. Uí Chanainn (1962) defined *sméirdhris* as 'blackberry, bramble'.
246. Uí Chanainn (1962) defined *fordhris* as 'sweet briar, bramble'.

R. ulmifolius ('elm-leaved bramble'), is easily recognisable within this complex. It has very white felty undersides to its leaves and is common on roadsides throughout Ireland. Its fruits are smaller and less fleshy than many other hedgerow brambles.

Blackberries (in Irish, *smear*, *sméra* or *smear dubh*) were one of the most important wild food sources of ancient Ireland and were widely gathered and used as a food and sugar source throughout history (Kelly, 1997). Vickery (2010) suggested that in some parts of Britain and Ireland, blackberries were not widely eaten or considered good to eat. For example, he recorded that in Co. Monaghan during World War I blackberries were gathered by children, presumably for export to the United Kingdom, but very few local people ate them. In Viking Dublin the excavated contents of drains and cesspits included seeds of strawberries, blackberries, apples, rowan berries, sloes and plums (Anonymous, 1973).

Blackberries were also used as famine food in Ireland. Woodham-Smith (1962) mentioned that by 25 September 1846 at Clashmore, Co. Waterford, people were living on blackberries after the failure of the potato crop in the area. Blackberries would normally have been at their most abundant at that time, after the so-called mean months of June, July and August, when the previous year's potato crop would have been exhausted. During these summer months of shortage, herrings were also eaten widely in Ireland. They were caught from late July until the end of August (Kennedy et al., 1999). Oatmeal was also an important food eaten in Ireland especially during the summer months, in particular in the north-east of the country.

Some authors have commented that it is remarkable that wild foods, particularly wild berries, have not been more widely collected and used in Ireland over the last century or more. In part, this is probably due to the increasing disconnect between the Irish population and nature and the increasing proportion of Irish people living in urban settings. Furthermore, changing agricultural practices have resulted in the loss of hedgerows throughout the country so that some species (e.g., *Malus sylvestris*, crab-apple) that provide a wild harvest have become much rarer. Since the rise of the motorcar, collecting wild fruits along roadsides was also less accepted than might be expected, with a belief that the fruits would be covered in dust and dirt spread by passing cars or contaminated with lead from petrol fumes (before its replacement by lead-free fuel in the 1980s). Evans (1957) has suggested that a bitter folk memory of the Great Famine may also have led to wild foods being shunned. During the Famine wild foods were one of the only available sources of food in many parts of the country, and this may have resulted in their being neglected and despised afterwards, so that today there is by and large, little on-going tradition of the collection of wild berries, cresses, edible seaweeds and wild fungi.

Today, blackberries are nevertheless the most widely collected wild fruits and are used for making jams, jellies, fool, wine, liqueurs, syrups, pie fillings and other purposes, sometimes mixed with other fruits such as apples, elderberries, rosehips and crab-apples. Sharkey (2000) noted that in former times in rural Ireland autumn fruits (such as blackberries) would be mixed with porridge and curds and eaten as part of the midday meal, adding some sweetness to an otherwise bland-tasting meal. Blackberry picking generally ends on or around 10 October in Ireland, mainly because after this the fruits are damaged by frosts, but also because that is traditionally known as the day on which the Devil (or the 'pooka' or *púca na sméar*, the 'blackberry sprite'; Dinneen, 1927) contaminates or makes them inedible by spitting (or urinating?) on them. October 11th was old Michaelmas Day. Country people also suggested that around this time blackberries became infested with maggots or became mildewed and unwholesome. Hart (1898) mentioned the tradition in Co. Donegal that at All Hallows' Eve the Devil shakes his *bratogs* (rags) over the blackberries and they are then unfit to eat. Hart suggests this is probably a play on words, as *bratogs* means both 'rags' and 'maggots' in Irish.

Blackberry jam and blackberry jelly are probably amongst the most popular wild plant-derived foods used in Ireland today. Some wild blackberries are collected and used on a commercial scale by companies in Ireland, such as Follain Teo, of Cúil Aodha, Maghchromtha, Co. Cork, which makes jam from wild collected fruits. Blackberry jam is easily made and has a short cooking time. However, blackberries contain only low amounts of natural pectin, which is important for setting the jam. For that reason, mixing blackberries with apples (or wild-collected crab-apples) will produce a jam that sets well. To make blackberry jam the berries are washed and stewed before an equal amount of sugar is added and stirred until dissolved. After that, the mixture is boiled rapidly until the setting point is reached. The jam is then spooned into hot jam jars (sterilised in the oven—don't forget to sterilise the lids, too) and sealed. If blackberry jam is made without apples some lemon juice can be added to help setting, or special jam sugar (with added pectin) can be used. To make blackberry jelly, the stewed blackberries are strained through a jelly bag (don't squeeze the bag or the resulting jelly will be cloudy) and then the resulting juice is mixed with sugar and then continue the recipe in the same way as for jam. My grandmother, Maud Phair, had a recipe for bramble jelly that included 6 lb of blackberries stewed in 2.5–3 pints of water and 3 lb of tart cooking apples. She recommended that the blackberries should be picked and used under-ripe. The apples were stewed without peeling or coring them. After simmering the fruit for 45 minutes the mixture is mashed in the pan and dripped overnight through a muslin or jelly bag. The following day each pint of liquid is boiled with 1 lb of sugar until setting point is reached. Blackberries can also be made into fruit cheeses, for example, mixed with crab-apples (see *Malus syvlestris*), or made into chutney or ketchup, pickled or used to flavour vinegar.

Mears and Hillman (2007) noted that the crisp shoot tips of bramble harvested in spring are delicious when steamed or blended into nettle (*Urtica dioica*) soup. They suggested using thick tips that snap off crisply and whose spines are still soft. These same authors remind us that the berries combine well with other wild autumn fruits, bullaces, sloes, elderberries and crab-apples to make autumn puddings and provide a sweet contrast to other more bitter autumn fruits.

The roots of brambles were formerly used to prepare a black dye (Colgan, 1911; Moloney, 1919; Mahon, 1983) and if treated variously, blackberry fruits can be used to prepare blue, grey or rose pink dyes (Edlin, 1973). Uí Chonchubhair (1995) noted that a yellow dye is made from the roots and black hair dye from stems and leaves. Rutty (1772: 124) wrote that the 'common bramble, or blackberry bush' could be used for tanning: 'An ooze of the tops and stalks tans leather'.

Bramble stems have been used for weaving baskets. The bark and thorns are stripped off leaving fine pliable stems that are good for weaving. Lucas (1957) recorded that they were used to make baskets (back baskets, or *cliabh*), especially in the treeless, bushless moors of Erris, Co. Mayo. Bramble stems gathered in woodlands may be better for basketry because they can be longer and more flexible than the tougher compact bramble shoots that grow in hedgerows and more open locations (Joe Hogan, pers. comm., 2010). For basketry, the stems were drawn through a leather pad held in the palm to remove the thorns. Bramble stems were also used to bind straw to make horse collars (Williams, 1993). This same author noted that bramble stems were used to make tobacco pipes and also used to pack *sliotars* (hurley balls). Brambles were also used to make scallops or spars to hold roof thatch in place (O'Hare, 1993). McGlinchey (2007) mentioned a place in Donegal called Drom na Scolb, where it was said that monks would dry briars used in thatching the monastery.

Blackberry leaves, roots and fruits have been used for traditional medicinal practices as well because of their astringent, antiseptic and antifungal properties. Crushed dried leaves have been used to make tea to treat colds, coughs and influenza (O'Regan, 1997; Reader's Digest Association, 2006). In Ireland, it was formerly widely used to treat diarrhoea and other purposes, such as cuts, ulcers, burns, indigestion, swellings, sore feet and kidney problems, less frequently (Allen & Hatfield, 2004). K'Eogh (1735: 16) noted that boiled blackberry roots 'are good against the stone and provoke urine'. A soothing drink was made from the roots, and bramble juice squeezed from the plant was mixed with butter as an ointment for swellings (Williams, 1993). A vinegar made from blackberries has been used to treat fevers and colds, as well as gout and arthritis (O'Regan, 1997).

Allen and Hatfield (2004) recorded that this species was used to treat diarrhoea in cattle in various Irish counties. The roots were also boiled and used to treat scour in cattle (Munnelly, 2005).

Brambles were often encouraged to grow in hedgerows and along field margins to make them an impenetrable barrier for stock. Evans (1957) suggested that present-day barbed wire owes its origins to a rustic prototype, a hay-rope strung with thorns, for which brambles would have been the ideal 'thorn-source'.

Rubus idaeus

- **Wild raspberry (Scannell & Synnott, 1987)**
- ***Sú craobh* (Scannell & Synnott, 1987); *Súghchraobh* (de Vere Coneys, 1849; Moloney, 1919); *Sobha craobh*[247] (de Vere Coneys, 1849); *Subh-chraobh* (Dinneen, 1927)**

ROSACEAE—THE ROSE FAMILY
Native

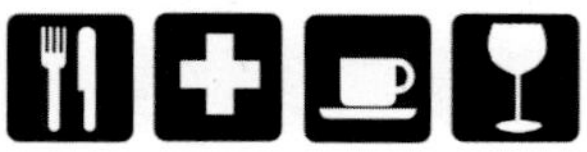

Rubus idaeus is frequently distributed in the north in Ireland and occasionally in the south in woods, hedges and shady places. This plant is not found in sufficient abundance for its delicious bright red or yellowish fruits to be picked in the same way as are those of the bramble or blackberry (*R. fruticosus*). Wild raspberry jam can be prepared in the same way as blackberry jam (see *Rubus fruictosus*). Raspberry wine and raspberry vinegar can also be made.

Mears and Hillman (2007) suggested that the wild raspberry was an important food for early people living in Britain and Ireland and noted that raspberry pips have been found in archaeological sites from a range of different periods, turning up in faecal deposits in ancient medieval cesspits, sewers, latrines and garderobes.

Wild raspberry leaf tea is said to be a useful medicinal tea. Raspberry leaves have been used medicinally for their astringent properties, to treat diarrhoea and as a gargle for throat and mouth infections (van Wyk & Wink, 2004). Wild raspberry wine can also be made. It is or was probably little gathered or used in Ireland, although Allen and Hatfield (2004) noted a record of the use for the plant to reduce pain in childbirth in Antrim and used as a tea in Limerick for sore throats. Shaw (1998) suggested that raspberry tea drunk during the final months of pregnancy helps to relax the womb and makes delivery easier. Vickery (2010) included raspberry leaves as an ancient cure for menstrual pain, with a record from Ballymote, Co. Sligo, in the Plant-lore Archive.

Rubus saxatilis

- **Stone bramble (Scannell & Synnott, 1987); Stoneberry (Hart, 1898)**
- ***Sú na mban mín* (Scannell & Synnott, 1987); *Crúibíní* (O'Rourke, 2006); *Súgh na mban mín* (Moloney, 1919)**

ROSACEAE—THE ROSE FAMILY

247. De Vere Coneys (1849) defined *Sobha craobh* as 'raspberries'.

Native

Rubus saxatilis is a widespread plant in Ireland but not abundant anywhere. It occurs in rocky habitats, such as limestone pavements (it is common in the Burren in Co. Clare) and on sand dunes.

It was said by K'Eogh (1735) to be useful against scurvy. The fruits of this species were traditionally gathered in the Aran Islands on St Mac Dara's Day, 16 July; they are wholesome to eat but somewhat tart (O'Rourke, 2006).

Rumex acetosa

- Sorrel (Scannell & Synnott, 1987); Cow sorrel (Colgan, 1911); Dog sorrel (Colgan, 1904); Bog sorrel (Hart, 1898); Sorrel dock (Hart, 1898); Crow-sorrel (Hart, 1898); Sourlick (Praeger, 1938); Sour'k (Praeger, 1938)
- *Samhadh bó* (Scannell & Synnott, 1987); *Samhadh* (Dinneen, 1927; Uí Chonchubhair, 1995); *Bileoga milse* (Ua Maoileoin, 2001); *Ribleachán*[248] (Moloney, 1919); *Puinneóg*[249] (Moloney, 1919; Dinneen, 1927); *Puineoga* (Cameron, 1883); *Sealgan* (Cameron, 1883; Dinneen, 1927; Ó Dónaill, 1977); *Samh* (de Vere Coneys, 1849); *Sobha* (de Vere Coneys, 1849); *Sámha bo* (Colgan, 1911); *Sámha* (Colgan, 1911)

POLYGONACEAE—THE KNOTWEED FAMILY

Native

Rumex acetosa.

Rumex acetosa is a very common perennial herb of grasslands throughout Ireland.

This plant was probably extensively used as a vegetable and salad in ancient Ireland (see *Urtica dioica* also). It was used in soups, sauces and salads, often to accompany fish. I enjoy a sauce of chopped sorrel leaves added to mayonnaise (home-made or otherwise), which makes an excellent accompaniment to fish such as salmon and trout, or chopped in an omelette. Sorrel leaves can also be boiled and used as a vegetable a bit like spinach. Drury (1984) highlighted its importance as a famine food in Ireland. Other important famine foods in Ireland were charlock (*Sinapis arvensis*), nettles, watercress and probably hazelnuts.

Sorrel soup is made by cooking a quantity of chopped sorrel leaves in butter for about 10 minutes, together with a chopped onion, some herbs and a spoonful of flour. Boiling water is added to this mixture, then it is liquidised (or not, according to taste), and breadcrumbs and seasoning are added and the mixture simmered for about an hour. Before serving, two egg yolks and a generous amount of cream is added into the mixture (Mabey, 1972). This soup is delicious and through the cooking loses much of the sharp acidic taste that is characteristic of raw sorrel leaves.

Mrs Beeton (1861: 207) gave an interesting recipe for sorrel and gooseberry sauce to go with goose or ducklings. Half a pint of gooseberries are boiled in water until tender then mashed and passed through a sieve. Then, 1/4 of a pint of sorrel-juice, a glass of sherry, a

248. Dinneen (1927) defined *ribleachán* as 'a kind of sorrel'.

249. Uí Chanainn (1962) gave the word '*puinneóg*' for 'sorrel, dock'.

teaspoon of sugar and 1 oz of butter are added and simmered for 3 or 4 minutes before being served very hot. I've not tried it, but it sounds an interesting and possibly delicious sauce for poultry, despite Mrs Beeton's reservation that 'at fashionable tables it is now seldom or never served'.

The roots of this species were used to obtain a red dye, and it has also been used to make green dyes (Mahon, 1983). It may also have been formerly used as a mordant to set colour in dyeing clothes.

It was used in Ireland in traditional medicine to 'cleanse the blood', treat kidney ailments, reduce fevers, stop bleeding, heal sores and bruises, for jaundice, cancer, burns, and sore throats and was mixed with thistle tops and plantain heads to cure consumption (Allen & Hatfield, 2004). Ui Sheasnáin (1982) recorded that in Cavan, chewed sorrel leaves were mixed with unsalted butter and applied to a wound to halt bleeding, haemorrhages and treat cuts. Egan (1887) recorded its use in Ireland to treat cancer where a decoction of dried blossoms was drunk. According to Uí Chonchubhair (1995), it was also used as a vermifuge. Mears and Hillman (2007) pointed out that *Rumex acetosa* and *R. acetosella* should not be eaten by people suffering from kidney or bladder stones.

Rumex acetosella

- Sheep's sorrel (Scannell & Synnott, 1987); Serrag (Hart, 1898); Soollick (Hart, 1898); Wee sour'k (Praeger, 1938); Red sour'k (Praeger, 1938)
- *Samhadh caorach* (Scannell & Synnott, 1987); *Ruanaidh* (Cameron, 1883); *Pluirín seangan* (de Vere Coneys, 1849; Dinneen, 1927)

POLYGONACEAE—THE KNOTWEED FAMILY

Native

Rumex acetosella is a common perennial herb of acid grasslands, heaths, peaty soils and cultivated ground; it is much rarer on calcareous soils. It is found commonly throughout Ireland but particularly in upland and mountain areas.

This plant was sometimes used medicinally as a diuretic (Moloney, 1919; Uí Chonchubhair, 1995). Threlkeld (1726: unnumbered [4]) noted that the leaves of this and the previous species are 'cooling, quench thirst, repress bile [and] very good in fevers and Cordial' and were eaten during the summer months in salads and good for the treatment of scurvy. This species can be used in very similar ways as *Rumex acetosa*. To make a cold refreshing drink, the leaves are steeped in hot water for 10 to 15 minutes then sweetened and served chilled (Peterson, 1977).

Remains of *Rumex acetosella* have been found in excavations from Fishamble Street that date back more than 1000 years, which may suggest that it was used for some purpose in Dublin in those times because it would not have been a common weed at that time on the calcareous land surrounding Dublin (Geraghty, 1996).

**Rumex conglomeratus*

- Clustered dock (Scannell & Synnott, 1987); Sour dock (Hart, 1898); Sharpleaved dock (Brunker, 1950)

Rumex conglomeratus.

- *Copóg thriopallach* (Scannell & Synnott, 1987); *Copóg ghéar* (Moloney, 1919)

POLYGONACEAE—THE KNOTWEED FAMILY

Native

Rumex conglomeratus is common in most parts of Ireland in damp grasslands.

*No ethnobotanical uses of this plant have been noted in Ireland.

Rumex crispus.

Rumex crispus

- Curled dock (Scannell & Synnott, 1987); Docken (Colgan, 1911); Dogweed (Hart, 1898)
- *Copóg chatach* (Scannell & Synnott, 1987); *Copóg sráide* (Colgan, 1911)

POLYGONACEAE—THE KNOTWEED FAMILY

Native

Rumex crispus is a very common plant throughout Ireland of grasslands, roadsides, waste places, cultivated ground and shingle beaches. A tall variety (var. *uliginosus*) is found also on estuarine mud.

According to van Wyk and Wink (2004) *Rumex crispus* has been used in Europe in traditional medicine to treat internal bleeding and as a diuretic and laxative and against internal parasites (such as tapeworm and roundworm). These same authors noted it has also been used externally for skin complaints, such as ulcers, boils and tumours (van Wyk & Wink, 2004). This plant was also used medicinally in Ireland; juice squeezed from the leaves was applied to bruises, and decoctions of the roots were drunk as a purgative to treat constipation, liver complaints, skin diseases and rheumatism (Allen & Hatfield, 2004). It was also used as a veterinary remedy to treat various ailments, including coughs in cattle and horses and farcy (Allen & Hatfield, 2004).

Mears and Hillman (2007) suggested that this is the most palatable dock with edible leaves; however, they warn that it should be eaten in moderation only because of the oxalic acid it contains and then in combination with dairy products. If the leaves are cooked as a vegetable, their bitterness may be reduced by changing the water a few times during boiling. In North America where the species is naturalised, Native Americans and early settlers boiled the leaves as food (Gilmore, 1991).

Rumex hydrolapathum

- Water dock (Scannell & Synnott, 1987); Sour dock (Hart, 1898); Great water dock (Brunker, 1950)
- *Copóg uisce* (Scannell & Synnott, 1987); *Copóg mhór uisce* (Moloney, 1919); *Bior-chopóg*[250] (Dinneen, 1927; Ó Dónaill, 1977)

POLYGONACEAE—THE KNOTWEED FAMILY

Native

250. Dinneen (1927) defined *bior-chopóg* as 'water dockleaf'. Ó Dónaill (1977) listed as '*biorchopóg*'.

Rumex hydrolapathum is a tall perennial dock of wet drains, shallow water, marshes and ditches that is uncommon in Ireland but distributed throughout.

According to Allen and Hatfield *Rumex hydrolapathum* was used in Ireland to treat scurvy, rheumatism, mouth ulcers and as a blood purifier. In Co. Cavan it was used to treat asthma (Maloney, 1972). Wade (1804: 27) remarked that this species is a useful and powerful astringent, and that its pulverized root 'is perhaps the safest and best dentifrice [toothpaste] we have'.

*Rumex maritimus

- Golden dock (Scannell & Synnott, 1987)
- *Copóg bhuí* (Scannell & Synnott, 1987)

POLYGONACEAE—THE KNOTWEED FAMILY

Native

Rumex maritimus is a very rare annual dock in Ireland of marshy places.

*No ethnobotanical uses of this plant have been noted in Ireland.

Rumex obtusifolius

- Broad-leaved dock (Scannell & Synnott, 1987); Butter dock (Colgan, 1904); Dock (Colgan, 1904); Docken (Hart, 1898[251]; Colgan, 1911); Coppag[252] (Hart, 1898)
- *Copóg shráide* (Scannell & Synnott, 1987); *Bileóg sráide* (Moloney, 1919); *Cupóg sráide* (Ó Criomhthain, MSS, n.d., ca. 1908; Colgan, 1911)

POLYGONACEAE—THE KNOTWEED FAMILY

Native

Rumex obtusifolius.

Rumex obtusifolius is an abundant plant throughout Ireland of grasslands, roadsides and waste places.

According to Allen and Hatfield (2004) *Rumex obtusifolius* was formerly used in Ireland to treat ringworm, stings and rheumatism and to staunch bleeding (sometimes mixed with donkey's dung), and a decoction of the seeds was used to treat coughs, colds and bronchitis, and the roots were used to treat liver ailments, jaundice, cancerous sores, corns and heart troubles. To treat ringworm the roots were made into a paste mixed with equal parts of sulphur and lard (McCarthy & Hawkes, 2007). Uí Chonchubhair (1995) noted that its roots boiled in vinegar were used to wash scabs or broken or itchy skin. The same author recorded that the leaves cooked with meat were believed to be high in iron and therefore useful to strengthen the liver and blood and to treat anaemia. Ní Ghaoithín (1978) wrote that in the Great Blasket Island the roots of dock (probably this species, or perhaps *R. crispus*) were peeled,

251. Praeger (1938) gave this as a general common name for *Rumex* species.
252. Hart (1898) gave this name as a general term for *Rumex* species.

diced and boiled with fresh mutton grease and unsalted butter to make an ointment for healing burns or any wounds. The ointment could be stored for later use as required.

Lewis-Stempel (2012) said that *Rumex obtusifolius* (and *R. crispus*) are edible and suggested picking very young leaves as a substitute for cabbage. The leaves should either be boiled twice or cooked in slightly sugared water to reduce their bitterness. The leaves have a high oxalic acid content. Lewis-Stempel also gave a recipe for stuffed dock leaves, prepared in the same way as are stuffed vine leaves (with a filling of cooked rice, herbs, seasoning and grated cheese). Rolled up with their filling in the dock leaves, they are kept together with a toothpick, brushed with olive oil and baked in an oven for about half an hour.

Dock leaves, usually of this species, are still much used, particularly by children, to rub onto and reduce the pain of nettle stings (*Urtica dioica*). This is probably the most widely known medicinal use and practice involving any Irish wild plant today. The leaves of this species were good for sore feet (Williams, 1993).

Colgan (1904) noted that in parts of Co. Dublin, its leaves were used as a cool, clean wrapping for rolls of fresh butter, giving it a local common name of 'butter dock'.

Lucas (1957) recorded that dock stems were used to make baskets (back baskets, or *cliabh*), especially in the treeless, bushless moors of Erris, Co. Mayo. The flower stems of dock were cut close to the ground when mature and left to dry for a couple of days but used while they were still flexible. If dried they were sometimes soaked in seawater to make them more pliable before use.

Greenish-white, yellow and black dyes were extracted from the roots (Mahon, 1983; Williams, 1993).

**Rumex pulcher*

- Fiddle dock (Scannell & Synnott, 1987)
- *Copóg chumtha* (Scannell & Synnott, 1987)

POLYGONACEAE—THE KNOTWEED FAMILY

Possibly Introduced

Rumex pulcher is found only rarely in Ireland, mainly near the south coast in dry sandy places.

*No ethnobotanical uses of this plant have been noted in Ireland.

**Rumex sanguineus*

- Wood dock (Scannell & Synnott, 1987)
- *Copóg choille* (Scannell & Synnott, 1987)

POLYGONACEAE—THE KNOTWEED FAMILY

Native

Rumex sanguineus is a common dock throughout Ireland of woodlands, hedgerows and damp shady places.

*No ethnobotanical uses of this plant have been noted in Ireland.

**Rumex tenuifolius*

- Narrow-leaved sorrel (Scannell & Synnott, 1987)

- *Samhadh caorach caol* (Scannell & Synnott, 1987)

POLYGONACEAE—THE KNOTWEED FAMILY
Native

Rumex tenuifolius is known only from Co. Wexford.

*No ethnobotanical uses of this plant have been noted in Ireland.

---

*Ruppia cirrhosa

- Spiral tasselweed (Scannell & Synnott, 1987)
- *Scothóga uisce* (Scannell & Synnott, 1987)

RUPPIACEAE—THE TASSELWEED FAMILY
Native

Ruppia cirrhosa is a perennial submerged aquatic plant found on coastal mudflats and in brackish drains, ditches and pools close to the sea. This species is rare in Ireland and is found mainly on the north-east and south-east coasts.

*No ethnobotanical uses of this plant have been noted in Ireland.

---

*Ruppia maritima

- Beaked tasselweed (Scannell & Synnott, 1987)
- *Scothóga mara* (Scannell & Synnott, 1987)

RUPPIACEAE—THE TASSELWEED FAMILY
Native

Ruppia maritima occurs in similar habitats as *R. cirrhosa*. It can be found all around the Irish coasts but is common only in the south-west.

*No ethnobotanical uses of this plant have been noted in Ireland.

*Sagina apetala

- Annual pearlwort (Scannell & Synnott, 1987); Common pearlwort (Anonymous, 1978)
- *Mongán lom* (Scannell & Synnott, 1987); *Lus na bpéarlaí*[253] (Dinneen, 1927)

CARYOPHYLLACEAE—THE CARNATION OR PINK FAMILY
Native

Sagina apetala is a tiny annual plant found on paths, walls and bare ground throughout Ireland.

*No ethnobotanical uses of this plant have been noted in Ireland.

---

*Sagina maritima

- Sea pearlwort (Scannell & Synnott, 1987)
- *Mongán mara* (Scannell & Synnott, 1987)

CARYOPHYLLACEAE—THE CARNATION OR PINK FAMILY
Native

Sagina maritima is found all around the Irish coastline on walls, pavement cracks, piers and wasteground near the sea.

*No ethnobotanical uses of this plant have been noted in Ireland.

---

*Sagina nodosa

- Knotted pearlwort (Scannell & Synnott, 1987); Knotted spurrey (Brunker, 1950)
- *Mongán gluineach* (Scannell & Synnott, 1987)

CARYOPHYLLACEAE—THE CARNATION OR PINK FAMILY
Native

Sagina nodosa is a perennial herb particularly in the west and centre of Ireland of damp, sandy and peaty places, such as lake-shores and marshes.

*No ethnobotanical uses of this plant have been noted in Ireland.

---

*Sagina procumbens

- Procumbent pearlwort (Scannell & Synnott, 1987)
- *Mongán sinte* (Scannell & Synnott, 1987)

CARYOPHYLLACEAE—THE CARNATION OR PINK FAMILY
Native

Sagina procumbens is a tiny perennial found commonly throughout Ireland in damp lawns, grassy places and banks, paths and streamsides.

*No ethnobotanical uses of this plant have been noted in Ireland.

---

*Sagina subulata

- Heath pearlwort (Scannell & Synnott, 1987)

253. Dinneen gave this Irish name for 'pearl-wort'.

- *Mongán móna* (Scannell & Synnott, 1987)

CARYOPHYLLACEAE—THE CARNATION OR PINK FAMILY
Native

Sagina subulata is a plant of dry, sandy and rocky habitats that is occasionally found only near the west and north coasts of Ireland.

*No ethnobotanical uses of this plant have been noted in Ireland.

Sagittaria sagittifolia

- Arrowhead (Scannell & Synnott, 1987)
- *Rinn saighde* (Scannell & Synnott, 1987)

ALISMATACEAE—THE WATER-PLANTAIN FAMILY
Native

Sagittaria sagittifolia is a rare aquatic plant of pools, ponds, slow-moving streams, lakes and canals in the centre and east of Ireland only.

Although arrowhead has been widely used as a food source in parts of its range, such as by Native North Americans, there are no records of such use in Ireland. Its corms are edible (Heywood, 1978).

In Ireland it may have been used occasionally as an ornamental plant for aquaria and garden ponds.

Salicornia species

- Glasswort (Scannell & Synnott, 1987); Samphire
- *Lus na gloine* (Scannell & Synnott, 1987); *Praiseach na mara* (Cameron, 1883)

AMARANTHACEAE (Chenopodiaceae)—THE GOOSEFOOT FAMILY
Native

Including:

Salicornia dolichostachya
- Long-spiked glasswort

Salicornia europaea
- Common glasswort

Salicornia fragilis
- Yellow glasswort

Salicornia nitens
- Shiny glasswort

Salicornia perennis (Arthrocnemum perenne)
- Perennial glasswort

Salicornia pusilla
- One-flowered glasswort

Salicornia products in France.

Salicornia ramosissima

- Purple glasswort

These are saltmarsh plants, represented in Ireland by the different species listed above. *Salicornia dolichostachya* and *S. fragilis* are found growing in mud on the lower parts of salt marshes and are quite common throughout the Irish coastline. *Salicornia europaea* is more generally distributed throughout many Irish salt marshes and is common. *Salicornia nitens* is a rare species, mainly found on the south and south-east coasts. It occurs in the upper and middle parts of salt marshes. *Salicornia pusilla* occurs on the upper parts of salt marshes and is quite rare in Ireland, occurring mainly on the south coast. The sixth species in Ireland is *S. perennis*, a perennial species (all the others are annuals) sometimes separated into a different genus, *Arthrocnemum*. It is rare in Ireland and occurs only on tidal mud on the south coasts of Co. Wexford.

Salicornia is picked and used as a vegetable from mainland European coasts, but is not very commonly collected in Ireland. It was sold occasionally in the market in Galway in the mid-1990s (M. Sheehy Skeffington, pers. comm., 2007). I have also seen it sold as a fresh vegetable in a fish shop on Howth Pier, labelled as imported from France. Prendergast and Sanderson (2004) described its use in Britain, harvested from the salt marshes on Norfolk and Lincolnshire coasts, particularly during June to September. Glasswort can be eaten raw as a crispy snack or boiled or steamed as a vegetable, like asparagus (Mears & Hillman, 2007; Hilton, 2007). It is also pickled or can be made into purée. *Salicornia* soup is also good to eat. Mabey (1998) mentioned that it featured as a dish served at the wedding breakfast of Prince Charles and Lady Diana Spencer in 1981, sent from the Royal Sandringham Estate in Norfolk. I have been served *Salicornia* as a 'gourmet' vegetable in more than one Irish hotel.

The name glasswort comes from its former use to provide a source of alkali for glass manufacture.

Salix species

- Willow (Scannell & Synnott, 1987); Sallow (Mackay, 1836)
- *Saileach* (Scannell & Synnott, 1987); *Cran-sailigh* (de Vere Coneys, 1849; Dinneen, 1927); *Suil* (de Vere Coneys, 1849); *Soileóg* (de Vere Coneys, 1849); *Crann saileoige* (Dinneen, 1927); *Sail* (Dinneen, 1927; Uí Chanainn, 1962; Ó Dónaill, 1977); *Saileog* (Uí Chanainn, 1962; Ó Dónaill, 1977)

SALICACEAE—THE WILLOW FAMILY

Including:

Salix alba

- White willow (Scannell & Synnott, 1987)
- *Saileach bhán* (Scannell & Synnott, 1987); *Saileóg* (de Vere Coneys, 1849; Moloney, 1919; Dinneen, 1927)

Possibly Introduced (possible archaeophyte)

Salix aurita

- Eared willow (Scannell & Synnott, 1987)
- *Crann sníofa* (Scannell & Synnott, 1987); *Sliotharnach* (Ua Maoileoin, 2001); *Crann svíomhtha* (Colgan, 1911; Dinneen, 1927); *Saileán* (Colgan, 1911)

Native

Salix caprea

- Goat willow (Scannell & Synnott, 1987); Great sallow (Anonymous, 1978); Pussy willow (Anonymous, 1978); Wild rod (McCarthy & Hawkes, 2007); Goat sallow (Praeger, 1938)
- *Sailchearnach* (Scannell & Synnott, 1987); *Saileog* (Cameron, 1883)

Native

Salix cinerea (Salix atrocinerea)

- Rusty willow (Scannell & Synnott, 1987); Sally (Anonymous, 1978; Webb et al., 1996); Common sallow (Brunker, 1950; Anonymous, 1978); Grey willow (Anonymous, 1978); Thillag (Hart, 1898); Grey sallow (Praeger, 1938); Salliagh (Harron, 1986)
- *Saileach liath* (Scannell & Synnott, 1987); *Crann svíomtha* (Moloney, 1919); *Crann sníomhtha* (Moloney, 1919)

Native

Salix fragilis

- Crack willow (Scannell & Synnott, 1987)
- *Saileach bhriosc* (Scannell & Synnott, 1987); *Gad sailí*[254] (Ó Dónaill, 1977)

Introduced

Salix pentandra

- Bay willow (Scannell & Synnott, 1987)
- *Saileach labhrais* (Scannell & Synnott, 1987)

Native

Salix purpurea

- Purple willow (Scannell & Synnott, 1987)
- *Saileach chorcra* (Scannell & Synnott, 1987)

Possibly Introduced

Salix triandra

- Almond willow (Scannell & Synnott, 1987); Almond-leaved willow; French willow
- *Saileach na dtrí bhall* (Scannell & Synnott, 1987)

Possibly Introduced

Salix viminalis

- Osier (Scannell & Synnott, 1987); Twig (McCarthy & Hawkes, 2007); Basket osier (Brunker, 1950)
- *Saileánach* (Scannell & Synnott, 1987); *Gallshaileach* (Moloney, 1919; Dinneen, 1927); *Maothán* (Moloney, 1919); *Fineamhain* (Cameron, 1883); *Bior-ghas* (Dinneen, 1927); *Suileasc* (Dinneen, 1927); *Tuigíos*[255] (Dinneen, 1927); *Fualascach*[256] (Ó Dónaill, 1977); *Gallsaileach* (Ó Dónaill, 1977); *Slat sailí* (Ó Dónaill, 1977); *Biorraide*[257] (Uí Canainn, 1962); *Féith*[258] (Uí Chanainn, 1962)

Introduced

254. Ó Dónaill (1977) defined *gad sailí* as 'osier withe'.
255. Dinneen (1927) defined *tuigíos* as 'twigs, osiers'.
256. Ó Dónaill (1977) defined *fualascach* as 'twigs, osiers'.
257. Uí Chanainn (1962) defined *biorraide* as 'twig, osier sapling'.
258. Uí Chanainn (1962) defined *féith* as 'fibre, twig, honeysuckle and osier'.

Salix aurita.

Salix caprea.

Salix alba is a common species throughout Ireland occurring in hedges, river-banks and other wet habitats. *Salix aurita* is a very common species found in mountain heaths, moors, bog-margins, ditches, drains and stream-sides in the mountains. *Salix caprea* is quite often seen growing in hedgerows and woodlands. It can grow to be quite a substantial tree. *Salix cinerea* is very common in wet woodlands, scrub and hedgerows. *Salix fragilis* and *S. viminalis* are widespread in hedgerows and waste places. *Salix pentandra* is widespread in Ireland but only common in the north and occurs in hedgerows, scrub and river-sides. *Salix purpurea* is a possible introduction, but it may be native in the midlands. It is found relatively commonly in hedges, roadsides, bog-margins and lake-shores. *Salix triandra* is a possible introduction, but it may be native in the south-east. It is a rare species found in wet habitats in hedgerows and scrub.

It is often difficult, if not impossible, to tell whether one or other of the introduced willow species have spread naturally or been planted in hedgerows. Today many specimens probably survive where they were originally planted as cuttings in a hedgerow either to supply rods for basketry or to fill a gap in the hedge. Webb and Scannell (1983) mention that *Salix alba* and *S. fragilis* were widely planted in the Connemara and Burren regions for ornamental purposes; that *S. pentandra* was planted for hedging and *S. viminalis* for basket making. In the same regions, *S. purpurea* occurs mainly in hedges and the native species *S. cinerea* (*S. atrocinerea*), *S. repens*, *S. aurita*, *S. caprea* and *S. herbacea* are probably rarely, if ever, planted.

Willows have traditionally been one of the most widely used and useful of native woody plants, used for craftwork, furniture, basketry, currach construction and many other purposes. Willow can be cleft into long, thin bands that are flexible and easily bent while fresh and green or after steaming. Most willows used in basketry, wickerwork and other craft work are non-native species such as *Salix viminalis* (osier), *S. purpurea* (purple osier), *S. fragilis* (withy), *S. triandra* (almond-leaved willow or French willow), *S. alba* (white willow) and various hybrids, which have been planted and grown, often on roadsides, ditches, river-banks and hedgerows, and used in Ireland for centuries.

Willow wood is lightweight, pliable and flexible. Its timber is pinkish brown, very soft and fine grained. Although willow wood is not of high quality, its importance is partly because of its ability to grow very quickly and its easy availability. Rutty (1772: 125) wrote that its fine-grained wood was used for a large number of utensils, particularly large cups, cans and vessels 'valued for their keeping milk sweet'. Willows have traditionally been used for basket making and wickerwork. O'Hare (1993) noted that (as recorded by Reverend J. Hall when he travelled through Ireland in 1813) doors made from wicker or sally rods, in a frame of bog deal (*Pinus sylvestris*) were widely used for simple houses. Wicker was used for a wide range of other purposes in Ireland, such as for furniture and wicker canopies over the hearth of simple houses and was often made from willow rods, although other species such as hazel (*Corylus avellana*), elder (*Sambucus nigra*) and bramble (*Rubus fruticosus*) would also have been used.

Rutty (1772: 127) made specific mention of a species of osier, called 'The golden Ozier' that was particularly 'worthy of notice by reason of the singular flexibility of its twigs, which is such that it is used not only for tying up bundles of Turneps, but may also, in some sort, supply the place of thread of flax or hemp in mending broken dishes, brogues etc. and it makes good baskets'. Rutty said that it was imported and propagated by 'William Connolly' and grown on the banks of the river Liffey near Chapelizod, Co. Dublin. This was probably William Conolly (1662–1729), also known as 'Speaker Conolly', a lawyer and Irish politician, who built the magnificent Castletown House in Celbridge, Co. Kildare. O'Sullivan (1977) recorded that in the 1900s willows (called 'sally') were grown in several damp places in the Aran Islands, Co. Galway, and used for various purposes, particularly for basketry—for making panniers for donkeys, lobster pots, baskets to store potatoes, turf and a variety of other uses. Upright rods were planted in the ground and other lighter rods woven in and out of them to create the basket of varying shapes and sizes depending on the use. Osiers were normally harvested after one year's growth and used either with or without their bark.

A creel made from *Salix* species.

Some basket-makers maintained their own 'sally gardens' where they would grow the rods (*saighléanach*, 'a willow rod or osier', Dinneen, 1927) they needed for their trade. These gardens were generally stocked with osiers. Different varieties were grown which might have different colours or characteristics. Delaney (1983: 216) described one of these sally gardens from Co. Longford:

> There are eight or nine stools of osiers in this garden, which is an extension of the kitchen garden. Two kinds of sally are grown, the green sally and the golden osier, and though they differ in colour and name, they both belong to the same species, *Salix viminalis*. The rods are cut every year about November or December, made into neat bundles and placed against the wall of an outhouse, where they are left standing, exposed to the weather. In this way they mature slowly. Weak rods are not cut but left to grow for a second year, when they will be strong enough to use in the making of heavy baskets such as the 'cleeve', or back basket as it is sometimes called. . . . Sometimes before weaving the baskets, the rods were boiled and then peeled. The peelings came off in one piece, like taking off a stocking. Strong ropes were made of these peelings, in the same way as making a thumbrope of straw or hay, and that they would last for years.

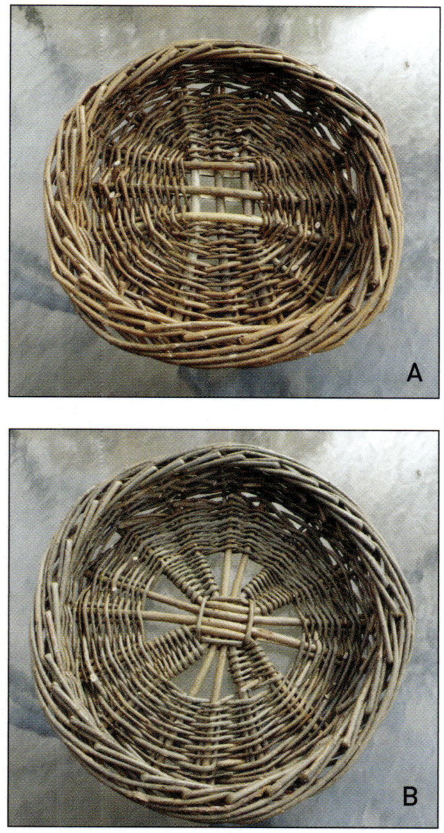

Traditional potato skib patterns: A, From the Aran Islands. B, From Joyce Country, Co. Galway.

Sally (or salley) gardens are probably best known from the William Butler Yeats poem 'Down by the Salley Gardens':

> Down by the salley gardens my love and I did meet;
> She passed the salley gardens with little snow-white feet.
> She bid me take love easy, as the leaves grow on the tree;
> But I, being young and foolish, with her would not agree.
> In a field by the river my love and I did stand,
> And on my leaning shoulder she laid her snow-white hand.
> She bid me take life easy, as the grass grows on the weirs;
> But I was young and foolish, and now am full of tears. (Yeats, 1949: p. 68)

Near Annascaul, Co. Kerry, I recently found what may have formerly been a sally garden. It consisted of more than a dozen specimens of osier (*Salix viminalis*), together with a handful of specimens of withy (or crack willow; *S. fragilis*) grown along a series of field margins and river-banks. They were all growing within easy distance of an old ruined

A former sally garden of willows near Annascaul, Co. Kerry.

farmhouse and looked as if they had not been harvested for 10 to 15 years. I like to imagine that this was the former home of a basket maker.

Cultivating willows around Irish country cottages for use in basketry was probably a common practice. Dutton (1808: 145) noted that in Co. Clare, 'Few cottages are without sallows for kishes or baskets, and which every labourer knows how to make'. Sheehy Skeffington (1984) records that there are old specimens of *Salix viminalis*, the osier, planted along the Grand Canal in Dublin city that may originally have been planted as a source of material for basketry. An ancient and rather sinister use of sally rods in the past has been noted by Friel (1997), who mentioned that the body of a man preserved in a bog near Castleblakeney in Co. Galway was found in 1821 and dated back to the Iron Age. The man was naked except for a deerskin leather cape that was tied around his neck with twisted sally rods. It is suggested that he was ritually strangled with these rods. The body is now preserved in the National Museum in Dublin. Gardey (1971) described a pre-famine coffin made from sally and hazel rods found in a graveyard in Co. Down. The frame was made from two hazel rods lashed together top and bottom and the pliable sally rods lapped around them. He also noted that there are records of similar rod and wicker coffins in Co. Westmeath where they were lined with rushes.

The National Museum of Ireland (Museum of Country Life) in Castlebar, Co. Mayo, includes a large range of objects made from willow, including many baskets. Willow rods were preferred for basketry, although when they were scarce, briars, docks and even heather were used. The Museum also includes a willow-made calf basket, like a muzzle, used to prevent young calves from suckling cows and eating bedding. One such example in the Museum was made by Thomas Loughnane from Clondelara, Co. Offaly (Ref. No: F1951.69).

Hogan (2001) described and illustrated an artefact called a 'cradle bird' made from willow rods (and hazel too). This was used to trap blackbirds and other birds of similar size. The cradle trap would drop onto the bird when it stood on a curved willow rod that acted as the trigger for the trap. Such an object would now be illegal in Ireland and their use had declined by the 1950s.

Rutty (1772: 126) said that osiers were used in the construction of fish weirs, and were also planted to support 'impetuous rivers' as the roots helped to bind the soil and prevented ditches, dykes and banks from being washed away by the water.

According to Delaney (1980: 241–242), in Co. Offaly, a flail was made from sally used in the threshing of corn: '. . . the striking stick was called the boolcheen [Irish, *builtin*] and was made of holly or crabtree; the handle was called the handstaff and was usually made of a fairly thick sally stick; the caps were made of black sally, cut in the summer or late harvest — "black sallies that you get in a swamp". They were dried and (then) cut in lengths of about sixteen inches. They had to be seasoned before they could be used. To season them they used to hang them in the house, near the fire' (from the reminiscences of Thomas Horan). The two parts of the flail were joined together with materials such as eelskin (Siggins, 2004).

Willow rods were used to make the ribs and laths forming the hulls of currachs, the traditional boats of the west and south-west of Ireland. In Co. Kerry, *naomhóg*-makers favoured and used goat willow (*Salix caprea*) to provide the bent wood needed for making the bows of these currachs (H. C. Lönze, in MacCárthaigh, 2008). The use of willow in the construction of currachs has now more or less disappeared, although hazel still continues to be used in some regions. Cameron (1883) mentioned that 'gads' or willow ropes were made in Ireland up to the 1880s at least. MacCoitir (2003) and Baker (1999) noted that willows were used in Ireland for making various household implements. Several old Irish harps are made from willow wood, including the famous Trinity College Dublin harp that dates from the 14th century (Rimmer, 1969). The oak staves of barrels were bound into shape by hoops made from hazel or willow up until the 19th century, after which they were replaced by metal hoops (McCracken, 1971). Willow rods (see also *Corylus avellana*) were traditionally used in the making of coracles, small oval boats used on the river Boyne.

In the west of Ireland up until the time of the famine, shovels (used for digging potato fields and for other crops) were made from a single piece of wood, usually willow, and strengthened with a strip of metal along the edge (Evans, 1957). Evans also mentioned that 'black sally' was used in the handles of turf-cutting slanes in the Antrim Glens, where a bent piece of this wood was spliced onto the handle because of its sweat-absorbing qualities. Upon reading this, I was uncertain what species of willow was suggested by the name 'black sally'. *Salix myrsinifolia* (*S. nigricans*), the dark-leaved willow is called *Saileach dhubh* in Irish, but it is relatively rare in Ireland and I was not convinced that this was the species meant here or in the earlier reference to 'black sally' use in Co. Offaly. Joe Hogan (pers. comm., 2010) told me, however, that this is a commonly used collective name for the native willows, that is, those not (or not much) used in basket making.

Willow rods derived from coppiced trees were widely used to make 'spars' or 'scollops', twisted hoops of varying shapes and sizes, sharpened at each end and used to peg down thatch on traditional cottages. The spars were pushed through the thatch and rooted into the sod scraws below, which line the undersides of the thatch. Many thatchers would maintain their own willow coppice, which they would harvest to make spars whenever needed (Sharkey, 2000). In the midlands, hazel was often used in preference to willow spars.

Salix alba var. *caerulea* is the cricket bat willow, a very fast-growing variety. It is regarded as the best timber for cricket bat production and is occasionally grown as an ornamental tree in gardens in Ireland. Willow rods were used by children to make bows and arrows as they were flexible, strong and straight. In suburban settings, though, the arrows were more usually made from non-native garden bamboos, which were prized by children for being very straight and easily split to insert a feather to make the arrow flights. A rough sieve made of cut shoots of osier (or alternatively heather or hazel) was used to filter the sugar-rich wort drawn off from a vat of malted barley used in poteen making in Ireland. Yeast was subsequently added to the wort for fermentation, which was later distilled to yield the alcohol (poteen) (Ó Curraidhín, n.d., ca. 2009).

Charcoal made from willow wood was a favourite for artists, and willows in general make good charcoal (Dinneen, 1927). Rutty (1772: 126) said that it was used by painters 'for delineating their design on papers' being fine and apt to split into pencils'. Milliken and Bridgewater (2004) mentioned that plantations of goat willow (*Salix caprea*) were grown for that purpose in Scotland. In the early 1800s charcoal made from willow (and alder) was used in the manufacture of gunpowder near Cork (Wakefield, 1812). Dinneen recorded that in an East Kerry custom on St Patrick's Day, a sally root was charred and crosses marked on the shoulder.

Willow bark has been used for tanning and for weaving coarse ropes, and willow roots were used in the preparation of a purplish red dye (Edlin, 1973). Rutty (1772: 126) said that the bark of the 'broad-leaved sally' [presumably *Salix caprea*] is used for tanning. The barks of *S. viminalis* and *S. caprea* have been the species of willow most widely used for tanning in northern Europe (Howes, 1974). They give a light yellow-brown leather that is soft and flexible and much used for gloves (Howes, 1974). Siggins (2004) noted that the manufacture of ropes and cord from the bark of trees seems to be a tradition in Co. Roscommon. Wade (1804) recorded that the bark of *S. cinerea* was made into coarse paper and 'paste board' (although not necessarily in Ireland). Willow bark was also smoked like tobacco (Lysaght, 1979).

The leaves of *Salix* species contain salicylic acid (the basis for the pain-killing medicine, aspirin) and have long been recognised as a cure for aches and pains. In Ireland, the ashes of willow were formerly mixed with a fatty substance such as lard to make an ointment to treat ringworm, baldness and other ailments (Allen & Hatfield, 2004). Sally rods were used to provide splints for broken bones (Lysaght, 1979). To cure baldness the ashes would be mixed with hog's lard and turpentine to form a thick paste and rubbed onto the head nightly (Logan, 1972; 1981). However, apparently the treatment does not work if the roots of the hair are dead (Logan, 1972). Allen and Hatfield (2004) also record the use of *Salix alba* in Ireland to treat pip, a disease of poultry.

Willows are also being experimented with as a potential biomass crop for Ireland, particularly for their cultivation and short-rotation coppicing on cut-over bogs.

The following are some of the willow hybrids (involving both native and non-native parents) that have been recorded in Ireland. Some of them are certainly naturally occurring hybrids, whereas others have probably been propagated, cultivated and used for their particular basket-making properties and other uses. It is not surprising that many willows found in the countryside are difficult to identify, even for an expert.

- *Salix fragilis* × *S. pentandra* (*S.* ×*meyerana*)
- *Salix fragilis* × *S. triandra* (*S.* ×*speciosa* (*S.* ×*alopecuroides*)
- *Salix alba* × *S. fragilis* (*S.* ×*rubens*)—Hybrid crack-willow
- *Salix alba* × *S. pentandra* (*S.* ×*ehrhartiana*)
- *Salix triandra* × *S. viminalis* (*S.* ×*mollissima*)—Sharp-stipuled willow
- *Salix purpurea* × *S. viminalis* (*S.* ×*rubra*)—Green-leaved willow
- *Salix purpurea* × *S. viminalis* × *S. cinerea* (*S.* ×*forbyana*)—Fine osier
- *Salix viminalis* × *S. caprea* (*S.* ×*sericans*)—Broad-leaved osier
- *Salix viminalis* × *S. caprea* × *S. aurita* (*S.* ×*stipularis*)—Eared osier (Reynolds, 2002)
- *Salix viminalis* × *S. cinerea* (*S.* ×*smithiana*)—Silky-leaved osier
- *Salix viminalis* × *S. aurita* (*S.* ×*fruticosa*)—Shrubby osier
- *Salix caprea* × *S. cinerea* (*S.* ×*reichardtii*)
- *Salix caprea* × *S. aurita* (*S.* ×*capreola*)
- *Salix cinerea* × *S. aurita* (*S.* ×*multinervis*)
- *Salix cinerea* × *S. myrsinifolia* (*S.* ×*strepida*)
- *Salix cinerea* × *S. phylicifolia* (*S.* ×*laurina*)—Laurel-leaved willow
- *Salix aurita* × *S. myrsinifolia* (*S.* ×*coriacea*)
- *Salix aurita* × *S. repens* (*S.* ×*ambigua*)
- *Salix caprea* × *S. cinerea* × *S. viminalis* (*S.* ×*calodendron*)

The weeping willow, *Salix* ×*sepulcralis* (*S. alba* × *S. babylonica*), is occasionally planted as an ornamental tree.

**Salix herbacea*

- **Dwarf willow (Scannell & Synnott, 1987)**
- ***Camshaileog* (Scannell & Synnott, 1987)**

SALICACEAE—THE WILLOW FAMILY

Native

Salix herbacea is a rare species in Ireland that is restricted to mountain tops and occurs on cliffs, ridges, ledges and summits.

Dinneen (1927) mentioned a little piece of Irish folklore about this species, which people said was once a large tree, but was cursed when its wood was used to make Our Lord's Cross.

*No ethnobotanical uses of this plant have been noted in Ireland.

**Salix myrsinifolia (Salix nigricans)*

- **Dark-leaved willow (Scannell & Synnott, 1987)**
- ***Saileach dhubh* (Scannell & Synnott, 1987)**

SALICACEAE—THE WILLOW FAMILY
Probably Native

Salix myrsinifolia is mainly confined to the north and north-east of Ireland where it is rare.

*No ethnobotanical uses of this plant have been noted in Ireland.

**Salix phylicifolia (Salix hibernica)*

- **Tea-leaved willow (Scannell & Synnott, 1987); Irish willow (Nelson, 1984)**
- ***Saileach ghaelach* (Scannell & Synnott, 1987)**

SALICACEAE—THE WILLOW FAMILY
Native

Salix phylicifolia is a very rare species in Ireland and is found only on the Ben Bulben range in Cos. Leitrim and Sligo.

*No ethnobotanical uses of this plant have been noted in Ireland.

Salix repens

- **Creeping willow (Scannell & Synnott, 1987); Sally (Colgan, 1904); Bogall (Hart, 1898); Cran commer (Hart, 1898)**
- ***Saileach reatha* (Scannell & Synnott, 1987)**

SALICACEAE—THE WILLOW FAMILY
Native

Salix repens is a common species throughout Ireland of sand dunes and other sandy, peaty and rocky habitats near the coast; it also occurs inland, although only rarely.

In the Glenalla region of Donegal a decoction (where the leaves are placed in boiling water and the resulting liquid drunk) from the leaves of *Salix repens* was made to treat 'pains in the head' (Hart, 1898: 370).

Salsola kali

- **Prickly saltwort (Scannell & Synnott, 1987); Saltwort (Brunker, 1950)**
- ***Lus an tsalainn* (Scannell & Synnott, 1987)**

AMARANTHACEAE (Chenopodiaceae)—THE GOOSEFOOT FAMILY
Native

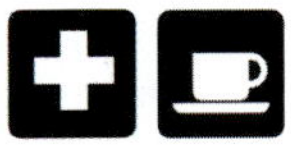

Salsola kali is an occasional plant of sandy seashores all around the Irish coastline.

Like *Suaeda* species, this plant was also burned to make soda ash for the production of sodium carbonate for soap manufacture. The caustic ash that resulted was used also for medicinal purposes, to remove old scabs and leprosy and for cleaning the skin. A drink made from the juice of *Salsola kali* was also recorded as being used to help against depression as well as to stimulate urine or expel the stillborn (Uí Chonchubhair, 1995).

Salvia horminioides (Salvia verbenaca)

- Wild clary (Wild sage) (Scannell & Synnott, 1987)
- *Tormán* (Scannell & Synnott, 1987); *Ceann cinnchoille* (Moloney, 1919); *Sáiste chuic* (Moloney, 1919); *Sáiste fhiadhain* (Moloney, 1919; Dinneen, 1927); *Athair liath* (de Vere Coneys, 1849; Moloney, 1919; Dinneen, 1927); *Sáiste cnuic* (Dinneen, 1927)

LAMIACEAE (Labiatae)—THE DEAD-NETTLE FAMILY
Native

Salvia horminioides is a very rare plant of coastal habitats on the south and east of Ireland that has been used in other countries as an aromatic herb and for wine making. I have not found records of its use in Ireland for such purposes.

Cameron (1883) mentioned that the mucilage produced by the seeds of this plant, when applied to the eye, had a reputation for clearing it of dust (hence its English name 'clear-eye' or 'clary', according to Cameron). However, I have not found records of its use in this way in Ireland. Dinneen (1927: 176) gave the Irish name *ceannbhán cinn coille* for 'clary'.

The common kitchen herb sage is *Salvia officinalis* (in Irish, *sáiste*, de Bhaldraithe, 1959; *saisde*, de Vere Coneys, 1849). It has been used for culinary and medicinal purposes for several centuries in Ireland but is not native.

Sambucus nigra

- Elder (Scannell & Synnott, 1987) (or also called in parts of Ireland, Boor [or Bour/Bore] tree, or Bootree); Bore-tree (Colgan, 1911); Thrumman (Hart, 1898)
- *Trom* (Scannell & Synnott, 1987); *Bualtraidh* (Moloney, 1919); *Ruis* (de Vere Coneys, 1849; Moloney, 1919; Dinneen, 1927; Ó Dónaill, 1977; Uí Chanainn, 1962); *Druman* (Cameron, 1883); *Droman* (Cameron, 1883); *Peith* (de Vere Coneys, 1849); *Peith-bhog* (*-buighe*) (de Vere Coneys, 1849); *Tromán*[259] (de Vere Coneys, 1849); *Tromm* (or *Truimm*) (de Vere Coneys, 1849); *Crann truim* (*druim*) (Dinneen, 1927); *Craobh tromáin* (Dinneen, 1927); *Ceireachán* (*dubh*) (Ó Dónaill, 1977); *Crann troim* (Ó Dónaill, 1977); *Troum* (Colgan, 1911)
- *Ellena ghou* (in Yola)

CAPRIFOLIACEAE—THE HONEYSUCKLE FAMILY
Native

---

259. Uí Chanainn (1962) gave the Irish word *tromán* for 'dwarf elder'.

Elder is a common shrub or small tree found everywhere in Ireland. Although *Sambucus nigra* is generally a small tree, some substantial specimens do occur occasionally. The current girth champion of elder in Ireland is a specimen growing at Greenside, Ballysax, Co. Kildare, that is 2.32 m in girth (and 6 m tall) (Anonymous, 2005). It occurs widely on roadsides and in woodlands, hedgerows, ditches and waste places and is particularly frequent in disturbed nitrogen-rich habitats, such as the corners of fields where cattle or sheep congregate. It may have been widely planted in some parts of Ireland but is now completely naturalised throughout the country.

Elder has been an extremely important source of food and medicine in Europe since the earliest times. In medicine it is recorded as having been used to treat a wide variety of ailments, such as respiratory troubles, swellings, burns and scalds and skin problems, and an ointment made from the leaves was a popular Dublin remedy for piles (Moloney, 1919). Lysaght (1979) reported that clay from under an elder tree was applied to an aching tooth (this was a more pleasant treatment than another Irish treatment alternative, putting a live frog in one's mouth), and elder bark or leaves were used in Ireland to make a poultice for ulcers. Logan (1981) noted that the soft pith of elder stems was formerly made into a poultice with cream and used for the treatment of ringworm. Mackay (1836) remarked that its bark was used medicinally by country practitioners. In the Mizen region of West Cork, elderberry juice made of boiled berries was used to 'clear up the yellow cast of jaundice in old people' (McCarthy & Hawkes, 2007: 80). In the Middle Ages, it was burnt and the smoke inhaled for hernia. According to Williams (1993), tea made from the inner bark was used as a cure for epilepsy or the inner bark was mixed with sheep fat as an ointment for burns. The same author noted that it was used as a cure for warts in various parts of the country; in Clare the leaves were rubbed on the warts, but in Longford it was necessary to bury pieces of elder wood to affect a cure. Lady Wilde (1890) also mentioned that elder was used to treat epilepsy. A twig was cut into nine parts and strung around the neck as a necklace. Lady Wilde (1887: 91) also mentioned the use of elder bark to treat a burn. 'There is a pretty secret to cure a burn without a scar: Take sheep's suet and the rind of the elder-tree, boil both together, and the ointment will cure a burn without leaving a mark.' In Cavan, the berries were mixed with wine and rubbed on the forehead to cure pimples. Elderberries are strongly laxative if eaten raw. Dried berries have been used for a variety of medicinal purposes such as bronchitis, colds, sore throat, influenza and infections (Ciesla, 2002).

Sambucus nigra was used for various veterinary remedies in Ireland: Allen and Hatfield (2004) recorded its use to treat pigs; Williams (1993) noted that the flower buds and shoots were used as a cure for rot in sheep and boiled leaves used to cure blackleg in Wicklow. McGlinchey (2007) wrote that the elder was also used to make a type of water-gun in Co. Donegal to give an enema to cattle suffering from gas (a hollowed out bootree [elder] stick was used for this purpose).

Elder flowers are used for culinary purposes, such as elderflower 'champagne', cordial, pickles or fritters. Elder flowers can also be added to fruit cheeses and add interest and flavour to these (see *Malus sylvestris*, crab-apple, for the method by which fruit cheeses are made). Elderberry wine was formerly used as a treatment for sciatica and is still an occasionally made country recipe. I have never made elderberry wine, but I have tried a bottle that I bought in Hampton Court in London and kept for 10 years before drinking. I found it very acidic (perhaps I kept it too long) and after two glasses I awoke the next morning with a bad hangover! I used the rest of the bottle to stew

Sambucus nigra.

Elderberries.

some pears, which were delicious. Elderberries can be used to flavour vinegar too, or used in chutney, or made into ketchup or syrup. Ó Céirín and Ó Céirín (1978) provided a recipe for elderberry soup, which can be served hot or chilled, made with 500 g/1 lb elderberries, grated lemon rind, 15 g/0.75 oz of corn flour or potato flour and a little sugar. Ciesla (2002) noted that the dried berries were used like currants in cooking. Some people are allergic to raw elderberries so it is probably advisable always to cook them. Williams (1993) suggested that the berries should not be fed to chickens as it makes their eggs taste odd. Elderberries can be combined with blackberries to make a delicious jam: 450 g/1 lb of elderberries and 450 g/1 lb of blackberries are simmered together until soft; 900 g/2 lb of sugar and 4 tablespoons of lemon juice are stirred in until the sugar has dissolved in the mixture; after this, the jam is boiled rapidly until setting point has been reached when it can be spooned into sterilised jam jars and sealed. To know when the setting point has been reached a small amount of the jam mixture is spooned onto a cold saucer. When it has cooled, it is pushed with a finger and if it wrinkles well then the jam is ready. If not, or if the mixture wrinkles only slightly, return the pan to the heat and boil rapidly for a couple more minutes. Jam can also be made from elderberries alone, which has a delicate flavour and wonderful colour (I serve it with roast lamb and chicken), but it has a stronger taste if mixed with blackberries.

The Romans used elderberries as a hair dye (Phillips, 1983), and in Ireland, Sharkey (2000) noted that dark hair washed with elderberries takes on a rich glossy shine. The bark and leaves have been used to prepare black and green dyes, respectively (Edlin, 1973). A blue dye can be made from elderberries if salt is added to the mix. Mahon (1983) listed *Sambucus nigra* as providing blue, purple and green dyes. In his book, *The Last of the Name*, Charles McGlinchey (2007: 108) said that 'they made some sort of ink, as far as I remember hearing, from the seeds of the elder tree'. Elder flower water (made by distilling the flowers) was said to be a favourite in the 18th century for whitening the skin and removing freckles (Ody, 1997a). Gallagher (2012: 25) provided a recipe for elder flower face cream. Lots of flowers are placed in a jar with white wax and almond oil, then the jar is boiled in a pan of water and the mixture is stirred from time to time. After simmering for an hour or so, when cool, the face cream is ready to use. Gallagher said that some people use clarified lard instead of oil and wax.

Elder timber from the trunk is hard and light-coloured and has been used as a substitute for boxwood, dogwood or even ivory, or dyed black, as ebony (Edlin, 1973). Elder wood had other uses too, such as for rough basketry. For example, Larner (2005) included a useful description of the construction of poorer mud cabins in early Co. Kerry. The walls were made of mud and the roof of timber and scraws (grassy or sedgy sods, such as the top layer of the bog). The chimneys were made of elder rods woven like a basket and plastered with clay. Williams (1993) noted that its wood was used to make meat skewers and tacks for shoe-making. Rutty (1772: 127) wrote that the wood of elder cut into splinters made very good toothpicks, 'not wounding the gums as those of Quills sometimes do'.

Elder twigs were sometimes hung in farm dairies as a talisman to keep away evil spirits (Sharkey, 2000). The elder tree was considered unlucky because it was supposed to be the tree on which Judas hanged himself and it was not used for making fuel or boats (Lysaght, 1979). *Cercis siliquastrum*, the Judas tree, is more usually regarded as the tree on which Judas hanged himself, which is not native to Ireland, and occurs instead in southern Europe and western Asia.

The branches of elder contain a central soft pith that is easily removed to make a hollow tube. For that reason they have been used in Ireland in the past to make whistles, pea-shooters, pop guns and cases for needles (Lysaght, 1979). Vickery (2010) noted its use in Co. Longford where the pith is removed and filled with molten lead to 'make a good weapon for protection on a journey or out walking at night'. When the pith is removed from stems they can be used to make musical instruments, blow-pipes and the pea-shooters of children (through which haws were often shot). Edlin (1973) reported that in Britain elder piths have been used by makers of fine instruments such as watches to remove oil and dirt from delicate instruments.

Williams (1993) noted that in Ros Muc, Co. Galway, in the 1940s, boys would use the branches to make guns (like blow-pipes), shooting paper stoppers several feet from the cleaned out barrel. Records in the Irish Folklore Commission indicate that elder was used to make pea-shooters and guns elsewhere in Ireland also. Once the pith had been cleaned out, another stick was cut to fit the barrel and used as a ramrod to shoot out pellets usually of chewed paper. McGlinchey (2007) described water guns made in his childhood during the mid-1800s in Co. Donegal: 'We used to make water guns out of a bootree stick. The inside is soft and can be cleaned out. Then we got a thin stick and tied a piece of cloth on to the end of it so that it fitted the hole well. At the knot we made a small hole to take the water. We could suck up the full of it of water and shoot it out again, and young people had great fun with it.'

Elder pits have been widely used in Ireland in universities by botany students, as a medium in which soft plant materials are imbedded and held to enable them to be thinly sectioned easily for plant anatomy studies through the microscope. As a student of botany in the 1970s elder piths were a regular part of my botany practical classes at Trinity College Dublin.

In Scotland, a traditional spout or tap (*bour*) was often made from an elder twig, which was used when drawing off the sap from a birch tree when making birch wine (Milliken & Bridgewater, 2004). I have not found records of such usage in Ireland, although the alternative common name, 'boor tree', in parts of Ireland suggests that this usage may have been known here too.

The roots of elder are reported to have insecticidal properties; badgers rub themselves vigorously against them to kill lice in their fur (Reader's Digest Association, 2006). An elder bush growing near the vegetable patch was also welcomed in the countryside as its heavy smell was thought to discourage butterflies, which would otherwise lay their eggs on brassicas (cabbage, cauliflowers, kale, Brussels sprouts, etc.) (Sharkey, 2000).

After giving a talk on Irish ethnobotany in Dingle, Co. Kerry, in January 2010, a local man told me that at one time elder twigs were commonly used in the region to castrate lambs. Flexible two-year-old twigs would be selected and twisted around the testicles, effectively cutting off their blood supply. Ten days later the testicles could then be removed without causing the lamb too much distress.

Dinneen (1927) gave the Irish word *starragán*, meaning 'a frame or rack composed of ropes and elder rods for supporting layers of flax in kiln-drying'. De Vere Coneys (1849) gave the Irish names *cran-fearna*, *fearn* (*-na*) and *fearnóg* (*-óige*, *-óga*) for the elder tree, but I suspect this is an error as these names are applied to the alder tree.

Another species of elder, *Sambucus ebulus* (dwarf elder) (Scannell & Synnott, 1987), danesweed (Webb & Scannell, 1983), dane-wort (Colgan, 1904) or dane's blood (Praeger, 1938), is introduced in Ireland and occasionally established on roadsides and in hedgerows and waste places and around old ruins. In Irish its name is *tromán*. Moloney (1919) listed the following alternative Irish names for dwarf elder: *péith*[260], *péith bhog*, *moluirt*,

260. This Irish word was given by Dinneen (1927), Ó Dónaill (1977) and Uí Chanainn (1962) to mean 'dwarf elder'.

fliodh an bhalla[261], *ballfhuirt* and *lus na ndanar*[262]. Moloney (1919) wrote that this species is said to grow where battles were fought against the Danes.

A dark purple-leaved cultivated form, sometimes with pink flowers, are occasionally grown in Irish gardens (Nelson, 2000b).

The leaves and stems of elder are poisonous and should not be eaten. Raw berries may also cause gastrointestinal distress.

Samolus valerandi (brookweed). Watercolour by Lydia Shackleton. Courtesy of the National Botanic Gardens, Glasnevin.

Samolus valerandi

- Brookweed (Scannell & Synnott, 1987)
- *Falcaire uisce* (Scannell & Synnott, 1987)

PRIMULACEAE—THE PRIMROSE FAMILY

Native

Samolus valerandi is an occasional plant of wet habitats, such as ditches, marshes and muddy and wet, rocky habitats by the sea. In Ireland it frequently occurs close to the coast but is also found on the shores of limestone lakes farther inland.

Borlase (1769) and Wood (1821) mentioned that *Samolus valerandi* was an important plant that was probably used in rituals by the Druids in Britain and Ireland for religious purposes. The reference to the Druids' use of this species probably originally comes from Pliny's *Natural History*, in which he mentioned that it was gathered by the Druids of Gaul using the left hand only, by a person who had been fasting. It was thought to provide important protection against disease.

Samolus valerandi.

Sanguisorba minor (Poterium sanguisorba)

- Salad burnet (Scannell & Synnott, 1987)
- *Lus an uille* (Scannell & Synnott, 1987); *Bileach lorcáin* (Moloney, 1919); *Namhneigh fladhain* (Wade, 1808b)

ROSACEAE—THE ROSE FAMILY

Native

Sanguisorba minor is an occasional plant seen in the south of Ireland on dry banks, in grasslands and gravelly habitats. It is very rare in the north.

Salad burnet has traditionally been used as a medicinal plant and is included in many European herbals; it can be used in salads but has a bitter taste. Mackay (1836) said that the leaves smell and taste like cucumber. Mears and Hillman (2007) suggested that young leaves can be chopped as salads and older leaves steamed as a vegetable. I have not tried it.

261. Dinneen (1927) defined *fliodh an bhalla* as 'dwarf alder', but he was presumably referring to dwarf elder.
262. *Lus na ndanar* was also given by Dinneen (1927).

Rutty (1772: 77) said that *Sanguisorba minor* and *S. officinalis* (listed by him as '*Pimpinella sanguisorba* and *P. minor*') were 'put green into Wine as a cordial, and to give it a grateful taste and smell like Melon: also fresh infused, they mend stale drink. They are cultivated for salads, the young leaves, immediately succeeding the seedlings, being used for this purpose, and giving a grateful flavour to the other ingredients.'

Moloney (1919) mentioned that in Ireland *Sanguisorba minor* and *S. officinalis* were said to increase the milk of cows.

Sanguisorba officinalis (Poterium officinale)

- Great burnet (Scannell & Synnott, 1987)
- *Lus an uille mór* (Scannell & Synnott, 1987); *Clabhrus mór* (Moloney, 1919)

ROSACEAE—THE ROSE FAMILY

Native

Sanguisorba officinalis.

Sanguisorba officinalis is very rare in Ireland and is known from only a few sites, on lake-shores in Co. Mayo (Lough Conn and Lough Cullin) and on dry banks in Cos. Down (Donaghadee) and Antrim (Carnlough).

It is native to Europe and Asia; the great burnet has been used for traditional medicine for a variety of purposes, including to stop bleeding, treat diarrhoea, haemorrhoids, wounds, ulcers, and burns and eczema (van Wyk & Wink, 2004).

Mears and Hillman (2007) suggested its use in salads and that it can be steamed as a vegetable.

Due to its rarity in Ireland, wild collected great burnet has probably been little used for ethnobotanical purposes, although it and *Sanguisorba minor* have been commonly grown in Irish herb gardens.

Moloney (1919) mentioned that in Ireland it was said to increase the milk of cows.

Sanicula europaea

- Sanicle (Scannell & Synnott, 1987); Wood sanicle (Anonymous, 1978); Sinicle (Hart, 1898)
- *Bodán coille* (Scannell & Synnott, 1987); *Bodán coilleadh* (Moloney, 1919); *Caogma* (Cameron, 1883); *Buine* (Cameron, 1883); *Cadhgamad* (Dinneen, 1927)

APIACEAE (Umbelliferae)—THE CARROT FAMILY

Native

Sanicula europaea is frequently found in woodlands and other shady habitats, such as old hedgerows.

Sanicle, or wood sanicle, has a long history of use in traditional medicine, to treat respiratory problems, diarrhoea, stomach upset, flatulence, excessive bleeding and as a gargle for sore throats and mouth infections (van Wyk & Wink, 2004). In Ireland, wood sanicle was used internally and externally to stop severe bleeding (Mackay, 1836; Moloney, 1919). It was also used to stop bleeding piles and for the treatment of consumption (Hart,

1898; Allen & Hatfield, 2004). It was also noted as being used to treat gapes (gape worms) in young pheasants in Co. Donegal. Cameron (1883) gave the meaning of the Irish name, *buine*, as an 'ulcer', which perhaps refers to its former use in traditional medicine in Ireland.

Saponaria officinalis

- Soapwort (Scannell & Synnott, 1987); Bruisewort (Dinneen, 1927)
- *Garbhán creagach* (Scannell & Synnott, 1987); *Lus an tsiabuinn* (Moloney, 1919); *Lus úchta* (Moloney, 1919; Dinneen, 1927); *Gairbhin creugach* (Cameron, 1883); *Gairbhín-cregach*[263] (de Vere Coneys, 1849); *Lus an tsiabainn* (Dinneen, 1927)

CARYOPHYLLACEAE—THE CARNATION OR PINK FAMILY
Introduced (possible archaeophyte)

Saponaria officinalis.

Saponaria officinalis is a herbaceous perennial species that can be frequently found throughout Ireland, although it is more common in the southern half. It occurs in waste places and on roadsides and river-banks, often as a relict of former cultivation.

It has also been used in beer-making to help create a good head for the beer. In Ireland it is usually represented by a double-flowered (*flore pleno*) variety in which the male parts of the flower (the stamens) have been replaced by extra petals. It is sometimes grown as a decorative ornamental ground-cover plant in cottage gardens, but spreads vigorously.

It was formerly widely used in country areas to make home-made soap. The plant is boiled and yields a lathery liquid used for washing clothes (Sharkey, 2000). The name *Saponaria* comes from the Latin for soap, *sapo*. The soap produced is gentle and mild and appropriate for use with delicate fabrics. It is reported that it was used to clean the Bayeaux tapestry (Pfaf.org, 2008) and other old and delicate fabrics (Lewington, 1990).

Soap can be regarded as a Celtic invention (according to Clyne, 1989). The word *sapo* was used by Pliny the Elder in his *Natural History*, in which the making of soap from ashes and tallow[264] was mentioned for use as a greasy mixture (pomade) to style the hair. Pliny mentioned its use amongst the Gauls in northern Europe.

Saponaria officinalis has also been used medicinally for inflammations of the lungs (Moloney, 1919), as an expectorant for coughs and bronchitis and for a variety of other folk remedy purposes. Moloney said that on the Continent it was used to treat syphilis.

The plant contains saponins, which are complex organic compounds called glycosides. Some authors (e.g., Peterson, 1977) listed this as a poisonous plant, and it causes severe irritation to the digestive tract if used internally.

**Saussurea alpina*

- Alpine saw-wort (Scannell & Synnott, 1987)

263. This word was given also by Dinneen (1927) to mean 'a bruisewort, a plant growing on rocks by the shore, good for bruises' and by Ó Dónaill (1977) to mean 'bruiswort'.

264. Tallow is a rendered form of animal fat, mainly from beef and mutton.

- *Sábhlus sléibhe* (Scannell & Synnott, 1987)

ASTERACEAE (Compositae)—THE DAISY FAMILY
Native

Saussurea alpina is a very rare alpine plant found on higher mountain cliffs in Ireland, mainly in scattered locations along the west, for example, in the mountains of Kerry, Connemara and Donegal. Elsewhere, this species can be found extending throughout Eurasia, in arctic or montane habitats. Global warming at higher latitudes is likely to impact the distribution of this alpine taxon, perhaps threatening the survival of *S. alpina* in Ireland.

*No ethnobotanical uses of this plant have been noted in Ireland.

*Saxifraga aizoides

- Yellow (mountain) saxifrage (Scannell & Synnott, 1987)
- *Mórán buí* (Scannell & Synnott, 1987); *Mionán Mhuire* (Moloney, 1919)

SAXIFRAGACEAE—THE SAXIFRAGE FAMILY
Native

Saxifraga aizoides is very rare in Ireland and generally found in the mountains. It is found only in wet rocky habitats in the north, from Cos. Sligo to Donegal, and in Northern Ireland in Co. Antrim.

*No ethnobotanical uses of this plant have been noted in Ireland.

Saxifraga granulata

- Meadow saxifrage (Scannell & Synnott, 1987)
- *Mórán léana* (Scannell & Synnott, 1987); *Mionán Mhuire* (Moloney, 1919; Dinneen, 1927)

SAXIFRAGACEAE—THE SAXIFRAGE FAMILY
Native (or possible archaeophyte)

Saxifraga granulata.

Saxifraga granulata is a rare and declining plant found mainly near the east coast of Ireland, in lowland grasslands and sand dunes. There is a large, long-naturalised population growing in woodland at the National Botanic Gardens in Glasnevin, Dublin.

The meadow saxifrage is an attractive creamy white-flowered perennial plant sometimes grown as an ornamental plant in horticulture, including a double-flowered (*flore pleno*) form. An older common English name of the double form is 'Pretty Maids', well known for the old English nursery rhyme:

> Mary, Mary, quite contrary,
> How does your garden grow?
> With silver bells, and cockle shells,
> And pretty maids all in a row.

However, interpretation of the meanings of this rhyme are obscure and somewhat disputed; some have argued that it may be a mocking reference to Queen Mary I of England and her inability to produce an heir.

Saxifraga granulata has also been used in traditional medicine in Europe, but in Ireland it has probably always been too rare to have had much, if any, usage. It has traditionally been used to treat gravel and kidney and bladder stones (Allen & Hatfield, 2004), a use perhaps reflected in its scientific epithet (*saxifraga*, meaning 'rock-breaker'). The scientific name *Saxifraga*, meaning 'breaker of rocks or stones', dates back to Dioscorides, the Greek writer on medical botany in the 1st century AD. This use most likely relates to the medicinal use of *Saxifraga* species in the treatment of kidney or bladder stones (Webb & Gornall, 1989), and not to the fact that many saxifrages grow in the crevices of rocks. The specific saxifrage referred to by Dioscorides was *Saxifraga granulata*, a species of grasslands.

**Saxifraga hartii (Saxifraga rosacea* subsp. *hartii)*

- Hart's saxifrage (Scannell & Synnott, 1987)
- *Mórán creige* (Scannell & Synnott, 1987)

SAXIFRAGACEAE—THE SAXIFRAGE FAMILY
Native

Saxifraga hartii is known only from cliffs on Arranmore Island off the west coast of Co. Donegal. This endemic saxifrage was named by Professor David A. Webb in 1950 to honour one of Ireland's most distinguished botanists, Henry Chichester Hart (1847–1908), author of the *Flora of County Donegal* (1898). Hart's saxifrage was later considered by Webb (1987) to be an endemic subspecies of the Irish saxifrage *S. rosacea*, because intermediate forms between it and this more widespread and variable mossy saxifrage are known also on Arranmore.

*No ethnobotanical uses of this plant have been noted in Ireland.

**Saxifraga hirculus*

- Marsh saxifrage (Scannell & Synnott, 1987); Yellow marsh saxifrage (Praeger, 1938)
- *Mórán réisc* (Scannell & Synnott, 1987)

SAXIFRAGACEAE—THE SAXIFRAGE FAMILY
Native

Saxifraga hirculus is a very rare saxifrage found only in marshes and wet bogs and largely now confined to Co. Mayo in western Ireland and Co. Antrim in Northern Ireland.

*No ethnobotanical uses of this plant have been noted in Ireland.

**Saxifraga hirsuta*

- Kidney saxifrage (Scannell & Synnott, 1987)
- *Mórán giobach* (Scannell & Synnott, 1987)

SAXIFRAGACEAE—THE SAXIFRAGE FAMILY
Native

Saxifraga hirsuta is quite frequent in woodlands and shady places only in south-west Ireland, Cos. Kerry and West Cork. Elsewhere in Ireland it occurs only as an introduction. Hybrids of this species with *S. spathularis* occur in Cos. Galway and Mayo, although the parent species is not known there.

*No ethnobotanical uses of this plant have been noted in Ireland.

*Saxifraga hypnoides

- Mossy saxifrage (Scannell & Synnott, 1987)
- *Mórán caonaigh* (Scannell & Synnott, 1987)

SAXIFRAGACEAE—THE SAXIFRAGE FAMILY
Native

Saxifraga hypnoides is rare except in the Burren in Co. Clare. Mossy saxifrage occurs on bare rock, in the grikes of limestone pavements, as well as in rocky grasslands in Ireland. Elsewhere, *S. hypnoides* is a boreal montane element in the European flora.

*No ethnobotanical uses of this plant have been noted in Ireland.

---

*Saxifraga nivalis

- Alpine saxifrage (Scannell & Synnott, 1987)
- *Mórán an tsneachta* (Scannell & Synnott, 1987)

SAXIFRAGACEAE—THE SAXIFRAGE FAMILY
Native

Saxifraga nivalis is a very rare species in Ireland, known only from cliffs on the Ben Bulben range, Co. Sligo, and it is now listed as a threatened species. Elsewhere, *S. nivalis* exists as Arctic populations at higher elevations or latitudes.

*No ethnobotanical uses of this plant have been noted in Ireland.

Saxifraga oppositifolia

- Purple saxifrage (Scannell & Synnott, 1987)
- *Mórán sléibhe* (Scannell & Synnott, 1987); *Clabhrus* (de Vere Coneys, 1849)

SAXIFRAGACEAE—THE SAXIFRAGE FAMILY
Native

Purple saxifrage is a rare plant found in rocky mountain habitats in the north and west only, extending from Co. Derry in Northern Ireland to Co. Galway in western Ireland. The flowers have a pleasant fresh taste and are edible, but because the plant is confined to populations with restricted distributions on some of the higher mountains in Ireland, it was probably never used significantly for food, if at all.

Saxifraga oppositifolia is the official county flower of Co. Derry in Northern Ireland.

Saxifraga rosacea

- Irish saxifrage (Scannell & Synnott, 1987)
- *Mórán gaelach* (Scannell & Synnott, 1987)

SAXIFRAGACEAE—THE SAXIFRAGE FAMILY
Native

Saxifraga rosacea in the Burren, Co. Clare.

Saxifraga rosacea is a rare species that is found most commonly in Co. Kerry in the mountains on cliffs and screes. It also occurs elsewhere in the west of Ireland, such as in the Burren in Co. Clare, on limestone pavements. *Saxifraga rosacea* is also common on the Aran Islands in Galway Bay, which have a karst topography similar to that of the nearby Burren. It is also recorded from the Galtee mountains in Co. Tipperary.

Saxifraga rosacea is occasionally cultivated as an ornamental plant in Irish gardens, grown in rockeries and as a pot plant in an alpine house. Nelson (2000b) listed a number of cultivated varieties, hybrids and forms of this species raised in Ireland and grown in gardens to a greater or lesser extent.

Saxifraga spathularis (St Patrick's-cabbage). Watercolour by Lydia Shackleton. Courtesy of the National Botanic Gardens, Glasnevin.

Saxifraga spathularis

- St Patrick's-cabbage (Scannell & Synnott, 1987)
- *Cabáiste an mhadra rua* (Scannell & Synnott, 1987); *Cabáiste Phádhraic* (Webb et al., 1996)

SAXIFRAGACEAE—THE SAXIFRAGE FAMILY

Native

Saxifraga spathularis is frequently found in woodlands and shady habitats in the west (Cos. Galway and Mayo) and south-west (Cos. Kerry and Cork). It is also found rarely in other mountain habitats elsewhere in Ireland.

St. Patrick's-cabbage is occasionally grown in gardens in Ireland as an ornamental, although more commonly seen in gardens is the hybrid, *Saxifraga ×urbium* 'London pride'. *Saxifraga ×urbium* can be confused with a species native to the Pyrenees, *S. umbrosa* (Webb & Gornall, 1989). *Saxifraga ×urbium* is in fact thought to be a hybrid between *Saxifraga spathularis* and *Saxifraga umbrosa* (Webb et al., 1996).

Saxifraga spathularis also forms hybrids with *S. hirsuta* (called *Saxifraga ×geum*) wherever they meet in the south-west of Ireland, and this hybrid also occurs in Cos. Galway and Mayo where pure *S. hirsuta* (kidney saxifrage) does not now occur.

Scannell and Synnott (1987) gave *mórán cathrach* as the Irish name for London Pride. Moloney (1919) also listed the following Irish names for London Pride: *Gabáiste (Cabáiste) daoine maithe* and *Gabáiste madraí ruaidh*. Ó Dónaill (1977) gave the Irish word *clabhras* for 'saxifrage'. Other authors apply this name to *Pimpinella*.

Munnelly (2005), quoting a local farmer in Miltown Malbay, Co. Clare, suggested that a plant called 'Saint Patrick's Leaf', which may be *Saxifraga spathularis*, was used to treat young cattle and calves, especially if they had dysentery and were passing blood. It was boiled in new milk and then given to the cattle.

**Saxifraga stellaris*

- Starry saxifrage (Scannell & Synnott, 1987)
- *Mórán réaltach* (Scannell & Synnott, 1987)

SAXIFRAGACEAE—THE SAXIFRAGE FAMILY
Native

Saxifraga stellaris is an occasional plant of wet rocky habitats in the mountains, such as on cliffs, screes and by mountain streams, in scattered locations around Ireland.

*No ethnobotanical uses of this plant have been noted in Ireland.

**Saxifraga tridactylites*

- Rue-leaved saxifrage (Scannell & Synnott, 1987)
- *Mórán balla* (Scannell & Synnott, 1987)

SAXIFRAGACEAE—THE SAXIFRAGE FAMILY
Native

Saxifraga tridactylites is an annual plant of dry walls, rocks, sand dunes, gravel pits and quarries, and other open habitats throughout Ireland. It is commonest in the west and parts of the south and centre of Ireland but is rarely distributed in the north.

*No ethnobotanical uses of this plant have been noted in Ireland.

Scandix pecten-veneris

- Shepherd's needle (Scannell & Synnott, 1987); Venus' comb (Mackay, 1836)
- *Gob an ghoirt* (Scannell & Synnott, 1987); *Creagach* (Moloney, 1919)

APIACEAE (Umbelliferae)—THE CARROT FAMILY
Introduced

Scandix pecten-veneris is a very rare or extinct annual weed of waste places and arable land. Apparently, some parts of this plant are edible and have been used for food, cooked as a potherb or eaten raw in salads. I do not know whether this plant was ever eaten in Ireland, but as a weed of arable fields it may well have been used as part of the wild crop that was sometimes utilised during years when the field was left fallow.

Schoenoplectus lacustris (Scirpus lacustris)

- Common club-rush (Scannell & Synnott, 1987); Bulrush (Anonymous, 1978)
- *Bogshifín* (Scannell & Synnott, 1987); *Bog-bhuinne* (Moloney, 1919); *Curcais* (Moloney, 1919); *Féitheacha fiadhaine* (Moloney, 1919); *Bois-shithbhín* (Webb et al., 1996); *Gibiun* (Cameron, 1883); *Bóg mhuine* (Cameron, 1883); *Boigean* (Cameron, 1883); *Bóg luchair* (Cameron, 1883; Dinneen, 1927); *Min-fheur* (Cameron, 1883); *Bogbhuin* (de Vere Coneys, 1849; Dinneen, 1927); *Boic-shithbhín* (Dinneen, 1927); *Curchais*[265] (Dinneen, 1927; Uí Chanainn, 1962); *Bocshífín* (Uí Chanainn, 1962)

265. Dinneen (1927) defined *curchais* as 'bulrush' and Uí Chanainn (1962) as 'bulrush, reed and flax'.

Schoenoplectus lacustris.

CYPERACEAE—THE SEDGE FAMILY
Native

Schoenoplectus lacustris is a common plant throughout Ireland of shallow water at the margins of lakes, rivers, canals, fens and turloughs. A separate and distinct subspecies, subsp. *tabernaemontani* grows in brackish water all around the Irish coast.

In Ireland when weavers spoke about weaving with rushes this was often the species used (Hogan, 2001). In Britain, *Schoenoplectus lacustris* has been traditionally used up to the present time to make matting. The stems are up to 2 m long and are woven or plaited to form mats, baskets, seats of chairs, bags, etc. (Simpson & Inglis, 2001; Prendergast & Sanderson, 2004). Excavations at Fishamble Street, Dublin, have yielded samples of this species, which was probably used as bedding (along with straw, which would have been more commonly used) (Geraghty, 1996). It has also been used in Ireland in similar ways, and coopers use it for filling and sealing the spaces between the ribs of barrels (Mackay, 1836). McGlinchey (2007) also mentioned that coopers in Donegal would put bulrushes between the staves of the barrels and they would never leak. The rushes were cut from a simple raft made from planks, boards and ladders with a long pole with a hook on the end. The rushes were cut when in seed and tied in bundles before being brought to Derry for sale to the coopers working there. The price for a sheaf of these bulrushes in the mid-1800s was 6 d. At that time, there were 240 d (pence) in a pound, 20 shillings in a pound and 12 d in 1 shilling.

Hogan (2001) noted that this species is widely used to make traditional St Brigid's crosses today but that in the past *Juncus* species were more usual.

Séamas MacPhilib (2000) and S. MacPhilib in MacCárthaigh (2008) documented the use of club-rush to make rush river rafts in Co. Roscommon, using materials harvested from the banks of the River Suck. A rush raft made from this species by Patrick Gately of Dysart, Co. Roscommon, is included in the collections of the National Museum of Ireland (Museum of Country Life) in Castlebar, Co. Mayo (Ref. No: F1962.124). MacPhilib also recorded that this species was used in thatching. The rushes were harvested in late July or August and left for a few days to become pliable before being used. The rush rafts were used for fishing and fowling and to help people get around the area of Dysart in the south of the county during times when the 'callows' were flooded. There are also records of similar crafts made from rushes elsewhere in Ireland, for example, in Cos. Longford, Leitrim, Clare and Mayo. MacPhilib also noted an 1866 account of rush rafts on Lough Erne in Ulster, where a timber frame, generally of ash (*Fraxinus excelsior*), would be stuffed with bundles of rushes. Rushes tied in bundles were also used as swimming aids, made into a belt or v-shaped bow placed under the arms, or simply as bundles of rushes tied together, sat upon and paddled. These have been recorded in Cos. Mayo, Meath, Cavan, Cork and Kerry (MacPhilib, 2000).

The underwater rhizomes contain starch and may have been used as a food source in early times in Britain and Ireland (Mears & Hillman, 2007) and are edible if roasted. Seeds are also edible.

Schoenoplectus lacustris is planted for use in natural reed-bed water-treatment purification systems.

Its use for medicinal purposes in parts of Europe, as an astringent and to treat genitourinary disorders, has been noted by Simpson and Inglis (2001), but I can find no records of its use medicinally in Ireland.

This species has sometimes been called 'bulrush', but this name in Ireland is more

A rush raft. B shows the underside of the raft, covered with rushes. Photographs courtesy of the National Museum of Ireland (Museum of Country Life), Castlebar.

frequently applied to *Typha* spp. It is therefore unclear if Irish names translated in English given in various dictionaries refer to this species, or to *Typha*, for example, *buighin* (*-ine*) or *buigshbhin* (*-bhne*) (de Vere Coneys, 1849). Ó Dónaill (1977) gave the Irish word *bogshifín* (or any of the following alternatives: *boigean*, *biocsheimhin*, *boigeimhin*, *boicshimhín*, *boicshithbhín*, *boicshifín*, *bogeimhin*, *bogbhuinne*, *bogluachair*[266], *buigiún*, *buigrín*) for 'bulrush'.

Schoenoplectus triqueter (Scirpus triqueter)

- Triangular club-rush (Scannell & Synnott, 1987)
- *Bogshifín tríchúinneach* (Scannell & Synnott, 1987)

CYPERACEAE—THE SEDGE FAMILY
Native

Schoenoplectus triqueter is a very rare species in Ireland and is known only from the river Shannon estuary in Cos. Limerick and Clare.

Simpson and Inglis (2001) recorded that it is used as a fibre source for matting in China and cultivated for 'taiko' mats in Taiwan, but there are no records of its use in Ireland.

Schoenus nigricans

- Black bog-rush (Scannell & Synnott, 1987)
- *Sifín* (Scannell & Synnott, 1987); *Seimhin* (*-mhne*) (*-mhneacha*) (de Vere Coneys, 1849; Cameron, 1883); *Seimhin luachra* (de Vere Coneys, 1849)

CYPERACEAE—THE SEDGE FAMILY
Native

266. This name was also given by Uí Chanainn (1962) for 'bulrush'.

Schoenus nigricans.

Schoenus nigricans is a common rush of fens, lake-shores, marshes and limestone springs and also acid bogs in western and central Ireland but rare elsewhere.

This plant was occasionally used for thatching, at least in Menlo, on Lough Corrib, Co. Galway, in the 1990s (M. Sheehy Skeffington, pers. comm., 2007).

Mackay (1806) reported that it was used for cattle grazing in seasons when fodder was short.

Simpson and Inglis (2001) recorded that it has been used in making ropes in Italy, but there are no specific records that I have found of its use in Ireland for such purposes.

Scilla verna

- Spring squill (Scannell & Synnott, 1987); Vernal squill (Anonymous, 1978)
- *Sciolla earraigh* (Scannell & Synnott, 1987)

ASPARAGACEAE (Hyacinthaceae)—ASPARAGUS FAMILY

Native

Scilla verna is a pretty spring-flowering, bulbous plant that occurs in coastal habitats all along the eastern coast of Ireland, from Derry to Wexford, in coastal grasslands and on rocky and sandy habitats. It is a characteristic plant for maritime heaths in eastern Ireland. Spring squill can be locally common. *Scilla verna* was chosen as the official flower of Co. Down in 2002.

No ethnobotanical uses of this plant have been noted particularly for Ireland. However, *Scilla verna* is native to similar coastal heath lands in Britain and found throughout western Europe. The bulbs, as squill powder, have been long been used as an expectorant in traditional cough medicines and remedies. The Reverend John M. Wilson (1849: 318) observed the mucilage of the bulbs, when separated from the acrid, poisonous portion, were 'extensively employed by calico-printers'. Writing about spring squill in southern England, Anne Pratt (1857: 154) noted that 'the bulbs of the genus, though not reckoned among our active poisons, are of doubtful wholesomeness.'

**Scirpus sylvaticus*

- Wood club-rush (Scannell & Synnott, 1987)
- *Bogshifín coille* (Scannell & Synnott, 1987)

CYPERACEAE—THE SEDGE FAMILY

Native

Scirpus sylvaticus is found in wet moorlands and riversides mainly in the north-east of Ireland and very rarely elsewhere.

*No ethnobotanical uses of this plant have been noted in Ireland.

Scleranthus annuus

- Annual knawel (Scannell & Synnott, 1987)
- *Cabhair Mhuire* (Scannell & Synnott, 1987); *Mionán Mhuire* (Moloney, 1919); *Lus Órtha* (Moloney, 1919; Dinneen, 1927)

CARYOPHYLLACEAE—THE CARNATION OR PINK FAMILY
Native

Scleranthus annuus is a very rare annual weed of wasteground and roadsides in dry sandy habitats that is mainly distributed in the north-east and is even rarer elsewhere in Ireland.

According to Moloney (1919), it was used in the treatment of urinary ailments and said to be a favourite herb with herbalists at the beginning of the last century for all diseases accompanied by a 'disordered urinary function'.

Scorzoneroides autumnalis (Leontodon autumnalis)

- Autumn hawkbit (Scannell & Synnott, 1987); She lion (Hart, 1898); Hawkbit (Hart, 1898)
- *Crág phortáin* (Scannell & Synnott, 1987); *Caisearbhán caol dearg* (Uí Chonchubhair, 1995)

ASTERACEAE (Compositae)—THE DAISY FAMILY
Native

Scorzoneroides autumnalis is a fairly frequent plant of damp grasslands such as meadows, pastures and roadsides throughout Ireland.

Uí Chonchubhair (1995) recorded that on the Dingle Peninsula, Co. Kerry, it was used in an infusion with other species (*Teucrium scorodonia* and *Arctium minus*) to treat tuberculosis, as a laxative and for 'decay'.

Mears and Hillman (2007) reported that this plant is bitter and not edible.

Scrophularia auriculata (Scrophularia aquatica)

- Water figwort (Scannell & Synnott, 1987); Water betony; Rose noble (Hart, 1898)
- *Donnlus uisce* (Scannell & Synnott, 1987); *Fothram* (Ua Maoileoin, 2001); *Dúnlus uisce* (Moloney, 1919; Dinneen, 1927); *Súgh fearna* (Moloney, 1919); *Biadhtas uisce* (Moloney, 1919); *Donn uisce* (Dinneen, 1927); *Samhadh curraigh* (Dinneen, 1927); *Samhadh feárna* (Dinneen, 1927)

SCROPHULARIACEAE—THE FIGWORT FAMILY
Native

Scrophularia auriculata is a widespread plant throughout Ireland of wet habitats, such as ditches and stream banks, and is most frequently distributed in the south and west, although it is nowhere common.

It was used in Ireland in traditional folk medicine to clean wounds (Moloney, 1919). K'Eogh (1735: 46) said that the roots are good for 'scrophulous tumours, the King's evil, the piles, cancerous ulcers and the itch' and can be used to treat hard swellings.

Scrophularia nodosa

- Figwort (Scannell & Synnott, 1987); Knotted figwort (Brunker, 1950); Rose noble (Hart, 1898; Praeger, 1938)
- *Donnlus* (Scannell & Synnott, 1987); *Lus na gcnapán* (Moloney, 1919; Dinneen, 1927); *Fothrum* (Moloney, 1919); *Farach dubh*[267] (Moloney, 1919; Dinneen, 1927); *Dúnlus* (de Vere Coneys, 1849; Moloney, 1919; Ó Dónaill, 1977); *Fotrum* (de Vere Coneys, 1849; Cameron, 1883); *Farach-duibh* (de Vere Coneys, 1849); *Fothram* (Tóibín, 1965; Ó Dónaill, 1977)

SCROPHULARIACEAE—THE FIGWORT FAMILY
Native

Scrophularia nodosa is a relatively common plant of hedges, ditches, roadsides and waste places.

Allen and Hatfield (2004) and Uí Chonchubhair (1995) noted that figwort was formerly a highly prized medicinal herb in Ireland. It was widely used externally to treat a variety of skin problems, as well as used in the treatment of tuberculosis and for liver, kidney and stomach ailments. It was used to make an ointment for piles and various skin problems and also used to treat farcy, an infectious disease of horses (Moloney, 1919). In Donegal *Scrophularia nodosa* and *S. auriculata* were pounded and bruised and used to treat 'splays or sprains' (Hart, 1898: 385). Lady Wilde (1890) listed it (as 'rose-noble') as an effective treatment for scrofula. Allen and Hatfield (2004) also recorded its use in veterinary medicine in Ireland to treat red-water fever and to help in the expulsion of afterbirth in cattle. It was also used to treat blasts (swellings) on cow's nipples (Williams, 1993). 'Figs' is an old English name for piles (Vickery, 2010), so its common name suggests the use of this species to treat haemorrhoids.

According to Hart (1898), both *Scrophularia nodosa* and *S. auriculata* were used as a treatment for ringworm; the roots were ground and stewed, then strained and added to lard or butter to make an ointment. It was also boiled in milk and drunk to treat lung complaints, or mixed with lard or goose grease to make an ointment for burns (Williams, 1993).

**Scrophularia umbrosa*

- Green figwort (Scannell & Synnott, 1987)
- *Donnlus glas* (Scannell & Synnott, 1987)

SCROPHULARIACEAE—THE FIGWORT FAMILY
Native

Scrophularia umbrosa is a very rare plant known only from the river-banks of the rivers Liffey and Bann.

*No ethnobotanical uses of this plant have been noted in Ireland.

267. Dinneen (1927) defined *farach dubh* as 'great fig-wort'.

*Scutellaria galericulata

- Skullcap (Scannell & Synnott, 1987)
- *Cochall* (Scannell & Synnott, 1987)

LAMIACEAE (Labiatae)—THE DEAD-NETTLE FAMILY
Native

Scutellaria galericulata is found only occasionally and mainly in the north in marshes and on lake-shores. It can also be found in wet places close to the sea, although it is rare in this habitat.

*No ethnobotanical uses of this plant have been noted in Ireland.

*Scutellaria minor

- Lesser skullcap (Scannell & Synnott, 1987)
- *Cochall beag* (Scannell & Synnott, 1987)

LAMIACEAE (Labiatae)—THE DEAD-NETTLE FAMILY
Native

Scutellaria minor is found mainly in the south-west and west in bog-margins, marshes, river-banks and lake-shores; it is rare elsewhere in Ireland.

*No ethnobotanical uses of this plant have been noted in Ireland.

Sedum acre

- Biting stonecrop (Scannell & Synnott, 1987); Yellow stonecrop (Anonymous, 1978); Wall-pepper (Brunker, 1950; Anonymous, 1978)
- *Grafán na gcloch* (Scannell & Synnott, 1987); *Púirín seangán* (O'Rourke, 2006); *Lusaire na seangán* (Moloney, 1919); *Lusradh an tsionnaigh* (Dinneen, 1927); *Lusradh na seangán* (Dinneen, 1927); *Póirín seangán* (Dinneen, 1927); *Lus na seangán* (Ó Dónaill, 1977)

CRASSULACEAE—THE STONECROP FAMILY
Native

Sedum acre is a common plant throughout Ireland of dry habitats, such as rocky areas, walls and sand dunes, and occurs mainly on limestone rock or calcareous soils.

According to Allen and Hatfield (2004), *Sedum acre* was used medicinally in Ireland to treat worms in humans and animals and for kidney problems. It has also been used to treat warts (O'Rourke, 2006). Harrop (1977) said that the leaves are too bitter to use fresh but dried make a useful substitute for pepper. Moloney (1919) said that this plant 'produces vomiting'.

Sedum acre (biting stonecrop). Watercolour by Lydia Shackleton. Courtesy of the National Botanic Gardens, Glasnevin.

**Sedum anglicum*

- English stonecrop (Scannell & Synnott, 1987)
- *Póiríní seangán* (Scannell & Synnott, 1987); *Púiríní seangán* (Uí Chonchubhair, 1995)

CRASSULACEAE—THE STONECROP FAMILY
Native

Sedum anglicum is a common plant throughout Ireland, although is much rarer inland, of dry habitats close to the coast, such as rocky shores, walls and dry heaths. It occurs mainly on granite. It is much rarer inland.

Another species of *Sedum*, *S. album*, is an introduced perennial plant found throughout Ireland, and is established on walls. Uí Chonchubhair (1995) gave *S. album* the Irish name *grafán bán na gcloch*.

*No ethnobotanical uses of this plant have been noted in Ireland.

Sedum telephium

- Orpine (Scannell & Synnott, 1987); Heal-all (Praeger, 1938); Livelong (Harron, 1986)
- *Tóirpín* (Scannell & Synnott, 1987); *Lus na laogh* (de Vere Coneys, 1849; Moloney, 1919; Dinneen, 1927); *Tóirpíneach* (Ó Dónaill, 1977)

CRASSULACEAE—THE STONECROP FAMILY
Possibly Introduced

Sedum telephium is a possible introduction, although it may be native in some habitats, such as in woodlands in the north-centre part of Ireland (Webb et al., 1996). It is more frequent as a garden escape.

It was noted by Wade (1804: 36) as a medicinal plant: 'A decoction of the leaves is said to be a very serviceable and forcible diuretic'.

Mears and Hillman (2007) suggested that the leaves and tubers of this species are edible but bitter.

It is occasionally grown as a garden plant in Ireland.

Sempervivum tectorum

- House-leek (Reynolds, 2002)
- *Teinne eagla* (Reilly, 2001); *Toirpín* (de Vere Coneys, 1849; Moloney, 1919; Dinneen, 1927); *Lus an tóiteá(i)n* (Moloney, 1919; Ó Dónaill, 1977); *Orp* (de Vere Coneys, 1849; Moloney, 1919; Dinneen, 1927; Uí Chanainn, 1962); *Sincín*[268] (Moloney, 1919; Dinneen, 1927); *Glan-eagla* (Moloney, 1919); *Tir-pin* (Cameron, 1883); *Norp* (*Nuirp*) (de Vere Coneys, 1849; Dinneen, 1927); *Biadh an tsionnaigh* (Dinneen, 1927); *Buachaill tighe*[269] (Colgan, 1914; Dinneen, 1927); *Luibh* (or *Lus*) *an tóiteain* (Dinneen, 1927); *Tein-eaglach* (Dinneen, 1927); *Tinicín* (or *Rinicín*) (Dinneen, 1927); *Teineagal* (Ó Dónaill, 1977)

CRASSULACEAE—THE STONECROP FAMILY
Introduced

---

268. The variant spelling '*sinicín*' was given by Dinneen (1927).
269. A variant spelling of this word was given by Vickery (1985) as *buachaill a'tighe*.

House-leeks were traditionally planted on the roofs of houses and in Ireland on thatched cottages as they were believed to protect them as a charm against fire (Moloney, 1919; Reynolds, 2002). Colgan (1914) reported that this plant was reputed to have no effect in protecting a house from fire or lightning unless it had been stolen from its previous owner, or at least taken without his knowledge. Howes (1974) suggested that they were planted on roofs to help keep tiles and slates in position. Records of it being planted on roofs in Europe go back more than 2000 years (Grigson, 1958). It is rarely if ever naturalized in Ireland but is occasionally seen as a relict of cultivation, especially on walls and ruins. It is also frequently cultivated as a garden plant in rockeries.

Its alternative common name 'Welcome home husband, however drunk you be' (Clapham et al., 1962) comes from the tradition that the leaves chewed can help to disguise the aroma of alcohol on the breath. I've not tried it so cannot vouch for its effectiveness!

David Moore (1834–35) recorded that in Co. Londonderry 'country people would apply it to burns as well as external inflammations'. The juice of the leaves were used to cure sore eyes in Co. Donegal (Hart, 1898). Such a use is also recorded by Logan (1972). Egan (1887) mentioned that the juice of the house-leek was applied to cure corns and warts, and Kinahan (1881) referred to their use as a poultice for sore legs. Vickery (2010: 47) quoted a use of the house-leek to induce abortions, recorded from Kiltimagh, Co. Mayo, when 'a young girl got in "trouble", her mother would take some of these plants, boil them, and give the water to her daughter to drink. Later on she would tell the girl to climb up on a high wall and jump down. That would make the girl all right.'

It was used in veterinary medicine to treat diarrhoea in cattle in Co. Wicklow and to help in the expulsion of afterbirth in Co. Mayo (Vickery, 1985; Allen & Hatfield, 2004).

House-leeks are grown as ornamental plants in some Irish gardens and are particularly useful as rockery plants for dry places.

Senecio aquaticus

- **Marsh ragwort (Scannell & Synnott, 1987); Marsh ragweed; She benweed (Hart, 1898)**
- ***Buachalán corraigh* (Scannell & Synnott, 1987); *Samhadh curraigh* (Moloney, 1919); *Samhadh corraigh* (Uí Chonchubhair, 1995); *Boglus* (Colgan, 1911); *Bothalán* (Colgan, 1911); *Bothalán buidhe* (Colgan, 1911)**

ASTERACEAE (Compositae)—THE DAISY FAMILY
Native

Senecio aquaticus is distributed throughout Ireland but is commonest in the west in wet fields, marshes and ditches, and on lake-shores and roadsides.

The pounded leaves were formerly used in Ireland to make poultices for suppurating wounds (Moloney, 1919). Colgan (1911) gave the Irish name *boglus* for this species in Clare Island and Achillbeg, Co. Mayo, used for an early state of the plant when it is just a rosette of succulent leaves. There, the leaves were pounded and used as a poultice for sores. However, it is also possible that this name is applied to the poultice, rather than the plant itself, as it was also used for a poultice made from at least one other species.

Senecio aquaticus and *S. erucifolius* were used in veterinary treatments for a range of animal ailments in various parts of Ireland, including the treatment of wounds, red-water fever and gapes (Allen & Hatfield, 2004).

Mahon (1983) listed this species as the source of a yellow dye.

**Senecio erucifolius*

- Hoary ragwort (Scannell & Synnott, 1987); Wormwood-leaved ragwort (Brunker, 1950)
- *Buachalán liath* (Scannell & Synnott, 1987)

ASTERACEAE (Compositae)—THE DAISY FAMILY
Native (or possible archaeophyte)

Senecio erucifolius is a relatively rare ragwort found mainly in the east, particularly in the Dublin region, on roadsides and dry banks and in grasslands and waste places.

*No ethnobotanical uses of this plant have been noted in Ireland.

Senecio jacobaea.

Senecio jacobaea

- Ragwort (Scannell & Synnott, 1987); St James's wort (Moloney, 1919); Ragweed (Colgan, 1904); Bunkishawn (Colgan, 1904); Cushawn (Colgan, 1904); Bohalawn (Colgan, 1904); Boolyan (Colgan, 1904); Birdseed (Colgan, 1904); Benweed (Hart, 1898; Praeger, 1938; Harron, 1986); Bogluss bwee (Hart, 1898); Buckles bwee (Hart, 1898); Boughaleen bwee (Hart, 1898); Buckieshaun (Hart, 1898); Fairy's horse (Hart, 1898); Yellow boy (Hart, 1898)
- *Buachalán buí* (Scannell & Synnott, 1987); *Meacan an dáthabha* (Ua Maoileoin, 2001); *Buachalán buidhe* (Moloney, 1919); *Fear talmhan* (Moloney, 1919); *Boltán buidhe* (Moloney, 1919; Dinneen, 1927); *Cuiseóg bhuidhe* (Moloney, 1919; Dinneen, 1927); *Buadhghallan* (Cameron, 1883); *Guiseag (or Cuiseag) bhuidhe* (Cameron, 1883); *Buafanán na h-easgarán* (de Vere Coneys, 1849); *Balcaiseán (Bolcaiseán)*[270] (Dinneen, 1927); *Buabhall (-allán)* (Dinneen, 1927); *Fean talmhan* (Dinneen, 1927); *Buachaltán* (Ó Dónaill, 1977); *Bothalán* (Colgan, 1911); *Bothalán buidhe* (Colgan, 1911); *Bualtán* (Colgan, 1911); *Bualterán* (Colgan, 1911); *Buchrillán bhuide* (McCarthy & Hawkes, 2007)
- *Boouchelawn* (in Yola)

ASTERACEAE (Compositae)—THE DAISY FAMILY
Native

Senecio jacobaea is an abundant and persistent weed of roadsides, fields, sand dunes and waste places in every part of Ireland.

If eaten, this plant is poisonous to cattle and horses, although it seems to be harmless to sheep. The Control of Noxious Weeds Act came into effect in Ireland in 1936 in the Republic, and ragwort was declared a noxious weed under it in a 1937 Order (along with thistle and dock), which required landowners to control its growth. Ragwort is also controlled under a 1977 Noxious Weeds Order in Northern Ireland. Kelly (1997) said that it causes cirrhosis of the liver in cattle and horses. It most often causes poisoning in cattle when eaten dried in hay or when it is incorporated into silage. When it is growing in the fields the cattle avoid it.

270. Dinneen (1927) said that this name is 'usual as an English word [for rag-weed, rag-wort] in Fingal [Co. Dublin]'.

The species has been used quite extensively in Ireland as a medicinal plant. The leaves were applied to animal wounds as a poultice. It was used for humans in the same way for the treatment of cancers (Moloney, 1919). It has been used medicinally in Ireland to treat jaundice, boils and abscesses, cuts, sores and inflammations, coughs, colds and sore throats, rheumatism, sprains and swollen joints, measles, bowel hives, warts and nettle stings (Egan, 1887; Allen & Hatfield, 2004). K'Eogh (1735) described it as good for healing wounds and ulcers, and, according to him, gargling the juice cures inflammations, tumours and abscesses of the throat. Also, he suggested that a plaster made of ragwort, with oatmeal and butter, is useful for sciatica as well as for a variety of inflammations and tumours.

Colgan (1914) described being told in north Co. Dublin that ragweed (called 'She-Bulkishawn') was an ingredient in a famous horse medicine, which contained a total of 12 plant ingredients. Apparently, this plant was regarded as having a male equivalent (He-Bulkishawn), which was an entirely different species, common tansy (*Tanacetum vulgare*). Colgan reported a similar folk botany 'sexual system' for plants elsewhere in the country, where different forms of the same species or different species were regarded as being male and female elements of a single entity.

In the Cooley Mountains in Co. Louth it was apparently traditional to allow goats to graze fields where *Senecio jacobaea* was growing, with the aim of removing a plant that is poisonous to other stock. The goats would then be milked and this milk was then renowned as an excellent cure for asthma (Helen Rock, pers. comm., 2010).

Drury (1985) told of the use of ragwort in Britain as a cure for an ailment of horses called 'staggers'. The same author said that its alternative English name, 'St James's wort', derives from St James being the patron saint of horses in English folklore.

According to a project shown at the Young Scientist Exhibition in 2008 from Bandon Grammar School, Co. Cork, ragwort has been used in parts of Ireland as a fish poison. Ragwort is collected and put into a bag and floated in a river. Fish in the river swim downstream to escape the poisons released from the plant and are then caught in a net placed there by the poachers.

The flowers of ragwort were used to produce a brown dye (Sharkey, 2000), although it is not a long-lasting dye unless mixed with alum (Williams, 1993). Writing in 1758 about a visit to Co. Kerry, Richard Pococke, an English clergyman and later Bishop of Ossory, wrote: 'On the green sod near the strand, I saw a plant very common in Hamps. with stalks of yellow flowers on it in clusters, the flowers not so big as a pin's head, they call it *Dah divill* (sand bank dye) the root of it dying a fine red & is used by the common people for dying stockins &c.' (McVeagh, 1995). Presumably, this is the sand dune variety, *Senecia jacobaea* var. *dunense*.

In the islands of Scotland, ragwort stems were used to make baskets, which Milliken and Bridgewater (2004) said must have been fairly rough and ready. I have not found records of ragwort baskets made in Ireland, but I have no doubt that any such plant material would have been utilized in Ireland to make simple baskets whenever needed, particularly if 'better' plants such as willow rods were not readily available.

Senecio jacobaea var. *dunense*.

*Senecio sylvaticus

- Heath groundsel (Scannell & Synnott, 1987)
- *Grúnlas móna* (Scannell & Synnott, 1987)

ASTERACEAE (Compositae)—THE DAISY FAMILY

Native

Senecio sylvaticus is not frequently found in Ireland, but it is nevertheless widespread throughout in dry, heathy habitats, such as sandy banks and boulder-clay cliffs.

*No ethnobotanical uses of this plant have been noted in Ireland.

Senecio vulgaris

- Groundsel (Scannell & Synnott, 1987)
- *Grúnlas* (Scannell & Synnott, 1987); *Gronnlus* (Moloney, 1919; Dinneen, 1927); *Grúlus* (Moloney, 1919); *Buachalán na neascarán* (Moloney, 1919); *Crann lus* (Cameron, 1883; Dinneen, 1927; Uí Chanainn, 1962); *Buafanan na h'easgaran* (Cameron, 1883); *Grúnlus* (Dinneen, 1927); *Grán-lus* (Dinneen, 1927); *Buachalán nach n-eascaraíonn* (Ó Dónaill, 1977); *Clár boinn* (Ó Dónaill, 1977); *Gronnlus* (Colgan, 1911)

ASTERACEAE (Compositae)—THE DAISY FAMILY
Native

Senecio vulgaris.

Senecio vulgaris is an extremely abundant annual weed of waste places, gardens, sand dunes and cultivated ground.

Groundsel was formerly widely used in Ireland medicinally for a variety of aliments: to absorb pus from wounds, boils and blisters; to ease toothache (as a hot poultice on the face); as a blood-cleaning tonic, to staunch nosebleeds; as a diuretic; and for chapped hands (MacMahon, 1994; Allen & Hatfield, 2004). It was also used as a purgative for children. Moloney (1919) mentioned that it was widely used for such purposes in Ireland as a safe option for children before the introduction of castor oil. A sprig of groundsel was added to milk that was boiled and then strained and given to constipated babies (Moloney, 1919). Cameron (1883) said that one of its Irish names, *buafanan na h'easgaran*, can be translated as 'a remedy to overcome the plague', a name also given to *Senecio jacobaea* (ragwort). McGlinchey (2007) said that groundsel heated over the fire and applied as a poultice gave great ease in the cases of gravel.

This species was used in veterinary treatments for a variety of animal ailments in various parts of Ireland, including the treatment of red-water fever, blackleg and pip (Allen & Hatfield, 2004).

It has been a popular food for hens and canaries in Ireland (Williams, 1993).

**Sesleria caerulea (Sesleria albicans)*

- Blue moor-grass (Scannell & Synnott, 1987); One-pointed grass (Hart, 1898)
- *Féar boirne* (Scannell & Synnott, 1987); *Síoda móna* (Moloney, 1919); *Béitean* (Moloney, 1919)

POACEAE (Gramineae)—THE GRASS FAMILY
Native

Sesleria caerulea is the dominant grass species seen among the karst topography of the Burren, in Co. Clare. *Sesleria caerulea* is found in a variety of other limestone habitats that include dry grasslands, sand dunes, limestone pavements and rocky places. It is found in similar calcareous habitats in the western half of Ireland as far north as Donegal but is absent from the east.

*No ethnobotanical uses of this plant have been noted in Ireland.

Sherardia arvensis

- Field madder (Scannell & Synnott, 1987)
- *Dearg faille* (Scannell & Synnott, 1987); *Balla cnis chonchulainn* (Moloney, 1919); *Dearg faille* (Moloney, 1919)

RUBIACEAE—THE BEDSTRAW FAMILY

Native

Sherardia arvensis is a little annual plant throughout Ireland that can occur as a weed of cultivated ground and lawns, as well as in a variety of dry habitats, such as sand dunes and dry grasslands.

It was used as a dye (Moloney, 1919) yielding a red colour (Mahon, 1983).

**Sibthorpia europaea*

- Cornish moneywort (Scannell & Synnott, 1987)
- *Pingin Dhuibhneach* (Scannell & Synnott, 1987)

VERONICACEAE (Scrophulariaceae)—THE SPEEDWELL FAMILY

Native

Sibthorpia europaea is a creeping plant of damp places, with small, round leaves with tiny pink-white flowers, that is found in Ireland only in the Dingle Peninsula, Co. Kerry. It grows on stream banks and the bases of walls and boggy places, mainly on the north side of the peninsula.

*No ethnobotanical uses of this plant have been noted in Ireland.

Sibthorpia europaea.

Silene species

- Campion (Scannell & Synnott, 1987)
- *Coireán*[271] (Scannell & Synnott, 1987)

CARYOPHYLLACEAE—THE CARNATION OR PINK FAMILY

**Silene acaulis*

- Moss campion (Scannell & Synnott, 1987); Cushion pink (Praeger, 1938)
- *Coireán caonaigh* (Scannell & Synnott, 1987)

CARYOPHYLLACEAE—THE CARNATION OR PINK FAMILY

Native

271. Other Irish names for *Silene* species include *cireán* and *coiligh* (Moloney, 1919).

Silene acaulis is a rare plant in Ireland of high mountain habitats in the north-west, from Cos. Derry to Mayo, that grows on rocky ledges and cliffs.

*No ethnobotanical uses of this plant have been noted in Ireland.

Silene alba (Silene latifolia)

- White campion (Scannell & Synnott, 1987)
- *Coireán bán* (Scannell & Synnott, 1987)

CARYOPHYLLACEAE—THE CARNATION OR PINK FAMILY
Native (or possible archaeophyte)

Silene alba is an occasional weed of cultivated ground, tilled fields and disturbed waste places that is distributed mainly in the south-east and centre of Ireland.

*No ethnobotanical uses of this plant have been noted in Ireland.

Silene dioica

- Red campion (Scannell & Synnott, 1987)
- *Coireán coilleach* (Scannell & Synnott, 1987)

CARYOPHYLLACEAE—THE CARNATION OR PINK FAMILY
Native

Silene dioica is a plant of woodlands, hedgerows and shady places and occasionally found in parts of the north and east; it is rare elsewhere in Ireland.

*No ethnobotanical uses of this plant have been noted in Ireland.

Silene uniflora (Silene vulgaris subsp. *maritima)*

- Sea campion (Scannell & Synnott, 1987)
- *Coireán mara* (Scannell & Synnott, 1987); *Púr faille* (Uí Chonchubhair, 1995); *Coireán mara* (Ó Dónaill, 1977); *Púir fhaille* (Ó Criomhthain, MSS, n.d., ca. 1908)

CARYOPHYLLACEAE—THE CARNATION OR PINK FAMILY
Native

Silene uniflora is a common plant of the Irish coasts where it grows on rocky and shingle shores, harbour walls and cliffs. It can also be found inland, although rarely, on mountain tops.

According to Uí Chonchubhair (1995), a decoction of this plant was used in wine to stop bleeding, treat urinary infections and kidney stones and heal sores and ulcers. The same author recorded that a drink made from the seeds in wine was used as an antidote against stings, bites and bad humour.

Silene vulgaris

- Bladder campion (Scannell & Synnott, 1987)

- *Coireán na gcuach* (Scannell & Synnott, 1987); *Coilleán coille* (Dinneen, 1927)

CARYOPHYLLACEAE—THE CARNATION OR PINK FAMILY
Native

Silene vulgaris is an occasional plant throughout Ireland of roadsides and disturbed soils; it is more common in the south-east.

Mears and Hillman (2007) reported this species to be edible. The shoots apparently have a taste and texture similar to that of mangetouts (snow peas or snap peas). Rutty (1772: 69) said that 'the leaves boiled are said to eat very pleasant, like pease, and proved of great use as an esculent, in the year 1685, when a swarm of Locusts had destroyed the harvest'.

The Irish word *coireán* means 'a small cauldron' (Dinneen, 1927), presumably because of the shape of the inflated tubular calyx.

Silybum marianum

- Milk thistle (Scannell & Synnott, 1987)
- *Feochadán Muire* (Scannell & Synnott, 1987); *Beárnán breac* (Moloney, 1919); *Fothannan beanduighte* (Cameron, 1883); *Bleachtán (-áin)* (Dinneen, 1927); *Bleicht-phiseán* (Dinneen, 1927)

ASTERACEAE (Compositae)—THE DAISY FAMILY
Introduced (possible archaeophyte)

Silybum marianum is now a rare thistle in Ireland of sandy banks and waste places that are generally close to human habitation.

Dónal O'Mathuna, writing in *The Irish Times* (*Healthplus Supplement*, Tuesday, 5 August 2008), described how milk thistle was widely used in Mediterranean regions since ancient times to treat a variety of liver diseases. Extracts from the plant are used to make a compound called 'silymarin' that is said to protect the liver from poisons that cause liver damage. The compound is used to treat jaundice, cirrhosis of the liver (most commonly caused by alcohol damage), hepatitis and other chronic liver diseases. Medical science has not confirmed whether its use in this regard is effective or not, and the results of some studies are inconclusive one way or the other. O'Mathuna reported cases where it was given to patients suffering from poisoning from the death cap fungus, *Amanita phalloides* (in Irish, *caidhp bháis*, Ó Dónaill, 1977). Some of the patients recovered, but others died, and it is unknown whether milk thistle treatment helped. It was used in Ireland as a medicinal plant for various purposes, including the treatment of whooping cough and warts (Allen & Hatfield, 2004). Its leaves have white veins which were said to contain the milk of the Virgin Mary, which gave it its common name. Its name in Irish, *feochadán Muire*, has similar origins. It was also said to help increase the flow of nursing mother's milk.

Gerard's *Herball* from 1597 suggested that milk thistle can be eaten. Its leaves were used as a pot herb, as a vegetable or in salads (the leaves were trimmed of prickles first) and its stalks peeled, soaked to remove the bitterness and cooked like rhubarb. The bracts surrounding the flowerheads can also be eaten like those of the globe artichoke. The seeds, called 'kenguel seeds', were roasted as a substitute for coffee (Howes, 1974). There are no records of its use for food or as a beverage in Ireland.

**Simethis mattiazzii (Simethis planifolia)*

- Kerry lily (Scannell & Synnott, 1987)
- *Lile Fhíonáin* (Scannell & Synnott, 1987)

XANTHORRHOEACEAE (Asphodelaceae)—THE ASPHODEL FAMILY
Native

Simethis mattiazzii is an attractive plant with long grass-like leaves and small white flowers that are purple on the outsides. It can be found in Ireland only in coastal heathy places in the extreme south-west corner of Co. Kerry, mainly around Derrynane, and at Ardgroom in Co. Cork nearby.

*No ethnobotanical uses of this plant have been noted in Ireland.

Sinapis arvensis

- Charlock (Scannell & Synnott, 1987); Preshaugh (Tóibín, 1965); Preshagh (Praeger, 1938); Yellow weed (Colgan, 1904)
- *Praiseach bhuí* (Scannell & Synnott, 1987); *Praiseach gharbh* (de Vere Coneys, 1849; Moloney, 1919); *Lus na súl mbuidhe* (Moloney, 1919; Dinneen, 1927); *Currán buidhe* (Moloney, 1919); *Nóinín buidhe* (Moloney, 1919; Dinneen, 1927); *Garabhóg (-óige, -óga)* (de Vere Coneys, 1849); *Corrán buidhe* (Dinneen, 1927); *Praiseach bhuidhe* (Dinneen, 1927); *Lus na súl buí* (Ó Dónaill, 1977)

BRASSICACEAE (Cruciferae)—THE CABBAGE AND CRESS FAMILY
Probably Introduced (probable archaeophyte)

Sinapis arvensis is probably an introduced species throughout Ireland and is found as a weed in waste places and on roadsides and agricultural land.

In the past it was used as an occasional food source, and the seeds are edible when boiled or baked. According to Milliken and Bridgewater (2004), in Orkney, Scotland, in the 1880s during the three months before harvest when grain was scarce, 'reuthie bread' was made from the peppery seeds, which was the only bread available at that time. I have not found records of such usage in Ireland.

Williams (1993) noted that in times of hardship or famine it was used to make 'vegetable potage' and was sold in the streets of Dublin in the 18th century as a vegetable. The same author recorded that in West Galway it was eaten mixed with nettles and the yellow maize meal that was distributed as a famine relief food in the 1840s; however, eating too much charlock causes the skin to turn yellow. Charlock was also used to make gruel, porridge or 'stirabout'.

Lucas (1959), Drury (1984) and Vickery (1995) highlighted its importance as a famine food in Ireland, where it was called 'boiled prushia'. Lucas (1959), quoting from Samuel Burdy's *Life of the Late Rev. Philip Skelton, With Some Curious Anecdotes* (Dublin 1792), said: 'In 1757, a remarkable dearth prevailed in Ireland; the effects of which were felt most severely in the rough and barren lands of Pettigo. Mr Skelton went out into the country to discover the real state of his poor, and travelled from cottage to cottage over mountains, rocks and heath. He was then a witness to many scenes of sorrow, to which the gay world were insensible, and which could only be felt by a soul so sympathetic as his. In one cabin he found the people eating boiled prushia by itself for their breakfast, and tasted this sorry food which seemed nauseous to him. Next morning he gave orders to have prushia gathered and boiled for his own breakfast, that he might live on the same sort of food with the poor. He eat [sic] this for one or two days; but at last his stomach turned against it, he set off immediately for Ballyshannon to buy oatmeal for them, and brought thence with all speed as much as appeased the hunger of some of them'. McGlinchey (2007) also wrote about charlock being collected by people during times of famine, where *praiseach* was being pulled to make soup from amongst the landlord's cornfields in Co. Donegal in about 1770 or 1780. Writing about the effects of the famine in the 1840s in Donegal, he also wrote that 'the women pulled *praiseach* round the fields and made it into a kind of broth with turnips or whatever else they could get'.

Lucas (1960) mentioned that a plant bearing the name 'praiseach' was in widespread use in Ireland for food, but he was uncertain whether this referred to charlock or to some cultivated form of kale or cabbage.

A saying of the time quoted in Ireland until recent times was:

Praiseach bhuidhe na ngort, biadh na ndaoine mbocht.
'Charlock of the fields, food of poor people'.

Other important wild famine foods in Ireland were nettles, sorrel, dandelions (roots and leaves), water-cress and probably hazelnuts.

Allen and Hatfield (2004) noted that a preparation of charlock flowers was used in parts of Ireland as a cure for jaundice. Logan (1972) mentioned that yellow-flowered plants such as charlock were often believed to be useful in the treatment of jaundice. K'Eogh (1735) recorded that it was made into a poultice and used to treat convulsions and cramps. It was boiled in water and strained through linen, then taken with added honey to treat coughs (Williams, 1993).

Sisymbrium officinale

- Hedge mustard (Scannell & Synnott, 1987)
- *Lus an óir* (Scannell & Synnott, 1987); *Meilise* (Moloney, 1919)

BRASSICACEAE (Cruciferae)—THE CABBAGE AND CRESS FAMILY
Probably Introduced (probable archaeophyte)

Sisymbrium officinale is a common weed of waste places, roadsides and cultivated ground. It is uncertain whether this is a native species, and authorities on the matter are divided.

Uí Chonchubhair (1995) noted that this plant was used as a pot herb in the preparation of a sauce to go with salt fish and was also used medicinally; the seeds were used to treat bleeding and incontinence and were given to children as a vermifuge. The plant was also reportedly used to treat swellings of the breast or testicles, to treat sciatica, and as an antidote to poison. Moloney (1919) listed its use in traditional medicine for hoarse throats and chest complaints.

**Sisyrinchium bermudiana*

- Blue-eyed-grass (Scannell & Synnott, 1987)
- *Feilistrín gorm* (Scannell & Synnott, 1987)

IRIDACEAE—THE IRIS FAMILY
Native

Sisyrinchium bermudiana is a perennial iris relative of short wet grasslands and lake-shores that is quite common in Co. Kerry; it is rare in other parts of the midlands and the west. Beyond Ireland it is native to North America only.

*No ethnobotanical uses of this plant have been noted in Ireland.

Sium latifolium

- Greater water-parsnip (Scannell & Synnott, 1987)
- *Meacan uisce*; *Cosadh dubhadh* (Cameron, 1883); *Fochlacht* (Dinneen, 1927); *Ráib uisce* (Dinneen, 1927)

APIACEAE (Umbelliferae)—THE CARROT FAMILY
Native

Sium latifolium is a rare perennial plant in Ireland of wet habitats, such as marshes and river-banks, and is found particularly in the midlands in the river basins of the Erne and Shannon.

It is regarded as a poisonous plant but has been used medicinally in Ireland (Moloney, 1919) 'for scrofulous swellings in the neck in combination with garlic and butter'. In other parts of its range, such as in the Altai Mountains of Siberia, it has been used to treat various ailments, including bronchitis, asthma, tetanus and cancer tumours. It is also said to have hallucinogenic properties.

According to Wade (1804: 20), it is 'highly noxious to cattle'.

Smyrnium olusatrum.

Smyrnium olusatrum

- Alexanders (Scannell & Synnott, 1987)
- *Lusrán grándubh* (Scannell & Synnott, 1987); *Lus na ngrán dubh* (Moloney, 1919; Dinneen, 1927; Uí Chonchubhair, 1995)

APIACEAE (Umbelliferae)—THE CARROT FAMILY
Introduced (possible archaeophyte)

Smyrnium olusatrum is widely established throughout the coastal counties of Ireland on roadsides and waste places; it is much rarer inland.

It was formerly used as a vegetable and pot herb (e.g., Mackay, 1836; Moloney, 1919), but it is now more or less entirely replaced by celery. It is one of my favourite wild foods, although some people tell me they find it revolting! It is clearly an acquired taste. The flowering shoots should be picked early in the season just as the flowerheads are forming. These shoots, or stalks, can then be cut into lengths, peeled and boiled or steamed and served with melted butter, as one would serve asparagus. The delicate flavour is somewhat reminiscent of fennel or celery. March is the ideal month to collect and use it. Later in the season the shoots become woody and stringy, and the flavour becomes too strong for my taste. The leaves and flowerheads are also useful in soups and salads, and the roots can be boiled and eaten as a vegetable, like parsnips. When ripe, the seeds can be ground and used instead of pepper (Hilton, 2007).

Hugh Fearnley-Whittingstall, the British TV chef, recently included a recipe for a liqueur made from Alexanders in his television programme, 'A River Cottage Christmas'. Alexanders shoots (500 g) and a few leaves were liquidised, squeezed, and strained and the juice boiled and added to 75 g of sugar. Then, when the sugar was melted, the liquid was added 1 part to 4–5 parts vodka (or white rum) to make the liqueur. A few seeds can also be added when it is being bottled.

Smyrnium olusatrum.

Solanum dulcamara

- Bittersweet (Scannell & Synnott, 1987); Woody nightshade (Colgan, 1904; Brunker, 1950; Anonymous, 1978; Scannell & Synnott, 1987)
- *Fuath gorm* (Scannell & Synnott, 1987); *Searbhóg mhilis* (Moloney, 1919; Dinneen, 1927); *Lus na hoidhce* (Moloney, 1919; Dinneen, 1927; Ó Dónaill, 1977); *Slat ghorm* (de Vere Coneys, 1849; Moloney, 1919); *Fuathghorm* (Moloney, 1919; Ó Dónaill, 1977); *Dréimire gorm* (de Vere Coneys, 1849; Ó Dónaill, 1977; Webb & Scannell, 1983); *Miathog buidhe* (de Vere Coneys, 1849; Cameron, 1883; Dinneen, 1927); *Míogbhuidhe* (de Vere Coneys, 1849; Dinneen, 1927); *Míogaidhe* (Dinneen, 1927)

SOLANACEAE—THE NIGHTSHADE FAMILY
Native

Solanum dulcamara is a common perennial plant throughout Ireland of a variety of habitats such as hedges, roadsides, waste places, seashores and woodland edges.

Solanum species are members of the nightshade family and include the powerfully narcotic and poisonous species *Atropa belladonna* (deadly nightshade) and *Hyoscyamus niger* (henbane), as well as the edible species *S. tuberosum* (potato). Mackay (1836: 189) commented casually in his *Flora Hibernica*, which was published during the early years of the Irish potato famine, that the poisons present in the potato are 'dissipated by heat, and the tubers are rendered quite wholesome'.

Solanum dulcamara.

Bittersweet is native to Europe and Asia and has been used in traditional medicine to treat chronic eczema, skin conditions and rheumatism, as well as respiratory problems such as bronchitis and asthma (van Wyk & Wink, 2004). Moloney (1919) noted that it is a narcotic and formerly used as a sedative to promote sleep (hence one of its Irish names: *lus na hoidhce*, meaning 'nightshade', Ó Dónaill, 1977). In Ireland, it was also used medicinally to treat cuts (Allen & Hatfield, 2004).

The attractive-looking and succulent berries of this species are bitter and very poisonous, so they should not be used. Rutty (1772: 175) wrote that 'Sir John Floyer gave thirty berries to a dog, and that in three days it died mad'.

Solanum tuberosum

- Potato (Reynolds, 2002)

- *Prátaí* (pl) (Dinneen, 1927); *Práta* (s) (Dinneen, 1927); *Póirín* (s)[272] (Dinneen, 1927); *Póirínidhe* (pl)[273] (de Vere Coneys, 1849); *Pruisteal* (*-téil*)[274] (de Vere Coneys, 1849); *Potáta* (Dinneen, 1927); *Fata* (Dinneen, 1927); *Préata* (Dinneen, 1927); *Preáta* (Dinneen, 1927); *Buntáta* (Dinneen, 1927); *Fadhbán*[275] (Uí Chanainn, 1962)
- *Bidaades* (in Yola) (Dolan & Ó Muirithe, 1979)
- *Kun'el*; *Molem*; *Cullions* (MacGréine, 1934: *sup cullions*—a few potatoes; *chelp cullions*—boiled potatoes) (in Shelta)
- *Cnoc*; *Criabhōg* (in Bēarlagair na Sāer)

SOLANACEAE—THE NIGHTSHADE FAMILY
Introduced

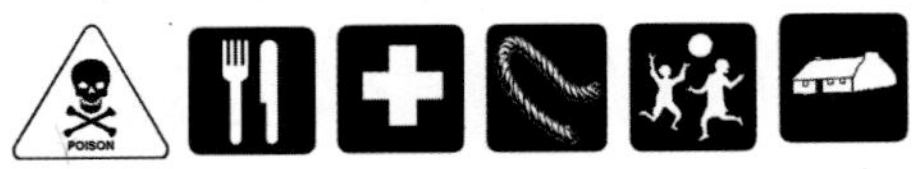

Potatoes are well known as the archetypal Irish vegetable, introduced to Europe from South America several centuries ago, probably as early as 1536.[276] The potato is native to the high Andean mountains, where it was a staple food for the Incas for several hundred years before the Spanish conquistadors arrived. The early Spanish explorer and conquistador, Pedro de Cieza de León, saw potatoes in cultivation in Quito (the present-day capital of Ecuador), Popayan and Pasto (both in present-day Colombia), which he documented in his *La Crónica del Perú*, published in 1553. The Inca civilization was built not on the treasures for which they are best known, gold and silver, but instead on two natural treasures that they have passed on to the modern world: potatoes and corn (maize). In the Andes, the Incas used as many as eight potato species as food, but only one of them, *Solanum tuberosum*, has become established as a western food (Lewington, 1990). At one time the Incas ruled a huge area of Andean South America, north from Quito in what is now Ecuador, through Peru, south to where Santiago, the capital city of Chile, is today. The introduction of the potato to Europe is often attributed to Sir Frances Drake or Sir Walter Raleigh, but much of what has been written about their route to Ireland from the Andes is fanciful (Hobhouse, 1986). It has even been suggested that potatoes floating ashore from the wrecked ships of the Spanish Armada in 1588 in western Ireland caused the country to be colonised by this crop and an (unreferenced) Wikipedia entry in 2010 said that Basque fishermen used potatoes as ships' stores for their voyages across the Atlantic in the 15th century and introduced the tuber to western Ireland where they landed to dry their cod. Other writers have said that the potato was probably first carried to Europe in the 16th century by the Spanish, probably as early at the 1530s, where they were first cultivated in Spain and then gradually spread from there into other European countries (Gilbert, 1917). Potatoes were certainly being grown by the first settlers in Virginia in North America before 1585, and Sir Frances Drake brought potatoes to England when he returned from his voyages in 1580.

Whatever their route of introduction to Ireland, potatoes had become a common vegetable crop in Ireland by the late 17th century, and by the early part of the 18th century they were well on their way to becoming the staple food supply for a significant proportion of the Irish population. During the height of their popularity amongst the rural poor in Ireland it was not unusual for each person to consume as much as 14 lb of potatoes daily (Woodham-Smith, 1962). Ironically, in view of what happened in the 1800s, Rutty wrote in

272. Dinneen (1927) defined *póirín* (*s*) as 'a small potato'.
273. This translates as 'small seed potatoes'.
274. This translates as 'a batch of potatoes for roasting'.
275. Uí Chanainn (1962) defined *fadhbán* as 'a large potato'.
276. Exactly when they were introduced is a subject of some controversy as many early introductions to Spain may have been of the sweet potato (*Ipomoea batatas*, a member of the Dioscoreaceae), a different plant entirely.

An Essay Towards a Natural History of the County of Dublin (1772: 82) that the potato is 'a most excellent substitute, for the farinaceous grains, and which has very often preserved us from famine'.

Potatoes in La Paz market, Bolivia.

Fourteen pounds of potatoes.

During the century leading up to the Great Potato Famine in the 1840s the potato crop failed on several occasions, but these early failures did not have the devastating impact on the population that they did later because potatoes had not yet grown to totally dominate agriculture and the diets of the majority of poorer tenant farmers throughout the land.

In the 1840s, Ireland had a population of about eight million people, of whom at least one third relied on the potato as their staple food supply. In parts of Ireland potatoes were more or less the only significant crop grown for food. The potato is also highly productive, and it is recorded that an acre and a half (0.6 ha) of potatoes could provide a family of five or six with food for a year, while to grow enough food from a cereal crop required four to six times the amount of land, and considerably better skills in tillage were necessary to achieve a productive cereal crop (Woodham-Smith, 1962). Therefore, when the potato crop was hit by the pathogenic fungus, *Phytophthora infestans* (in Irish, *an dúchan*, or 'potato disease'), in 1845, the country was devastated by an increasingly serious famine that year and in 1846 and 1847. In 1845, in about 10 weeks the 'potato murrain' (as the disease was sometimes called) spread through Belgium, the Netherlands, France, Britain and Ireland, causing the loss of about half of the crop that year (Nelson, 1995). In 1845 potato blight resulted in the loss of about one third of the total crop grown in Ireland. The following year about three-quarters of the harvest were lost and in 1847 the potato yields were also very low, mainly because a smaller area was planted, as seed potatoes were scarce (Kennedy et al., 1999).

The Great Famine (or in Irish, *an Gorta Mór*, or 'the Great Hunger') changed Ireland forever. It was a period of mass starvation during which approximately one million people died of hunger or disease. A further million people emigrated, resulting in the Irish population dropping by 20% to 25% between 1845 and 1852. The impact of the Famine in Ireland during this period cannot be overstated. Cecil Woodham-Smith (1962) included accounts of the horrific suffering caused by the loss of this staple food in many parts of the country. One of these contemporary accounts, written by a Captain Wynne on Christmas Eve, 24 December 1846, read: 'I ventured through the parish [Clare Abbey, Co. Clare] this day to ascertain the condition of the inhabitants, and, altho' a man not easily moved, I confess myself unmanned by the intensity and extent of the suffering I witnessed more especially among the women and children, crowds of whom were to be seen scattered over the turnip fields like a flock of famished crows, devouring the raw turnips, mothers half naked, shivering in the snow and sleet, uttering exclamations of despair while their children were screaming with hunger. I am a match for anything else I may meet with here, but this I cannot stand'.

Charles McGlinchey (2007) wrote about the effects of the famine in Co. Donegal in the 1840s: 'There was a lot of travelling people on the roads, coming from places where the famine was worse. I heard of a travelling man being found dead along the road and his mouth full of grass and weeds. One day at dinner-time a man and a boy called at a house in Gortnahinson and asked them for God's sake to give them the skins of the potatoes in the basket. They made balls of the skins and ate them and moved on up the Owenirk road. Some days after that the boy was found dead on the old road below Minamrua'.

Irish traditional ways of life changed dramatically during this period, and it is regarded as a watershed in Ireland's history. The potato had a significant impact on life in Ireland before the famine, allowing a significant growth of population in a few generations and bringing into cultivation much land that had previously been uncultivated. In 1700 the Irish population was about one million people and it climbed steadily until reaching its peak of more than eight million in the mid-1840s, before continuously declining (mainly through emigration in the post-famine period) until the mid-1940s to 1950 when it stood at just over four million.

It is remarkable the impact that a single plant species, the potato, had on so many aspects of Ireland's life, landscape, economy, history, traditions (including its ethnobotany) as well as on the history of many other countries to which the Irish emigrated, including Britain and, in particular, the United States. The Great Famine in Ireland was probably the world's first well-documented environmental disaster, which has left scars on Ireland and a significant international legacy that has lasted right up to the present day.

In 1917 A. W. Gilbert wrote that the potato accounted for about 25% of the food of European and English-speaking people, when 161 million tonnes of potatoes were produced (averaged over five years from 1908 to 1912), compared to 128.38 million tonnes of corn (maize), 106 million tonnes of wheat, 65.6 million tonnes of oats, 57 million tonnes of rye and 33.41 million tonnes of barley. Accurate figures for rice production at that time were lacking. Today potatoes are still one of the largest world crops, ranked fourth after rice, wheat and maize.

Worldwide there are more than 5000 different potato varieties known, of which some 3000 are grown in the Andes. Nevertheless, more than 99% of all cultivated potatoes are today derived from a subspecies (*Solanum tuberosum* subsp. *tuberosum*) that is indigenous to south-west Chile, where it has been cultivated in the Chiloé Archipelago for as long as 10,000 years (Miller, 2008). The Andean subspecies is *S. tuberosum* subsp. *andigena*.

Even after the famine, and despite the continuing effects of blight and the memory of the tragic events of the 1840s, the potato remained the most important food item for most of Ireland's rural farming communities. Michael MacGowan (2003: 3) wrote in his autobiography, *The Hard Road to Klondike* (first published in Irish in 1959 as *Rotha Mór an tSaoil*), that in the 1860s and 1870s when he was a young lad in Co. Donegal:

> Mostly they survived on potatoes and they worked hard to raise that particular crop. Spades were used in our area all the time. There was an odd wooden plough here and there but there were so many rocks and stones around the place that even that could hardly be used. There was no talk then about artificial fertilizers and, as every family had only one cow, that left natural manure in short supply. People had to go down to the sea-shore to watch the tides early and late, collecting sea-weed and wrack to put on the land.... When I was young, there were households that ate potatoes four times a day — on two occasions they were boiled, and on the other two they were in the form of potato-cakes. They lay heavy on the stomach but, in those days, fresh fish were as plentiful as grass—and as cheap. Fresh ling and cod from Tory [Island, off the Donegal coast] sold at twopence apice. They had these and other good food as well: limpets, periwinkles, cockles, dulse, laver, mussels. Many is the time when there would be real hardship that they would have to make do with that food alone. I often heard the old people say it was this food saved the people in the 'Bad' Year (1847) and in another poor year when I was a child.

The potato remains a hugely popular food source in Ireland. Indeed most families in

Ireland still eat potatoes, if not every day, then at least several times a week. When I was growing up in Limerick and Kerry, a meal was not really regarded as complete unless potatoes were included in one form or another. As a student in the 1970s I remember visiting my local T. D. (public representative) in Dáil Éireann. While there, he very kindly took me to lunch in the canteen, where I was given a plate of spaghetti bolognese, with potatoes and cabbage!

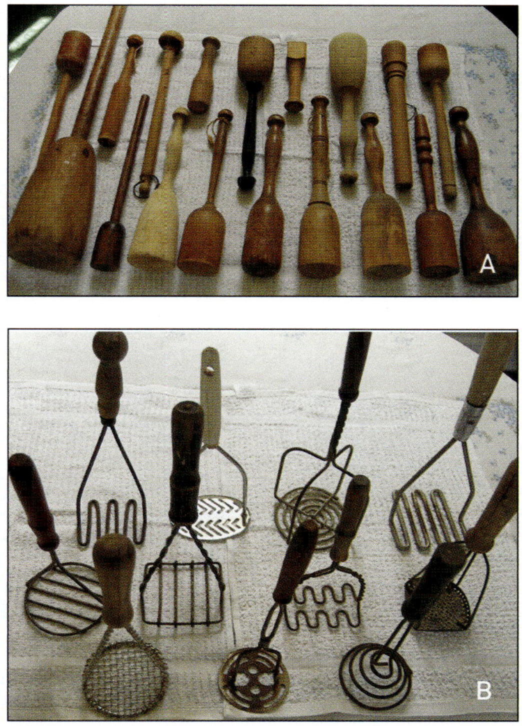

Various types of potato mashers (author's collection).

A 'spud gun'.

Traditional methods of cooking potatoes in Ireland have included boiling, roasting, steaming, baking and frying. A multitude of recipes including potatoes are widely known in Ireland, including potato cakes, chips, mashed potatoes, dumplings, potato crisps and so on. Potatoes are also the main ingredient of colcannon, which is composed of mashed potatoes mixed with shredded cabbage (or kale) and onions (or leeks). Colcannon (in Irish, *cál ceannann*[277], meaning 'white-headed cabbage', Dinneen, 1927) has long been a traditional dish for Halloween in Ireland when small coins would be mixed in as a surprise, in much the same way as small treats, charms or coins are traditionally added to Christmas plum puddings.

Potatoes are used in many other traditional Irish recipes, such as Irish stew (lamb, onions and potatoes mainly) and champ. Champ is made by boiling and mashing and then seasoning potatoes well with salt and pepper, then adding boiling milk in which chopped onions have been cooked and mixed well before serving with melted butter. In parts of Northern Ireland nettle tops (*Urtica dioica*) or parsley or peas were also added. An old Irish rhyme celebrates champ: 'There was an old woman who lived in a lamp; she had no room to beetle her champ'. The beetle was a traditional wooded pestle used to mash potatoes, often made from beech or sycamore wood.

There is a wide diversity of folklore surrounding the potato in Ireland, including many different Irish names of potatoes and their different varieties (e.g., *fataí*, *prátaí*, *potátaí*, praties, *creatháin*, *seacaidí*, *cailleacha*, 'stags', *craoileacáin*, *fadhbháin*, *geinidí*, *prátaí odhruighthe*, etc.) (O'Súilleabháin, 1942). Uí Chanainn (1962) gave the Irish word *bruthnóg* (*-óige*) for 'roast potatoes'. In Scots Gaelic the potato is *buntata*, an adaptation of the Spanish word *batata* (Cameron, 1883). Customs and practices associated with planting, harvesting and storing potatoes are also widely remembered in Ireland. It is still widely regarded in Ireland that St Patrick's Day is the best day to plant a potato crop.

For children in 1960s Ireland, a 'spud gun' was a common and popular toy. An ordinary potato was used to provide ammunition for the gun. When the end of the barrel was plunged into the potato a small plug of potato would be left in it, which would then be shot out by means of compressed air when the trigger was pulled.

Many people in Ireland today also know that a cut potato rubbed on a car windscreen is a good way to prevent ice from forming on the glass, a good example of modern-day plant lore.

My mother would often add a small piece of coal to a pot of peeled potatoes in water, if they were to be left for a while, as she was certain that the coal prevented the potatoes from turning brown.

277. Dinneen (1927) also gave the word *ceallaigh* for the Irish dish of colcannon.

Potatoes have been widely used in Ireland and elsewhere as food for farm animals (fed to them both raw and cooked), although of course in Ireland they were more commonly used for food for human consumption. Starch was extracted from potatoes, for use in the ironing of clothes (O'Súilleabháin, 1942). Potato starch was also used industrially to make sizing (sizing is a glaze or filler used to smooth or waterproof the surface of paper or textiles) for paper and textiles and in the manufacture of industrial alcohol. Potato flour has been made too (Gilbert, 1917). Rutty (1772: 127) wrote potato starch 'is exquisitely white, and not hurtful to Linen, for which purpose, the white Spanish Potatoe, being washed very clean, and grated very fine into a crock of clean water, is to be rubbed well with the hands, then strained through a hair sieve, throwing away the skins, and let what runs through the sieve settle at the bottom of the crock, then decant the water, and pour clean water on, and so shift by repeated assusions of water until it be quite clean, or the pulp white: then put the sediment into a clean linen cloth, and dry it in the sun or by the fire: thus five pounds of Potatoe are said to have yielded half a pound of clear Starch'.

The collections of the National Museum of Ireland (Museum of Country Life) in Castlebar, Co. Mayo, contain an example of a St Brigid's cross in which a potato was incorporated, as part of the ritualistic blessing of the potato crop for the forthcoming season. This potato, now blessed by St Brigid, was often the first seed potato planted for that year. The collections also include a 'Ghost potato lamp' from Keenagh More, Co. Mayo (Ref. No: F1941.282). It is a hollowed-out potato in which a ghostly face has been carved, which may have been used at Halloween, like the more commonly used turnip lamp.

Egan (1887: 11) recorded the use in Ireland of potatoes to treat sprains: 'Bathe in hot water in which they have been boiled. All the "mercury" of the potato supposed to be in the water', and Gilbert (1917) mentioned that a number of 'tinctures' have also been made from potatoes. In the Aran Islands they were used to treat warts, chilblains and corns (Dirrane, 1997); the raw potato was sliced and rubbed directly on and around the affected area.

Rutty (1772: 82) wrote that the potato was also used 'to supply us with drinks, as first being malted and managed as Barley: I am assured that a Liquor like Ale, has been prepared from it'. Poitín ('moonshine') has traditionally been made from potatoes or malted barley in Ireland. Poitín (or poteen, or potcheen) is an Irish distilled spirit that is often very strong with a high alcohol by volume (ABV), generally 60% to 95% (whiskey is usually 40%–43% and beer 4%–6% ABV). Because it was an illegally distilled spirit on which duty had not been paid, poitín was generally made in remote rural places in Ireland, and sometimes even on offshore islands. It was banned in Ireland in 1661 along with all private distillation, and only recently, on 7 March 1997, did it cease to be illegal. Poitín is made from a fermented mash made from potatoes (or malted barley), sugar and yeast to which water has been added (a Wikipedia recipe suggests 100 gallons of fresh water to 6 stone of potatoes and 6 stone of sugar, together with some yeast). As a special Irish product it has recently, in July 2007, been declared a protected European Union G. I. (geographical indicative) Status Spirit. Several legal poitíns are now made in Ireland, such as triple-distilled Knockeen Hills Poteen, which is produced in Co. Waterford, and Bunratty Potcheen, from Co. Clare. Illegal poitín is apparently still made in several parts of Ireland. The Bunratty Potcheen makers recommend mixing it with mead (the ancient honey-based drink) as a cocktail to make a 'Moonshiner'.

McGlinchey (2007) said that in Donegal in the 1800s poitín was generally made from barley. The barley was prepared by steeping it in water, then dried near a fire and turned so that it was all warmed a bit. After that, it was spread on the floor until it started to sprout, then kiln dried and ground in a mill before being steeped in barrels to 'ripen' (I presume, to ferment) before it 'would be ready for running [distilling]'. He said that Revenue Police were appointed after the Napoleonic wars to control illegal poitín production, and when the old people were worried about raids they would place a plank of wood on the road with an iron rod sticking up on it. When they put their ear to the end of the rod they could hear the police galloping their horses even a long way off.

Potatoes are frequently found growing in the wild in Ireland, mainly on roadsides and in waste places and landfill sites. These plants are derived from discarded tubers. In 2008 only 12,000 ha of land were cultivated with potatoes in the Republic of Ireland (data from the Central Statistics Office of Ireland).

Apart from the edible tubers, the rest of the plant is poisonous, including the yellowish green fruits, and should not be eaten. Green-sprouting tubers are also poisonous and contain a poisonous alkaloid, solanine, which can be fatal if eaten (Woodward, 1985).

Solidago virgaurea

- Goldenrod (Scannell & Synnott, 1987); Aaron's rod (Hart, 1898)
- *Slat óir* (Scannell & Synnott, 1987); *Luibh an fhallsaora* (Moloney, 1919; Dinneen, 1927); *Luibh an easa* (Moloney, 1919); *Luibh naoimh seán* (Moloney, 1919); *Lus an fhalsaora* (Uí Chonchubhair, 1995)

ASTERACEAE (Compositae)—THE DAISY FAMILY
Native

Solidago virgaurea.

Solidago virgaurea is an occasional plant in Ireland of heathy grasslands and rocky ground, particularly in mountainous areas, but also of lake-shores. It can also be found among the limestone pavements of the Burren, Co. Clare. It is widespread throughout except for the north-east and adjacent areas of the centre of Ireland, where the plants appear to be absent.

Goldenrod is known to herbalists as the 'herb of the palsy' as a treatment for apathy, depression and nervous exhaustion (Moloney, 1919). According to Allen and Hatfield (2004), it was used in Ireland for heart and kidney problems, stomach upsets, flatulence, colds and pulmonary tuberculosis (Allen & Hatfield, 2004). In some parts of its native range it has also been used as a diuretic, for rheumatism, and to treat kidney and bladder problems as well as included in cough medicines (van Wyk & Wink, 2004).

Sonchus species

- Sow-thistle (Scannell & Synnott, 1987); Swine's thistle (Hart, 1898)
- *Bleachtán* (Scannell & Synnott, 1987); *Bainne muice* (Dinneen, 1927; Ó Dónaill, 1977; Webb et al., 1996); *Fóthannán mín* (Dinneen, 1927); *Leitís girr-fhiaidh* (Dinneen, 1927)

ASTERACEAE (Compositae)—THE DAISY FAMILY
Native

Sonchus arvensis.

Including:

Sonchus arvensis

- Perennial sow-thistle (Scannell & Synnott, 1987); Corn sowthistle (Colgan,

1904; Brunker, 1950); Melsh weed (Colgan, 1904); Milk weed (Colgan, 1904); Milk thistle (Hart, 1898); Milk thrissel (Hart, 1898)
- *Bleachtán léana* (Scannell & Synnott, 1987)

Sonchus asper
- Prickly sow-thistle (Scannell & Synnott, 1987); Spiny milk-thistle or sow-thistle (Clapham et al., 1962); Melsh weed (Colgan, 1904); Milk weed (Colgan, 1904); Rough sowthistle (Brunker, 1950)
- *Bleachtán colgach* (Scannell & Synnott, 1987)

Sonchus asper.

Sonchus oleraceus
- Smooth sow-thistle (Scannell & Synnott, 1987); Common smooth sow-thistle (Brunker, 1950; Anonymous, 1978); Melsh weed (Colgan, 1904); Milk weed (Colgan, 1904)
- *Bleachtán mín* (Scannell & Synnott, 1987); *Bleach(t) fheochadán* (Uí Chanainn, 1962; Ua Maoileoin, 2001); *Bainne muice* (Moloney, 1919); *Leitís geirrfhiadh* (Moloney, 1919); *Bliocht fheóchadán* (Moloney, 1919); *Bog ghioghan* (Cameron, 1883); *Bliochdfochadán (-áin)* (de Vere Coneys, 1849); *Fothanán mín* (de Vere Coneys, 1849)

Sonchus arvensis is a reasonably common plant throughout Ireland of roadsides, waste places, seashores, river-banks and bogs and as a weed of cultivated field margins. *Sonchus oleraceus* is a very common weed of agricultural land throughout Ireland, particularly disturbed ground, farm-yards, roadsides and waste places. *Sonchus asper* is also common but not quite as abundant; it occurs in similar habitats as those of *S. arvensis* and *S. oleraceus* but also on sand dunes and walls.

Sonchus oleraceus.

According to Phillips (1983), sow-thistles (*Sonchus oleraceus*) have been eaten as a vegetable since ancient times. The leaves can be boiled or steamed like spinach, cooked in soup and omelettes, or used as a pot herb. Young leaves can be used in salads. The young inner non-spiney leaves of over-wintering rosettes of *S. asper* are also edible and can be used in ways similar to those of the previous species. Peterson (1977) suggested that the leaves of sow-thistles can be used in most of the same ways as those of dandelions (*Taraxacum officinale*), but that the spines of *S. asper* should be removed before cooking.

Sow-thistles were formerly used in Ireland for the treatment of warts and burns (Allen & Hatfield, 2004).

Hart (1898) said that *Sonchus* species were regarded as being good food for pigs in Donegal and given the common name 'swine's thistle'.

*Sorbus anglica

- English whitebeam (Rich et al., 2010)
- *Fionncholl gallda*

ROSACEAE—THE ROSE FAMILY

Native

Sorbus anglica is a small woodland tree, usually no more than 10 m tall, that is very rare in Ireland where it is only known from Killarney, Co. Kerry (Rich et al., 2010). The English whitebeam is found only on carboniferous limestone, with its distribution restricted to similar wooded habitats over limestone in Great Britain.

*No ethnobotanical uses of this plant have been noted in Ireland.

Sorbus anglica.

Sorbus aria

- Common whitebeam (Scannell & Synnott, 1987)
- *Fionncholl* (Scannell & Synnott, 1987)

ROSACEAE—THE ROSE FAMILY
Probably Introduced

Sorbus aria is a small shrubby tree, typically no more than 15 m tall, that is distributed throughout much of Ireland in woodlands, scrub and on the karst limestone pavements of the Burren, Co. Clare. The largest known recorded tree is Ireland is 20.2 m in height, from Cahervilla House, Golden, Co. Tipperary (Anonymous, 2005). Its status as a native tree in Ireland is in doubt (Webb & Scannell, 1983; Rich et al., 2010). *Sorbus aria* is grown as an ornamental tree in Irish gardens and is an escape from cultivation in most of its widespread Irish locations.

Wade (1804: 37) said of this species (under the name *Crataegus aria*): 'From the fruit of the beam tree an ardent spirit may be distilled, and … the wood of this tree affords an excellent charcoal for making gun-powder'.

Vickery (1995: 393) quoted a letter from Drumshanbo, Co. Leitrim, to the magazine *Ireland's Own* in 1993 which tells of this species being used to predict the weather: 'the leaves of the whitebeam trees slightly turned and giving glimpses of their silvery underside predict rain'.

Sorbus aucuparia (Pyrus aucuparia)

- Rowan (Scannell & Synnott, 1987); Mountain ash (Scannell & Synnott, 1987); Quicken tree (Hart, 1898; Colgan, 1904; Webb, 1977); Kwayreen (Colgan, 1904); Caorran (Hart, 1898); Creaghan (Hart, 1898)
- *Caorthann* (Scannell & Synnott, 1987); *Fuinnseóg coilleadh* (Moloney, 1919); *Luis* (de Vere Coneys, 1849; Moloney, 1919; Dinneen, 1927; Uí Chanainn, 1962; Ó Dónaill, 1977); *Pairtainn dearg* (Cameron, 1883); *Caor* (*-ra*) (de Vere Coneys, 1849); *Caorthan* (de Vere Coneys, 1849); *Fuinseóg coille* (de Vere Coneys, 1849); *Crann caortainn* (Dinneen, 1927; Ó Dónaill, 1977)

ROSACEAE—THE ROSE FAMILY
Native

Sorbus aucuparia is an attractive, small tree of rocky places, glens and mountain riversides that is found frequently in the mountains throughout Ireland. It is known by diverse names in Ireland. *S. aucuparia* is grown as an ornamental tree in parks and gardens (Wyse Jackson, 2006). The largest known specimen in Ireland is 18.5 m in height and grows at

Sorbus aucuparia.

Headfort Preparatory School, Kells, Co. Meath (Anonymous, 2005). It grows at higher altitudes in Ireland than almost any other tree and can tolerate exposed and cold situations. Elsewhere in Europe, rowan is a well-known temperate element of boreal forests. It is a sexually producing tree within a genus known for its apomict (asexual) populations, and its bright red to yellow berries are attractive to and dispersed by birds.

Rowan timber is dense, pale-brown in colour and used for various purposes, including carving, wood turning, tool handles and, formerly, to make long bows (Hickie, 2002). In Britain its wood has been used like that of hazel to make crates, hoops, creels and baskets, and its bark was used for tanning (Edlin, 1973), but in Ireland I have not recorded such uses. Howkins (1996) mentioned that rowan wood was incorporated into household objects, such as butter churns and milk stirrers in Britain, to protect the contents from bewitchment or the 'evil eye'. Hart (1898) mentioned similar uses for rowan wood in Ireland, in Cos. Derry and Donegal. Rowan wood is tough and used for a variety of implements (MacCoitir, 2003), such as tool handles.

McGlinchey (2007) wrote that children made whistles from a branch of the rowan tree: 'We cut a straight piece without knots and tapped the bark round and round with the handle of a knife or stick until the bark got loose and came off in a piece. Then we cut a bit for a mouthpiece and stopped up the other end. Some would put a pea inside the whistle.'

Rowan twigs were sometimes hung in farm dairies as a talisman to keep away evil spirits (Sharkey, 2000). O'Dowd (2011) referred to a custom in Athea, west Co. Limerick, where rushes, ferns and mountain ash were spread on floors and yards on May Eve for protection against evil.

Rowan berries (*caora*) were one of the most significant wild fruits of ancient Ireland and play an important part in folklore. Their remains have been found in Viking Dublin excavations dating back more than 1000 years (Geraghty, 1996). Although bitter to eat, they may have been sweetened with honey. In medieval Dublin the excavated contents of drains and cesspits included seeds of strawberries, blackberries, apples, rowan berries, sloes and plums (Anonymous, 1973). The fruits of this tree have been used for a wide variety of culinary purposes, such as jams and jellies, or fermented and distilled to make an alcoholic spirit. Mears and Hillman (2007) suggested that rowan fruits may not have been widely used as food in ancient times in Britain and Ireland; a range of processing methods were tried but always resulted in unpalatable outcomes. They also reported that the berries can cause poisoning with symptoms including convulsions and respiratory distress. However, the berries are also known as a natural source of sorbitol that is widely used in the pharmaceutical and food industries (although it is now produced synthetically from glucose). Sorbitol is used as a sugar substitute for diabetics; it is about half as sweet as sucrose (van Wyk & Wink, 2004).

Rowan jelly is easily made and is useful to accompany roasted meat and other dishes (used in the same way as one would use red currant jelly). To do this 2 lb (900 g) of rowan berries are added to 300 mL (1/2 pint) of water and simmered gently until the fruits are very soft. The fruits may be pressed a little during cooking to ensure that most of the juice is extracted. The resulting mixture is strained through a jelly bag or muslin. Then, 450 g (1 lb) of sugar is added for each 200 mL (1 pint) of liquid and heated gently until the sugar has dissolved. The mixture is then boiled rapidly until setting point is reached. After that the jelly is poured into sterilised jam jars and sealed. Rowan jelly can have quite a strong or distinctive bitter flavour that not everyone will like, so rowan berries may be added to apples (or wild-collected crab-apples) to make a milder jelly. I've made jelly from other closely related non-native *Sorbus* species (such as *S. cashmiriana*, from Kashmir), but they are not my favourite fruits for use as their jellies have a strong aftertaste.

Phillips (1983) and Ó Céirín and Ó Céirín (1978) provided recipes for rowan wine, and the latter authors noted that in Ireland the berries were used to flavour alcoholic drinks, especially mead (wine made from honey and barley). In Northern Europe, the berries are used to make cider, and in Wales they are used to make ale. Scully (1916: 94) noted that early writers such as Drs Smith (ca. 1756) and Wade (1804) considered this species and other *Sorbus* species in Co. Kerry very useful, and said, 'Thus, if dried and ground, the berries made a wholesome bread, they could also be made to afford a "not unpleasant jelly", while last, though probably not least, an ardent spirit of a "fine flavour" might be distilled from them'.

In traditional medicine, *Sorbus aucuparia* has been used to treat diabetes, scurvy, kidney problems, arthritis, diarrhoea and haemorrhoids (van Wyk & Wink, 2004), although not necessarily for these purposes in Ireland. Egan (1887) mentioned its use in traditional medicine in Ireland, to treat worms. A few rowan berries were eaten before breakfast for a few days. Allen and Hatfiled (2004) noted its use in veterinary medicine to treat strangles; the smoke of burnt wood was inhaled. An infusion of the leaves was used to treat rheumatism, and the leaves when burnt and inhaled were said to be useful as a treatment for asthma (Moloney, 1919). In Ireland its bark was boiled to make a cough treatment; leaves were used to make a poultice for sore eyes and the berries were eaten raw to cleanse the blood and cure worms (Allen & Hatfield, 2004). In Co. Cavan, the berries were eaten raw as a tonic for weak blood (Maloney, 1972).

The berries of *Sorbus aucuparia* can be toxic, but this is dosage-dependent. Poisoning can happen with over-consumption of the berries or seeds by a child or animal. Like other seeded fruit in the Rosaceae family (bitter almonds, apricots, even apples), the seeds contain cyanogenic glycosides (Campbell & Chapman, 2000). The enzyme β-glucosidase is found in the human small intestine and releases hydrogen cyanide from these compounds, which has resulted in lethal poisonings. But few of us would consume enough seeds to be so affected. Rowan berries also contain small amounts of parascorbic acid, which can irritate mucous membranes, if the berries are consumed raw.

Sorbus devoniensis

- Devon whitebeam (Rich et al., 2010)
- *Fionncholl Domhnann* (Scannell & Synnott, 1987)

ROSACEAE—THE ROSE FAMILY

Native

The Devon whitebeam is a small tree, growing up to about 20 m, that is named after one of its endemic localities in south-west England. This species is quite rare in Ireland and distributed in the south and south-east, mainly in Cos. Waterford, Kilkenny and Wexford, where it occurs in hedgerows, woodland edges and scrub. There are scattered records of this tree in Northern Ireland, where at least some of them have seeded from trees planted in gardens (Rich et al., 2010).

The only use of this tree that I have found is in Co. Tyrone, from Northern Ireland, where McNeill (2010: 216) cited records of its occurrence in Moghan Moss in that county: 'When I was at the Moghan Moss site I met a local man who recalled the whitebeam trees from his childhood, 50 years earlier, and he recalled eating the fruit, which he called 'French apples''. According to Rich et al. (2010), its fruits have been reputedly picked, sold and eaten in Devon, and have been made into wine. I have not found any records of its use to make wine in Ireland. MacCoitir (2003) noted that in England the fruits of this whitebeam were eaten.

Sorbus hibernica

- Irish whitebeam (Scannell & Synnott, 1987)
- *Fionncholl gaelach* (Scannell & Synnott, 1987); *Bíoma bán*[278] (Ó Dónaill, 1977)

ROSACEAE—THE ROSE FAMILY
Native

Sorbus hibernica is distributed occasionally throughout Ireland, but in particular in the midlands, on carboniferous limestone, and is conspicuous in hedgerows, thickets, scrubs, woods, cliffs, lake-shores and rocky places. It is a small, shrubby tree, growing up to at least 7 m. Its name *S. hibernica* honours its restricted distribution (i.e., it is endemic to Ireland). It is probably an apomictic species (Rich et al., 2010).

Whitebeam provides a hard and long-lasting wood but has probably been little used for timber in Ireland. Few trees attain a significant-sized trunk to make them valuable for their timber. In Britain, *Sorbus aria* is recorded as having been used to make handles of knives and other tools, cogwheels, billiard cues and wooden spoons (Edlin, 1973); it and related species were probably used similarly in Ireland.

The fruits of whitebeams are said to be edible and can be used to make jellies, similar to rowan berry jelly. Phillips (1983) suggested that they are best used, like medlars (the fruits of *Vangueria infausta*, in the Rubiaceae family), when they begin to go rotten.

**Sorbus rupicola*

- Rock whitebeam (Scannell & Synnott, 1987)
- *Fionncholl creige* (Scannell & Synnott, 1987)

ROSACEAE—THE ROSE FAMILY
Native

Sorbus rupicola is quite a rare tree in Ireland that is found at scattered locations, mainly in the north and west of Ireland. This shrubby tree to 10 m tall can be found in habitats such as rocky woodlands and hedgerows and rocky lake islands in the Connemara region and is often associated with limestone.

*No ethnobotanical uses of this plant have been noted in Ireland.

**Sorbus scannelliana*

- Scannell's whitebeam (Rich et al., 2010)
- *Fionncholl Scannell* (Rich et al., 2010)

ROSACEAE—THE ROSE FAMILY
Native

Sorbus scannelliana is a very rare endemic Irish tree, described in 2009 and found only on a wooded limestone ridge east of Blue Pool, Ross Island, Co. Kerry. Five plants of

278. Ó Dónaill (1977) defined *bíoma bán* as 'whitebeam'.

Scannell's whitebeam are known. It is named in honour of the late Maura Scannell, former head of the Irish National Herbarium at the National Botanic Gardens, Glasnevin. It is a small tree, first found in 1988 (Rich et al., 2010). It is not known to be in cultivation.

*No ethnobotanical uses of this plant have been noted in Ireland.

Sparganium angustifolium (Sparganium affine)

- Floating bur-reed (Scannell & Synnott, 1987)
- *Rísheisc chaol* (Scannell & Synnott, 1987); *Rígh-sheisc* (*Ar snámh*) (Moloney, 1919)

TYPHACEAE (Sparganiaceae)—THE BULRUSH FAMILY
Native

Sparganium angustifolium is an aquatic plant of relatively deep water in lakes and pools, and more rarely, in slow-moving rivers, that is mainly distributed in the west; it is much rarer elsewhere in Ireland.

*No ethnobotanical uses of this plant have been noted in Ireland.

Sparganium emersum (Sparganium simplex)

- Unbranched bur-reed (Scannell & Synnott, 1987)
- *Rísheisc lom* (Scannell & Synnott, 1987); *Rígh-sheisc* (*Lom*) (Moloney, 1919); *Righ-sheisg* (*-ge*) (de Vere Coneys, 1849)

TYPHACEAE (Sparganiaceae)—THE BULRUSH FAMILY
Native

Sparganium emersum is an aquatic plant of ditches, drains, lake margins and slow streams that is found occasionally throughout Ireland.

*No ethnobotanical uses of this plant have been noted in Ireland.

Sparganium erectum (Sparganium ramosum)

- Branched bur-reed (Scannell & Synnott, 1987); Flag, Flaggers (or Flaggon[279]) (Hart, 1898); Wild shellistring (Colgan, 1914); Flagger (Colgan, 1914; Dinneen, 1927)
- *Rísheisc* (Scannell & Synnott, 1987); *Rígh-sheisc* (*Craobhach*) (Moloney, 1919); *Rísheisc chraobhach* (Uí Chonchubhair, 1995); *Seilistring* (*-e*) (Dinneen, 1927); *Sellistring* (Colgan, 1911)

TYPHACEAE (Sparganiaceae)—THE BULRUSH FAMILY
Native

Sparganium erectum is a common aquatic plant throughout Ireland of slow-moving rivers and streams, canals, ditches, pools, ponds and marshes.

Dinneen (1927) gave the Irish name *seisc mhadraidh* for 'burr-reed' but did not indicate if this name applied to all *Sparganium* species.

*No ethnobotanical uses of this plant have been noted in Ireland.

---

279. These names apply to the leaves, not to the flowers

**Sparganium minimum (Sparganium natans)*

- Least bur-reed (Scannell & Synnott, 1987)
- *Rísheisc mhion* (Scannell & Synnott, 1987); *Rígh-sheisc mhín* (Moloney, 1919)

TYPHACEAE (Sparganiaceae)—THE BULRUSH FAMILY
Native

Sparganium minimum is an occasional plant throughout Ireland of bogs, drains, ditches, marshes, lake margins and canals.

*No ethnobotanical uses of this plant have been noted in Ireland.

Spartina ×townsendii.

Spartina ×townsendii

- Townsend's cord-grass (Scannell & Synnott, 1987); Rice grass (Anonymous, 1978)
- *Spairtíneach aimrid* (Scannell & Synnott, 1987)

POACEAE (Gramineae)—THE GRASS FAMILY
Introduced

Spartina ×townsendii is a hybrid grass (a cross between two non-native species, *S. alterniflora* and *S. maritima*) that has been planted in tidal mud flats and muddy estuaries in various parts of Ireland in land reclamation efforts and to prevent coastal erosion. It is now spreading and invasive in some sites where it has become well established. It was not known in Ireland before 1925 when it was introduced from the south of England (Boyle, 1977). It was imported largely through purchases of the Irish Government for coastal protection reasons. Efforts to control it as an invasive alien plant have been put in place in locations such as the North Bull Island in Dublin Bay where it threatened natural ecosystems and the feeding areas for wading bird populations (Jeffrey, 1977).

Several other species of *Spartina* are also found planted in the wild in Ireland: *S. maritima* (small cord-grass; *spairtíneach mhara*), *S. anglica* (common cord-grass; *spairtíneach ghalida*) and *S. pectinata* (pectinate cord-grass; *spairtíneach chíre*) (Scannell & Synnott, 1987). These species are not as vigorous in growth as *S. ×townsendii*, but they may have been imported inadvertently when *S. ×townsendii* was brought in (Boyle, 1977).

Spergula arvensis

- Corn spurrey (Scannell & Synnott, 1987); Cloonbawn (Colgan, 1904); Wild broom (Colgan, 1904); Granyagh (Hart, 1898); Yagh (Hart, 1898)
- *Corrán lín* (Dinneen, 1927; Ó Dónaill, 1977; Scannell & Synnott, 1987); *Cluain lín* (Moloney, 1919; Dinneen, 1927); *Currán lín* (de Vere Coneys, 1849; Moloney, 1919); *Clúbhán* (Colgan, 1911; Moloney, 1919); *Cabrois* (Cameron, 1883); *Cabróise* (de Vere Coneys, 1849; Dinneen, 1927); *Clamhán (-áin)* (Dinneen, 1927)

CARYOPHYLLACEAE—THE CARNATION OR PINK FAMILY
Introduced (possible archaeophyte)

Spergula arvensis is a small, annual agricultural weed found throughout Ireland in cultivated ground and on disturbed soils.

Milliken and Bridgewater (2004) mentioned that the seeds of this weed are rich in oils and were formerly used as food by Shetland islanders, to make bread. I have not found records of such usage in Ireland.

**Spergularia marina*

- Lesser sea-spurrey (Scannell & Synnott, 1987)
- *Cabróis mhara bheag* (Scannell & Synnott, 1987)

CARYOPHYLLACEAE—THE CARNATION OR PINK FAMILY
Native

Spergularia marina is quite common throughout Ireland on seashores and in salt marshes.

*No ethnobotanical uses of this plant have been noted in Ireland.

Spergularia media

- Greater sea-spurrey (Scannell & Synnott, 1987)
- *Cabróis mhara mhór* (Scannell & Synnott, 1987); *Corrán mara mór* (Uí Chonchubhair, 1995)

CARYOPHYLLACEAE—THE CARNATION OR PINK FAMILY
Native

Spergularia media is quite common throughout Ireland on seashores and in salt marshes.

Mears and Hillman (2007) suggested that this plant is edible and useful in salads.

Spergularia media.

**Spergularia rubra*

- Sand spurrey (Scannell & Synnott, 1987)
- *Cabróis duimhche* (Scannell & Synnott, 1987)

CARYOPHYLLACEAE—THE CARNATION OR PINK FAMILY
Native

Spergularia rubra is a rare annual plant throughout Ireland of dry sandy and gravelly places.

*No ethnobotanical uses of this plant have been noted in Ireland.

**Spergularia rupicola*

- Rock sea-spurrey (Scannell & Synnott, 1987)
- *Cabróis na gcloch* (Scannell & Synnott, 1987); *Corrán mara* (Uí Chonchubhair, 1995)

Spergularia rupicola.

CARYOPHYLLACEAE—THE CARNATION OR PINK FAMILY
Native

Spergularia rupicola is a coastal plant of walls and rocks that is distributed all around the Irish coast.

*No ethnobotanical uses of this plant have been noted in Ireland.

*Spiranthes romanzoffiana

- Irish lady's-tresses (Scannell & Synnott, 1987)
- *Cúilín gaelach* (Scannell & Synnott, 1987)

ORCHIDACEAE—THE ORCHID FAMILY
Native

Spiranthes romanzoffiana is a rare orchid in Ireland that is distributed in the south-west, west and north-east only; it occurs in a variety of wet and boggy grassland habitats and on lake-shores.

*No ethnobotanical uses of this plant have been noted in Ireland.

*Spiranthes spiralis (Spiranthes autumnalis)

- Autumn lady's-tresses (Scannell & Synnott, 1987)
- *Cúilín Muire* (Scannell & Synnott, 1987)

ORCHIDACEAE—THE ORCHID FAMILY
Native

Spiranthes spiralis is an orchid of dry coastal sandy grasslands and sand dunes; it is commoner in the south than in the north.

*No ethnobotanical uses of this plant have been noted in Ireland.

*Spirodela polyrhiza (Lemna polyrhiza)

- Greater duckweed (Scannell & Synnott, 1987)
- *Ros lachan mór* (Scannell & Synnott, 1987)

LEMNACEAE—THE DUCKWEED FAMILY
Native

Spirodela polyrhiza is a tiny aquatic plant of streams, ditches, pools, ponds, lakes and slow-moving rivers that is rare throughout Ireland except in the Lough Neagh region of Northern Ireland where it is common.

*No ethnobotanical uses of this plant have been noted in Ireland.

Stachys officinalis (Betonica officinalis; Stachys betonica)

- Betony (Scannell & Synnott, 1987); Wood betony (K'Eogh, 1735)
- *Lus beatha* (Scannell & Synnott, 1987); *Lus beathaigh* (Moloney, 1919; Dinneen, 1927); *Lus mhic bheathaigh* (de Vere Coneys, 1849; Moloney, 1919; Dinneen, 1927); *Lus biotaine* (De Bhaldraithe, 1959); *Biotaine* (Ó Dónaill, 1977)

LAMIACEAE (Labiatae)—THE DEAD-NETTLE FAMILY
Native

Stachys officinalis is a very rare plant in Ireland of woodlands.

Betony has a long history of use in traditional medicine, as a sedative and bitter tonic, to treat many different diseases and ailments, such as diarrhoea, indigestion, catarrh, mouth ulcers, headaches and neuralgia, as well as to relieve stress and tension (van Wyk & Wink, 2004), although due to its rarity in Ireland, wild plant material has probably been little used for such purposes. Small doses of the root were used as an emetic, and powdered dried leaves were said to be a good 'sternutatory' (a medicine that causes or promotes sneezing) (Wade, 1804). Harrop (1977) said that tea made from it is aromatic and astringent.

Clyne (1989) said that it was eaten raw as a salad in the Highlands of Scotland, but I have not found records of such usage in Ireland.

Betony is occasionally grown as an ornamental plant in gardens and herb gardens in Ireland.

Stachys species

LAMIACEAE (Labiatae)—THE DEAD-NETTLE FAMILY
Native

Including:

Stachys palustris

- Marsh woundwort (Scannell & Synnott, 1987); Clown's all-heal (Anonymous, 1978); Sheep's brisken (Hart, 1898)
- *Cabhsadán* (Scannell & Synnott, 1987); *Duilleóg na saor* (Moloney, 1919; Dinneen, 1927; Ó Dónaill, 1977); *Lus na scor* (de Vere Coneys, 1849)

Stachys sylvatica

- Hedge woundwort (Scannell & Synnott, 1987); Day nettle (Hart, 1898)
- *Créachtlus* (Scannell & Synnott, 1987); *Caubsadan* (Cameron, 1883)

Stachys palustris is a common plant in Ireland of marshy habitats, such as ditches and drains, river-banks and marshy fields. *Stachys sylvatica* is also common but occurs in shady woodland habitats and hedges.

Both species have been used as medicinals and applied to fresh wounds to help healing (Moloney, 1919; Uí Chonchubhair, 1995; Allen & Hatfield, 2004). K'Eogh (1735) regarded it as an excellent herb to treat wounds, when mixed with hog's lard and applied as a poultice and the leaves were used to 'cure the bite of a mad dog'.

Hart (1898) mentioned that *Stachys palustris* may have been dried and used in Co. Donegal as a tobacco substitute.

According to Mahon (1983), a yellow dye was obtained from this plant in Ireland.

Mears and Hillman (2007) mentioned that *Stachys sylvatica* is used in soups and cheeses in France. In North America, Peterson (1977) noted that the tubers of *S. palustris* have been used for food, eaten raw, pickled or cooked. No records of the use of either species in Ireland have been noted though.

Stellaria graminea

- Lesser stitchwort (Scannell & Synnott, 1987)
- *Tursarraing bheag* (Scannell & Synnott, 1987); *Tursairrinín* (Moloney, 1919); *Tursáinín* (Dinneen, 1927)

CARYOPHYLLACEAE—THE CARNATION OR PINK FAMILY
Native

Stellaria graminea is a common plant throughout Ireland of roadside verges, hedgerows, grasslands, ditches, marshes, bog-margins and meadows.

Lesser stitchwort and greater stitchwort, as their common names suggest, were formerly used in the treatment of stitches (the flowers were chewed) and rheumatism (Moloney, 1919).

Mears and Hillman (2007) reported that young vegetative shoots can be harvested throughout the season and used in salads, boiled or steamed.

Ó Dónaill (1977) gave the words *teanga éanáin* or *turasáinín* to mean 'stitch-wort'.

Stellaria holostea

- Greater stitchwort (Scannell & Synnott, 1987); Star of Bethlehem (Hart, 1898)
- *Tursarraing mhór* (Scannell & Synnott, 1987); *Tuirseach* (Moloney, 1919; Ó Dónaill, 1977); *Tursarrain* (Moloney, 1919); *Tursainín mhór* (Uí Chonchubhair, 1995)

CARYOPHYLLACEAE—THE CARNATION OR PINK FAMILY
Native

Stellaria holostea is a common plant in the east of Ireland of woodlands, hedgerows and scrub; it is much rarer in the west.

In southern Ireland *Stellaria holostea* was boiled with sugar to prepare a remedy for whooping cough (Allen & Hatfield, 2004).

Roy Vickery included a record from Clonlara, Co. Clare, in the Plant-lore Archive (Vickery, 2010) of children squeezing the seedpods of stitchwort between the fingers to make them 'pop'.

Stellaria media

- Chickweed (Anonymous, 1978; Scannell & Synnott, 1987); Chicken weed (or Chickenweed) (Hart, 1898; Colgan, 1904)
- *Fliodh* (Scannell & Synnott, 1987); *Fuillig* (Ó Criomhthain, MSS, n.d., ca. 1908); *Fligh* (*-ghe*) (de Vere Coneys, 1849); *Fleadha* (Dinneen, 1927; Uí Chanainn, 1962); *Luibh na bhFranncach* (Dinneen, 1927); *Fliach* (Tóibín, 1965); *Flith* (or *Flic*) (Colgan, 1911)

CARYOPHYLLACEAE—THE CARNATION OR PINK FAMILY
Native

Stellaria media is extremely widespread and common throughout Ireland in waste places, roadsides, farmyards, cultivated fields, gardens and sand dunes.

It is native to and widespread in Europe, Asia and North America, and is now a cosmopolitan weed. This species has been widely and extensively used in traditional medicine, mainly for the treatment of skin problems (van Wyk & Wink, 2004). Moloney (1919) said it was formerly used in the treatment of rheumatism. According to Allen and Hatfield (2004), it was used in the form of a poultice to reduce swellings from sprains or mumps and other inflammations, as well as internally for coughs, sore throats, colic and jaundice. In the Belleek area of Co. Donegal, Hart (1898) recorded that it was regarded as good for sprains and bruises and could be taken as an infusion. Lady Wilde (1890) also listed this as a popular Irish 'cure'. It is said to contain potassium salts that provide it with these medicinal properties (Uí Chonchubhair, 1995). O'Regan (1997) recommended an ointment or lotion made from chickweed to treat itching. Chickweed cream is said to be a useful treatment to relieve the irritation of eczema and also to sooth minor burns, sunburn, insect stings and to help draw out splinters and boils (Ody, 1997b). Chickweed has also been used as an ingredient in cosmetics in Ireland too.

It was used as a veterinary remedy to treat red-water fever (combined with *Scrophularia nodosa*), blackleg, convulsions in pigs, gapes and pip (Allen & Hatfield, 2004).

Stellaria media (chickweed). Watercolour by Lydia Shackleton. Courtesy of the National Botanic Gardens, Glasnevin.

It is recommended by Phillips (1983) and Mears and Hillman (2007) as an edible winter and spring herb for salads and soup. MacMahon (1994) said that it was eaten during the famine in Ireland. It was eaten during the Siege of Derry in 1689 when it was much in demand (Rutty, 1772; Williams, 1993). I find it rather bitter to use as a salad, but if it is mixed with other leaves then it is more tasty. It can also be cooked or steamed like spinach. The seeds are edible, although tedious to collect. They can be ground up and added to flour, used in baking bread or to thicken soups (Hilton, 2007). Houston and Milne (2008) provided a recipe for a chickweed sandwich mix: lemon juice, Worcestershire sauce, sea salt and ground black pepper are mixed with a bunch of chickweed, then combined with chicken, ham, tomatoes or other ingredients to make sandwich fillings. Róise Rua, in her memoir of life in Arranmore Island, Co. Donegal (Ua Cnáimhsí, 2009: 136), said that during the famine years of the 1840s in the island, people collected chickweed and 'boiled it into some sort of broth', with water added and a bowl of oatmeal too. She said 'That sort of "food" was difficult to stomach; it was said it sometimes brought on disease'.

It is much used as a green vegetable for caged birds (e.g., for canaries and budgerigars) especially when they are moulting as it is apparently good for their feathers. Mackay (1836) noted that it is frequently given to birds, which are fond of its seeds.

*Stellaria neglecta

- Greater chickweed (Scannell & Synnott, 1987)
- *Fliodh mhór* (Scannell & Synnott, 1987)

CARYOPHYLLACEAE—THE CARNATION OR PINK FAMILY

Native

Stellaria neglecta is a rare plant in Ireland that is found mainly in damp shady habitats.

*No ethnobotanical uses of this plant have been noted in Ireland.

**Stellaria pallida*

- Lesser chickweed (Scannell & Synnott, 1987)
- *Fliodh bheag* (Scannell & Synnott, 1987)

CARYOPHYLLACEAE—THE CARNATION OR PINK FAMILY
Native

Stellaria pallida is an occasional plant throughout Ireland of old walls, bare ground and sand-hills.

*No ethnobotanical uses of this plant have been noted in Ireland.

**Stellaria palustris*

- Marsh stitchwort (Scannell & Synnott, 1987)
- *Tursarraing chorraigh* (Scannell & Synnott, 1987)

CARYOPHYLLACEAE—THE CARNATION OR PINK FAMILY
Native

Stellaria palustris is a rare plant in Ireland that is mainly in the midlands in calcareous marshes and fens.

*No ethnobotanical uses of this plant have been noted in Ireland.

**Stellaria uliginosa (Stellaria alsine)*

- Bog stitchwort (Scannell & Synnott, 1987)
- *Tursarraing mhóna* (Scannell & Synnott, 1987)

CARYOPHYLLACEAE—THE CARNATION OR PINK FAMILY
Native

Stellaria uliginosa is a frequent plant throughout most of Ireland in stream banks, marshes, ditches and other wet habitats.

*No ethnobotanical uses of this plant have been noted in Ireland.

Suaeda maritima

- Annual sea-blite (Scannell & Synnott, 1987)
- *Blide mhara* (Scannell & Synnott, 1987); *Praiseach na mara* (Cameron, 1883)

AMARANTHACEAE (Chenopodiaceae)—THE GOOSEFOOT FAMILY
Native

Suaeda maritima is found all around the Irish coastline on stony, muddy and rocky shores, and occasionally, on sandy shores.

Mears and Hillman (2007) said this plant to be a useful salad and vegetable species.

It was formerly burned to form an ash containing an impure form of washing soda (sodium carbonate). It was also used to make soap and in the manufacture of glass (Uí Chonchubhair, 1995) (see also *Salsola kali*).

**Subularia aquatica*

- Awlwort (Scannell & Synnott, 1987); Water awlwort (Brunker, 1950)
- *Briogadán uisce* (Scannell & Synnott, 1987)

BRASSICACEAE (Cruciferae)—THE CABBAGE AND CRESS FAMILY

Native

Subularia aquatica is a very rare, tiny annual plant in Ireland that is found mainly in the west and north-east on lake-shores.

*No ethnobotanical uses of this plant have been noted in Ireland.

Succisa pratensis (Scabiosa succisa)

- Devil's-bit scabious (Scannell & Synnott, 1987); Devil's-bit (Hart, 1898; Anonymous, 1978); Biting billy (Hart, 1898); Carl doddy (Hart, 1898); Curl doddy (Praeger, 1938); Lough shule (Hart, 1898)
- *Odhrach bhallach* (Scannell & Synnott, 1987); *Greim an diabhail* (Moloney, 1919; Dinneen, 1927); *Odharach mullaigh* (de Vere Coneys, 1849; Moloney, 1919); *Urach bhallach* (Webb et al., 1996); *Ura bhallach* (Cameron, 1883); *Caisearbhán beag* (Dinneen, 1927); *Úrach mullaigh* (Dinneen, 1927); *Úrach balla* (Dinneen, 1927)

DIPSACACEAE—THE TEASEL FAMILY

Native

Succisa pratensis is a common plant throughout Ireland of many different habitats, such as hedgerows, marshes, heaths, grasslands and cut-away bogs.

Apparently, it received its common name because it contained so many cures that 'it angered the Devil', so he cut its root short. It was used to treat bad eyes and running sores, and it was mixed with butter in Galway and rubbed onto the skin to treat 'eruptions' (Williams, 1993). The same author noted that the leaves were boiled to provide a cure for coughs and liver disease. Its roots were formerly used in Ireland to treat sores, wounds and as an antiseptic (Allen & Hatfield, 2004). Mackay (1836) mentioned that the name 'scabious' is derived from the former use of this species (and probably *Knautia arvensis*—field scabious too) to treat skin diseases such as leprosy.

Williams (1993) noted that it was used for a childhood game where the shoots were thrown slowly to see if they returned, like a boomerang. Wood-Martin (1902) and Thiselton Dyer (1889) mentioned a different children's game with this species where the shoots were twisted and, as they slowly untwisted in the hand, the following rhyme was spoken:

'Curl-doddy on the midden,
Turn round, an' take my bidden'

Curl-doddy is an old English name for this species, as the head of flowers resembles the curly head of hair of a small boy (Thiselton Dyer, 1889).

A blue dye can be obtained from its leaves (Mahon, 1983; Williams, 1993).

Symphytum officinale

- Comfrey (Scannell & Synnott, 1987)
- *Compar* (Scannell & Synnott, 1987); *Meacan an chomparaí* (Ua Maoileoin, 2001); *Meacan an chompanaí* (Anonymous, 1978); *Lus na gcnámh mbriste* (Moloney, 1919; Dinneen, 1927); *Meacan dubh* (Moloney, 1919; Dinneen, 1927); *Lus na gcnámh briste* (Uí Chonchubhair, 1995); *Meacan an chompair* (de Bhaldraithe, 1959); *Glaisín coille* (Dinneen, 1927); *Meacan an chompraidhe* (Dinneen, 1927); *Meacan compare* (Ó Dónaill, 1977); *Comprái* (Ó Criomhthain, MSS, n.d., ca. 1908)

BORAGINACEAE—THE BORAGE FAMILY
Possibly Introduced (possible archaeophyte)

Symphytum officinale is a perennial herb of damp, shady habitats, such as river-banks and woodlands. It is occasionally found throughout Ireland but is commonest in the north.

Comfrey has a long history of usage in traditional medicine in Europe and western Asia, where it is native (van Wyk & Wink, 2004). In Ireland and elsewhere, it was formerly used by herbalists to promote the setting of bone fractures (hence the Irish name *lus na gcnámh briste*, 'the herb of the broken bones') and used both internally and externally for that purpose (Moloney, 1919; Mabey, 1972; Synnott, 1979). The root was grated and made into a mash. It was then put around the limb and would set hard in a short time, like plaster of Paris is used to set broken bones (McDaeid, 1991). It was used medicinally in Ireland to treat cuts, wounds and boils (the roots were boiled and used as a poultice in Co. Kerry) (Logan, 1981). Other medicinal use in Ireland has included treatments for colds, rheumatism, toothache, kidney ailments, skin complaints and as a cosmetic, to 'improve the complexion' (Allen & Hatfield, 2004). O'Regan (1997) mentioned that a poultice of mashed comfrey leaves has been used to treat the severe burns, rashes, blistering and swellings caused by giant hogweed (*Heracleum mantegazzianum*). Extracts of comfrey have also been used as an ingredient of cosmetics in recent years in Ireland.

It was used as a veterinary remedy to treat sprains and broken legs, swellings in horses, swollen cows' udders, sore feet and diarrhoea (Allen & Hatfield, 2004) and in Co. Cavan, the leaves were boiled and broken up and given in a bran mash to animals to treat scour (Uí Sheasnáin, 1982). Moloney (1919) said that it was formerly also used for fattening pigs. In Co. Limerick, the roots were used to make plasters for the broken limbs of poultry (Vickery, 2010).

The young leaves of comfrey can be cooked and eaten as spinach or in soup (Synnott, 1979; Phillips, 1983), but they are not to everyone's taste, being hairy and rather glue-like! The bitterness can be reduced by boiling the leaves in several changes of water. Young shoots can be eaten as one would asparagus (Hilton, 2007) or the leaves dipped into batter and made into fritters (Mabey, 1972). Dried leaves can be used to make tea. However, Mears and Hillman (2007) suggested that despite comfrey being said to be edible, they would not advise eating comfrey roots. Some warnings and reservations on the use of comfrey internally have been given as they contain pyrrolizidine alkaloids and have been said to cause liver cancers. The shoots and leaves, especially, should not be eaten too often or in large quantities.

Comfrey has been cultivated in gardens in Ireland as an ornamental and elsewhere as a fodder crop for stock (Howes, 1974).

The leaves of this species are quite similar in appearance to those of foxglove, which are highly poisonous, so comfrey should be harvested and used with great care.

Tamus communis

- Black bryony (Scannell & Synnott, 1987)
- *Crotharnach* (Scannell & Synnott, 1987); *Crotharnaighe* (Dinneen, 1927); *Unach (-aighe) dubh*[280] (Dinneen, 1927; Ó Dónaill, 1977)

DIOSCOREACEAE—THE BLACK BRYONY FAMILY
Introduced

Tamus communis is a perennial climbing plant that is naturalised, although rare, in woodlands and hedgerows in various parts of Ireland (Reynolds, 2002). It is known to be poisonous but was formerly used for medicinal purposes, particularly as a diuretic. The roots have also been made into a poultice to treat the discolouration in bruises and black eyes. It is not known to what extent black bryony has been used in traditional medicine in Ireland and whether it was originally introduced for its medicinal properties, although I suspect it may have been.

Dinneen (1927) recorded a use of this plant in veterinary medicine: 'I saw the plant and roots boiled for a drink for a wasting cow, and the cow fully recovered'.

White bryony (*Bryonia dioica*) is an unrelated plant, a member of the Cucurbitaceae, the cucumber family. It is an introduced plant that has escaped into the wild in Ireland from gardens on several occasions (Reynolds, 2002). Dinneen (1927) and Ó Dónaill (1977) gave the Irish name *unach gheal* for 'white bryony'. This plant is also poisonous and if the orange berries are eaten can cause vomiting, abdominal pain and diarrhoea (Woodward, 1985). Juice from white bryony can cause irritation to the skin too. Drury (1985) recorded that in Britain, white bryony was used as a fertility stimulant in horses. I have not found records of such usage in Ireland.

The berries of *Tamus communis* are poisonous and should not be eaten. Woodward (1985: 104) gave the symptoms of poisoning with black bryony as 'burning and blistering of the mouth, abdominal pain, diarrhoea; can be fatal'.

Tanacetum parthenium (Chrysanthemum parthenium)

- Feverfew (Scannell & Synnott, 1987); Featherfew (Hart, 1898)
- *Lus deartán* (Scannell & Synnott, 1987)

ASTERACEAE (Compositae)—THE DAISY FAMILY
Introduced (possible archaeophyte)

Tanacetum parthenium is an occasional weed throughout Ireland of wasteground, roadsides and ditches, and is usually found near houses.

It is native to south-east Europe and Asia Minor and has been widely used in traditional

280. Dinneen (1927) defined *unach (-aighe) dubh* as 'black briony'.

medicine both in these regions and elsewhere to treat migraine headache, fever, rheumatic and skin complaints and gynaecological problems (van Wyk & Wink, 2004). In Ireland it has also been well regarded as a valuable medicinal plant and used as an effective treatment for headaches, and in particular, for migraines. It has also been used as an ingredient in cosmetics in Ireland in recent years.

A double-flowered variety is occasionally grown in gardens.

Tanacetum vulgare (Chrysanthemum vulgare)

- Tansy (Scannell & Synnott, 1987); Ginger plant (Hart, 1898)
- *Franclus* (Scannell & Synnott, 1987); *Leamhach bhuí* (Ua Maoileoin, 2001); *Lus na bhfran(n)cach* (Moloney, 1919; Dinneen, 1927; Ó Dónaill, 1977); *Lus na bhfeacóg*[281] (Moloney, 1919; Dinneen, 1927); *Tamhsae* (Cameron, 1883); *Lus na fraince*[282] (de Vere Coneys, 1849; Dinneen, 1927); *Frannc-lus* (Dinneen, 1927); *Tannsae* (Dinneen, 1927); *Tansae* (Ó Dónaill, 1977)

ASTERACEAE (Compositae)—THE DAISY FAMILY

Introduced (possible archaeophyte)

Tanacetum vulgare is an occasional weed throughout Ireland of wasteground, roadsides and ditches, and is usually found near houses.

Moloney (1919) noted its use to expel intestinal worms. It was also formerly used as a vermifuge in Co. Louth (Synnott, 1979).

In the Aran Islands the juice was drunk for fevers; elsewhere in Ireland it has been used for a variety of ailments, including jaundice, cuts, indigestion, flatulence, gout and pains in the joints, and in veterinary medicine to treat red-water fever, worms in horses, and pip (Egan, 1887; Allen & Hatfield, 2004). In Co. Kerry it was boiled with unsalted butter, strained and allowed to cool, and then used as a poultice for cuts (O'Sullivan, 1994). In Yorkshire during the 17th century, it was chopped finely and made into a ball with fresh butter and used to heal the wound of castrated lambs. The butter helped to heal the wound and the tansy kept the flies away (Drury, 1985). Elsewhere, outside of Ireland, it has been used to treat migraines, liver problems and loss of appetite and externally for neuralgia, rheumatism and various skin complaints (van Wyk & Wink, 2004).

It was used to flavour drisheen (made from sheep's blood, milk and breadcrumbs) (Moloney, 1919) in Co. Cork (Akeroyd, 1996) and as a flavouring for milk puddings and egg dishes (Hilton, 2007). The flowers and fresh, young aromatic leaves can also be used as a seasoning for fish, in lieu of sage, but used sparingly as it has a very strong flavour (Peterson, 1977). Rutty (1772: 86) said that it was 'fried brownish in the pan, and being qualified with the juices of other fresh herbs, as Spinage, green Corn, Violet and Primrose leaves, at the entrance of Spring, and eaten hot, with the juice of Orange and Sugar, makes one of the most agreeable of the herbaceous dishes; the tender leaves are put into pancakes; and puddings are greened with the juice, to both which they give a not unpleasant taste, and are very wholesome, serving to dissipate the flatulencies, contracted in time of Lent, by eating fish, legumes, etc'.

Although used in Ireland for culinary purposes, Peterson (1977) provided a warning that although tansy is quite safe if used in tiny quantities needed for seasoning, it contains

281. Dinneen (1927) gave this name for 'black tansy'.
282. Dinneen (1927) gave as '*lus na Frainnce*'.

an oil that can be fatal when consumed in large quantities, and as such he listed it as a poisonous plant.

Taraxacum officinale agg.

- Dandelion (Scannell & Synnott, 1987); Casharawan (Hart, 1898); Glaslarwan (Hart, 1898)
- *Caisearbhán* (Scannell & Synnott, 1987); *Caistsearbhán* (Moloney, 1919); *Caisearbhán caol dearg* (Moloney, 1919); *Caisearbhán cos dearg* (Moloney, 1919); *Bláth buidhe* (Dinneen, 1927); *Searbhán* (Dinneen, 1927); *Bainne-cith-na n'éun*[283]; *Caisre bhán*[284] (Ó Criomhthain, MSS, n.d., ca. 1908); *Castsearbhán* (Tóibín, 1965)

ASTERACEAE (Compositae)—THE DAISY FAMILY
Native

Dandelions are extremely abundant throughout Ireland and are found in a wide variety of habitats, such as meadows and pastures, roadsides and waste places, sand dunes, fens, walls and banks, lawns and gardens. Most dandelions reproduce asexually, which gives rise to a large number of different forms and 'micro-species' that are often lumped together under the aggregate name *Taraxacum officinale*. Scannell and Synnott (1987) recorded more than 70 such micro-species in Ireland, most of which can be identified only by a small number of experts who specialise in this group. Some micro-species are characteristic of specific habitats, such as marshlands and sand dunes.

Dandelions are well recognised in Ireland today as useful edible plants, rich in vitamin A, even if most people rarely gather or use them. Nevertheless, the leaves are occasionally collected in Ireland and used as a salad or vegetable. Young leaves are best when gathered before the flowers develop, if they are to be eaten raw. As a vegetable, the leaves need to be boiled for 10 to 15 minutes. The leaves can be blanched (covered or earthed-up) so that the green colour does not develop, in which case the leaves are even tastier. Young flowerbuds can be boiled and served with butter, or pickled (Peterson, 1977). The flowers can be made into fritters, dipped in batter and fried.

Roasted roots can be eaten or used as a coffee substitute. Ó Céirín and Ó Céirín (1978) suggested that the roots are best gathered in autumn when they are long and fat. They can be left to dry (unpeeled) before being slow-roasted and ground to make 'coffee'. I have made delicious dandelion coffee by washing and roughly chopping the roots and then dry roasting them in the oven for an hour or two. To make the coffee, place one teaspoon of the ground roots into half a pint of water, bring to the boil and simmer for about 5 minutes. During World War II ('The Emergency') dandelion coffee was often made in Co. Cork, according to O'Regan (1997). Phillips (1983) also provided recipes for dandelion flower

Taraxacum officinale.

283. Mairín Uí Chonchubhair (pers. comm., 2011) tells me that this name is applied to any plant with white milky latex, such as *Euphorbia helioscopia* (sun spurge). In the early 1900s the Islandman Tomás Ó Criomhthain, from Kerry's Great Blasket Island, gave the Irish name *bainne-cith-na n'éun* for *Conium maculatum* (hemlock) in a manuscript list of Irish plant names compiled for the Norwegian scholar Carl Marstrander.

284. Mairín Uí Chonchubhair (pers. comm., 2011) tells me that this name is pronounced 'cais-re-bhán' in West Kerry.

wine, dandelion beer and coffee. In the United Kingdom, it is used to make a traditional commercially available fizzy soft drink with burdock (*Arctium minus*).

Dandelion flower syrup can also be made, which goes well with pancakes, ice-cream and other sweet desserts and puddings. The taste is delicious, like that of a wildflower honey. I have made this syrup by simmering about 300 flowers in half a litre of water for an hour, adding juice of a lemon and leaving to steep overnight. The next day the mixture is strained and then boiled up gently with a kilo of white sugar for about two hours until a nice thick syrupy consistency is achieved. The syrup should be bottled in sterilised jars and stored in a refrigerator until used.

Dandelions have been widely used medicinally for a wide variety of ailments, including as a diuretic (Logan, 1972, 1981), to help dissolve gall stones, kidney stones, and for obesity, cellulite and water retention (Reader's Digest Association, 2006). In French, its name is *pissenlit* (or *pisse au lit*) (meaning 'bed wetter') referring to its diuretic effect. Allen and Hatfield (2004) documented a large number of records of the use of *Taraxacum officinale* for folk medicine in Ireland, particularly to treat warts, but also for use in the treatment of coughs, colds and respiratory problems. Lady Wilde (1890) mentioned its use as a treatment for asthma. It has also been used for 'consumption', jaundice and liver problems, stomach pains and upsets, rheumatism, heart problems, toothache and as a tonic. There is also a very wide range of ailments for which it has been used less commonly in some parts of Ireland, such as thrush, sprains and swellings, weak or broken bones, headaches, diabetes, sore eyes, external cancers, anaemia, cuts and nervousness. Logan (1972, 1981) noted that the milky juice of dandelions was used for the treatment of a stye in the eye. Logan (1972) also mentioned that a sandwich of bread and butter, with a filling of fresh dandelion leaves was used as a treatment for tuberculosis, and that chewing, sucking and swallowing the juice of dandelion leaves was used to treat bronchitis. The same recipe of dandelion leaf sandwiches was used in Donegal to treat indigestion (McClintock, 1912). Akeroyd (1996) reported that an infusion of the roots was used in the West Cork islands (such as Sherkin Island) to treat rheumatism. Dandelion tea was used to treat asthma in Co. Cavan (Maloney, 1972).

Allen and Hatfield (2004) noted the use of *Taraxacum officinale* in veterinary remedies to treat liver flukes and diarrhoea in calves, blackleg, cowpox, pip and pains in horses. It was also fed to animals, particularly poultry and pigs in Cork (Williams, 1993).

The leaves and roots have also been dried and smoked as a tobacco substitute (Williams, 1993).

The seed heads of the dandelion continue to provide the basis for a popular childhood game—the seed head, sometimes called a 'Dandelion Clock', is blown and the number of puffs it takes to scatter all the seeds indicates the time of day. It has also been used for another children's game, 'She loves me, she loves me not', where the seed head is blown and when all the seeds have dispersed the result is known.

A magenta dye is derived from its roots (Mahon, 1983; Williams, 1993).

**Teesdalia nudicaulis*

- **Shepherd's cress (Scannell & Synnott, 1987)**
- ***Lus an bhuachalla* (Scannell & Synnott, 1987)**

BRASSICACEAE (Cruciferae)—THE CABBAGE AND CRESS FAMILY
Native

Teesdalia nudicaulis is a very rare plant found only in a few places in Ireland, on sandy shores of Lough Neagh in the north and on acid sand dunes near Dundrum in Co. Down and Coleraine, Co. Derry.

*No ethnobotanical uses of this plant have been noted in Ireland.

Teucrium scordium

- Water germander (Scannell & Synnott, 1987)
- *Niolúch uisce* (Scannell & Synnott, 1987)

LAMIACEAE (Labiatae)—THE DEAD-NETTLE FAMILY
Native

Teucrium scordium is a rare plant in Ireland of lake-shores in Lough Derg and Lough Ree and in turloughs along the river Shannon; it is very rarely distributed elsewhere at sites around Galway Bay. Wade (1804) noted *Teucrium scordium* as a medicinal plant, and its powdered leaves were used to treat worms.

Teucrium scorodonia

- Wood sage (Scannell & Synnott, 1987); Sage (Colgan, 1904)
- *Iúr sléibhe* (Scannell & Synnott, 1987); *Sáise cnoic* (Ua Maoileoin, 2001); *Sáiste choilleadh* (Moloney, 1919); *Iubhar sléibhe* (Dinneen, 1927); *Sáiste chnuic*[285] (Dinneen, 1927; Ó Dónaill, 1977); *Sáiste choille*[286] (Dinneen, 1927; Ó Dónaill, 1977)

LAMIACEAE (Labiatae)—THE DEAD-NETTLE FAMILY
Native

Teucrium scorodonia is a common plant of mountainous areas of Ireland and is found on rocky slopes and in heaths, hedges, open woodland and scrub. It is very rarely distributed in the lowland parts of Ireland.

It has been used medicinally in Ireland in the treatment of intestinal colic (Moloney, 1919) and for rheumatism, coughs and colds, tuberculosis, colic, gripe, indigestion, palpitations, sprains and heart pains (Allen & Hatfield, 2004). Dinneen (1927) said that it was widely used in Ireland as a cough remedy. Colgan (1911) said that it was grown in gardens in Clare Island, Co. Mayo, as a cure for coughs. Uí Chonchubhair (1995) recorded that on the Dingle Peninsula, Co. Kerry, it was used in an infusion with other species (*Scorzoneroides autumnalis* and *Arctium minus*) to treat tuberculosis, as a laxative and for 'decay'. It is reported to have been used (under the name 'wild sage') in the islands of West Cork to treat internal injuries (Akeroyd, 1996).

According to Mackay (1836), the plant is extremely bitter and was sometimes used as a substitute for hops, presumably for beer-making. Certainly, if one chews the leaves they leave a strong, bitter and lingering taste in the mouth.

**Thalictrum alpinum*

- Alpine meadow-rue (Scannell & Synnott, 1987)
- *Rú léana Alpach* (Scannell & Synnott, 1987); *Ruibh (Rubha) Ailpeac* (Moloney, 1919)

RANUNCULACEAE—THE BUTTERCUP AND CROWFOOT FAMILY

285. Dinneen (1927) and Ó Dónaill (1977) defined *sáiste chnuic* as 'mountain sage'.
286. Dinneen (1927) and Ó Dónaill (1977) defined *sáiste choille* as 'wood sage'.

Native

Thalictrum alpinum is a very rare plant in Ireland of high mountains in the west.

*No ethnobotanical uses of this plant have been noted in Ireland.

Thalictrum flavum

- Common meadow-rue (Scannell & Synnott, 1987)
- *Rú léana* (Scannell & Synnott, 1987)

RANUNCULACEAE—THE BUTTERCUP AND CROWFOOT FAMILY

Native

Thalictrum flavum is a rare plant throughout Ireland of marshes, wet fields and fens, as well as rocky ground by turloughs in limestone regions.

Moloney (1919) listed it as having been taken internally as an antispasmodic, and it was used to treat 'falling sickness' (epilepsy).

Mahon (1983) recorded that it was used in dyeing textiles in Ireland and provides a yellow dye.

**Thalictrum minus*

- Lesser meadow-rue (Scannell & Synnott, 1987)
- *Rú léana beag* (Scannell & Synnott, 1987); *Ruibh bheag* (Moloney, 1919)

RANUNCULACEAE—THE BUTTERCUP AND CROWFOOT FAMILY

Native

Thalictrum minus is a rare plant in Ireland that is found mainly near the coast on lake-shores, mountains, rocky and stony habitats and sand dunes.

*No ethnobotanical uses of this plant have been noted in Ireland.

Thlaspi arvense

- Field penny-cress (Scannell & Synnott, 1987)
- *Praiseach fhia* (Scannell & Synnott, 1987); *Praiseach na gcaorach* (de Vere Coneys, 1849; Moloney, 1919; Dinneen, 1927); *Praiseach na fhiaidh* (or *fhiadh*) (de Vere Coneys, 1849; Moloney, 1919; Dinneen, 1927)

BRASSICACEAE (Cruciferae) THE CABBAGE AND CRESS FAMILY

Introduced (Archaeophyte)

Thlaspi arvense is an annual plant throughout Ireland of weedy habitats, such as roadsides, waste places, disturbed ground, landfill sites and cultivated fields. Today it is rare in most parts of Ireland but has been recorded nevertheless from most parts of the island. It has been found in Ireland for centuries with the earliest records dating back to Viking times (Geraghty, 1996) when it would have been a weed of cultivated ground, crops and waste places.

I have not found records of its use in Ireland, although it has been employed elsewhere for food (e.g., Peterson, 1977). Nevertheless, I suspect that it was probably gathered in Ireland for food, particularly during times of famine or shortage. The young leaves can be used in salads or cooked (boiled or steamed) as a green vegetable. Dried seed heads can be ground and used as a seasoning.

Its Irish name, *praiseach fhia*, means a 'deer's pot-herb'.

Thymus polytrichus (Thymus drucei; Thymus praecox; Thymus serpyllum)

- Wild thyme
- *Tím chreige*; *Luibh an fhraoigh* (Ua Maoileoin, 2001); *Lus mhic righ breatain* (Moloney, 1919); *Lus mhich righ breattuin* (de Vere Coneys, 1849); *Lus mhic ríogh breatan* (or *na mbrat*) (Dinneen, 1927; Ó Dónaill, 1977); *Tím órdha* (Dinneen, 1927)

LAMIACEAE (Labiatae)—THE DEAD-NETTLE FAMILY
Native

Thymus polytrichus is a frequent and widely distributed plant in many coastal habitats throughout Ireland such as cliffs, sand dunes, dry banks, seaside grasslands and heaths and is much less common inland.

Wild thyme species have been widely used in traditional medicine especially in southern and south-eastern Europe, to treat coughs, colds, bronchitis, throat infections and gastrointestinal problems (van Wyk & Wink, 2004). In Ireland, it was formerly used in the treatment of whooping cough, as an anti-spasmodic and to relieve flatulence (Moloney, 1919). Allen and Hatfield (2004) noted its use to counter respiratory troubles and as a sedative to induce sleep and calm the nerves, as well as its use as a cure for tuberculosis when mixed with honeysuckle and wild sage and taken as an infusion. It was used medicinally to treat trembling limbs and palsies, to settle the stomach, against spasms and twitches and used in various preparations against tuberculosis (Williams, 1993). Thyme has also been used as an antiseptic to treat a variety of ailments (Shaw, 1998). Jude (1933), however, said that it [*Thymus serpyllum*] was 'of very little medicinal value' but yields commercially useful oil.

Thymus polytrichus.

It is used widely as a herb for flavouring food. The wild thyme is not as strongly flavoured as the cultivated species (*Thymus vulgaris*) but still an excellent herb for culinary purposes, such as with roasted meat, omelettes, scrambled eggs, soup and stews. The leaves can be used to make tea. Ironically, most thyme used for culinary purposes today in Ireland is imported, not wild-gathered. A tasty thyme jelly can be made by adding finely chopped thyme leaves to apple jelly or crab-apple jelly during cooking, together with several tablespoons of malt or wine vinegar. It is delicious with roasted meat and cheeses.

Uí Chonchubhair (1995) noted that like other aromatic herbs, thyme was believed to 'influence the course of true love'.

Tilia species

- Lime (Scannell & Synnott, 1987)
- *Crann teile* (Dinneen, 1927); *Teile* (Dinneen, 1927)

MALVACEAE (Tiliaceae)—THE MALLOW FAMILY
Introduced

Including:

Tilia cordata

- Small-leaved lime (Stace, 2010; Forbes & Northridge, 2012)

Tilia platyphyllos

- Large-leaved lime (Stace, 2010)

Tilia cordata × *Tilia platyphyllos* (*Tilia* ×*vulgaris*)

- Lime (Stace, 2010)

Limes are occasionally grown as ornamental trees in gardens and demesnes and are also used for municipal and a small amount of forestry planting.

In other countries the bark of *Tilia* species was used in rope-making, but I know of no records of such usage in Ireland.

Tea may be made from dried flowers of lime and the flowers have also been used to flavour liqueurs, lemonade and cordials (Lewis-Stempel, 2012). Gallagher (2012) said that tea made from dried lime flowers was used as an effective treatment for headache in Co. Donegal. Wine may also be made from the sap of lime trees.

Torilis japonica.

**Torilis japonica*

- Upright hedge-parsley (Scannell & Synnott, 1987)
- *Fionnas fáil* (Scannell & Synnott, 1987)

APIACEAE (Umbelliferae)—THE CARROT FAMILY
Native

Torilis japonica is an occasional plant throughout Ireland of hedges, roadsides, waste places and woodland margins, and rather rarely of meadows.

*No ethnobotanical uses of this plant have been noted in Ireland.

**Torilis nodosa*

- Knotted hedge-parsley (Scannell & Synnott, 1987)
- *Lus na gcloch fuail* (Scannell & Synnott, 1987)

APIACEAE (Umbelliferae)—THE CARROT FAMILY
Native

Torilis nodosa is generally a rare plant in Ireland, except in parts of the south and south-east, and is found on roadsides, dry banks, hedgerows, stony places and wasteground.

*No ethnobotanical uses of this plant have been noted in Ireland.

Tragopogon pratensis

- Goat's-beard (Scannell & Synnott, 1987)
- *Finidí na muc* (Scannell & Synnott, 1987); *Feasóg ghabhair* (Moloney, 1919; Dinneen, 1927); *Gabharulcha* (Moloney, 1919); *Finidigh na muc* (Moloney, 1919)

ASTERACEAE (Compositae)—THE DAISY FAMILY
Native

Tragopogon pratensis is an uncommon plant in Ireland that is found mainly in the centre of the country in dry grasslands, roadsides and waste places.

This species has an edible taproot, which should be eaten only before the flowering shoots develop, at which point the plant becomes rather woody (Mears & Hillman, 2007). They can be eaten raw but are better roasted or lightly steamed. Young flowerbuds are also edible if pickled, steamed or boiled. The young leaves can be used in salads or boiled as a green vegetable (Peterson, 1977). Rutty (1772: 87) said that the root is delicious in salads and boiled, if used before the stalk develops, and it is 'very nourishing, in tabid[287] and emaciated bodies'.

A close relative, *Tragopogon porrifolius*, salsify, is an excellent but rather little used vegetable, cultivated occasionally in Irish gardens. Its roots are peeled and boiled and steamed and can be served as a strongly flavoured vegetable with butter. Its roots can also be roasted and used as a coffee substitute (Peterson, 1977).

**Trichophorum cespitosum (Scirpus caespitosus)*

- Deergrass (Scannell & Synnott, 1987)
- *Cíb cheannghеal* (Scannell & Synnott, 1987); *Luchair mhéiscreach* (Moloney, 1919; Dinneen, 1927); *Ciabh chean dubh* (de Vere Coneys, 1849); *Ciabh chin dubh* (de Vere Coneys, 1849); *Cruach-luachar* (*Cruaiche-luachra*) (de Vere Coneys, 1849; Dinneen, 1927)

CYPERACEAE—THE SEDGE FAMILY
Native

Trichophorum cespitosum is a widespread and frequent plant throughout Ireland of bogs, wet heaths and moors.

Clyne (1989) said this plant is the principal food for sheep and cattle in the Highlands of Scotland from March until the end of May each year.

*No ethnobotanical uses of this plant have been noted in Ireland.

**Trifolium arvense*

- Hare's-foot clover (Scannell & Synnott, 1987)
- *Cos mhaideach* (Scannell & Synnott, 1987)

FABACEAE (Leguminosae)—THE PEA AND CLOVER FAMILY
Native

Trifolium arvense is occasionally distributed along the eastern coasts of Ireland on dry and sandy ground and rarely in the west.

*No ethnobotanical uses of this plant have been noted in Ireland.

287. A tabid person is one suffering from 'tebes', a progressive bodily wasting or emaciation.

Trifolium campestre.

**Trifolium campestre (Trifolium agrarium)*

- Hop trefoil (Scannell & Synnott, 1987)
- *Seamair dhuimhche* (Scannell & Synnott, 1987); *Seamar obcheannach* (Wade, 1808b)

FABACEAE (Leguminosae)—THE PEA AND CLOVER FAMILY

Native

Trifolium campestre is quite frequent on dry ground and in waste places, mainly in the southern half of Ireland.

*No ethnobotanical uses of this plant have been noted in Ireland.

Trifolium dubium.

Trifolium dubium

- Lesser trefoil (Scannell & Synnott, 1987); Yellow clover (Anonymous, 1978); Shamrock (Praeger, 1938; Anonymous, 1978); Lesser hop trefoil (Brunker, 1950)
- *Seamair bhuí* (Scannell & Synnott, 1987); *Seamair Mhuire* (Ua Maoileoin, 2001); *Seamróg* (Moloney, 1919; Dinneen, 1927)

FABACEAE (Leguminosae)—THE PEA AND CLOVER FAMILY

Native

Trifolium dubium is an annual plant found very commonly throughout Ireland in open grasslands, sandy and gravely habitats, waste places and sand dunes.

According to Nelson (1991), this clover is the species most widely used as shamrock (in Irish, *seamróg*). Uí Chonchubhair (1995) and Williams (1993) noted that clovers were eaten in Ireland during times of food shortage and famine.

Baker (1999) mentioned an Irish cure for lumbago: pounded 'dog-fern' (I am unsure what plant this is). Roots and shamrock were mixed together with butter made on May morning and 'holy salt'. This ointment was rubbed on the back and left untouched until the cure was complete. This treatment is given by Lady Wilde (1890).

A vintage Irish postcard containing a small packet of shamrock seed.

**Trifolium fragiferum*

- Strawberry clover (Scannell & Synnott, 1987)
- *Seamair mhogallach* (Scannell & Synnott, 1987); *Seamar suthlar* (Wade, 1808b)

FABACEAE (Leguminosae)—THE PEA AND CLOVER FAMILY

Native

Trifolium fragiferum is a rare perennial clover of damp grasslands, such as sand-dune slacks close to the sea, that is distributed mainly in the east; it is rarely found elsewhere in Ireland.

*No ethnobotanical uses of this plant have been noted in Ireland.

**Trifolium glomeratum*

- Clustered clover (Scannell & Synnott, 1987); Roundheaded clover (Brunker, 1950)
- *Seamair chlibíneach* (Scannell & Synnott, 1987)

FABACEAE (Leguminosae)—THE PEA AND CLOVER FAMILY
Native

Trifolium glomeratum is a very rare annual clover in Ireland of habitats close to the sea and sandy and gravelly ground; it is primarily distributed in the south and east, from Cos. Waterford and Wexford only.

*No ethnobotanical uses of this plant have been noted in Ireland.

**Trifolium medium*

- Zigzag clover (Scannell & Synnott, 1987)
- *Seamair sceabhach* (Scannell & Synnott, 1987); *Seamar seangan* (Wade, 1808b)

FABACEAE (Leguminosae)—THE PEA AND CLOVER FAMILY
Native

Trifolium medium is an uncommon perennial clover throughout Ireland of grassy and rocky habitats, such as roadsides and wasteground.

*No ethnobotanical uses of this plant have been noted in Ireland.

**Trifolium micranthum (Trifolium filiforme)*

- Slender trefoil (Scannell & Synnott, 1987); Least clover (Brunker, 1950)
- *Seamair bheag* (Scannell & Synnott, 1987)

FABACEAE (Leguminosae)—THE PEA AND CLOVER FAMILY
Native

Trifolium micranthum is a rare clover scattered across Ireland, mainly in coastal and sub-maritime areas, although it is largely absent in the west. Slender trefoil can be found along roadsides, in grasslands and as a weed in lawns.

*No ethnobotanical uses of this plant have been noted in Ireland.

**Trifolium occidentale*

- Western clover (Scannell & Synnott, 1987)
- *Seamair mhara* (Scannell & Synnott, 1987)

FABACEAE (Leguminosae)—THE PEA AND CLOVER FAMILY
Native

Trifolium occidentale is a rare clover very similar to *T. repens* and is found in Ireland only on shallow soils overlying glacial boulder clay along the east, southeast and southern coasts.

*No ethnobotanical uses of this plant have been noted in Ireland.

**Trifolium ornithopodioides*

- Bird's-foot clover (Scannell & Synnott, 1987)
- *Seamair éin* (Scannell & Synnott, 1987)

FABACEAE (Leguminosae)—THE PEA AND CLOVER FAMILY

Native

Trifolium ornithopodioides is a rare annual clover found on the south and east coasts in sandy and gravelly places. It has been found from Co. Down in northern Ireland south to Co. Cork in Ireland.

*No ethnobotanical uses of this plant have been noted in Ireland.

Trifolium pratense

- Red clover (Scannell & Synnott, 1987); Cow clover (Dinneen, 1927); Honeysuckle (Hart, 1898)
- *Seamair dhearg* (Scannell & Synnott, 1987); *Clóbhar dearg* (Ua Maoileoin, 2001); *Seamar chapaill* (de Vere Coneys, 1849; Moloney, 1919); *Seamar capull* (Wade, 1808b); *Seamar bó* (Dinneen, 1927)
- *Meelough* or *Milagh* (meaning clover or any trefoil in Yola)

FABACEAE (Leguminosae)—THE PEA AND CLOVER FAMILY

Native

Trifolium pratense is extremely common and widespread in Ireland in grassy habitats, such as meadows, pastures and roadsides. In many places it has been planted as a fodder crop and persists as a relict of such cultivation but is native in most areas too.

Clovers such as this extremely abundant species and *Trifolium repens* are important fodder plants in Ireland, often forming the dominant species of rich pastures used for grazing cattle and sheep. As such they are amongst Ireland's most important renewable plant resources. Clovers have the ability to fix nitrogen from the atmosphere through root nodules (containing nitrogen-fixing bacteria) and, in effect, create their own fertilizer. For that reason they are an extremely important green manure, helping to enhance the fertility of the places where they grow. Various large-leaved and large-flowered non-native varieties of red clover are often grown.

Trifolium pratense.

Phillips (1983) suggested that red clover leaves and flowers can be eaten in salad and the flowers used to make wine. Colgan (1904) recorded that the flowers of red clover were sucked by children for their sweet nectar. Patrick Weston Joyce (1875) recorded that in the *Annals of the Four Masters* (which date from the 1600s but were largely compiled from early medieval manuscripts) mention is made of the use of flowers to make wine: 'these flowers moreover were full of wine, so that the wine was pressed into bright vessels'. Weston Joyce interpreted this as a bardic way of saying that the flowers were made into wine. Perhaps these wine-making flowers were clovers.

Allen and Hatfield (2004) noted its use as a cough remedy in Ireland. They also recorded that its leaves were applied to bee stings in Co. Offaly and as an infusion of flowers to treat cancer in Wicklow and Meath. The roots were used to treat colds, and cancer was treated with an infusion of buds or shoots (Williams, 1993). The same author noted that the leaves and flowers of clovers were boiled with *Agrimonia* for heartburn and hay fever. Gallagher (2012) recorded that in Co. Donegal, red clover has been used to reduce angina. It is boiled in water and when cool the juice is drunk. Red clover is known as a medicinal plant in other parts of the world. It contains isoflavones that can mimic the effect of oestrogen. It also contains anti-inflammatory salicylates and it has been used to treat menopausal symptoms, because of their oestrogenic effect (Reader's Digest Association, 2006). It is not recommended for those with a history of tumours, which can be stimulated by the hormone oestrogen. The flowerheads are dried and used to prepare infusions, liquid extracts, ointments and tinctures.

Red clover has been used as a veterinary remedy to treat sore tongue in cattle in Co. Tipperary (Allen & Hatfield, 2004).

Red clover is worn occasionally as shamrock (Nelson, 1991).

Trifolium repens

- White clover (Scannell & Synnott, 1987); Dutch clover (Brunker, 1950); Shamrock (Praeger, 1938)
- *Seamair bhán* (Scannell & Synnott, 1987); *Clóbhar bán* (Ua Maoileoin, 2001); *Seamar seamrog* (Wade, 1808b)
- *Meelough* or *Milagh* (clover or any trefoil, in Yola)

FABACEAE (Leguminosae)—THE PEA AND CLOVER FAMILY
Native

White clover is an extremely common and often abundant plant in all types of grassland habitats in Ireland, along roadsides, in meadows, pastures and sand dunes.

Although a native plant, it has been widely cultivated as a forage crop for several centuries. Wade (1808a) wrote, 'Seed has been usually imported from Holland—the quantity astonishing—one seed-shop has been known to sell 50 tons annually'.

Trifolium repens.

Allen and Hatfield (2004) record the use of shamrock, possibly this species, to treat coughs in Kerry and liver ailments in Cavan. It is worn occasionally as shamrock (Nelson, 1991). Clover flowers are an important nectar source for foraging bees.

Mears and Hillman (2007) noted that white clover heads with developing seed pods can be eaten while green and tender, dried and ground down to make a flour. If gathered too late though the heads become dangerous to eat as toxins develop in them. These authors suggested that it is entirely possible that clover may have been used as food during the Mesolithic period in these islands. Peterson (1977) mentioned that the flowerheads of clovers are difficult to digest raw but can be eaten in quantity if soaked for several hours in salty water or boiled for five to ten minutes. Rutty (1772: 88) said that white clover is very nourishing and was much eaten by the ancient Irish 'before the introduction of potatoes' and that it gave them strength and firmness.

A four-lobed leaf of a clover (a so-called four-leafed clover) is regarded as lucky in Ireland.

Trifolium scabrum

- Rough clover (Scannell & Synnott, 1987)

- *Seamair gharbh* (Scannell & Synnott, 1987)

FABACEAE (Leguminosae)—THE PEA AND CLOVER FAMILY
Native

Trifolium scabrum is a very rare plant in Ireland that is found only in sandy habitats near the south and east coasts, extending from Cos. Dublin to Waterford.

*No ethnobotanical uses of this plant have been noted in Ireland.

*Trifolium striatum

- Knotted clover (Scannell & Synnott, 1987)
- *Seamair stríocach* (Scannell & Synnott, 1987)

FABACEAE (Leguminosae)—THE PEA AND CLOVER FAMILY
Native

Trifolium striatum is a rare plant in Ireland of dry grasslands and rocky habitats mainly near the south and east coasts. It has been sporadically recorded from Cos. Antrim and Down in Northern Ireland, extending south and west into Cos. Cork and Kerry.

*No ethnobotanical uses of this plant have been noted in Ireland.

*Trifolium subterraneum

- Subterranean clover (Scannell & Synnott, 1987)
- *Seamair faoi thalamh* (Scannell & Synnott, 1987)

FABACEAE (Leguminosae)—THE PEA AND CLOVER FAMILY
Native

Trifolium subterraneum is a very rare clover in Ireland that was originally known from only one gravel river-bank in Co. Wicklow on the east coast. Additional populations are now known from Cos. Wicklow and from Co. Waterford. Elsewhere, this small, procumbent clover has widely naturalised outside of its coastal distribution around Europe and Britain.

*No ethnobotanical uses of this plant have been noted in Ireland.

**Triglochin* species

- Arrowgrass (Scannell & Synnott, 1987)
- *Barr an mhilltigh* (Scannell & Synnott, 1987)

JUNCAGINACEAE—THE ARROWGRASS FAMILY
Native

There are two species of these rush-like plants in Ireland found in marshy habitats. I can find no recorded use for either, although elsewhere in the world other species are used as food. For instance, *Triglochin procera*, the creek lily or water ribbon, has small tubers which form on the roots, and in Arnhem Land (Northern Territory) and Victoria, Australia, they have been used as a food source, eaten raw or cooked, by Aboriginal people (Isaacs, 1987). *Triglochin dubium* has been used in the same way by Aboriginal people in the Northern Territory of Australia (Lindsay et al., 2001).

*Triglochin maritima

- Sea arrowgrass (Scannell & Synnott, 1987)
- *Barr an mhilltigh mara* (Scannell & Synnott, 1987)

JUNCAGINACEAE—THE ARROWGRASS FAMILY
Native

Triglochin maritima is found in salt marshes and coastal muddy habitats all around Ireland.

*No ethnobotanical uses of this plant have been noted in Ireland.

*Triglochin palustris

- Marsh arrowgrass (Scannell & Synnott, 1987)
- *Barr an mhilltigh* (Scannell & Synnott, 1987)

JUNCAGINACEAE—THE ARROWGRASS FAMILY
Native

Triglochin palustris is found in wet fields and marshes throughout Ireland.

Clyne (1989) said that sheep and cattle are fond of eating this species of the hillsides in Scotland.

*No ethnobotanical uses of this plant have been noted in Ireland.

*Trisetum flavescens (Avena flavescens)

- Yellow oat-grass (Scannell & Synnott, 1987)
- *Coirce buí* (Scannell & Synnott, 1987); *Coirce orfolltach*[288] (White, 1803; Wade, 1808a)

POACEAE (Gramineae)—THE GRASS FAMILY
Native

Trisetum flavescens is widespread in Ireland in limestone grasslands, sand dunes and dry banks.

*No ethnobotanical uses of this plant have been noted in Ireland.

Trollius europaeus

- Globeflower (Scannell & Synnott, 1987); Golden ball (Hart, 1898)
- *Leolach* (Scannell & Synnott, 1987)

RANUNCULACEAE—THE BUTTERCUP AND CROWFOOT FAMILY
Native

Trollius europaeus is a very rare plant in Ireland that occurs mainly on lake-shores and river-banks in the north-west (two sites in Co. Donegal and two sites in Co. Fermanagh). It was also discovered in one location in Co. Cavan in 1995 (Reilly, 2001).

288. Wade (1804, 1808a) applied this name to *Avena flavescens*.

It is occasionally grown as an ornamental garden plant for the herbaceous border (Wyse Jackson, 2006).

Trollius europaeus is the official county flower of Co. Fermanagh.

Tuberaria guttata (Helianthemum guttatum)

- Spotted rock-rose (Scannell & Synnott, 1987)
- *Grianrós breac* (Scannell & Synnott, 1987); *Plúr na gréine* (Moloney, 1919); *Rós na carraige* (Moloney, 1919)

CISTACEAE—THE ROCK-ROSE FAMILY
Native

Tuberaria guttata is a very rare plant in Ireland of dry rocky heaths near the west and south-west coasts only.

According to Moloney (1919), this plant is 'said to have been used in the treatment of shingles or wild fire (Herpes Zoster)'.

Tussilago farfara

- Colt's foot (Scannell & Synnott, 1987); Cold-foot (Hart, 1898); Could-foot (Hart, 1898)
- *Sponc* (Scannell & Synnott, 1987); *Duilleabhar spuinnc* (Moloney, 1919; Dinneen, 1927); *Bileóg spuinnc* (Moloney, 1919); *Gallán greannchair* (Moloney, 1919; Dinneen, 1927); *Luibh i n-aghaidh cosachtach* (*-taighe*) (Moloney, 1919; Dinneen, 1927); *Fathan* (Cameron, 1883); *Athan* (Cameron, 1883); *Cassachdaighe* (Cameron, 1883); *Adhain* (de Vere Coneys, 1849); *Aidhne* (de Vere Coneys, 1849); *Cluaisliath* (de Vere Coneys, 1849; Dinneen, 1927); *Duilliur-spuinc* (*-ce*) (de Vere Coneys, 1849); *Gallán-mór* (*-áin-mhóir*) (de Vere Coneys, 1849); *Adhann* (Dinneen, 1927; Ó Dónaill, 1977); *Fanann* (McGlinchey, 2007); *Comann* (*-ainn*) (Dinneen, 1927); *Copóg shleamhain* (Dinneen, 1927); *Sponclus* (Ó Dónaill, 1977)

ASTERACEAE (Compositae)—THE DAISY FAMILY
Native

Tussilago farfara is an extremely common plant throughout Ireland in damp heavy soils on roadsides, wet fields, clay banks and waste places.

Colt's foot has been used in traditional medicine for centuries, in particularly as a cough medicine; however, because it contains a range of carcinogenic alkaloids, its usage is not recommended today (van Wyk & Wink, 2004). In Ireland, its leaves were used in the treatment of asthma and spasmodic bronchial infections (Moloney, 1919, Logan, 1981; Allen & Hatfield, 2004). Its name is derived from the Latin word *tussis*, meaning 'a cough'. Cameron (1883) said that one of its Irish names, *cassachdaighe*, is derived from *casachd*, 'a cough', and *aighe*, 'a remedy'. Peterson (1977) provided a recipe for making colt's foot cough sweets or throat lozenges. The fresh leaves are boiled and sugar is added to the resulting liquid extract. To make the hard sweets, two cups of sugar are added to every cup of the extract and boiled until the rich syrup forms a hard ball when dropped into cold water ('cough drops'). Mackay (1836) recorded that its leaves were used medicinally as an infusion, or smoked like tobacco, to relieve the symptoms of asthma. The leaves were also mixed with unsalted butter for eczema and its juice used to treat earache (Williams, 1993).

Sharkey (2000) noted that colt's foot was used as a substitute for expensive tobacco mixed with wild rose petals and other aromatic herbs (known as 'sponc'). A recipe for herbal smoking mixture is given by Howes (1974) as colt's foot (50%), clover flowers (30%), lavender flowers (10%) and rose petals (10%). Because of the hairy undersides of the leaves, the colt's foot helps to bind the mixture together and may also assist it to burn well (Howes, 1974).

The petals have been used to make wine (Phillips, 1983) but I have not been able to find any records of its traditional use in this way in Ireland, although it may have been used as an ingredient of beer and jellies in Ireland (Williams, 1993). Peterson (1977) mentioned that in North America the leaves have been burnt and the residue used as a salt-like seasoning.

Uí Chonchubhair (1995) noted that the white cottony hairs on the undersurface of the leaves were collected and used as tinder in times past. Rutty (1772) mentioned such usage. Cameron (1883) said that in Scotland the leaves were dipped in saltpetre and then dried as tinder or touchwood. One of its Irish names, *fathan* (or *athan*), means 'fire' (Cameron, 1883). Dinneen (1927) said that it was used for 'tinder, as tobacco and as a specific, and good for many ills'; altogether a useful wild plant.

Typha species

- Bulrush (Scannell & Synnott, 1987); Reed mace (Anonymous, 1978); Cat's tail; Blackhead (Hart, 1898); Blackheads (Praeger, 1938)
- *Coigeal* (Scannell & Synnott, 1987); *Bodán dubh*[289] (Ó Dónaill, 1977); *Eireaball cait* (Ó Dónaill, 1977)

TYPHACEAE—THE BULRUSH FAMILY

Native

Including:

Typha angustifolia

- Lesser bulrush (Scannell & Synnott, 1987)
- *Coigeal chaol* (Scannell & Synnott, 1987); *Bodan* (Cameron, 1883)

Typha latifolia

- Bulrush (Scannell & Synnott, 1987); Reed mace (Anonymous, 1978)
- *Coigeal na mban sí* (Scannell & Synnott, 1987); *Cuigeal na mban sidhe* (Moloney, 1919); *Bodán dubh* (Moloney, 1919); *Earball cuit* (Moloney, 1919; Dinneen, 1927); *Coigeál na m-ban sighe* (de Vere Coneys, 1849); *Coiseog* (*-oige*) (Dinneen, 1927); *Curchais*[290] (Dinneen, 1927); *Sifín*[291] (Dinneen, 1927); *Simhean* (*-mhin*) (Dinneen, 1927); *Buigiún* (Dinneen, 1927); *Bansidhe coigeal* (Colgan, 1904)

Typha latifolia is a widespread plant in Ireland of ditches, canals, river and lake margins, and marshes. *Typha angustifolia* is much rarer and occurs mainly in the northern half of Ireland in pools, marshes and along the margins of lakes.

289. Ó Dónaill (1977) defined *bodán dubh* as 'great reed-mace'.
290. See also *Schoenoplectus lacustris*.
291. Dinneen (1927) defined *sifín* as 'a bulrush'.

Typha latifolia.

The English names of *Typha* include reed mace or cat's tail. Some authors, such as Cameron (1883), suggested that it is incorrect to call this plant 'bulrush' as this name should be applied to *Schoenoplectus lacustris*. However, as Webb (1977) pointed out, the name 'bulrush' is almost always applied to this genus when used in popular speech in Ireland.

According to Allen and Hatfield (2004), *Typha latifolia* was used as a toothache remedy in Ireland. It was formerly regarded as a talisman against harm. Colgan (1904) reported that in north-west Dublin a couple of shoots of this plant would be placed in the cow-shed to keep it safe and protect it against fire.

Typha latifolia.

The rhizomes of *Typha latifolia* are edible and were probably a useful starch source for early people in Ireland and can be harvested throughout the winter months. The young shoots and immature flower spikes are also edible and can be steamed or boiled as a vegetable or added to soups. Roasted seeds can be ground down to make flour for baking (Hilton, 2007). Other species of *Typha* are used elsewhere in the world for food. For example, throughout Australia the roots of *T. orientalis* and *T. domingensis* (neither species occur in Ireland) are eaten raw or sometimes cooked by Aboriginal people (Maiden, 1899; Isaacs, 1987). Peterson (1977) described *Typha* spp. as 'possibly one of the best and certainly the most versatile of our native edible plants' in Eastern and Central North America and noted its use as an asparagus-like vegetable, for flour, pickles and as a potato-like starch source. Green immature flower spikes can be collected, just before they emerge from the leaf sheaths, then boiled for about 15 minutes and eaten like corn on the cob, spread with melted butter. The flowerheads produce large quantities of bright yellow pollen, which can be shaken out of the heads into a bag, sifted and used as flour. Peterson (1977) suggested it is best mixed 50/50 with wheat flour. The same author described the way flour can be obtained from the rootstock. The roots are washed and peeled of the outer skin to reveal the starchy inner core. This is then crushed in cold water to separate the starch from the fibres. Once the fibres have been removed and the starch has settled, the water is poured off. After washing again, a pure white flour is left that can be used immediately or dried and stored.

Cameron (1883) and Clyne (1989) mentioned the use of the downy seeds produced by *Typha* for stuffing pillows and the leaves for making chair-bottoms, mats, thatch and sometimes hats or bonnets. In North America, various indigenous groups used *Typha latifolia* down to make dressings for burns and scalds, as well as to prevent chafing on infants and newborn babies. In a family in which the birth of a child was expected the women collected a large quantity of *Typha* down onto which the newborn baby was placed. The down that stuck to the baby was later rubbed off after moistening with milk from the mother's breasts (Gilmore, 1991).

Typha latifolia is planted for use in natural reed-bed water-treatment purification systems. Both species of *Typha* may also have been used for thatching.

Typha species were used in Ireland as an alternative to *Juncus* species in the making of rush lights (Ashley, 2001), although Drury (1925) would dispute this and said that bulrushes were not used for such purposes.

Ulex species

- Gorse (Scannell & Synnott, 1987)
- *Aiteann* (Scannell & Synnott, 1987); *Aitinn* (Lucas, 1958); *Conasg* (Cameron, 1883); *Teine* (de

Vere Coneys, 1849); *Praip aitinn*[292] (Dinneen, 1927); *Teithne* (Dinneen, 1927; Uí Chanainn, 1962); *Tor aitinn* (Dinneen, 1927; Ó Dónaill, 1977); *Tom aitinn*[293] (Ó Dónaill, 1977)
- *Goss* (in Yola) (Dolan & Ó Muirithe, 1979)

FABACEAE (Leguminosae)—THE PEA AND CLOVER FAMILY
Native

Including:

Ulex europaeus
- Gorse (Scannell & Synnott, 1987); Furze (Scannell & Synnott, 1987); Whin (Scannell & Synnott, 1987); French furze (Colgan, 1904); Atchenagh (Hart, 1898); Common furze (Brunker, 1950)
- *Aiteann gallda* (Scannell & Synnott, 1987); *Aiteann gallach* (Ua Maoileoin, 2001); *Aitean (-tine)* (de Vere Coneys, 1849); *Oir*[294] (de Vere Coneys, 1849; Dinneen, 1927); *Scothán aitinn*[295] (Ó Dónaill, 1977); *Aiteann uaibhreach* (Lucas, 1958); *Aiteann franncach* (Lucas, 1958)
- *Wyddeer* (in Yola)

Ulex gallii
- Western gorse (Scannell & Synnott, 1987); Autumn gorse (Scannell & Synnott, 1987); Furze (Colgan, 1904); Irish furze (McCarthy & Hawkes, 2007); Dwarf furze (Brunker, 1950)
- *Aiteann gaelach* (Scannell & Synnott, 1987); *Aiteann mín* (Lucas, 1958); *Aiteann chaiteach* (Lucas, 1958); *Aitinne bheag* (Lucas, 1958); *Aitinnin* (or *Aitnín*) *gaedhealach*[296] (Lucas, 1958)

Ulex europaeus is an extremely common plant throughout Ireland of heaths, rough grassland, bog-margins, hedges, rocky places, scrub and open woods. On higher ground it is sometimes replaced by *U. gallii*, which is found on acid (lime-free) soils, in heaths, particularly in the mountains, and in exposed places. It is common in most parts of Ireland except the centre.

In Ireland, gorse is known by two alternative common (English) names: furze and whin. The name gorse, which is the name used in England, has become much more widespread and widely used over the last few decades. Lucas (1958) said that a line can be drawn through Ireland from Drogheda, Co. Louth, in the east to Westport, Co. Mayo, in the west, where the name 'whin' is used north of the line, and 'furze' is used south of it. The name 'furze' crops up in place names in various counties, such as The Furry Glen in Dublin's Phoenix Park and Furzyhill in Co. Meath. A. T. Lucas (1958) wrote what is probably the most comprehensive account of the history and uses of any wild plant in Ireland in his article, 'Furze: A Survey and History of Its Uses in Ireland', published by the Folklore of Ireland Society (*An Cumann Le Béaloideas Éireann*) in the journal *Béaloideas*, Vol. 26: 1–204. I have drawn heavily on his work in my account here.

Gorse has a long history of ethnobotanical use in Ireland and has been used for a wide range of purposes. Land covered in gorse ('furzy pasture') was regarded as of considerable economic value in the 1600s (from evidence included in the Civil Survey of 1654–56) and

292. Dinneen (1927) defined *praip aitinn* as 'a furze bush'.

293. Ó Dónaill (1977) defined *tom aitinn* as 'whin-bush'.

294. These authors also apply this name to broom (*Cytisus scoparius*) and the spindle-tree (*Euonymus europaeus*).

295. Ó Dónaill (1977) gives the meaning 'furze bush' for *Scothán aitinn*.

296. A diminutive form found in usage in Co. Cork (Lucas, 1958).

Ulex europaeus (gorse, furze). Watercolour by Lydia Shackleton. Courtesy of the National Botanic Gardens, Glasnevin.

Ulex europaeus.

made up a substantial acreage of land that was described as 'profitable' in many parts of the island (Lucas, 1958). Lucas points out that there are records of furze being included in ecclesiastical tithes as an important commodity, further suggesting its value in economic terms, at least in the 16th, 17th and 18th centuries. It was even grown as a crop, sown in rows a few ridges apart, and in a few years when the plants had grown to 8 to 10 ft tall, they were harvested for fuel wood, roofing materials and fodder for horses, sheep and goats. For example, Lucas (1958) provided information on its cultivation as a crop in the barony of Forth in 1683.

Lysaght (1979) noted that it was used in fences, for closing doors and gaps, as domestic fuel, fuel for limekilns, as horse and cattle fodder, in bedding for animals and in sheds and animal shelters, in manure, as roofing for dwelling houses, in reed fences and in foundations for paths and roads, for harrowing, for making hurleys, walking sticks and umbrella handles and even for cleaning chimneys.

There is also a word in Irish for a 'furze-gatherer', which is *cnocadóir* (Dinneen, 1927). 'Furze gathering' is *cnocadóireacht*. A small two-pronged fork made from the limbs of a shrub or furze bush, used in the left hand when cutting the tops off furze and other shrubs, is called a *gabhlóg* in Irish, and an axe for grubbing or cutting furze is called a *grafán* (Dinneen, 1927).

Allen and Hatfield (2004) noted its widespread use in Ireland for medicinal purposes to treat coughs, colds, sore throats and hoarseness, consumption, heartburn, hiccups, jaundice, heart problems, dermatitis and ringworm, and to reduce swellings and as a tonic. In Co. Cavan it was used to treat asthma (Maloney, 1972). Lucas (1958: 185) said that in Aghowle, Co. Wicklow, it was 'thought to cure hiccough, the soft green tops of the dwarf kind being chewed and the juice swallowed', and in Carn, Co. Wexford, the tops of gorse were eaten or the bark was sucked to cure heartburn. Lucas (1958: 185) also provided a gorse-based cure for coughs: 'Furze was sometimes used to make a cough cure. The flowers were packed tightly into a jam-crock. Brown sugar (Demerara) was put on top and the crock covered and put in a saucepan of water and stewed slowly. I think a very little water must have been added to the furze blossom to make the syrup'. To cure worms in children, the flowers were boiled in milk and the mixture strained and given to the child on an empty stomach in the morning. In Derry, hoarseness was relieved by swallowing the flowers and a tonic was also made from the flowers, after stewing them in water for several hours. In Co. Meath, the young tops were boiled and used to make a potion to reduce swellings in people and animals.

Gorse was used as a veterinary medicine to eliminate worms in livestock (and children), coughs in cattle and swellings and pains in horses. Ui Sheasnáin (1982) records that the tips of furze bushes were mashed with a hammer and mixed with linseed oil and given to horses to treat an infestation of 'bots'. Bots are the larvae of bot flies, which resemble honeybees but do not sting (*Gasterophilus nasalis*, *G. hemorrhoidalis* and *G. intestinalis* are the most common species of bots in horses). Bot flies lay their eggs in the horse's hair and are ingested when the horse licks itself. The larvae then attach themselves to the stomach and intestinal walls, even within the nasal cavity, to feed and develop overwinter. The pupae are then passed by the horse to develop into adult bot flies. A 'dose' of furze fodder for horses was widely regarded as a treatment for this problem throughout Ireland. Sometimes gorse was boiled or steeped in boiling water to extract the juice and given to the horse, or sometimes gorse tips and flowers were mixed with bran, or even occasionally the flowers alone were given.

Ulex gallii in the autumn.

Gorse burning was a traditional practice that continues today to encourage the growth of new shoots, which are eaten by livestock (MacCoitir, 2003). *Ulex europaeus* seems to have been the species mainly used as fodder. Lucas (1958: 75) included information from someone in Kilgrant, south Co. Tipperary, who said, 'About sixty years ago I saw furze being used as horse and mule feed. Young *aiteann gallda* [*U. europaeus*] was used for this. I never saw *aiteann gaolach* [*U. gallii*] used as feed for animals'. O'Hare (1993) noted that as fodder, gorse was generally fed to horses. Lucas (1958) records many references to it being cut for fodder and bedding for cattle and horses throughout Ireland. Smith ([1756] 1979) recorded that gorse was used for fodder in Co. Kerry. Lucas (1958) documented several records of it being fed to cattle too, and in some instances it was used to milch cows; however, it has been said to taint the milk. When grown as a fodder crop, gorse was cut with a scythe when it reached 6 to 8 in. tall. It was particularly useful in winter when other fodder crops were scarce. Furze was cut in Ireland with a short, curved billhook that was held in one hand while the other hand gathered and collected the cut branches. To save the gathering hand (usually the left hand) from enduring all the prickles, it was wrapped in straw rope (*súgán*), like a mitten (called a *dornóg* or a *mútóg*) (Lucas, 1958). The mitten was made by winding a length of *súgán* around the hand while holding a piece of stick about 10 in. long. The stick served the dual purpose of providing a foundation for the mitten and a grip to hold it in place while it was being worn. A leather glove was also sometimes used. The handle of the billhook was often shielded by a leather (generally of horsehide) sheath or cuff, to protect the cutting hand. The cutter generally wore leggings too, or else a *súgán* would be wrapped around his legs, right down to his boots.

Once cut, bundles of gorse were trampled and then transported home, either in back-baskets or on carts. At home, the gorse was pounded to get rid of its prickles to prepare for its use as fodder. The shoots were finely chopped, pounded or crushed to get rid of the spines. A furze chopping machine was sometimes used. Lucas (1958: 80) included an account of its preparation in Co. Cork, provided by the Rev Horatio Townsend in the *Munster Farmer's Magazine* (1812), dating from 1811: 'The mode of preparing it, is by procuring a large wooden trough, or I believe more frequently inclosing a small area with flags, and chopping it in one of these with a sharp edged spade until it is completely divested of prickles. Even with this simple management one man will easily manufacture enough to feed four horses a day'. A hatchet or a heavy blade, like a butcher's chopper, was also used to chop gorse shoots. A stone mortar was a common method to prepare gorse as fodder by crushing the shoots. They were crushed with a stout wooden mallet or pounder on a flat stone that was sometimes surrounded in a wicker basket, or by a ring of other stones. Lucas (1958) records that in Clonbroney, Co. Longford, the mallet was made from a section of a 'fir tree' (presumably a piece of bog pine, *Pinus sylvestris*) and was very heavy. Lucas also records that in Tullyfern, Co. Donegal, wood from ash and holly were both used to make the mallet, and beech wood in Clonallan, Co. Down. A stone trough or basin was also used, into which the gorse was put and then pounded by the heavy mallet or 'beetle'.

Two types of furze choppers.

The mallet sometimes had blades set into its face that would chop the gorse at the same time as it was being pounded.

O'Hare (1993) mentioned that the mud floors of houses in the Brosna area of Co. Kerry had special holes in them, which were used for holding branches of gorse when it was being cut and prepared as fodder. (O'Hare gives an example in which gorse was cut with a *sleán* and weighed down with a 'box' [the central part] of a cart-wheel). Lucas (1958) described these from Co. Kerry as rectangular pits sunk into the ground, lined with boards, or more frequently with stone flags set on edge. The bottom of the trough was lined with a thick plank or board. In many old Kerry houses the trough was situated in the kitchen, generally near the door. Presumably this was to enable the fodder to be prepared in the evenings and at night, and especially during the winter. Some early houses had the horses stabled at one end of the house so the trough could also act as a simple manger. Knocking stones, flat slabs or stone basins were where gorse was crushed (or melled) before being fed to horses. These could sometimes be seen in farm-yards, and there were also a few water-driven 'whin-mills' known (Evans, 1957). Grigson (1958) said that in Ireland gorse cut for fodder was crushed by wooden mauls on a block of stone.

In the 19th century new machinery was introduced that was used to chop gorse for fodder, making the process of preparing it much less time consuming and strenuous. One such machine survives at my cottage in Lougher, Annascaul, Co. Kerry, where I found it in the ruins of an old shed, and it is now immobilised as a monument to old Irish agricultural practices by a sycamore tree that has grown around it. Such machines were also used to slice turnips and hay and were sometimes called 'chaff-cutters'.

When used as bedding for stock on the farm, gorse was spread on the floor of the shed where the animals were housed. A layer, sometimes more than a foot in depth, was spread and trampled to make it compact. The surface was then covered with hay on top (or if that was in short supply, with rushes, bracken, sedges or coarse grass). Gorse provided a warm insulating layer for the animals and allowed urine to flow through it so that the shed remained warm and dry (Lucas, 1958). When the gorse and dung from the animals were cleared out from time to time it was all added to the dunghill to rot down and made excellent manure. Gorse and dung were sometimes mixed with sea sand, and also with seaweed to make manure and used to fertilise crops of potatoes and turnips.

Lucas (1958) described the use of furze for hedging, often grown on the tops of earthen banks surrounding the fields in the 1700s and 1800s. He included references to it being grown from seed, right up to the 19th century and said that 'the general consensus of opinion being that the practice ceased about 20–60 years ago' (i.e., 1900–1940) (Lucas, 1958: 33). He also included a description of a method by which a gorse fence was made:

> In Kerry when a new fence was made there were stones and sods in alternate rows on both sides with earth in the middle. When the fence was almost completed a pound of furze seed was bought in the shop (it was always dear) and was 'fed' into a rope made of hay, a few grains at a time, while a second person did the twisting with a *crúicín* [an implement for twisting hay or straw ropes]. The *súgán* was then wound into a ball and was later unwound and laid on top of the fence, all along from end to end. The rope was then covered with earth to a depth of one or two inches and trampled on firmly when covered (Lucas, 1958: 34).

Lucas (1958) also mentioned a variation on this method from Co. Donegal where the straw used to make the *súgán* would be steeped in horse manure and dried before being made into a rope. Haw seeds (from hawthorn, *Crataegus monogyna*) were also sown in this way. Smith ([1756]; 1979) also mentioned the use of gorse as an effective hedging plant in Co. Kerry, even in bleak mountain situations. The hedging gorse plants would variously have been either raised from seed or transplanted as young plants from the surrounding countryside. Gorse seed was collected as a source to grow it as a cultivated crop. It is even recorded that its seed was exported from Cork to France in the early 1800s (Wakefield, 1812).

Gorse played an important part in drain-making too in agricultural fields. A trench would be dug and lined with flat stones. To prevent loose soil washing into the new drain a covering of gorse would be placed over it to act as a filter. Lucas (1958) has documented the use of gorse in such drain construction in many parts of Ireland. Another simple type of field drain also used gorse. A trench was dug and lined with gorse, then covered over with an inverted sod and filled with earth again. When the gorse decayed, it left an underground channel with the clay having set above it. In water-logged conditions the gorse might not decay, and in these drains it remained there to help keep the watercourse open.

Gorse was also used in road-making, particularly in mountain, marshy or boggy regions. The ancient material used for making roads in Ireland was logs which 'floated' in the bog and onto which the roadway was laid. With the destruction of Ireland's woodlands, particularly in the 16th and 17th centuries, people turned to using bog timber, and when that depleted too, gorse provided a useful alternative as the foundation for roads. Lucas (1958) described the road-making process: The first task was to dig a drainage ditch on each side. The road itself was then levelled and graded and covered with grassy sods, grass-side up. Sometime later a layer of gorse was then spread on top of the sods. A layer of mud, gravel or marl was then spread on top of the gorse, to help prevent it from rooting and coming up through the road. The next layer on top of this was of large stones, and then the road was finished off with a final covering of gravel.

Gorse has also been burnt as a fuel in Ireland, in traditional open fires and ovens. It burns quickly and gives out great heat (Sharkey, 2000). It was used for kindling and as an alternative fuel to peat, particularly if higher temperatures were needed for purposes such as baking, or if wood and peat were in short supply (Lucas, 1958; Rotherham, 2009). In parts of Ireland where turf was scarce, furze was a particularly important fuel source. Other fuels used in times of scarcity were twigs and branches from hedgerows, straw and stubble, dried horse and cow dung and dried grass. As a fuel source, furze was superior to any of these. In Dublin, bundles of furze (or 'faggots', *feagóid*[297]) or sheaves, were brought into the city for sale as fuel. Lucas (1958: 42) provided records of such usage in Dublin in the 1500s and 1600s and mentioned a complaint made in 1619 that 'great firr and faggott reekes' built close to the city walls 'doe overtop the said walles in height', indicating that at that time great quantities of furze must have been used for fuel. Large quantities of furze were brought into the city to supply bakers' ovens. Although the earliest reference quoted by Lucas (1958) of such usage dates to 1568, it was well established as a fuel source by that time. When cut for fuel the furze was generally left to season before being transported home (or for sale) in a cart and stacked before use. On the farm, the fuel was often used in the preparation of food for pigs and poultry, being used to heat large pots or boilers. According to Lucas (1958) griddle bread was invariably baked on a stand over furze embers. In Wexford, during ploughing matches, there were also faggot cutting competitions.

Bakers' ovens were often heated using furze, which was freely available, burned easily, provided good heat and left little ash. Furze was also used to some extent to provide fuel for limekilns too. Rutty, in 1772, said that in Dublin the bakers would sell the gorse ashes to make a lye (a strongly alkaline liquid made by leaching wood ashes), used for scouring pewter and bleaching linen (quoted in Lucas, 1958).

297. Lucas (1958). A faggot was defined as 'long furze tied in bundles by briar bands'.

Sharkey (2000) also mentioned that gorse bushes were often grown (or at the very least, not removed) near the house as they made a convenient place to dry clothes. It was also used as a primitive harrow (Akeroyd, 1996), particularly for covering grass seed and for sweeping weeds and earth from potato drills before they are earthed up (Lucas, 1958). It was also used to make rudders in traditional boats (Williams, 1993). Evans (1957) mentioned that the roots of gorse (whin) were used to make the tiller. Gorse was also widely used to provide an insulating base for hay or corn stacks.

Gorse was used to help bind the mud used in the walls of early mud-built houses (cabins) in Ireland (Lucas, 1958; O'Hare, 1993). Gorse was also used in roofing, particularly for thatch on rough cabins, sheds and temporary shelters. It was also used as a base layer spread over the rafters of cottages onto which sods were laid, with straw, rushes or reeds used for thatching. The layer of gorse may well have acted as insulation and also to catch dust and debris falling from the sods or scraws used on the roof. Lucas (1958) described gorse timber used to make wickerwork, onto which thatch was fastened. Its use as a foundation for thatch was widespread in Ireland. Lucas (1958: 134) also described another thatching technique (recorded in 1946) with gorse, where '[s]hort Irish furze is tied together in small bundles and placed on the rafters and sewn on to them with thatching twine, one man being inside the house, another outside during the process'.

Records of the use of gorse to make u-shaped scallops for holding down thatch are known. Willow or hazel rods were more commonly used though. In some places, two-year-old pieces of the gorse stem about 2 ft in length were selected. These were heated in the middle over a turf fire so that they would be pliable and bendable without breaking. In other parts of Ireland older wood was preferred.

Gorse rods were sometimes used for the ribs of lobster pots, and Lucas (1958) mentioned records of this use in the south of the Mullet Peninsula, Co. Mayo. Its use here was probably due to the absence of any other more suitable material for rods. But in Galway, the rods are recorded as having been used for the base of the lobster pot, and the rest of the pot made with willow (osier) rods. Other miscellaneous similar uses described by this same author include its use to make the hole pins for currachs in Connemara; fitments for straddles and panniers in the Mullet and in North Kerry; stone-breakers' hammer handles in Co. Mayo; tool handles (in general) in Co. Carlow and whip handles in Co. Cavan.

Clamps or pits in which potatoes were stored over the winter were sometimes fenced with gorse or roofed with hay and then a layer of gorse, both to keep the potatoes free of frost and also to prevent damage from browsing animals. Sometimes apples were stored in a similar way (Lucas, 1958).

The cleaning of chimneys with gorse was a widespread practice throughout Ireland. A dense-headed shoot of gorse was selected, or sometimes several tied together. The gorse was then either drawn through the chimney in one direction or moved backwards and forwards with a rope tied to it, with one end coming out through the chimney top and the other out of the fireplace. Another method was to tie the gorse onto a stick which would then be inserted up into the chimney, and if it was not long enough, another piece was tied onto that, and so on, until the whole chimney was cleaned, rather like the screw-on sections of a more modern chimney sweep's brush. Gorse was also used to clean the scum out of wells and to capture any weeds or debris floating in the water (Lucas, 1958).

Gorse timber was used to make hurleys and walking sticks. In Irish a 'furze hurley' is *spaic aitinn*, *spealóg* (Ó Dónaill, 1977) or *caimín* sticks (Lucas, 1958). Lucas (1958) said that furze was quite commonly used to make hurleys all over Ireland. He reported many records where a piece of furze timber with a bent end was selected and carved to make a hurley; however, finding a piece with a suitably straight shaft was often a problem. Usually, timber pieces that were 30 to 36 in. long were selected, cleaned and planed. Sometimes pieces were used that ended with a stout stump or root, which were also considered useful. Gorse hurleys were predominantly used by young boys, but in earlier times they may have been widely used by older players too.

The museum collections at the National Botanic Gardens include several examples of gorse walking sticks. These would have been steamed to make them flexible so that they could be straightened. The examples at Glasnevin were made in the United Kingdom by Henry Howell and Company, who operated from 1832 to 1934 at a shop at 76 Aldersgate, London. To make a walking stick, a straight rod was selected and the branches trimmed off flush with the stem. The bark was stripped and then the thicker end was turned to make a crook. This was achieved by boiling or steaming the end until it was soft and could be bent into shape. Then the bent end was tied with a string to keep this shape until it had set hard again. These sticks were also used for driving animals on the farm and to and from market (Lucas, 1958).

A related use of *Ulex* species is the making of umbrella handles that Lucas also described as a minor industry in Ireland in the late 1800s. Curved crooks were fashioned from gorse wood and fastened to the metal or wooden shafts of umbrellas. Lucas reported that in Rathdrum, Co. Wicklow, an Englishman named Booker employed seven to 10 men in preparing umbrella crooks from gorse wood. Apparently, in Fore, Co. Westmeath, for about two years during the early part of the 1900s, a group of people from Scotland cut extensive crops of gorse for such umbrella handles, and there are also records of the harvesting of gorse timber for the same purposes in Cloondoorney, north of Tulla, Co. Clare, in Tracton and Fermoyle, Co. Cork, as well as in Moynalty and Rathmore, Co. Meath and Carn, Co. Wexford. In Fermoyle the collectors used a kind of boiler mounted on wheels to boil the sticks and help strip off the bark.

The long roots of gorse were used as rough threads in shoe-making according to Williams (1993). Lucas (1958) mentioned such usage, traced to Rutty (1772: 115), who stated that 'the long fibrous roots running sometimes the length of a yard or more under ground, being twisted together, were formerly used by the Irish, as threads to sew their brogues with...' However, Lucas thinks this is somewhat fanciful and found no other evidence for such usage.

Although there are not many records of gorse being eaten, the flowers have been used to make wine and teas. Phillips (1983) provided recipes for both. The beautiful, bright yellow gorse flowers were also used to make a yellow dye (Moloney, 1919; Uí Chonchubhair, 1995). Mahon (1983) mentioned that a yellow dye can be obtained from the bark and young shoots as well as the flowers. In Dublin, gorse flowers were used to colour Easter eggs, *cúbóga*, up to the 1930s (Williams, 1993). It was also commonly used in Ulster for colouring Easter eggs (Lucas, 1958). Lucas said that the custom of dyeing Easter eggs with gorse was chiefly confined to Antrim, Armagh, Down and Tyrone, and also Donegal. The eggs were hard boiled and taken by youngsters to play with in the mountains, rolled (called 'trundling') or thrown as far as they could, and if broken, then eaten. Apparently, this sport was only practiced by Presbyterians, which would explain its predominance in the northern counties.

Lucas (1958) mentioned that in Kilcoole, Co. Wicklow, and in Killarney, Co. Kerry, there are records of gorse being used to shake the frost off the foliage of young potato plants, so that when the sun rose, frost would not burn the tender young stalks.

In horse-racing, the Irish national hunt steeplechase rules specified that in steeplechase courses there had to be fences constructed that would include a hedge, sometimes made from dead brushwood or gorse, 2 ft in width and not less than 4 ft 6 in. in height (Lucas, 1958).

In the garden, a cultivated variety of gorse, *Ulex europaeus* (strictus), is an attractive ornamental shrub, sometimes called 'Irish gorse' or 'Irish furze'. Mackay (1925a: 166) said that this had been discovered in 'the Marquis of Londonderry's park, [Mount

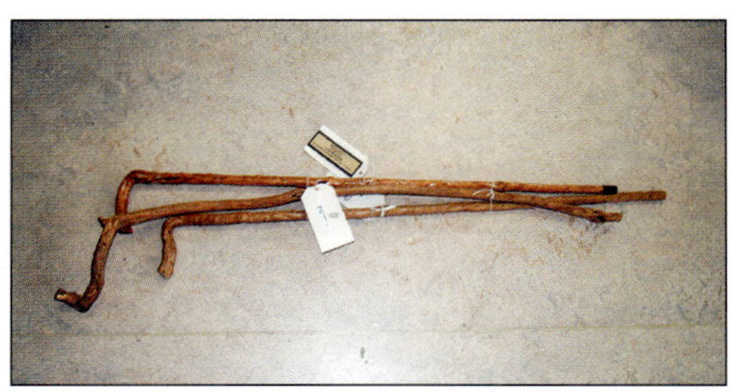

Walking sticks made from *Ulex* species. From the Economic Botany collection of the National Botanic Gardens.

Stewart] county of Down [by] Mr. John White'. He considered it to be a new species (*U. strictus*), but it has subsequently been recognised as a variety of *U. europaeus*. *Ulex gallii*, 'Mizen Head', is also a cultivated gorse of Irish origin, a prostrate, mat-forming shrub that was collected at Mizen Head, Co. Cork, around 1970 by Peter and Patricia Cox (Nelson, 2000b).

Ulex europaeus is the official flower of Belfast city.

Ulmus species

- Elm (Scannell & Synnott, 1987)
- *Leamhán* (Scannell & Synnott, 1987); *Ailm*[298] (Dinneen, 1927); *Sleamhán* (Dinneen, 1927; Uí Chanainn, 1962); *Tuilm*[299] (Dinneen, 1927); *Crann leamhán*[300] (Ó Dónaill, 1977)

ULMACEAE—THE ELM FAMILY

Including:

Ulmus glabra (Ulmus montana)

- Wych elm (Scannell & Synnott, 1987); Irish elm; Ellum (Colgan, 1904)
- *Leamhán sléibhe* (Scannell & Synnott, 1987); *Leamh* (Moloney, 1919); *Fionn-choll* (Dinneen, 1927)

Native

Ulmus procera (Ulmus campestris)

- English elm (Scannell & Synnott, 1987); Common elm (Anonymous, 1978)
- *Leamhán gallda* (Scannell & Synnott, 1987)

Introduced

Wych elm was once an important, widespread and common tree on the better fertile soils of the Irish lowlands but declined in ancient times, probably largely due to clearance for agriculture. Elsewhere, *Ulmus glabra* is native to temperate Europe. In recent decades most mature elms in Ireland have been killed by Dutch elm disease, and today native trees survive mainly as isolated specimens in mountains and glens of the west and north-west in Ireland. It has been planted in other locations throughout Ireland, but most of these survive as scrubby individuals that have regrown from the stumps of trees killed by the disease. The largest known surviving tree of this species is one growing at Killegar House, Co. Leitrim, that is 35 m tall (and 3.5 m in girth), and the Irish girth champion of this species from Kilcarrig Lodge, Co. Carlow, is 4.53 m in girth (and 21 m tall) (Anonymous, 2005).

Elm wood is hard, heavy, very resistant to water and is very tough, which made it a valuable timber for many purposes. Elm wood was often used for canal lock gates, harbour-works and the keels of ships (Arnold, 1968). Harry Madill, writing in MacCárthaigh (2008), described the use of elm planking combined with larch timber for the construction of a yawl, or traditional boat, built in 1924 in Killough, Co. Down. Local native elm trees were sometimes harvested for their timber for boat building and apparently some boat

---

298. Dinneen (1927) also gave this same Irish word for palm tree. *Ailm* is also indicated by Uí Chanainn (1962) to mean 'palm tree'.

299. Dinneen (1927) defined *tuilm* as 'vagina, or oak or elm (?)'.

300. Ó Dónaill (1977) defined *crann leamhán* as 'elm-tree'.

builders could judge the suitability of a particular elm tree for boat building by chewing its leaves (C. Levis, in MacCárthaigh, 2008).

Drainage and water pipes made from *Ulmus* trunks were a standard feature of plumbing systems in Georgian Ireland in the 18th century, and for example, elms planted along the Royal Canal in Dublin were originally planted for such purposes (McCracken, 1971). The trunks were bored while still green and unseasoned. Each length was attached to its neighbour by having one end tapered down, rather like a lead pencil, and fitted into the hollowed-out end of the next tree, with the joint between them sealed with tallow (Edlin, 1973).

A bowl made from *Ulmus* species.

Elm wood was also used in the manufacture of wooden pumps (O'Sullivan, 1969). Pumps made from wood for raising water from wells were once common throughout Ireland until they were replaced by iron pumps and piped water schemes. Wooden pumps were manufactured and installed in Ireland as recently as 1960 when the last wooden pump makers in Ireland ceased production. Larch logs (*Larix* spp.), a non-native species, were used in recent years but elm wood was used for some of the parts, including the 'sucker' and the spout. Rods to draw the sucker up and down the shaft of the pump were generally made of 'red pine' (a name that can refer to the timber of a number of imported conifer species), which did not become water-logged (O'Sullivan, 1969).

MacCoitir (2003) noted that elm has been used for laying piles under bridges and buildings in damp places as well as for ship-building, water troughs and furniture. He also records that it was used to make coffins and, due to its flexibility, wych elm was popular in medieval times for longbows. Divining rods have also been made from wych elm. Elmwood has a twisting grain, which helps to make it resistant to splintering and makes it ideal for purposes such as the seats of chairs. The seats of Windsor chairs, a classic piece of furniture for traditional cottages in Ireland, were invariably made of elm wood and the other parts made from beech or yew (Sharkey, 2000).

Other uses for elm timber included beds, coffins, wheel stocks, rigging blocks and pulleys (Lysaght, 1979). Apparently, elm wood has been popular to make library shelves for books as it is one of the few native timbers in these islands that does not mark or stain the books (Matthew Jebb, pers. comm., 2010).

Elm rods have been found during excavations in medieval Dublin city sites that had been used in the construction of walls of house (ash and hazel rods were also used), onto which mud was plastered to make traditional wattle-and-daub houses (Anonymous, 1973).

Ropes were also made from elm bark. Some fine reins made from elm bark in Co. Roscommon are included in the collections of the National Museum of Ireland at its Museum of Country Life in Castlebar, Co. Mayo. Bootlaces made from tree bark are also known in Ireland (such as a two-ply pleated example made by James Grady of St John's, Athlone, Co. Roscommon, that is in the collections of the National Museum of Ireland in Castlebar, Co. Mayo (Ref. No: F1959.72), as well as plough traces and goat spancels (a rope used for tying or hobbling the legs of the goat to prevent it from wandering). This elm bark can be very strong indeed and was ideally suited for such purposes. I have seen the bark of North American elms used for basketry but have not found records of its use for such purposes in Ireland.

Elm was used as a remedy for scalds and burns. A slimy decoction derived from mucilage of the inner bark of the elm was used as a salve (Threlkeld, 1726) and used to reduce swellings and inflammation; also used to staunch bleeding, treat jaundice and other ailments in Ireland (Allen & Hatfield, 2004). Ody (1997b) mentioned its use against stomach inflammations, gastritis, indigestion and for coughs, but I have not found records of its use for these ailments in Ireland. However, Gallagher (2012: 26) said that elm leaves are one of the best cosmetics in the world: 'Go to any elm tree after heavy rain or in the early morning when the dew is on the leaves. Gather the youngest and greenest leaves and draw gently over the face leaving the dampness to dry on, repeat daily. That's all is required. The skin will be softer and more beautiful and freckles will almost disappear'.

Elm foliage was used to feed tethered cattle in ancient Ireland, which may have contributed in part to its decline in Ireland (McCracken, 1971).

McCarthy and Hawkes (2007: 84) said that in the Mizen region of West Cork a small branch of elm was used 'to beat the caterpillars off cabbages'.

The new leaves are edible and young developing fruits can also be eaten raw, cooked or used in stews or soups. They are best while still soft in April and May (Mears & Hillman, 2007). Drury (1984) mentioned that meal made from the bark of elms was used as a famine food in Ireland. The bark of other trees was also used in a similar way, such as Scots pine, birch, beech and lime, but not oak or willow, which contain too much tannin.

Wych elm was formerly a common tree, as was the non-native *Ulmus procera*, the English elm, and both were grown as ornamental trees of parks, parkland and large gardens but now much reduced due to Dutch elm disease. *Ulmus minor* (*U. carpinifolia*), the small-leaved elm, also occurs in Ireland as an introduction. It is difficult to be certain which species of elm may occur in different localities in Ireland, as specimens are not easy to distinguish, particularly specimens growing in hedgerows that have sprouted from suckers. As well, a hybrid elm, *U. ×hollandica* (Dutch elm), has also been recorded and may be widespread.

The English elm is prone to dropping limbs without warning, even in still weather (Anonymous, 1941), which makes it a health and safety concern for managers of public parks and gardens.

In Irish a name for elm is *leamhán*, as in '*Dinc den leamhán a scoilteann í féin*' (Only a wedge of its own wood will succeed in splitting the elm) (Ua Maoileoin, 2001).

Umbilicus rupestris

- Navelwort (Scannell & Synnott, 1987); Wall Pennywort (Anonymous, 1978; Doogue & Krieger, 2010)
- *Cornán caisil* (Scannell & Synnott, 1987); *Arán glas* (Ó Criomhthain, MSS, n.d., ca. 1908; Ua Maoileoin, 2001); *Lus na pingne*, or *pinginne* (Dinneen, 1927) or *pingine* (Dónaill, 1977); (Moloney, 1919); *Cornán leacáin* (Moloney, 1919); *Carnán caisil* (de Vere Coneys, 1849; Dinneen, 1927; Uí Chonchubhair, 1995; Webb et al., 1996); *Lamhainn cat leacain* (Dinneen, 1927)

CRASSULACEAE—THE STONECROP FAMILY

Native

Umbilicus rupestris is a common plant throughout most of Ireland of walls, banks and non-calcareous rocks. It can also be found growing as an epiphyte on the trunks and boughs of trees, especially in wet woodlands.

It was formerly used medicinally in Ireland for a variety of different ailments, such as to treat lumps and swellings, sore eyes, tuberculosis, jaundice, headaches, worms, kidney trouble and earaches as well as to induce abortions (Allen & Hatfield, 2004). It was also mixed with ivy to treat warts; used to treat corns in Waterford and, as it grows on walls (out of stones), it was regarded as a useful treatment for bladder stones (Williams, 1993). In the Mizen region of West Cork, McCarthy and Hawkes (2007) recorded that the sap was put onto bee stings. In his book on the folklore of Co. Clare, Westropp (2000) said that it was reputed in that county that it would only cure Protestants! McNeill (2010) speculated that it may have been planted as a medicinal herb in several locations in Co. Tyrone.

It was also used as a veterinary remedy to treat sick cattle in Kerry and scruff on the legs of horses in Wicklow (Allen & Hatfield, 2004). In Dunquin, Co. Kerry, a decoction of the plant was given to cows after calving to expel the afterbirth, and it was also used medicinally to treat high fevers, inflammations, piles, chilblains, liver complaints, kidney stones and as a diuretic (Uí Chonchubhair, 1995).

The fleshy leaves can be used in salads or sandwiches (Phillips, 1983; Uí Chonchubhair, 1995; Mears & Hillman, 2007).

According to Mahon (1983) and Williams (1993), this plant yields yellow and red dyes.

Urtica dioica

- Nettle (Scannell & Synnott, 1987); Coolfaugh (Hart, 1898)
- *Neantóg* (Scannell & Synnott, 1987); *Leantóg* (Ua Maoileoin, 2001); *Neannta* (Moloney, 1919; Dinneen, 1927); *Caolfhail* (Moloney, 1919); *Feanntag* (Cameron, 1883); *Deanntag* (Cameron, 1883); *Iontag* (Cameron, 1883); *Iuntag* (Cameron, 1883); *Caolaigh* (de Vere Coneys, 1849); *Cúlfáidh* (Dinneen, 1927); *Neanntóg* (Dinneen, 1927); *Neanntóg loiscneach*[301] (Dinneen, 1927); *Neannta buidhe* (Dinneen, 1927); *Cál faiche*[302] (Ó Dónaill, 1977); *Neantán* (Ó Dónaill, 1977); *Cúlfáich* (Uí Chanainn, 1962); *Leanntóg* (Ó Criomhthain, MSS, n.d., ca. 1908); *Lanntóg* (Colgan, 1911)

URTICACEAE—THE NETTLE FAMILY
Native

Urtica dioica is an abundant perennial plant that grows to about 1 m in height and can be found throughout Ireland growing in waste places, grassy places, hedgerows, woodlands, fens and close to houses. It is especially common in places where the soils have been enriched with nutrients (notably phosphates) from the dung of cattle and sheep.

Nettles are well known for their painful stinging hairs[303]. Nevertheless, they have been widely used for a wide variety of medicinal purposes in Ireland, particularly as a spring tonic to cleanse the blood, to treat sores, rashes, colds and coughs, as a counter-irritant for rheumatic complaints and as a tea to treat measles (Allen & Hatfield, 2004). Dirrane (1997) wrote that nettles were the best cure for measles used in Inishmore in the Aran Islands when she was growing up in the 1890s and in the early part of the 20th century. Her mother would pick wild nettles and boil and strain them to create a juice, which was most effective once the spots appeared on the skin. Elsewhere, as a treatment for worms in children, the roots of nettles were boiled to make a powerful purgative (Logan, 1972, 1981). A tea was made or the plant was eaten to treat dropsy, jaundice and as a general defence against ill health, and according to general homeopathic principles, it was said to be good against rashes, eczema and measles (Williams, 1993). Lady Wilde (1887) said that the nettles need to be gathered in a churchyard to be effective against dropsy. In Co.

301. Dinneen (1927) gave the meaning 'common stinging nettle' for *neanntóg loiscneach*. Dinneen also listed the Irish word *áinleog*, meaning 'a weed of the nettle kind'.

302. Ó Dónaill (1977) gave the meaning 'nettles' for *cál faiche*.

303. *Urtica dioica* subsp. *galeopsifolia*, the fen nettle, a subspecies without stinging hairs, is rare in Ireland.

Urtica dioica (nettle). Watercolour by Lydia Shackleton. Courtesy of the National Botanic Gardens, Glasnevin.

Cavan, boiled nettles eaten for dinner three times during Lent were said to be useful to prevent boils and infections (Ui Sheasnáin, 1982).

Nettles have been used since ancient times for sexual stimulation; as an aphrodisiac that would have been prepared and used for genital flagellation (Rätsch, 1997). However, I know of no records of their use for such purposes in Ireland. Nevertheless, scourging or flogging with nettles was recommended against paralysis, weak blood flow, neuritis and rheumatism (Williams, 1993; Vickery, 2010). Milliken and Bridgewater (2004) reported that the Nitinat people of Vancouver Island in British Columbia, western Canada, would rub their hands with nettle leaves before handling their fishing gear, and they also used the stems to make twine. Formerly, in the south of Ireland, the day before May Day was known as 'Nettlemas Day' when schoolboys would rush about with bunches of nettles stinging their friends and foes alike (O'Regan, 1997). Cameron (1883), however, said that it was generally the girls who would run amuck with branches of nettles, stinging everyone they met.

Nettles are used as a source of fibre in several countries and can be woven into cloth. Uí Chonchubhair (1995) reported that cloth made from nettle fibres has been in use since the Bronze Age, and nettles have long been used in the manufacture of dyes, inks and paper. Nettles are listed by Mahon (1983) as having provided green and yellow dyes. The nutlets of nettles have been found in pits excavated from Fishamble Street, Dublin, dating back more than 1000 years. Geraghty (1996) suggested that these pits may have been used for retting nettles as well as to ret flax to obtain their fibres for textiles. *Urtica*-based textiles have been found in numerous Scandinavian Viking sites. Mabey (1998) reported that in Britain a flint arrowhead has been found attached by nettle fibres to the remains of its shaft and that nettle cloth was manufactured in Scotland and Scandinavia into the early 19th century. They were used to make military clothing in Germany during World War I. Fabrics made from nettle fibres can be bleached to be as white as the best linen (Lewington, 1990). Rutty (1772: 132) described how ashes derived from burnt nettles are good for bleaching linen and said that 'in the country they make up the ashes of nettles and other herbs into balls, with stale urine, and burn these balls, till from an ash-grey colour they become white, and for this purpose these balls undergo two burnings, and then are used as Soap for cleaning and whitening linen cloth'.

In Ireland, it has occasionally been used as a vegetable when the nettle tips are picked and boiled, as an ingredient of soup, or made into gruel with oatmeal. Nettles are rich in proteins (anything up to 20% of the leaf dry weight), fats (2%) and vitamins (100 mg of vitamin C per 100 g as well as vitamins B, A, and several minerals, such as calcium, potassium and iron); altogether an extremely nutritious food (Mears & Hillman, 2007)! Nettle soup made with onions, potatoes and garlic is delicious, as is nettle soup flavoured with a little nutmeg. Mackay (1836: 233) noted that nettles were eaten by the poor in Northern Ireland 'during the late years of scarcity, owing to the failure of the potato crops'. Nettles tops were often included in champ, a traditional Irish recipe (made from potatoes, onions and milk) in Northern Ireland. John Lewis-Stempel (2012) said that nettle purée can be spread on toast, with a poached egg on top. Kelly (1997) reported that nettles (and sorrel, *Rumex acetosa*) were used as food by the ancient Irish. He mentioned, for example, that it was claimed that St Cóemgen lived for seven years with no food but nettle and sorrel.

Urtica dioica.

Drury (1984) highlighted the importance of nettles as a famine food in Ireland. Dinneen (1927) gave the Irish phrase *ag ithe na neanntóg* ('eating the nettles'), with reference to their common usage during the famine years as food, which he said survives in some places. Other important famine foods in Ireland were charlock (*Sinapis arvensis*), sorrel, water-cress and probably hazelnuts. Nettle soup, called *brotchán neanntóg* in Irish, was probably widely eaten in Ireland since early times (Ó Céirín & Ó Céirín, 1978). It was recorded as an ingredient of haggis (made from chopped bacon, nettles, leeks, cabbage, suet and oatmeal) in Ireland and a gruel made from oatmeal, chopped nettles, butter and milk or stock (Ó Céirín & Ó Céirín, 1978). Deep-fried nettle leaves can be eaten like green crisps (Mabey, 1998). Today, nettles are used to flavour farmhouse cheeses made by several producers around Ireland, such as 'Dingle Peninsula Cheese' made by Maja Binder near Kilcummin, Castlegregory, Co. Kerry.

Milliken and Bridgewater (2004) mention that, in Scotland, nettles were sometimes forced as a spring vegetable and even grown under glass to obtain an early season vegetable, and also 'earthed-up' like rhubarb to blanch the stalks. I have not found records of these practices in Ireland, though.

Phillips (1983) provided a recipe for nettle beer. The nettles are boiled in water (100 stalks with 12 litres of water), then sugar (1.5 kg) and cream of tartar (50 g) are added and boiled up until the sugar has melted. When the mixture has cooled, yeast is added and allowed to ferment. After a few days the scum on top is skimmed off, the liquid is decanted (leaving the sediment behind) and bottled.

An unusual use for nettles was described by Rutty (1772: 132); that is, to turn green gooseberries a reddish colour. He said that this was well known and practiced by gardeners near Dublin. Alternate layers of gooseberries and nettles were added to a tub, 'by which means they acquire a red colour, superior to what they have upon the trees'. I've not tried it to see if it works.

Nettles were used as a veterinary remedy to treat red-water fever in Co. Kerry and pip in Co. Tipperary (Allen & Hatfield, 2004).

O'Regan (1997) mentioned the use of nettles in Ireland to make a spray against greenfly. One part nettles is added to 10 parts of water in a bucket. The mixture is left for some days until it starts to smell, at which time it can be used.

Urtica urens

- Small nettle (Scannell & Synnott, 1987); Annual nettle (Anonymous, 1978)
- *Neantóg bheag* (Scannell & Synnott, 1987); *Neannta* (Moloney, 1919); *Caolfhail* (Moloney, 1919)

URTICACEAE—THE NETTLE FAMILY

Probably Introduced (possible archaeophyte)

Urtica urens.

Urtica urens is annual and grows much smaller in size, to 60 cm, than its perennial relative. It is occasionally found as a weed of waste places and cultivated fields. It may be an introduction in Ireland, but if so, it was brought to

Ireland in ancient times, perhaps with the first farmers.

It is edible and can be used for food similarly to the previous species.

**Utricularia australis (Utricularia neglecta; Utricularia major)*

- Bladderwort (Scannell & Synnott, 1987)
- *Lus borraigh mór* (Scannell & Synnott, 1987); *Lus an bhorraigh* (Dinneen, 1927; Uí Chanainn, 1962; Ó Dónaill, 1977)

LENTIBULARIACEAE—THE BLADDERWORT FAMILY

Native

Utricularia australis is a rare aquatic insectivorous plant of bog-holes and pools, ditches and drains and lake margins found mainly in the west of Ireland.

*No ethnobotanical uses of this plant have been noted in Ireland.

**Utricularia intermedia*

- Intermediate bladderwort (Scannell & Synnott, 1987); Irish bladderwort (Anonymous, 1978)
- *Lus borraigh gaelach* (Scannell & Synnott, 1987)

LENTIBULARIACEAE—THE BLADDERWORT FAMILY

Native

Utricularia intermedia is an aquatic insectivorous plant of bog-holes and pools, ditches and drains and lake margins found mainly in the west of Ireland.

*No ethnobotanical uses of this plant have been noted in Ireland.

**Utricularia minor*

- Lesser bladderwort (Scannell & Synnott, 1987)
- *Lus borraigh beag* (Scannell & Synnott, 1987)

LENTIBULARIACEAE—THE BLADDERWORT FAMILY

Native

Utricularia minor is an aquatic insectivorous plant found mainly in the west of Ireland in bare peat as well as in bog-holes and pools, ditches and drains and lake margins.

*No ethnobotanical uses of this plant have been noted in Ireland.

**Utricularia vulgaris*

- Greater bladderwort (Scannell & Synnott, 1987)
- *Lus an bhorraigh* (Scannell & Synnott, 1987)

LENTIBULARIACEAE—THE BLADDERWORT FAMILY

Native

Utricularia vulgaris is a rare aquatic insectivorous plant found in bog-holes and pools, ditches and drains and lake margins mainly in the west of Ireland.

*No ethnobotanical uses of this plant have been noted in Ireland.

Vaccinium myrtillus

- Bilberry (Scannell & Synnott, 1987); Whortleberry (Colgan, 1904); Fraughan (Frauchan) (Conry, 2012); Frochan (Webb & Scannell, 1983); Frocken (Colgan, 1904); Blaeberry (Hart, 1898); Blueberry (Hart, 1898); Coraseena (Hart, 1898); Bogberry (Uí Chanainn, 1962); Heatherberry (Conry, 2012); Mulberry (Conry, 2012); Whinberry (Conry, 2012); Whorts[304] (Conry, 2012); Black-whorts (Conry, 2012); Huckleberry (Harron, 1986)
- *Fraochán* (Scannell & Synnott, 1987); *Fraochóg* (de Vere Coneys, 1849; Moloney, 1919; Dinneen, 1927; Uí Chanainn, 1962); *Broileóg*[305] (de Vere Coneys, 1849; Dinneen, 1927); *Cran-fraocháin* (de Vere Coneys, 1849; Dinneen, 1927); *Lus na gorm dhearg* (de Vere Coneys, 1849); *Dearc fraoigh* (Dinneen, 1927); *Lusradh na bhfraochán* (Dinneen, 1927); *Lusradh na geire boirnighe* (Dinneen, 1927); *Breileog*[306] (Uí Chanainn, 1962); *Mónadán* (Uí Chanainn, 1962); *Mónóg*[307] (Uí Chanainn, 1962)

ERICACEAE—THE HEATHER FAMILY
Native

Vaccinium myrtillus is a small sub-shrub in Ireland found commonly in all areas on mountains, heathy woods, bogs and peaty habitats. In late summer it produces black berries tinged with a bluish bloom. Bilberries (*fraochóga*), or whortleberries, were one of the most important wild fruits of early Ireland and were widely gathered as part of the diet for the first human populations. The fruits are sweet and tasty when ripe. Remains of bilberries have been found in Viking-era excavations of Dublin dating back more than 1000 years (Geraghty, 1996) when they were probably imported into the city as food.

The fruits have been widely used in Ireland to make jams, tarts, drinks and were also eaten raw. They are excellent as pie fillings and in cheesecakes. Unlike cowberries (*Vaccinium vitis-idaea*), they do not contain the toxic benzoic or oxalic acid but instead are rich in vitamins and anthocyanins (Mears & Hillman, 2007).

To make bilberry jam, simmer 1 lb/450 g of bilberries over low heat until they have softened. Then add 1 lb/450 g of sugar. After the sugar has dissolved, boil rapidly until the setting point has been reached (see *Sambucus nigra*, elder, on how to determine the setting point of jam). Then spoon the jam into sterilised jars and seal. Bilberry jam may also be made from slightly unripe (red) berries, in which case a little water must be added to the mixture when cooking. The berries become almost black when ripe.

Ó Céirín and Ó Céirín (1978) provided a recipe for 'fraughan soup', made in a similar way to elderberry soup (see *Sambucus nigra*), as well as recipes for the use of bilberries, or fraughans, in puddings, buns, jellies, pies, wine and other drinks.

Bilberry seeds are known from archaeological excavations and early written documents (Kelly, 1997). The seeds have been found in excavations of Viking and Anglo-Norman sites in Dublin (Mitchell, 1987). Rutty (1772: 89–90) said that 'they are cried about in our streets [presumably in Dublin] in autumn, and eaten with sugar, and in cream and milk: they are also put in brandy and whiskey, to which they give a bright tincture, and an agreeable taste'. Hart (1898) recorded that at the end of July a 'Bilberry Day' was held

304. Pronounced 'hurts'.

305. Dinneen (1927) defined *broileóg* as 'whortle berry', *lusradh na bhfraochán* as 'whortleberry plant' and *lusradh na geire boirnighe* as 'bear whortleberry plant'.

306. Uí Chanainn (1962) defined *breileog* as 'whortleberry' and *mónadán* as 'bogberry'.

307. Uí Chanainn (1962) defined *mónóg* as 'cranberry, bogberry'. Dinneen (1927: 711) also gives the word *maonóg* for 'a bogberry', which may be this species, but is more likely to be used generally for *Vaccinium oxycoccos*, cranberry.

in Co. Tyrone when people would go to the woods to collect bilberries. Elsewhere in Ireland, it was called 'Fraughan Sunday' (generally the last Sunday in July) when people would have a day out in the hills gathering bilberries and courting. McGlinchey (2007) referred to it as 'Heather-berry Sunday' in Co. Donegal. In Kilkenny the fruits were made into fraughan cakes and specially eaten by young girls at a bonfire dance, and in Longford they were made into a 'wine' with sugar (MacCoitir, 2006). Bilberry picking was traditional in parts of Ireland (particularly in the north) on Garland Sunday or Lammas Sunday, the last Sunday of July (Evans, 1957).

The archives of the Irish Folklore Commission's Schools' Folklore Scheme details whortleberry picking in Co. Tipperary (quoted here from Vickery (1995: 31):

> In the woods on the slopes of the Galtee mountain there are always a good supply of whorts [they are sometimes called whortleberries]. Every summer people go out into the wood picking these whorts; this has been done for about 200 years. They start picking on 29 June and continue to the end of August. Pickers, consisting sometimes of the whole family, leave early in the morning for the wood, and after picking for some hours one of their number will light a fire and prepare some tea, which they all sit down and enjoy about mid-day. Sometimes the fruit is plentiful and pickers have no difficulty in filling their cans, but very often they are not so fortunate. ... Some people pick from three to seven gallons in the day and get from nine pence to one shilling per gallon. The whorts are sent across to England where they are used to make DYE or to mix with the cheaper kinds of jam. Buyers meet the pickers every evening to buy their day's picking.'

Michael J. Conry (2011) has written and published a comprehensive account of the collecting and use of this species in Ireland. In his book he described the large-scale picking of bilberries in Ireland, particularly in the first half of the 20th century, much of which was exported to England and Wales. For example, in 1941 more than 220,000 four-pound baskets of fruit were exported to Britain. In other years the quantities exported fluctuated, in part because of less productive fruiting seasons and with changing fruit prices. Conry also highlights the important contribution that bilberry harvesting made to the local economies in several parts of Ireland, particularly providing an income for poorer families. Conry (2011) also highlighted the important link between traditional bilberry collecting and the Lughnasa festival in Ireland. Lughnasa was one of the quarterly feasts of the old Irish year when people gathered to celebrate the harvesting of the first crops of the new season. As Conry reported (2011: 19), 'Lughnasa, which is the Gaelic for August, got its name from the mythical god Lugh'. This festival included dancing and music as well as social and sporting events. Bilberry collecting was also an important part of the festivities, which were often moved into July to correspond with the best bilberry picking season. Conry documents bilberry collecting in a wide range of counties through Ireland, including Kildare, Wicklow, Carlow, Tipperary, Wexford, Leitrim, Sligo, Cavan, Down, Mayo, Kerry as well as other counties.

Collecting berries still continues today in some parts of Ireland but to a very limited extent. In part this is due to the loss of good woodland and mountainside habitats for the wild bilberries. Many formerly productive mountain bilberry habitats have been afforested with conifer plantations, and suitable old woodland habitats are now few and far between.

The bilberry has had important traditional medicinal uses in various parts of its range in many of the cold northern temperate parts of the world, including in Europe, Asia and North America, where it has been used to treat arthritis, gout, diarrhoea (especially in children), dermatitis, haemorrhoids and other skin complaints, mouth and throat inflammations, poor circulation, and digestive, kidney and urinary problems (van Wyk & Wink, 2004). The Reader's Digest Association (2006) noted also its use for vein, lymphatic and eyesight disorders. In Ireland, it was used as a diuretic for kidney problems, to treat jaundice, asthma, measles and general pains (Allen & Hatfield, 2004). Williams (1993) recorded its use to treat bladder stones. K'Eogh (1735) recorded the value of its berries to treat fevers, liver complaints and scurvy.

MacCoitir (2006) noted that their fruits have been used as a dye for clothes and as a spray for potatoes, and that they were also collected for commercial sale until recent times. Mahon (1983) listed it as providing purple and blue dyes. This species has also been used to make brooms.

Vaccinium oxycoccos

- Cranberry (Scannell & Synnott, 1987)
- *Mónóg* (Scannell & Synnott, 1987); *Cornán* (Moloney, 1919); *Muileóg* (Moloney, 1919); *Mionóg* (Moloney, 1919); *Crúibín* (Moloney, 1919); *Mónóg sléibhe* (Dinneen, 1927)
- *Kraanberry* (in Yola; this name was also used for gooseberry)

ERICACEAE—THE HEATHER FAMILY
Native

Vaccinium oxycoccos is a creeping, slender wiry sub-shrub of wet lowland bogs that is rare in Ireland and is found only occasionally except in parts of the midlands.

It was widely gathered in former times, particularly in the midland counties where it is frequent in bogland habitats. It would have been more abundant in the past before the midland raised bogs were extensively harvested for fuel. Mackay (1836) said that its berries were used in Dublin and elsewhere for making tarts. Smith (1815: 358) described this species under various common names: 'moss-berries', 'moor-berries' and 'bog-berries', and mentioned that it was used in Co. Cork in pickles and tarts. According to Wade (1804: 30), the berries were used in Sweden to 'boil silver plate to its proper whiteness'.

Vaccinium oxycoccos has been used in traditional medicine to treat urinary problems, including cystitis and kidney disorders, as well as against scurvy (van Wyk & Wink, 2004), although I have not found records for its present or past use in medicine in Ireland.

Juices and wines can also be made from cranberries.

Vaccinium vitis-idaea

- Cowberry (Scannell & Synnott, 1987); Red whortleberry (Harron, 1986)
- *Bódhearc* (Scannell & Synnott, 1987); *Braighleog* (Cameron, 1883); *Bó-dhearc* (Cameron, 1883)

ERICACEAE—THE HEATHER FAMILY
Native

Vaccinium vitis-idaea is a very low-growing shrub found only rarely in Ireland, mainly in the east and north of the country, amongst heather on peaty high mountain slopes.

It produces sour red berries in late summer and autumn. They contain benzoic and oxalic acid and are mildly toxic if eaten in quantity (Mears & Hillman, 2007). Nevertheless, they have been an important food source for ancient people in several northern countries. They can be made into fruit leathers, jellies, jams and their juice added to spirits (in the same way that sloes are added to gin to make sloe gin). There are no records of its use for food in ancient Ireland, probably due to its rarity, but as it is an important component of the heath flora of *Pinus sylvestris* woods in some countries, it may have been more widespread when pine was common in Ireland. At that time, it is likely that it would have been collected and used by early Irish settlers who were hunter–gatherers. This species occurs commonly throughout the circumboreal forests of Europe, Asia and North America, right up to the subarctic, and in Scandinavia they are an important part of indigenous cuisine and are called 'lingonberries'; they are used to make a variety of foods including jams, compote, jellies, juices and liqueurs. I presume that they can also be used to make wine.

Valeriana officinalis.

Valeriana officinalis

- Common valerian (Scannell & Synnott, 1987); Wild valerian (Anonymous, 1978); Bolaira (Hart, 1898); Maid's conceit (Hart, 1898); Malara (Hart, 1898)
- *Caorthann corraigh* (Scannell & Synnott, 1987); *Lus na dtrí mbille* (Moloney, 1919; Dinneen, 1927; O'Regan, 1997); *Caorthann* (Moloney, 1919); *Lus na dtrí mballán* (de Vere Coneys, 1849; Moloney, 1919; Dinneen, 1927); *Carthan curaigh* (de Vere Coneys, 1849)

VALERIANACEAE—THE VALERIAN FAMILY
Native

Valeriana officinalis is a widespread and quite common herb of woodlands, roadsides and wet marshy ground by rivers and streams.

The roots have been widely used in Ireland as a sedative and nerve relaxant and to treat a variety of ailments, including wounds, sore eyes, liver ailments and tuberculosis (Mackay, 1836; Allen & Hatfield, 2004). According to Moloney (1919), the root was used in the treatment of 'nervy and hypersensitive folk'. I have read that this plant is a powerful sedative and can be habit-forming, so it must be used with very great caution.

In the Aran Islands the leaves of wild valerian soaked in whiskey were a 'well known relief for aches and pains' (Roden, 1994). In Donegal, it was used 'in old times as tea before tea was known'.

Some people have believed that there is a connection between this plant and the modern drug Valium used to treat tension, anxiety, insomnia and other conditions. However, they are not related despite the similarity of their names. Valium is a synthetic drug, also known as diazepam.

Cameron (1883) pointed out the intriguing fact that Irish names for another species of wild valerian, *Valeriana dioica*, the marsh or dwarf valerian, are *carthan arraigh* and *caoirin leana*, even though this plant is not recorded in Ireland (it is widespread in England, southern Scotland and eastern Wales). Even though it occurs in Scotland, there is no name for this species in Scots Gaelic. Is it possible that it was formerly present in Ireland but became extinct, perhaps because of its medicinal value? Of course it is also possible that this Irish name has been misapplied. The Irish name, *caoirin leana*, translates as 'that which gleams in the marsh', so it may have been a conspicuous plant.

Valerianella locusta

- Common cornsalad (Scannell & Synnott, 1987); Lamb's lettuce (Brunker, 1950; Webb, 1977)
- *Ceathrú uain* (Scannell & Synnott, 1987); *Ceathramh-uain* (de Vere Coneys, 1849; Dinneen, 1927); *Leitís uain* (Ó Dónaill, 1977)

VALERIANACEAE—THE VALERIAN FAMILY
Native

Valerianella locusta is an annual plant found growing as a weed of cultivated fields, gardens, sandy places and dry banks, especially near the coast.

The leaves are edible and useful in salads (Phillips, 1983; Mears & Hillman, 2007). Rutty (1772: 67) said that 'it may also be eaten boiled as Cale, when young. It is sold in our market', presumably in Dublin.

Verbascum thapsus

- Mullein (Scannell & Synnott, 1987); Poor man's blanket (Hart, 1898); High-taper (Praeger, 1938); Hag's taper (Praeger, 1938); Aaron's rod (Harron, 1986)
- *Coinnle Muire* (Scannell & Synnott, 1987); *Coinneal Mhuire* (Moloney, 1919; Tóibín, 1965; Uí Chonchubhair, 1995); *Lus mó* (Ó Criomhthain, MSS, n.d., ca. 1908; Moloney, 1919; Dinneen, 1927); *Cuineal (or Cuingeal) Mhuire* (Cameron, 1883)

SCROPHULARIACEAE—THE FIGWORT FAMILY
Native

Verbascum thapsus is found particularly in the southern half of Ireland in dry gravel and sandy habitats, seashores and waste places.

This plant was used in traditional medicine in Ireland as a cure for constipation and diarrhoea, and a poultice made from the leaves was used to treat weeping sores and boils (Egan, 1887; Moloney, 1919). To treat boils a leaf was roasted between dock leaves and moistened with spittle and then applied to the boil—the spittle must come from an Irishman, at least in Co. Dublin, but in Ulster anyone's spittle will do! (Egan, 1887). In Rathmullan, Co. Donegal, the woolly leaves were made into a warm drink and used to treat an old cough (Hart, 1898). It was also used in the Aran Islands to treat a bad cough; the leaves were boiled and strained, and the juice then taken hot or cold, sometimes with added sugar (Dirrane, 1997). Wood-Martin (1902) and Allen and Hatfield (2004) noted that it was a favourite remedy for pulmonary tuberculosis throughout Ireland. The leaves would be boiled in milk and then strained. The thick mucilaginous liquid would be drunk warm twice daily. It was also used to treat coughs, colds, sore throats, catarrh, bronchitis and asthma. The same authors noted that it had other minor medicinal uses in Ireland too: as a decoction for diarrhoea, for cramp and liver and kidney problems, to treat boils, as a poultice on 'running sores', and for bee stings. In Kerry water in which it was boiled was rubbed on the body to ease pains (probably rheumatism). Mackay (1836: 190) noted that 'the leaves boiled in milk are much used by the peasantry in haemorrhoidal complaints'. Howes (1974) mentioned that its leaves were used in asthma cigarettes, although probably not in Ireland.

Mullein was used as a veterinary remedy to treat sores on animals in Co. Dublin (Allen & Hatfield, 2004) and has been used to treat pulmonary diseases of cattle in Scotland (Cameron, 1883). Drury (1985) reported that in Kent in England, country people gave a decoction of mullein leaves to cattle to cure various lung diseases.

Dried leaves can be used to make a refreshing tea.

Verbena officinalis (vervain). Watercolour by Lydia Shackleton. Courtesy of the National Botanic Gardens, Glasnevin.

Verbena officinalis

- Vervain (Scannell & Synnott, 1987)
- *Beirbhéine* (Scannell & Synnott, 1987); *Crúba leómhain* (Moloney, 1919); *Buidheach* (Moloney, 1919); *Trombhod* (Cameron, 1883)

VERBENACEAE—THE VERVAIN FAMILY

Introduced

Verbena officinalis has been recorded throughout much of Ireland but is now rare. It occurs on roadsides, in rough grassland and waste places, on calcareous soils.

Allen and Hatfield (2004) noted that vervain has been highly regarded for exceptional magico-religious potency for centuries, and in parts of Britain it has been used for medicinal purposes, for example, to treat fevers, as a tonic, for wounds, as an eye lotion and as a vermifuge. In Ireland, Moloney (1919) said that it was worn round the neck for scrofula. Allen and Hatfield (2004) also noted a record of the use of this plant to treat fevers in Co. Cavan. Elsewhere in the world vervain has been used for a wide variety of ailments and conditions, including headaches, nervous exhaustion, depression and gall bladder problems.

Cameron (1883) mentioned that vervain is said to have been used in the religious ceremonies of the Druids.

Veronica species

- Speedwell (Scannell & Synnott, 1987)
- *Lus na banaltra(n)* (Moloney, 1919; Dinneen, 1927); *Seamar chré* (de Vere Coneys, 1849; Moloney, 1919; Ó Dónaill, 1977); *Lus cré* (Moloney, 1919)

VERONICACEAE (Scrophulariaceae)—THE SPEEDWELL FAMILY

Native

Including:

Veronica anagallis-aquatica

- Blue water-speedwell (Scannell & Synnott, 1987)

- *Biolar grá* (Scannell & Synnott, 1987); *Fualachtair* (de Vere Coneys, 1849; Moloney, 1919; Dinneen, 1927)

Veronica arvensis

- Wall speedwell (Scannell & Synnott, 1987)
- *Lus cré balla* (Scannell & Synnott, 1987); *Luibh an fhiabhrais* (Ua Maoileoin, 2001)

Veronica catenata

- Pink water-speedwell (Scannell & Synnott, 1987)
- *Biolar grá dearg* (Scannell & Synnott, 1987)

Veronica hederifolia

- Ivy-leaved speedwell (Scannell & Synnott, 1987)
- *Lus cré eidhneach* (Scannell & Synnott, 1987)

Possibly Introduced

Veronica montana

- Wood speedwell (Scannell & Synnott, 1987)
- *Lus cré coille* (Scannell & Synnott, 1987)

Veronica officinalis

- Heath speedwell (Scannell & Synnott, 1987); Common speedwell (Brunker, 1950; Anonymous, 1978); Clairtha (Hart, 1898)
- *Lus cré* (Scannell & Synnott, 1987); *Seamar chré* (Dinneen, 1927)

Veronica scutellata

- Marsh speedwell (Scannell & Synnott, 1987)
- *Lus cré corraigh* (Scannell & Synnott, 1987)

Veronica serpyllifolia

- Thyme-leaved speedwell (Scannell & Synnott, 1987)
- *Lus an treacha* (Scannell & Synnott, 1987); *Luibh a treatha*[308] (Corrigan, 1984); *Lus na treatha* (Dinneen, 1927)

About 15 species of *Veronica* occur in Ireland. All but four or five of them are native and most are widespread throughout the country. *Veronica anagallis-aquatica* is a common plant of streams, canals and river-banks. *Veronica catenata* is very similar and occurs in wet habitats. *Veronica arvensis* is an annual plant of dry habitats and common on walls, arable land, wasteground and sand-hills. *Veronica montana* is similar but is found mainly in shady habitats, such as in woodlands and hedgerows. *Veronica hederifolia* is mainly found in the east in waste places and as a weed of cultivation and is rarely distributed in the west. *Veronica officinalis* is a common species of heaths and heathy grasslands, as well as walls and some open woodlands. *Veronica scutellata* is common in marshes, ditches and lake margins. *Veronica serpyllifolia* is a common species in most areas on roadsides, grasslands, wasteground and cultivated land and as a weed of lawns. There are several introduced species of *Veronica*, all of them considered weeds: *V. persica*[309], *V. filiformis* and *V. peregrina*.

Various *Veronica* species have been used by nursing mothers for sore breasts. *Veronica* species were also used to treat coughs, colds and chest problems (Hart, 1898; Moloney, 1919). *Veronica serpyllifolia* was used as a treatment for whooping cough (Moloney, 1919). It (and probably *V. officinalis* too) was formerly used as the classic Irish herbal remedy for whooping cough (Allen & Hatfield, 2004). Mackay (1836) noted that *V. officinalis* has astringent leaves, hence its use medicinally for tea.

308. Corrigan defined *luibh a treatha* as 'herb of the whooping-cough'.
309. Uí Chonchubhair (1995) gave the Irish name for *Veronica persica* as *lus cré garraí*.

MacMahon (1994) noted 'speedwell was sewn on [a] garment to protect the wearer' in Ballyheigue, Co. Kerry.

Veronica beccabunga and *Nasturtium* species.

Veronica beccabunga

- Brooklime (Scannell & Synnott, 1987); Jellymanders (Hart, 1898); Well ink (Hart, 1898; Praeger, 1938)
- *Lochall* (Scannell & Synnott, 1987); *Lochal mothair* (de Vere Coneys, 1849; Moloney, 1919; Dinneen, 1927); *Biolar Mhuire* (de Vere Coneys, 1849; Moloney, 1919; Dinneen, 1927; Ó Dónaill, 1977); *Biolar mhuin* (Cameron, 1883); *Falcaire an uisce* (Dinneen, 1927); *Fochlacht* (Dinneen, 1927; Ó Dónaill, 1977); *Fualachtair*[310] (Ó Dónaill, 1977)

VERONICACEAE (Scrophulariaceae)—THE SPEEDWELL FAMILY
Native

Veronica beccabunga is a common species of ditches, drains, marshes and slow-moving streams throughout Ireland.

Brooklime was probably an important edible wild food plant in ancient Ireland. It can be used in the same way as water-cress (*Nasturtium* species) but has a more bitter taste. It is eaten in salads or cooked—boiled or steamed. Mears and Hillman (2007) warn that care should be taken in eating this plant raw or when cooking it lightly as it often grows in nitrogen-rich habitats where liver fluke is present.

It was used medicinally as a 'Diet Drink in the Spring against the Scurvy; it cleanseth the Kidneys of Gravel, and slimy Humours' (Threlkeld, 1726: unnumbered [9]). Allen and Hatfield (2004) noted its use as a diuretic to treat jaundice, urinary and kidney ailments in Ireland, to ease coughs and colds, as an expectorant, and to cure scurvy. In Dunquin, Co. Kerry, it was 'knotted and used as a poultice to ease labour pains' (Uí Chonchubhair, 1995: 196). K'Eogh (1735) recommended it to dissolve hard swellings in the liver and spleen, and it was boiled with water-cress to cure a variety of kidney complaints (Williams, 1993).

A brown dye was obtained from this plant for dyeing textiles (Mahon, 1983).

Veronica chamaedrys

- Germander speedwell (Scannell & Synnott, 1987); Billy bright-eye (Hart, 1898); Germander (Hart, 1898); Jellymander (Hart, 1898); Jerrymander (Hart, 1898)
- *Anuallach* (Scannell & Synnott, 1987); *Lus cré* (Uí Chonchubhair, 1995)

VERONICACEAE (Scrophulariaceae)—THE SPEEDWELL FAMILY
Native

Veronica chamaedrys is a very common species throughout Ireland of hedgerows, roadsides and meadows.

310. Ó Dónaill (1977) defined *fualachtair* as a '(kind of) brooklime'.

According to Allen and Hatfield (2004), it was used in Ireland to treat jaundice; the leaves and stems were boiled, and the resulting liquid was drunk, sometimes with milk and sugar.

The dried leaves of *Veronica chamaedrys* provide an excellent substitute for tea (Wade, 1804).

Ó Dónaill (1977) gave the Irish word *dairín* (or *dairghin*) for germander, but this probably refers to a species of *Teucrium*, rather than *Veronica chamaedrys*.

Viburnum opulus

- Guelder rose (Scannell & Synnott, 1987); Water elder; Dogberry (Hart, 1898)
- *Caor chon* (Scannell & Synnott, 1987); *An rós holóndach* (Moloney, 1919); *Ceiriocán* (de Vere Coneys, 1849; Dinneen, 1927)

CAPRIFOLIACEAE—THE HONEYSUCKLE FAMILY

Native

Viburnum opulus.

Viburnum opulus berries.

Viburnum opulus is a widespread shrub throughout Ireland but is only occasionally seen. It occurs in hedgerows, scrub, woodland edges and on riverbanks and lake-shores.

Moloney (1919) said that Guelder rose was originally introduced from Gelderland in Holland, hence its Irish name. However, there is no evidence that this is the case, and the species is regarded as native in Ireland.

Guelder rose is used in northern Europe to flavour a paste of honey and flour; in Siberia the fruits are distilled to yield a spirit, and in Canada they are used as a substitute for cranberries and to make jelly (Phillips, 1983). Its fruits should only be used when cooked—they are not safe to eat raw. I have made jelly from Guelder rose, which is very good to eat and has an interesting flavour and is a useful to accompany roasted meat. However, I have read suggestions that it is regarded as a toxic plant in some countries, so I use it only sparingly. Woodward (1985: 49) included it in her book of poisonous plants and gives the symptoms of poisoning as 'gastrointestinal irritation, vomiting, diarrhoea, collapse; can be fatal'.

This is occasionally grown in Irish gardens as an ornamental plant, particularly as the form where all the flowers are sterile and are borne in large, white globular heads, called the 'Snowball Tree'.

**Vicia cracca*

- Tufted vetch (Scannell & Synnott, 1987); Fitches (Colgan, 1904); Wild fitches (Colgan, 1904)
- *Peasair na luch* (Scannell & Synnott, 1987); *Peasair luch na coillte* (Moloney, 1919); *Piseánach tiugh* (Moloney, 1919; Dinneen, 1927); *Fiseánach thiubh* (Uí Chonchubhair, 1995); *Pesseir badanach* (Wade, 1808b)

FABACEAE (Leguminosae)—THE PEA AND CLOVER FAMILY

Native

Vicia cracca is a very common plant throughout Ireland of hedges, roadsides and wasteground.

*No ethnobotanical uses of this plant have been noted in Ireland.

**Vicia hirsuta*

- Hairy tare (Scannell & Synnott, 1987); Hairy vetch
- *Peasair arbhair* (Scannell & Synnott, 1987)

FABACEAE (Leguminosae)—THE PEA AND CLOVER FAMILY
Native

Vicia hirsuta is a trailing annual found occasionally in waste places, dry banks and the margins of fields, particularly in the eastern half of Ireland.

*No ethnobotanical uses of this plant have been noted in Ireland.

**Vicia lathyroides*

- Spring vetch (Scannell & Synnott, 1987)
- *Peasair earraigh* (Scannell & Synnott, 1987)

FABACEAE (Leguminosae)—THE PEA AND CLOVER FAMILY
Native

Vicia lathyroides is a rare annual plant in Ireland of sandy places near the east and north coasts.

*No ethnobotanical uses of this plant have been noted in Ireland.

**Vicia orobus*

- Wood bitter-vetch (Scannell & Synnott, 1987)
- *Peasair shearbh* (Scannell & Synnott, 1987)

FABACEAE (Leguminosae)—THE PEA AND CLOVER FAMILY
Native

Vicia orobus is a very rare plant of scrub and rocky habitats in the west, midlands and north-east of Ireland.

*No ethnobotanical uses of this plant have been noted in Ireland.

Vicia sativa

- Common vetch (Scannell & Synnott, 1987)
- *Peasair chapaill* (Scannell & Synnott, 1987); *Peasair fiadhain* (Cameron, 1883); *Pis chapuill* (de Vere Coneys, 1849); *Pesseir capull* (Wade, 1808b); *Pis chapaill*[311] (Dinneen, 1927)

FABACEAE (Leguminosae)—THE PEA AND CLOVER FAMILY
Native and Introduced

Vicia sativa occurs in Ireland as two distinct subspecies: *V. sativa* subsp. *sativa*, which is introduced to Ireland and occasionally survives in the wild as a remnant where it has been cultivated for fodder, and *V. sativa* subsp. *nigra*, which is a native plant that

311. Dinneen (1927) defined *pis chapaill* as 'horse pea, vetch tare (*Vicia sativa*)'.

is rather rarer and occurs only occasionally in dry, gravel and sandy habitats, particularly in the east and south of Ireland and is much rarer in other parts of the island.

Mackay (1836) noted that *Vicia sativa* was a valuable fodder plant for feeding horses and cattle early in the season and could be harvested several times each year.

Mears and Hillman (2007) suggested that spring shoots can be eaten as a salad or steamed as a vegetable and that young green seedpods are edible. Vickery (1995) included a record from Clonmel, Co. Tipperary, of children eating the small, black seeds and green pods as a roadside snack on the way to school in the 1950s.

In Irish a general name for *Vicia* is *fiseánach* (Ua Maoileoin, 2001), or *fiailteacha* (Uí Chanainn, 1962).

Vicia sepium

- Bush vetch (Scannell & Synnott, 1987); Fitches (Colgan, 1904); Wild fitches (Colgan, 1904)
- *Peasair fhiáin* (Scannell & Synnott, 1987); *Peasair fhiadhain* (Moloney, 1919); *Pesseir dubh* (Wade, 1808b); *Piseánach coit-cheannta*[312] (Dinneen, 1927); *Piseánach na mbreas* (Dinneen, 1927); *Pis phréacháin*[313] (Dinneen, 1927; Ó Dónaill, 1977)

FABACEAE (Leguminosae)—THE PEA AND CLOVER FAMILY
Native

Vicia sepium is an extremely common plant throughout Ireland of roadsides, hedges, rough meadows, scrub and waste places.

Like *Vicia sativa*, Mears and Hillman (2007) suggested that young shoots can be eaten in salads or as a vegetable.

**Vicia sylvatica*

- Wood vetch (Scannell & Synnott, 1987)
- *Peasair choille* (Scannell & Synnott, 1987); *Pis éanáin* (Dinneen, 1927)

FABACEAE (Leguminosae)—THE PEA AND CLOVER FAMILY
Native

Vicia sylvatica is a rare plant found in woodlands, scrub and rocky places mainly in the eastern half of Ireland.

*No ethnobotanical uses of this plant have been noted in Ireland.

Viola arvensis

- Field pansy; Heartsease (Scannell & Synnott, 1987)
- *Lus croí* (Scannell & Synnott, 1987); *Salchuach* (Moloney, 1919; Dinneen, 1927); *Cogal gorm* (Moloney, 1919)

VIOLACEAE—THE VIOLET FAMILY
Probably Introduced (possible archaeophyte)

312. Dinneen (1927) defined *piseánach coit-cheannta* as 'common or bush vetch'.
313. Dinneen (1927) defined *pis phréacháin* as 'common vetch' and Ó Dónaill (1977) as 'wild vetch'.

Viola arvensis is a widely distributed annual herb found on cultivated ground throughout Ireland.

William Allingham (1824–1889), a Donegal man born in Ballyshannon, wrote in his diary: 'Walked over Carntullagh to Bruckless [in Co. Donegal]… some women gathering wild heartease 'for tea', they use many herbs thus' (Allingham, 1985: 39). Allingham is best known for his poem 'The Fairies'.

Up the airy mountain,
 Down the rushy glen,
We daren't go a-hunting,
 For fear of little men;
 Wee folk, good folk,
 Trooping all together;
 Green jacket, red cap,
 And white owl's feather!

(Allingham, 'The Fairies', stanza 1)

Hart (1898) suggested that the name 'Hearts' Ease' was used in Co. Donegal for *Prunella vulgaris*, selfheal, where an infusion of that plant was valued for the heart. Hart suggested that *Viola* was never an esteemed herb in Donegal.

Moloney (1919) mentioned that *Viola* species were little used medicinally in Ireland at that time.

**Viola canina*

- Heath dog-violet (Scannell & Synnott, 1987); Dog violet (Brunker, 1950)
- *Sailchuach mhóna* (Scannell & Synnott, 1987); *Fanaigse* (Moloney, 1919; Dinneen, 1927); *Bíodh a leithéid*[314] (Moloney, 1919; Dinneen, 1927); *Sailchuach (-aiche) (-acha)* (de Vere Coneys, 1849)

VIOLACEAE—THE VIOLET FAMILY
Native

Viola canina is a widespread plant in Ireland of stony heaths, lake-shores and sand-hills.

*No ethnobotanical uses of this plant have been noted in Ireland.

**Viola hirta*

- Hairy violet (Scannell & Synnott, 1987)
- *Sailchuach ghiobach* (Scannell & Synnott, 1987); *Salchuach* (Moloney, 1919); *Cogal gorm* (Moloney, 1919)

VIOLACEAE—THE VIOLET FAMILY
Native

314. Dinneen (1927) listed this name for 'violet'.

Viola hirta is a rare plant in Ireland of limestone grasslands and pavements.

*No ethnobotanical uses of this plant have been noted in Ireland.

**Viola lactea*

- Pale dog-violet (Scannell & Synnott, 1987)
- *Sailchuach liath* (Scannell & Synnott, 1987)

VIOLACEAE—THE VIOLET FAMILY
Native

Viola lactea is a very rare violet in Ireland of heathy places that is found mainly in the south.

*No ethnobotanical uses of this plant have been noted in Ireland.

Viola lutea

- Mountain pansy (Scannell & Synnott, 1987); Yellow mountain pansy (Brunker, 1950)
- *Sailchuach shléibhe* (Scannell & Synnott, 1987); *Goirmín sléidhe* (Moloney, 1919)

VIOLACEAE—THE VIOLET FAMILY
Native

Viola lutea is a beautiful perennial herb that is similar to the dune-pansy but with larger flowers. It is found in upland grasslands, cliff-tops and some sand dunes in southern parts of Ireland but only quite rarely.

According to Moloney (1919), a laxative syrup was made from *Viola* species and taken internally (Moloney, 1919).

Mahon (1983) noted that a blue dye was obtained from this plant.

Viola odorata

- Sweet violet (Scannell & Synnott, 1987)
- *Sailchuach chumhra* (Scannell & Synnott, 1987); *Salchuach* (Moloney, 1919); *Cogal gorm* (Moloney, 1919; O'Regan, 1997)

VIOLACEAE—THE VIOLET FAMILY
Native

Viola odorata is found mainly in woodlands and hedges particularly in the west, where it is probably native, and as a garden weed throughout the country. It is undoubtedly an escape from gardens in many places that it occurs throughout Ireland.

Viola odorata is probably the most economically important species of violet in the world, of which there are about 400 known. It is grown commercially in the south of France for the production of essential oils used in the manufacture of perfume, flavouring, cosmetics and toiletries and a sweet liqueur called *parfait amour*. It has had only limited use in Ireland.

According to Allen and Hatfield (2004), it was formerly used in Ireland to treat tumours and headaches.

Crystallized flowers of this species are traditionally used as edible decorations.

**Viola palustris*

- Marsh violet (Scannell & Synnott, 1987)
- *Sailchuach chorraigh* (Scannell & Synnott, 1987); *Salchuach* (Moloney, 1919); *Cogal gorm* (Moloney, 1919); *Badán meascáin*[315] (Dinneen, 1927)

VIOLACEAE—THE VIOLET FAMILY
Native

Viola palustris is found in wet bogs, marshes and other swampy acid habitats, particularly in the western half of Ireland and in the south-east.

Drury (1985) mentioned the use of this plant in Britain to make an ointment to cure wounds and bites in cattle, but there are no records of such usage in Ireland.

*No ethnobotanical uses of this plant have been noted in Ireland.

**Viola persicifolia*

- Fen violet (Scannell & Synnott, 1987); Turlough violet (Webb & Scannell, 1983)
- *Sailchuach uisce* (Scannell & Synnott, 1987)

VIOLACEAE—THE VIOLET FAMILY
Native

Viola persicifolia is a pretty light-blue flowered violet that is found only rarely in Ireland, except in the flood zones of turloughs (disappearing lakes) in limestone areas such as those in Cos. Clare and Galway.

*No ethnobotanical uses of this plant have been noted in Ireland.

**Viola reichenbachiana*

- Early dog-violet (Scannell & Synnott, 1987); Early wood violet (Brunker, 1950); Pale wood violet (Harron, 1986)
- *Sailchuach luath* (Scannell & Synnott, 1987)

VIOLACEAE—THE VIOLET FAMILY
Native

Viola reichenbachiana is found throughout Ireland in woodland habitats. It is not always easy to distinguish from *V. riviniana*, but it is generally an indicator of old woods, on limestone soils.

*No ethnobotanical uses of this plant have been noted in Ireland.

---

315. Dinneen (1927) listed the name *badán meascáin* for a plant called 'bog violet', which probably refers to this species.

*Viola riviniana

- Common dog-violet (Scannell & Synnott, 1987); Wood violet (Brunker, 1950)
- *Fanaigse* (Scannell & Synnott, 1987); *Sailchuach* (Uí Chonchubhair, 1995); *Sailchuach chon*[316] (Ó Dónaill, 1977)

VIOLACEAE—THE VIOLET FAMILY
Native

Viola riviniana is an extremely common violet throughout Ireland of woods, hedges and other shady habitats as well as short grassland, road-sides, banks and sand dunes.

*No ethnobotanical uses of this plant have been noted in Ireland.

*Viola tricolor subsp. curtisii

- Sandhill pansy (Webb & Scannell, 1983); Sand pansy (Brunker, 1950); Sea pansy (Praeger, 1938)
- *Goirmín duimhche* (Scannell & Synnott, 1987); *Goirmín searraigh* (Uí Chonchubhair, 1995)

VIOLACEAE—THE VIOLET FAMILY
Native

Viola tricolor subsp. *curtisii* is a perennial herb that occurs commonly throughout Ireland on sand dunes although it is rare on lake-shores. Its flowers are either violet or yellow, or can be a mixture of both.

Ody (1997b) mentioned the use of *Viola tricolor*, heartsease or wild pansy, to treat skin conditions and inflammations, including nappy rash, as well as internally for coughs and chest problems. However, there are no records of its use in traditional medicine in Ireland that I have been able to locate.

*No ethnobotanical uses of this plant have been noted in Ireland.

Viola tricolor subsp. *curtisii.*

*Viola tricolor subsp. tricolor

- Wild pansy (Scannell & Synnott, 1987)
- *Goirmín searraigh* (Scannell & Synnott, 1987); *Gormán searraigh* (Moloney, 1919); *Goirmín* (Moloney, 1919); *Salchuach* (?) (Moloney, 1919)

VIOLACEAE—THE VIOLET FAMILY
Native

Viola tricolor subsp. *tricolor* is a delicate annual herb with violet flowers that is found mainly in the north on cultivated ground.

*No ethnobotanical uses of this plant have been noted in Ireland.

Viscum album

- Mistletoe (Reynolds, 2002)
- *Druadhlus* (Moloney, 1919; Dinneen, 1927); *Uile-íce* (Moloney, 1919; Dinneen, 1927); *Uile iceach* (Cameron, 1883); *Guis* (Cameron, 1883); *Súgh-dharach* (*-ruigh*) (de Vere Coneys,

316. Ó Dónaill (1977) defined *sailchuach chon* as 'dog violet'.

Viscum album growing on a poplar tree at the National Botanic Gardens, Glasnevin.

1849); *Subh darach*[317] (Dinneen, 1927); *Drualus* (Ó Dónaill, 1977)
SANTALACEAE (Viscaceae)—THE BASTARD-TOADFLAX FAMILY
Introduced

Viscum album is nowhere a widespread plant in the wild in Ireland. Allen and Hatfield (2004) noted both its general use as a medicinal plant to soothe nerves, and its specific use in palliating epilepsy and hysteria. K'Eogh (1735) too documented such uses of the plant in Ireland. *Viscum* tea has traditionally been used to treat high blood pressure (van Wyk & Wink, 2004), although probably not in Ireland. Cameron (1883) reported that Pliny (AD 23–79) wrote that it was given as a potion by the druids as an antidote to all poisons.

In Britain, it was used as a source of birdlime, obtained from the sticky berries (Edlin, 1973). The birdlime was spread on twigs on which birds would land and be stuck to it until caught. Rutty (1772) said that mistletoe birdlime was better than that made from holly bark.

Wood (1821: 163) wrote that Pliny described the druidic priests, covered with a white vestment, climbing up the oak tree to harvest the mistletoe with a golden hook. They used the mistletoe to treat various diseases. Wood said that according to Pliny, they also 'administered in disorders of cows and swine, the *Lycopodium selago*, or the fir-club-moss; which they plucked with the right, and the *Samolus valerandi*, brook-weed, which they gathered with the left hand'.

It is a common plant used today for Christmas decorations in Ireland, generally from imported stock; it is now traditional for people in Ireland to 'kiss under the mistletoe'. Many trees in the National Botanic Gardens in Glasnevin have mistletoe naturalised on them, probably with a greater density of mistletoe than anywhere else in Ireland. In recent years it has become a tradition to prune the mistletoe just before Christmas and provide it to visitors at a Santa Claus grotto in the Gardens.

Mistletoe is a poisonous plant and the berries should not be eaten.

**Vulpia bromoides (Festuca bromoides)*

- Squirreltail fescue (Scannell & Synnott, 1987)
- *Feisciú aimrid* (Scannell & Synnott, 1987); *Fescufhér aimrid* (White, 1803)

POACEAE (Gramineae)—THE GRASS FAMILY
Native

Vulpia bromoides is an occasional grass of walls, rocky outcrops, gravel and sand dunes.

*No ethnobotanical uses of this plant have been noted in Ireland.

**Vulpia fasciculata*

- Dune fescue (Scannell & Synnott, 1987)
- *Feisciú duimhche* (Scannell & Synnott, 1987)

POACEAE (Gramineae)—THE GRASS FAMILY
Native

317. Dinneen (1927) defined *subh darach* as 'a mistletoe-berry, or mistletoe'.

Vulpia fasciculata is a rare sand dune grass found along the east coast of Ireland.

*No ethnobotanical uses of this plant have been noted in Ireland.

**Vulpia myuros*

- Rat's-tail fescue (Scannell & Synnott, 1987)
- *Feisciú balla* (Scannell & Synnott, 1987)

POACEAE (Gramineae)—THE GRASS FAMILY
Native or Possibly Introduced (possible archaeophyte)

Vulpia myuros is found mainly in the southern half of Ireland on walls and in dry gravel habitats.

*No ethnobotanical uses of this plant have been noted in Ireland.

**Wahlenbergia hederacea*

- Ivy-leaved bellflower (Scannell & Synnott, 1987)
- *Milleanna móna* (Scannell & Synnott, 1987)

CAMPANULACEAE—THE BELLFLOWER FAMILY
Native

Wahlenbergia hederacea is a beautiful little perennial plant with pale blue flowers that is rare in Ireland and found mainly in the south-west and south-east growing on river-banks and moist grassy places.

*No ethnobotanical uses of this plant have been noted in Ireland.

**Zannichellia palustris*

- Horned pondweed (Scannell & Synnott, 1987)
- *Lus na linne* (Scannell & Synnott, 1987)

POTAMOGETONACEAE (Zannichelliaceae)—THE PONDWEED FAMILY
Native

Zannichellia palustris is a submerged aquatic of both fresh and brackish water that is found particularly in coastal areas and in ponds, lakes, wet ditches and bog pools.

*No ethnobotanical uses of this plant have been noted in Ireland.

Zostera species

- Eelgrass (Scannell & Synnott, 1987); Grasswrack (Praeger, 1938; Brunker, 1950); Sweet grass (Hart, 1898); Slishgrass (Brunker, 1950); Sleechgrass (Praeger, 1938)
- *Miléarach* (Scannell & Synnott, 1987); *Meilsceánach* (De Bhaldraithe, 1959; Ó Dónaill, 1977)

ZOSTERACEAE—THE EELGRASS FAMILY

Including:

Zostera angustifolia
- Narrow-leaved eelgrass (Scannell & Synnott, 1987)
- *Miléarach chaol* (Scannell & Synnott, 1987)

Zostera marina
- Common eelgrass (Scannell & Synnott, 1987); Millgrass (Colgan, 1911); Sweet grass (Hart, 1898)
- *Miléarach* (Scannell & Synnott, 1987); *Mil-fhéurach* (Moloney, 1919); *Miliurach* (Colgan, 1911)

Zostera noltei
- Dwarf eelgrass (Scannell & Synnott, 1987)
- *Miléarach bheag* (Scannell & Synnott, 1987)

Native

Zostera species are perennial marine plants that grow from underground creeping stems. They are found in muddy and sandy shores, generally below the low water mark. *Zostera angustifolia* and *Z. noltei* are both relatively rare in Ireland, found mainly around tidal estuaries and on mud. *Zostera marina* is more widespread.

Eelgrass was collected on sea shores after storms and used as a decay-resistant underlay for roof thatching in the north of Ireland (Letts, 1997). Clyne (1989) said that it was extensively used for thatching in Scotland as it is more long-lasting than straw. I have not found records of eelgrass being used in a similar way in other parts of Ireland. For example, Bernie Goggin, a local naturalist in Dingle, Co. Kerry, told me he had never heard of *Zostera* being used in this way in Co. Kerry, despite the presence of extensive *Zostera* beds offshore in parts of the Dingle Peninsula.

In Scotland it was used for bedding cattle and to stuff mattresses as it was believed to prevent flea infestations (Vickery, 1995, 2010; Milliken & Bridgewater, 2004). In other parts of the world (but not in Ireland, as far as I am aware) it has been used as a packing material and for sound insulation; called 'alva marina' (Howes, 1974).

Amongst the islands of Scotland, the roots of eelgrass were gathered and chewed for their sweet-tasting juice (Milliken & Bridgewater, 2004), but I have not found any records of such usage in Ireland.

CONIFERS

Abies species

- Fir
- *Giúis* (Anonymous, 1978)

PINACEAE—THE PINE FAMILY
Introduced

A number of *Abies* species are grown for forestry purposes in Ireland, used both for timber and as Christmas trees. These include the noble fir (*A. procera*) that is native to western North America (from north-west California to Oregon and Washington State) and the silver fir[318] (*A. alba*), a tree native to Europe (Mitchell, 1974). *Abies alba* self-seeds in or near plantations and in demesnes, but is rarely naturalised elsewhere in Ireland (Reynolds, 2002). Korean fir (*A. koreana*) is native to the higher mountains of South Korea and is a very popular ornamental tree for gardens (Mitchell, 1974). It is grown for its foliage but also for the abundant cone production even on young trees that are only 1–2 m tall. Nordmann fir (*A. nordmanniana*) is also grown in Ireland, particularly for Christmas tree production. It is native to the mountains south and east of the Black Sea, in Turkey, Georgia and the Russian Caucasus, as well as in northern Armenia (Mitchell, 1974).

Timber from the silver fir, *Abies alba*, is very similar to that of Norway spruce *Picea abies*, pale cream in colour and with a fine texture. This wood is used for a variety of purposes in Ireland including joinery, construction and pole production.

The custom of erecting and decorating a Christmas tree is now universal in Ireland but can be traced back to 16th-century northern Germany. The tradition spread rapidly throughout Germany and then abroad. In the early 19th century, the custom became popular among the nobility and spread to royal courts as far as Russia. In Britain, the Christmas tree was introduced in the early 1800s, but the custom hadn't yet spread much beyond the royal family. As a child Queen Victoria was familiar with the custom. She was married to her German cousin Prince Albert in early 1840, and the custom became even more widespread in Britain and Ireland.

Cedrus species

- Cedar
- *Céadar* (Anonymous, 1978); *Crann sheudar* (Cameron, 1883); *Cran-cédair* (de Vere Coneys, 1849); *Céadras* (Dinneen, 1927; Ó Dónaill, 1977)

PINACEAE—THE PINE FAMILY
Introduced

318. Anonymous (1978) and Ó Dónaill (1977) gave the Irish *giúis gheal* for 'silver fir' and Dinneen (1927) lists *crann aibiéid*.

Cedars are popular ornamental trees in Ireland, grown widely in larger gardens, parks, demesnes and churchyards, even in towns and cities, despite the large size of the trees. Three cedars are frequently seen in Ireland: *Cedrus libani* (cedar of Lebanon), native to the Mediterranean region, notably in the mountains of Lebanon itself; *C. deodara* (Deodar cedar), native to western Himalaya and *C. atlantica* (Atlantic cedar), which is generally grown as a blue form, *C. atlantica* var. *glauca*, and is native to the Atlas Mountains of Algeria and Morocco in North Africa (Mitchell, 1974).

Cedar wood is a useful timber for various purposes. The term 'cedar' has been used to describe the wood of other species in commerce that have a natural fragrance resembling that of the true cedars. The heartwood of cedars is light brown in colour, strongly scented and somewhat resinous and distinct from the paler sapwood (Anonymous, 1941). Former uses have included garden furniture, gates and fences and other outdoor objects. Good-quality timber was also used in floor joinery, as well as for doors and window frames. However, cedar wood is very rarely available or used in Ireland today.

Chamaecyparis lawsoniana

- Lawson's cypress[319] (Anonymous, 2005)
- *Cufróg Lawson* (Anonymous, 1978)

CUPRESSACEAE—THE JUNIPER FAMILY
Introduced

Chamaecyparis lawsoniana is widely grown in Ireland both as an ornamental tree and as a vigorous and fast-growing hedge. Lawson's cypress is native to western North America. Numerous cultivars are grown with variously coloured foliage and growth forms. *Chamaecyparis lawsoniana* occasionally self-sows in the wild in Ireland, usually close to a parent tree and even onto walls (Reynolds, 2002).

The wood of Lawson's cypress provided a useful timber that has a fragrant spicy odour. It is even textured and rather similar to that of cedar. Its heartwood is pale pinkish brown in colour and the sapwood only marginally lighter in colour (Anonymous, 1941).

Sprigs of Lawson's cypress are sometimes used in Ireland, distributed to parishioners and blessed during Palm Sunday church services as a palm substitute.

Chamaecyparis nootkatensis (Nootka cypress), from north-western North America, is also grown in Ireland, but much less frequently, as is *C. pisifera* (Sawara cypress), from Japan.

Cupressus species

- Cypress
- *Cufróg* (Anonymous, 1978); *Cupraisc* (Anonymous, 1978); *Cuphair* (Cameron, 1883)

CUPRESSACEAE—THE JUNIPER FAMILY
Introduced

319. De Vere Coneys (1849) gave the Irish word *cran-chuphair* for 'cypress', and Dinneen (1927) gave the words *crann copair*, *cufróg*, *cupraisc* and *crann cupnaisce*. Ó Dónaill (1977) gave the Irish name *crann cufróige* for 'cypress-tree' and *sípris* for 'cypress'. These names may perhaps be applied more generally to cypress trees, rather than to this species alone.

A range of exotic species of *Cupressus* are grown in Ireland, mainly as ornamental trees in demesnes, parks and large gardens and also occasionally for forestry purposes. These include *C. macrocarpa* (Monterey cypress), from California, used as a hedging plant in Ireland but now largely replaced by Leyland cypress. The other species that is most commonly grown is *C. sempervirens* (Italian cypress), native to the Mediterranean region and southern Europe. Monterey cypress has been found self-seeded in the wild in several parts of Ireland (Reynolds, 2002) but is nowhere well established as a wild plant.

A hybrid cross between *Cypress* and *Chamaecyparis*, called × *Cupressocyparis leylandii* (Leyland cypress) is abundantly grown in Ireland. This hybrid cypress occasionally develops as large trees but it is much more commonly utilised as a fast-growing and very hardy hedging plant.

Juniperus communis

- Juniper (Scannell & Synnott, 1987); Caorragh aitiyean[320] (Hart, 1898)
- *Aiteal* (Scannell & Synnott, 1987); *Beárnán brighde* (Moloney, 1919); *Lubhar beinne* (Moloney, 1919); *Biora leacra* (Moloney, 1919; Ó Dónaill, 1977); *Iubhar talamh*[321] (Cameron, 1883); *Iubhar beinne* (Cameron, 1883; Dinneen, 1927); *Ubhar chraige* (Cameron, 1883); *Gibire gearr* (McGlinchey, 2007); *Iubhar craige* (Dinneen, 1927); *Iúr binne* (Ó Dónaill, 1977); *Iúr creige* (Ó Dónaill, 1977)

CUPRESSACEAE—THE JUNIPER FAMILY
Native

Juniperus communis is a shrub that is found in the wild in Ireland mainly in the west and north in rocky ground, heaths, limestone pavement and lake-shores. Two different subspecies occur in Ireland, subsp. *communis*, which is either erect or prostrate in growth form with prickly branches, and subsp. *nana* (also known as subsp. *alpina*), which has a prostrate and spreading habit with branches that are not prickly to touch. Juniper is grown as an ornamental plant in parks and gardens although usually an exotic species rather than *J. communis*. De Vere Coneys (1849) gives the Irish name *samhán* (*-áin*) for the non-native *J. sabina*, the savin tree.

Juniperus communis.

People in ancient Ireland probably gathered juniper berries quite extensively. Juniper berries (in Irish, *dearc fiodha*, Dinneen, 1927) are used to flavour gin (The word gin comes from *genever*, a Dutch word meaning juniper). They are highly aromatic and rich in sugars. They are used for culinary purposes, especially associated with cooking game and in marinades. The berries should be picked in autumn when they have turned black. A delicious traditional recipe including juniper berries is potted goose (shredded goose meat mixed with goose fat, seasoning and juniper berries). Corned beef, flavoured with saltpetre and juniper berries, was a traditional dish for Easter Sunday in some parts of Ireland

320. Hart (1898: 369) gave this name for *Juniperus nana* from Fanet, Co. Donegal, mentioning that *caor* is a 'berry' and *aitinn* is 'gorse or whin'.

321. Dinneen (1927) gave this same name for 'rough spleenwort' (*Polystichum aculeatum*).

Juniperus communis.

(Sharkey, 2000). Saltpetre is a common name for potassium nitrate, which was often used as a preservative (and as an anaphrodisiac, to quell the libido). Dried juniper berries are available in Irish supermarkets but are apparently not of Irish origin. A peppery spice can be obtained by grinding up the seeds (Edlin, 1973). In Germany, juniper berries are used to flavour sauerkraut and, in other parts of Europe, the berries are used as an ingredient of vegetable dishes, pies and stuffings (Hilton, 2007). Hilton (2007) also noted that tea can be made from juniper leaves and stems and the flavour enhanced with the addition of a few berries.

Juniper berries were also used a diuretic (Logan, 1972, 1981; Allen & Hatfield, 2004; van Wyk & Wink, 2004). K'Eogh (1735: 67) stated that were used to 'provoke urine, and cure old coughs, flatulence and colic pains. The gum of the tree expels worms from the body and stops excessive menstrual flow'. Juniper branches were used to make poultices for wounds and when burned were used to 'purify' the air or for fumigation (Wyse Jackson, 1994; MacCoitir, 2003). McGlinchey (2007: 83) wrote that 'there's a small bush called the gibire gearr [probably a type of juniper] that grows in dry hilly places. It burns like fir in the fire when it's withered. The she-kind has green berries all over it but the he-kind has no berries at all. They say if the berries and leaves are boiled the water is good for people that are run down or have bad blood. They mend well after it'. Grigson (1958) noted that juniper berries were used as an abortifacient and to assist in childbirth, but does not say whether this usage applied to Ireland. Mabey (1998) suggested that this may 'have echoes in the Victorian belief in the effectiveness of gin for the same purpose', that is, to promote abortions and miscarriages. Vickery (2010) mentioned that it was widely used in Britain as an abortifacient and gives the name 'bastard-killer' as a Somerset name for the species. A large number of plants worldwide have been used as abortifacients, including species of *Euphorbia*, *Arum* and *Rubia* (Howes, 1974). One of the best known in northern Europe was *Aristolochia clematitis* (birthwort) that is found in Britain as a wild plant but has not been recorded in the wild in Ireland (Reynolds, 2002), although it has been grown in some gardens.

O'Rourke (2006) reported that on the Aran Islands juniper branches were used as a substitute for yew branches on Palm Sunday (the yew doesn't occur on the Aran Islands, and juniper is now extinct in Inisheer). Juniper wood has been used in the British Isles for fine carving work and woodturning. It is durable, can be well polished and has a delicate and lasting fragrance (Edlin, 1973). An oil may be distilled from the wood too (Howes, 1974). In the Faroe Islands, juniper wood was used for smoking meat (Milliken & Bridgewater, 2004) but no records of its use for such purposes in Ireland have been found. It seems to have been little used in Ireland for its timber. Few trees are found today that reach a size that would be worth harvesting for its wood in any case. It burns well and can be used for fuel, producing very little smoke. For that reason Warren (2006) said that it was used in Scotland as fuel for illicit stills.

Juniper was also used in Britain as an abortifacient and contraceptive for horses (Drury, 1985).

Juniper has been used to provide a brown dye (Kelly, 1997). Juniper oil has been used in aromatherapy.

Larix species

- Larch
- *Learóg* (Anonymous, 1978); *Crann laireoige* (*learóige*) (Dinneen, 1927; Ó Dónaill, 1977); *Laireog* (Dinneen, 1927)

PINACEAE—THE PINE FAMILY
Introduced

Including:

Larix decidua

- European larch
- *Learóg Eorpach* (Anonymous, 1978)

Larix kaempferi (Larix leptolepis)

- Japanese larch
- *Learóg Sheapánach* (Anonymous, 1978)

Larix ×marschlinsii (Larix ×eurolepis)

- Hybrid larch

Although introduced, larch has a long history of cultivation in Ireland for forestry purposes as a shelter-belt tree. The European larch *Larix decidua* has self-seeded in many locations in Ireland, generally close to existing plantations. The Japanese larch *L. kaempferi* has been widely planted and can also self-seed. Larch timber is moderately hard and heavy, brownish white in colour and striped with broad dark brown resinous bands. It is quite similar to that of Scots pine.

Larix ×eurolepis is a hybrid tree that results from a cross between the European and Japanese larch. Unlike most conifers in Ireland, the larch is deciduous and will lose its leaves in winter. Japanese larch was introduced to Britain in 1861 and was more widely planted as a forestry plantation species in the early 1900s. The hybrid arose as a chance crossing on the estate of the Duke of Atholl at Dunkeld, Tayside, Scotland, ca. 1904 (Mitchell, 1985). This hybrid tree is quite vigorous and grows faster than either of its two parents. It also survives better on poor infertile soils than the two parent species and so is now the more usual choice for forestry plantations.

McCracken (1988–1989) recorded that larch was a common component of Irish timber plantations in Co. Meath in the early 1800s, for example. Amongst a total of 1.45 million conifers planted by tenants in that county from 1800 to 1849, about half were larch while other species included Scots pine, silver fir and spruces. I have seen larch poles used as major structural timbers and rafters in a traditional cottage in south Galway built probably as early as 1830 (Micheline Sheehy Skeffington, pers. comm., 2009), indicating that even then it was a valuable timber source.

Picea species

- Spruce (Anonymous, 2005)
- *Sprús* (Ó Dónaill, 1977; Anonymous, 1978)

PINACEAE—THE PINE FAMILY
Introduced

Picea species are widely and commonly planted in Ireland, particularly for forestry purposes. *Picea sitchensis* (Sitka spruce; in Irish, *sprús Sitceach*, Ó Dónaill, 1977; Anonymous, 1978) is the most widely planted forestry species in Ireland and regenerates and self-seeds

Picea sitchensis.

Picea syrup.

very commonly. *Picea sitchensis* is native to the west coast of North America from Alaska to California (Mitchell, 1974). *Picea abies* (Norway spruce; in Irish, *sprús Lochlannach*, Ó Dónaill, 1977; Anonymous, 1978) is also grown for forestry in Ireland. It is much more rarely planted but is nevertheless widespread. It is a European native tree, found in Scandinavia and western Russia, extending south as far as the Alps and the Balkans (Mitchell, 1974). Remarkably, it was native to Ireland in interglacial warm periods, and notably recorded during a period called the Gortian Warm Stage, between 250,000 and 200,000 years ago (Mitchell, 1986).

Some outstanding *Picea sitchensis* specimens rank amongst the tallest trees currently known in Ireland. One specimen of Sitka spruce (measured in 2004) growing in Curraghmore, Portlaw, Co. Waterford, was measured at 55 m tall and 6.7 m in girth (Anonymous, 2005). Of the top 24 tallest trees recorded in Ireland in the Tree Council of Ireland's book, *Champion Trees* (Anonymous, 2005), 13 are specimens of this species.

Spruce timber, especially Sitka spruce, has been used in Ireland for a variety of purposes, including joinery, building and construction work, manufactured boards and paper (Neeson, 1991). Spruce timber is pale yellow brown to almost white in colour and fine textured.

Spruces are used in Ireland as Christmas trees but have become less popular in recent years. They have been replaced by 'dropless' species, particularly firs (*Abies* species), because spruces do not hold onto their leaves once they dry out. *Picea* species are grown as ornamentals in larger gardens, parks and tree collections throughout Ireland.

Pinus sylvestris

- Scots pine (Scannell & Synnott, 1987); Fir (Harron, 1986)
- *Péine Albanach* (Scannell & Synnott, 1987); *Giuis* (Dinneen, 1927; Webb et al., 1996); *Giumhas* (de Vere Coneys, 1849; Colgan, 1911); *Giús* (Dinneen, 1927; Uí Chanainn, 1962); *Crann giuis* (Dinneen, 1927); *Giúsach*[322] (Uí Chanainn, 1962)

PINACEAE—THE PINE FAMILY

Formerly Native, probably Extinct and Reintroduced

Pinus sylvestris is one of the most widely distributed species of pine in the world and despite the common name, Scots pine, by which it is generally known in Ireland, it also occurs through much of northern Europe, including Scandinavia and Russia, extending as far as Siberia in Asia, as well as occurring in Scotland, of course, where, as a native plant, it is largely found in the Highlands. It is widely grown and is well naturalised in North America, in parts of Canada and the United States.

322. Uí Chanainn (1962) defined *giúsach* as 'pine, fir'.

De Vere Coneys (1849) gave the Irish name *cran-giumhuis* for a 'fir tree' or a 'pine tree'.

In post-glacial times in Ireland, the Scots pine was an abundant tree throughout Ireland. However, it is thought that the Scots pine became extinct in Ireland before it was reintroduced perhaps about 300 years ago. Remarkably, three Scots pine cones have been found in excavations from Fishamble Street, Dublin, dating to ca. 1025, even though it is generally considered to have become extinct in Ireland in AD 300. At the least, it was very rare and mainly confined to pockets in the extreme west of Ireland (Geraghty, 1996). Of course these cones may have been imported rather than have come from locally grown trees. Nevertheless, this is an interesting anomaly. Today, *Pinus sylvestris* is common again. It is planted in most parts of the island and frequently self-seeds, particularly in the west. The Scots pine is also grown in gardens.

Pinus sylvestris (Scots pine). Watercolour by Lydia Shackleton. Courtesy of the National Botanic Gardens, Glasnevin.

A range of pine species are grown for forestry purposes in Ireland, including *Pinus sylvestris*, and some of these have become established in the wild. One such naturalised pine is *P. contorta*, lodgepole pine (or shore pine; Anonymous, 1978) (in Irish, *péine contórtach*), a common tree in western North America occurring in the wild from Alaska south to California. Lodgepole pine was named for its preferred use in the Native American tepee lodge. The long, straight and lightweight timber made it ideal for horse transport in nomadic cultures. *Pinus contorta* is the second most important forestry species in Ireland, second only to *Picea sitchensis*, Sitka spruce. *Pinus radiata*, Monterey pine, native to two small areas of southern California in the United States, has been planted occasionally in Ireland, particularly in the west and south-west, near the coast. Some exceptionally large trees of the Monterey pine are known, such as one on the Queen's Drive, Muckross, Killarney, Co. Kerry, that is 45 m tall (Anonymous, 2005).

A pine cone from another species, *Pinus pinea*, stone pine, has also been found in Viking Dublin, from near the waterfront at Wood Quay, dated to the 13th century. This was never native in Ireland and is a Mediterranean species whose seeds are edible and have commonly been used for food. It would have been imported into Dublin (Geraghty, 1996).

The wood of *Pinus sylvestris* is rather soft whitish to whitish brown, striped prominently with resinous-brown belts. The sapwood colour varies from, rarely, almost white to dark brown and very resinous. Scots pine was traditionally used in ancient Ireland for its timber (red deal). Pine wood or 'deal' was used for a variety of purposes, such as to make the gunwales of currachs, the traditional boat of western Ireland. Scots pine timber has had other industrial and local uses in Britain and Ireland, such as for telegraph poles, pit-props, railway sleepers, masts of ships, boat-building and ladders (Arnold, 1968).

Pinus sylvestris.

The National Museum of Ireland (Museum of Country Life) in Castlebar, Co. Mayo, contains a rich collection of vernacular Irish furniture, much of it made from pine (probably much of it from *Pinus sylvestris*). A study that is needed in Ireland is research into which timber was used for what purposes. Timbers are not always easy to identify and need to be reviewed by an expert, often using microscopic

A 'goat ladder' made from pine.

characteristics to differentiate the woods of various tree species. Although, as wood was scarce in many parts of Ireland, particularly during the last few centuries, objects were often fashioned from whatever wood was available at the time, sometimes even from driftwood gathered on the sea-shores, and also from bog timber (notably pine, oak and yew).

An interesting object in the Museum's collection made from pine is a 'goat ladder'. This is made like a small ladder through which the heads of two goats would be placed, to keep them together and discourage them from wandering and to make it difficult for them to jump ditches and fences.

Scots pine stump below a bog in Co. Mayo.

Resin was collected from *Pinus sylvestris* for waterproofing boats and for preserving timber (Lucas, 1954). Pine was also the source of an important weatherproofing material, 'Stockholm tar', a high-grade tar or 'pitch' that has been widely used to treat ropes and timber for outdoor usage. It was much used in ship building. The pine wood was burnt slowly and steadily at a low heat in a special kiln. As the wood burned, a dense tar would accumulate and be collected in a chamber below the kiln. Stockholm tar has been an important product in Scandinavia for many centuries. Rutty (1772: 123) wrote that tar was said to have been made 'with good success, from some of our subterranean Pines [bog pine] in Ireland'.

Pine wood (or bog deal) dug from the bogs was used to provide domestic light (Mackay, 1836; MacCoitir, 2003) and for torches, splinters (in Irish, *birín* [pl: *-ní*], *geataire*, *sciotachán* or *meathán*; Dinneen, 1927) and firewood. An account from 1937 (included in Lysaght, 1979: 29) described such usage: 'There were no candles in those times but it was how everyone would go to the bog and root up a good thick block of bog deal timber and bring it home with him. He would get his hatchet then and split that block up into *sliseogs*, nice thin long sliseogs. The sliseogs were then dipped in tallow from the cattle or sheep and whenever they wanted light someone would hold them in his hand or maybe they would put it standing in the middle of a heap of potatoes while they would be eating them for supper. The sliseogs that were burnt that way, were called *caisnes* or *caisninis*'. During World War II in Co. Kerry when there was a shortage of paraffin oil for domestic lighting (rural electrification was still 10 years or more away), splinters of bog pine were used instead for lighting (O'Shea, 2009). The resin-rich heartwood was often chosen to make fire splinters too, sometimes from bog stumps (Edlin, 1973; Pilcher & Hall, 2001). Mackay (1836: 259) noted that the resinous roots of bog pine were sold in Dublin as firewood at that time and used 'by the peasantry in the west of Ireland in lieu of candles'. Crane (1907) said that it was widely used as firewood in Kerry homes. Lucas (1957) reported that the thin strips of resinous pine were far superior and cheaper than the rush lights used previously. The collections of the National Museum of Ireland (Museum of Country Life) in Castlebar, Co. Mayo, include an impressive torch made from bog deal (*Pinus sylvestris*). McGlinchey (2007: 15) mentioned that in Co. Donegal in the 1800s, 'All the old houses had a couple of holes built in the wall at the side of the fireplace. These were called *poll-cloiche* or bowils. They would hold a pipe or fir-splits or things like that'. McGlinchey (2007: 20) also said that the fir-splits were used to provide light when going around the house. He said that 'they were nearly as good as a box of matches or a candle. When they went to the byre to milk, someone took a coal and two or three fir-splits to show them light'. Milliken and Bridgewater (2004) mention the use of bog pine for similar purposes in Scotland and describe tapers used as 'rustic candles', which were about 2 ft long and three quarters of an inch by a quarter of an inch wide. Rutty included bog wood in *An Essay Towards a Natural History of the County of Dublin*

(1772: 99) under the name '*Abies subterranean*, Bog Fir or Subterraneous Firr or Deal'; however, what he actually referred to was *P. sylvestris*, not *Abies*. Rutty stated that it was 'dug, at a depth of three or four feet, in the bogs of Glancullen [Glencullen] mountain, and supplies Dublin with an excellent material to light their fires with, burning like a candle, and of a grateful smell: and in some places slips of it, cut in pieces, serve instead of candles, as it burns with a clear and durable flame'. In Donegal, Campbell (1974) described how torches made from dried pieces of bog pine were used in poaching salmon. A torch would be lit by the riverside to dazzle the fish, which could then be easily gaffed and hauled out.

Fibres from bog pine were used to make ropes in ancient Ireland (Lysaght, 1979), a practice that continued up to the beginning of the 20th century. The timber was soaked in hot water, then beaten with wooden mallets to separate the fibres before being woven or twisted together into ropes. Pine ropes were used to tie down thatch, corn- and hay-stacks, to tie up bundles for carrying, in cording beds and for use in sailing boats (Lucas, 1957). They were twisted into two- or three-ply ropes and were extremely strong and durable. Mackay (1825a) noted that this was carried out in Queen's County (Co. Laois). Amongst the uses of such ropes were as cording for beds, and according to Wakefield (1812: 524), 'in damp places they will last much longer than hempen ones. The usual length for a bed is twenty yards, which are commonly bought for ten-pence'. Evans (1957) mentioned that it was the root fibres of bog pine that were twisted into ropes to make mattresses, and bog timber supplied the bed frame. Evans (1967 [1942]: 63) wrote that 'scraws were generally sewn to the roof timbers [themselves often made from bog pine] with loops of sugán or fir-rope'. Bog pine ropes are recorded as having been used to bind bundles of rushes together to make bundle rafts in the Urlaur lakes in east Co. Mayo (S. MacPhilib in MacCárthaigh, 2008). Rutty (1772: 99) said that 'This wood, moreover, is sometimes beaten into strings, and combed and spun, and several strings are twisted together to make ropes, which are sold at some of our fairs, having this peculiar excellence, that being used for cording on beds on damp earthen floors, they are not subject to rot, as the hempen cords are, by moisture'.

Bog deal (or bog pine, *Pinus sylvestris*) was also used to make scallops or spars to hold roof thatch in place (O'Hare, 1993). They were said to be the best material for such purposes, as they did not rot readily. Scallops made from this material and other species were often prepared and sold at markets and fairs. P. J. Magnier, an 88-year-old man I met in August 2011, at the Bridges of Ross, near Loop Head, Co. Clare, told me that his father used what they called 'bog pitch pine' to make spars for thatching. He also told me of a ladder being made from bog pine which was very strong and 'will last forever'. He said that bog pine was much preferred by craftsmen to bog oak as it was much stronger and could be found in good lengths.

Bog pine timbers were a useful source of beams for houses, fence posts, carts, looms, boats, furniture, spinning wheels and other purposes in Ireland. Evans (1967 [1942]: 63), said that 'In most Irish cottages the frame of the roof is of bog-fir, but its use has been superseded by imported fir'. Reverend J. Hall recorded, when he travelled through Ireland in 1813, doors (called *scolpáns*) made from wicker of sally rods[323], in a frame of bog deal, were widely used for simple houses (O'Hare, 1993). Ua Cnáimhsí (2009: 131) described the use of bog deal to build a house in Arranmore Island, off the west coast of Co. Donegal in the late 1800s: 'There was plenty wood around at the time: you'd find fir in the bog and, as well as that, planks were often washed ashore. It wasn't easy hiding those planks from the eyes of the revenue men or the police or the bailiffs'. The 'plenty wood' does not seem to have included timber from growing trees though.

To find the bog timbers, farmers would probe the bog with augers to search for it (Edlin, 1973) (see also *Quercus* spp.). Dutton (1808: 285) described the method used to

323. Ó Dónaill (1977) gave the Irish word *brilleog, -oige* for sally rods.

Pinus sylvestris on Lough Leane, Co. Kerry.

discover these bog pine trees in Co. Clare: 'Very early in the morning, before the dew has evaporated, a man with a long sharp spear goes out into the bog, and as the dew never lies on the part over the trees, be they ever so deep, he can ascertain their length, and by thrusting down his spear he easily discovers whether they are sound or rotten; if sound he marks with a spade the spot where they lie, and at his leisure proceeds to extricate them from their bed'. Dinneen (1927: 89) gave the Irish word *bearadóir* as 'a prober, one who probes for bog-wood by means of an iron bar or spike with a wooden handle, the prober having stuck the bar into the bog applies his teeth to the wooden handle to detect the bog-wood'. The Irish word *bior-mhaide* was given by Dinneen (1927) as a long iron rod used for searching for bog oak. Bog pine (as well as bog oak and yew) provided an extremely important raw material for many rural communities who relied on it for many purposes by the 18th and 19th centuries when the woodlands in their region had been felled (Lucas, 1957). Evans (1957) mentioned that in Kerry whole boats were sometimes built from bog fir and had the advantage of being impermeable to water, and bog fir ropes could be used for the rigging. McGlinchey (2007) mentioned that in Co. Donegal in the 1800s people made three-legged stools out of bog fir.

John McEvoy's *Statistical Survey of the County of Tyrone* . . . (1802: 188) indicated the importance of bog pine to Ireland in 1802: 'The chief reliance this county has, is upon the bog-fir, which many of the bogs produce in great plenty.... This timber is sometimes sold at half a crown the cubical foot, for the purposes of loom-timber and mill-shafts.... Even grained bog-fir is looked upon to be full as good for roofing and lofting stables, &c. as foreign fir or pine; it makes excellent laths, and when beaten out into small filaments, is found to answer for ropes, which are principally used as cording of beds, and, in damp places, will last considerably longer than hempen ropes'.

Pieces of bog timber, including bog pine, were also burnt to provide ash to fertilise rough infertile fields. Róise Rua's memoir of life in Arranmore Island, Co. Donegal, in the early 1900s mentioned that 'They collected lumps of roughage they called stumps, pieces of bog fir and so on and set fire to them [in March or April]. These fires burnt for a week or so, and the ashes were then raken into the soil and the field dug up once again. You could only do this sort of burning in moorland or bogland' (Ua Cnáimhsí, 2009: 164).

Mears and Hillman (2007) noted the use of pine pollen as a foodstuff when cooked. It is rich in oils and protein. The pollen is gathered just as the cones are beginning to shed their pollen. They also noted that young pine needles (leaves) may be eaten and have a sweet lemony flavour. Older needles may be chopped and used to make tea. Drury (1984) mentioned that meal made from the bark of Scots pine bark was used as a famine food in Ireland. The bark of other trees was also used in a similar way, such as birch, beech, elm and lime, but not oak or willow, which contained too much tannin.

Fresh young branches have been used to provide an essential oil used medicinally in parts of Europe, as an expectorant and to treat respiratory problems (van Wyk & Wink, 2004).

Pseudotsuga menziesii

- Douglas fir (Anonymous, 2005)
- *Giúis Dhúghlais* (Ó Dónaill, 1977; Anonymous, 1978)

PINACEAE—THE PINE FAMILY

Introduced

Pseudotsuga menziesii is native to western North America and extends along the Rocky Mountains north from British Columbia in Canada, south through California into Mexico (Mitchell, 1974). *Pseudotsuga menziesii* is grown in Ireland occasionally for forestry and also as a tall ornamental conifer in demesnes, larger parks and some private gardens. It has been widely used as a shelter-belt tree and occasionally self-sows in the wild.

A specimen of *Pseudotsuga menziesii* at 56 m tall (in 2004) and 5 m girth is currently the tallest tree in Ireland and grows in Powerscourt Gardens, Enniskerry, Co. Wicklow. This is closely followed by another specimen of the same species growing in Avondale Forest Park in Rathdrum, Co. Wicklow, that is 55 m tall and 3.35 in girth (Anonymous, 2005). Of the top 24 tallest trees recorded in Ireland in the Tree Council of Ireland's book, *Champion Trees* (Anonymous, 2005), seven of them are specimens of Douglas fir.

Douglas fir timber is quite heavy, moderately durable and reddish brown or brownish white in colour with broad resinous-brown bands. It is quite similar to the timber of larch but generally darker than the timber of Scots pine. It has been used for a wide variety of purposes, including joinery, furniture, building and construction, railway sleepers, outdoor cladding and plywood. In Ireland it has been the favoured choice for transmission poles, and especially for high-tension electricity lines, due to the long lengths commercially available.

Taxus baccata

- Yew (Scannell & Synnott, 1987)
- *Iúr* (Scannell & Synnott, 1987); *Lubhar* (Moloney, 1919); *Eó* (Moloney, 1919; Ó Dónaill, 1977); *Iuthar* (Cameron, 1883); *Iubhar* (de Vere Coneys, 1849; Cameron, 1883; Dinneen, 1927); *Iughar* (Cameron, 1883); *Cran-iuthair* (de Vere Coneys, 1849); *Iogha* (de Vere Coneys, 1849); *Eo-chrann*[324] (Dinneen, 1927); *Íodha*[325] (Dinneen, 1927); Uí Chanainn, 1962); *Crann iúir* (Ó Dónaill, 1977); *Iúir coiteann*[326] (Ó Dónaill, 1977)

TAXACEAE—THE YEW FAMILY

Native

Taxus baccata is widely grown in Ireland but is much rarer now as a native plant. In the wild, as a native tree, it is found in woodlands, cliffs, lake-islands and rocky places, mainly in the west and north of Ireland. It commonly self-sows elsewhere as an escape from cultivation, and seedlings arising from bird-sown seeds are also frequent in many parts of Ireland. Yews are probably the longest living trees in Ireland and some celebrated ancient trees are more than 6 m in girth. Current estimates give a yew of 5 m girth an age of more than 400 years (Anonymous, 2005), and several specimens in Cos.

Taxus baccata on Innisfallen Island in Killarney, Co. Kerry.

324. Dinneen (1927) defined *eo-chrann* as 'a yew-tree'.
325. Dinneen (1927) also used this as a name for ivy.
326. Ó Dónaill (1977) defined *iúir coiteann* as 'yew-tree'.

Taxus baccata fruits. The green seeds are surrounded by a red, fleshy aril.

Cork, Kildare, Wexford and Wicklow have been estimated as 700 to 1200 years old. The specimen with the largest recorded girth size is from Bunclody, Co. Carlow, that is 6.78 m in girth and 16.5 m tall (Anonymous, 2005).

Yew timber is close-grained, fine veined and hard. The heartwood is a deep reddish brown colour. The sapwood is a yellowish white in colour and is softer than the heartwood. Yew is an extremely important tree of folklore in Ireland and was much valued traditionally for its hard timber, which was used to make a variety of household objects, such as bowls, handles, pipes, barrel hoops and shields. Rutty (1772) said that it was used by wood turners, engravers, cabinet makers, mathematical instrument makers and comb and pipe makers. Its rarity today must be, in part, due to its importance as a prized timber resource, and in part, due to the general clearance of native Irish woodlands and its use for smelting iron-ore. Smith ([1756] 1979: 382) recorded that the yew was formerly abundant in parts of Co. Kerry 'until it was destroyed for making coals [charcoal] for the ironworks'. In early times, spears and longbows were made from yew wood (Grigson, 1958; MacCoitir, 2003), which is very flexible. However, Lowe (1897: 115) noted that 'The Irish do not seem to have been much acquainted with the use of the bow until the time of Edward IV, who enforced the practice amongst the English residents and their Irish servants'. Bog yew was much used during the 19th century for making fine furniture. It was said to be superior to the tropical timber, rosewood, and at least one manufacturer in Ireland, Arthur Jones, who operated in Dublin from 1820 to 1860 specialised in making bog yew furniture (Teahan, 1994). Rutty (1772: 129) said that bog yew was sometimes so hard as to 'give fire at the stroke of a hatchet' and was used to make rolling pins, punch ladles, chests and tables.

The use of yew wood has an extensive history in Ireland. John Evelyn in his *Sylva, or a Discourse of Forest-trees and the Propagation of Timber in His Majesties Dominions* (1664: 297) said that yew wood was used for the 'cogs of mills, posts to be set in moist grounds, and everlasting axel-trees, there is none to be compared with it; likewise for the bodies of lutes, theorbo's [a plucked string instrument], bowles, wheels, and pins for pullies; yea, and for tankards to drink out of …'. In addition, the Vikings used yew timber to make nails to secure the planks on their clinker-built boats (Arnold, 1968). Yew wood has been identified in at least one old Irish harp (Rimmer, 1969). The 'Domhnach Airgid', a celebrated silver decorated church shrine, dating from approximately AD 1350 and used to hold religious relics, was carved from a solid block of yew wood (Teahan, 1994). A bucket made from yew wood with elaborate bronze bindings probably dating back to the 9th century was found near the Kinegad river in Co. Meath and is now in the National Museum of Ireland. This bucket was probably used for ecclesiastical purposes, to carry wine or holy water (Ryan, 2000). Indeed, even a block cartwheel dated to about 400 BC and made of alder and with yew dowels, was found in a bog in Co. Roscommon in 1968–69 (Teahan, 1994).

The Muckross Abbey yew tree depicted on a picture postcard from the early 1900s.

References to the planting of yew trees in Ireland represent some of the earliest records of tree planting. For example, according to Forbes (1933: 168–199), *The Annals of the Four Masters*, which chronicles medieval history in Ireland, mentions that in 1162, 'the Monastery of the monks at Iubhar chin trechta was burned, with all its furniture and books, and also the yew tree which St Patrick himself had planted'. In Ireland some trees, such as individual yew trees, are associated with ecclesiastical sites and have been regarded as sacred. As a result, people have been reluctant to interfere with them or cut them down. Lysaght (1979) mentioned a yew tree in the grounds of Muckross Abbey in Killarney, Co. Kerry, that the Office of Public Works wanted to have pruned. However, nobody in the parish would touch it as it was recorded in tradition that a soldier from Cromwell's army tried to cut the tree for fuel, but his hatchet slipped and cut him, and he bled to death.

In about AD 1185 Geraldus Cambrensis wrote that the yew grows in Ireland in cemeteries and holy places (O'Meara, 1982). Mackay (1836) said that the yew was extensively planted in churchyards in Ireland for eventual use in cabinet-making and for making bows, but whether that was really the purpose of such plants is a matter of conjecture. It is also said that some ancient churchyard yews survive from (or at least the tradition of planting them in religious sites does) pre-Christian days when the yew was sacred in these islands. In 1307 Edward I decreed that yews should be planted in churchyards to protect their churches from gale damage (Vickery, 2010). Another suggestion is that yew trees were planted in churchyards to discourage their grazing by stock, or that they were there to ward off evil spirits. However, Vickery (2010) said that the reason yews have been planted in churchyards has never been satisfactorily explained.

O'Rourke (2006) wrote that yew branches were used to represent palm fronds on Palm Sunday, which to Irish speakers was known as *Domhnach an Iuir* (Yew Sunday) (Danaher, 1972). It was also often the species that was burnt to produce ash for Ash Wednesday ceremonies (Vickery, 1995). In his book on the yew trees of Great Britain and Ireland, John Lowe (1897: 96) wrote that 'the Yew was a funereal tree, the companion of the grave among the Celtic tribes' and noted that 'in Ireland … it is still the custom for the peasants to wear in their hats or buttonholes sprigs of yew from Palm Sunday until Easter Day' (p. 100).

The Irish yew *Taxus baccata* 'Fastigiata' (in Irish, *Iúr Éireannach*), a columnar variety, is a ubiquitous feature of graveyards and originated from a wild native tree growing near Florence Court, Co. Fermanagh[327], in the 18th century. The 'Fastigiata' cultivar is so well regarded that it was featured on an Irish stamp, in 1984.

All parts of the yew (apparently except the succulent aril, typically red, that surrounds the seed) are said to be highly poisonous and so accordingly, it has been little used in folk medicine. (I have found only one record of its use: to treat ringworm in Co. Kildare, mentioned by Allen and Hatfield, 2004.) The red arils resemble berries and their flavour is said to be quite sickly sweet and gelatinous. The colourful arils are attractive to birds, which disperse the hard seeds. As to their flavour, I've certainly not tried them! *The Irish Times* on Saturday, 12 April 2008, included a story of an unfortunate 37-year-old woman found on Christmas Day in 2007 who had died from drinking tea made from yew needles, killed by the toxic alkaloid, taxane, which they contain. Woodward (1985: 105) gives the symptoms of yew poisoning as 'vomiting, abdominal pain, diarrhoea, convulsions, delirium, cardiovascular paralysis; death very common and survival not common'.

Clyne (1989: 113) said that 'juice' of the yew was applied to the tips of arrows as a poison, but I have never heard of such usage connected with the tree in Ireland.

Yew clippings have been gathered in Ireland in recent years as a source for the extraction of the chemotherapy drug, taxol, used in the treatment of cancers, particularly ovarian and breast cancers. The anti-cancer properties of taxol were discovered when it was extracted from the bark of Pacific yew, *Taxus brevifolia*. This type of yew occurs only in the Pacific coast mountain ranges of North America and has been under considerable threat, harvested as a commercial source of taxol. Now that synthetic taxol and the similar drug Paclitaxel are available, the conservation assessment of Pacific yew has improved to less threatened levels. Taxol and various taxol derivatives act to inhibit cell division by interfering with mitotic spindles, thus stopping the rapid growth of tumours (van Wyk & Wink, 2004).

327. It is sometimes referred to as the 'Florence Court yew' (Anonymous, 1978)

Addison's Walk, a canopy and avenue of ancient yew trees at the National Botanic Gardens, Glasnevin.

Yew trees are frequent trees for parks and gardens, and yew hedges are popular as they respond well to trimming and regular clipping. Yew is therefore one of the most popular plants for topiary. Yew wood was extensively used (along with the strawberry-tree *Arbutus unedo*) for 'Killarney work' marquetry—elaborately inlaid furniture and boxes depicting scenes from Killarney and Irish symbols such as shamrocks and harps. In Britain yew wood was also used in a similar way for the marquetry furniture and objects known as 'Tunbridge ware' (Lowe, 1897).

Thuja plicata

- Western red cedar (Anonymous, 2005)

CUPRESSACEAE—THE JUNIPER FAMILY

Introduced

Thuja plicata is native to north-western North America and is grown in Ireland as an ornamental tree in demesnes, parks and large gardens. It is also occasionally grown for forestry purposes.

The wood of *Thuja plicata* is useful for various outdoor purposes and joinery, such as weather-boards, shingles, greenhouse frames and garden furniture as it is very durable and resistant to fungal decay (Anonymous, 1941). The heartwood varies in colour from pale pinkish brown to chocolate and the sapwood is paler in colour.

Tsuga heterophylla

- Western hemlock
- *Himlic iartharach* (Ó Dónaill, 1977)

PINACEAE—THE PINE FAMILY

Introduced

Tsuga heterophylla is native to the west coast of North America (Mitchell, 1974) and is grown in Ireland as a tall ornamental conifer in demesnes, larger parks and some private gardens. It has been widely used as a shelter-belt tree and occasionally self-sows in the Irish landscape. *Tsuga heterophylla* produces a useful timber for construction and other purposes.

FERNS AND FERN ALLIES

Adiantum capillus-veneris

- Maidenhair fern (Scannell & Synnott, 1987)
- *Dúchosach* (Scannell & Synnott, 1987); *Tae scailpeach* (Dinneen, 1927; O'Rourke, 2006); *Dubhchosach* (Moloney, 1919; Dinneen, 1927); *Failteann fionn* (Moloney, 1919); *Cos dubh* (Dinneen, 1927)

PTERIDACEAE (Adiantaceae)—THE RIBBON FERN FAMILY
Native

The Maidenhair fern is a beautiful fern with delicate leaves on long slender shiny black petioles. *Adiantum capillus-veneris* thrives in damp, rocky places, mostly in western Ireland most commonly amongst the limestone pavements and rocks of the Burren, in Co. Clare, and on the nearby Aran Islands. Where it occurs elsewhere in the wild in Ireland it is often an introduction.

Synnott (1979) noted that John K'Eogh in *Botanologia Universalis Hibernica* (K'Eogh, 1735: 74) reported that the maidenhair fern was brought in sacks from the Burren, Co. Clare, to be sold in Dublin. K'Eogh said that it was 'pulmonic, lithontriptic, dieuretic, styptic and deopulative … and wonderfully helps those afflicted with asthmas, shortness of breath and coughs, occasioning a free expectoration [and] it is good against the jaundice, dropsy, diarrhaea, haemoptysis, and the bitings of mad dogs'. Moloney (1919) mentioned also that it was sometimes used as an expectorant. McGlinchey (2007) also wrote about the medicinal use of this plant, as '*dubhchosach*' to treat heart ailments. Lady Wilde (1890) gave a recipe of 'Woodbine' (*Lonicera periclymenum*) and 'Maiden-hair' (*Adiantum capillus-veneris*) in milk with oatmeal, to treat dysentery.

Mabey (1972) described how maidenhair fern was used as the basis of *capillaire*, formerly a popular flavouring. The fern was simmered for several hours, then orange-water and sugar were added to make a syrup. *Capillaire* was mixed with fruit juices to make soft drinks. I have not found records of its use as such a flavouring in Ireland.

Dinneen (1927) mentioned that it was sometimes used as a substitute for tea in Ireland.

In the early 1900s, J. M. Synge visited and wrote about the Aran Islands (Synge, 1979: 8) and mentioned that 'two little girls … told me with a sort of chant how they guide 'ladies and gintlemins' in the summer to all that is worth seeing in their neighbourhood, and sell them pampooties[328] and maidenhair ferns, which are common among the rocks'. This is an early indication of the popularity of this fern, collected from the wild, as an ornamental plant. It is occasionally grown in Ireland on rockeries and rock walls and as an indoor or conservatory pot plant.

Asplenium adiantum-nigrum

- Black spleenwort (Scannell & Synnott, 1987)
- *Fionncha dubh* (Scannell & Synnott, 1987); *Lubhar talmhan* (Moloney, 1919); *Lus na seilge*

328. Pampooties were traditional footwear worn in the Aran Islands until the 1950s, made from a single piece of uncured hide, tied around the foot with twine or a leather string.

(Moloney, 1919); *Lus an corráin (dubh)*[329] (Moloney, 1919; Dinneen, 1927); *Craobh mac fiadh* (Cameron, 1883)

ASPLENIACEAE—THE SPLEENWORT FAMILY

Native

Asplenium adiantum-nigrum, or black spleenwort, is very common in Ireland. Black spleenwort occurs on rock walls and shady banks particularly in the south and is less frequently seen on the central plain of Ireland than elsewhere. This fern is widely distributed in Europe, ranging into central Asia and North America.

Logan (1980: 87) mentioned that this fern used to grow on the walls of the ruined St Finnian's church at Killemlagh, Co. Kerry, 'where pilgrims would go seeking the cure of 'diseases of a scrofulous nature' on 16 March. The story is that St Finnian was a leper and used a fern which grew on the walls to treat this leprosy. Logan (1980) said that the fern was also used to treat other skin diseases.

Asplenium ceterach.

Asplenium ceterach (Ceterach officinarum)

- Rustyback (Scannell & Synnott, 1987); Scalefern (Praeger, 1938; Brunker, 1950)
- *Raithneach rua* (Scannell & Synnott, 1987); *Raithneach mhéiscreach* (Moloney, 1919)

ASPLENIACEAE—THE SPLEENWORT FAMILY

Native

Asplenium ceterach is a common, small fern of limestone rocks and mortared walls that extends through Europe to central Asia. Although this fern is found throughout Ireland, it is most common in the south. This is an attractive fern preferring calcium-rich, basic habitats, whose distribution has probably been greatly extended by the existence of mortared walls and historic human construction.

According to Threlkeld (1726: unnumbered [20]) it was used medicinally for 'Swellings of the Milt' [male reproductive organs] and 'Fluxes of the Belly'.

**Asplenium cuneatum (Asplenium cuneifolium)*

- Black spleenwort (Scannell & Synnott, 1987)
- *Fionncha dubh* (Scannell & Synnott, 1987)

ASPLENIACEAE—THE SPLEENWORT FAMILY

Native

Asplenium cuneatum is native to mainland Europe. Specimens referred to this species

329. Ó Dónaill (1977) defined this word as 'spleenwort'.

from Ireland are probably not correct but instead are a minor variant of *A. adiantum-nigrum,* and in Ireland they are found only on serpentine rocks in Connemara (Parnell & Curtis, 2012).

*No ethnobotanical uses of this plant have been noted in Ireland.

**Asplenium marinum*

- Sea spleenwort (Scannell & Synnott, 1987)
- *Fionncha mara* (Scannell & Synnott, 1987)

ASPLENIACEAE—THE SPLEENWORT FAMILY
Native

Asplenium marinum is an attractive fern of exposed cliffs and cracks and crevices of rocks by the sea that is found throughout the coasts.

*No ethnobotanical uses of this plant have been noted in Ireland.

**Asplenium obovatum (Asplenium billotii, Asplenium lanceolatum)*

- Lanceolate spleenwort (Scannell & Synnott, 1987)
- *Fionncha lansach* (Scannell & Synnott, 1987)

ASPLENIACEAE—THE SPLEENWORT FAMILY
Native

Asplenium obovatum is a very rare fern in Ireland and is now found mainly near the east and south coasts, growing on walls and shady banks.

*No ethnobotanical uses of this plant have been noted in Ireland.

**Asplenium onopteris*

- Irish spleenwort (Scannell & Synnott, 1987)
- *Fionncha biorach* (Scannell & Synnott, 1987)

ASPLENIACEAE—THE SPLEENWORT FAMILY
Native

Asplenium onopteris is similar to *A. adiantum-nigrum* but is less commonly seen in Ireland. It is found in drier habitats, such as on rocks, walls and banks, and is often associated with limestone.

*No ethnobotanical uses of this plant have been noted in Ireland.

Asplenium ruta-muraria

- Wall-rue (Scannell & Synnott, 1987)
- *Luibh na seacht ngábh* (Scannell & Synnott, 1987); *Luibh na seacht ngádha* (Moloney, 1919)

ASPLENIACEAE—THE SPLEENWORT FAMILY
Native

Asplenium ruta-muraria.

Asplenium ruta-muraria is a common fern throughout Ireland. As both its common and Latin names indicate, this fern thrives on walls and in rocky crevices, with a preference for basic substrates.

Threlkeld (1726: unnumbered [4–5]) noted wall-rue was used in the treatment of rickets and convulsions in children and as a remedy 'against narrowness of children's breaths' (presumably asthma). It was used as a decoction or powder. Epilepsy was said to have been cured by taking nine pieces of the cranium (skull) of a dead man, grinding them down to make a fine powder and dissolving it in a decoction of wall-rue. While fasting, the patient swallows a portion of the liquid each morning until it is all gone. If he fails to finish the treatment, the owner of the skull was said to return to look for the unswallowed portion (Wilde, 1887; Wood-Martin, 1902).

Asplenium scolopendrium.

Asplenium scolopendrium (Phyllitis scolopendrium)

- Hart's-tongue (Scannell & Synnott, 1987); Fox's tongue (Colgan, 1904); Cow tongue (Hart, 1898); Fox tongue fern (Hart, 1898)
- *Creamh na muice fia* (Scannell & Synnott, 1987); *Leadhb gadhair* (Uí Chonchubhair, 1995; Ua Maoileoin, 2001); *Teanga chapaill* (O'Rourke, 2006); *Teanga fiaidh* (Moloney, 1919; Dinneen, 1927); *Teanga fia* (Ó Dónaill, 1977); *Craoibh-muice-fiadh* (de Vere Coneys, 1849; Dinneen, 1927); *Creamh muice fiadha* (Dinneen, 1927)

ASPLENIACEAE—THE SPLEENWORT FAMILY

Native

Hart's-tongue fern is a common fern of hedgerows, shady banks, walls, woodlands and moist cliffs throughout Ireland.

Vickery (1995) noted a record from Co. Limerick in the manuscript collection of the Irish Folklore Commission of Hart's-tongue fern used medicinally. The leaf was burned and applied to a burn.

Various cultivated forms of this attractive fern are grown as ornamental plants in Irish gardens, such as forms with wavy, curled or divided edges to their fronds.

**Asplenium septentrionale*

- Forked spleenwort (Scannell & Synnott, 1987)
- *Fionncha ladhrach* (Scannell & Synnott, 1987)

ASPLENIACEAE—THE SPLEENWORT FAMILY

Native

Forked spleenwort is a very rare fern found now only on volcanic rock cliffs near Roundstone, Co. Galway.

Although no medicinal uses of this plant are known in Ireland, elsewhere, for example, in the Jammu Province of India, the leaves were burnt and the smoke inhaled to treat chest and head colds (Pennacchio et al., 2010).

*No ethnobotanical uses of this plant have been noted in Ireland.

Asplenium trichomanes

- Maidenhair spleenwort (Scannell & Synnott, 1987); Black maidenhair (Praeger, 1938)
- *Lus na seilge* (Scannell & Synnott, 1987); *Lus an corráin* (Moloney, 1919; Dinneen, 1927; Uí Chonchubhair, 1995); *Urthalmhan*[330] (de Vere Coneys, 1849; Cameron, 1883)

ASPLENIACEAE—THE SPLEENWORT FAMILY

Native

Asplenium trichomanes is a common fern throughout Ireland and Great Britain of walls and dry rocky habitats.

Cameron (1883) said that this plant was formerly used as a remedy for ailments associated with the spleen, but no ethnobotanical uses have been noted in Ireland, despite its English and Irish names, which suggest such usage (Cameron said that *sealg* means 'spleen' in Scots Gaelic). However, as it is a common plant in Ireland, I suspect that it was probably used for medicinal purposes. Elsewhere in the world this species has been used for medicinal purposes. For example, in Lesotho, Africa, head and chest colds were treated by inhaling smoke produced by burning the leaves of this species (Pennacchio et al., 2010).

Asplenium trichomanes.

**Asplenium viride*

- Green spleenwort (Scannell & Synnott, 1987)
- *Fionncha glas* (Scannell & Synnott, 1987); *Lus an chorráin* (*glas*) (Moloney, 1919; Dinneen, 1927)

ASPLENIACEAE—THE SPLEENWORT FAMILY

Native

Green spleenwort is an alpine fern that is found in high mountain rocky habitats, on basic substrates, in the west of Ireland from Co. Donegal down to the south-west, and also in the Galtee Mountains of Co. Tipperary.

*No ethnobotanical uses of this plant have been noted in Ireland.

330. Cameron (1883) applied this name to *Asplenium* in general, not to a specific species. Dinneen (1927) also gave this name for 'spleenwort'.

Athyrium filix-femina

- Lady-fern (Scannell & Synnott, 1987); Fox's fern (Colgan, 1911)
- *Raithneach Mhuire* (Scannell & Synnott, 1987); *Raithneach mhadra mhaith* (Moloney, 1919); *Raithnighe* (de Vere Coneys, 1849); *Raithneach mhada ruaidh* (Dinneen, 1927); *Raithneach mhadra*[331] (Colgan, 1911)

WOODSIACEAE (Athyriaceae)—THE LADY-FERN FAMILY
Native

Athyrium filix-femina is a common fern of woodlands, hedges, bog-margins and shady ditches throughout Ireland and Britain, indeed throughout temperate boreal regions in the Northern hemisphere. Lady fern is preferentially found on moist, acidic substrates.

According to Uí Chonchubhair (1995), lady-fern has been used medicinally in Ireland; its rhizome was crushed, sweetened and taken with water as a vermifuge and as a decoction in wine to help remove obstructions of the liver and spleen. In North America this species has also been used in traditional medicines and as food. For example, the Native American Ojibwa and Potawatomi tribes used the roots to make a medicinal beverage, and the Meskwaki used it to treat breast pains (King, 1984).

Moerman (1998: 115) cites multiple uses by Native Americans for this plant, such as a diuretic, to treat cancer, for sores, and to help breast milk flow and ease labour in childbirth.

K'Eogh (1735) recorded that the plant (rhizome and fronds) was burnt in midsummer and the ash used to bleach linen. Rutty (1772: 113) said that 'scarce any herb produces so much salt by burning as this, and therefore the most profitable application of it is to burn it for its ashes for making soap, and making green bottles and glasses'. Rutty also recorded that it made the clearest crystal glass. Rutty suggested that it should be cut between Midsummer and St James's tide (i.e., late June) when full of sap. He said that ferns which grow on 'gritty land' are preferred. Apparently, balls of the ash were made of lady-fern and the male-fern (*Dryopteris filix-mas*) and water, which were used as an alternative for washing linen instead of soap. Before use, the balls were put in a fire until they went white, 'by which means they easily go into powder'.

Blechnum spicant

- Hard fern (Scannell & Synnott, 1987); Black-footed canavawn (Colgan, 1911)
- *Raithneach chrua* (Scannell & Synnott, 1987); *Ceannabhán dubhcosach* (Moloney, 1919); *Raithneach cruaidh* (Moloney, 1919); *Raithneach (an) mhadra* (Uí Chonchubhair, 1995); *Ceanabhán dubh chosach* (Colgan, 1911)

BLECHNACEAE—THE HARD-FERN FAMILY
Native

Blechnum spicant is a common fern of woodlands, shady banks, bogs, mountains and moorlands, with a preference for calcium-rich subtrates. This fern can be found

331. Colgan (1911: 4.13) listed this word for 'dog fern' in Kerry.

throughout Ireland and also occurs throughout temperate Europe, eastern Asia and western North America.

Its only use as a medicinal plant in Ireland was noted by Allen and Hatfield (2004) as one of eight ingredients in a drink used to treat a cough after a fever in Co. Mayo.

Botrychium lunaria

- Moonwort (Scannell & Synnott, 1987)
- *Lus na míosa* (Scannell & Synnott, 1987); *Luanlus* (Moloney, 1919; Dinneen, 1927); *Dealtlus* (Moloney, 1919; Dinneen, 1927; Ó Dónaill, 1977); *Éasglur* (Moloney, 1919); *Easc-lus* (Dinneen, 1927)

OPHIOGLOSSACEAE—THE ADDER'S-TONGUE FAMILY
Native

Botrychium lunaria is a widespread but rare fern in Ireland of acidic substrates in dry grasslands, open woodlands and high elevations on rocky ledges in the mountains.

Cameron (1883) referred to an old Limerick story that a man in Clonmel jail (in Co. Tipperary) could open all the locks by means of this plant, but he does not explain how this was done, or where the man obtained the plant!

Cryptogramma crispa

- Parsley fern (Scannell & Synnott, 1987)
- *Raithneach chas* (Scannell & Synnott, 1987)

PTERIDACEAE (Adiantaceae)—THE RIBBON FERN FAMILY
Native

This is a very rare fern in Ireland that is mainly in the north and east in rocky habitats. It is a montane fern with an arctic-alpine distribution across Europe into eastern Asia.

*No ethnobotanical uses of this plant have been noted in Ireland.

Cystopteris fragilis

- Brittle bladder-fern (Scannell & Synnott, 1987)
- *Raithneach bhriosc* (Scannell & Synnott, 1987)

WOODSIACEAE—THE LADY-FERN FAMILY
Native

This fern is most frequent in the north and west in Ireland where it grows in moist rocky lime-rich habitats, including limed walls, on limestone pavements and in damp shady rocky ravines. In the east and south of Ireland it is much rarer.

*No ethnobotanical uses of this plant have been noted in Ireland.

Diphasiastrum alpinum (Lycopodium alpinum)

- Alpine clubmoss (Scannell & Synnott, 1987)
- *Garbhógach shléibhe* (Scannell & Synnott, 1987); *Garbhógach an tsléibhe* (Moloney, 1919)

LYCOPODIACEAE—THE CLUBMOSS FAMILY
Native

Alpine clubmoss is restricted to high mountain moors and grassy heaths on peaty soils. It is rare in Ireland, but particularly known from Cos. Galway and Wicklow and formerly in Kerry where *Diphasiastrum alpinum* is probably now extinct.

*No ethnobotanical uses of this plant have been noted in Ireland.

**Dryopteris aemula*

- Hay-scented buckler-fern (Scannell & Synnott, 1987)
- *Raithneach chumhra* (Scannell & Synnott, 1987)

DRYOPTERIDACEAE—THE BUCKLER-FERN FAMILY

Native

Hay-scented buckler-fern has a broadly oceanic distribution in Ireland, inhabiting acidic to neutral soils in lowland woods, shady banks, and sea cliffs, mainly in the west. *Dryopteris aemula* is less frequently seen in the central plain and east of Ireland. *Dryopteris aemula* has a north Atlantic distribution and is found on the Atlantic coasts of France and the northern Iberian peninsula, as well as in the Azores, Madeira and the Canary Islands.

*No ethnobotanical uses of this plant have been noted in Ireland.

Dryopteris affinis (Dryopteris assimilis; Dryopteris borreri; Dryopteris pseudomas)

- Scaly male-fern (Scannell & Synnott, 1987)
- *Raithneach ghainneach* (Scannell & Synnott, 1987); *Raithneach na ngainní órga* (Uí Chonchubhair, 1995)

DRYOPTERIDACEAE—THE BUCKLER-FERN FAMILY

Native

Dryopteris affinis is a frequent fern of acidic woodlands and hedgerows and on stream banks in the mountains and amongst rocks. *Dryopteris affinis* is particularly common in the west in Ireland but is somewhat less frequent in the east.

Uí Chonchubhair (1995) recorded that this species and *Dryopteris filix-mas* were used medicinally in Ireland; their rhizomes were crushed, boiled, sweetened and taken with water to treat obstructions of the liver and spleen and to treat rickets in children. They were apparently also effective against tapeworm.

**Dryopteris carthusiana (Dryopteris spinulosa)*

- Narrow buckler-fern (Scannell & Synnott, 1987)
- *Raithneach chaol* (Scannell & Synnott, 1987)

DRYOPTERIDACEAE—THE BUCKLER-FERN FAMILY

Native

This fern is seen in bogs and moist heaths in Ireland, amongst rocks, in damp woods and streamsides, particularly in the north and is less common in the south.

*No ethnobotanical uses of this plant have been noted in Ireland.

Dryopteris dilatata (Dryopteris austriaca)

- Broad buckler-fern (Scannell & Synnott, 1987)
- *Raithneach leathan* (Scannell & Synnott, 1987)

DRYOPTERIDACEAE—THE BUCKLER-FERN FAMILY
Native

Dryopteris dilatata is a common fern throughout Ireland of woodlands, hedgerows, shady banks, streamsides, heaths and moorland, and acidic substrates. In damp situations, this fern can even be epiphytic on tree branches.

*No ethnobotanical uses of this plant have been noted in Ireland.

Dryopteris dilatata.

Dryopteris filix-mas (Lastrea filix-mas)

- Male-fern (Scannell & Synnott, 1987); Basket fern (Hart, 1898)
- *Raithneach mhadra* (Scannell & Synnott, 1987); *Marc raithneach* (Moloney, 1919; Dinneen, 1927; Ó Dónaill, 1977); *Raithneach madraidh* (Dinneen, 1927); *Marcraithneach* (Uí Chanainn, 1962)

DRYOPTERIDACEAE—THE BUCKLER-FERN FAMILY
Native

Dryopteris filix-mas is one of the most common ferns that is widespread not only throughout Ireland but temperate circumpolar regions in Eurasia and North America. This deciduous fern can be found in woodlands, hedgerows, mountain slopes, river and stream banks and a variety of shady habitats.

It has been used in the treatment of intestinal worms, and in particular, tapeworm (Moloney, 1919).

It was used as a veterinary remedy to treat various ailments including red-water fever, liver fluke, kidney failure and white scour in various parts of Ireland (Allen & Hatfield, 2004). The roots were boiled and strained and given to animals to cure liver fluke in Co. Roscommon (Vickery, 1995).

Rutty (1772: 113) recorded that the ash made from burning *Dryopteris filix-mas* was used to wash linen, instead of soap: balls of the ash were made of this and *Athyrium filix-femina* (Lady fern) and water, and before use, they were put in a fire until becoming white, 'by which means they easily go into powder'.

**Dryopteris oreades (Dryopteris abbreviata)*

- Mountain male-fern (Scannell & Synnott, 1987)
- *Raithneach shléibhe* (Scannell & Synnott, 1987)

DRYOPTERIDACEAE—THE BUCKLER-FERN FAMILY
Native

Mountain male-fern is a very rare fern known only from a few scattered sites in Ireland. Its seasonally deciduous colonies occur generally on inaccessible rock ledges and slopes on acidic substrates. In Ireland it is exclusively a montane plant, occurring on well-drained rocky ledges and open slopes on mountain scree. This fern has been found in Co. Down in Northern Ireland where colonies may no longer persist. Elsewhere in Ireland

Dryopteris oreades has been found in Cos. Cork and Kerry. *Dryopteris oreades* is a sub-oceanic boreal fern in Wales, Scotland and eastern England and other coastal montane sites in mainland Europe where generally it has an arctic-alpine distribution.

*No ethnobotanical uses of this plant have been noted in Ireland.

Equisetum species

- Horsetails (Scannell & Synnott, 1987); Mares-tail (Hart, 1898)
- *Gliogán* (Webb et al., 1996); *Cuiridín* (Uí Chonchubhair, 1995)

Equisetum arvense

- Field horsetail (Scannell & Synnott, 1987); Black scutch (Colgan, 1904); Horse pipes (Hart, 1898)
- *Scuab eich ghoirt* (Scannell & Synnott, 1987); *Eireaball capaill* (Ó Dónaill, 1977; Ua Maoileoin, 2001)

EQUISETACEAE—THE HORSETAIL FAMILY
Native

Equisetum arvense is a common plant throughout Ireland of roadsides, wasteground, sand dunes, banks and dry, stony and grassy places. It is also a common garden weed.

Shaw (1998) noted its herbal use to treat cystitis, rheumatoid arthritis, diarrhoea and incontinence in elderly people, but there are no records of such usage in Ireland. The same author recorded that young shoots can be eaten as a vegetable, like asparagus, and that the plant was used as a scour for pots and pans (the stems are rich in silica). In parts of North America this species has also been used in traditional medicine by Native Americans as a diuretic as well as to treat dropsy, kidney and bladder ailments (King, 1984).

Uí Chonchubhair (1995) mentioned that horsetails are poisonous to stock because of the presence of an enzyme that destroys vitamin B1, which results in symptoms of B1 deficiency (staggers[332]) in animals, mainly horses, and that this toxicity remains even when it is dried and stored, so it should not be included in hay or bedding.

In Co. Clare, horsetails have been used to treat 'hoose' in cattle. Hoose is a bronchial infestation by nematodes, seen in cattle, sheep, goats and pigs. Cattle with hoose were moved to a place where *Equisetum* grew, and after grazing on horsetails, the cattle were said to be cured (Munnelly, 2005).

Equisetum fluviatile (Equisetum limosum)

- Water horsetail (Scannell & Synnott, 1987); Feadoge (Hart, 1898); Pipes (Hart, 1898); Water pipes (Brunker, 1950)
- *Scuab eich uisce* (Scannell & Synnott, 1987)

EQUISETACEAE—THE HORSETAIL FAMILY

332. Holden (2009) reported that in 1747 one of the most infamous Irish highwaymen of the 16th century, 'Captain' James Freney, lost his beloved horse 'Beefsteaks' to staggers.

Native

Equisetum fluviatile is widespread throughout Ireland in wet habitats, such as marshes, ditches, lake margins and canals.

Hart (1898) recorded that this plant (listed by him as *Equisetum limosum*) was used in Co. Donegal by boys to make pipes or whistles called a *feadoge*.

Equisetum hyemale

- Rough horsetail (Scannell & Synnott, 1987); Dutch rush (Colgan, 1904); Shavegrass (Brunker, 1950)
- *Bioróg* (Scannell & Synnott, 1987); *Gadhar* (Cameron, 1883); *Liobhag* (Cameron, 1883); *Cuiridin* (Cameron, 1883)

EQUISETACEAE—THE HORSETAIL FAMILY

Native

A table mat made of *Equisetum hyemale*.

Equisetum hyemale is a rare plant of damp, shady habitats, such as open woodlands and river-banks, mainly in the east of Ireland.

Due to the high silica content of their stems, *Equisetum* species are, as Mackay (1836: 351) noted, 'admirably suited for the polishing of hard woods, ivory, brass etc. This species is by far the best kind for such purposes and is imported from Holland in considerable quantity, under the name of Dutch Rushes'. Cameron (1883) said that they were used for scouring pewter and wooden things in the kitchen. The Irish name *liobhag* comes from the Irish word *liobh*, meaning to 'smooth' or 'polish'. In North America this species is called 'scouring rush'. The Native American Ojibwa tribe used it to scour kettles (King, 1984). Clyne (1989) said that it was used in Scotland for scouring pewter and kitchen utensils. Rutty (1772) also mentioned that it was used 'like a file' for polishing wood, bones and metal. I have also seen its shoots made into attractive table-mats from Latvia, but not yet in Ireland (M. Sheehy Skeffington, pers. comm., 2009).

In North America it was also used in traditional medicine to treat gonorrhoea and kidney ailments (King, 1984), but I have not found records of similar usage in Ireland.

**Equisetum* ×*moorei*

- Moore's horsetail (Scannell & Synnott, 1987)
- *Scuab eich an Mhóraigh* (Scannell & Synnott, 1987)

EQUISETACEAE—THE HORSETAIL FAMILY

Native

Equisetum ×*moorei* is an interesting hybrid between *E. hyemale* and *E. ramosissimum*, although the latter has never been found in Ireland. It is found occasionally on the east coast of Ireland in Cos. Wexford and Wicklow and grows on sand dunes and clay banks.

*No ethnobotanical uses of this plant have been noted in Ireland.

Equisetum palustre

- Marsh horsetail (Scannell & Synnott, 1987)
- *Scuab eich chorraigh* (Scannell & Synnott, 1987); *Cuiristín* (Moloney, 1919)

EQUISETACEAE—THE HORSETAIL FAMILY
Native

Equisetum palustre is commonest in the north in Ireland, although it can be found throughout in a range of wet habitats, such as marshes, ditches and damp grasslands.

Mears and Hillman (2007) noted that the rhizome buds are edible and can be gathered for food between November and the following spring and cooked by boiling or steaming. Again, I have not found records of its use for food in Ireland. Other authors regard it as poisonous.

Moloney (1919) gave the Irish name *cuairidín* for *Equisetum nudum* (naked horsetail), which Colgan and Scully (1898) listed as a variety of *E. palustre*.

**Equisetum pratense*

- Shade horsetail (Scannell & Synnott, 1987)
- *Scuab eich mhóinéir* (Scannell & Synnott, 1987)

EQUISETACEAE—THE HORSETAIL FAMILY
Native

Equisetum pratense is found in damp, shady habitats and is distributed only in the north in Ireland.

*No ethnobotanical uses of this plant have been noted in Ireland.

Equisetum sylvaticum

- Wood horsetail (Scannell & Synnott, 1987); Fox's tail (Colgan, 1904)
- *Scuab eich choille* (Scannell & Synnott, 1987)

EQUISETACEAE—THE HORSETAIL FAMILY
Native

Equisetum sylvaticum is a relatively rare plant in Ireland, commonest in the north. It grows in wet woodlands and on stream-banks, particularly in mountain areas.

Mears and Hillman (2007) noted that although this species is generally poisonous, the buds at the ends of the rhizome are edible. These buds swell in September and October and can be cooked by steaming or boiling. They have a soft, floury texture and taste of potato with a hazelnut aftertaste. I have not found records of *Equisetum sylvaticum* used for food in Ireland.

Equisetum telmateia (Equisetum maximum)

- Great horsetail (Scannell & Synnott, 1987); Fox's tail (Colgan, 1904)
- *Feadóg* (Scannell & Synnott, 1987); *Earball capaill* (Moloney, 1919); *Gliogán* (Dinneen, 1927); *Gliogar* (Colgan, 1911)

EQUISETACEAE—THE HORSETAIL FAMILY
Native

Equisetum telmateia is the largest native horsetail of Ireland, and although it grows in shady hedgerows throughout, it is nowhere common.

Mahon (1983) listed this plant as providing a green dye.

An interesting horticultural use of this plant was described to me in 2010 by Seamus O'Brien, Head Gardener of the National Botanic Gardens, Kilmacurragh, in Co. Wicklow, as told to him by a visitor to the Garden. Shoots of the horsetail are boiled until a thick juice is made; the juice is then diluted and applied as a fungicide spray to treat mildew, blackspot and other fungal diseases of garden plants. Apparently, it is very effective.

**Equisetum ×trachyodon*

- Mackay's horsetail (Scannell & Synnott, 1987)
- *Scuab eich Mhic Aoidh* (Scannell & Synnott, 1987)

EQUISETACEAE—THE HORSETAIL FAMILY
Native

Equisetum ×trachyodon is a hybrid between *E. hyemale* and *E. variegatum* that occurs on lake-shores, river-banks and in wet woods, often in the absence of both parents. It is found mainly in the north and west.

*No ethnobotanical uses of this plant have been noted in Ireland.

**Equisetum variegatum*

- Variegated horsetail (Scannell & Synnott, 1987)
- *Scuab eich bhreac* (Scannell & Synnott, 1987)

EQUISETACEAE—THE HORSETAIL FAMILY
Native

Equisetum variegatum is rare in Ireland except in parts of the east. It is found on dune-slacks, along river-banks and in fens and canals.

*No ethnobotanical uses of this plant have been noted in Ireland.

**Gymnocarpium dryopteris*

- Oak fern (Scannell & Synnott, 1987)
- *Raithneach dharach* (Scannell & Synnott, 1987); *Sceamh darach*[333] (Moloney, 1919); *Scím d(h)arach* (Moloney, 1919; Ó Dónaill, 1977)

WOODSIACEAE—THE LADY-FERN FAMILY
Native

Gymnocarpium dryopteris is a very rare fern in Ireland that is found only on shady rocks in the Antrim Mountains.

*No ethnobotanical uses of this plant have been noted in Ireland.

**Gymnocarpium robertianum*

- Limestone fern (Scannell & Synnott, 1987)
- *Raithneach aolchloiche* (Scannell & Synnott, 1987)

WOODSIACEAE—THE LADY-FERN FAMILY
Native

Gymnocarpium robertianum is a very rare fern in Ireland that is found only amongst limestone rocks on a hill near Headford, East Mayo.

*No ethnobotanical uses of this plant have been noted in Ireland.

Huperzia selago (Lycopodium selago)

- Fir clubmoss (Scannell & Synnott, 1987); Stag-moss (Hart, 1898); Blessed heather (a name given for 'clubmoss') (McCarthy & Hawkes, 2007)
- *Aiteann Muire* (Scannell & Synnott, 1987); *Crúibíní sionnaigh* (Dinneen, 1927; Webb et al., 1996); *Garbogach* (*Iosóipe*) (Moloney, 1919); *Garbhógach* (Dinneen, 1927)

LYCOPODIACEAE—THE CLUBMOSS FAMILY
Native

Huperzia selago is an occasional plant of bogs and moorlands, particularly at higher altitudes.

Borlase (1769) and Wood (1821) mentioned that this was an important plant and probably used in rituals by the druids in Britain and Ireland for religious purposes. It was also

333. This name was also given by Dinneen (1927); however, I suspect that it should probably be applied to *Polypodium vulgare* growing on trees such as oaks, rather than to *Gymnocarpium dryopteris*.

thought good for eye ailments. Reference to the druids' use of this species probably originally came from the Pliny's *Natural History*, in which he mentioned this plant was gathered by the druids of Gaul after offerings had been made of bread and wine, cut without the use of iron, by someone dressed in white, with washed bare feet. It was thought to be an important talisman against all harm.

According to Mackay (1836) and Cameron (1883), it was used in the Scottish Highlands instead of alum to fix colours in dyeing. I am not aware whether it was used in Ireland for that purpose.

**Hymenophyllum tunbrigense*

- Tunbridge filmy-fern (Scannell & Synnott, 1987)
- *Dallán coille* (Scannell & Synnott, 1987); *Raithneach scanánach* (Moloney, 1919)

HYMENOPHYLLACEAE—THE FILMY FERN FAMILY
Native

Hymenophyllum tunbrigense is a beautiful and delicate little fern that is found particularly in the south-west of Ireland. It grows on tree trunks, over moist rocks, and in sheltered wet and humid cracks and gullies. It is much rarer elsewhere in Ireland.

*No ethnobotanical uses of this plant have been noted in Ireland.

**Hymenophyllum wilsonii*

- Wilson's filmy-fern (Scannell & Synnott, 1987)
- *Dallán sléibhe* (Scannell & Synnott, 1987)

HYMENOPHYLLACEAE—THE FILMY FERN FAMILY
Native

Hymenophyllum wilsonii is very similar to *H. tunbrigense* and is found in more or less the same habitats, although it tends to occur at higher elevations and in the mountains.

*No ethnobotanical uses of this plant have been noted in Ireland.

**Isoetes echinospora*

- Spring quillwort (Scannell & Synnott, 1987)
- *Lus an cleite an earraigh* (Scannell & Synnott, 1987)

ISOETACEAE—THE QUILLWORT FAMILY
Native

Isoetes echinospora is a rare fern-relative of shallow water at the margins of acid mountain and corrie lakes in western Ireland only.

*No ethnobotanical uses of this plant have been noted in Ireland.

**Isoetes lacustris*

- Quillwort (Scannell & Synnott, 1987)
- *Lus an chleite* (Scannell & Synnott, 1987)

ISOETACEAE—THE QUILLWORT FAMILY
Native

Isoetes lacustris occurs in habitats similar to that of *I. echinospora* but grows in deeper water and along the margins of lakes. It is more common than *I. echinospora* and is found mainly in the west and north of Ireland.

*No ethnobotanical uses of this plant have been noted in Ireland.

**Lycopodiella inundata (Lycopodium inundatum)*

- Marsh clubmoss (Scannell & Synnott, 1987)
- *Garbhógach chorraigh* (Scannell & Synnott, 1987)

LYCOPODIACEAE—THE CLUBMOSS FAMILY
Native

Lycopodiella inundata is commonest in West Galway and is very rare elsewhere in Ireland. It grows in wet bogs and lake-shores.

In North America the spores of this species have been used by the Native Americans as a dusting powder (King, 1984), but no similar usage is known in Ireland, where it has probably always been too rare to have been used.

*No ethnobotanical uses of this plant have been noted in Ireland.

Lycopodium clavatum

- Stag's-horn clubmoss (Scannell & Synnott, 1987); Wolf's claw (Wolfes-claw) (K'Eogh, 1735); Wandering fog (Hart, 1898); Lion's paw (Hart, 1898)
- *Garbhógach na mbeann* (Scannell & Synnott, 1987); *Garbhóg* (*-óige*) (or *Garbhógach*) *sléibhe*[334] (Dinneen, 1927)

LYCOPODIACEAE—THE CLUBMOSS FAMILY
Native

Lycopodium clavatum is a rare plant in Ireland of mountain grasslands and heaths.

It was said to 'restore ropy wine in a few days'[335] (Rutty, 1772: 121). Wade (1804: 94) also reported this use, but did not quote Rutty or any other particular source for this information.

O'Súilleabháin (1942) mentioned that clubmoss was used for food during times of famine but does not specify which species was used or what parts of the plant were eaten.

**Ophioglossum azoricum*

- Small adder's-tongue (Scannell & Synnott, 1987)
- *Lus na teanga beag* (Scannell & Synnott, 1987)

OPHIOGLOSSACEAE—THE ADDER'S-TONGUE FAMILY
Native

334. Dinneen (1927) gave this name for 'a club moss, or wolves' claws'.

335. Restoring 'ropy' wine suggests that wine that is beginning to spoil or to go musty, or that needs to have excess acidity removed, can be improved with the addition of the stag's-horn clubmoss.

Ophioglossum azoricum is rare in Ireland and is found only in the west, in sand dunes and coastal grasslands.

*No ethnobotanical uses of this plant have been noted in Ireland.

Ophioglossum vulgatum

- Adder's-tongue (Scannell & Synnott, 1987)
- *Lus na teanga* (Scannell & Synnott, 1987); *Lus na teangan* (Moloney, 1919); *Lus na nathrach* (Moloney, 1919)

OPHIOGLOSSACEAE—THE ADDER'S TONGUE FAMILY
Native

Ophioglossum vulgatum is an occasional plant throughout Ireland of rough and rocky grasslands.

It was recorded by K'Eogh (1735) as an excellent herb for healing wounds and bruises and for treating inflammations of the liver. An oil made from it can be applied to inflammations, burns, wounds, St Anthony's fire (a skin disease caused by ergot infection) and for the treatment of all 'hot tumours'.

Osmunda regalis

- Royal fern (Scannell & Synnott, 1987); Flowering fern (Hart, 1898); King fern (Hart, 1898)
- *Raithneach riúil* (Scannell & Synnott, 1987); *Raithneach an ríogh* (Moloney, 1919; Dinneen, 1927); *Bior-raithneach* (Moloney, 1919; Uí Chonchubhair, 1995); *Gall raithneach* (Colgan, 1911; Moloney, 1919; Dinneen, 1927; Ó Dónaill, 1977); *Inniúin-móm* (Moloney, 1919); *Raithneach ríoghamhail* (Dinneen, 1927); *Ríogh-raithneach* (Dinneen, 1927); *Raithneach Ghallda* (Dinneen, 1927; Ó Dónaill, 1977); *Raithneach capaill* (Dinneen, 1927); *Uinniún múin* (Dinneen, 1927)

OSMUNDACEAE—THE ROYAL FERN FAMILY
Native

Osmunda regalis.

Osmunda regalis can be found throughout Ireland, although it is only common in the north-west, west and south-west. It grows in bogs, river-sides, peaty banks, fens, bog-drains, lake-shores, ditches and damp woods.

This plant was formerly collected and harvested in several parts of the west of Ireland during the late 20th century for export and use in orchid cultivation.

Vickery (1995) noted a record from the Irish Folklore Commission of royal fern being used medicinally in Co. Clare to treat sprains; the fern was pounded and made into a poultice and applied to the sprain covered with a bandage. This same author in a later publication (2010) mentioned that in Co. Galway, it was called 'bog onion', and its roots were used as a treatment for sciatica and rheumatism. The roots were cut into slices and then pounded into a mash and bottled with some water and left until it formed a thick, white juicy substance before being rubbed on the aching part of the body. Threlkeld (1726: unnumbered [58]) noted that 'the roots are accounted good for obstructions of liver, and spleen, and particularly esteemed a specifick for rickets in children, as also for ruptures, wounds, and bruises'.

Wade (1804) reported that its roots were used in the 'North of Europe' to stiffen linen instead of starch, but it is not known whether it was used in this way in Ireland. Rutty (1772: 121) wrote that the root boiled in water is said to be very slimy, and 'to be used in the North for size, to stiffen their linen yarn'.

Cameron (1883) said that in Ireland it is called 'bog-onion'.

**Pilularia globulifera*

- Pillwort (Scannell & Synnott, 1987)
- *Lus an phiollaire* (Scannell & Synnott, 1987)

MARSILEACEAE—THE PILLWORT FAMILY
Native

Pilularia globulifera is a very rare fern relative in Ireland that is found only in the north and west. It grows in shallow water on lake-shores, ponds and in slow-moving streams.

*No ethnobotanical uses of this plant have been noted in Ireland.

Polypodium vulgare.

Polypodium vulgare (sensu lato)

- Common polypody (Scannell & Synnott, 1987); Checkleweed (Hart, 1898); Ladies' comb (Hart, 1898)
- *Scim chaol* (Scannell & Synnott, 1987); *Raithneach mhadaruaidh* (Ua Maoileoin, 2001); *Sceamh* (Dinneen, 1927); *Sceamh na gcloch* (Moloney, 1919; Dinneen, 1927; Uí Chonchubhair, 1995); *Mearlag* (Moloney, 1919); *Céis-chrann* (Cameron, 1883); *Sgeamh* (*-eimh*) (*-eamha*) (de Vere Coneys, 1849); *Sgeamh chroin* (de Vere Coneys, 1849); *Sceamh crainn* or *Sceam darach*[336] (Dinneen, 1927); *Scim na gcloch* (Ó Dónaill, 1977)

POLYPODIACEAE—THE POLYPODY FAMILY
Native

Including:

Polypodium australe (Polypodium cambricum)

- Southern polypody (Scannell & Synnott, 1987)
- *Scim leathan* (Scannell & Synnott, 1987)

336. Dinneen (1927) defined *sceamh crainn* or *sceam darach* as 'polypody of the oak (*Polypodium quercinum*)'. This name has been applied to the 'oak fern' also, *Gymnocarpium dryopteris*, but I suspect it should also be applied to *Polypodium* growing as an epiphyte on mature trees.

Polypodium interjectum

- Intermediate polypody (Scannell & Synnott, 1987)
- *Scim mheánach* (Scannell & Synnott, 1987)

Polypodium vulgare (sensu stricto)

- Polypody (Scannell & Synnott, 1987)
- *Scim chaol*; *Raithneach mhadaruaidh* (Ua Maoileoin, 2001)

Polypodium vulgare is an aggregate species including *P. vulgare* (sensu stricto), *P. interjectum* and *P. australe*. There are no known distinctions between the folk uses of different species of this aggregate in Ireland, and they can really only be distinguished effectively and accurately by an expert. Polypody ferns are common throughout Ireland on walls and rocky habitats and in shady hedgerows and banks and growing (as epiphytes) on trees.

According to Moloney (1919), the roots were formerly used in an infusion to treat rheumatism and powdered as snuff. Uí Chonchubhair (1995) noted that the snuff was used to cure nasal polyps and that a decoction of the crushed rhizome, sweetened with honey and taken with a pinch of aniseed, was used to remove obstructions of the spleen and to ease fevers. Hart (1898: 380) mentioned that 'the root boiled is good for a bad stomach'.

A range of cultivated forms of these ferns have been grown in Irish gardens as ornamentals, with variously shaped and twisted fronds (Nelson, 2000b).

---

*Polystichum aculeatum

- Hard shield-fern (Scannell & Synnott, 1987); Rough spleenwort
- *Ibheag chrua* (Scannell & Synnott, 1987); *Ibheag gharbh* (Moloney, 1919); *Iubhar talmhan*[337] (Dinneen, 1927)

DRYOPTERIDACEAE—THE BUCKLER-FERN FAMILY
Native

Polystichum aculeatum is occasionally found throughout Ireland in shady habitats and on banks, rocky places, hedgerows and woodlands.

Cameron (1883) and Clyne (1989) indicated that its powdered roots were made into a medicinal drink in Scotland that is taken in water as a specific cure for worms. I have not found records of its use for this purpose in Ireland.

*No ethnobotanical uses of this plant have been noted in Ireland.

---

*Polystichum lonchitis (Aspidium lonchitis)

- Holly fern (Scannell & Synnott, 1987)
- *Ibheag dheilgneach* (Scannell & Synnott, 1987); *Raithneach chuilinn* (Moloney, 1919); *Ibheag gharbh*[338] (Dinneen, 1927)

DRYOPTERIDACEAE—THE BUCKLER-FERN FAMILY
Native

Polystichum lonchitis is a very rare fern in Ireland, found only in mountains in the west, growing in rocky crevices.

*No ethnobotanical uses of this plant have been noted in Ireland.

337. Dinneen (1927) defined *iubhar talmhan* as 'rough spleenwort'. Cameron (1883) listed this as the name for *Juniperus*.

338. Dinneen (1927) defined *ibheag gharbh* as 'prickly shield fern'.

**Polystichum setiferum (Aspidium angulare)*

- Soft shield-fern (Scannell & Synnott, 1987)
- *Ibheag bhog* (Scannell & Synnott, 1987); *Ibheag mhín* (Moloney, 1919; Dinneen, 1927)

DRYOPTERIDACEAE—THE BUCKLER-FERN FAMILY

Native

Polystichum setiferum is a common fern of shady habitats, such as woodlands and hedgerows. It is one of the commonest ferns found throughout Ireland, except in the west where it is rare.

Hart (1898: 371) gave a local name for 'shieldfern' from Derry as 'crock-fern'.

*No ethnobotanical uses of this plant have been noted in Ireland.

Pteridium aquilinum.

Pteridium aquilinum

- Bracken (Scannell & Synnott, 1987); Fern (Colgan, 1904); Fairen (Colgan, 1904)
- *Raithneach mhór* (Scannell & Synnott, 1987); *Raithean-mhagh*[339] (Dinneen, 1927); *Raithín* (Dinneen, 1927); *Raithneach*[340] (Colgan, 1911)

DENNSTAEDTIACEAE—THE BRACKEN FAMILY

Native

Pteridium aquilinum is a common, and in some places, abundant plant throughout Ireland of rough acid grasslands, heaths and neglected or abandoned pastures. In a review of the history and ethnobotany of bracken worldwide, Rymer (1976) outlined the use of this species as thatch, compost, food, fuel, medicine, in the manufacture of glass, litter and bedding for animals, as a source of potash, for making soap and for bleaching and dyeing, and as a covering for tender horticultural plants from winter frosts. Some of these are known uses for bracken in Ireland.

It is the source of a green, yellow-green or yellow dye (Mahon, 1983; Kelly, 1997). In Britain it has been used in soap-making and as a source of potash fertiliser (Prendergast & Sanderson, 2004). The ashes were used in the manufacture of soap and glass (Mackay, 1836) and to bleach linen (Williams, 1993). Danaher (1964a) mentioned that soap made from the ashes of ferns (probably bracken) were said to be good for the complexion. Bracken was also burnt as a source of potash in Scotland; about 'five acres of fully stocked bracken land could yield a ton of ash, although this varied according to the productivity of the soil' (Milliken & Bridgewater, 2004: 96). In 1917 one ton of bracken ash could be sold for £17, a high price during World War I when imported supplies of potash were hard to come by.

In Ireland it was used to treat rickets, worms, burns, scalds and made into soup to treat Bright's disease (a kidney disorder). It was also used in veterinary medicine, for example, in Co. Cavan, where it was made into tea and fed to cows with a cough (Williams, 1993).

Milliken and Bridgewater (2004) indicate that livestock that feed on its rhizomes develop a serious disorder known as 'trembles' in Scotland.

339. Dinneen (1927) defined *raithean-mhagh* as 'a fern-covered plain or field, a bracken'.

340. While the Irish word *raithneach* means 'fern'; this name is often applied to *Pteridium aquilinum* specifically.

Pteridium aquilinum.

Bracken leaves contain a carcinogenic compound, ptaquiloside, which is broken down when the leaves are composted. Mears and Hillman (2007) noted that in several parts of the world, bracken rhizomes are or have been used as a source of starch for hunter–gatherer peoples (in New Zealand and in the Pacific Northwest of the United States), but they failed to find plants in Britain with sufficiently thick rhizomes to merit harvesting as a food source. In parts of Australia, Aboriginal people gather rhizomes, beat them into a paste and roast the paste in hot ashes. Then the skin is peeled off and it is eaten with meat (Maiden, 1899; Isaacs, 1987). Maiden (1899) also mentioned that in Japan, starch was extracted from bracken roots and made into cakes. However, Peterson (1977) warned that in Japan there have been reports of stomach cancer attributable to excessive consumption of bracken over a long period of time. In China, I have been served noodles made from bracken.

Williams (1993) noted that bracken spores were found during the Wood Quay excavations in Dublin, where it was probably used for bedding or animal litter. The same author noted that bracken fronds were placed over holes dug in the ground to store potatoes. Geraghty (1996) noted that fronds of bracken were commonly found in house floors in excavations from Fishamble Street, Dublin, that date back more than 1000 years, and stems seem to have been used as 'cavity wall fill'.

In parts of Ireland stone wall-beds were built into the sides of traditional cottages (half-ruined examples may be seen in the deserted houses of Slievemore on Achill Island, Co. Mayo) (Evans, 1957). Dried bracken (or heather) would have been spread on them to provide soft bedding. Bracken was probably widely used in Ireland for bedding up until the 18th century, simply spread on the floor in a heap or in bracken-filled beds made from a simple frame into which the bracken would have been packed tightly. They would be quite comfortable, and I have made a bracken mattress on one occasion when camping, after my airbed sprung a leak. I slept very well. In Scotland, Milliken and Bridgewater (2004) recorded that a bed made with fresh green fern fronds was a traditional cure for rickets, and, in parts of Scotland, newly married couples were encouraged to spend their wedding night on a hard barn floor covered only with bracken. I have not found records of these practices in Ireland.

Until quite recently, bracken was used for various purposes around Irish farms, such as under haycocks (to help keep rising damp from the hay) and for animal bedding (John McLoughlin, pers. comm., 2008) and even for rough thatching. McCarthy and Hawkes (2007: 5) described its use for fuel, particularly when baking bread over an open fire: 'If the turf store was running low, dry fern (bracken) was thrown into the fire when it was opened, as this gave a strong heat'. They also said that in the Northside region of Mizen, Co. Cork, it was used for wrapping cured fish and bacon. Pennant (1776) described how ferns (probably bracken) were used as the general thatching material on the Isle of Skye in the 1700s. Dutton (1808) mentioned that 'fern' (almost certainly bracken) was used for thatching in Co. Clare. However, we have little information on the use of bracken for thatching in Ireland. Nevertheless, in Scotland it was certainly widely used for that purpose. Milliken and Bridgewater (2004) mentioned how a good bracken thatched roof could last for up to 30 years, at least on the sunny side of the house. There the stems were pulled up by the roots, stripped of their fronds and generally tied in place with birch bark or heather. One cart load of bracken could be harvested during a day's pulling, which would thatch six square yards of roof. They noted that the average crofter's roof covered some 115 square yards, so thatching with bracken was a significant task.

Bracken would also have been used as fuel, particularly when other sources were scarce. It burns strongly and gives out lots of heat but burns too quickly to make it an ideal fuel. I have used it though as kindling to get the fire started.

Clyne (1989) (in *Gaelic Names for Flowers and Plants*) said that this fern has commonly been called 'the fern of God' in Ireland because if its stem is cut into three pieces, a pattern in the shape of 'G' will be seen in the first cut, 'O' in the second and 'D' in the third. I had not heard this story before in Ireland, nor have I tried cutting it to see if it is true.

**Selaginella selaginoides*

- Lesser clubmoss (Scannell & Synnott, 1987)
- *Garbhógach bheag* (Scannell & Synnott, 1987)

SELAGINELLACEAE—THE LESSER CLUBMOSS FAMILY
Native

Selaginella selaginoides is a fern-relative of bogs, marshes, fens, lake-shores, heathy grassland, wet places in sand dunes and mountain ledges and cliffs that is most commonly found in the north. It is also occasionally found in the centre of the country and is rare elsewhere.

*No ethnobotanical uses of this plant have been noted in Ireland.

**Thelypteris limbosperma (Oreopteris limbosperma, Lastrea thelypteris)*

- Lemon-scented fern (Scannell & Synnott, 1987); Mountain fern (Webb et al., 1996)
- *Raithneach bhuí* (Scannell & Synnott, 1987); *Raithneach mhilis an tsléibhe* (Moloney, 1919); *Crim-raithneach* (Moloney, 1919)

THELYPTERIDACEAE—THE MARSH FERN FAMILY
Native

Thelypteris limbosperma is a rare fern in Ireland. It can be found throughout on rocky mountain slopes and cliffs and in stream-banks, ditches and drains and open woodlands.

*No ethnobotanical uses of this plant have been noted in Ireland.

**Thelypteris palustris (Lastrea thelypteris)*

- Marsh fern (Scannell & Synnott, 1987)
- *Raithneach chorraigh* (Scannell & Synnott, 1987)

THELYPTERIDACEAE—THE MARSH FERN FAMILY
Native

Thelypteris palustris is a rare fern in Ireland. It can be found throughout in marshes.

*No ethnobotanical uses of this plant have been noted in Ireland.

**Thelypteris phegopteris (Phegopteris connectilis)*

- Beech fern (Scannell & Synnott, 1987)
- *Raithneach feá* (Scannell & Synnott, 1987)

THELYPTERIDACEAE—THE MARSH FERN FAMILY
Native

Thelypteris phegopteris is found mainly in the north in rocky mountainous habitats, such as cliffs and rocky slopes.

*No ethnobotanical uses of this plant have been noted in Ireland.

Trichomanes speciosum

- Killarney fern (Scannell & Synnott, 1987)
- *Raithneach Chill Airne* (Scannell & Synnott, 1987); *Raithneach ruainneach* (Moloney, 1919)

HYMENOPHYLLACEAE—THE FILMY-FERN FAMILY
Native

Trichomanes speciosum is rare in Ireland and is found mainly in the south-west in Cos. Cork and Kerry. This delicate and beautiful fern grows close to waterfalls and in wet shady crevices or under rocks and in caves. It was formerly relatively frequent in some parts of those counties, particularly around Killarney, where it gained its common name.

Scully (1916: 360) wrote in the *Flora of County Kerry*: 'few plants have suffered more from the depredations of unscrupulous collectors, botanical and mercenary, than has the Killarney Fern'. There are accounts of 'basket fulls' of the fern being removed from the wild from places such as Torc waterfall by local people and carried around to various hotels in Killarney where they were offered for sale to tourists. Scully recorded that he saw a large box of fronds and roots being sold in Killarney by someone who had come from Sneem and offered them for £5. When the seller could not get any takers at that price the lot was split and sold off in pieces at costs varying from 2s 6d to 7s 6d. The plant was a popular species to include in sealed bottle gardens, particularly during the 'fern craze' of the Victorian era.

Coprinus comatus. Watercolour by Lydia Shackleton. Courtesy of the National Botanic Gardens, Glasnevin.

ALGAE AND MISCELLANEOUS

Algae

- *Algaí*

Seaweed (marine algae)

- *Feamainn* (Anonymous, 1978); *Feamuin* (*-mna*) (de Vere Coneys, 1849); *Fadhairt* (*-artha*) (Dinneen, 1927); *Faimleach* (*-ligh*) (Dinneen, 1927); *Fraileach* (*-lighe*) (Dinneen, 1927)

Seaweeds have been used in Ireland from ancient times up to the present day for food, fodder for animals and as a soil conditioner or fertiliser. Seaweeds are rich in iodine and minerals and would formerly have been valuable dietary supplements. A small number of people still make their living as professional seaweed harvesters in parts of the west of Ireland. Seaweed is commercially harvested in Ireland today for use in seaweed baths and in the preparation of liquid seaweed extracts and industrial alginates, agars and carrageenans. Alginates are used for many purposes, such as in dentistry in the making of impressions of the jaw before fitting orthodontic appliances.

Seaweed also provided an important famine food in Ireland. In Cecil Woodham-Smith's book about the Irish potato famine, *The Great Hunger: Ireland 1845–1849* (1962), she described a letter written to the British authorities in February 1847, during the height of the potato famine, which said that not a piece of seaweed existed west of Skibbereen in Co. Cork because it had all been picked for food. However, Róise Rua, in her memoir of life on Arranmore, wrote that her step-father who lived through the famine years said, 'You'll hear people suggesting the hunger wasn't as severe in Arranmore as it was throughout the mainland for there was plenty of sea-food there for people to lay their hands on—dulse, sloak, carrageen [all seaweeds], barnacles, winkles, and so on. The sea-food wasn't really a great help' (Ua Cnáimhsí, 2009: 137). Instead, they survived on the cargoes of imported meal brought into the region by sea.

Rutty (1772: 36) mentioned that seaweed (*Alga marina*) 'was an article of food at the siege of Londonderry, and purchased at a considerable price'. The siege took place in 1689 and lasted 105 days, from 8 April to 28 July, during which it is said that 8000 of the 30,000 population died, mostly from disease.

In coastal areas, cattle were often fed on seaweed from the earliest times in Ireland (Kelly, 1997). The seaweed is rich in nutrients and salt and provides a good substitute for grass. Kelly also suggested that being rich in salt, seaweed may have been fed to cattle to increase their thirst and therefore their milk yield.

Seaweed harvested from the rocks or washed up on the shores of parts of the west of Ireland has been used as a soil conditioner, mixed with sand, for the cultivation of potatoes and other crops, particularly in the Aran Islands, Co. Galway (O'Sullivan, 1977).

The distinguished Limerick-born botanist, William Henry Harvey (1811–1866), in his book *A Manual of the British Marine Algae* (1841: xliii–xliv) wrote:

> On many shores the harvest of the deep is anxiously looked after, and as carefully attended to as that of the land, and indeed the last is often dependant on the former

A load of seaweed, harvested for use as manure. Reprinted with permission. From Mason, Thomas H. 1936. *The Islands of Ireland*. Batsford, London.

for its abundance ... On many of our coasts, as along the west of Ireland, the poorer classes are almost entirely dependant for the cultivation of their potatoes on the manure offered by their rocky shores, and frequent gales of wind. After a storm they may be seen congregating in numbers from the surrounding county, with horses and carts, or with panniers; and the poorest, who cannot afford the assistance of a donkey, are themselves bearers of burdens, eagerly collecting what is thrown up and carrying it beyond the reach of the tide. The kinds preferred for potatoes are the large and succulent Laminarieae, which rapidly melt into the ground, and when these are abundant other kinds are neglected.

William Henry Harvey (1811–1866).

O'Sullivan included a detailed description of the use of seaweed in the Aran Islands in the 1900s where it was used both as a fertilizer and 'cash crop'. Alternating layers of sand and seaweed would be laid on top of the bare limestone rock to create new fields for cultivation of potatoes, rye and vegetables. Seaweed was used widely as a fertilizer in coastal areas but was also carted inland for use. In Achill Island, Co. Mayo, for example, a traditional boat, the 'Achill Yawl' was used to harvest seaweed along the coast or it was gathered as driftweed in the sea. The seaweed was then brought home to fertilise the field crops of potato, oats or rye (J. Cunnane, D. Ó Gallchobhair & J. Kilbane, in MacCárthaigh, 2008). Until the middle of the 20th century transporting seaweed was one of the major tasks of the Galway hookers, the large traditional sailing boats, which also transported peat from Connemara to the Aran Islands and north Clare (MacCárthaigh, 2008). At the height of this trade there were up to 100 boats involved in this trade moored at Kinvara. The largest boats could carry up to 8 tonnes of seaweed, which was brought from Connemara to Galway and sold at the small ports along the south Galway and north Clare piers and harbours to the local farmers who came down to these ports with the horse and cart to collect the seaweed (P. Barry & D. Scott, in MacCárthaigh, 2008). Evans (1957) records that in the Mourne Mountains wrack and shell sand were carted inland as much as 30 miles from the coast. In the Aran Islands, the islanders also produced 'kelp', an iodine-rich compound derived from burning seaweed, which was sold onto the mainland and used in the manufacture of chemicals and drugs. The seaweed was cut at low tide and carried onto the shore, often in donkey panniers, where it was dried. Twenty-five tonnes of seaweed were needed to make 1 tonne of kelp. In June the seaweed was burned in a carefully controlled fire in a specially constructed 'kelp kiln'. When the weed melted it was stirred with special wooden rakes until it formed into a liquid molten mass, which became solid when cool. Later it was discovered that if not stirred the kelp became a powder that was equally rich in iodine and was easier to transport in bags. The cash obtained from selling the kelp was often used to buy turf for fuel or timber (there were no peat deposits and few trees on Aran). Kelp burning was practiced on the mainland too. Dutton (1808: 265) records that on the shores of Co. Clare a 'considerable quantity of kelp is manufactured on the extensive shores this county possesses; it is generally made in so careless and dishonest a manner, that the value is considerably less than that of Scotland or other countries; when it sells in Scotland for 6/- per tonne, it only brings in Ireland 4/- per ton so far are they from clearing the sea-weed from sand, shells, or other extraneous substances'.

In Arranmore, the island off the west coast of mainland Co. Donegal, kelp-burning was common also in the 1800s. Róise Rua, in her memoir of life on Arranmore in the late 1800s, described kelp-making:

> Kelp-making was in full swing in my seasons on the island. A ton of kelp made between three and four pounds then [in the 1890s], and they reckoned it took about five tons of seaweed to make one ton of kelp. Seaweed was washed ashore at times, especially while the sea was rough after a gale, but most of the time it had to be cut with hooks. It was cut with a big hook, or kelp hook as they call it in places. It was especially made for the job. It had a long shaft so that the man using it could dip it down in the water and cut the seaweed growing down on the seabed. They usually did this at low tide, for it was then easy to cut; the best time of all was low water in the spring tides. Seaweed floats to the surface as soon as it's cut and then it has to be loaded aboard the boat. After it was brought ashore, it was spread out on the stone fences nearby; it was turned a few times to make sure it was thoroughly dry. It was then heaped in cocks and left like that for three weeks before it was burned in the kiln. It was a rare day you wouldn't see smoke rising from the kelp kilns throughout the Rosses at that time. When the kelp was made, it had to be taken by boat as far as Burtonport or Rutland Harbour, and sold there. Later on, a ship would arrive to take a cargo of kelp to Scotland. Kelp-making was always done during the summer; you couldn't do that sort of work any other time of the year (Ua Cnáimhsí, 2009: 56).

Harvey (1841) mentioned that all the large *Fucus* and *Laminaria* species, as well as *H. lorea* [its accepted name is now *Himanthalia elongata*] and *Chorda filum* were used for making kelp.

Despite the bountiful harvest of seaweed along the coast, it was not a free resource. Róise Rua also remembered that 'the tenants had to pay a rent of fifty pounds: twenty-six for the use of the land and twenty-four for the use of the seashore—making kelp, picking winkles or shellfish, dulse and the like' (Ua Cnáimhsí, 2009: 58–59). Ní Shúilleabháin (1988: 59) wrote that in the 1930s on the Blasket Islands in Co. Kerry on Good Friday 'we have a custom here to go to the strand for sea grass and other things that's growing on the rocks… All the women and children go to the strand that day. It is an old custom I think'.

Rutty (1772: 101) said of '*Alga marina*. Sea Wreck': 'Burnt, it yields Kelp useful for Soap boilers, Bleachers, and makers of Glass, tho' little of Kelp is made on our coasts, but brought us from the Co. of Galway. The Sea weed is also used as a rich Manure, like Marl, by some of our people on our Coasts, as by the Scotch and French, and is said to be particularly good for Asparagus'.

Uí Sheasnáin (1982) recorded the use of seaweed in traditional medicine in Ireland to treat asthma: 'hold a handful of seaweed to the mouth and breathe through it'.

Kelp, an iodine-rich compound derived from burning seaweed. From the Economic Botany collections of the National Botanic Gardens.

Dried seaweed for kelp. Reprinted with permission. From Mason, Thomas H. 1936. *The Islands of Ireland.* Batsford, London.

A postcard from the early 1900s showing a harvest of seaweed.

Seaweed tea was occasionally made in Ireland. MacConghail (1987) mentioned that on the Blasket Islands in Co. Kerry, a drink called '*sleadaí*' was made from seaweed.

Dinneen (1927) gave the Irish word *flocas* for a spongy seaweed used when dried for stuffing mattresses. It is not clear to which species of seaweed this name is applied.

In recent years the possible use of seaweed as a biofuel has been suggested and this may become a productive future crop in Ireland. Seaweed is a potential source of bioethanol that is typically made from crops such as maize (corn), sugarcane, and oil from rapeseed (canola), peanut and palm oil. Brown algae such as *Laminaria* species, which grow quickly and can be sustainably harvested, are the most likely future source for use as a biofuel crop, although microalgae living in the seas around Ireland would also be a possibility for exploitation. Research in this field is being carried out at the Irish Seaweed Centre at the National University of Ireland in Galway.

Alga unspecified

Allen and Hatfield (2004) noted that the scum of stagnant water, presumably species of blue-green algae, has been applied to burns in Co. Meath.

Alaria esculenta.

Alaria esculenta

- Dapperlocks; Bladderlocks
- *Láracha* (Anonymous, 1978); *Muiririn* (Cameron, 1883)

Alaria esculenta is an edible brown alga that has been collected (commercially and locally) in Ireland, particularly in parts of the south-west, west and north-west. It can be used as food, thinly sliced in salads. It can also be boiled in water to make soup or dried and powdered to make seasoning.

Cameron (1883) said that, according to Dr Drummond, in some parts of Ireland it is called 'murlins', probably a corruption of its Irish name, and also *sparain* (or *sporain*), because the pinnate leaves were thought to resemble a Scottish Highlander's sporran.

Ascophyllum nodosum

- Knotted wrack; Yellow wrack[341] (Colgan, 1911)
- *Feamainn bhuí* (Anonymous, 1978); *Trioscar dubh na ruadhog* (Moloney, 1919); *Feamain bhuidhe* (Dinneen, 1927)

Ascophyllum nodosum is a brown alga of which some 16,000 tonnes are harvested annually in Ireland. Lorna Siggins (writing in *The Irish Times*, 28 August 2007) documented

341. This English name was reported by Colgan (1911) from Belclare, Clew Bay, Co. Mayo.

the work of Johnny Clochartaigh, a professional seaweed harvester in southern Connemara, Co. Galway, harvesting seaweed (apparently mainly *Ascophyllum nodosum*) for industrial use. Clochartaigh cuts the seaweed by hand with a knife and uses a pitchfork to work it into bales (called *climín*) for which he is paid €38 per tonne by Arramara Teo, a state-sponsored body and manufacturing concern (established in 1947). Once cut and tied, the bales are towed inshore by currach (a traditional small boat with a canvas-covered hull). The material is processed to yield alginates, used as a gelling agent for ice cream, a binding agent for tablets, wound dressing and dental impressions and in textile and electrode production. It is also used in the manufacture of horticultural products (liquid feeds) and as an animal-feed supplement. In 2007 Arramara Teo estimated that about 300 people work in seaweed harvesting in Cos. Galway and Mayo, some of them harvesting throughout the year. In 2000 seaweed production was estimated to be of value €8.8 million to the national economy. On the west coast some families and landowners hold traditional seaweed harvesting rights, known as *Spiteanna feamainne*. One Irish company is producing a seaweed powder (made mainly from *Ascophyllum nodosum*) for the cosmetic and algotherapy market.

'A wrack harvest. Connemara'.

Chondrus crispus

- **Carrageen (Carrigeen) moss; Irish moss; Sea moss**
- ***Cosáinín* (Ó Dónaill, 1977; Anonymous, 1978); *Clúimhín cait* (Ua Maoileoin, 2001); *Carraigín* (Moloney, 1919); *Cosáinín carriage* (Moloney, 1919); *Máthair an duilisc* (Moloney, 1919); *Ceann donn* (Moloney, 1919; Dinneen, 1927; Uí Chanainn, 1962)**

Chondrus crispus is a popular edible red alga that is still frequently gathered on Irish coasts. An extract of this species, carrageenan, is used in processed foods for stabilisation, thickening and gelation and has been widely used in the food industry worldwide as an additive for ice-cream, chocolate milk, custards, cheeses, jellies, confectionary, meats and protein drinks. It is also used to thicken cosmetics (Lewington, 1990).

During my childhood in the 1960s in Camp, on the Dingle Peninsula in Co. Kerry, large quantities of carrageen moss gathered from the lower rocky shores were collected, dried and sold by local people during the time of low spring tides. It is used to make a mucilaginous *blanc mange* soup and is said to have medicinal properties.

To prepare the carrageen it is boiled in water or milk (about 1:3 by volume), with sugar or other flavourings (such as lemon juice, cinnamon, nutmeg, coffee or chocolate) added to taste until the seaweed has more or less dissolved. After the remaining undissolved bits have been removed the liquid is poured into a mould and left to cool and set. Carrageen soup can also be made or it can be added to other soups to give them body. Carrageen soup is made by boiling 200 mL of dried carrageen in 1 litre of water to which are added slices of bacon, chopped celery, carrots, onions, herbs and seasoning. After the mixture has simmered for close on an hour it can be blended and served. During wartime shortages in Ireland my grandmother used carrageen instead of gelatine in cooking (Lois Wyse Jackson, pers. comm., 2007). White Lennon (2002) provided a recipe for a traditional-style Irish seafood chowder made from potatoes, onions, various fish and two seaweed ingredients, carrageen and dillisk (*Palmaria palmata*—dulse) which help to thicken the soup as well as to give it flavour.

Carrageen moss has been used over many centuries in traditional medicine, mainly to treat coughs and for bronchitis (Jude, 1933; van Wyk & Wink, 2004). Logan (1981) records that carrageen, to which nasturtium seeds have been added, boiled and drunk, was used in Ireland as a cure for lung diseases in general and tuberculosis in particular. This species and *Mastocarpus stellatus* (also called carrageen moss or Irish moss) were widely used in Ireland for medicinal purposes, to treat coughs, colds, sore throats, chest and lung ailments (including tuberculosis), burns and kidney trouble (Allen & Hatfield, 2004). The carrageen was boiled in water or milk, strained and drunk hot. It was used in cough mixtures (Moloney, 1919) and for shortness of breath in the Aran Islands (Doolan, 1994).

MacMahon (1994) provided details of the collecting of carrageen in his book on Ballyheigue in north Kerry. There it was only collected at spring tides and spread out on the grass to bleach. When dried, a sackfull would be sold for 7 shillings and 6 pence a stone.

Extracts of carrageen have also been used in cosmetics in recent years in Ireland.

Chorda filum

- Sea laces; Mermaid's tresses; Dead man's rope
- *Ríseach* (Colgan, 1911); *Driamlaichean*[342] Clyne (1989)

Chorda filum is a brown alga that occurs around much of the Irish coastline. It is an annual plant that has long string-line fronds, which can grow up to 8 m long.

Clyne (1989) said that this species becomes so tough that it was formerly frequently used as a fishing line and in the manufacture of fishing nets.

Corallina officinalis

- Coraline
- *Cúnach trágha* (Moloney, 1919); *Cúnach trá*

Corallina officinalis is a red alga collected (commercially and for local use) in Ireland, particularly in parts of the south-west, west and north-west and is used as a soil fertiliser and industrially in modern-day skin care products.

Desmarestia aculeata

- Witch's hair
- *Gruaig trágha* (Moloney, 1919)

Desmarestia aculeata is a brown seaweed that is probably not edible (as I have heard that this species can give off sulphuric acid as it breaks down). It would have been collected and used as a soil conditioner and fertilizer along with other seaweeds with which it is associated.

342. This is the Scottish Gaelic name for this species.

Fucus serratus

- Serrated wrack
- *Míoránach* (Ó Dónaill, 1977; Anonymous, 1978); *Trioscar dubh ladharthach* (Moloney, 1919); *Turscar dubh ladharach* (Dinneen, 1927); *Bolgán* (or *Boilgín*) (Colgan, 1911)

Fucus vesiculosus

- Bladder wrack
- *Feamainn bhoilgíneach* (Anonymous, 1978); *Dúlamán na gclog* (Anonymous, 1978); *Trioscar dubh nag clog* (Moloney, 1919); *Feamain bhuilgíneach* (Moloney, 1919; Dinneen, 1927); *Feamain cháilíneach* (Moloney, 1919); *Leathach*[343] (Dinneen, 1927); *Turscar (dubh) na gclog* (Dinneen, 1927); *Bolgán* (Colgan, 1911); *Boilgín* (Colgan, 1911); *Feamainn dubh* (Colgan, 1911)

Fucus vesiculosus is a brown alga collected (commercially and for local use) in Ireland, particularly in parts of the south-west, west and north-west of Ireland and used as a soil fertiliser and industrially in modern-day skin care products. Seaweed baths were a frequent feature of Irish seaside resorts in the past in the west and south of Ireland in the late 19th and early 20th centuries, such as at Kilkee, Co. Clare, and Salthill, Co. Galway. Only one fully operational seaweed bath centre continues to operate today, in Killala Bay at Enniscrone, Co. Sligo. *Fucus serratus* is the species used, which is steamed and placed in baths of seawater to help ease aches and pains, such as arthritis and rheumatism. Harvey (1841: lii) mentioned that mucus derived from *Fucus vesiculosus* 'and other species' was used medicinally to treat 'diseases of the glands', gaining healing benefits from the iodine that the mucus contains.

Smith ([1756] 1979) documented the use of 'seawrack' mixed with sea sand as a manure to fertilise land for crops on the Magharees Peninsula north of Castlegregory in Co. Kerry. Hart (1898: 367) mentioned that bladder wrack was used for manure in Co. Donegal, where 'stones are regularly set in the slob in some places, and a crop cut off'.

It was formerly used in Northern Ireland for fattening pigs and by herbalists to treat obesity (Moloney, 1919) and marketed under the name 'Anti-fat' (Jude, 1933). The same author mentioned that it was used to treat scrofula and joint rheumatism in Ireland.

Cameron (1883) mentioned that in the Hebrides of Scotland cheeses were dried without salt by covering them with ashes of this seaweed.

Moloney (1919) gave an Irish name for *Fucus canaliculatus* (channelled wrack; see *Pelvetia canaliculata*) as *dualamán*, and Dinneen (1927) gave *dubh-lomán*. Colgan (1911) also gave an Irish name for this alga, *cos sionnaigh* (fox's paw), reported from Louisburgh, Co. Mayo. Cameron (1883: 100) said that this species was a favourite feed for cattle in Scotland, which farmers would give to 'counteract the injurious effects of sapless food, such as straw and hay'. Dinneen (1927: 1279) gave the Irish name *turscar dubh na ruadhóg* for 'knobbed wrack'.

343. Dinneen (1927) defined *leathach* as 'wrack'. *Buinne leathaigh* is a pile of wrack thrown up on the shore (Co. Mayo).

Fucus species growing on the seashore have been used as winter grazing for cattle, sheep and deer in Scotland when grass is in short supply and to counter the effects of old dry straw and hay (Clyne, 1989). Harvey (1841) also mentioned that this species was used as winter food for horses and cattle in the west of Ireland.

Himanthalia elongata (Himanthalia lorea)

- **Thongweed; Buttonweed; Sea spaghetti**
- *Ríseach* (Anonymous, 1978; Colgan, 1911); *Imleacán cloch* (Anonymous, 1978); *Scaoileach*[344] (Colgan, 1911); *Corrleach* (Colgan, 1911)

Himanthalia elongata is a brown alga that is eaten in France after drying or pickling. Some collecting (commercially and for local use) in Ireland, particularly in parts of the south-west, west and north-west. In the north of Scotland this species was used to make a kind of sauce for fish or fowl, resembling ketchup (Clyne, 1989).

I have come across (but not yet tried) a recipe with this species of seaweed: 'Spicy garlic tomato and sea spaghetti'. Some chopped onions and mushrooms are fried until soft, then the sea spaghetti (50 g), a few crushed cloves of garlic and a tin of chopped tomatoes are added, and the mixture is seasoned with a little turmeric, pepper and chili powder. It is then simmered for a while until soft and served, perhaps with baked potatoes.

Laminaria species

Including:

Laminaria digitata

- **Kelp, Oarweed; Sea rod (Colgan, 1911)**
- *Coirleach* (Anonymous, 1978); *Ribíní* (Anonymous, 1978); *Stamh* (Cameron, 1883); *Slát-mhára* (Cameron, 1883); *Liach* (*Léiche*)[345] (Dinneen, 1927); *Slat mhara*[346] (Colgan, 1911); *Feamain dearg* (Colgan, 1911)

Laminaria hyperborea

- **Kelp; May weed**

Laminaria saccharina (Saccharina latissima)

- **Sea belt; Sugar kelp; Leafy ware**
- *Rufa* (Anonymous, 1978); *Madraí rua* (Anonymous, 1978); *Copógach woher*[347] (Colgan, 1911)

---

344. Colgan (1911) said that this Irish name is applied to *Himanthalia* at 'Rinvyle' [Renvyle, Co. Galway] and to *Laminaria* at Cloghmore [Co. Mayo]. He said also that the word is derived from the verb *scaoilim*, 'to scàtter or spread loosely', in allusion to the waving fronds.

345. Dinneen (1927) defined *liach* (*léiche*) as 'kelp-weed (*Laminaria*) (Clare I.)'.

346. According to Colgan (1911: 4.15), this name 'applied to the thick stems of the Tangle in Clare Island and on Achill Sound as it is in Donegal, in the South Isles of Aran, and even in Co. Dublin, where the shortened form *slots* or *slocks* is in use: one of the chief materials for kelp-making'.

347. Colgan (1911) said this is 'a compound Irish–English name used for the species at Cloghmore, Achill Sound [Co. Mayo], but perhaps generic for the *Laminarias*'.

Laminaria drying in Inisheer, Aran Islands.

Laminaria hyperborea is a brown algae that is collected (commercially and for local use) in Ireland, particularly in parts of the south-west, west and north-west. It is used as a soil fertiliser and industrially in modern-day skin care products.

MacMahon (1994: 133) recorded that a 'long ribbon seaweed' called *múr buí* [probably *Laminaria* spp.] was collected at the Black Rock near Ballyheigue, north Kerry, and dried and sold for £30 a ton in the 1950s or 1960s and 'eventually taken to Liverpool for iodine and liniment'.

Phillips (1983) suggested that the fronds of *Laminaria saccharina* can be used to make a salty snack chip when dried, cut into squares and fried. I have also tried *L. hyperborea* this way, cut into squares and dried in an oven. I enjoy them like this, as a low-calorie snack to go with a drink. However, the younger members of my family think they taste pretty disgusting and say that they will stick with potato crisps. One source mentioned to me that *L. saccharina* is a useful vegetable that can be added to almost anything that is boiled. To prepare the seaweed, it is simply cut into strips and boiled. In one recipe I found, it is cooked with honey, crushed garlic and soy sauce. I've come across kelp used in a home-made energy drink called 'Iron Crazy' served in the Cúl Gairdín Restaurant in Dingle, Co. Kerry, in 2010, made with spinach, apricots, carrots, pumpkin seeds and kelp powder.

Laminaria digitata.

Hart (1898: 388) mentioned that *Laminaria* stems (called 'slack marrow', 'slab marrow' or 'slock marrow'; *slacan* is a club and *mara* is the sea) were sometimes used as cudgels when a row takes place amongst the kelp burners. When the bailiffs went to Tory Island, Co. Donegal, for taxes, 'the women beat them out of it with slack marrow [*Laminaria* stems]'. Harvey (1841: l) said that cut sections of *L. digitata* were used to make knife handles in Scotland, but did not mention such usage in Ireland. He wrote: 'A pretty thick stem is selected, and cut into pieces about four inches long. Into those, when fresh, are stuck blades of knives, such as gardeners use for pruning or grafting. As the stem dries it contracts and hardens closely and firmly embracing the hilt of the blade. In the course of some months the handles become quite firm, and very hard and shrivelled, so that when tipped with metal they are hardly distinguished from hartshorn [the horns of male red deer]'.

As a child growing up in the Dingle Peninsula, Co. Kerry, in the 1960s, occasionally a piece of seaweed (usually *Laminaria,* locally called 'ribbons') was hung by the door of the house (indoors) and if moist, it was said to foretell rain, operating like a primitive barometer.

Moloney (1919) gave a range of Irish names for the various species of *Laminaria* in Ireland: *trioscar, trioscar an aithchinn, fuip, slat mara, stamh, burrscaimh, bárlóg, feamnach, feam* and *múrach*[348]. Cameron (1883) gave other names: *duidhean, liaghag, leathagan, barr stamh* and *bragair*. Dinneen (1927: 73) gave the Irish name *báirleog* (*-oige*) for 'a ribbon-like sea-weed growing in clusters at end of a thick stem (feam)'.

Mastocarpus stellatus

- Carrageen moss; Irish moss

348. This name was also given by Dinneen (1927).

Mastocarpus stellatus is a red alga that is gathered (to a limited extent) on Irish coasts for local and commercial use, and like *Chondrus crispus,* it is also known as carrageen moss or Irish moss.

This species and *Chondrus crispus* were widely used in Ireland for a variety of medicinal purposes (see *Chondrus crispus*) (Allen & Hatfield, 2004).

Nostoc species

Nostoc species are dark-coloured jelly-like blue-green algae (cyanobacteria) found in solution hollows in limestone rocks, such as in the Burren, Co. Clare, and in the Aran Islands. They have been used medicinally in Ireland to treat burns (O'Rourke, 2006).

Hart (1898: 387) gave the common name 'shot star' for *Nostoc commune.*

Palmaria palmata dried and ready to eat as a snack.

Palmaria palmata (Rhodymenia palmata)

- Dulse; Dillisk
- *Duileasc* (Anonymous, 1978); *Creathnach* (Anonymous, 1978); *Duileasc garbh* (Moloney, 1919); *Duiliasg* (Cameron, 1883); *Rapán* (Dinneen, 1927); *Dileascach na g'cloch* (Ó Criomhthain, MSS, n.d., ca. 1908)

Palmaria palmata, a red alga, was the most prized edible seaweed in ancient Ireland (Kelly, 1997). It is still gathered locally in Ireland today and is generally eaten dried, sold in small packets. I have come across a variety of recipes in which it is used for food in Ireland, such as mixed with wholemeal flour and used in the baking of brown soda bread, or mixed with lentils and various vegetables to make soup. Dulse can also be cooked by simmering in water until tender or chopped and eaten raw. Maja Binder makes a semi-hard cheese called 'Dilliskus' at Kilcummin, Dingle Peninsula, Co. Kerry, which is flavoured with dulse seaweed (Anderson & McLaughlin, 2011).

Harvey (1841: 63) recorded that it was much eaten in Ireland and Scotland as well as in most of the northern countries of Europe 'by the poor along the shores, and is transmitted as an article of humble luxury over most parts of the country'. He noted that it was generally eaten raw, either fresh from the sea or having been dried and was sometimes cooked. Specimens growing near low-water mark are shorter, sweeter and less leathery to eat. Harvey (1841: 63) also noted that 'cattle, especially sheep, are fond of it'. Hart (1898: 372) reported that it was 'eaten by the poor people' in Donegal and exported to Liverpool and Glasgow, dried in the sun for use but also eaten fresh off the rocks. Rutty (1772: 57–58) said that this species (listed as '*Fucus membranaceus ceranoides*') was gathered on the rocks about Dalkey Island in Co. Dublin, 'and being dried and infused in water is said to smell like a violet. It is sold in the streets of Dublin, being dried, and is eaten, causing a sweet breath and is said to kill worms'.

Packaged *Palmaria palmata.*

It was widely used in folk medicine to protect against general ill health as well as to treat worms in children (in the Aran Islands), sore throats and scurvy (Allen & Hatfield, 2004).

Mahon (1983) noted that dulse was used to obtain a brown dye for dyeing textiles.

Palmaria palmata (dulse), on sale in Kilkee, Co. Clare.

Pelvetia canaliculata (Fucus canaliculatus)

- Channelled wrack
- *Dualamán* (Moloney, 1919); *Cos sionnaigh* (Colgan, 1911); *Dubh-lomán* (Dinneen, 1927)

In Ireland this brown seaweed was used as a famine food and probably also sometimes for fodder for cattle and sheep. It has also been used as a soil fertiliser and conditioner.

Porphyra umbilicalis

- Laver
- *Sleabhac* (Anonymous, 1978); *Sleaidí* (Ua Maoileoin, 2001); *Sleabhacán* (Moloney, 1919); *Slabhach* (Moloney, 1919); *Sloucan* (Cameron, 1883); *Slabhacán* (*-cáin*) (de Vere Coneys, 1849)

Porphyra umbilicalis is a red alga used in cooking with oatmeal, bread and potatoes (called 'sloucawn' or 'slukane') and is very nutritious. Phillips (1983) provides a recipe for laver bread. The seaweed is torn up into pieces and simmered in lightly salted water for at least four hours. When it is cooked it will resemble a coarse purée. After the water is drained off, the laver bread can be cooled and eaten, ideally with pasta and seafood (Hilton, 2007). Another recipe I found involves frying the seaweed in hot bacon fat then sprinkling with rolled oats and frying again on both sides for about 10 minutes until well done. If boiled for hours until tender and dressed with butter, vinegar and pepper, it can be served as a vegetable (Moloney, 1919). It is best collected in the spring when it is at its most tender. Cameron (1883) said that it was eaten stewed with leeks and onions.

MacMahon (1994) provided details of its traditional means of collection and preparation 'the picking season is short—from January to early March. *Sliúchán* is prepared by boiling, pounding, boiling again, and thickening. Add butter and pepper for a delicious meal—and moreover, some say it is an aphrodisiac!' Laver can be dried and stored and, indeed, in this state it is used to wrap Japanese sushi.

Edible seaweeds were used as famine food in Ireland during the years of the Great Famine. For example, Graham (1997: 86) wrote that in the 1840s on the Dingle Peninsula of Co. Kerry, 'cholera and fever forced people to the cliffs collecting lever (sic) and seagulls' eggs'.

Rutty (1772: 58) said that it was 'well known … in Dublin, being gathered from the rocks about Portrane; boiled in Sea-water, and brought to us in little earthen pots, and tho' not of an inviting look, is of a very pleasant flavour and relish, and a little salt, and eaten with bread, as a dessert, or otherwise'.

It was formerly used as packing for lobsters and oysters (Moloney, 1919).
It was used in the Aran Islands to ease indigestion (Allen & Hatfield, 2004).
Dinneen (1927) gave the Irish name *coirnín* (or *cuirnín*) for *Porphyra laciniata.*

Ulva lactuca

- Sea lettuce
- *Glasán*[349] (Dinneen, 1927; Anonymous, 1978) ; *Sleaidí* (Anonymous, 1978)

Ulva lactuca is a green alga that is edible, although according to Kelly (1997) it is inferior in taste to *Porphyra umbilicalis*. It can be chopped and used in salads and dried and powdered to make a salty seasoning.

Moloney (1919) listed the following Irish names for another species, *Ulva intestinalis* (under the name *Enteromorpha intestinalis*): *líonáin uaithne* and *lineáil uaithne.*

Fungi

- *Fungais* (Anonymous, 1978); *Múscáin* (Anonymous, 1978)

Fungi are widely collected and used for food in various parts of mainland Europe, as well as elsewhere in the world, and edible forms include both Ascomycota (fungi such as morels) and Basidomycota (true mushrooms). In Ireland cultivated shop-bought mushrooms are popular and widely used for a wide variety of culinary purposes. For some unexplained reason, wild-collected fungi do not seem to have been popularly gathered and used much in Ireland. It may be that other foods were abundant and so fungi were not as needed. As well, folk knowledge on which fungi are good to eat and which ones are poisonous was apparently never well-developed and so all or most fungi were avoided. While it is tempting to suggest that knowledge of edible fungi has simply disappeared in Ireland, early writings such as those by Philip O'Sullivan Beare in the early 1600s seem to indicate that they were never popular. O'Sullivan Beare wrote in Latin on Ireland's natural history in his *Zoilomastix*: 'Concerning fungi or mushrooms which the Irish make nothing of, there is no reason why we should discuss further' (O'Sullivan, 2009: 217). Nevertheless, Rutty (1772: 59) said that truffles (*Fungi subterranei*) were sent to Dublin from Co. Meath, and were a very ancient food.

The Irish are today very reticent to collect and eat wild fungi, fearing that they will eat poisonous species, and this is entirely a reasonable concern. Nevertheless, many immigrants from central Europe in Ireland are happy to collect and use wild fungi, having been brought up in parts of Europe, such as Poland and the Baltic countries, where collecting wild fungi is a traditional countryside pastime.

The National Poisons Information Centre in Beaumont Hospital, Dublin, has reviewed the cases of mushroom poisoning in Ireland. During the period between 1997 and 2000, there were 114 incidents reported to the Centre: 0.22% of all poisoning enquiries (Cassidy et al., 2002: 380). Sixty-two of these were related to the taking of the so-called magic mushroom *Psilocybe semilanceata*, known for its hallucinogenic properties. The average

349. Dinneen (1927) gave this name for '*Ulva latissima*'.

age of these patients was 17.5 years old and they were predominantly male. Forty-three of the incidents were due to unknown mushrooms having been eaten. In a few cases, keen fungi collectors were also poisoned, having made errors in identification.

Wood-Martin (1902) included a delightful Irish name for mushrooms, *fas-na-haon-oidhche* (or *fas-na-hannahy*). Anonymous (1978) gave the term *fas na haon oíche* as the 'growth of one night'.

Mildew is a fungus, or mould, and this is generally called *mil-cheo* or *ceo mhillteach* in Irish (Dinneen, 1927).

Agaricus campestris

- Field mushroom
- *Beacán* (Anonymous, 1978); *Agairg* (Clyne, 1989); *Bacán baeraigh*[350] (Uí Chanainn, 1962)

AGARICACEAE—A MUSHROOM FAMILY

Native

Field mushrooms are widespread and are the most popularly collected fungi (division Basidomycota) in Ireland. It is certainly the most commonly collected wild fungus used for food, and for most people, the only fungus they will trust themselves to collect and eat. Although there are poisonous mushrooms similar to their field cousins, such as the yellow-stained mushroom *Agaricus xanthoderma*, these are generally not lethal and will only cause sickness and stomach upsets. The field mushroom is found in meadows, lawns and other grasslands throughout Ireland, and indeed throughout temperate Europe, Asia and North America. *Agaricus campestris* is related to the commercial mushroom (*A. bisporus*) ubiquitously sold in shops and supermarkets in Ireland. Both mushrooms are white-coloured and grow on a short stem. They can be eaten raw or cooked, but I find the best way to enjoy their full flavour is to slice them and fry them quickly in butter. Another close relative of *A. campestris* is *A. arvensis*, or horse mushroom, which is similar in appearance but simply much larger and grows in grassy habitats. The horse mushroom is also good to eat and sometimes just one large specimen can be enough for a meal.

Amanita muscaria

- Fly agaric

AMANITACEAE—THE FLY AGARIC FAMILY

Native

Amanita muscaria is the infamous scarlet-red or orange-capped fungus that is found occasionally growing in birch woods and other woodland habitats in Ireland. It can also be found worldwide in the cooler climates of the Northern Hemisphere. This mushroom is easy to recognise by its bright colour and the distinctive white warts that often cover the cap and have been widely depicted in folklore and popular culture and which are

350. Uí Chanainn (1962) defined *bacán baeraigh* as 'mushroom'.

often associated with Christmas scenes. The name fly agaric derives from its former use as a fly control. Broken pieces of the cap would be placed in milk and used to stupefy flies (Phillips, 1981). This basidomycete fungus is both **poisonous** and hallucinogenic and should **not** be eaten. Woodward (1985: 110) described the symptoms of fly agaric poisoning as 'rapid; pain, diarrhoea, vomiting; excessive salivation and perspiration. Hallucinations, delirium, convulsions, coma and respiratory failure precede death. Mild cases recover in a few hours. Rarely fatal'.

Amanita muscaria is strongly hallucinogenic and has been used in northern Europe by local tribes, such as in Scandinavia and Siberia, for ritualistic and religious purposes. It may have been the species used as the ritual dream-inducing drink 'Soma'. Phillips (1981) suggested that its effect on individuals can differ—some people are strongly affected, whereas it has little impact on others. Emboden (1972) described its use by people in north-eastern Europe. The women of the tribe would chew the dried fungus and mould it into a sausage, which was then eaten by the men. Two fungi were enough to cause a significant hallucinogenic effect. The urine passed by those who ate the fungus was collected and drunk and apparently had an even stronger effect. As early as the 18th century, the Irish writer Oliver Goldsmith (1762: 122) wrote that the poorer Tartars of Koraki (in northern Russia) would gather around the huts of richer people and when they emerged they would collect their urine, which he described as 'delicious fluid', and 'of this they drink with the utmost satisfaction, and thus they get as drunk and as jovial as their betters'.

Despite its wide use in northern parts of Europe and the extensive Scandinavian links with Ireland from the earliest of times, I have been unable to find any documentation on its use in Ireland, although my feeling is that it could well have been a source of narcotic drugs in ancient Ireland. However, further research will be necessary to clarify if that was really the case.

Moerman (1998: 64) noted that Goodrich et al. (1980) recorded that the Kashaya Pomo tribe of the American Southwest use the fungus as a drug, referring to it as 'poison plant'.

Auricularia auricularia-judae

- Jew's ear

AGARICACEAE—THE MUSHROOM FAMILY
Native

Auricularia auricularia-judae grows in the autumn on dead or dying tree branches, particularly on elder (*Sambucus nigra*) on which it is common. It is edible but needs to be cooked well. It is best sliced and simmered or fried and used in a variety of recipes. To what extent it was formerly collected and used as food in Ireland is not known.

Boletus edulis

- Cep; Penny bun

AGARICACEAE—THE MUSHROOM FAMILY
Native

Boletus edulis is an extremely tasty, nutty-flavoured fungus found in beech or oak woods, particularly in the autumn months. In Italian it is known as 'porcini' or 'porcino' and sold under that name in Ireland when imported from mainland Europe. It can be eaten fried, raw, pickled or as an ingredient in various dishes. It can also be eaten fresh or dried, preserved and stored for later use. It is not known to what extent it was formerly collected and used as food in Ireland.

Mabey (1972) suggested that a good way to eat this fungus is in cep and beetroot soup, which is traditionally served in Poland on Christmas Eve.

Cameron (1883) gave *bonaid* as an Irish name for *Boletus*.

Cantharellus cibarius

- Chanterelle
- *Cantarnaid* (Anonymous, 1978)

AGARICACEAE—THE MUSHROOM FAMILY
Native

Cantharellus cibarius is a beautiful yellow- or orange-coloured edible fungus that is extremely sought after and desirable. It is collected mainly in the autumn months of September and October but can be found from July to December, mainly in broad-leaved oak and beech woodlands. Chanterelle are best eaten fresh, fried and used to accompany eggs, or in omelettes. To what extent it was formerly collected and used as food in Ireland is not known, but it is an extremely popular and widely used edible fungus in continental Europe (Mabey, 1972), and I have recently seen them on sale in markets in France and the Netherlands.

A similar species, the false chanterelle, *Hygrophoropsis aurantiaca*, can be poisonous, so before you use it, be absolutely sure of its identification.

Coprinus comatus

- Shaggy ink cap; Lawyer's wig

AGARICACEAE—THE MUSHROOM FAMILY
Native

Coprinus comatus is commonly found on wasteground, roadsides, fields and lawns and is a valuable edible wild food, particularly since it is widely distributed in grasslands across temperate Europe and North America. The fungi should be picked and eaten while young and are best prepared within a few hours of harvest. Older specimens can still be eaten but will colour the foods to which they are added in an alarming way! This true mushroom (Basidomycota) has a mild flavour and can be cooked like ordinary culinary mushrooms, ideally fried in butter. I don't recommend eating it raw. This species does not contain the mushroom compound coprine, which can cause adverse reactions if eaten with alcohol. Coprine does occur in the related *Coprinopsis atramentaria*, so beware its cousin! Specimens harvested after a few hours will begin to break down and dramatically dissolve to a black ink-like liquid filled with the spores. To what extent this mushroom was formerly collected and used as food in Ireland is not known, but its edibility has been tested by humans and animals for a very long time.

Mabey (1972) provides a recipe to make ketchup from this species. Young shaggy ink caps are packed closely in an earthenware jar, with salt spread on each layer. When the jar is full it is placed in the oven to simmer gently for a couple of hours and when ready strained through muslin. A little pepper and nutmeg is added to the liquid, which is boiled again, strained and bottled in sterilised containers. The ketchup keeps for a long time but should be used quickly once it has been opened.

Laetiporus sulphureus

- Chicken of the woods

AGARICACEAE—THE MUSHROOM FAMILY

Native

Laetiporus sulphureus is a large bracket fungus that occurs on a range of different trees, especially on dead or dying branches. Young specimens are edible and quite good to eat, but only when cooked (they should not to be eaten raw). It is best parboiled, then fried. To what extent this fungus was formerly collected and used as food in Ireland is not known.

Langermannia gigantea (Lycoperdon giganteum; Calvatia gigantea)

- Giant puffball; Blindman's buff[351] (Hart, 1898)
- *Caise an phúca* (Anonymous, 1978); Puffball: *Balbhán béice* (Ua Maoileoin, 2001); *Bolgán béice* (Anonymous, 1978); *Caochag* (Cameron, 1883); *Bolg béice* (Dinneen, 1927)

AGARICACEAE—THE MUSHROOM FAMILY

Native

Langermannia gigantea is occasionally found in meadows, gardens and woodlands throughout Ireland and can grow to a substantial size. According to Logan (1981), it was used medicinally to stop bleeding when applied to a wound; the spores are applied to the wound and quickly stop the bleeding. Pressing the flesh of a puffball onto a wound seems to have been a well-known way of treating them in Ireland until relatively recent times. Allen and Hatfield (2004) noted that puffballs were used to staunch bleeding in cattle when polled in Co. Wicklow. A large piece of fungus skin, possibly of this species, was found during excavations in Viking Dublin, and Geraghty (1996) speculates that it might have been used as a wound dressing.

It is also edible and very tasty. This puffball can be sliced and fried like a steak or cut up into pieces and used in the various ways that one would use ordinary mushrooms, such as in stews, sauces, soups and so on. It should only be used if the flesh is fresh and pure

351. Hart (1898: 367) said that this name was applied to species of *Lycoperdon* in Fanet, Co. Donegal.

white. It becomes bitter and yellows with age. It is easily recognised, and for that reason it is probably reasonable to say it is frequently collected and used for food in Ireland. The common puffball (*Lycoperdon perlatum*) is also edible.

Cameron (1883) said that the giant puffball was used for smothering bees. Milliken and Bridgewater (2004) also mentioned this use for a smouldering puffball, to smoke out bees, or to move a fire from one location to another.

Leccinum versipelle

- Orange birch bolete

AGARICACEAE—THE MUSHROOM FAMILY
Native

Leccinum versipelle is a large, distinctive orange-topped fungus found in birch woodlands and scrub. It is edible and quite good to eat. The caps can be fried whole or cut up and used in various dishes. It is a good species to stuff with garlic and breadcrumbs and various fillings. It can be collected during the summer months up to midwinter but is commonest in the early autumn. To what extent it was formerly collected and used for food in Ireland is not known.

Lepiota procera

- Parasol mushroom

AGARICACEAE—THE MUSHROOM FAMILY
Native

Lepiota procera is found in woodlands, grasslands and roadsides. It is large and when fully open, the cap can be as much as 25 cm across, borne high on a stem up to 30 cm tall. It is relatively common and generally found during the late summer months and into autumn. It is edible and very good to eat. To what extent it was formerly collected and used for food in Ireland is not known.

Lepista saeva

- Field blewit; Blue-leg

TRICHOLOMATACEAE
Native

Lepista saeva is a fungus that occurs in woodlands, woodland margins and some grasslands, such as roadsides and waste places. Its cap is brown in colour, and it has a lilac-streaked stem. It can be collected in the autumn and is edible and good to eat, but it should be cooked. To what extent it was formerly collected and used for food in Ireland is not known.

Morchella esculenta (common morel). Watercolour by Lydia Shackleton. Courtesy of the National Botanic Gardens, Glasnevin.

Morchella esculenta

- Common morel
- *Morchal* (Anonymous, 1978)

MORCHELLACEAE—THE MOREL FAMILY

Native

Morchella esculenta is found in deciduous woodlands, especially those of ash and elm on limestone soils. This Ascomycete fungus is recognisable by its honeycomb-like cap (although in this way it is similar to the false morels, which are variably toxic) and its hollow stems. The common morel is collected in the spring and is one of the most highly prized and tasty edible fungi in Ireland. It can be dried, stored and cooked in a multitude of different ways, similar to ordinary mushrooms, but only much more delicious! To what extent it was formerly collected and used for food in Ireland is not known.

Pleurotus ostreatus

- Oyster mushroom

PLEUROTACEAE

Native

Pleurotus ostreatus is a bracket fungus that grows from the dead branches and trunks of trees and is found mainly on broad-leaved species, such as beech. It is pale in colour, fan-shaped and good to eat if cooked well, as it can be rather tough. Oyster mushrooms can be collected during much of the year, but to what extent they were formerly collected and used as food in Ireland is not known.

Polyporus species

- Bracket fungi
- *Caise*

POLYPORACEAE

Native

There are several *Polyporus* species known in Ireland, and they belong to the same order, Basidomycota, as the familiar field and culinary mushrooms. Cameron (1883) gave the Irish name *caise* for *Polyporus*, meaning 'cheese'. The widespread *P. squamosus* occurs throughout Europe and North America where it is imaginatively known as the dryad's saddle or pheasant's back mushroom.

Some species of *Polyporus* have been used in the Highlands of Scotland for making tinder and for sharpening razors (Cameron, 1883). Only the young forms can be eaten, as bracket fungi become quite tough with age.

Psilocybe semilanceata

- Liberty cap; Magic mushrooms

STROPHARIACEAE
Native

There are several species of *Psilocybe* found in Ireland, but the best known, used as a hallucinogen, is *Psilocybe semilanceata*. Liberty caps are named for the historic, soft felt caps they resemble and occur in pastures, lawns and on roadsides during the late summer and autumn. Perhaps the best known and commonly encountered magic mushroom is *P. semilanceata*, which is occasionally gathered in Ireland and used for the psychedelic compounds psilocybin, baeocystin and psilocin (Stamets, 1996). This species is known as perhaps the most potent, psychoactive mushroom. Other species of *Psilocybe* are collected for similar purposes in other parts of the world. They all have psychoactive compounds and their collection is illegal throughout Europe and North America.

Lichens

- *Léicin* (Anonymous, 1978); *Losa liag* (Dinneen, 1927)

There are few recorded ethnobotanical uses of lichens in Ireland, other than their use as dyes (*dathú* and *ruamadh* are Irish words for dyeing). I have not found records of the use of lichens for food, although in other parts of the world several species are regarded as edible. For example, in North America, species of *Cladonia*, *Umbilicaria*, *Gyrophora* and *Cetraria* have been used to prepare flour, cooked as vegetables or added to soup and made into blancmange (Peterson, 1977).

Lichens have been used in Ireland as dyestuffs for centuries. According to Mairet (1916) and Mahon (1983) *Parmelia saxatilis* and *P. omphalodes*, often referred to as 'crottle', are the most commonly used species by traditional weavers in Ireland and western Scotland for dyeing wool used in making tweed. Other lichens listed by Mairet (1916) as the sources of dyes in Britain and Ireland include *Ramalina scopulorum* (yellow brown), *Flavoparmelia* (formerly *Parmelia*) *caperata* (stone crottle) (yellow), *Parmelia ceratophylla* (dark crottle) (this is probably *Allantoparmelia alpicola*), *Parmelia parietina* (= *Xanthoria parietina*) (yellowish brown) and *Lecanora tartarea* (red or crimson). Sue Redican, the Blasket Island weaver in Co. Kerry, uses *Parmelia* species today too.

Dyes called *orchil* and *cudbear* (in Scotland) were commonly imported into Britain and used mainly to dye wool (Mairet, 1916). These were made from tropical and subtropical lichens and from species of *Roccella* growing on rocks in the Canary Islands. They were also made from native lichens and produced beautiful red and purple dyes. The dye was formed by the action of ammonia and oxygen on the crushed lichen. In early times, the lichen was treated with stale urine and slaked lime and then boiled with the wool in a slightly acid solution (made acid by the addition of weak solutions of sulphuric acid, alum or tartar). In the Shetlands and Western Highlands of Scotland the dye made in this way was called *cudbear*, generally from *Lecanora tartarea* (= *Ochrolechia tartarea*) and *Urceolaria calcarea* (= *Aspicilia calcarea*), but also less commonly from *Evernia prunastri*, *Lecanora pallescens* (= *Ochrolechia parella* or *O. szatalaensis*), *Umbilicaria vellea*, *U. pustulata* (= *Lasallia pustulata*) and *Parmelia perlata* (= *Parmotrema perlatum*). In Ireland, this dye was sometimes called *arcel* (Mairet, 1916).

Other records of lichens used for dyeing include those listed by Smith (1756: 378) using pre-Linnean polynomial Latin names: *Lichenoides rigidum, Eryngii folio referns, Muscus Islandicus purgans.* Smith reported that this was mixed with whey and used in Co. Kerry as a purgative. Howard Fox (pers. comm., 2008) suggested that his probably refers to *Cetraria islandica,* so-called Island moss, a lichen that is only known in Ireland from the summit of Torc Mountain in Killarney.

Regarding *Lichenoides saxatile, tinctorium foliis pilosis purpureis, Muscus tinctorius crustae modo petris adnascens,* Smith (1756: 378) said that in Kerry the 'country people use it for dyeing purple, for which purpose they steep it in stale urine, and make it up into balls with lime. I have seen considerable quantities of it thus prepared, sold in Dingle Market.' Howard Fox (pers. comm., 2008) suggested that this probably refers to *Ochrolechia tartarea,* a lichen known from high altitude mountain rocks, for example in Co. Kerry, or the previously mentioned *Parmelia omphalodes* or *P. saxatilis.*

The lichen material used for dyeing was scraped from the rocks, dried and made into a powder, then mixed with urine for some weeks before use. Purple and crimson dyes were obtained from such species. The lichens were usually gathered off the rocks in July or August, preferably after rainfall (Mahon, 1983), when presumably it would have been easier to scrape off.

Wood-Martin (1902) said that in Co. Sligo the common lichen *Stricta pulmonaria* (= *Lobaria pulmonaria*), also called the tree lung-wort, hazel rag and crottles, was applied and used as a cure for piles.

Up to the present day a number of Irish lichens have been used to make very realistic and effective architectural models and in model railway building. Dyed various colours, different species have been used to make miniature model hedges, bushes, shrubberies, tree canopies and other natural features.

Paul Whelan (Whelan, 2011) recently prepared and published an excellent and well-illustrated guide to more than 250 lichen species in Ireland, which makes identification of the commonest species in Ireland so much easier.

Cladonia chlorophaea

- **Mealy pixie-cup**

This lichen was boiled in new milk in Co. Waterford as a remedy for whooping cough (Allen & Hatfield, 2004).

Parmelia species

Parmelia species, called *crotal* in Irish, were used medicinally in Ireland as a cure for a bad sore under the chin, as well as for burns and cuts (Allen & Hatfield, 2004). They were also added to soup made for invalids, which was probably made from carrageen

(*Chondrus crispus* and *Mastocarpus stellatus*) (Allen & Hatfield, 2004). They are also used for dyeing. McGlinchey (2007: 85) tells the story of the use of *crotal* to cure a sore: 'Paddy Óg Roddy one time had a sore under the chin that was getting worse and spreading. A travelling woman called in Roddys one day to roast a meal of potatoes for herself and noticed Paddy with his chin bandaged. She went out and gathered a handful of *crotal* [lichen] off the rocks and made a poultice of it. That one poultice left him alright. Travelling people knew a lot that way'.

Colgan (1911) gave the Irish name *scra cloch*, 'stone scurf', for *Parmelia saxatilis* in Clare Island, Co. Mayo, but said that in the Carna District of Galway the same name is given to another lichen, *Ramalina scopulorum* (= *R. siliquosa*).

Cameron (1883) said that *Parmelia* species provide a reddish brown colour for dyeing.

See also Lichens (unspecified).

Ramalina siliquosa

- Sea ivory

Ramalina siliquosa is a common species on coastal rocks. It is used to produce a pale peach colour for dyeing by the Blasket Islands weaver Sue Redican in Co. Kerry (Redican, pers. comm., 2010).

Colgan (1911: 4.11) gave the Irish name *féitóg liath* for '*Ramelina scopulorum*', used in Clare Island to produce a yellow, or saffron, dye.

Usnea species

- Beard lichen
- *Léicean meigeallach* (Anonymous, 1978)

According to Allen and Hatfield (2004), beard lichen was used in Leitrim to treat sore eyes (probably *Usnea subfloridana*).

See also Lichens (unspecified).

Xanthoria parietina

- Orange lichen; Yellow scale; Maritime sunburst lichen; Shore lichen

Xanthoria parietina is a common species on coastal rocks. The lichen is used by the Blasket Islands weaver Sue Redican, in Co. Kerry, to produce a pale brown colour for dyeing (Redican, pers. comm., 2010).

Mosses

- *Caonacha* (Anonymous, 1978); *Cúnlach* (Ua Maoileoin, 2001)

Remarkably few uses for mosses have been described in the literature in Ireland, other than the uses for *Sphagnum* moss.

O'Hare (1993) recorded that moss (unspecified species) was mixed with mud in the construction of mud houses to help bind the walls. Mud houses were the commonest form of housing in much of rural Ireland up to the 19th century.

An Irish medicinal remedy for inflammation was nine handfuls of mountain moss, dried to powder. Nine pinches of this were added to nine pinches of ashes from the hearth, mixed in whey and taken every Tuesday and Thursday (Wilde, 1890; Wood-Martin, 1902; Baker, 1999).

Large amounts of mosses, some identified to species level, have been excavated in cess-pits from medieval sites in Dublin, at Winetavern Street and Fishamble Street (Geraghty, 1996), where mosses would have been used as toilet paper.

Dried mosses were also used as fuel for household fires (O'Súilleabháin, 1942), perhaps as tinder and probably only to a limited extent, and used especially when other fuels were limited.

Wild collected mats of moss have been used in Ireland in recent decades to provide a decorative outer lining for hanging baskets filled with annual flowers and for use in floristry (for example, in decorative wreaths at Christmas and for naturalistic landscapes).

Moss has been used in Scotland in the smoking of malt in the production of whiskey and was perhaps used in a similar way in Ireland, although I have not found any records of such use. Mosses have also been used to make bedding.

Fontinalis antipyretica

- Willow moss
- *Sailchaonach* (Anonymous, 1978)

Allen and Hatfield (2004) noted a record of the use of this moss in Co. Limerick to staunch bleeding.

Sphagnum species

- Sphagnum moss; Bog moss[352]; Fog[353] (Hart, 1898)
- *Sfagnam* (Anonymous, 1978); *Súsán*[354] (Ó Dónaill, 1977); *Fionnamhóin*[355] (Uí Chanainn, 1962)

Sphagnum species.

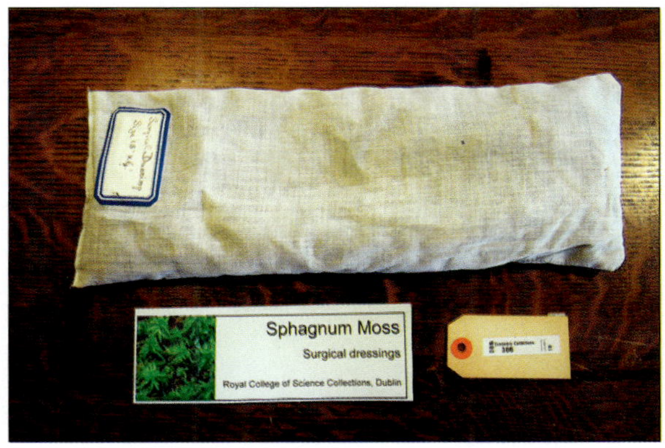

Sphagnum species surgical dressing. From the Museum collections of the National Botanic Gardens at Glasnevin.

Sphagnum mosses are the dominant species of Ireland's bogs. *Sphagnum* moss is well known for its use as an absorbent material, and can absorb up to 20 times its own weight. It is also known to have useful antibacterial and antiseptic properties. For that reason it was used in medieval times as toilet paper and for dressing wounds (e.g., in Co. Cavan; Maloney, 1972). It may also have been used by women as sanitary towels (Geraghty, 1996). McCarthy and Hawkes (2007) said that *Sphagnum* was used as nappies for babies in the Mizen region of West Cork.

There are records of *Sphagnum* species having being collected during World War II in Co. Wicklow by Capuchin nuns for the preparation of medicinal wound dressings. Several examples of *Sphagnum* dressings are included in the Museum collections of the National Botanic Gardens at Glasnevin. MacMahon (1994) recorded that in Ballyheigue, North Kerry, *Sphagnum* species (including a species called 'drowned kittens') were used as a dressing for wounds. Baker (1999) mentioned that *Sphagnum* was used by the Irish to treat the wounded after the Battle of Clontarf in 1014. Bog moss (presumably *Sphagnum* spp.) was also used in a poultice to treat a sprain (Logan, 1972). O'Regan (1997) mentioned that in the 1940s a rector's wife in Co. Sligo organised the collection of *Sphagnum* for Dr Steeven's Hospital in Dublin, where it was sterilised and used to treat wounds associated with tuberculosis.

A project by David Burns and Donna Noonan included in the BT Young Scientist and Technology Exhibition held at the Royal Dublin Society in Ballsbridge, Dublin, in 2008, suggested that *Sphagnum* could be used as an effective means of packing fruit and vegetables and could also be useful for preventing bruising and extending the shelf life of many fruits and vegetables.

Sphagnum mosses make up a high proportion of most peat deposits in Ireland. The ethnobotanical significance and uses of peat are described in the introductory chapters.

352. Dinneen (1927) gave the Irish name *cadach* (*-aighe*) for 'bog moss'
353. Hart (1898: 375) said that this name referred to dry mosses, including *Sphagnum* and other dense mosses.
354. Ó Dónaill (1977) defined *súsán* as 'sphagnum, peat moss'.
355. Uí Chanainn (1962) defined *fionnamhóin* as 'bogmoss, sphagnum'.

Liverworts

- *Aelusanna*[356] (Ó Dónaill, 1977; Anonymous, 1978); *Ae abhann* (Dinneen, 1927)

Liverworts have few recorded uses in Ireland, although elsewhere their anti-bacterial properties have encouraged their use medicinally, as dressings for wounds and other purposes. K'Eogh (1735) said that common liverwort was useful against inflammations and liver ailments, as well as for jaundice, dropsy and gonorrhoea.

Cameron (1883) gave the Irish name *cuisle aibheach* for *Marchantia polymorpha*.

356. Ó Dónaill (1977) gave the singular form of this word as '*Aelus*'.

APPENDIX I
PLANT NAMES IN IRISH LOCAL PLACE NAMES

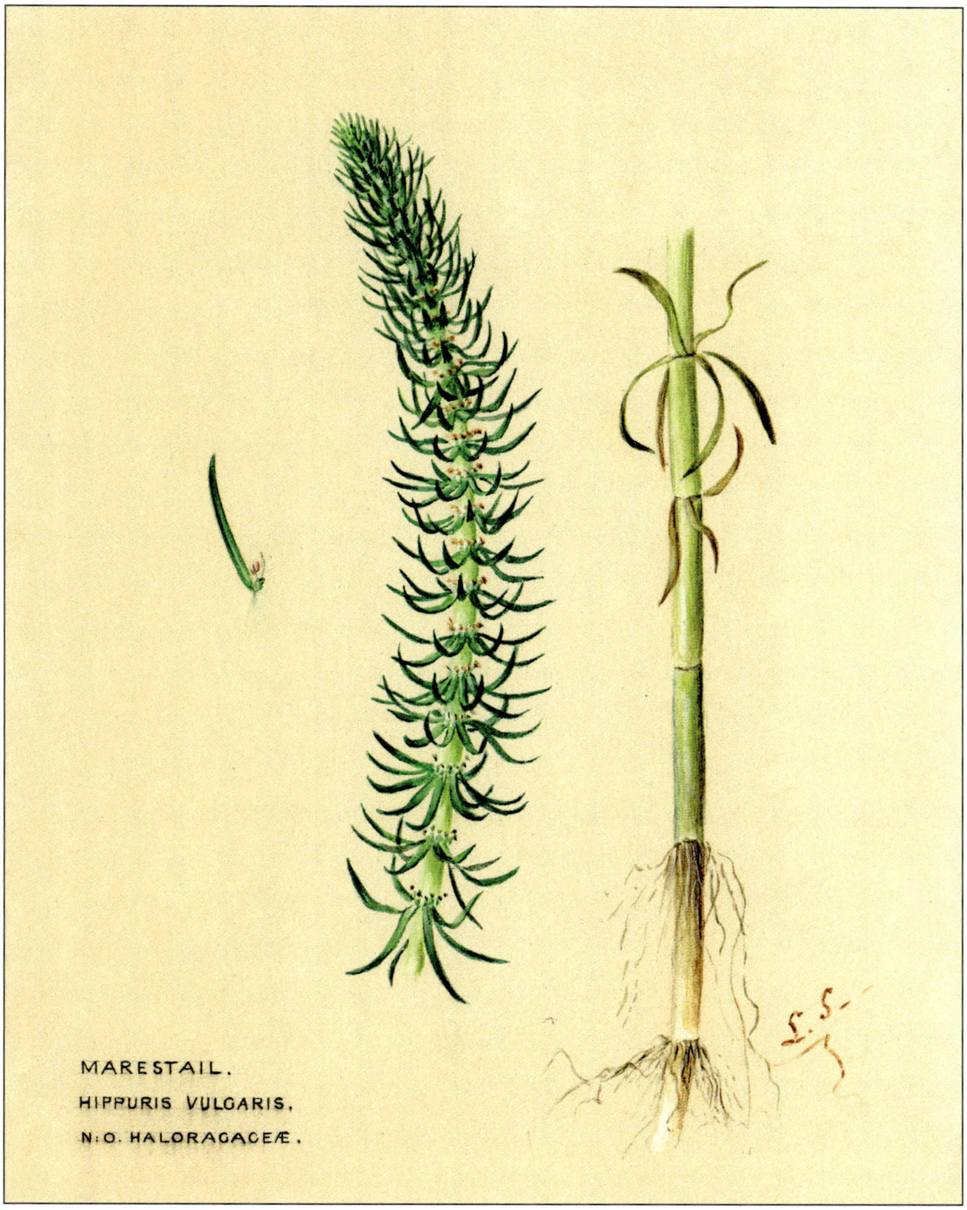

Hippuris vulgaris (mare's-tail). Watercolour by Lydia Shackleton. Courtesy of the National Botanic Gardens, Glasnevin.

PLANT NAMES IN IRISH LOCAL PLACE NAMES[1]

Alder

Burrisnafarney in Tipperary; the *burris* or borough of the alder-plain.

Farnagh, Farnane, Farnoge, a place of *Fearns* [Farns], or alders.

Farney in Monaghan; *Fearnmhagh* [Farnvah], from *The Book of Rights* (*Leabhar-na-gCeart*), the alder plain (*fearn* and *magh*). Also Fernagh and Ferney.

Ferns; *Fearna* [Farna], from *The Annals of the Four Masters* [2], alders, a place abounding in alders.

Ash

Corrinshigo, Corrinshigagh; the round hill of the ash trees.

Funcheon; *Fuinnseann* [Funshin], the ash tree: the ash-producing river.

Funshin, Funshinagh, Funshog, Funshoge; a place producing ash trees (*fuinnse*).

Unshinagh, Inshinagh; Uinseannach, a place producing ash trees (*uinnse* and *fuinnse*).

Apple

Aghowle in Wicklow; *Achadh-abhla*, the field of the apple trees.

Annahavil; *Eanach-abhaill*, the marsh of the orchard (*abhall*).

Oola in Limerick and Waterford; *Ubhla* [Oola], a place of apples, an orchard (from *ubhall* or *abhall*).

Oulart in Wexford; *abhall-ghort* [oulort], an orchard, compounded of *abhall* and *gort*.

Arbutus**

Ardnaconnia; *Ard-na-caithne*, the Irish name of the village of Smerwick, Dingle, Co. Kerry; the height of the arbutus.[3]

Cahnicaun, the wood at the back of the Eagle's Nest, near Killarney, Co. Kerry (*caithne*, arbutus or strawberry tree).

Isknagahiny; *Eisc-na-gcaithne*, a lake near Lough Currane, Co. Kerry; the stream-track of the arbutus trees.

Owenacahina, a stream that flows from Barley Lake down to Glengarriff, Co. Kerry.

Quin, Co. Clare (*cuinche*, another form of the Irish name of *Arbutus*).

Quinsheen, a small island in Clew Bay, Co. Mayo; little arbutus island.

Barley

Cappaghnahoran, west of Mountrath, Co. Laois** (*ceapach*, barley field).

Coolnahorna, a placename found both in Cos. Wexford and Wicklow** (*cuil*, angle or corner).

Craignahorn, near Portrush, Co. Derry**; the rock of the barley.

Glennyhorn [or Cloonnahorn], Clontibert, Co. Monaghan**; meadow of the barley.

Knocknahorna; the hill of the barley (*eórna*).

Loughorne, near Newry, Co. Down**; barley lake.

Lough Ourna, north of Nenagh, Co. Tipperary and another one near Glengarriff, Co. Kerry**[4]

Mulnahorn, townlands in Cos. Fermanagh and Tyrone**; barley hill.

Tavnaghorna, a small stream near Cushendall, Co. Antrim**; barley field.

Bean**

Ardnaponra; *Ard-na-bponaire*, near Moate, Co. Westmeath; (*pónaire*, bean) the height of the beans.

Coolpowra, near Portumna, Co. Galway; the hill-back (*cúl*) of the beans.

Gorteennaphoria, Moyaliff, Co. Tipperary; bean field.

Gortaphoria, near Dingle Bay, west of Drung Hill, Co. Kerry; bean field.

Bilberry*

Freehans, Co. Waterford.

Froghanstown, *Baile Fraochán;* Co. Westmeath.

Gortnavreaghaun, Co. Clare; field of the bilberries.

Birch

Aghavea in Fermanagh; *Achadh-beithe* [Ahabehy], from *Four Masters*, the field of the birch trees.

Ballaghbehy; the road of the birch (*beith*).

Ballybay in Monaghan; *Bel-atha-beithe* [Bellabehy], the ford mouth of the birch.

PLANT NAMES IN IRISH LOCAL PLACE NAMES *(continued)*

Birch *(continued)*

Beagh; *Beitheach* [Beha], a place of birches.

Beheenagh, Behernagh; a place of birches.

Corbeagh; round hill of the birch.

Currabaha, Currabeha; the marsh of the birch.

Glanbehy; birchy glen.

Kilbaha; *Coill-beithe*, birch wood.

Kilbeheny; *Coill-beithne* [Kilbehena], from *The Annals of the Four Masters*[2], birch wood.

Rossbegh or Rossbehy west of Killarney; the peninsula of birches.

Bluebell*

Cloonboo, Co. Galway; meadow of bluebells.

Coolbooa, Co. Waterford; nook of bluebells.

Bramble, Blackberry

Ardrass, Cos. Mayo and Kildare, and Ardress, near Loughgall, Co. Armagh; *Ard-dreas***; the height of brambles.

Dressoge, Dressogagh, Dresternagh, Dresternan, Dristernan, Drishaghaun, Drishane, Drishoge; a briery or bushy place.

Drinagh, Drinaghan; a place producing *dreens* or blackthorns.

Smear, Columkille, Co. Longford** (*sméar*, a blackberry).

Drisheen, west of Skibbereen, Co. Cork**.

Drishoge, a name used in several places in Dublin, Roscommon and Tipperary**.

Drissoge in Meath and Carlow, and Dressoge in Fermanagh and Tyrone**.

Gortnadrass, near Achonry, Sligo**; the briar field.

Kildress, Co. Tyrone**; the church of the brambles.

Cappanasmear, near Borrisokane, Co. Tipperary**; the plot (*ceapagh*) of the blackberries.

Coolnasmear, near Dungarvan, Co. Waterford**; blackberry corner.

Creenasmear, by Muckish Mountain in Co. Donegal** (*crioch*, a district).

Drumnasmear, Layd, Co. Antrim**; the ridge of the blackberries.

Smearlagh, a stream flowing into the river Feale, near Listowel, Co. Kerry**.

Broom

Garranakilka, Co. Tipperary**; the broom garden.

Giltagh, Co. Fermanagh**.

Guilcagh; a place producing broom (*giolcach*).[5]

Guileagh, Co. Waterford**.

Guilkagh, Listerlin, Co. Kilkenny**.

Gilkagh, Moylough, and also near Ballymoe, all in Co. Galway**.

Bulrush**

Cloonshivna, Co. Galway; the meadow of the bulrushes.

Shiven River; *Simhne*, near Mountbellew, Co. Galway (the bulrush, *sibhin* or *simhin* [shiveen], and the plural *simhne* [shivna].[6]

Shimna River, near Tollymore Park, south of Newcastle, Co. Down.

Tawnanasheffin, Co. Mayo; the field of the bulrushes.

Loughfea, a lake near Carrickmacross, Co. Monaghan; the lake of the bulrushes (*feadh*, 'any strong rush from which they make lights' [Joyce, 1875]).

Loughaunnavaag, near Kilconnell, Co. Galway; the lake of the bulrushes.

Clover*

Knocknashamroge, Co. Wicklow; hill of the shamrock.

Lugnashammer, Co. Roscommon; hollow of the clover.

PLANT NAMES IN IRISH LOCAL PLACE NAMES *(continued)*

Corn**[7]

Clonarrow, near Philipstown, Co. Offaly; corn meadow.

Coolanarroo, Tuosist, south-west of Kenmare, Co. Kerry (*cuil*, a corner).

Derryarrow, near Mountrath, Co. Laois; the oak grove of the corn.

Knockanarroor, near Likkarney, Co. Kerry; *Cnoc-an-arbhair*, the hill of the corn.

Lissanarroor, near Galbally, Co. Limerick (*lis*, a fort).

Meenanarwa, near Lough Finn, Co. Donegal; the *meen* or mountain flat of the corn.

Dandelion

Gortnasharvoge, near Ballinasloe, Co. Roscommon**; the field of the dandelions.

Moneysharvan; *Moin-na-searbhan*, Maghera, Co. Derry**; the bog of the dandelion (*searbhán* or *searbhóg*, dandelion).

Toberataravan; *Tobar-a'-tsearbháin*, Tumna, east of Boyle, Co. Roscommon**; well of the dandelion.

Pollsharvoge, Meelick, south-east of Foxford, Co. Mayo**; the hole of the dandelion.

Dock**

Cloonnagoppoge, Co. Mayo; dock-leaved meadow (*copóg*, a dock).

Coppanagh, a name that occurs in Ulster, Connaught and Leinster, Copney and Copany, in Cos. Tyrone, Armagh and Donegal.

Glencoppogagh, Upper Bodoney, Co. Tyrone; the glen of the dock leaves.

Lagnagoppoge, south of Strangford, Co. Down (*lag*, a hollow).

Elder

Trim in Meath; full name *Ath-truim* [Ah-trim], the ford of the elder bushes.

Tromaun, Trumman, Trummery; a place producing elder bushes (*trom*).

Elm

Laghil, Laghile, Laughil, Loughill, Loughil; *Leamhchoill* [Lavwhill], elm wood.

Laune River at Killarney; *Leamhain*, from *Four Masters*, elm; the elm-producing river.

Lavagh; *Leamhach* [Lavagh], a place producing elms.

Lislevane in Cork; *Lios-leamhain*, elm fort.

Longfield; in almost all cases a corruption of *Leamh-choill* [Lavwhill], elm wood.

Lorum in Carlow; *Leamh-dhruim* [Lavrum], elm ridge.

Ferns

Ardrahan, Co. Galway; *Ard-rathain*, the height of the ferns.

Drumraine, Drumrainy; ferny ridge (*rathain*).

Inishraher, a small island in Westport Boy, Co. Mayo**; ferny island.

Knockranny; *Cnoc-raithnigh* [rahnee], ferny hill.

Lisrenny, north of Ardee, Co. Louth**; ferny fort.

Rahan, Co. Offaly; *Raithin*, a ferny place.

Ranaghan, Rannagh; a ferny place (*raithne*, a fern).

Tavnaghranny, Grange of Layd, Co. Antrim** (*tavnagh*, a field).

Flag iris

Allistragh, Co. Armagh; *Aileastrach**.

Ellistrim, Co. Galway; *Eileastraim**.

Ellistrin, near Letterkenny, Co. Donegal**.

Ellistron, near Ballinrobe, Co. Mayo**.

Lisatilister; *Lios-a'-tsiolastair*, near Carrickmacross, Co. Monaghan**; fort of the flaggers or yellow iris.

Flax**[8]

Coolaleen, Killeedy, near Broadford, Co. Limerick; *Cúil-a'-lin*, the corner of the flax.

Crockaleen, near Enniskillen, Co. Fermanagh; flax-hill.

PLANT NAMES IN IRISH LOCAL PLACE NAMES *(continued)*

Flax** *(continued)*

Curraghaleen, west of Athlone (*curragh*, a marsh).

Gortaleen, a name found in Cos. Cork and Kerry; the field of the flax.

Killaleen, near Drumahaire, Co. Leitrim, and Killyleen, near the town of Monaghan; the wood (*coill*) of the flax.

Lisheenaleen, a name found in Cos. Cork, Galway and Tipperary and Rathleen, near Inistioge, Co. Kilkenny; all perhaps derived from forts places where flax was processed.

Monaleen, near Newtownmountkennedy, Co. Wicklow; perhaps derived from a place where flax was steeped as part of its preparation for fibre.

Foxglove*

Carrickatane, Co. Tyrone*; rock of the foxglove.

Carrickateane and Carrickatane, both in Co. Cavan**; the rock of the foxglove.

Cloomameragaun, Co. Galway*; meadow of the foxglove.

Gortnamearacaun, Inchicronan, Co. Clare* & **; field of the foxglove.

Lugateane, Co. Roscommon*; hollow of the foxglove.

Drumantine, north of Newry, Co. Down**; the foxglove ridge.

Gortateean; *Gort-a'-tsiain***; the field of the foxglove.

Mullantain, near Stewartsrown, Co. Tyrone** and Mullantine, townlands in Cos. Kildare and Armagh**; the hill (*mul*) of the fairy finger.

Garlic**

Clooncraff; *Cluain-creamha*, Co. Roscommon (*creamh*, garlic), the meadow of wild garlic.

Clooncrave, Killucan, Co. Westmeath and Cloncrew, Co. Limerick.

Craffield, Co. Wicklow; Clooncraffield, Castlerea, Co. Roscommon; the meadow of the wild garlic wood.

Derrycramph, Cavan and Derrycraff, Co. Mayo; the oak wood or the wild garlic.

Inishcraff; *Inis-creamha*, Lough Corrib.

Knavagh, Tiranascragh, near the Shannon, north of Portumna, Co. Galway (*cneamh*, a variant of the Irish name).

Gooseberry**

Killaspeenan, near Newtown Butler, Co. Fermanagh; the wood (*coill*) of the gooseberries (*spionán*, gooseberry or gooseberry bush).

Spinans, Donaghmore, Co. Wicklow. Also Speenoge in Co. Donegal.

Carrickspringan, near Moynalty, Co. Meath and Lisnasprunane, a fort in the townland of Garranroe, near Adare, Co. Limerick—both corruptions of this word.

Gorse

Coolattin; the corner of the furze (*aiteann*).

Grass**

Fearglass, Co. Leitrim (*fér* or *féur*, grass), green grass.

Ferbane and Fearboy, both in Co. Offaly; whitish grass and yellowish grass, respectively.

Lissanair; *Lios-an-fhéir*, Kilmihil, Co. Clare; the fort of the grass.

Monairmore and Monearmore, several townlands in Munster and Connaught (*moinfhéur* [monair], a mountain meadow; literally bog-grass, *móin*, bog).

Ballinvonear; *Baile-an-mhóinfhéir*, near Doneraile, Co. Cork; the town of the mountain meadow.

Hawthorn, Whitethorn

Aghnaskea, Aghnaskeagh, Aghnaskew; *Achadh-na-sceach*, the field of the white-thorn bushes.

Ardskeagh; the height of the *skeaghs* or bushes.

Bawnskeha; the green field of the bush (*seeach*).

Clonskeagh; *Cluain-sceach*, the meadow of the white thorns.

Corskeagh; the round hill of the white thorns.

PLANT NAMES IN IRISH LOCAL PLACE NAMES *(continued)*

Hawthorn, Whitethorn *(continued)*

Gortnaskea, Gortnaskeagh, Gortnaskeha, Gortnaskey; the field of the *sceachs* or whitethorn bushes.

Knocknaskagh, Knocknaskeagh; the hill of the *sceachs* or whitethorn bushes.

Lisnaskea in Fermanagh; the fort of the *sceach* or whitethorn tree.

Skagh, Skeagh, Skea; *Sceach*, a white thorn bush.

Skahanagh, Skehanagh; a place full of *sceachs* or white thorns.

Skeheen; a little *sceach* or bush.

Tullynaskeagh; the hill of the white thorns.

Hazel

Callowhill; *Collchoill*, hazel wood (*coll* and *coil*).

Colehill; *Coll-choill*, hazel wood.

Collon, Cullan, Cullane, Cullaun; a place of hazels (*coll*).

Dromcolliher in Limerick; a corruption of *Druim-Coll-choille* [Drum-collohill], the ridge of the hazel wood.

Heather

Ballinree; sometimes *Baile-an-fhraeigh*, the town of the heath (*fraech*); sometimes *Baile-an-righ*, the town of the king.

Freagh, Freugh; *Fraech*, heath, a heathy place.

Freaghduff, Freeduff; black heath.

Freaghillan, Freaghillaun; heathy island (*oilean*).

Freughmore, Freaghmore; great heath.

Letterfrack; *Leitir Fraic*, hillside of heather.

Ummerafree; the ridge of the heath (*fraech*).

Herb[9]**

Clonliff, near Dublin City and another place of the same name in Kinawly, Co. Fermanagh; herb meadow.

Drumliff; *Druim-luibh*, the name of three townlands in Cos. Cavan and Fermanagh; the ridge of the herbs.

Drumliffin, near Carrick-on-Shannon, Co. Leitrim; the ridge of the herbs.

Knockalohert; *Cnoc-a'-lubhghuirt*, Kilbrin, west of Doneraile, Co. Cork; the hill of the garden (*lubh-gort* or *lubgort* [looart, loovart], a herb garden or herb plot).

Faslowart, near Lough Gill, Co. Leitrim; *fás*, a wilderness.

Holly

Cloncullen; holly meadow.

Cullen; *Cuillionn* [Cullen], holly, holly land.

Cullenagh, Cullentra, Cullentragh; a place producing holly.

Drumcullen, Drumcullion; the ridge of holly.

Kilcullen; *Cill-cuillinn*, the church of the holly.

Knockacullen; the hill of the holly.

Moycullen in Galway; the plain of holly.

Honeysuckle*

Aghnaveiloge, Co. Longford; field of honeysuckle.

Stranafeley, Co. Fermanagh; river bottom of honeysuckle.

Ivy

Clonenagh in Co. Laois; *Cluain-eidhnech* [enagh]; the meadow of ivy.

Marsh-mallow

Joyce (1875) suggested that several places in Ireland take their name from marsh-mallows, but because its Irish name (*leamh* or *leamhach*) is shared with elm, it is not always clear which species is meant. The accepted Irish name for elm is *eamhán*. Examples include places called Leugh, Lewagh, Leo, Leoh, Luogh and Rathnaleugh, which Joyce interprets as the fort of the marsh-mallows. I'm not convinced. The marsh-mallow, *Althaea officinalis*, is probably an introduced plant in Ireland.

PLANT NAMES IN IRISH LOCAL PLACE NAMES *(continued)*

Moss**

Keenagh, Keenaghan, Keenoge, all occurring in several places in Ireland, for example, the village of Keenagh near Ballymahon in Co. Longford; a mossy place (*caonach* [keenagh], moss).

Drumkeenagh, Cleenish, Co. Fermanagh; the ridge of the moss.

Caherakeeny, west of Tuam, Co. Galway; the *caher* or stone fort of the moss.

Carrivekeeny, near Newry, Co. Armagh.

Carrowkeeny, Kiltown, Co. Roscommon**; mossy quarter.

Nettle

Ballynant, near Kesh, Co. Fermanagh**.

Ballynanty, Co. Limerick and Ballinanty, Co. Wicklow**; the townland of the nettles.

Cappananty, Corcomohide, Co. Limerick**.

Knockananty, south-east of Limerick, Co. Limerick**. The hill of the nettles.

Nantinan in Limerick; *Neantanán*, a place of nettles (*neanta*).

Oak

Adare; *Ath Dara*, the ford of the oak tree.

Aghindarragh in Tyrone; the field of the oak.

Ballaghaderreen in Mayo; the road of the *derreen* or little oak wood.

Ballinderry; the town of the *derry* or oak wood.

Cordarragh; round hill of the oak (*dair*).

Darragh, Derragh; a place producing oaks.

Darraragh, Darrery; *Dairbhreach*, an oak forest, a place abounding in oaks.

Dernish, Derinch, Derinish; oak island.

Derrada, Derradd; *Doire-fhada*, long oak grove.

Derreen; little derry or oak grove or wood.

Derreens, Derries; oak groves.

Derreennacno, near Dromdaleague, Co. Cork**.

Derry; *Doire* [Derry], an oak grove or wood.

Derryad, Derryadda; *Doire-fhada*, long oak wood.

Derrybane, Derrybawn; whitish oak wood.

Derrybeg; little oak wood.

Derrycnaw, Feakle, Co. Clare**; the oak wood of the nuts.

Derrycreevy; the oak wood of the branchy tree.

Derrydorragh, Derrydorraghy; dark oak wood (*dorcha*).

Derryduff; black oak wood.

Derryfadda; long oak wood.

Derrygarriff, Derrygarve; rough oak wood (*garbh*).

Derrylahan, Derrylane; broad oak wood (*leathan*).

Derrylea; grey oak wood.

Derrylough, Derryloughan; the oak wood of the lake.

Derrymore; great oak wood.

Derrynahinch; the oak wood of the island or river meadow (*inis*).

Derrynane in Kerry; *Doire-Fhionain* [Derry-Eenane: *Fh* silent], the oak grove of St Finan Cam, a native of Corkaguiny, who flourished in the 6th century.

Derrynaseer; the oak grove of the *saers* or carpenters.

Derryvullan in Fermanagh; *Doire-Maelain* [Derry-Velan: *M* aspirated], from *Four Masters*, Maelan's oak grove.

Doogary; *Dubhdhoire* [Dooary], black derry or oak wood.

Edenderry; the hill brow of the oak wood.

PLANT NAMES IN IRISH LOCAL PLACE NAMES *(continued)*

Oak *(continued)*

Kildare. According to Animosus, St Brigid built her little cell here under a very high oak tree; and hence it was called *Cill-dara*, which the same writer translates as *Cella quercûs*, the cell or church of the oak.

Kildorrery in Cork; *Cill-dairbhre* [Kildarrery], the church of the oaks.

Knockaderry; the hill of the oak wood.

Lackandarragh, Lackendarragh; the hill side of the oaks.

Lough Derravara in Westmeath; *Loch-Dairbhreach* [Darravara], from *Four Masters*, the lake of the oaks.

Portumna in Galway; *Port-omna*, from *Four Masters*, the landing place of the oak.

Oats

Coolacork, Dunganstown, Co. Wicklow** (*cuil*, angle or corner; *coerce*, oats).

Farranacurky, near Lisnaskea, Co. Fermanagh**; oats bearing land.

Gortachurk, near Bellananagh, Co. Cavan** (*gort*, field)

Gortahork, Gortahurk; the field of the oats.

Lissacurkin, in two places in Co. Roscommon**; the fort of the oats.

Parsnip[10]**

Mackan, Macknan, Mackanagh, Macknagh, Mackney; a place producing parsnips (*meacan*, a parsnip or carrot).

Cloonmackan, Clonmackan; parsnip meadow.

Gortnamackan and Gortnamackanee; the field of the parsnips.

Killygorden in Donegal; *Coill-na-gcuiridin* [Kilnagurridin], from *Four Masters*, the wood of the parsnips.

Pea**

Coolnapish and Coolnapisha, in Cos. Carlow, Kilkenny and Limerick.

Pishanagh, two townlands in Co. Wesmeath.

Pine[11]**

Cappayuse; *Ceapach-ghiumhais*, Moore, Co. Roscommon; the garden plot of the fir.

Meenaguse, near Inver, Co. Donegal (*meen*, a mountain meadow).

Monagoush, near Ardmore, Co. Waterford (*moin*, bog).

Ragwort*

Coolaboghlan, Co. Laois; the nook of the ragwort.

Knocknamoghalaun, Co. Mayo; the hill of the ragwort.

Rowan, Mountain ash, Quicken tree

Ballaghkeeran; the road of the *keerans* or quicken trees.

Corkeeran, Corrakeeran; the round hill of the *keerans* or quicken trees (*caerthainn*).

Drumkeeran; the ridge of the quicken trees.

Rush

Ardlougher; *Ard-luachra*, rushy height.

Cloonlougher; the meadow of the rushes (*luachra*).

Drumlougher; the ridge of the rushes.

Greaghnaloughry, north-east of Ballinamore, Co. Leitrim**; the *greagh* or mountain flat of the rushes.

Letterlougher, Upper Cumber, Co. Derry**; the rushy *letter* or wet hillside.

Lougher, Co. Kerry[12], and Co. Meath** and Loughry, in some places in Co. Tyrone**.

Loughlohery; *Loch-luchra*, near Cahir, Co. Tipperary**; the lake of the rushes.

Rye

Ardataggle, Ardateggle; *Ard-a'-tseagail*, the height of the rye (*seagal*).

Coolataggle, near Borrisoleigh, Co. Tipperary** (*cuil*, a corner).

PLANT NAMES IN IRISH LOCAL PLACE NAMES *(continued)*

Rye *(continued)*

Lissataggle; *Lios-a'-tseagail*, Currans, near Castleisland, Co. Kerry** (the fort of the rye).

Pollataggle, near Gort, Co. Galway** (the hole or pool of the rye).

Sedge

Muing; a sedgy place.

Silverweed*

Caherbriscaun, Co. Galway; fort of the silverweed.

Coolnamrisklawn, Co. Kilkenny; nook of the silverweed.

Lugbriscan, Co. Louth; hollow of the silverweed.

Sloe, Blackthorn

Aghadreen, Aghadreenagh, Aghadreenan, Aghadrinagh; the field of the *dreens* or sloe bushes (*draeighean*).

Aglishdrinagh in Cork; *Eaglais-draeighneach*, the church of the *dreens* or sloe bushes.

Clonarney; *Cluan-airne*, the meadow of sloes.

Dreen; *Draeighean* [Dreean], the blackthorn.

Dreenagh; a place producing blackthorns.

Dreenan, Drinan, Drinaun; blackthorn, a place of blackthorns.

Killarney; *Cill-airne*, the church of the sloes.

Meenadreen; the mountain meadow of blackthorns.

Sorrel

Aghintamy; *Achadh-an-tsamhaidh*, near Monaghan**; the field of the sorrel (*samhadh*, sorrel, pronounced 'saua', 'sawva' and 'sow').

Carrigathou, near Macroom, Co. Cork**; the rock of the sorrel.

Carrowntawa and Carrowntawy in Co. Sligo** (*carrow*, a quarter-land).

Curraghatawy; *Currach-a'tsamhaidh*, Drumreilly, near Ballinamore, Co. Leitrim**; the marsh of the sorrel.

Derrintawy, in Co. Leitrim and Derreenatawy in Co. Roscommon** (*derreen*, oak wood); oak wood of the sorrel.

Garshooey; *Garrdha-samhaidh*, west of Derry**; sorrel garden.

Illaunatoo, Kilmihil, Co. Clare**; sorrel island.

Inishmacsaint, a parish in Fermanagh, taking its name from an island in Lough Erne, which is called in the annals *Inis-muighe-samh* [moy-sauv], the island of the plain of sorrel, from which the present name has been formed by a corrupt pronunciation.

Kilsough; *Coill-samhach*, near Skerries, Co. Dublin**; sorrel wood.

Knockatoo, Co. Galway**; sorrel hill.

Knockatavy, Co. Louth**; sorrel hill.

The river Sow, which enters the estuary of the river Slaney at Castlebridge just north of Wexford**.

Sooey, Ballynakill, near Rivertown, Co. Sligo**; sorrel-bearing land.

Thistle**

Barrafohona, Co. Cork (*barr*, hill-top).

Fofannybane, Kilcoo, Co. Down; white thistle land.

Fofannyreagh, Kilcoo, Co. Down; striped thistle land.

Foffanagh, a townland just north of Buncrana, Co. Donegal; a place full of thistles (*fofannán*, *fothannán*, thistles).

Foghahagh, near Ballymoe, Co. Roscommon.

Fohanagh, Co. Galway.

Glen Fofanny, a river flowing down Slieve Donard just south of Newcastle, Co. Down; thistle glen.

Tonyfohanan, Co. Monaghan (*tonnagh*, a mound).

Vetch*

Pishanagh, Co. Westmeath; place of the vetches.

PLANT NAMES IN IRISH LOCAL PLACE NAMES *(continued)*

Wall fern[13]

Carrigskeewaun, Kilgeever, Co. Mayo; the rock of the wall ferns.

Drumnascamph; *Druim-na-sceamh*, Clonduff, Co. Down; the ridge of the wall ferns (*sgeamh*, wall fern).

Meenscovane, Duagh, Co. Kerry; the smooth plain of the wall ferns.

Water-cress

Aghaviller in Kilkenny; *Achadh-biorair* [Ahabirrer], from *Four Masters*; the field of the water-cresses (*r* changed to *l*).

Askinvillar; *Easc-an-bhiolair*, a stream running down the Blackstairs Mountains at Killann, Co. Wexford**; the water-course of cresses.

Billeragh; *Biolcrach*, a name that occurs in several places in Cos. Cork and Kerry, Billary, in Cos. Donegal, Monaghan and Tyrone, and Bellary in Co. Wexford**; a place that produces cresses.

Toberaviller, near the town of Wicklow**; water-cress well.

Watergrasshill in Cork; a translation of the Irish name *Cnocan-na-biolraighe* [Knockaun-na-billery], the little hill of the water-cresses.

Wheat**

Crinnaghtane, near Kilworth, Co. Cork.

Crinnaghtaun, near Cappoquin, Co. Waterford.

Crunagh, Loughgilly, Co. Armagh.

Crunaght, near Markethill, Co. Armagh.

Portnacrinnaght, Kilnamanagh, Co. Roscommon; the port or landing place of wheat.

Tullanacrunat; *Talamh-na-cruithneachta*, near Castleblaney, Co. Monaghan; the land of the wheat.

Tullycreenaght, near the town of Antrim; wheat hill.

Wild garlic

Annacramph in Armagh and Monaghan; *Eanach-creamha*, the marsh of the wild garlic.

Cloncrew in Limerick; *Cluain-creamha* [Crawa], the meadow of wild garlic.

Cranfield; a corruption of *Creamh-choill* [Cravwhill], the wood (*coill*) of wild garlic (*creamh*).

Craughwell; *Creamh-choill*, wild garlic wood.

Willow

Annakisha; the ford of the *kish* or wickerwork causeway.

Clonsilla; *Cluain-saileach*, the meadow of sallows.

Cloonsillagh or Cloonsellan; the meadow of sallows.

Corsillagh, near Newtown Mountkennedy, Co. Wicklow, and Corsilloga, Aghnamullen, Co. Monaghan**; the round hill of the sallows.

Cratloe, Crataloe; sallow wood.

Currasilla, in Cos. Tipperary and Kilkenny**; the *curragh* or marsh of the osiers.

Gortnasillagh; the field of the sallows.

Lisnasallagh, in two places in Co. Cork and one near Saintfield in Co. Down**; the fort of the sallows.

Lough Sillan, near Shercock, Cavan**; the osier-producing lake.

Pollsallagh, Pollsillagh; the hole of the sallows.

Seltan; a place of sallows.

Solloghod in Tipperary; *Sulchoid* [Sollohed], from *Four Masters*, sallow wood.

Sylaun; a place of sallows.

Tooreennasillane, near Skibbereen, Co. Cork**; the little bleach-field of the osiers (a bleach-field, or *tuar*, was a place that something was spread out to dry).

Yew

Altinure; *Alt-an-iubhair* [yure]; the glen side of the yew tree.

Aughall in Tipperary and Aughil in Derry; *Eochaill*, the yew wood (*eo* and *coill*).

Aughnanure near Oughterard in Galway; *Achadh-na-niúbhar* [Ahananure]; the field of the yew trees.

Ballinure; the town of the yew tree (*iubhar*).

Ballynure; *Baile-an-iubhair*, the town of the yew.

PLANT NAMES IN IRISH LOCAL PLACE NAMES *(continued)*

Yew *(continued)*

Clonoghil; the meadow of the yew wood (*eóchaill*).

Crockanure; *Cnoc-an-iúbhair*, the hill of the yew.

Donohill; the fortress of the yew wood.

Drumanure; the ridge of the yew tree.

Gortanure, Gortinure; the field of the yew.

Killoe; *Cill-eó*, from *O'Clery's Calendar of Irish Saints, or, The Martyrology of Donegal*, the church of the yews.

Killure; *Cill-iubhair*, the church of the yew.

Kinure; *Ceann-iubhair*, the head of the yew.

Knockanure; *Cnoc-an-iubhair*, yew hill.

Mayo; *Maigh-Eo* or *Muigh-Eo*[14], the plain of the yews.

Moynure; the plain of the yew (*iubhiar*).

Newrath; *'n-Iubhrach* [nuragh], the yew land; by the incorporation of the article.

Newry; Ancient name (*Iubhar-cinn-tragha*) [yure-kin-traw]; the yew tree at the head of the strand. In after ages this was shortened to *Iulhar*, which, with the article prefixed, and *y* added, became the present form Newry.

Oghill; *Eóchaill* [Oghill], yew wood (*eó* and *coill*).

Rush in Dublin; *Ros-eo* [Rush-o], from *Four Masters*, the peninsula of the yew trees.

Terenure; *Tir-an-iubhair*, the land of the yew.

Tullynure; *Tulach-an-iubhair*, the hill of the yew.

Ture; the yew. The word *iubhar* [yure] has incorporated the *t* of the article.

Uragh; *Iulhrach* [yuragh], yew land.

Uregare in Limerick; *Lubhar-ghearr* [yure-yar], short yew tree.

Wood of O near Tullamore in King's County; the Irish name is *Eóchaill*, yew wood, the same as Youghal; modern name an attempted translation: Wood of O, i.e., the wood of the *eó* or yew.

Yewer near Killashandra in Cavan; an anglicised form of *Iubhar* [yure], the yew tree.

Youghal in Cork. A yew wood grew anciently on the hill slope now occupied by the town, hence it was called *Eóchaill* [Oghill], from *Four Masters*, i.e., yew wood.

Source: Adapted from Joyce (1922), MacCoitir (2006) and Joyce (1875). The source of most names given is Joyce (1922), as well as those included in other works (marked * for MacCoitir, 2006 and ** Joyce, 1875.

1. Irish forms are in italics. Pronunciations are given in brackets.

2. *The Annals of the Four Masters* dates from the 1600s but was largely compiled from early medieval manuscripts and chronicals medieval history in Ireland.

3. Arbutus has never been recorded at this locality and has not been known on the Dingle Peninsula in historic times.

4. Joyce (1875) suggested that these lakes may have gained their name from having been used to steep and malt barley around their margins.

5. Joyce (1875) says that this word is generally applied to the common broom in the south and east of Ireland but to a reed in the north and west.

6. Joyce (1875) says that this word occurs frequently in local names.

7. Corn of any kind is represented by the Irish word *arbhar* (Joyce, 1875).

8. Joyce (1875) suggested that some places that incorporate flax in their name may be derived from the wild or fairy flax, rather than the cultivated crop plant used for linen making.

9. Joyce (1875: 317) said that when the Irish word for herb, *luibh*, occurs in place names in Ireland, as it frequently does, it suggests that these were 'originally to designate places which were particularly rich in the smaller vegetable productions'. *Luibh* is usually anglicised to 'liff' in placenames.

10. Joyce (1875) interpreted the Irish word *meacan* as parsnip, although it is better understood as the name given to any tap-rooted plants such as burdock (*Arctium minus*), horse-radish (*Armoracia rusticana*), wild carrot (*Daucus carota*) and turnip (*Brassica rapa*).

11. Joyce (1875: 337) included this under the heading of 'fir', *giumhas*, a fir tree. However, firs are not native to Ireland and this name can be applied instead to 'bog deal', i.e., pieces of the timber of the Scots pine (*Pinus sylvestris*) buried in bogs in many parts of Ireland. He said that the tree has not given names to many places, 'which would appear to show that in former times it was not very abundant'. Where it does occur, it is not always certain whether it refers to a live tree or to bog deal.

12. This is a townland on the south side of the Dingle Peninsula, Co. Kerry, east of Annascaul, where I have a cottage and where a considerable portion of this book was written.

13. According to Joyce (1875), *Polypodium vulgare* is the plant that is intended in these plant names.

14. Uí Chanainn (1962).

Dryas octopetala (mountain avens). Watercolour by Lydia Shackleton. Courtesy of the National Botanic Gardens, Glasnevin.

APPENDIX II
K'EOGH AND THRELKELD NAMES

ASSIGNING up-to-date scientific names to the polynomial Latin names used by K'Eogh is not always easy, straightforward or certain. In some cases I have included in the Table that follows what I think is probably the species meant by K'Eogh (K). In other cases I have followed the interpretation of polynomial names used by Threlkeld (1726) (T) suggested by Charles Nelson (1988). Sometimes I cannot be sure what plant is intended by K'Eogh, so I have omitted these. I have generally also left out plants that were purely grown in gardens. In each case I have tried to reproduce the names as given in K'Eogh, even though in some cases there are obvious printer's errors and the printing is difficult to read in other instances. Letters in square brackets are my best guess when occasionally a word is badly printed and difficult to interpret.

The following Table also includes the Irish names of plant species included in Threlkeld's *Synopsis Stirpium Hibernicarum* (1726). I have followed Nelson's interpretation in most cases on what modern scientific names correspond with the polynomial Latin names used by Threlkeld. The work undertaken by Donal Synnott (in Nelson, 1988) in extracting an annotated list of the Irish names used by Threlkeld has been most useful and helpful, and I have followed his interpretations as well. For example, I have given an Irish name in brackets when the spelling of this name in Threlkeld's index differs from the spelling for the same plant in the main text. It is clear that either Threlkeld and K'Eogh used the same sources in determining the Irish names for many plant species included in their books, or K'Eogh used Threlkeld's work as the basis for many of the Irish names he included. The latter seems more likely and this view has been expressed by Williams (1987) too, who noted that some misprints included in Threlkeld's work have been copied by K'Eogh.

Threlkeld provided Irish names for more than 260 plant species, which is more than half of the plants he included in his book (Mitchell, 1974). As to the sources of these names, Threlkeld (1726: 17–18) said that he copied them from a manuscript 'written sometime before the civil Wars in 1641 and probably by that Revd. Irish Divine, Mr. Heaton . . .' [1] (Walsh, 1978). Mitchell (1974) has pointed out that Threlkeld incorporated names from other sources too and perhaps also some names that he collected personally.

1. Richard Heaton (ca. 1610–1666) was a native of Yorkshire who came to Ireland as a chaplain in 1633 (Mitchell, 1974) and is regarded as Ireland's first botanist.

Some Irish names of native and other wild Irish plants included in K'Eogh (1735) (K) and Threlkeld (1726) (T).

Scientific name		Irish names[1]
Abies alba	K	*Crann Giumhais; Giuhass*
Abies sp.	T	*Crann Giumhais; Crann Guivhais*
Achillea millefolium	K	*Ahair Talhum; Ahirhallune*
	T	*Ahair Talhum*
Achillea ptarmica	T	*Luss Corrain*
	K	*Luss Corrain*
Adiantum capillus-veneris	K	*Chossa dub*
Agrimonia eupatoria	K	*Murgrine; Murgrahin*
	T	*Murgraihin; (Murgrachin)*
Agrostemma githago	K	*Cogal*
	T	*Cogal*
Ajuga reptans	K	*Gl[a]ssair Heile; Makkindagh fian*
	T	*Glasair heile*
Alchemilla vulgaris	K	*Leauhagh buih; Crub Lehion; Dearna Muire; Fallinbantierna*
	T	*Leauhagh buih; Crub lehoin; Dearna Muire*
Alisma plantago-aquatica	K	*Cor Chapog; Endilloge*
	T	*Cor Chapog*
Alliaria petiolata	K	*Bocuineall; Coillah*
	T	*Bo cuineall; Gairleog Coillah; (Gairleog coilleagh)*
Allium ursinum	T	*Creaugh coilleah*
Allium vineale	K	*Creoaugh Coilleah*
	K	*Gairleog muire; Knave*
	T	*Gairleog muire*
Alnus glutinosa	K	*Fearnoge; Farnoge*
	T	*Fearnog*
Althaea officinalis	K	*Lavaghvye*
Ammophila arenaria	K	*Muiriunagh; Mehin Muirneagh; Mehain*
	T	*Muiriunagh; Muirneagh; (Muirniagh); Mehini*
	T	*Mehain*
Anagallis arvensis	K	*Falcaire fihain; Ruinn Rusk; Falcuir fuar*
	T	*Falcaire fihain; Ruinn ruisk; Falcuire fuar*
Anemone nemorosa	K	*Nead coilleah*
	T	*Neal coilleah*
Angelica sylvestris	K	*Gah laubran; Bonan lane*
	T	*Gah lauhren; Bonan leane; Cuinneog meig; Coahrain curraih*
Anthemis cotula	K	*Finell moidur; Finell madra*
	T	*Finell; (Finel); Moidiur madra; Moidiur mairhes*
Antirrhinum majus	K	*Sriumh na laugh*
	T	*Srium na laugh; (Srium na laog)*
Apium graveolens	T	*Mer[f]i*
Aquilegia vulgaris	K	*Choffacolumbe; Crubaleone*
Arbutus unedo	K	*Keeora caihne; Uhla caihne* (fruit)
	T	*Keeora caihne; (Keora caihne); Uhla caihne* (fruit)

Some Irish names of native and other wild Irish plants included in K'Eogh (1735) (K) and Threlkeld (1726) (T). *(continued)*

Scientific name		Irish names[1]
Arctium lappa	T	*Meacan tovach; Meacan tuam*
Arctium spp.	K	*Meacan tuam; Meacantovach; Mackantuel*
Armoracia rusticana	K	*Meacain Ragum; Usquebah na Mrahir; Iskybagh na Bragher*
	T	*Meacain Rugum; (Meacan Ragum)*
Artemisia absinthium	K	*Murmonta*
Artemisia maritima	K	*Lyanetragh*
Artemisia vulgaris	K	*Bofullan Ban; Bofullan Liah; Liaghluss*
	T	*Murmonta; Bofullan ban; Bofullan liah; Liagh luss*
	T	*Mongagh measga; (Mongagh measgah)*
Arum maculatum	K	*Maccough; Clovas a Ga[e]bir; Clovis a Geauhur*
	T	*Clovas a gachir; Geahur; Gachrin cuihih; Gachrin gairin keehih*
Asparagus officinalis	K	*Lussprakky; Lussugagh; Lussprakkyfian*
Asplenium ruta-muraria	K	*Riffebulla*
Asplenium scolopendrium (Phyllitis scolopendrium)	K	*Creauh muckefiah; Cnafmuckafiah; Grammuckfiah*
	T	*Creauh Muck e fiah*
Asplenium trichomanes	K	*Dugh chossa*
	T	*Duh Chossa*
Astragalus glycyphyllos	T	*Maddai Milis*
Atriplex patula	K	*Keahruha Lebain; Eirele[h]og; Praisseagh min*
	T	*Keahruhe luhain; Eirelehog; Praisseagh min*
Atropa belladonna	K	*Luss mor; Luss na die mor; Luss un Mongair*
	T	*Lus na dih mor; (Lus dih Mor); Luss an Mongair*
Avena sativa	T	*Praisseagh earle*
Avena sp.	K	*K[u]erke*
Ballota nigra	K	*Breneluss*
Bellis perennis	T	*Noinin; Nonin*
Beta spp.	K	*Bethesbee*
Betula pubescens	T	*Beihe*
Betula sp.	K	*Beighy; Begh*
Blackstonia perfoliata	K	*Dedgha buihe*
	T	*Dedgha buihe*
Borago officinalis	K	*Borastegorrum*
Botrychium lunaria	K	*Lussna missa; Lussina[moe]s*
	T	*Luss na Missa; Lus na Mees; (Luss na Misia)*
Brassica nigra	K	*Garrivoge*
Bryonia dioica	K	*Urinagh Gal*
Calluna vulgaris	K	*Friogh*
	T	*Fraogh*
Caltha palustris	K	*Bearnan Bealtaine; Luss buih Bealtaine*
	T	*Plubairsin; (Plubarsin); Bearnan Bealtain; Luss buih Bealtaine*
Cannabis sativa	K	*Knaib*
	T	*Knaib*

Some Irish names of native and other wild Irish plants included in K'Eogh (1735) (K) and Threlkeld (1726) (T). ***(continued)***

Scientific name		Irish names[1]
Capsella bursa-pastoris	K	*Sradine; Luss Spairain; Luss na f[o]l[n]*
	T	*Sradin; Luss a spairair; Luss na fola*
Cardamine pratensis	K	*Glourane; Gleoran; Lenymurry*
	T	*Gleoran; Billar gragan*
Carlina vulgaris	K	*Fohenanemine*
Carpinus betulus	K	*Crann Sleauhain*
	T	*Crann sleauhain*
Centaurea cyanus	K	*Gorman; Lossgor-rumane*
	T	*Gorman*
Centaurea erythraea	K	*Dremire muire; Dreamrymurry; Deaga Dearg*
	T	*Dremite Muire; Deaga dearg*
Asplenium ceterach (Ceterach officinarum)	K	*Lussnasballiga*
Chelidonium majus	K	*Garevye*
Chenopodium bonus-henricus	K	*Prisseagh; Prahar; Praisseagh Earla; Hali*
	T	*Praisseagh braher; Hali*
Chrysanthemum segetum	K	*Buihainbuihe; Lehan; Uyclaragh*
	T	*Buihain buihe; Lehan*
Chrysosplenium oppositifolium	K	*Gloris; Lussnaliuh*
	T	*Gloris; Luss na liuh*
Cicorium intybus	K	*Cassrevane more gorm*
Circaea lutetiana	T	*Fuinn sheagagh; (Fuinn seagah)*
Cochlearia officinalis	K	*Billar-traihe; Ulanagh; Creaghta*
	T	*Billar traihe*
Colchicum autumnale	T	*Crogh*
Comarum palustre (Potentilla palustris)	T	*Kna Leana*
Conium maculatum	K	*Minvear; Muinmhear; Mungvair*
	T	*Minvear; Munimhear*
Conopodium majus	K	*Currhlane; Clorane*
	T	*Currlan*
Cornus sanguinea	K	*Brieniel*
Corylus avellana	K	*Coul*
	T	*Coll*
Crataegus monogyna	K	*Sgeagh; Skeaghgal*
	T	*Sgeagh*
Crithmum maritimum	K	*Greirig; Greiuhrigin; Greilig; Creuhirg*
	T	*Greiuhrigin; Creuhirg; Greirig; Greilig*
	T	*(Grelig)*
Cuscuta spp.	K	*Cluhan Dearg; Clavane line; Clavane thyeme*
Cuscuta europaea	T	*Cluhan dearg*
Cynoglossum officinale	K	*Tanga gohow*
	T	*Tanga gohow*

Some Irish names of native and other wild Irish plants included in K'Eogh (1735) (K) and Threlkeld (1726) (T). ***(continued)***

Scientific name		Irish names[1]
Cytisus scoparius	K	*Gileagh; Gilkagh*
	T	*Gileagh*
Daucus carota	K	*Meacan buigh; Miohlucan; Hurgoman; Mellughcane*
	T	*Miohlucan; (Mioh lucan); Durgoman; (Murgoman)*
	T	*Milbogan; Meacan buih*
Descurainia sophia	K	*Fiuelmuire; Gorruliummine*
	T	*Finel Muire*
Digitalis purpurea	K	*Lussmore; Moerecan; Sihane Sleivhe*
	T	*Bearnan Bearnagh; Moerecan; (Mearecan); Meregan na m'na Sioh*
	T	*Sihan; Sihan Sleibhe*
Dipsacus fullonum	K	*Leadan fiane; Leadan Cardeen*
	T	*Leadan*
Drosera sp.	K	*Druigdin monah; Drughtine mona; Eildriugd*
Drosera rotundifolia	T	*Druighdin Monah; Eil driugd*
Dryopteris filix-mas	K	*Raihneagh muir; Raigneagh Madderavah*
	T	*Raihneagh Muire; Raigneagh Madde ravah*
Elytrigia repens (*Elymus repens*)	K	*Brimere; Bruimfher; Urin*
	T	*Bruim fher; (Bruimfher); Urin; (Urin)*
Empetrum nigrum	T	*Froeghan Fhih; (Froeghan fihih); Brallan duh*
Equisetum sp.	K	*Irebull Copil Cait I Callagan; Scob Eigh*
Equisetum ?fluviatile	T	*Irebull Cait I Callagan; Scob Eich*
Equisetum hyemale	K	*Cuirridin Ganbir*
	T	*Cuirridin gauhair*
Equisetum palustre	T	*Cuirristin*
Eriophorum sp.	K	*Keannan ban; Fiodha monah; Feahog*
	T	*Keanuan ban; Siodha monah; Keannah na Monah; Scahog finae*
Erophila verna	K	*Illrugh grugan*
Eryngium maritimum	K	*Holimtragh; Ci[u]illen Traihe*
	T	*Cuillin Traihe; (Cuillin trahe)*
Euonymus europaeus	K	*Feorus*
	T	*Feorus*
Eupatorium cannabinum	K	*Scahogmuire*
	T	*Scahog Muire*
Euphorbia hyberna	K	*Makinboy; Makinbeigh au thle*
	T	*Makinboy*
Euphorbia peplus	K	*Gearneve*
	T	*Gear neve*
Euphrasia sp.	K	*Raihairkin; Dreaeryenrick; Glanrosk*
	T	*Kaihairkin; (Raihairkin); Keeimin Kaehairk; (Reimin Raihairk)*
	T	*Rensnin Raihairk; (Ruisnin Raihairk); Linn Raihairk; Luss a Bainne*
	T	*Luss na bainne*
Ficaria verna (*Ranunculus ficaria*)	K	*Lussnagnappane; Crann Arcain*
Filago minima	T	*Liah Luss beug; Loghall Keeol; (Loghall Keoll)*

Some Irish names of native and other wild Irish plants included in K'Eogh (1735) (K) and Threlkeld (1726) (T). ***(continued)***

Scientific name		Irish names[1]
Filago vulgaris	K	*Liablussroid*
	T	*Liah Luss roid*
Filipendula ulmaria	K	*Airgid Lovaghra; Airgidlugher*
	T	*Airgid Lovaghra*
Filipendula vulgaris	K	*Lussbrynagh*
Foeniculum vulgare	K	*Fenelecurba*
Fragaria vesca	T	*Suh Talhum; (Sauh talhum)*
Fraxinus excelsior	K	*Crann Funshog; Fuinnshop*
	T	*Crann Fuinnseog*
Fumaria officinalis	K	*Mantenaigh; Fumeterre; Seragh taihum; Çummansbarry*
	T	*Cuman Scarraigh; (Cuman searraigh); Detagh-talhum; Mantenaigh*
	T	*(Maintenagh); Fumeterre*
Galeopsis angustifolia	K	*Cruelife*
Galium aparine	K	*Garuh luss; Luss Garuh; Airmeireigh; Sophthilan*
	T	*Airmeirigh; Garuh luss; Luss garuh*
Galium odoratum	K	*Luss moleas*
	T	*Luss moleas*
Galium verum	K	*Balah chnise; Chouquilon; Madra Freaugh*
	T	*Balah Chnise; Chongulloin; Madra Freaugh*
Geranium columbinum	K	*Creghlagh; Creghdarig; Cuberanagh*
Geranium robertianum	K	*Rubel Rib; Rialcuiall; Rian Rib*
	T	*Ruhel Rih; (Ruhell Rih); Rial Guiall; Rian Rih*
Geum urbanum	K	*Mockell; Macull; Maghall*
	T	*Macull; (Macul); Maghal; (Maghall)*
Glechoma hederacea	K	*Aherluss; Abairlussa; Fientalloon*
	T	*Ahair lussa*
Halimione portulacoides	K	*Brenluss*
Hedera helix	K	*Oihin; Jen; Ihnane*
(or *Hedera hibernica*)	T	*Oihin*
Helleborus foetidus	K	*Dahouh Duh; Crubmahuin; Makkin sleve; Crubbear*
	T	*Dahouh duh; (Dahow Duh); Crub Mahuin*
Helleborus sp.	T	*Dahow duh*
Heracleum sphondylium	K	*Phoraan; Gouran; Louran*
Hieracium sp.	K	*Scaruhnamuck*
	T	*Searuhan na Muck*
Hieracium pilosella	K	*Clovas Liah; Clovas Lugh; Cluesine lugboge; Cluose Lugh*
	T	*Clovas Liah; Clovas Lugh*
Hordeum murinum	T	*Cuiseog Fhinn; (Cuiseog Fhin)*
Hordeum vulgare	T	*Orr na mor*
Humulus lupulus	K	*Hope*
Hyacinthoides non-scriptus	K	*Ruba-Muc; Lussna Ginle Gorah*
	T	*Buha Muck; Luss na Ginle Gorah*
Hydrocotyle vulgaris	K	*I[u]ss[n]a Pinin; Ivil Isle*
	T	*Luss na pininn*

Some Irish names of native and other wild Irish plants included in K'Eogh (1735) (K) and Threlkeld (1726) (T). ***(continued)***

Scientific name		Irish names[1]
Hyoscymus niger	K	*Gafainn; Keoghna Geark; Crann Goffen*
	T	*Gafainn; (Gefain); Keogh na Geark*
Hypericum androsaemum	K	*Meastork keeil; Masaturkalta*
	T	*Meastork keeil; (Meastook keel); Meas aturk alta*
Hypericum perforatum	T	*Allais Muire; Beahnova; Taed Coluim Kille; (Ted Colum kill)*
	T	*Beahnua Bicinionn; (Beannua buinnion); Beahnua firionn; (B. firionn* and *B. firion)*
Hypericum sp.	K	*Benhuir Allais muire; Ted Collumkiil; Beahnore; Beahnua Tirionu*
Hypericum tetrapterum	K	*Beahnua Firionn; Grinanagh*
	T	*Beahnua firiann; (Beahnua firion)*
Ilex aquifolium	K	*Cullin*
	T	*Cullin*
Inula helenium	K	*Makkinalen*
	T	*Meacan Eillin; (Meacan Eclin)*
Iris foetidissima	K	*Gloricum*
	T	*Gloricum*
Iris pseudacorus	K	*Felister; Silistar; Sillid drinn*
	T	*Sillistar; (Sillistir); Sillis drinn; (Sillis Arin)*
Juglans sp.	K	*Cran Knuhg Francugh*
Juniperus communis	K	*Uhar Greige; Uhar Traihe; Savinagh*
	T	*Uhar Creige; (Uhar Greige); Beeora Leacra*
Kickxia elatine	K	*Luss cre; Luss mide*
Knautia arvensis	K	*Cabon Guisain*
	T	*Cabon Guisain*
Lamium album	K	*Neantog maruh; Keighnantogebane*
	T	*Neantog Maruh*
Lamium amplexicaule	K	*Neantog keeogh*
	T	*Neantog keeogh*
Lamium purpureum	K	*Nantog-muire; Keigh nantoge Daurig*
	T	*Neantog Muire*
Lemna sp.	K	*Gran Laghan; Russ Laghan; Granrunog; Mackgan Aihir*
Lemna minor or *L. polyrhiza*	T	*Gran Laghan; (Gran Lagan); Gran tunnog; (Gran Cunnog)*
	T	*Mackgan Aihir; Rus Laghan*
Leonurus cardiaca	K	*Lussnacline*
Lepidium (coronopus) sp.	K	*Clybaghvye*
Lepidium latifolium	K	*Bochunell*
Lepidium sativum	K	*Pibargass; Billarfrancugh; Gall-Billar*
	T	*Pibargass; Billar Francagh; Gail (Gall) Billar; (Gall Billar)*
Lapsana communis	K	*Ston na Duilleog; Duilleoge Brideog; Duilleog mah; Duilleog min*
	T	*Ston n'duilleog; Duilleog brideog; Duilleog Mah; Duilleog min*
Leucanthemum vulgare	K	*Aisbeochan; Easbeaig-beeah*
	T	*Aisbeoc ban; Easbeaiig boeah; Luss na Frank*
Ligustrum vulgare	K	*Crandlu*
Linum catharticum	K	*Keaolagh; Millvoot; Eribleminugh; Misugh*
	T	*Keoeolagh; (Keaolagh)*

Some Irish names of native and other wild Irish plants included in K'Eogh (1735) (K) and Threlkeld (1726) (T). ***(continued)***

Scientific name		Irish names[1]
Linum usitatissimum	T	*Lion na m'ban Sioth; (Lion na mban sioh)*
Lithospermum officinale	K	*Couragh*
Lolium multiflorum (or *L. temulentum*)	T	*Ruinhelais; Kuinhelean Darnel*
Lolium perenne	T	*Braillan*
Lolium temulentum	K	*Brailan*
Lonicera periclymenum	K	*Duillear Fehlin; Feileog Facrann; Dahelin*
	T	*Duillear Fehlin; (Duillir fehlin); Feileog fa chrann; (Feilog fa chran)*
	T	*Keis Fa chrann; Faheline*
Lychnis flos-cuculi	K	*Keeorrogagh Lenu; Luss na Gile*
	T	*Keeorogagh lena; Luss na gile*
Lycopodium clavatum	T	*Garvogagh Sleive*
	K	*Garbogagh flieve*
Lycopus europaeus	K	*Foohran Curroih*
	T	*Eeohran Curroih; (Eeohran Curroigh)*
Lysimachia nemorum	T	*Seamar muire; Lus na Colum-kill; Lus no Colum-kill*
Lysimachia nummularia	K	*Lussna dawpinin*
Lythrum salicaria	K	*Beala[n] Leane; Irrebul Caitin; Conaire*
	T	*Bralan Leane; Irebul Caitin; (Irebull Caitin); Conaire*
Malva neglecta	K	*Hiccus Francagh*
	T	*Hocus Francagh*
Malus sylvestris	K	*Crann owle; Uhla; Crann owle a fain*
	K	*Hocus fihain*
	T	*Hocus Tihain; (Hocus fihain); Luss na Meal Muire; (Luss na Meal Muirre)*
Marrubium vulgare	K	*Orafunt*
	T	*Orafunt*
Melilotus sp.	K	*Melodagh; Crubinkitt; Phlurinshingan*
Melittis melissophyllum	T	*Cuslin-gann-dauri; (Cuslin gan Dauri); Causadan*
Mentha aquatica	K	*Cortline; Cartlein; Pisba[ne]en*
	T	*Pismin; Cartloinn; (Cartloin)*
Mentha arvensis	K	*Dunmindas Isky*
Mentha longifolia	T	*Luss na beag*
Mentha pulegium	K	*Voragagh*
Menyanthes trifoliata	K	*Ponair capuil; Porahappil; Bagharane; Pacharan*
	T	*Ponair Capuil; Pacharan; (Pachran)*
Mycelis muralis	K	*Nunsinugh*
	T	*Bliutan*
Myosotis arvensis	K	*Lussmide; Lussminle*
	T	*Luss Mide; Luss Minnle*
Myosotis scorpioides	T	*Coharagh*
Myrica gale	K	*Rindeog; Rileogagh*
	T	*Riudeog; (Ruideog); Raodagh; (Rodagagh)*
	T	*Rileogagh*

Some Irish names of native and other wild Irish plants included in K'Eogh (1735) (K) and Threlkeld (1726) (T). *(continued)*

Scientific name		Irish names[1]
Nasturtium officinale	K	*Billar; Billrugh*
	T	*Billar*
Nuphar lutea	K	*Duillog baite buigh*
Nymphaea alba	K	*Duikeogbaile; Cohinih aubun; Liagh loghar*
	T	*Cohinih Auhun; Cuirrinin ba'n; Duilleog baite; (Curririn ban)*
	T	*Liagh Loghar*
Oenanthe crocata	K	*Dahow ban; Takow*
	T	*Fealla bog; Dahow ban; Tahow*
Ononis repens	K	*Chuinvalegh; Strang boh; Strang Terrain; Trian Terrain*
	T	*Srang boh; Srang terrain; (Srangterrain); Tri an terrain*
Ophioglossum vulgatum	K	*Tongueabowbally; Tongueabowlaw; Lusnateanagh*
	T	*Luss na teangah*
Orchis mascula	K	*Keark brea[c]k*
Origanum vulgare	K	*Marhine fiane*
Orobanche rapum-genistae	K	*Muchog*
	T	*Muchog*
Osmunda regalis	K	*Ranagh Rihule*
Oxalis acetosella	K	*Samsogy; Shamsoge*
Papaver rhoeas	K	*Blah na bodagh; Cailleagh Dearg; Papin; Snannagh*
	T	*Cailleagh Dearg; Papin*
Papaver somniferum	T	*Colladin Poipin*
Parietaria judaica (Parietaria diffusa)	K	*Mintus Caissil; Mindas balla*
	T	*Mintus Caissil*
Pastinaca sativa	K	*Meacan Rih; Meacan Rih fian*
	T	*Meacan buih; Cuiridin ba'n; (Cuirridin ban)*
Pedicularis sylvatica	K	*Moahlin monoh; Luss na Giolla; Milshan monah*
	T	*Moahlan Monah; (Moahlin Monah); Luss Rihagh; (Luss Riah)*
	T	*Luss an Giolla; Rieauhaih; Milshan Monah*
Persicaria bistorta	K	*Stonisi; Cabpoge nieve*
	T	*Stonsi*
Persicaria hydropiper	K	*Glunedarig; Gluneaghdearg*
	T	*Gluneagh Dearg; (Glunnegh Dearg)*
Persicaria maculosa	K	*Gluneagh mo'r; Gloonagh vye*
	T	*Gluneagh Mo'r*
Petasites hybridus	K	*Pubboll gallan; Oubbnel; Burrane*
	T	*Puboll; Gallan*
Petroselinum crispum	K	*Persill*
Phragmites australis	K	*Birrah Laghan; Cuisgirnag*
	T	*Birrah; Laghan; (Lagan); Cuisgirnah*
Pimpinella saxifraga	K	*Lussinnuille fiane*
Pinguicula vulgaris	K	*Bodan measgan; Woghter go Toan; Iem go Illinygh*
	T	*Bodan Measgan; (Bodan Measgar); Ouaghdar go toin*

Some Irish names of native and other wild Irish plants included in K'Eogh (1735) (K) and Threlkeld (1726) (T). ***(continued)***

Scientific name		Irish names[1]
Pinus sylvestris	T	*Giusagh; Guisagh*
Plantago lanceolata	K	*Slan Luss*
	T	*Slan Luss*
Plantago major	K	*Croabh-Phadruig; Crugh-Phatrick*
	T	*Cor Chapog*
Polygala vulgaris	K	*Luss Bain*
	T	*Luss Baine; (Luss bainne); Bo gorm*
Polygonum aviculare	K	*Gluneagh beug; Fereglunagh*
	T	*Glunneagh beug*
Polypodium vulgare	K	*Sgim-na Clogh; Skave-chrine; Sgim na Darah*
	T	*Sgim na Clogh; Sgim na Darah*
Polystichum aculeatum	K	*Uhartalhum*
	T	*Uhar talum; (Uhar talvan)*
Populus tremula	K	*Crann na Crih; Creigh*
	T	*Crann na Crih*
Potamogeton natans	K	*Duileasg na auhun; Liagh Roda*
	T	*Duilleasg na auhun; Liagh Roda*
Potentilla anserina	K	*Brisklane; Briclane*
	T	*Brisclan*
Potentilla erecta	K	*Leanarthy; Neaubnadis; Menedin; Lievenet*
	K	*Nauniddy; Makkinnefin*
	T	*Neauhnadis; (Neaunadis); Menedin; Neaubnuid*
	T	*Livenet*
Potentilla reptans	K	*Cuigiagh; Lussernacoaghmore*
	T	*Cuigvear Muirre; (Cuigvear Muire); Meagigh*
Primula veris	K	*Seuerine*
Primula vulgaris	K	*Banne bo Bliughtain; Buiheacan bo bliught; Seiuherin; Seiuhirgin*
	T	*Bainne bo bliughtain; Buiheacan bo bliught; Seiuherin; Seiuhirgin*
	T	*(Seihirgin); Seiuheirclan*
Prunella vulgaris	K	*Duban Kean Cassagh; Duanmebeg; Keanavan beug; Duih muih*
	T	*Keannavan beug; Duhan Kean Cassagh; Duih Muih*
Prunus avium	K	*Crann Silin darug*
	T	*Crann silin*
Prunus padus	K	*Cran Silindugh; Dunrisk*
	T	*Dunreisk*
Prunus spinosa	K	*Droihean; Drine*
	T	*Droihean*
Pteridium aquilinum	K	*Ranugh*
Pulicaria dysenterica	K	*Lasnacorscart*
Quercus sp.	K	*Crann Darrah; Daerr*
Quercus robur or *Q. petraea*	T	*Crann Darrah; Farran; Furtan*
Radiola linoides	K	*Lussnamakkave*

Some Irish names of native and other wild Irish plants included in K'Eogh (1735) (K) and Threlkeld (1726) (T). ***(continued)***

Scientific name		Irish names[1]
Ranunculus aquatilis or other spp.	T	*Lionan Auhun; (Lionan Auhuin); Flingh visge; (flingh uisge)*
	T	*Niul Uisge; Linaneagh; Snaihe bahih*
Ranunculus bulbosus	T	*Tile talvan*
	T	*Grann arcain; (Gran arcain)*
Ranunculus flammula	T	*Lassair Lena; (Laissar Lena); Inniagh*
Ranunculus peltatus	T	*Turkis fihain*
Ranunculus repens	K	*Fearban; Gairgain*
	T	*Fearban; Gairgin*
Reseda luteola	K	*Buigh more; Boigh-more*
	T	*Buigh Mor*
Rhamnus cathartica	K	*Brenedrain; Maide Bren*
	T	*Bren uhall; Maide bren*
Rhinanthus minor	K	*Bodan Chloigin*
	T	*Bodan Chloigin*
Ribes uva-crispa	K	*Crann Spinan*
Rorippa amphibia	K	*Makkin Ragum Isky; Raboon Isky*
Rosa canina	T	*Feirdriss*
Rosa rubiginosa	T	*Feirrdriss cuhra; (Feirdriss Cuhra)*
Rubus fruticosus	K	*Driseog; Smerugh Dugh (berries)*
	T	*Driseog*
Rubus idaeus	K	*Maohan Conaire; Sugh cruh (fruit); Sugh creeve (fruit)*
	T	*Maohan Conaire; Suh Cruh*
Rubus saxatilis	K	*Sou-a-meen*
	T	*Soon-a-man-meene; (soon-a-man meene)*
Rumex acetosa	T	*Saugh boh; Sealgan*
Rumex acetosella	K	*Sauha Keeoragh; Keirogah; Savachyragh*
	T	*Sauha keeoragh; (Sauha keeoragagh); Keirogah*
Rumex conglomeratus	K	*Cuppogegere*
Rumex hydrolapathum	K	*Cuppoge more Isky*
Rumex obtusifolius	T	*Copog*
Ruscus aculeatus	T	*Brusglagh*
Salix alba	K	*Soileog; Saileagh; Sillioge*
	T	*Soileog; Saileach*
Salix ?viminalis	K	*Soileog; Saileagh Isky*
Salvia verbenaca	K	*Nartruiel fian*
Sambucus ebulus	K	*Bollfurt; Lusluh-volvaird; Mallavourt*
	T	*Ballfurt; Luss luhvolvaird; (Luss luh volvaird)*
Sambucus nigra	K	*Crann Tromain; Thrum*
	T	*Crann Tromain*
Sanguisorba minor	K	*Lusinuila*
Sanicula europaea	K	*Reagum; keagamuid Buir; Keama; Makkil*
	K	*Chimud*
	T	*Reagum; Buir; Keama; Keaga muid*

Some Irish names of native and other wild Irish plants included in K'Eogh (1735) (K) and Threlkeld (1726) (T). ***(continued)***

Scientific name		Irish names[1]
Saxifraga granulata	K	*Eigreim; Minane Murry; M[o]ran*
Schoenoplectus lacustris (Scirpus lacustris)	K	*Buigown; Buigh an buin; Beugsheinhin*
	T	*Buigoun; Buigh Auhuinn; (Beug Auhuin); Beug Sheinhin*
Scleranthus annuus	K	*Minam muire*
	T	*Minan Muire*
Scrophularia auriculata	K	*Sugh farnugh; Bogluss; Savacurha*
Scrophularia nodosa	T	*Forum Dunlass; (Forum Dun luss); Faruh duh; Luss na Knapan*
Secale cereale	K	*Segael; Saggil*
	T	*Segael*
Sedum acre	K	*Grifan na Clogh; Garuanclogh*
	T	*Grifan na Glogh; (Gri fan na clogh)*
Sedum telephium	K	*Torpine*
Sempervivum tectorum	K	*Feinneeagla; Finnikin; Tirpin; Torepen*
	T	*Teinne Eagla; (Tanne Eagla); Tinnikin; Tirpin*
Senecio aquaticus	T	*Bogluss; (Bog luss)*
?Senecio fluviatilis	K	*Jeglante; (Saracens Consound)*
Senecio jacobaea	K	*Pfulian bugih balkisan; Booellan buigh*
var. *flosculosus*	T	*Pfullen Buih Balkisan*
Senecio vulgaris	K	*Groondluss*
Silaum silaus	T	*Eigreim; Moran*
Silene dioica	T	*Cororan Coilleah*
Silene vulgaris	K	*Cororan coilleah*
Silybum marianum	K	*Fofannan breack; Fofannan muire; Bernan breack; Fohenane Banny*
	T	*Fofannan breack; Fofannan Muire; Bearnan breack; Fosannan beannaight*
Sinapis arvensis	K	*Preaisseagh Garugh; Prasseagh Buigh; Borastebye*
	T	*Praisseagh buigh; Preaisseagh Garuh; (Praisseagh Garuh); Gas an Chunnagta*
	T	*Garuhog*
Sisymbrium officinale	K	*Melisse; Lusanore*
	T	*Melissa*
Sium latifolium	K	*Folaghter; Folaght; Cossa-duh*
	T	*Folaght; Cossa duh*
Smyrnium olusatrum	K	*Lussarangrandugh; Livenamuea*
	T	*Alistrin; Lussaran gran duh*
Solanum dulcamara	K	*Slat gorm; Dremire gorm; Migbuih*
	T	*Stat Gorm; Mihagh Uisge; Mig buih; Dremire Go m*
	T	*(Dremire Gorm)*
Solanum tuberosum	K	*Potaty; Photy*
Solidago virgaurea	K	*Stal Ore*
Sonchus asper	K	*Sharravane muck*
Sonchus oleraceus	K	*Bainne muck; Blinght Fofannan; Fofannan min*
	T	*Bainne Muck; Bliught Fofannan; Fofannan Min*
Sorbus aucuparia	K	*Keora Cubra; Keeora Cahrain; Kouragh [?]un*

Some Irish names of native and other wild Irish plants included in K'Eogh (1735) (K) and Threlkeld (1726) (T). *(continued)*

Scientific name		Irish names[1]
	T	*Keeora Cahrain; (Keora Cahran)*
Sorbus cf. *hibernica*	T	*Keora Cuhra*
Sparganium erectum	K	*Seisg madrah; Rih seisg*
	T	*Seisg Madrah; Rih Seisg*
Sparganium ramosum	K	*Seisg Madrah; Rih Seisg*
Spergula arvensis	T	*Curran lin; Cluan lin; Cabrois; (Cabruis)*
Stachys palustris	K	*Furrhum; Forrhum*
Stachys sylvatica	K	*Uhar sliab (ure slieve); Ahair Liah; Lussna fiah; ure cleagh*
	T	*Ahair liah; Uhar Sliah; Luss na fiah; Sai[?]c Fihain*
	T	*Ebar Slaiuhe*
Stellaria holostea	K	*Tursarrin*
	T	*Tursarrin; (Tursacrin)*
Stellaria media	K	*Fliagh; Blihagh*
	T	*Flium*
Succisa pratensis	K	*Uragh-ballagh; Oir Ballagh; Urugh vallugh*
	T	*Uragh Ballagh; Oir Ballagh*
Symphytum sp.	K	*Makkinduff; Lussnaknauh; Briodi; Coomfry*
Symphytum officinale	T	*Luss na Knauh Brisdi; (Lus na Knau brisdi)*
Tamus communis	K	*Urinaghdugh*
Tanacetum parthenium	K	*Meadhdruah; Midrua*
	T	*Meadh druah*
Tanacetum vulgare	K	*Lussna Frank; Lussfrancugh*
Taraxacum officinale	K	*Caisearuhan; Caprevane*
	T	*Caissearuhan; (Cais searuhan)*
Taxus baccata	K	*Uhar*
	T	*Uhar*
Teesdalia nudicaulis	T	*Gass an Cailleagh*
Teucrium scordium	K	*Nonlough; Nilluegh Isky*
Teucrium scorodonia	K	*Shaste Kil[t]y; Shaste fhain*
Thlaspi arvense	K	*Praisseagh fiach; Praisseagh na Keeoragh*
	T	*Praisseach fiach; (Praisseach fiaeh); Praisseagh na keeoragh*
Thymus polytrichus	K	*Thiemcregge*
Trifolium pratense	K	*Seamar capuil; Seamar Leanne*
	T	*Seamar Leanne; (Seamar Leanna); Seamar Capuil; (Seamar Capul)*
Trofolium repens	K	*Shamrock*
	T	*Shamrocks; (Shamrock)*
Triticum aestivum	K	*Crinugth*
	T	*Cruih Neaght*
Tussilago farfara	K	*Lussernaspunck; Ahain fahain; Duiluir Spunck*
	T	*Ahain; Fahain; Duilluir Spuink; (Duilliur Spuink)*
	T	*Fohannan*
Typha latifolia or *Typha angustifolia*	T	*Bodan Duh; Gah Bodan; Cogillah na M'ban Sioih*
Ulex europaeus	K	*Aitinn*

Some Irish names of native and other wild Irish plants included in K'Eogh (1735) (K) and Threlkeld (1726) (T). ***(continued)***

Scientific name		Irish names[1]
	T	*Aitinn; (Aitin)*
Ulmus sp.	K	*Ailim; Slievane; Livane*
Ulmus glabra	T	*Ailim*
Umbilicus rupestris	K	*Cornnan Caisil; Loun Cait; Leacan; Imlekkanagh*
	T	*Cornnan Caisil; Loun Cait; Leacan*
Urtica dioica	K	*Neantog Loisgneath*
	T	*Neantog loisgneah; (Neantog Loisgneagh)*
Urtica urens	T	*Caol Fail*
Vaccinium myrtillus	K	*Fraghan; Feadog*
	T	*Frahog; Fraghan*
Vaccinium oxycoccos	K	*Monog*
	T	*Monog*
Valeriana officinalis	K	*Keerin Leanna; Carencurrugh*
	T	*Keerin Leanna; (Keerin Leana)*
Verbascum thapsus	K	*Cuineail muire; Kinnil vurry; Cunellmurry*
	T	*Cuineal Muire*
Verbena officinalis	K	*Vuegagh*
	T	*Luss Cre; Seamar Cre*
Veronica anagallis-aquatica	T	*Billar ika; Loghal muhar*
Veronica beccabunga	K	*Loghal more; Loghal muhar; Bilre murry*
	T	*Billar muire*
Veronica chamaedrys	K	*Nilluegh; Nonlough Gardeen; Nonlough Elough; Niuluragh*
	T	*Noulough; Elough; Niuluragh; (Niulurah)*
Veronica officinalis	K	*Seamearcre*
Viburnum opulus	T	*Keora Con*
Vicia cracca	K	*Pis Capuil; Fiarghail; Pissman*
	T	*Pis capuil; (Pis capul); Fiarghail*
Vicia faba	T	*Ponair*
Vicia sativa	K	*Piss Capuil; Piss duh*
	T	*Piss Capuil; (Pis capul); Pis duh*
Viola sp.	K	*Sail Covagh; Sa[i]chugh*
Viola odorata	T	*Sail Covagh; (Sail Covah)*
Viola tricolor	T	*Gorman-seareagh*
Viscum album	K	*Kynaghdarny*
Chara hispida	T	*Cloish; (Clo Isk)*
Slauke, Sea Lettuce	K	*Slukane*
Rhodymenia palmata	T	*Duilleasg; Faminkiren*
Agaricus campestris	T	*Fas na heanaih; Bocan Beragh; Leurin Sugagh*
Jew's ear fungus	K	*Cluesudy*
Tuber aestivum (fungus)	T	*Bolgan beike; Bal Beike; Bucaille boh; Buile baka*
	T	*Trimane*
Common liverwort	K	*Duil[?]eag na Cruichneaghta; Jouin I[i]n; Hebow*

Some Irish names of native and other wild Irish plants included in K'Eogh (1735) (K) and Threlkeld (1726) (T). *(continued)*

Scientific name		Irish names[1]
Unidentified liverwort	T	*Duilleasg; Duilleasc na Cruichneaghte; Oihe auhunn; Oihe Auhuin*
Lichens including *Parmelia saxatilis*	T	*Crotall Coilleagh*
Polytrichum commune	K	*Noergagh*
	T	*Garuh luss na monah*
Polytrichum sp.	K	*Garub luss na Monah*

1. Spelling and capitalization of Irish names have been duplicated from the original sources.

DISPLAYS AND EXHIBITIONS OF TRADITIONAL FOLKLIFE

The following are some of the best places in Ireland to see objects of ethnobotanical interest, including reconstructed vernacular buildings constructed with a range of plant and other natural materials. Many have excellent exhibitions, information resources and good collections of artefacts made from plants (such as vernacular furniture) and objects used in agriculture.

Bunratty Castle and Folk Park, Bunratty, Co. Clare

Restored and recreated traditional buildings, furniture, artefacts and exhibitions related to Irish folklife, as well as the restored Bunratty Castle itself. The Folk Park includes a range of traditional rural thatched houses and a village street with houses, pubs and shops. Visitors can also see Bunratty House, a large Georgian house built in 1804.

Craggaunowen—The Living Past, near Quin, Co. Clare

A restored fortified Tower House, originally built ca. 1550, and a reconstructed crannog, ringfort and other features illustrating life in ancient Ireland. Also on display is 'The Brendan', a leather-hulled boat. The boat was built in 1976 by British explorer Tim Severin, who sailed in it from Ireland to Newfoundland, re-enacting the voyage of St Brendan the Navigator (ca. 484–577). Some believe St Brendan was the first European to reach North America.

Kerry Bog Village Museum, Glenbeigh, Co. Kerry

Various traditional recreated houses, such as a turf-cutter's house, blacksmith, labourer's dwelling and thatcher.

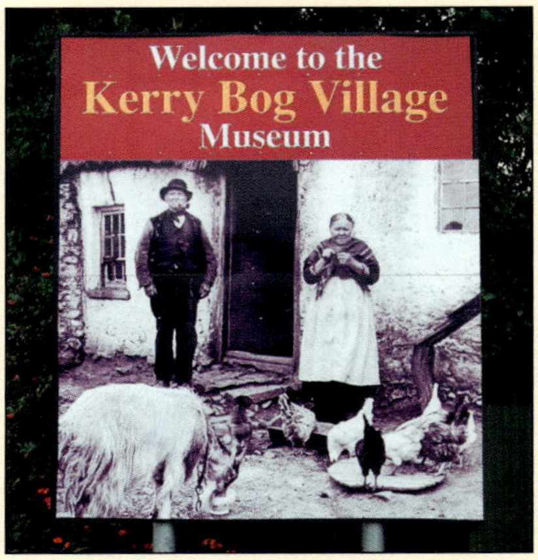

Muckross House, Gardens and Traditional Farms, The National Park, Killarney, Co. Kerry

Restored and recreated traditional buildings, furniture and artefacts mainly related to farming and traditional life in Ireland.

National Botanic Gardens, Glasnevin, Dublin 9

A rich collection of artefacts and products made from plants from around the world, including from Irish plants. Although the collection is not on public display, arrangements can be made to see specific objects that are housed in the National Herbarium there. Occasional ethnobotanical exhibitions are mounted.

National Museum of Ireland—Country Life, Turlough Park, Castlebar, Co. Mayo

Exhibitions and displays of artefacts and information related to traditional life in Ireland.

National Museum of Ireland—Archaeology, Kildare Street, Dublin 2

Archaeological exhibitions and artefacts, such as the remarkable Lurgan logboat and an exhibit and interpretation of Irish bog bodies.

National Museum of Ireland—Decorative Arts & History, Collins Barracks, Benburb Street, Dublin 7

Vernacular and antique furniture and other decorative objects.

Sligo Folk Park, Millview House, Riverstown, Co. Sligo

Displays of traditional buildings and agricultural artefacts.

The Irish National Heritage Park, Ferrycarrig, Co. Wexford

Displays, recreated buildings (such as a crannog) and sites of early Irish life from the Stone age, Bronze age, Celtic and Early Christian times, up to the Viking and Norman invasions.

Ulster American Folk Park, National Museums of Northern Ireland, Castletown, Omagh, Co. Tyrone

Restored and recreated traditional buildings, furniture, artefacts and exhibitions mainly related to Ulster folklife.

Ulster Folk and Transport Museum, National Museums of Northern Ireland, Cultra, Holywood, Co. Down

Restored and recreated traditional buildings, furniture, artefacts and exhibitions mainly related to 19th-century Ulster and emigration to North America.

GLOSSARY

ague—A fever, marked by periods of chills, sweating and other flu-like or malaria-like symptoms.

alginate—A salt of algenic acid, such as sodium or ammonium alginate, extracted from brown algae.

alum—An aluminium sulphate compound used as a chemical fixative.

angina—A sharp and spasmodic pain in various parts of the body, such as the chest or abdomen, usually associated with a lack of oxygen reaching the heart.

annual—A plant that grows from seed and then flowers, seeds and generally dies all within one year (or one growing season).

anti-inflammatory—Something that reduces the inflammation of tissues.

antispasmodic—Something that prevents or relieves spasms or convulsions.

aphrodisiac—A drug or other substance that arouses sexual desire.

aromatic—Something that has a spicy, pungent or fragrant odor that is generally agreeable.

astringent—A sharp or pungent medicine that tends to draw together tissues, for example, to help reduce blood flow.

berry—A fleshy fruit that has no hard parts other than the seed. The term berry is often and typically applied to other fleshy fruits that are not botanically 'berries'.

biennial—A plant that takes two years to come to flowering and to set seed, after which it often dies. It usually does not flower during its first year.

blackleg—An acute infectious disease of cattle where the beast generally has a high fever and a very swollen and painful thigh and there the skin becomes gangrenous and death often occurs quickly.

blasts—Facial swellings.

bulb—A short swollen underground structure consisting of a stem surrounded by scale-like modified leaves.

cathartic—A powerful agent used to relieve constipation (also called a purgative).

cauterize—To burn or sear and deaden tissue, for example, helping to eliminate disease of the skin through the formation of a callous.

cellulite—A deposit of fat, usually found on the thighs and buttocks, giving an uneven or dimpled appearance.

chilblains—An inflammation of the feet, hands or ears, followed by itchiness, usually as a result of exposure to moisture and low temperatures.

cirrhosis—A chronic disease of the liver, often caused by long-term over-use of alcohol or infection by the hepatitis virus.

colic—Abdominal pains, which often occur in spasms, often associated with feeding babies and young children.

composites—Members of the Compositae (Asteraceae) family; daisy family.

conjunctivitis—An eye infection causing inflammation of the conjunctiva.

consumption—An archaic word used to describe pulmonary tuberculosis.

copperas—Ferrous sulphate crystals, often used in the manufacture of iron pigments of various colours.

cystitis—A urinary infection which may be accompanied by blood in the urine and pain during urination.

dampers—A simple unleavened bread.

decoction—A substance, such as bark, leaves, roots or berries, that is steeped in water so as to extract soluble compounds to be drunk like tea, which is itself a decoction.

dermatitis—Inflammation of the skin.

diuretic—A substance that promotes the excretion of urine.
drisheen—A form of black pudding made with milk, water, sheep, cow or pig blood, breadcrumbs and mutton suet. Recipes vary from place to place and so do cooking methods. Tansy (*Tanacetum vulgare*) has often been used to flavour drisheen.
drones—Usually three pipes that are part of uilleann pipes that accompany the melody being played with a constant background note.
dropsy—A disease associated with the accumulation of excess fluid in the body cavity or tissues.
dysentery—The inflammation of the intestine, especially the colon, accompanied by abdominal pain and frequent stools often containing mucus and blood.
eczema—A non-contagious inflammation of the skin, characterized by redness, itching and lesions that may become encrusted and scaly.
emetic—A substance that causes vomiting.
epiphyte—A plant that grows on another, such as ferns on an oak tree.
ergot—A fungus that affects cereals and grasses.
erysipelas—An acute disease of the skin, also called St Anthony's fire, caused by ergot infection.
ethnobotany—The study of the use of plants by people.
expectorant—A substance that helps to clear mucus and phlegm from the chest and respiratory tract.
falling sickness—Epilepsy.
farcy—An infectious disease of horses and cattle, usually including lesions on the legs.
felons—Inflamed sores on the fingers.
fen—A wet vegetation community on alkaline, often calcareous, neutral or only slightly acid soils.
flatulence—The accumulation of excessive gas in the digestive tract.
fluke—A parasitic flatworm (Platyhelminthes) that invades the bodies of animals, including humans, such as liver fluke.
frond—A fern leaf, or the leaf-like expanded part of a seaweed.
fumigant—A chemical compound that is used in a gaseous state as an insecticide.
gangrene—The death of tissues, resulting from the loss of nutrient flow to the affected part, followed by bacterial attach and putrefaction.
gapes—A disease of young domestic fowl, characterised by gaping or gasping for breath and caused by gapeworms.
garderobe—A private room used for toilet purposes.
gout—The painful inflammation of the joints, usually the legs, feet and hands, caused by raised levels of uric acid in the blood, which causes uric crystals to be deposited in the joints. Gout can become chronic and cause deformities.
gripe—Sharp and spasmodic pains in the bowels.
gruel—A food made from ground meal mixed with water, rather like a thin, watery porridge.
haemorrhoid—Painful swellings of the rectum and anus, often with bleeding.
haemostatic—A substance that stops bleeding and haemorrhaging.
heath—A vegetation usually dominated by ling (*Calluna vulgaris*), heathers and heaths (*Erica* species) on nutrient-poor, usually sandy soils often with a shallow layer of peat.
herb—Any vascular plant that is not woody. The term herb is often used for species that are used for medicinal and culinary purposes, but the term is not restricted to such plants only.
hoose—A lung disease of calves caused by worms.
infusion—Material, such as leaves, steeped in water to dissolve the active or required ingredients.
jaundice—A liver disease or disorder often causing a yellowing of the skin.
laxative—A substance that increases the movement of material through the digestive tract and from the bowel.
malt/malting—The process by which cereal grains are prepared for brewing, in which the grains are made to germinate by soaking in water and are then quickly halted from germinating further by drying/heating with hot air. Malting is therefore a combination of two processes: the sprouting and the drying.

marquetry—Material such as thin sheets of veneer that are inlaid decoratively and often intricately into other wood to make decorative furniture and other objects.
mastitis—An inflammation of the breast, usually caused by bacteria and generally painful, preventing or slowing the flow of milk in nursing mothers, both in humans and other animals.
meadow—Grassland that is mown for hay and silage. In contrast, pasture is grazed grassland.
migraine—Periodic sharp headaches, often accompanied by other symptoms such as nausea, vomiting and affected vision.
mordant—A substance used to fix colours in dyeing or used to treat leather or other materials before dyeing.
mucilage—A sticky or gelatinous substance produced by various plants.
mucus—A viscous or slippery substance.
palsy—Complete or partial loss of muscle function causing weakness, tremors or loss of sensation.
pasture—Grassland that is grazed by stock, including cattle and sheep. In contrast, meadow is grassland that is mown for hay and silage.
perennial—A plant that lives, flowers and sets seed repeatedly over a number of years.
pip—A disease of birds where a thick mucous discharge forms a crust around the mouth and throat.
pleurisy—Inflammation of the pleura (the membrane surrounding the lungs), usually occurring as a complication of a disease such as pneumonia. It is generally accompanied by accumulation of fluid in the body cavity, chills, fever, and painful breathing and coughing.
poultice—A soft, moist mass that is wrapped in cloth, such as linen, gauze or towels, and applied when hot to treat a skin complaint or other ailment.
prophylactic—Something that is used to prevent or protect against disease.
purgative/purge—A substance that causes the evacuation of the bowels.
ramrod—A rod used to clean the barrel of a gun or other firearm.
red-water fever—A cattle fever carried and transmitted by cattle ticks. The disease *Clostridium haemolyticum* is called babesiosis and infects the red blood cells and is often fatal. It occasionally infects sheep too.
rheumatoid arthritis—A chronic disease causing the inflammation of the joints and resulting in stiffness, pain, deformities and loss of mobility.
rickets—A disease caused by lack of vitamin D and/or calcium, and insufficient exposure to sunlight, causing bone deformities, often in children.
ringworm—A contagious skin disease caused by several fungi, giving rise to circular itchy and scaly patches of skin.
rubefacient—A substance that causes irritation (and therefore redness) of the skin.
salep—Starchy meal ground from the roots of dried orchids used as a food and to make a milky beverage.
saltpetre—A common name for potassium nitrate, which was often used as a preservative.
scallops—Flexible rods used to anchor thatch in place.
scrofula—A form of tuberculosis that affects the lymph nodes, particularly in children, and is spread by unpasteurised milk from infected cows.
scurvy—A disease caused by a vitamin C deficiency, with symptoms including weakness, soft and bleeding gums and bleeding under the skin.
shingles—A viral infection similar to chickenpox that affects the nerves and causes a rash on the body, most often in older people.
shrub—A woody plant that is generally not more than 5 m (15 ft) tall, often with several stems or trunks that arise near the base of the plant.
soporific—A substance that causes or helps to induce sleep.
St Anthony's fire—A skin disease caused by ergot infection.
St Vitus dance—Sydenham chorea; an acute neurologic disorder that can appear several months after a streptococcal infection.
staggers—A disease in animals caused by B_1 deficiency.
strangles—A bacterial infection of horses, characterised by swelling of the lymph glands around the neck.

sternutatory—Of a medicine that causes or promotes sneezing.

tallow—Hard fat from cattle and sheep.

tannin—A substance rich in tannic acid used as an astringent particularly in the preservation of leather.

tuber—A swollen, fleshy, usually underground stem, such as a potato.

tuberculosis—An infectious disease caused by a bacterium that affects the lungs. Symptoms include coughing up sputum and mucus, weight loss, fevers and chest pains.

urinary—Relating to the production and excretion of urine.

vermifuge—A substance that is used to expel worms from the intestine.

wattle—A wall, fence or roof constructed of poles intertwined with twigs, rods, branches or reeds.

white scour—A disease of young calves caused by bacteria including *E. coli* which causes acute diarrhoea.

whitlow—An infected tendon sheath on a finger.

whooping cough—A contagious disease of the respiratory system often affecting children, characterised by spasmodic coughing.

BIBLIOGRAPHY

Adriaens, Monik. 2006. Family Medicinal Plant Gardens in the Rwenzori Region. Marianum Press Ltd, Entebbe, Uganda.

Akerele, Olayiwola, Vernon Heywood & Hugh Synge (editors). 1991. The Conservation of Medicinal Plants. Cambridge University Press, Cambridge.

Akeroyd, John R. (editor). 1996. The Wild Plants of Sherkin, Cape Clear and Adjacent Islands of West Cork. Sherkin Island Marine Station, Sherkin Island, Co. Cork.

Allen, D. E. & G. Hatfield. 2004. Medicinal Plants in Folk Tradition: An Ethnobotany of Britain and Ireland. Timber Press, Portland and Cambridge.

Allingham, William. 1985. William Allingham, A Diary: 1824–1889 (Penguin Lives and Letters) (Introduction by John Julius Norwich). Penguin Books, London.

Anderson, Glynn and John McLaughlin. 2011. Farmhouse Cheeses of Ireland. A Celebration. Collins Press, Cork.

Angiosperm Phylogeny Group III. 2009. An update of the Angiosperm Phylogeny Group classification for the orders and families of flowering plants: APG III. Bot. J. Linnean Soc. 161: 105–121.

Anonymous. 1840. The Irish in 1644: As described by a Frenchman of that period. The Irish Penny Journal 1(1), 4 July 1840.

Anonymous. 1941. A Handbook of Home-Grown Timbers (War emergency edition). Department of Scientific and Industrial Research, Forest Products Research Board. His Majesty's Stationery Office, London.

Anonymous. 1973. Viking and Medieval Dublin: National Museum Excavations, 1962–1973. Catalogue of Exhibition. National Museum of Ireland, Dublin.

Anonymous. 1978. Ainmneacha Plandaí agus Ainmhithe (Flora and Fauna Nomenclature). An Roinn Oideachais, Oifig an tSoláthair, Baile Átha Cliath.

Anonymous. 1997. Farming Since the Famine: Irish Farm Statistics 1847–1996. Central Statistics Office, Government of Ireland, Dublin.

Anonymous. 2005. Champion Trees: A Selection of Ireland's Great Trees. Tree Council of Ireland, Dublin.

Anonymous. 2007. Forestry & Timber Yearbook 2007. Irish Timber Growers Association, Dublin.

Anonymous. 2008. Preserving a landscape, reviving a taste. Natura 2000, European Commission DG. Env Nature Newsletter, 24 July 2008.

Arnold, James. 1968. The Shell Book of Country Crafts. John Baker (Publishers) Ltd, London.

Ardron, P. A., I. D. Rotherham & A. Oswald. 2008. Interpreting archaeological features. Pp. 189–196 *in* I. D. Rotherham, M. Jones, L. Smith & C. Handley (editors), The Woodland Heritage Manual: A Guide to Investigating Wooded Landscapes. Wildtrack Publishing, Sheffield.

Ashley, Robert. 2001. The Rushlight and Related Holders: A Regional View. Ashley Publications, Marlborough, Wiltshire.

A Society of Gentlemen. 1780. Young's Tour in Ireland. The Critical Review: Or Annals of Literature. Volume 49. A. Hamilton, London.

Badr, A., K. Müller, R. Schäfer, H. El Rabey, S. Effgen, H. H. Ibrahim, C. Pozzi, W. Rohde & F. Salamini. 2000. On the origin and domestication history of barley (*Hordeum vulgare*). Molec. Biol. Evol. 17: 499–510.

Baker, Margaret. 1999. Discovering the Folklore of Plants. Shire Publications Ltd, Buckinghamshire.

Bannister, P. 1965. *Erica cinerea* L. Biological Flora of the British Isles. Journal of Ecology 53: 527–542.

Barrington, Richard M. 1881. Report on the flora of the Blasket Islands, County Kerry. Proceedings of the Royal Irish Academy, Ser. 2, Vol. III, Science, No. 6.

Barrington, Richard M. 1906. Names and uses of *Molinia caerulea*. The Irish Naturalist 15: 219.

Beecher, Seán. 2005. A Dictionary of Cork Slang. Collins Press, Cork. <http://www.corkslang.com>, accessed 02 October 2011.

Beeton, Isabella. 1861. Mrs Beeton's Book of Household Management (New edition published in 1982). Chancellor Press, London.

Behre, Karl-Ernst. 1992. The history of rye cultivation in Europe. Vegetation History and Archaeobotany 1(3): 141–156.

Bell, Jonathan & Mervyn Watson. 2008. A History of Irish Farming, 1750–1950. Four Courts Press, Dublin.

Benson, Eva H. 1934. The bull-roarer in Ireland. The Irish Naturalists' Journal 5: 24.

Booth, Evelyn M. 1979. The Flora of County Carlow. Royal Dublin Society, Dublin.

Bord na Móna. 2009. Annual Report and Sustainability Report 2008/2009. <http://bnm.annualreport08-09.com/>, accessed 28 August 2011.

Borland, Mrs John. 1920. Some Kerry notes. Folklore 31(3): 234–237.

Borlase, William. 1769. Antiquities, Historical and Monumental, of the County of Cornwall. W. Bowyer and J. Nichols, for S. Baker and G. Leigh [etc.], London.

Boyle, P. J. 1977. *Spartina* on Bull Island. Pp. 88–92 *in* D. W. Jeffrey (editor), North Bull Island Dublin Bay: A Modern Coastal Natural History. Royal Dublin Society, Dublin.

Brown, O. P. 1885. The Complete Herbalist: Or, the People Their Own Physicians, by the Use of Natural Remedies; Describing the Great Curative Properties Found in the Herbal Kingdom. Fredk. W. Hale, London.

Brunker, J. P. 1950. Flora of the County Wicklow. Dundalgan Press, Dundalk.

Bush, John. 1769. Hibernia Curiosa: A Letter [dated 1764] from a Gentleman in Dublin, to his Friend at Dover in Kent. Giving a General View of the Manners, Customs, Dispositions, & of the Inhabitants of Ireland. W. Flexney, London.

Byrne, E. 2004. Funny bumps in the woods! Using archaeology to identify historical changes and uses in our native woodlands. Pp. 60–63 *in* Cara Doyle (editor), Ireland's Native Woodlands, Abstracts of Conference Papers. Woodlands of Ireland, Dublin.

Byrne, Hugh James. 1907. All Hallows Eve and Other Festivals in Connaught. Folklore 18(4): 437–439.

Byrnes, Emmet & Declan Little. n.d. [ca. 1990–2010]. A History of Woodland Management in Ireland: An Overview. Native Woodland Scheme No. 2. Forest Service and Woodlands of Ireland, Dublin.

Cameron, John. 1883. The Gaelic Names of Plants (Scottish and Irish). William Blackwood and Sons, Edinburgh and London.

Campbell, A. & M. Chapman. 2000. Handbook of Poisoning in Dogs and Cats. Blackwell Science, Oxford.

Campbell, Patrick. 1974. Growing up in Donegal. Béaloideas 42/44: 62–87.

Carr, John. 1806. The Stranger in Ireland: Or, a Tour in the Southern and Western Parts of that Country in the Year 1805. Lincoln & Gleason, Hartford.

Cassidy, N., P. B. Casey & J. A. Tracey. 2002. Incidence of mushroom poisoning in Ireland 1997–2000. Pp. 241–399 *in* Abstracts of the European Association of Poisons Centres and Clinical Toxicologists XXII International Congress. Clinical Toxicology 40(3): 301.

Chevallier. A. 1996. The Encyclopedia of Medicinal Plants. Dorling Kindersley, London.

Ciesla, W. M. 2002. Non-wood Forest Products from Temperate Broad-leaved Trees. Non-wood Forest Products 15. Food and Agriculture Organization of the United Nations, Rome.

Clapham, A. R., T. G. Tutin & E. F. Warburg. 1962. Flora of the British Isles, 2nd ed. Cambridge University Press, Cambridge.

Clare Champion. 2002. Riches of Clare: The Clare currach. 30 Aug. <http://www.clarelibrary.ie/eolas/claremuseum/news_events/ccclare_currach.htm>.

Clarke, Donal. 2010. Brown Gold. A History of Bord an Móna and the Peat Industry in Ireland. Gill & Macmillan, Dublin.

Clyne, Douglas. 1989. Gaelic Names for Flowers and Plants. Cruisgean, Furnace, Argyll, Scotland.

Coillte. 2008. <http://www.coillte.ie>, accessed 28 August 2011.

Colgan, Nathaniel. 1904. Flora of the County Dublin. Hodges, Figgis & Co. Ltd, Dublin.

Colgan, Nathaniel. 1911. Gaelic Plant and Animal Names, and Associated Folk-Lore, part 4. A Biological Survey of Clare Island in the County of Mayo, Ireland and of the Adjoining District. Hodges, Figgis & Co., Ltd, Dublin.

Colgan, Nathaniel. 1914. Field notes on the folk-lore of Irish plants and animals. The Irish Naturalist 23: 53–64.

Colgan, Nathaniel. 1919. On the occurrence of tropical drift seeds on the Irish Atlantic coasts. Proceedings of the Royal Irish Academy 35, Sect. B, 3: 25–54.

Colgan, Nathaniel & Reginald Scully. 1898. Contributions Towards a Cybele Hibernica, 2nd ed. Edward Ponsonby, Dublin.

Coll, Marian (editor). 2000. Our Trees: A Guide to Growing Ireland's Native Trees in Celebration of a New Millennium. Coillte and Woodlands of Ireland, Dublin.

Conry, Michael J. 2011. Picking Bilberries, Fraochâns and Whorts in Ireland: The Human Story. Privately published. Chapelstown, Carlow.

Corrigan, Desmond. 1984. The scientific basis of folk medicine: The Irish dimension. P. 1042 *in* Roy Vickery (editor), Plant-Lore Studies. The Folklore Society Mistletoe Series, Vol. 18, London.

Coward, Noël. 1973. The Lyrics of Noël Coward. The Overlook Press, Woodstock, New York.

Crane, C. P. 1907. Kerry. Methuen & Co, London.

Cross, J. R. 1975. Biological Flora of the British Isles. *Rhododendron ponticum* L. Journal of Ecology 63(1): 345–364.

Daly, Willie. 2010. The Last Matchmaker. Sphere, London.

Danaher, Kevin. 1964a. Irish Customs and Beliefs. Mercier Press, Cork (reprinted 2004).

Danaher, Kevin. 1964b. In Ireland Long Ago. Mercier Press, Cork.

Danaher, Kevin. 1972. The Year in Ireland. Mercier Press, Cork.

Danaher, Kevin [Caoimhín Ó Danachair]. 1978. Ireland's Vernacular Architecture [Foirgneamh na nDaoine], 2nd ed. Published for the Cultural Relations Committee of Ireland by Mercier Press, Cork.

Davis, Wade. 1986. The Serpent and the Rainbow. William Collins & Co. Ltd, Glasgow.

De Bhaldraithe, Tomás (editor). 1959. English-Irish Dictionary. Oifig an tSoláthair, Baile Átha Cliath.

Delaney, J. G. 1980. The harvest in north Offaly. Pp. 241–242 *in* H. Murtagh (editor), Irish Midland Studies: Essays in Commemoration of N.W. English. Old Athlone Society, Athlone.

Delaney, J. G. 1983. Baskets and their uses in the Midlands. Pp. 215–227 *in* Alan Gailey & Dáithí Ó hÓgáin (editors), Gold Under the Furze: Studies in Folk Tradition. The Glendale Press, Dublin.

Densmore, Frances. 1974. How Indians Use Wild Plants for Food, Medicine & Crafts (formerly titled 'Uses of Plants by the Chippewa Indians'). Dover Publications, New York.

Department of Agriculture. 1933. Home Preservation of Fruit (Leaflet No. 88). An Roinn Talmhaíochta, Dublin.

Department of Agriculture and Technical Instruction. 1904. Irish Industrial Exhibition, World's Fair, St Louis 1904: Handbook and Catalogue of Exhibits, Vol. 1. Ireland.

Department of Agriculture, Fisheries and Food. 2007. National Forest Inventory field data collection, November 2004–November 2006. An Roinn Talmhaíochta, Dublin.

The Department of Health/An Roinn Sláinte. 2011. Smoking—Key Statistics and Points. <http://www.dohc.ie/issues/smoking_ban/smokekey.html>, accessed 4 September, 2011.

Dickie, G. 1864. A Flora of Ulster and Botanist's Guide to the North of Ireland. C. Aitchison, Belfast.

Dillon, Eamon. 2006. The Outsiders: Exposing the Secretive World of Ireland's Travellers. Merlin Publishing, Ireland.

Dinneen, Patrick S. 1927. Foclóir Gaedhilge agus Béarla: An Irish-English Dictionary. Educational Company of Ireland, Ltd for the Irish Texts Society, Dublin.

Dirrane, Bridget. 1997. A Woman of Aran. Blackwater Press, Dublin.

Doherty, Michael L. 2001. The folklore of cattle diseases: a veterinary perspective 1. Béaloideas 69: 41–75.

Dolan, T. P. & Diarmaid Ó Muirithe (editors). 1979. A Glossary, with Some Pieces of Verse of the Old Dialect of the English Colony in the Baronies of Forth and Bargy, County of Wexford, Formerly Collected by Jacob Poole. The Past: The Organ of the Uí Cinsealaigh Historical Society, No. 13.

Doogue, Declan & Carsten Krieger. 2010. The Wild Flowers of Ireland. Gill & Macmillan, Dublin.

Doogue, Declan, David Nash, John Parnell, Sylvia Reynolds & Peter Wyse Jackson. 1998. Flora of County Dublin. Dublin Naturalists' Field Club, Dublin.

Doolan, Lelia. 1994. The sound of your footfall is as welcome as yourself. Pp. 235–243 *in* John Feehan, Cilian Roden, Michael O'Connell, Gordon D'Arcy, J. W. O'Connell, John Waddell, Paul Walsh, John de Courcy Ireland, Anne O'Dowd, Clancy Ó Conaola, Doolan Pádraigín, Lelia James Duran, Anne Korff, Joe McMahon, Patrick F. Sheeran & Pádraig Standún, The Book of Aran: The Aran Islands, Co. Galway. Tír Eolas, Kinvara, Co. Galway.

Drury, H. C. 1925. The rush light and its associates. Journal of the Royal Society of Antiquaries of Ireland 55: 99–111.

Drury, Susan M. 1984. The use of wild plants as famine foods in eighteenth century Scotland and Ireland. Pp. 43–60 *in* Roy Vickery (editor), Plant-Lore Studies. The Folklore Society Mistletoe Series, Vol. 18, London.

Drury, Susan. 1985. Herbal remedies for livestock in seventeenth and eighteenth century England: Some examples. Folklore 96(2): 243–247.

Drury, Susan. 1987. Customs and beliefs associated with Christmas evergreens: A preliminary survey. Folklore 98(2): 194–199.

Duke, J. A. & E. S. Ayensu. 1985. Medicinal Plants of China, Vol. 1. Reference Publications, Algonac, Michigan.

Duke, J. A., M. J. Bogenschutz-Godwin, J. duCellier & Peggy-Ann K. Duke. 2002. Handbook of Medicinal Herbs, 2nd ed. CRC Press, Boca Raton, Florida.

Dunton, John. 2003. Teague Land, or, A Merry Rample to the Wild Irish: Letters from Ireland, 1698 (edited by Andrew Carpenter). Four Courts Press, Dublin.

Durand, Jack. 2008. An Fhuinseog: Ash for Hurleys and for Profit. Tree Council of Ireland, Dublin.

Dutton, Hely. 1808. Statistical Survey of the County of Clare, with Observations on the Means of Improvement; Drawn Up for the Consideration of the Dublin Society. Graisberry and Campbell, Dublin.

Edlin, H. L. 1973. Woodland Crafts in Britain. David & Charles, Newton Abbot, Devon, United Kingdom.

Egan, F. W. 1887. Irish Folk-Lore. Medicinal Plants. The Folk-Lore Journal 5(1): 11–13.

Emboden, William. 1972. Narcotic Plants. Studio Vista, London.

Engel, Cindy. 2007. Zoopharmacognosy. Pp. 7–16 *in* Susan G. Wynn & Barbara J. Fougère (editors), Veterinary Herbal Medicine. Mosby Elsevier, St Louis, Missouri.

Evans, E. Estyn. 1957. Irish Folk Ways. Routledge & Kegan Paul Ltd, London.

Evans, E. Estyn. 1967 [1942]. Irish Heritage. W. Tempest, Dundalgan Press, Dundalk.

Evelyn, John. 1664. Sylva, or a Discourse of Forest-trees and the Propagation of Timber in His Majesties Dominions. J. Martyn and J. Allestry, London.

Evelyn, John. 1699. Acetaria: A Discourse of Sallets. B. Tooke, London.

Fargher, D. C. 1969. The Manx Have a Word for It: Manx Gaelic Names for Flora. Port Erin, Isle of Man. Privately published.

Farren, Margaret. It's us they're talking about: Lily Ivors and Hughie Copen. <http://www.clonmany.com>, accessed 23 May 2010.

Feehan, John & Grace O'Donovan. 1996. The Bogs of Ireland: An Introduction to the Natural, Cultural and Industrial Heritage of Irish Peatlands. University College Dublin, The Environmental Institute, Dublin.

Flower, Robin. 1944. The Western Island or The Great Blasket. Clarendon Press, Oxford.

Forbes, A. C. 1933. Tree planting in Ireland during four centuries. Proceedings of the Royal Irish Academy XLI, Sect. C (6): 168–199.

Forbes, Ralph S. & Robert H. Northridge. 2012. The Flora of County Fermanagh. National Museums Northern Ireland, Belfast.

Forest Service of Northern Ireland. 2009. Annual Report 2008/09 of the Northern Ireland Forest Service. Department of Agriculture and Rural Development, Government of the United Kingdom, Belfast.

Friel, Maeve. 1997. Here Lies: A Guide to Irish Graves. Poolbeg Press Ltd, Dublin.

Gailey, Alan & G. B. Adams. 1977. The bonfire in north Irish tradition. Folklore 88(1): 3–38.

Gallagher, Michael. 2012. Remedies and Cures of Bygone Era. Browne Printers Ltd., Letterkenny, Co. Donegal.

Gardey, Alan. 1971. A wicker coffin from Ballyshiel Graveyard, Annaclone, County Down. Ulster Folklife 17: 89–90.

Garrad, Larch S. 1984. Some Manx plant-lore. Pp. 75–83 *in* Roy Vickery (editor), Plant-Lore Studies. The Folklore Society Mistletoe Series, Vol. 18.

Garrison, William Lloyd. 1891. When potatoes were gone. Starving women and children on the west coast of Ireland. Letter to the Editor, The New York Times, 29 January 1891.

Gehring, Ronette & Kelly Kindscher. 2010. The medicinal use of native north american plants in domestic animals. Pp. 213–229 *in* David R. Katerere & Luseba Dibungi (editors), Ethnoveterinary Botanical Medicine: Herbal Medicines for Animal Health. CRC Press, Boca Raton, Florida.

Geraghty, Sarah. 2009. Wine Consumer Behaviour: An Irish Wine Market Analysis. A Literature Review. Shannon College of Hotel Management, Limerick.

Geraghty, Siobhán. 1996. Viking Dublin: Botanical Evidence from Fishamble Street. National Museum of Ireland, Medieval Dublin Excavations 1962–81, Ser. C, Vol 2. Royal Irish Academy, Dublin.

Gerard, John. 1597. The Herball or Generall Historie of Plantes. John Norton, London.

Gilbert, Arthur W. 1917. The Potato. MacMillan, New York.

Gilmore, Melvin R. 1991. Uses of Plants by the Indians of the Missouri River Region. [An enlarged edition of a work first published in 1914.] University of Nebraska Press, Lincoln and London.

Goldsmith, Oliver. 1762. The Citizen of the World or Letters from a Chinese Philosopher, Residing in London, to His Friends in the East, Vol. I. J. Newbery and W. Bristow, London.

Goodrich, Jennie, Claudia Lawson & Vana Parrish Lawson. 1980. Kashaya Pomo Plants. American Indian Studies Center, University of California, Los Angeles.

Graham, Dan. 1997. Dingle, Corca Dhuibne, and The Great Famine. *In* Michael Costello (editor), The Famine in Kerry. Kerry Archaeological and Historical Society, Kerry.

Green, Paul. 2008. Flora of County Waterford. National Botanic Gardens of Ireland, Dublin.

Grieve, Maud. 1931. A Modern Herbal, electronic version. <http://www.botanical.com/botanical/mgmh/a/aconi007.html>, accessed 01 September 2012.

Grigson, Geoffrey. 1958. The Englishman's Flora. Paladin, Frogmore, St Albans.

Hackney, Paul (editor). 1992. Stewart & Corry's Flora of the North-East of Ireland (3rd ed.). Institute of Irish Studies, The Queen's University of Belfast, Belfast.

Haddon, Alfred C. 1893. A wedding dance-mask, Co. Mayo. Folk-lore 4: 123–124.

Hall, Mr and Mrs S. C. 1841. Ireland: Its Scenery, Character, etc. Volume 1. How and Parsons, London.

Hanley, Brian. 2002. The IRA, 1926–1936. Four Courts Press, Dublin.

Hardy, Frederick Gavin & Michael D. Guiry. 2003. A Check-list and Atlas of the Seaweeds of Britain and Ireland. British Phycological Society, London.

Harron, John. 1987 [1986]. Flora of Lough Neagh. Belfast: Irish Naturalists' Journal Committee, University of Ulster, Coleraine.

Harrop, Renny (editor). 1977. Encyclopedia of Herbs. Marshall Cavendish Books Ltd, London.

Hart, Henry Chichester. 1873. *Euphorbia hyberna*, *Equisetum trachyodon*, etc., in Co. Galway. Journal of Botany 11: 338–339.

Hart, Henry Chichester. 1898. The Flora of the County Donegal, or, List of the Flowering Plants and Ferns with Their Localities and Distribution. Sealy, Bryers and Walker, Dublin.

Harte, Aidan. 2008. A sweathouse at Caherlehillian, Co. Kerry. Journal of the Kerry Archaeological and Historical Society (Ser 2) 8: 114–125.
Hartman, George. 1696. Family Physitian, or a Collection of Choice, Approv'd and Experienc'd Remedies, for the Cure of Almost All Diseases. H. Hills, London.
Harvey, William Henry. 1841. A Manual of the British Marine Algae: Containing Generic and Specific Descriptions of All the Known British Species of Sea-weeds. John Van Voorst, London.
He, Shan-an & Zhong-ming Cheng. 1991. The role of Chinese botanical gardens in conservation of medicinal plants. Pp. 229–237 *in* Olayiwola Akerele, Vernon Heywood & Hugh Synge (editors), The Conservation of Medicinal Plants. Cambridge University Press, Cambridge.
Headland, F. W. 1856. On poisoning by the root of *Aconitum napellus*. The london Lancet II: 18–22.
Herder, Ronald (editor). 1998. 500 Best-loved Song Lyrics. Dover, Mineola, New York.
Heywood, Vernon H. (editor). 1978. Flowering Plants of the World. Oxford University Press, Oxford, United Kingdom.
Hickie, David. 2002. Native Trees and Forests of Ireland. Gill & McMillan, Dublin.
Hill, M. O., C. D. Preston & D. B. Roy. 2004. PLANTATT: Attributes of British and Irish Plants: Status, Size, Life History, Geography and Habitats. Centre for Ecology and Hydrology (CEH), Wallingford.
Hilton, Jonathan. 2007. Wild Food for Free. Hamlyn, London.
Hobhouse, Henry. 1986. Seeds of Change: Five Plants that Transformed Mankind. Harper & Row, New York.
Hogan, Joe. 2001. Basketmaking in Ireland. Wordwell Ltd, Bray, Co. Wicklow.
Hogan, J. J. & Patrick C. O'Neill. 1947. A North-County Dublin glossary. Béaloideas 17: 262–283.
Holden, Michael. 2009. Freney the Robber, the Noblest Highwayman in Ireland. Mercier Press, Cork.
Honeychurch, Penelope N. 1986. Caribbean Wild Plants and Their Uses. Macmillan, London.
Hora, B. 1981. The Oxford Encyclopedia to Trees of the World. Oxford University Press, Oxford.
Houston, Fiona & Xa Milne. 2008. Seaweed and Eat It: A Family Foraging and Cooking Adventure. Virgin Books, London.
Howes, F. N. 1946. Fence and barrier plants in warm climates. Kew Bulletin 1: 51–87.
Howes, F. N. 1974. A Dictionary of Useful and Everyday Plants and Their Common Names. Cambridge University Press, Cambridge.
Howkins, Chris. 1996. Rowan: Tree of Protection. Privately published, Addlestone, Surrey.
Huxley, A. 1992. The New Royal Horticultural Society Dictionary of Gardening. Royal Horticultural Society, Wisley, United Kingdom.
Isaacs, Jennifer. 1987. Bush Food: Aboriginal Food and Herbal Medicine. Lansdowne Publishing, Australia.
Jackson, Albert & David Day. 1991. Collins Good Wood Handbook. Harper Collins, London.
Jain, S. K., with Sumita Srivastava. 1999. Dictionary of Ethnoveterinary Plants of India. Deep Publications, New Delhi.
Jeffrey, D. W. (editor) 1977. North Bull Island, Dublin Bay: A Modern Coastal Natural History. Royal Dublin Society, Dublin.
Johnson, Samuel. 1775. A Journey to the Western Islands of Scotland. W. Strahan & T. Cadell, London.
Joyce, James. 1914. Dubliners. Penguin Books, New York.
Joyce, James. 1922. Ulysses. Shakespeare and Company, Paris.
Joyce, Patrick Weston. 1870. The Origin and History of Irish Names of Places. McGlashan & Gill, Dublin.
Joyce, Patrick Weston. 1875. The Origin and History of Irish Names of Places (2nd Ser.). McGlashan & Gill, Dublin.
Joyce, Patrick Weston. 1903. A Social History of Ancient Ireland, Volume II. Longmans, Green and Co., London.
Joyce, Patrick Weston. 1922. Irish Local Names Explained. Educational Company of Ireland, Dublin. [Information from this book also provided at (2011): <http://www.libraryireland.com>.
Jude [Foley of Limerick]. 1933. Medicinal and Perfumery Plants and Herbs of Ireland: A Possible

Great Natural Industry Suitable to the Gaeltacht. M. H. Gill, Dublin.[1]

Kavasch, E. Barrie. 2005. Native Harvests: American Indian Wild Food and Recipes. Dover Publications, Mineola, New York.

Keller, G. N. 1918. The Cultivation of Medicinal Plants in Ireland. Department of Agriculture and Technical Instruction in Ireland. Department's Journal Vol. XVIII., No. 2.

Kells, Arthur. 1929. On picking hawthorn. The Irish Naturalists' Journal 2: 189–191.

Kelly, D. L. & S. Fuller. 1988. Ancient woodland in central Ireland: Does it exist? Pp. 363–369 *in* Fabio Salbitano (editor), Human Influence on Forest Ecosystems Development in Europe. Proceedings of a Workshop Held in Trento, Italy, 26–29 September 1988. ESF FERN-CNR, Pitagora Editrice, Bologna, Italy.

Kelly, Eamonn P. 1981. A log boat from Derryco. Journal of the Kerry Archaeological and Historical Society 14: 10–13.

Kelly, Fergus. 1997. Early Irish Farming: A Study Based Mainly on the Law-texts of the 7th and 8th Centuries A.D. School of Celtic Studies, Dublin Institute for Advanced Studies, Dublin.

Kennedy, Liam, Paul S. Ell, S. M. Crawford & L. A. Clarkson. 1999. Mapping the Great Irish Famine: A Survey of the Famine Decades. Four Courts Press, Dublin.

K'Eogh (Keogh), J. 1735. Botanalogia Universalis Hibernica, Cork (also published in translation by M. Scott [editor], 1991, as An Irish Herbal, Anna Livia, Press Ltd, Dun Laoghaire).

Kinahan, G. H. 1881. Notes of Irish folk-lore. Folk-lore Record 4: 96–125.

King, Frances B. 1984. Plants, People, and Paleoecology: Biotic Communities and Aboriginal Plant Usage in Illinois. Illinois State Museum, Scientific Papers, Vol. XX. Illinois State Museum, Springfield.

King, Noel (editor). 2009. Crossroads: Folklore from Kerry. Doghouse, Tralee, Co. Kerry.

Kinmouth, Claudia. 1993. Irish Country Furniture 1700–1950. Yale University Press, New Haven and London.

Koch, John T. 2006. Celtic Culture: A Historical Encyclopedia. ABC-CLIO, Santa Barbara, California.

Kraus, Erin. 2011. Wise-woman of Kildare: Moll Anthony and Popular Tradition in the East of Ireland. Maynooth Studies in Local History. Four Courts Press, Dublin.

Lambeth, M. 2008. Discovering Corn Dollies. Shire Classics. Shire Publications, Oxford.

Larner, J. 2005. Killarney History & Heritage. Collins Press, Cork.

Leavy, S. 2007. Bog treasures. Science Spin 20: 12–14.

Lenihan, Edmund. 1987. In Search of Biddy Early. Mercier Press, Cork and Dublin.

Letts, J. B. 1997. Historic Thatch (Ireland) Study: An Interim Report. Institute of Historic Building Conservation. Turnbridge Wells, Kent. <http:www.ihbc.org.uk/context_archive/56/thatchireland/nireland.html>

Lewington, Anna. 1990. Plants for People. Natural History Museum Publications, Royal Botanic Gardens, Kew and World Wide Fund for Nature Education Department, United Kingdom.

Lewis-Stempel, John. 2012. Foraging. The Essential Guide to Free Wild Food. Constable & Robinson., London.

Lhuyd, Edward. 1707. Archæologia Britannica, Giving Some Account Additional to What Has Been Hitherto Publish'd, of the Languages, Histories and Customs of the Original Inhabitants of Great Britain: From Collections and Observations in Travels Through Wales, Cornwal, Bas-Bretagne, Ireland and Scotland. Oxford, Printed at the Theater for the author, and sold by Mr. Bateman, London

Lindsay, Biddy Yingguny, Kitty Waliwararra, Frances Milijat, Helen Kuwarda, Rita Pirak, Albert Muyung, Edwin Pambany, Jack Maruridj, Patricia Marrfurra, and Glenn Wightman (2001). Malakmalak and Matngala Plants and Animals: Aboriginal Flora and Fauna knowledge from the Daly River area, northern Australia. Northern Territory Bulletin No. 26. Parks and Wildlife Commission of the Northern Territory, Darwin, Australia.

1. In the copy of this book in the Library of the National Botanic Gardens in Glasnevin, a note dated 20 Sept. 34 on the first page of this book reads, 'This is a thoroughly unreliable book'!

Logan, Patrick. 1972. Making the Cure: A Look at Irish Folk Medicine. Talbot Press, Dublin.

Logan, Patrick. 1980. The Holy Wells of Ireland. Colin Smythe Limited, Gerrards Cross, Buckinghamshire.

Logan, Patrick. 1981. Irish Country Cures. Appletree Press, Belfast (Republished in 2009 under the title Irish Folk Medicine).

Louw, Alison. 1983. Do-it-yourself Wine and Beermaking. Marshall Cavendish Books, London.

Lowe, John. 1897. The Yew-Trees of Great Britain and Ireland. Macmillan, London.

Lucas, A. T. 1954. Bog wood: A study in rural economy. Béaloideas 23: 128.

Lucas, A. T. 1957. Folk life. Pp. 196–206 *in* J. Meenan & D. A.Webb (editors), A View of Ireland. British Association for the Advancement of Science, Dublin.

Lucas, A. T. 1958. Furze: A survey and history of its uses in Ireland. Béaloideas 26: 1–204.

Lucas, A. T. 1959. Nettles and charlock as famine food. Breifne: Journal of the Breifne Historical Society, 1959: 137–146.

Lucas, A. T. 1960. Irish food before the potato. Gwerin 3(1): 8–43.

Lyons, Patrick. 1943. Stick-fencing. Béaloideas 13(1/2): 269–272.

Lysaght, Patricia. 1979. Trees in Irish Folk Tradition. Irish Biological Records Centre Newsletter, 14 (May 1979): 22–39.

Lysaght, Patricia. 2002. Shore Food Use in West County Clare. Béaloideas 70: 196–198.

Lysaght, Patricia. 2003. Hospitality at wakes and funerals in Ireland from the seventeenth to the nineteenth century: Some evidence from the written record. Folklore 114(3): 403–426.

Mabey, Richard. 1972. Food for Free: A Guide to the Edible Wild Plants of Britain. William Collins Sons & Co., London.

Mabey, Richard. 1998. Flora Britannica Book of Wild Herbs. Chatto & Windus, London.

Macalister, R. A. Stewart. 1937. The Secret Languages of Ireland. Cambridge University Press, Cambridge (Reprinted by Craobh Rua Books, Armagh, 1997).

MacCárthaigh, Críostóir (editor). 2008. Traditional Boats of Ireland: History, Folklore and Construction. Collins Press, Cork.

MacCoitir, N. 2003. Irish Trees: Myths, Legends & Folklore. Collins Press, Cork.

MacCoitir, N. 2006. Irish Wild Plants: Myths, Legends & Folklore. Collins Press, Cork.

MacConghail, Muiris. 1987. The Blaskets: People and Literature. Country House, Dublin.

Macdonald Douglas, Ronald (compiler). 1982. Scottish Lore and Folklore. Beekman House, New York.

MacGowan, Michael. 2003. The Hard Road to Klondike (Translated from the original Irish by Valentine Iremonger). Collins Press, Cork.

MacGréine, Pádraig. 1934. Some Notes on Tinkers and Their 'Cant'. Béaloideas 4(3): 259–263.

Mackay, James Townsend. 1806. A systematic catalogue of rare plants found in Ireland. Transactions of the Dublin Society 5: 121–184.

Mackay, James Townsend. 1825a. Catalogue of the indigenous plants of Ireland. Transactions of the Royal Irish Academy, 14: 103–198.

Mackay, James Townsend. 1825b. Supplementary notices of the plants and minerals of the islands of Arran. Transactions of the Royal Irish Academy, 14: 227–228

Mackay, James Townsend. 1836. Flora Hibernica: Comprising the Flowering Plants, Ferns, Characeae, Musci, Hepaticae, Lichenes and Algae of Ireland: Arranged According to the Natural System with a Synopsis of the Genera According to the Linnean System. William Curry, Dublin.

MacMahon, Bryan. 1994. The Story of Ballyheigue. Oidhreacht, Baile Uí Thaidhg, Co. Kerry.

MacPhilib, Séamas. 2000. Rush rafts in Ireland. Ulster Folklife 46: 1–8.

Magner, Donal. 2007. The woods we use. Pp. 21–36 *in* Forestry & Timber Yearbook 2007. Irish Timber Growers Association, Dublin.

Mahon, Bríd. 1983. Traditional dyestuffs in Ireland. Pp. 115–128 *in* Alan Gailey & Dáithí Ó hÓgáin (editors), Gold Under the Furze: Studies in Folk Tradition. Glendale Press, Dublin.

Maiden, J. H. 1899. Native food-plants. Agricultural Gazette of N.S. Wales, Sydney, Australia.

Mairet, Ethel. 1916. Vegetable Dyes, Being a Book of Recipes and Other Information Useful to the Dyer. Faber and Faber, London.

Maloney, Beatrice. 1972. Traditional herbal cures in County Cavan: Part 1. Ulster Folklife 18: 66–79.

Maple, William. 1729. A Method of Tanning without Bark. A. Rhames, Dublin.

Markey, Patrick. 2011. Around Ireland troubled fairy tales. The Irish Echo Online. <http://www.Irishecho.com/?p=46957>, accessed 12 July 2012.

Mason, Thomas H. 1936. The Islands of Ireland. Batsford, London.

Mathews, Aidan Carl. 1983. Minding Ruth. Gallery Press, Dublin.

McCarthy, Patrick & Richard Hawkes. 2007. Northside of the Mizen: Tales, Customs & History from the South-west of County Cork in Ireland. Mizen Productions, Goleen, Co. Cork.

McClintock, Letitia. 1912. Donegal cures and charms. Folklore 23(4): 473–478.

McCracken, Eileen 1971. Irish Woods Since Tudor Times. Their Distribution and Exploitation. David & Charles, Newton Abbot, United Kingdom.

McCracken, Eileen. 1988–9. Tree planting by tenants in Meath 1800 to 1850. Ríocht na Midhe 8: 3–20.

McDaeid, Caroline. 1991. 'Avondale': A history in weeds. Pp. 89–97 *in* Patrick J. Nugent (editor), Natural Heritage. Meath, Cavan, Westmeath Nature Yearbook 1991. Turner's Printing Co., Longford.

McEvoy, John. 1802. Statistical Survey of the County of Tyrone, with Observations on the Means of Improvement; Drawn Up in the Years 1801, and 1802 for the Consideration, and Under the Direction of the Dublin Society. Graisberry & Campbell, Dublin.

McGlinchey, Charles. 2007. The Last of the Name. Collins Press, Cork.

McInerney, Jim. 2005. The Gandelow: A Shannon Estuary Fishing Boat. A. K. Ilen Company, Limerick.

McKean, E. J. 1933. The bull-roarer in Ireland. The Irish Naturalists' Journal 4: 216–217.

McKenna, Neil. 2005. The Secret Life of Oscar Wilde: An Intimate Biography. Basic Books, New York.

McKillop, Anne. Interview with Annie McKillop. The Glens of Antrim Historical Society, <http://www.antrimhistory.net/content.php?cid=71>, accessed 23 May 2010.

McNeill, Ian. 2010. The Flora of County Tyrone. National Museums Northern Ireland, Cultra, Holywood, Co. Down.

McVeagh, John (editor). 1995. Richard Pococke's Irish Tours. Irish Academic Press, Dublin.

Mears, Ray & Gordon Hillman. 2007. Wild Food. BBC and Hodder & Staughton, United Kingdom.

Meehan, Joseph. 1906. The cure of elf-shooting in the north-west of Ireland. Folklore 17(2): 200–210.

Merck. 2010. Merck Veterinary Manual 2010, 10th ed. Merck & Co., Whitehouse Station, New Jersey. <http://www.merckvetmanual.com>, accessed 1 August 2012.

Miller, Nicole. 2008. Using DNA, scientists hunt for the roots of the modern potato. University of Wisconsin–Madison News, <http://www.news.wisc.edu/14649>, accessed 28 October 2013.

Milliken, William & Sam Bridgewater. 2004. Flora Celtica: Plants and People in Scotland. Birlinn, Edinburgh.

Milne, R. I. & R. J. Abbott. 2000. Origin and evolution of invasive naturalized material of *Rhododendron ponticum* L. in the British Isles. Molecular Ecology 9: 541–556.

Mitchell, Alan. 1974. A Field Guide to the Trees of Britain and Northern Europe. Collins, London.

Mitchell, Alan F. 1985. Conifers. Forestry Commission Booklet no. 15. Stationery Office Books, London.

Mitchell, Frank. 1986. The Shell Guide to Reading the Irish Landscape. Country House, Dublin.

Mitchell, G. F. 1987. Archaeology and Environment in Early Dublin. Medieval Dublin Excavations 1962–81, Ser. C, Vol. 1. Royal Irish Academy, Dublin.

Mitchell, Frank. 1989. Man & Environment in Valencia Island. Royal Irish Academy, Dublin.

Mitchell, Frank & Michael Ryan. 1997. Reading the Irish Landscape. Town House, Dublin.

Mitchell, M. E. 1974. The sources of Threlkeld's *Synopsis Stirpium Hibernicarum*. Proceedings of the Royal Irish Academy 74, Sect. B(1): 1–6.

Moerman, Daniel E. 1998. Native American Ethnobotany. Timber Press, Portland, Oregon.

Moloney, Michael. F. 1919. Irish Ethno-botany and the Evolution of Medicine in Ireland. M. H. Gill & Son, Dublin.

Moore, David. 1834–1835. Botany of the County of Londonderry. Unpublished manuscript in the Herbarium Library of the National Botanic Gardens, Glasnevin, Dublin.

Moore, David & Alexander Goodman More. 1866. Contributions Towards a Cybele Hibernica. Hodges, Smith & Co., Dublin.

Moore, Fionnbarr. 1996. Ireland's oldest bridge: At Clonmacnoise. Archaeology Ireland 10/4: 24–7.

Moutray Read, D. H. 1916. Some Characteristics of Irish Folklore. Folklore 27(3): 250–278.

Munnelly, Tom. 2005. Plant lore from West Clare. Béaloideas 73: 164–169.

Murray, Griffin. 2009. An ancient paddle from near Knocknagashel. Journal of the Kerry Archaeological and Historical Society (Ser. 2) 9: 32–35.

Murray, M. A. 1953. The Puck Fair of Killorglin. Folklore 64(2): 351–354.

na gCopaleen, Myles. (Brian O'Nolan). 1941. An Béal Bocht (The Poor Mouth), An Press Náisiúnta, Baile Átha Cliath.

National Advisory Committee on Drugs and Public Health Information and Research Branch. 2008. Drug Use in Ireland and Northern Ireland. 2006/2007 Drug Prevalence Survey: Cannabis Results, Bulletin 3. National Advisory Committee on Drugs, Dublin.

Neeson, Eoin. 1991. A History of Irish Forestry. Lilliput Press, Dublin.

Nelson, E. C. 1978. Tropical drift fruits and seeds on coasts in the British Isles and western Europe, 1. Irish beaches. Watsonia 12: 103–112.

Nelson, E. Charles. 1982. Tropical drift fruits and seeds: A new Irish species. Irish Naturalists' Journal 20(10): 452.

Nelson, E. Charles. 1984. An Irish Flower Garden: The Histories of Some of Our Garden Plants. Boethius Press, Kilkenny.

Nelson, E. Charles. 1988. The First Irish Flora, Synopsis Stirpium Hibernicarum, Caleb Threlkeld. Boethius Press, Kilkenny.

Nelson, E. Charles. 1991. Shamrock: Botany and History of an Irish Myth. Boethius Press, Kilkenny.

Nelson, E. Charles. 1995. The Cause of the Calamity: Potato Blight in Ireland, 1846–1847, and the Role of the National Botanic Gardens, Glasnevin. Department of Arts, Culture and the Gaeltacht, Stationery Office, Government Publications, Dublin.

Nelson, E. Charles. 2000a. Sea Beans and Nickar Nuts. BSBI Handbook No. 10. Botanical Society of the British Isles, London.

Nelson, E. Charles. 2000b. A Heritage of Beauty: the Garden Plants of Ireland, an Illustrated Encyclopaedia. Irish Garden Plant Society, Dublin.

Nelson, E. Charles. 2009. An Irishman's Cuttings: Tales of Irish Gardens and Gardeners and Plants and Plant Hunters. Collins Press, Cork.

Nelson, E. C., D. Lidstone, A. Hathway, D. McNally & J. M. B. Smith. 1990. Nickars! or, the art of growing long-distance flotsam. Moorea: Journal of the Irish Garden Plant Society 8: 2–13.

Nelson, E. C. & W. F. Walsh. 1993. Trees of Ireland. Lilliput Press, Dublin.

Nelson, E. C. & E. M. McCracken. 1987. The Brightest Jewel: A History of the National Botanic Gardens, Glasnevin, Dublin. Boethius Press, Kilkenny.

Ní Dhálaigh-Uí Chonchubhair, Máirín. 1999. Flóra an Bhlascaoid. *In* Máire Ní Chéilleachair (editor), Ceiliúradh an Bhlascaoid 3. The Great Blasket Commemoration. Peig Sayers Scéalaí 1873–1958. Coiscéim, Baile Átha Cliath.

Ní Ghaoithín, Máire. 1978. An tOileán a Bhí. An Clóchomhar Tta., Baile Átha Cliath.

Ní Sheasnáin, Áine. 1984. Some local lore of fairs. The Heart of Breifne 2(3): 66–71.

Ní Shúilleabháin, Eibhlís. 1988. Letters from The Great Blasket. Mercier Press, Cork.

Ó Céirín, Cyril & Kit Ó Céirín. 1978. Wild and Free: Cooking from Nature. O'Brien Press, Dublin.

Ó Criomhthain Tomás. n.d., ca. 1908. A manuscript list of Irish plant names compiled for the Norwegian scholar, Carl Marstrander by the Islandman, Tomás Ó Criomhthain in the early 1900s, from Kerry's Great Blasket Island. Located in Oslo University Library, Norway.

Ó Curraidhín, Mac Dara. n.d., ca. 2009. Déantús an Phoitín [Poteen Making]. (Booklet and accompanying DVD). Mac Dara Ó Curraidhín, An Spidéal, Co. Galway.

O'Donovan, John (translated and annotated by). 1868. Sanas Chormaic: Cormac's Glossary. O. T. Cutter, for the Irish Archaeological Society, Calcutta.

O'Dowd, Anne. n.d., ca. 2009. Tuí, Féar & Luachra: Straw, Haw & Rushes. Museum of Country Life, National Museum of Ireland, Castlebar.

O'Dowd, Anne. 2011. Green rushes under your feet! Spreading rushes in folklore and history.

Béaloideas 79: 82–112.
Ó Dubhshláine, Mícheál. 1996. An Práta. (Pádraig Ó Fiannachta, ed.), An Sagart, An Daingean (Dingle), Co. Kerry.
Ó Dubhshláine, Mícheál. 2009. Inisvickillane. A Unique Portrait of the Blasket Island. Brandon, Dingle, Co. Kerry.
Ody, Penelope. 1997a. Herbs for First Aid. Keats Publishing, Los Angeles.
Ody, Penelope. 1997b. 100 Great Natural Remedies: Using Healing Plants at Home. Kyle Cathie, London.
O'Hare, Patricia. 1993. A brief survey of the typical vernacular housing of a portion of East Kerry. Journal of the Kerry Archaeological and Historical Society 26: 5–72.
O'Mahony, T. 2006. *Mibora minima* (L.) Desv., early sand-grass (Poaceae) in West Cork (H3): An addition to the Irish flora. Irish Botanical News 16: 5–13.
O'Mahony, Tony. 2009. Wildflowers of Cork City and County. Collins Press, Cork.
O'Mathuna, Dónal. 2008. Does it work? Milk thistle and liver problems. Irish Times, Healthplus Supplement, 5 Aug. 2008.
O'Meara, John, trans. 1982. Gerald of Wales: The History and Topography of Ireland. Penguin Books., Harmondsworth, Middlesex.
O'Regan, Paula. 1997. Healing Herbs in Ireland. Primrose Press, Blackrock, Co. Dublin.
O'Reilly, Barry. 2004. Living Under Thatch: Vernacular Architecture in Co. Offaly. Mercier Press, Cork.
O'Rourke, Con. 2006. Nature Guide to the Aran Islands. Lilliput Press, Dublin.
Ó Sé, Maidhe Dainín. 2007. House Don't Fall On Me (translated by Gabriel Fitzmaurice). Mercier Press, Cork (First published as A Thig Ná Tit Orm, 1987, by Coiscéé).
O'Shea, Pat. 2009. Interview by Kevin Griffin. *In* Noel King (editor), Crossroads: Folklore from Kerry. Doghouse, Tralee, Co. Kerry.
O'Síocháin, P. A. 1962. Aran, Islands of Legend. Foilsiúcháin Éireann, Dublin.
Ospina, Alison. 2009. Green Wood Chairs: Chairs and Chairmakers of Ireland. Stobart Davies, Ammanford, United Kingdom.
O'Súilleabháin, Séan. 1942. A Handbook of Irish Folklore. Educational Company of Ireland for the Folklore of Ireland Society, Dublin.
O'Sullivan, Denis C. (edited by and translated from the Latin). 2009. The Natural History of Ireland: Included in Book One of the Zoilomastix of Don Philip O'Sullivan Beare. Cork University Press, Cork.
O'Sullivan, John C. 1969. Wooden pumps. Folk Life 7: 101–116.
O'Sullivan, Patrick. 1994. A Country Diary: The Year in Kerry. Anvil Books, Dublin.
O'Sullivan, Paul (editor). 1977. Field and Shore: Daily Life and Traditions Aran Islands, 1900. O'Brien Educational, Dublin.
Parnell, John & Tom Curtis. 2012. Webb's An Irish Flora. Cork University Press, Cork.
Pennacchio, Marcello, Lara Jefferson & Kayri Havens. 2010. Uses and Abuses of Plant-derived Smoke: Its Ethnobotany as Hallucinogen, Perfume, Incense and Medicine. Oxford University Press, Oxford.
Pennant, T. 1776. A Tour in Scotland in the Eighteenth Century. London.
Perrin, P. M. & O. H. Daly. 2010. A Provisional Inventory of Ancient and Long-established Woodland in Ireland. Irish Wildlife Manuals, No. 46. National Parks and Wildlife Service, Department of the Environment, Heritage and Local Government, Dublin.
Peterson, Lee Allen. 1977. A Field Guide to Edible Wild Plants of Eastern and Central America. Houghton Mifflin, Boston and New York.
Phillips, Roger. 1981. Mushrooms and Other Fungi of Great Britain and Europe. Pan Books, London.
Phillips, Roger. 1983. Wild Food. Pan Macmillan, London.
Pilcher, Jonathan & Valerie Hall. 2001. Flora Hibernica: The Wild Flowers, Plants and Trees of Ireland. Collins Press, Cork.
Pim, Sheila. 1945. Common or Garden Crime. Hodder & Stoughton, London.

Praeger, Robert Lloyd. 1909. A Tourist's Flora of the West of Ireland. Hodges Figgis and Co., Dublin.

Praeger, Robert Lloyd. 1915. *Peucedanum ostruthium* Linn. Irish Naturalist 24: 135.

Praeger, Robert Lloyd. 1938. A Flora of the North-East of Ireland, 2nd ed. Quota Press, Belfast.

Praeger, Robert Lloyd. 1949. Some Irish Naturalists: A Biographical Note-book. W. Tempest, Dundalgan Press, Dundalk.

Pratt, Anne. 1857. The Poisonous, Noxious, and Suspected Plants of our Fields and Woods. Society for Promoting Christian Knowledge, London.

Prendergast, H. D. V. & H. Sanderson. 2004. Britain's Wild Harvest. Royal Botanic Gardens, Kew, United Kingdom.

Preston, C. D. 1995. Pondweeds of Great Britain and Ireland. BSBI Handbook No. 8. Botanical Sociaty of the British Isles, London.

Preston, C. D., D. A. Pearman & T. D. Dines (editors). 2002. New Atlas of the British Flora: An Atlas of the Vascular Plants of Britain, Ireland, the Isle of Man and the Channel Islands. Oxford University Press, Oxford.

Quiller-Couch, Arthur (editor). 1919. The Oxford Book of English Verse: 1250–1900. Clarendon, Oxford.

Rackham, Oliver. 1976. Trees and Woodland in the British Landscape. J. M. Dent & Sons, London.

Rackham, Oliver. 1985. Ancient woodland and hedges in Britain. Pp. 68–105 *in* S. R. J. Woodell (editor), The English Landscape: Past, Present and Future. Wolfson College Lectures. Oxford University Press, Oxford.

Rackham, Oliver. 1995. Looking for ancient woodland in Ireland. Pp. 1–12 *in* J. R. Pilcher & S. Mac an tSaoir (editors). Woods, Trees and Forests in Ireland. Royal Irish Academy, Dublin.

Rackham, Oliver. 2003. Ancient Woodland (revised ed). Castlepoint Press, Dalbeattie, United Kingdom.

Rätsch, C. 1997. Plants of Love: The History of Aphrodisiacs and a Guide to Their Identification and Use. Ten Speed Press, Berkeley, California.

Reader's Digest Association. 2006. Nature's Medicines: A Guide to Herbal Medicines and What They Can Do for You. Reader's Digest Association, London.

Reilly, Paddy A. 2001. The Flora of County Cavan. Occasional Papers No. 13. National Botanic Gardens, Glasnevin, Dublin.

Reynolds, Sylvia C. P. 2002. A Catalogue of Alien Plants in Ireland. Occasional Papers No. 14. National Botanic Gardens, Glasnevin, Dublin.

Rich, T. C. G., L. Houston, A. Robertson & M. C. F. Proctor. 2010. Whitebeams, Rowans and Service Trees of Britain and Ireland: A Monograph of British and Irish *Sorbus* L. B.S.B.I. Handbook No. 14. Botanical Society of the British Isles, London.

Rimmer, Joan. 1969. The Irish Harp. Published for the Cultural Relations Committee by Mercier Press, Cork.

Roden, Cilian. 1994. The Aran flora. Pp. 35–46 *in* John Waddell, J. W. O'Connell & Anne Koroff (editors), The Book of Aran: The Aran Islands, Co. Galway. Tír Eolas, Kinvara, Co. Galway.

Rotherham, Ian D. 2009. Peat and Peat Cutting. Shire Publications, Oxford.

Rutty, John. 1772. An Essay Towards a Natural History of the County of Dublin, Vol. 1. W. Sleater, Dublin.

Ryan, Hugh. 1907. Reports upon the Irish Peat Industries: Part I. Economic Proceedings, Royal Dublin Society, Vol. 1: 371–402.

Ryan, Hugh. 1908. Reports upon the Irish Peat Industries: Part II. Economic Proceedings, Royal Dublin Society, Vol. 1: 465–546.

Ryan, Meda. 1978. Biddy Early: The Wise Woman of Clare. Mercier Press, Dublin and Cork.

Ryan, Michael. 2000. Early Irish Communion Vessels. The Irish Treasures Series. Country House, Dublin, in association with The National Museum of Ireland.

Rymer, L. 1976. The history and ethnobotany of bracken. Botanical Journal of the Linnean Society 73: 151–176.

Scannell, Maura. 2007. 'The Rush Gatherers, Lough Corrib, Connemara, Ireland': A chromolithograph of a painting by J. J. Hill. Journal of the Galway Archaeological and Historical Society 59: 179–184.

Scannell, M. J. P. & D. M. Synnott, 1987. Census Catalogue of the Flora of Ireland, 2nd ed. Stationery Office, Dublin.

Scully, Reginald W. 1916. Flora of County Kerry. Hodges Figgis and Co., Dublin.

Sell, Peter. 1991. The cherries and plums of Cambridgeshire. Nature in Cambridgeshire 33: 29–39.

Sharkey, Olive. 2000. Ways of Old: Traditional Life in Ireland. O'Brien Press, Dublin.

Shaw, N. 1998. Herbalism: An Illustrated Guide. Element Books, Shaftesbury, Dorset.

Shaw-Smith, David. 1984. Ireland's Traditional Crafts. Thames and Hudson, London.

Sheehy Skeffington, Micheline. 1984. Cultivated plants. Pp. 34–38 *in* Peter Wyse Jackson & Micheline Sheehy Skeffington. The Flora of Inner Dublin. Royal Dublin Society, Dublin.

Siggins, Albert. 2004. Jim Delaney and James 'A' Grady helps the National Museum. Rindoon Historical Journal, Vol. 1. St John's Parish Heritage Group, Lecarrow, Co. Roscommon.

Siggins, Lorna. 2007. An Irish Harvest 1: Seaweed in Connemara. The Irish Times, 28th August, 2007: 13.

Simpson, David A. & Cecilia A. Inglis. 2001. Cyperaceae of economic, ethnobotanical and horticultural importance: a checklist. Kew Bulletin, 56(2): 257–360.

Singleton, A. H. 1904. Dairy folklore, and other notes from Meath and Tipperary. Folklore 15(4): 457–462.

Sloane, Sir Hans. 1725. Natural History of Jamaica. London.

Smith, Charles. (1756) 1979. The Ancient and Present State of the County of Kerry: Being a Natural, Civil, Ecclesiastical, Historical and Topographical Description Thereof. Mercier Press, Cork and Dublin.

Smith, Charles. 1815. The Ancient and Present State of the County and City of Cork, revised edition. John Connor, Patrick Street, Cork.

Smith, John. 1882. Dictionary of Popular Names of Economic Plants. Macmillan, London.

Somerville-Large, Peter. 1985. Cappaghglass. Hamish Hamilton, London.

Stace, Clive. 2010. New Flora of the British Isles, 3rd ed. Cambridge University Press, Cambridge.

Stamets, Paul. 1996. Psilocybe Mushrooms of the World: An Identification Guide. Ten Speed Press, Berkeley, California.

Staniforth, Arthur. 2008. Straw and Straw Craftsmen. Shire Publications, Oxford.

Sullivan, D. J. & A. W. Stelfox. 1943. Survey of Inishtrahull, II. A list of flowering plants and ferns. Irish Naturalists' Journal 8: 116–123.

Sundararajan, R., A. H. Nazeer, K. Venkatesan, K. Mukherjee, B. P. Saha, A. Bandyopadhyay & P. K. Mukherjee. 2006. *Cytisus scoparius* link: A Natural Antioxidant. BMC Complementary and Alternative Medicine 6: 8. <http://www.biomedcentral.com/1472-6882/6/8>, accessed 1 June 2012.

Synge, J. M. 1979 [First published in 1907]. The Aran Islands. Oxford University Press, Oxford.

Synge, J. M. 2009. Travels in Wicklow, West Kerry and Connemara. Serif, London [First published as In Wicklow, West Kerry and Connemara in 1911 by Maunsell & Co. Ltd, Dublin].

Synnott, Chris. 2008. A history of bog butter finds in Kerry since 1854 and some recent radiocarbon dating evidence. Journal of the Kerry Archaeological and Historical Society (Ser. 2) 8: 97–113.

Synnott, Donal M. 1979. Folk-lore, legend and Irish plants. Pp. 37–43 *in* C. Nelson & A. Brady (editors). Irish Gardening and Horticulture. Royal Horticultural Society of Ireland, Dublin.

Teagasc (Agriculture and Food Development Authority). 2011. <http://www.teagasc.ie/forestry/technical_info/forestry_history.asp>, accessed 21 August 2011.

Teahan, John. 1994. Irish Furniture and Woodcraft. National Museum of Ireland in association with Country House, Dublin.

Thiselton Dyer, T. F. 1889. The Folk-lore of Plants. D. Appleton and Company, New York.

Threlkeld, Caleb. 1726. Synopsis Stirpium Hibernicarum. F. Davys et al., Dublin.

Tóibín, Seán. 1965. Blátha an Bhóithrín. Oifig An tSoláthair, Baile Átha Cliath.

Toner, Niall. 2007. Voted top in a straw poll, The Sunday Times (Irish Property Section), 8 April 2007.

Tutin, T. G., V. H. Heywood, N. A. Burges, D. A. Moore, D. H. Valentine, S. M. Walters & D. A. Webb (editors). 1980. Flora Europaea, Volume 5. Cambridge University Press, Cambridge.

Ua Cnáimhsí, Pádraig. 2009. Róise Rua: An Island Memoir (Translated from the Irish by J. J. Keaveny). Mercier Press, Cork.

Ua Maoileoin, Pádraig. 2001. Dúlra Duibhneach, 2nd ed. Binn Éadair, Baile Átha Cliath.

Uí Chanainn, Eibhlín. 1962. Foclóir an Mhic Léinn: The Scholar's Irish–English Dictionary. An Press Náisiúnta, Baile Átha Cliath.

Uí Chonchubhair, M. 1995. Flóra Chorca Dhuibhne: Aspects of the Flora of Corca Dhuibhne. Oidhreacht Chorca Dhuibhne, Baile an Fheirtéaraigh, Co. Kerry.

Ui Sheasnáin, Áine. 1982. Traditional medicine. Pp. 93–108 *in* Anna Sexton (editor), The Heart of Breifne. Cumann Staire Leathrátha, Cavan.

van Asseldonk, Tedje. 2010. Ethnoveterinary medical practice in the European Union (EU): A case study of the Netherlands. Pp. 373–387 *in* David R. Katerere & Dibungi Luseba (editors), Ethnoveterinary Botanical Medicine: Herbal Medicines for Animal Health. CRC Press, Boca Raton, Florida.

van Wyk, Ben-Erik & Michael Wink. 2004. Medicinal Plants of the World. Timber Press, Portland, Oregon.

Vickery, Roy. 1985. *Sempervivum tectorum* as an abortifacient. Folklore 96 (2): 253.

Vickery, Roy. 1995. A Dictionary of Plant-Lore. Oxford University Press, Oxford.

Vickery, Roy. 2010a. Garlands, Conkers and Mother-Die; British and Irish Plant-lore. Continuum, London.

Vickery, Roy. 2010b. Plant-Lore.com, <http://www.plant-lore.com>, accessed 2 November 2011.

Virmond, Alice. 2009. Bonar, Wild & Delicious. All About Travellers' Food: Traditional Ingredients in Old and New Recipes. North Kerry Traveller Women's Group, Tralee, Co. Kerry.

Vogel, A./Bioforce AG. 2012. Fresh from the Nature: A. Vogel Plant Encyclopaedia. *Crataegus monogyna* Jacq. & *laevigata* (Poir.) DC Hawthorn, <http://www.avogel.ch/en/plant-encyclopaedia/crataegus.php>, accessed 23 August 2012.

von Pückler-Muskau, Hermann. 1940 [originally published in 1833]. Tour of England, Ireland, and France, in the Years 1826, 1827, 1828, and 1829 (translated by Sarah Austin). Massie Publishing Co., Zurich.

Wade, Walter. 1794. Catalogus Systematicus Plantarum Indigenarum in Comitatu Dublinensi Inventarum. Dublin Society, Dublin.

Wade, Walter. 1804. Plantae Rariores in Hibernia Inventae: Or Habitats of Some Plants, Rather Scarce and Valuable, Found in Ireland; With Concise Remarks on the Properties and Uses of Many of Them. Graisberry and Campbell, Dublin.

Wade, Walter. 1808a. Sketch of Lectures on Meadow and Pasture Grasse, Delivered in the Dublin Society's Botanical Garden, Glasnevin. Graisberry and Campbell, Dublin.

Wade, Walter. 1808b. Sketch of Lectures on Artificial or Sown Grasses, as Lucern, Saint-Foin, Clovers, Trefoils, Vetches, &c. &c. Graisberry and Campbell, Dublin.

Wagner, Kelly. 2006. Tapping the potential of medicinal and aromatic plants in northern Europe. Geneflow 6: 35.

Wakefield, Edward. 1812. An Account of Ireland, Statistical and Political. Longman, Hurst, Rees, Orme & Brown, London.

Wakeman, William F. 1891. Archaeologia Hibernica. A Hand-Book of Irish Antiquities. Hodges Figgis and Co., Dublin.

Walsh, Laurence. 1978. Dr Richard Heaton of Ballyskenagh, 1601–1666. Parkmore Press, Roscrea.

Warren, Piers. 2006. British Native Trees: Their Past and Present Uses, Including a Guide to Burning Wood in the Home. Wildeye, United Kingdom.

Webb, D. A. 1950. *Saxifraga hartii* D. A. Webb. Proc. Roy. Irish Acad. 53, sect. B: 219.

Webb, D. A. 1977. An Irish Flora, 6th ed. Dundalgan Press, Dundalk.

Webb, D. A. 1987. *Saxifraga rosacea* Moench subsp. *hartii* (D. A. Webb) D. A. Webb. Bot. J. Linn. Soc. 95(4): 246.

Webb, D. A. & R. J. Gornall. 1989. Saxifrages of Europe. Christopher Helm, London.

Webb, D. A., J. Parnell & D. Doogue. 1996. An Irish Flora, 7th ed. Dundalgan Press, Dundalk.

Webb, D. A. & M. J. P. Scannell. 1983. Flora of Connemara and the Burren. Cambridge University Press, Cambridge.

Westropp, T. J. 2000. Folklore of Clare. Clasp Press, Ennis, Co. Clare (originally published as a series of articles in the journal *Folklore* in the early part of the 20th century).

Whelan, Paul. 2011. Lichens of Ireland: An Illustrated Introduction to over 250 Species. Collins Press, Cork.

WHO, IUCN, WWF. 1993. Guidelines on The Conservation of Medicinal Plants. The International Union for Conservation of Nature and Natural Resources (IUCN), Gland, Switzerland, in partnership with The World Health Organization (WHO), Geneva, Switzerland, and WWF—World Wide Fund for Nature, Gland, Switzerland.

White, Gilbert. [1775] 1996. Natural History of Selborne (Wordsworth Classics). Wordsworth Editions, Herts, United Kingdom.

White, John. 1803. An Essay on the Indigenous Grasses of Ireland. Graisberry and Campbell, Dublin.

White Lennon, Biddy. 2002. Best of Irish Potato Recipes. O'Brien Press, Dublin.

Wilde, Lady. 1887. Ancient Legends, Mystic Charms and Superstitions of Ireland: With Sketches of the Irish Past. Ticknor and Co., London

Wilde, Lady. 1890. Ancient Cures, Charms, and Usages of Ireland: Contributions to Irish Lore. Ward and Downey, London.

Williams, N. J. A. 1987. A note on John K'Eogh's *Herbal.* Eighteenth-Century Ireland/Iris an dá chultúr 2: 198–202.

Williams, Nicholas. 1993. Díolaim Luibheanna. Sáirséal Ó Marcaigh, Baile Átha Cliath.

Wilson, John M. (Reverend). 1849. The Rural Cyclopedia, or a General Dictionary of Agriculture, and of the Arts, Sciences, Instruments, and Practice, Necessary to the Farmer, Stockfarmer, Gardner, Forester, Landsteward, Farrier, &c, Vol. IV, Q–Z. A. Fullerton & Co., Edinburgh.

Wiynjorrotj, Phyllis, Sara Flora, Nipper Daybilama Brown, Peter Jatbula, Judy Galmur, Margaret Katherine, Francesca Merlan & Glenn Wightman. 2005. Jawoyn Plants and Animals: Aboriginal Flora and Fauna Knowledge from Nitmiluk National Park and the Katherine Area, Northern Australia. Northern Territory Bulletin No. 29. Ethnobiology Project (NRETA) and Jawoyn Association, Darwin, Australia.

Wood, Thomas. 1821. An Inquiry Concerning the Primitive Inhabitants of Ireland. Edwards and Savage, Cork.

Woodham-Smith, Cecil. 1962. The Great Hunger: Ireland 1845–1849. Hamish Hamilton, London.

Woodman, P. C., E. Anderson & N. Finlay. 1999. Excavations at Ferriter's Cove, 1983–95: Last Foragers, First Farmers in the Dingle Peninsula. Wordwell, Bray, Co. Wicklow.

Wood-Martin, W. G. 1902. Traces of the Elder Faiths of Ireland: A Folklore Sketch. A Handbook of Irish Pre-Christian Traditions, Vols. 1, 2. Longmans, Green, and Co., London.

Woodward, Lucia. 1985. Poisonous Plants: A Colour Field Guide. David & Charles, Newton Abbot, London.

Wren, R. C. 1988. Potter's New Cyclopaedia of Botanical Drugs and Preparations. C. W. Daniel Company, London.

Wyse Jackson, John & Hector McDonnell. 2007. Ireland's Other Poetry: Anonymous to Zozimus. Lilliput Press, Dublin.

Wyse Jackson, Peter. 1994. Irish Trees and Shrubs. Appletree Press, Belfast.

Wyse Jackson, Peter. 2006. Gardening with native Irish plants. Moorea 15: 11–16.

Wyse Jackson, Peter S. 2007. The potential impact of climate change on native plant diversity in Ireland. BGjournal 4(2): 26–29.

Wyse Jackson, Peter & Micheline Sheehy Skeffington. 1984. The Flora of Inner Dublin. Royal Dublin Society, Dublin.

Xiao, Pei-gen. 1991. The Chinese approach to medicinal plants: Their utilization and conservation. Pp. 305–314 *in* Olayiwola Akerele, Vernon Heywood & Hugh Synge (editors), Conservation of Medicinal Plants. Cambridge University Press, Cambridge.

Yeats, William Butler. 1949. The Poems of W. B. Yeats. Volume II. Macmillan, London.

INDEX TO SCIENTIFIC NAMES

Main entries in the Systematic List (see pg. 151) are boldface. Wherever possible, authorities have been listed, using Stace (2010), Tropicos, and the International Plant Names Index as sources.

Q

R

INDEX TO COMMON ENGLISH NAMES

C

D

E